Antibiotic Drug Resistance

Antibiotic Drug Resistance

Edited by

José-Luis Capelo-Martínez
Department of Chemistry of the Faculty of Science and Technology
NOVA University Lisbon, Lisbon, Portugal

Gilberto Igrejas
Department of Genetics and Biotechnology
Functional Genomics and Proteomics Unit
University of Trás-os-Montes and Alto Douro, Vila Real, Portugal

WILEY

This edition first published 2020

Registered Office
John Wiley & Sons, Inc., 111 River Street, Hoboken, NJ 07030, USA

Editorial Office
111 River Street, Hoboken, NJ 07030, USA

For details of our global editorial offices, customer services, and more information about Wiley products visit us at www.wiley.com.

Wiley also publishes its books in a variety of electronic formats and by print-on-demand. Some content that appears in standard print versions of this book may not be available in other formats.

Library of Congress Cataloging-in-Publication Data
Names: Capelo-Martinez, Jose-Luis, editor. | Igrejas, Gilberto, editor.
Title: Antibiotic drug resistance / edited by Jose-Luis Capelo Martinez, Gilberto Igrejas.
Description: First edition. | Hoboken, NJ : Wiley, 2019. | Includes bibliographical references and index. |
Identifiers: LCCN 2019016039 (print) | LCCN 2019017013 (ebook) | ISBN 9781119282532 (Adobe PDF) | ISBN 9781119282556 (ePub) | ISBN 9781119282525 (hardback)
Subjects: | MESH: Anti-Bacterial Agents | Drug Resistance, Bacterial
Classification: LCC RM267 (ebook) | LCC RM267 (print) | NLM QV 350 | DDC 615.7/922–dc23
LC record available at https://lccn.loc.gov/2019016039

Cover image: Courtesy of Mathieu F. Chellat, Luka Raguž and Rainer Riedl
Cover design by Wiley

Set in 10/12pt Warnock by SPi Global, Pondicherry, India

Printed in the United States of America

V10012830_080919

Contents

List of Contributors

Teresa Alarcón
Department of Microbiology,
Hospital Universitario La Princesa
Instituto de Investigación Sanitaria
Princesa
Madrid, Spain

Department of Preventive Medicine,
Public Health and Microbiology,
Medical School
Autonomous University of Madrid
Madrid, Spain

Egorov Alexey
Department of Chemistry
M.V. Lomonosov Moscow State
University
Moscow, Russia

Rustam Aminov
School of Medicine, Medical
Sciences and Nutrition
University of Aberdeen
Aberdeen, UK

Institute of Fundamental Medicine
and Biology
Kazan Federal University
Kazan, Russia

Haotian Bai
Institute of Chemistry
Chinese Academy of Sciences
Beijing, P. R. China

Luz Balsalobre
Department of Microbiology,
Hospital Universitario
La Princesa
Instituto de Investigación Sanitaria
Princesa
Madrid, Spain

Johan Bengtsson-Palme
Department of Infectious
Diseases
Institute of Biomedicine,
The Sahlgrenska Academy
University of Gothenburg
Gothenburg, Sweden

Centre for Antibiotic Resistance
Research (CARe)
University of Gothenburg
Gothenburg, Sweden

Wisconsin Institute
of Discovery
University of Wisconsin-Madison
Madison, WI, USA

Ana Blanco
Department of Microbiology
Hospital Universitario
La Princesa
Instituto de Investigación Sanitaria
Princesa
Madrid, Spain

Martine Bonnaure-Mallet
Univ Rennes, INSERM, INRA, CHU Rennes
Institut NUMECAN (Nutrition Metabolisms and Cancer)
Rennes, France

Teaching Hospital of Rennes
Rennes, France

João Carrola
Centre for the Research and Technology of Agro-Environmental and Biological Sciences (CITAB), University of Trás-os-Montes and Alto Douro, Vila Real, Portugal

Ana Luísa Carvalho
UCIBIO-REQUIMTE,
Departamento de Química
Faculdade de Ciências e Tecnologia
Universidade NOVA de Lisboa
Caparica, Portugal

Isabel Carvalho
Department of Veterinary Sciences, Department of Genetics and Biotechnology,
Functional Genomics and Proteomics Unit,
MicroART- Antibiotic Resistance Team, University of Trás-os-Montes and Alto Douro, Vila Real, Portugal;
Laboratory Associated for Green Chemistry (LAQV-REQUIMTE), New University of Lisbon
Monte da Caparica, Portugal;

Viviana G. Correia
UCIBIO-REQUIMTE,
Departamento de Química
Faculdade de Ciências e Tecnologia
Universidade NOVA de Lisboa
Caparica, Portugal

Carol Currie
Moredun Research Institute
Pentlands Science Park
Penicuik, Scotland, UK

Bruna de Oliveira Costa
S-Inova Biotech, Programa de Pós-Graduação em Biotecnologia
Universidade Católica Dom Bosco
Campo Grande
MS, Brazil

Lucía Fernández
Instituto de Productos Lácteos de Asturias (IPLA-CSIC)
Villaviciosa, Spain

Catarina Ferreira
Universidade Católica Portuguesa
CBQF – Centro de Biotecnologia e Química Fina
Laboratório Associado, Escola Superior de Biotecnologia
Porto, Portugal

Octávio Luiz Franco
S-Inova Biotech, Programa de Pós-Graduação em Biotecnologia
Universidade Católica Dom Bosco
Campo Grande, MS, Brazil

Centro de Análises Proteômicas e Bioquímicas
Programa de Pós-Graduação em Ciências Genômicas e Biotecnologia
Universidade Católica de Brasília
Brasília, DF, Brazil
Faculdade de Medicina
Programa de Pós-Graduação em Patologia Molecular
Universidade de Brasília
Brasília, DF, Brazil

Ana Paula Guedes Frazzon
Federal University of Rio Grande do Sul
Porto Alegre, Brazil

Jeverson Frazzon
Federal University of Rio Grande do Sul
Porto Alegre, Brazil

Anton Gadelii
Division of Experimental Infection Medicine
Department of Translational Medicine
Lund University
Malmö, Sweden

Pilar García
Instituto de Productos Lácteos de Asturias (IPLA-CSIC)
Villaviciosa, Spain

Elisabeth Grohmann
Life Sciences and Technology
Beuth University of Applied Sciences
Berlin, Berlin, Germany

Diana Gutiérrez
Instituto de Productos Lácteos de Asturias (IPLA-CSIC)
Villaviciosa, Spain

Anders P. Hakansson
Division of Experimental Infection Medicine
Department of Translational Medicine
Lund University
Malmö, Sweden

Jayanta Haldar
Antimicrobial Research Laboratory, New Chemistry Unit and School of Advanced Materials, Jawaharlal Nehru Centre for Advanced Scientific Research, Bengaluru, Karnataka, India

Karl-Omar Hassan
Division of Experimental Infection Medicine
Department of Translational Medicine
Lund University
Malmö, Sweden

Stefanie Heß
Department of Microbiology
University of Helsinki
Helsinki, Finland

Gilberto Igrejas
Department of Genetics and Biotechnology,
Functional Genomics and Proteomics Unit,
University of Trás-os-Montes and Alto Douro
Vila Real, Portugal

Anne Jolivet-Gougeon
Univ Rennes, INSERM, INRA, CHU Rennes
Institut NUMECAN (Nutrition Metabolisms and Cancer)
Rennes, France
Teaching Hospital of Rennes
Rennes, France

Verena Kohler
Institute of Molecular Biosciences
University of Graz
Graz, Austria

Department of Molecular Biosciences,
The Wenner-Gren Institute
Stockholm University, Stockholm, Sweden

Jackson O. Lay, Jr.
Department of Chemistry and Biochemistry
University of Arkansas
Fayetteville, AR, USA

S. Lohsen
School of Medicine
Emory University
Atlanta, USA

Célia M. Manaia
Universidade Católica Portuguesa
CBQF - Centro de Biotecnologia e Química Fina
Laboratório Associado, Escola Superior de Biotecnologia
Porto, Portugal

Michele Bertoni Mann
Federal University of Rio Grande do Sul
Porto Alegre, Brazil

Ulyashova Mariya
Department of Chemistry
M.V. Lomonosov Moscow State University
Moscow, Russia

Beatriz Martínez
Instituto de Productos Lácteos de Asturias (IPLA-CSIC)
Villaviciosa, Spain

José L. Martínez
Departamento de Biotecnología Microbiana
Centro Nacional de Biotecnología,
Consejo Superior de Investigaciones Científicas (CSIC)
Madrid, Spain

Rubtsova Maya
Department of Chemistry
M.V. Lomonosov Moscow State University
Moscow, Russia

Luís D. R. Melo
Centre of Biological Engineering (CEB)
Laboratório de Investigação em Biofilmes Rosário Oliveira (LIBRO)
University of Minho
Braga, Portugal

David J. Newman
Newman Consulting LLC
Wayne, PA, USA

Olga C. Nunes
LEPABE – Laboratory for Process Engineering, Environment, Biotechnology and Energy
Faculdade de Engenharia
Universidade do Porto
Porto, Portugal

Hugo Oliveira
Centre of Biological Engineering (CEB)
Laboratório de Investigação em Biofilmes Rosário Oliveira (LIBRO)
University of Minho
Braga, Portugal

Angelina S. Palma
UCIBIO-REQUIMTE
Departamento de Química
Faculdade de Ciências e Tecnologia

Universidade NOVA de Lisboa
Caparica, Portugal

Benedita A. Pinheiro
UCIBIO-REQUIMTE
Departamento de Química
Faculdade de Ciências e Tecnologia
Universidade NOVA de Lisboa
Caparica, Portugal

Patrícia Poeta
Microbiology and Antibiotic Resistance Team (MicroART), Department of Veterinary Sciences, University of Trás-os-Montes and Alto Douro, Vila Real, Portugal; Laboratory Associated for Green Chemistry (LAQV-REQUIMTE), New University of Lisbon, Monte da Caparica, Portugal;
New University of Lisbon, Monte da Caparica, Portugal

Fatemeh Rafii
National Center for Toxicological Research
U.S. Food and Drug Administration
Jefferson, AR, USA

Ana Rodríguez
Instituto de Productos Lácteos de Asturias (IPLA-CSIC)
Villaviciosa, Spain

Sílvio B. Santos
Centre of Biological Engineering (CEB)
Laboratório de Investigação em Biofilmes Rosário Oliveira (LIBRO)
University of Minho
Braga, Portugal

Paramita Sarkar
Antimicrobial Research Laboratory, New Chemistry Unit, Jawaharlal Nehru Centre for Advanced Scientific
Research, Bengaluru, Karnataka, India

Frank Schweizer
Department of Chemistry
University of Manitoba
Winnipeg, Canada

Nuno Silva
Moredun Research Institute
Pentlands Science Park
Penicuik, Scotland, UK

Osmar Nascimento Silva
S-Inova Biotech, Programa de Pós-Graduação em Biotecnologia
Universidade Católica Dom Bosco
Campo Grande, MS, Brazil

Vanessa Silva
Microbiology and Antibiotic Resistance Team (MicroART), Department of Veterinary Sciences, Department of Genetics and Biotechnology,
Functional Genomics and Proteomics Unit, University of Trás-os-Montes and Alto Douro, Vila Real, Portugal; Laboratory Associated for Green Chemistry (LAQV-REQUIMTE), New University of Lisbon, Monte da Caparica, Portugal
New University of Lisbon,
Monte da Caparica, Portugal

Margarida Sousa
Department of Veterinary Sciences, Department of Genetic and Biotechnology,
Functional Genomics and Proteomics Unit

University of Trás-os-Montes and
Alto Douro
Vila Real, Portugal

D.S. Stephens
School of Medicine
Emory University
Atlanta, GA, USA

John B. Sutherland
National Center for Toxicological
Research
U.S. Food and Drug Administration
Jefferson, AR, USA

Ankita Vaishampayan
Life Sciences and Technology
Beuth University of Applied Sciences
Berlin
Berlin, Germany

Manuel F. Varela
Eastern New Mexico University
Portales, NM, USA

Ivone Vaz-Moreira
Universidade Católica Portuguesa
CBQF – Centro de Biotecnologia e
Química Fina
Laboratório Associado, Escola
Superior de Biotecnologia
Porto, Portugal

LEPABE – Laboratory for Process
Engineering, Environment,
Biotechnology and Energy
Faculdade de Engenharia
Universidade do Porto
Porto, Portugal

Shu Wang
Institute of Chemistry, Chinese
Academy of Sciences
Beijing, P. R. China

Anna J. Williams
National Center for Toxicological
Research
U.S. Food and Drug Administration
Jefferson, Arkansas, USA

Mire Zloh
UCL School of Pharmacy
University College London
London, UK

Preface

In the fight to survive, bacteria have been able to find their own path to succeed despite what may be considered their worst-case evolutionary scenario, the advent of the antibiotic era. Bacterial resistance to antibiotics has reached levels of success unexpectedly 20 years ago. Some bacteria can resist, literally, everything human kind has invented to fight them. The situation is worsening as the bacteria causing pneumonia, tuberculosis, gonorrhea, and salmonellosis are becoming multiresistant to all known antibiotics. For humans, this problem is having a tremendous effect on our society, regardless of gender, age, or country, because the cost of medical care is increasingly high due to longer hospital stays and the need of sophisticated and expensive new drugs. There is an additional threat for immunodepressed patients who cannot survive multiresistant bacteria. Though the causes of multiresistance are complex, it seems the misuse of antibiotics in human medicine and more conspicuously in animal medicine and husbandry is one of the primary causes. To complicate things further, the spread of multiresistance is a pressing issue, as poor hygiene, unsafe sexual relationships, and poor food preservation are key factors that help to magnify the problem (World Health Organization 2019). Biocides and other antimicrobials could co-select for bacteria resistant to clinically relevant antibiotics, and therefore the use of these chemicals must be revisited (Oniciuc et al. 2019).

There is collateral damage in the use of antibiotics. The large amount of antibiotics produced every year (c. 100 K tons) has environmental implications of great concern. Antibiotics are now ubiquitous in the environment. For example, they have been detected in freshwater and in fish at sublethal concentrations that can contribute to spreading bacterial resistance and changing the composition of single-celled communities (Danner et al. 2019). The way antibiotic resistance is acquired is still under debate, but it is certainly multifactorial and complex.

Some of today's problems with antibiotic resistance are clearly growing so much and so rapidly that they seem intractable. Therefore, now more than ever, a holistic view of the problem is necessary. This book is intended to fill this gap. The first part of the book addresses the mechanisms of action of the antibiotics most used nowadays, namely, aminoglycosides, quinolones, beta-lactams, glycopeptides, and macrolides. The mechanisms by which bacteria develop antibiotic resistance, including mutations and gene transfer, are also explained. As an issue that negatively affects the living conditions of many people, the socioeconomic impact of antibiotic resistance on public health is also discussed, with special emphasis in public policies aimed at reducing or eliminating pathogens in the environment. Special attention is given to strategies devoted to overcoming antibiotic resistance, with focus on (i) new strategies to design drugs, (ii) antibiotics from natural sources, (iii) strategies based on antimicrobials and bacteriophages, (iv) sensitizing agents to restore antibiotic activity, (v) nontraditional medicines, and (vi) therapeutic options to treat infections caused by pathogenic biofilms.

We believe this book offers a unique global perspective of the problem of antibiotic resistance, as it integrates current knowledge in all related areas from new antibiotics to the reuse of old ones, from new strategies to fight bacteria based on natural products or bacteriophages to new synthetic drugs, and from the strategies to prevent the spread of antibiotic resistance to public policies to reduce the impact of the problem.

We are in debt to everyone involved in bringing this project to fruition, especially to Professor Ramaiah who kindly proposed the idea of the book and to the Wiley Editorial Team who generously embraced the idea. We thank all the authors who generously gave their time and expertise and Gonçalo Martins for helping us to compile the contributions.

Gilberto Igrejas
Associate Professor
University of Tras-os-Montes and Alto Douro

José-Luis Capelo-Martínez
Associate Professor
NOVA University Lisbon

References

Danner, M.-C., Robertson, A., Behrends, V., and Reiss, J. (2019). Antibiotic pollution in surface fresh waters: occurrence and effects (review). *Sci. Total Environ.* 664: 793–804.

Oniciuc, E.-A., Likotrafiti, E., Alvarez-Molina, A. et al. (2019). Food processing as a risk factor for antimicrobial resistance spread along the food chain (review). *Curr. Opin. Food Sci.* 30: 21–26.

WHO (2019). Antimicrobial resistance. https://www.who.int/news-room/fact-sheets/detail/antibiotic-resistance (accessed 26 February 2019).

About the Editors

José-Luis Capelo-Martínez obtained his PhD in Analytical Chemistry from the University of Vigo in 2002 with Prof. Carlos Bendicho and his postdoc in IST in Lisbon with Prof. Ana Mota (2002–2005). He was appointed as researcher at REQUIMTE (FCT/UNL, 2005–2009) and returned to the University of Vigo to become a principal investigator for the Isidro Parga Pondal program and a researcher–lecturer (2009–2012). He was an assistant professor in FCT/UNL (2012–2018), and in 2018 he was appointed associate professor in the Department of Chemistry of the Faculty of Science and Technology in the same institution. In 2007 he obtained his Spanish habilitation in analytical chemistry and in 2017 the Portuguese habilitation in biochemistry (analytical proteomics). Dr. Capelo is a fellow of the Royal Society of Chemistry and member of the Portuguese Chemistry Society. He co-leads the BIOSCOPE Research Group (www.bioscopegroup.org). He is the co-CEO of the PROTEOMASS Scientific Society and founder/co-CEO of the Chemicals start-up Nan@rts. Dr. Capelo has been researching on the following topics: (i) quantification of metal and metal species in environmental and food samples, (ii) new methods to speed protein identification using mass spectrometry-based workflows, (iii) accurate bottom-up protein quantification, (iv) bacterial identification through mass spectrometry, (v) fast determination of steroids in human samples, (v) biomarker discovery, (vi) application of sensors and chemosensor to the detection/quantification of metals, and (vii) nanoproteomics and nanomedicine.

He is an author or co-author of more than 200 manuscripts, 2 patents, 12 book chapters, and 4 books. He has mentored 12 PhD theses and currently he is mentoring 6.

Gilberto Igrejas is a professor at the University of Trás-os-Montes and Alto Douro (UTAD). He completed his PhD in Genetics and Biotechnology at the University of Trás-os-Montes and Alto Douro in collaboration with the Institut National de la Recherche Agronomique (INRA) in 2001 and a post-graduate degree in legal medicine at the National Institute of Legal Medicine – Porto/Faculty of Medicine of Porto University in 2002. He had a postdoctoral training in molecular genetics, as Visiting Scientist, at Commonwealth Scientific and Industrial Research Organisation (CSIRO), Australia in 2004–2005. Currently he is the head of the Functional Genomics and Proteomics Unit and an integrated member of the (Bio)Chem & OMICS, LAQV/REQUIMTE of Nova University in Lisbon. His research is focused on the use of omics tools, particularly genomics and proteomics, at the molecular genetics and biotechnology level of various plant, animal, and microbial species. These are in the chronological involvement as priority research areas, based on these tools and their scope: (i) characterization of genetic resources of wheat, rye, and triticale; (ii) proteomics applied to the detection of genes responsible for the functionality and allergenicity of wheat grain, rye, and triticale; (iii) genomics and proteomics applied to antibiotic resistance; (iv) nutrigenomics and proteomics applied to the evaluation of protein species; and, finally, (v) probiotics in biotechnology and health. With regard to scientific production, he has published more than 150 articles, 12 book chapters, 20 oral presentations by invitation, 40 oral presentations, 20 articles in technical and scientific journals, and 10 educational series as well as 250 communications in scientific meetings and records in GenBank, UniProt, and MLST. Dr. Igrejas continues to work on research projects, teaches international courses for doctoral and master's degrees, and collaborates with several national and international groups. He currently supervises three postdoctoral researchers, four PhD students, seven master students, and four undergraduate students. He successfully mentored more than 120 students and participated as member of 150 academic degree evaluation panels.

Part I

Current Antibiotics and Their Mechanism of Action

1

Resistance to Aminoglycosides

Glycomics and the Link to the Human Gut Microbiome

Viviana G. Correia, Benedita A. Pinheiro, Ana Luísa Carvalho, and Angelina S. Palma

UCIBIO-REQUIMTE, Departamento de Química, Faculdade de Ciências e Tecnologia, Universidade NOVA de Lisboa, Caparica, Portugal

1.1 Aminoglycosides as Antimicrobial Drugs

The exponential appearance of antibiotic-resistant infections, in particular those caused by Gram-negative pathogens, is a major public health concern. The observed decrease in the emergence of new effective antimicrobial drugs is an inevitable consequence of the use of antibiotics, and new approaches to fight infection are a matter in need of attention from the scientific community (Magiorakos et al. 2012). In response to this challenge, the optimization of existing drugs with known mechanisms of action and resistance, such as aminoglycosides, is an attractive approach for the development of new antimicrobials.

Aminoglycosides or aminoglycoside antibiotics (AGAs) are secondary metabolites of bacteria used in the warfare against other microorganisms, which were repurposed in medicine as broad-spectrum antibiotics in both humans and animals. This class of antibiotics has activity against Gram-negative and Gram-positive bacteria by targeting ribosomal RNA (rRNA), leading to protein misfolding. AGAs have predictable pharmacokinetics and often act in synergy with other antibiotics, such as beta-lactams, making them powerful anti-infective drugs (Hanberger et al. 2013). Despite their potential renal toxicity and ototoxicity and known bacterial resistance, diverse molecules of this family of antibiotics have been used in clinical practice for several decades (Thamban Chandrika and Garneau-Tsodikova 2018).

AGAs are constituted by a carbohydrate residue moiety and possess several amino and hydroxyl group functionalities, determinants for the interaction

Antibiotic Drug Resistance, First Edition. Edited by José-Luis Capelo-Martínez and Gilberto Igrejas.

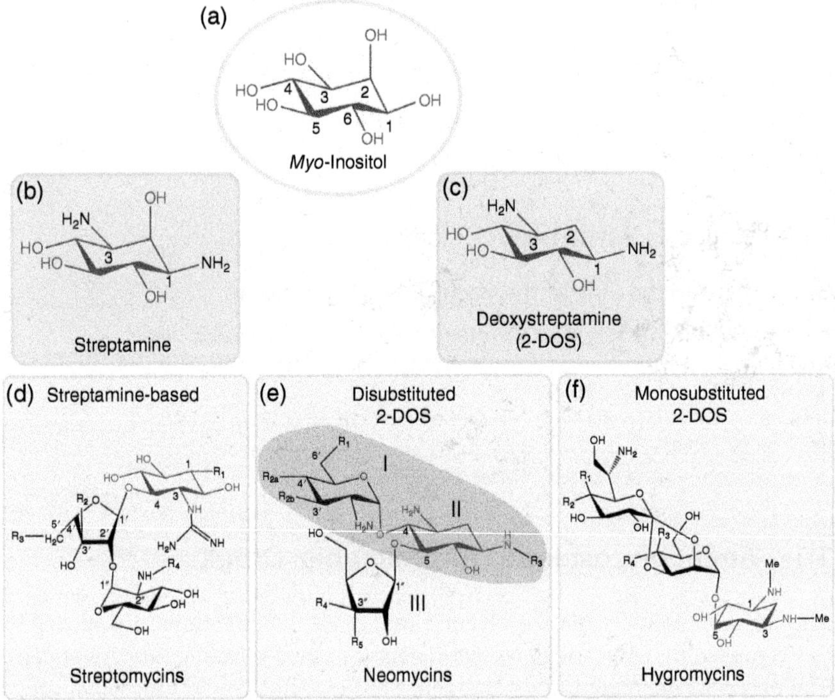

Figure 1.1 Core structural elements of the aminoglycosides and examples of clinically relevant AGA families. Each family has a primary structure with different substitutions (Rn) at hydroxyl and amino groups. The rings (I–III) of the neomycins' representative structure are numbered as usually observed in the literature for disubstituted 2-DOS AGAs. In panel (e), the pseudodisaccharide core structure paromamine, also termed as neamine, is demarked.

with target sequences on the rRNA and for impairing normal ribosomal function. Although natural AGAs share the same *myo*-inositol-based core (Figure 1.1), these molecules exhibit significant structural differences depending on the bacterial origin, which result in different biological activities. Importantly, the bacterial origin is also the driving force behind bacterial resistance, as it enables bacteria to alter the structure of AGAs by modifying their amino and hydroxyl groups.

Streptomycin was the first identified and characterized AGA and the first useful antibiotic obtained from a bacterial source (1944). This AGA was isolated from the soil-dwelling bacterial species *Streptomyces* and *Micromonospora* and successfully introduced into clinical practice in 1940 to treat tuberculosis. After the initial discovery of streptomycin and its streptamine-based relatives (Figure 1.1a), several others followed, and the development of bacterial resistance was largely overcome by introduction of AGAs derived from 2-deoxystreptamine (DOS) (Figure 1.1c), reviewed in (Davies 2007). These included neomycin (1949), kanamycin (1957), gentamycin (1963), tobramycin (1967), and sisomicin (1970). The acquisition of bacterial resistance for the DOS aminoglycosides prompted the development of novel and potent semisynthetic AGAs. These second-generation AGAs resulted from the insertion of a 4-hydroxy-2-aminobutyric acid (HABA) substituent of the C-1 amine group on the DOS ring of kanamycin and gentamycin-derived compounds. Examples are dibekacin (1971), amikacin (1972), arbekacin (1973), isepamicin (1975), and netilmicin (1976). However, because of their clinical usage, bacteria also developed resistance mechanisms against these semisynthetic antibiotics, almost leading to the abandon of AGAs.

Recently, the interest in AGAs research has resurged as consequence of the increasing number of strains resistant to other classes of antibiotics, such as the Gram-negative bacteria *Enterococcus faecium* responsible for serious invasive nosocomial infections (Buelow et al. 2017). New approaches have been used for developing semisynthetic AGAs using combined structure–activity relationship (SAR), in search for less toxic but effective AGAs (Thamban Chandrika and Garneau-Tsodikova 2018). The AGA plazomicin developed by Achaogen Inc. (ACHN-490) (Aggen et al. 2010), currently in phase III clinical trials, is an evidence of the renewed interest. Table 1.1 summarizes described AGAs and their distinctive features.

1.1.1 The Structure of Aminoglycosides

In this section, the basic structure of AGAs and the major differences between the various families and classes are overviewed. For a better understanding of the dissimilarities between them, a reference to the biosynthetic pathways is made. A more comprehensive review on the genetics and biosynthesis of AGAs is available in Piepersberg et al. (2007) and Becker and Cooper (2013).

As stated by their name, AGAs have in their composition amino-modified glycosides, which contain a carbohydrate linked to another functional group via a glycosidic bond. The common element to the core structure of the various AGA families is the *myo*-inositol (a cyclohexanehexol with six hydroxyl groups) (Figure 1.1a). AGA biosynthesis starts with a *myo*-inositol molecule and bifurcates into two distinct pathways that have (i) streptamine (Figure 1.1b) or (ii) 2-deoxystreptamine (2-DOS) (Figure 1.1c) as intermediates. Streptamine results from the introduction of an amino group at the 1,3-positions of the *myo*-inositol ring and is the intermediate of the streptamine-related AGA family. The few members of this family, including streptomycin, have the streptamine intermediate guanidinylated at the 3-position and a disaccharide unit linked to the 4-position (Figure 1.1d). For many AGA antibiotics, the core structure is the paromamine (Figure 1.1e), a pseudodisaccharide with the

Table 1.1 Overview of major aminoglycoside antibiotics (AGAs) and their distinctive features and effect on the human gut microbiome.

AGA	Core-derived structure	Common use	Effect on human gut microbiome	Related pathology or disease	Microbiome-related studies
Naturally occurring					
Apramycin (APR)	4-Monosubstituted 2-DOS	Veterinary	NA	NA	NA
Butirosin (BTR)	4,5-Disubstituted 2-DOS	Biochemical reagent	NA	NA	NA
Fortimicin (FOR)	6-Monosubstituted fortamine	Biochemical reagent	NA	NA	NA
Geneticin (G418)	4,6-Disubstituted 2-DOS	Medicine; biochemical reagent	NA	NA	NA
Gentamycin (GEN)	4,6-Disubstituted 2-DOS	Clinical; veterinary; agriculture	Decreased diversity; prevalence of *Enterobacter* spp.; reduced levels of SCFAs and other metabolites	Necrotizing enterocolitis; Crohn's; increased opportunistic infections	Zhao et al. (2013), Greenwood et al. (2014), and Shankar et al. (2015)
Hygromycin (HYG)	5-Monosubstituted 2-DOS	Veterinary; biochemical reagent	NA	NA	NA
Istamycin (IST)	Fortamine-related	Biochemical reagent	NA	NA	NA
Kanamycin (KAN)	4,6-Disubstituted 2-DOS	Clinical; veterinary; biochemical reagent	NA	NA	NA
Kasugamycin (KSG)	Streptamine-related	Agriculture	NA	NA	NA
Lividomycin (LIV)	4,5-Disubstituted 2-DOS	Biochemical reagent	NA	NA	NA
Neomycin (NEO)	4,5-Disubstituted 2-DOS	Clinical; veterinary	Long-term changes in diversity with oral administration	NA	Kim et al. (2017)
Paromomycin (PAR)	4,5-Disubstituted 2-DOS	Clinical; veterinary	Decreased abundance and diversity	NA	Heinsen et al. (2015)
Ribostamycin (RIB)	4,5-Disubstituted 2-DOS	Clinical	NA	NA	NA

Spectinomycin (SPC)	4,5-Disubstituted actinamine	Clinical; veterinary	NA	NA	NA
Streptomycin (STR)	4-Monosubstituted Streptamine	Clinical; veterinary; agriculture	Decreased diversity; abundance of *Ruminococcaceae* and *Bacteroidaceae*; bile acid metabolism and other pathways affected	Diabetes type 1; inhibition of cancer therapy; increased opportunistic infections	Sekirov et al. (2008), Antunes et al. (2011), Candon et al. (2015), Lichtman et al. (2016), and Routy et al. (2018)
Tobramycin (TOB)	4,6-Disubstituted 2-DOS	Clinical; veterinary	NA	NA	NA
Semisynthetic					
Amikacin (AMK)	HABA insertion; 4,6-disubstituted 2-DOS	Clinical; veterinary	NA	NA	NA
Arbekacin (ABK)	HABA insertion; 4,6-disubstituted 2-DOS	Clinical	NA	NA	NA
Dibekacin (DBK)	HABA insertion; 4,6-disubstituted 2-DOS	Clinical	NA	NA	NA
Isepamicin (ISP)	HABA insertion; 4,6-disubstituted 2-DOS	Clinical	NA	NA	NA
Netilmicin (NTM)	HABA insertion; 4,6-disubstituted 2-DOS	Clinical; veterinary	NA	NA	NA
Next-generation neoglycosides					
Plazomicin (PLZ)	Sisomicin derived; HABA insertion	*In phase III clinical trials*	NA	NA	NA

Source: Adapted from Piepersberg et al. (2007) and Becker and Cooper (2013).

Aminoglycosides are sorted by naturally occurring, semisynthetic, or next generation. The core-derived structures are included. AGAs' common use is highlighted to reflect on spread and constant contact with the human gut microbiome. Effect on human gut microbiome summarizes genomic, proteomic, and metabolic data. Pathologies and diseases known to be related with uptake of AGAs are also included.

2-DOS, 2-deoxystreptamine; AGA, aminoglycoside antibiotic; HABA, 4-hydroxy-2-aminobutyric acid; NA, information not available to our knowledge.

2-DOS intermediate derivatized with glucose at the 4-position, ring II and ring I, respectively. Paromamine-related AGAs are the 4,5-disubstituted neomycin (Figure 1.1e) and the 4,6-disubstituted kanamycins and gentamycins. These AGA classes are also named after the following intermediate in the biosynthetic pathway – neamine, substituted at the 6′-position of ring I with an amine group (R1 substitution; Figure 1.1e).

The AGA classes with pseudodisaccharides other than paromamine comprise the family of fortimicin-related pseudodisaccharides, which include the fortimicin and istamycin. These are produced in a distinct biosynthetic pathway, although sharing some steps with the general 2-DOS and streptomycin biosynthesis. Another AGA family is the monosubstituted 2-DOS AGA family, which includes the hygromycins (Figure 1.1f) and the apramycins, containing the 2-DOS unit as central core with substitutions in the 5- and 4-positions, respectively. Few members of these two last families have clinical importance, except for spectinomycin.

1.1.2 Mechanisms of Action

The bacterial ribosome is the final target of the AGAs, more precisely the A-site located on the 16S RNA of the 30S bacterial ribosomal subunit, which is the binding site of a cognate transference RNA (tRNA) (Figure 1.2).

The elucidation of this macromolecular machine was accomplished by studying the mode of action of AGAs and other ribosome-targeting antibiotics. The bacterial ribosome is composed of 3 rRNAs and 54 ribosomal proteins. The rRNA molecules are the main players, interacting with the messenger RNA (mRNA) and tRNA molecules during protein synthesis. The study of AGAs also enabled to understand the efficiency and fidelity of the translation process, decoding mRNA into protein. The AGAs target the elongation cycle, resulting in a decrease of the fidelity of the ribosome (Figure 1.2a) or in complete inhibition of protein synthesis (Figure 1.2b). During the elongation cycle, the ribosome alternates between an unlocked and a locked structure. This interchange is important for mRNA and tRNA translocation and ribosome recycling. In the unlocked ribosome, the deacylated tRNA in the P-site allows incoming of the elongation factor G, promoting the conformational change. In the locked state, the peptidyl-tRNA in the P-site prevents movements between ribosomal subunits and therefore provides a stable architecture required for the rapid and accurate reading of the mRNA molecules (Valle et al. 2003).

AGAs interfere with the elongation cycle in different ways. For example, streptomycin binds to RNA helices h27, h18 (the 530 loop), and h44 and to the ribosomal protein S12, interfering with the delivery of the aa-tRNA to the A-site of the ribosome by the elongation factor Tu. The antibiotic stabilizes a conformation of the A-site that has higher affinity to tRNAs (including non-cognate tRNAs) and is thus more error-prone (Carter et al. 2000). The 2-DOS

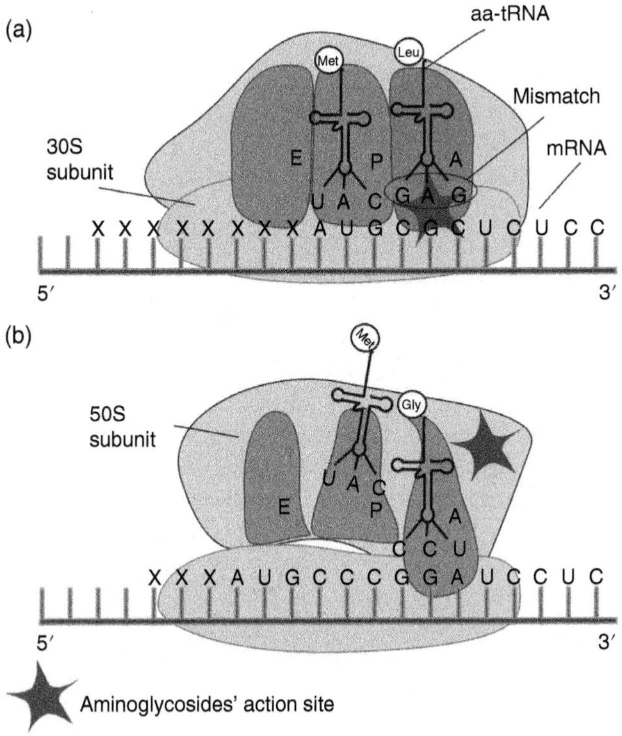

Figure 1.2 Schematic representation of the mode of action of AGAs in the bacterial ribosomes. During initiation, the small (30S) and large (50S) subunits bind, and the start codon of the mRNA is positioned along with the initiator aminoacylated tRNA (aa-tRNA) at the ribosomal P-site. In the elongation phase, an aa-tRNA is then delivered to the A-site, and the formation of a peptide bond between the amino acids attached to the tRNAs in the A- and P-sites occurs. The AGAs target the elongation cycle of protein synthesis by (a) increasing the error rate during the decoding process in the A-site and (b) inducing conformational changes in the 50S subunit, inhibiting the translocation of the tRNA molecules (from the A- and P-sites to the P- and E-sites).

AGAs such as hygromycin B, neomycin, and paromomycin inhibit the translocation of the tRNAs (catalyzed by the elongation factor G). These AGAs interact with the internal loop of h44 from the decoding center of the ribosome and stabilize a "flipped-out" conformation of the rRNA nucleotides A1492 and A1493. These local rearrangements of the highly conserved h44 impair the ribosome's ability to discriminate between cognate and non-cognate tRNAs, thus leading to an increase in the error rate (Carter et al. 2000; Borovinskaya et al. 2008; Wang et al. 2012). In addition to binding to h44 on the small subunit, neomycin has a second binding site within h69 of the 23S rRNA, which interacts with h44 of the 16S rRNA, creating an intersubunit bridge. In this second binding site, neomycin stabilizes a ribosomal intermediate hybrid state, between locked and unlocked state, impairing translocation and ribosome recycling (Wang et al. 2012).

The aberrant proteins produced are thought to interrupt the cell membrane by creating membrane channels, further allowing AGAs to penetrate the cell wall. This rapid uptake of AGAs leads eventually to lethality (Davis et al. 1986).

1.2 Mechanisms of Resistance

Bacterial resistance to AGAs is the result of a combined effect of three main causes (highlighted in the following subsections): the action of aminoglycoside-modifying enzymes (AMEs) (Section 1.2.1), the mutation and modification of AGA's target sequence in the ribosome (Section 1.2.2), and changes in uptake and efflux (Section 1.2.3). However, other causes may have a major influence in pathogenic organisms, such as in *Pseudomonas aeruginosa*: on the one hand, membrane proteases counteract the effect of aminoglycosides by eliminating the misfolded/mistranslated proteins produced in aminoglycoside-affected ribosomes (Kindrachuk et al. 2011; Krahn et al. 2012); on the other hand, biofilm production can impair the access of AGAs to the bacterial cells (Poole 2011). Moreover, *P. aeruginosa* has been found to produce periplasmic cyclic β-glucans, capable of directly binding kanamycin (Sadovskaya et al. 2010).

1.2.1 Aminoglycoside-Modifying Enzymes

The most clinically significant mechanism of aminoglycoside resistance is a consequence of enzymatic modification of specific chemical groups in the antibiotics. Interestingly, it is hypothesized that the enzymes originate from the organisms that produce the aminoglycoside itself (Perry et al. 2014). Over a hundred different enzymes have been identified, and the genes coding for AMEs are usually found on plasmids and transposons. Extensive reviews on the diverse AMEs, organisms, mechanisms of resistance to AGAs, strategies of inhibition, and 3D crystal structures have been done (Ramirez and Tolmasky 2010; Labby and Garneau-Tsodikova 2013; Garneau-Tsodikova and Labby 2016). There are three types of AMEs: (i) ATP (and/or GDP)-dependent aminoglycoside phosphotransferases (APHs), (ii) acetyl-CoA-dependent aminoglycoside acetyltransferases (AACs), and (iii) ATP-dependent aminoglycoside nucleotidyltransferases (ANTs). Figure 1.3 illustrates the action of the three types of enzymes, including the known crystal structures of representative members in complex with aminoglycosides and catalytic cofactors.

As AGAs target specific regions and sequences of RNA with particular orientations, the described enzymatic modifications in the basic structure of

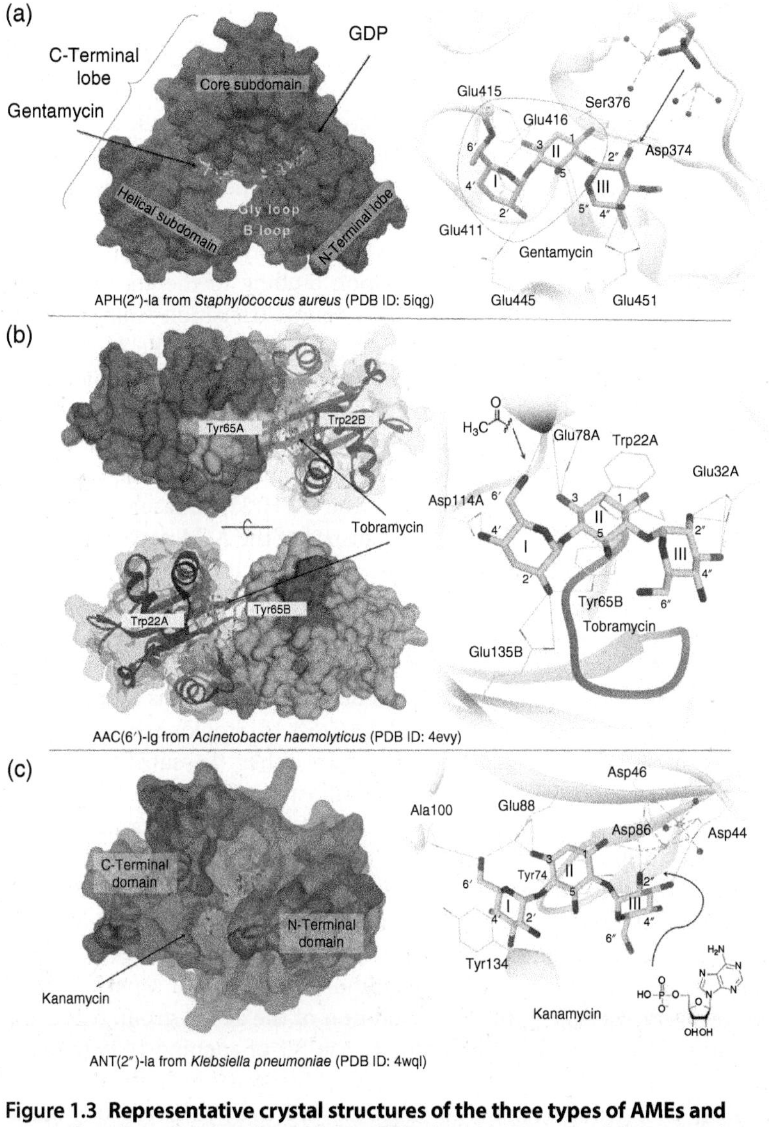

Figure 1.3 Representative crystal structures of the three types of AMEs and respective sites of aminoglycoside modification. (a) Structure of the aminoglycoside 2″-phosphotransferase APH(2″)-Ia from *Staphylococcus aureus*, in complex with gentamycin and GDP (Caldwell et al. 2016). The neamine-like moiety is circumscribed by a dashed blue line. (b) Dimeric structure of aminoglycoside 6′-*N*-acetyltransferase AAC(6′)-Ig from *Acinetobacter haemolyticus* in complex with tobramycin. Chains A and B are depicted in blue and red ribbons, respectively, while the yellow star marks the putative binding site of acetyl-CoA in each monomer (Stogios et al. 2017). (c) Structure of aminoglycoside nucleotidylyltransferase ANT(2″)-Ia from *Klebsiella pneumoniae* in complex with kanamycin (Cox et al. 2015). All protein structures are shown in surface representation. AGAs and GDP are represented as stick model; magnesium atoms and water molecules are represented as green and red spheres, respectively. Hydrogen bonds are represented by thin blue lines. Hydrogen atoms are omitted for simplicity. *Source:* All images were produced with program Chimera (Pettersen et al. 2004). *(See insert for color representation of the figure.)*

aminoglycosides will result in unfavorable steric and/or electrostatic interactions, causing a reduction in the ability of the antibiotic to bind to the RNA.

The phosphorylation of AGA hydroxyl groups upon the action of APHs introduces a negative charge into the molecule, which results in a dramatic change in its ability to bind to the A-site in the ribosome, due to electrostatic repulsion between the modified AGA and the RNA backbone. Examples are APH(2″) kinase enzymes, which phosphorylate the 2-hydroxyl in the C-ring in gentamycin-related AGAs (Figure 1.3a). Upon binding to the antibiotic, the kinase and its GTP cofactor (the phosphate donor) transition from a stable, inactive form to an activated form that triggers modification of the 2″-hydroxyl in 4,6-disubstituted AGAs. This transition, which Caldwell et al. named *catalytic triphosphate switch* (2016), is driven by conformational changes of the enzyme that bring into contact distinct regions of the protein, namely, the distal helical subdomain (of the C-terminal lobe) and the so-called Gly and B loops (of the N-terminal lobe), closing the APH(2″) kinase's active site (Figure 1.3a). In this site, the acceptor 2″-hydroxyl of the AGA stays directly in line to react with the GTP phosphate. The 4,6-disubstituted AGA is held in place, through its neamine-like moiety, by ionic, hydrogen, and ring stacking interactions with the protein residues. The second type of AMEs, acetyl-CoA-dependent AACs, promotes acetylation of amino groups in aminoglycosides (Figure 1.3b), affecting binding to rRNA. As observed by analysis of the crystal structures of AAC(6′)-Ig, AAC(6′)-Ih, and AAC(6′)-Iy (Stogios et al. 2017), the binding site of AAC(6′)-Ig can accommodate either 4,6-substituted (e.g. tobramycin; Figure 1.3b), 4,5-substituted (e.g. ribostamycin), or 4,1-substituted (e.g. amikacin) AGAs, which are held in place by stacking interactions with the central 2-deoxystreptamine ring (ring II). The protein residues that promote the stacking, a tryptophan and a tyrosine in alternate chains (Figure 1.3b), are also involved in the enzyme's dimerization. Further ionic and hydrogen contacts are done with rings I and III and the 6′-amine (to be acetylated). Although not visible in the crystal structures, the authors could assign a pocket to be occupied by acetyl-CoA (Figure 1.3b). Addition of the acetyl group to the 6′-amine position impairs hydrogen binding to the N1 of adenine 1408 (h44) in the 30S subunit of rRNA. The third type of AMEs, the ANTs, catalyzes the transfer of an AMP group from the co-substrate ATP to a hydroxyl group in the AGA (Figure 1.3c). The presence of a bulky group, such as AMP, prevents accommodation of the modified AGA in the 16S rRNA A-site, affected by steric clashes with guanine 1405. According to the work by Cox and colleagues (2015), ANT(2″)-Ia can accommodate and modify 4,6-disubstituted 2-deoxystreptamine-based AGAs, such as gentamycin, tobramycin, and kanamycin (Figure 1.3c), without major conformational changes of the polypeptide chain upon AGA binding. The antibiotic occupies a large cleft (in the frontier between the N- and C-terminal domains) with its central ring (ring II) facing the interior of the protein and rings I and III facing the solvent. Close to the

AGA, the crystal structure also reveals the presence of two pentacoordinated Mg^{2+} ions (Figure 1.3c). Two tyrosine residues form stacking interactions with the antibiotic. The authors hypothesize that the pocket is adequate in size to accommodate a nucleoside triphosphate. While *Mycobacterium tuberculosis* and *Mycobacterium abscessus* natively express AMEs, most pathogenic bacteria produce analogous enzymes that originate from horizontal gene transfer (Garneau-Tsodikova and Labby 2016), and this promiscuity is a major concern. This topic is addressed in Section 1.4.2. Furthermore, different activities can coexist in the same macromolecule, which then behaves as a bifunctional enzyme. An example of characterized bifunctional AME found in clinically relevant *Staphylococcus aureus* strains is the AAC(6′)-Ie/APH(2″)-Ia, which confers resistance to gentamycin kanamycin, and tobramycin (Smith et al. 2014).

1.2.2 Mutation or Modification of Ribosomal Target Sequences

As the main target for the action of aminoglycosides, bacterial ribosomes have a fundamental role in the mechanisms of resistance. These can take place either from specific mutations of the ribosome, mainly at the A-site, or from enzymatic modifications, performed by the 16S ribosomal RNA methyltransferases (RMTases). The latter mechanism is probably a resistance strategy of the aminoglycoside-producing bacteria, as observed for actinomycetes that produce RMTases to methylate their own 16S rRNA (Garneau-Tsodikova and Labby 2016).

In the bacterial ribosome, adenine 1408 and guanine 1491 (*Escherichia coli* numbering) of h44, located at the 16S RNA of the 30S subunit, are two of the residues directly involved in the inhibitory action of aminoglycosides (Carter et al. 2000). As in the human ribosomes these residues are replaced by guanine and adenine, respectively, AGAs cannot bind in the A-site and could be envisaged as ideal antibiotics. However, as a consequence of the probable prokaryotic origin of the mitochondrial ribosomes, residues adenine 1408 and guanine 1491 are present in mitochondrial ribosomes, making them susceptible to the inhibitory action of aminoglycosides (Han et al. 2005). This seems to be the cause for aminoglycoside toxicity in humans (Hobbie et al. 2008). In a different mechanism, translation is still enabled but, flipping of h44 helix (induced by binding of the aminoglycoside), results in mistranslated proteins (McCoy et al. 2011). Although uncommon (because usually lethal), mutations of the *rss* gene, which codes for this part of the rRNA, may confer resistance to aminoglycosides. Examples of mutations in h44 helix were found in resistant strains of *M. tuberculosis* (Georghiou et al. 2012) and *M. abscessus* (Nessar et al. 2011), where key hydrogen bonding interactions of the aminoglycosides with the rRNA are disrupted.

Bacterial resistance mechanisms that rely on modification of ribosomes by 16S RMTases are strongly related to plasmid transference among bacterial species. A list of known exogenously acquired 16S RMTases has been compiled and can be found in Garneau-Tsodikova and Labby (2016). Depending on the methylation position (which prevents the binding of the aminoglycoside), RMTases are classified into two families: (i) the methyl group is added in the N7 position of nucleotide G1405 (enzymes ArmA, RmtA, RmtB, RmtC, RmtD1, RmtD2 RmtE, RmtF, RmtG, and RmtH, which confer resistance to 4,6-disubstituted 2-DOS aminoglycosides), and (ii) the methyl group is added in the N1 position of A1408 (enzyme NpmA, which confers resistance to both 4,5- and 4,6-disubstituted 2-DOS aminoglycosides and to apramycin) (Garneau-Tsodikova and Labby 2016).

1.2.3 Changes in Uptake and Efflux

Other relevant causes of resistance to AGAs are related to the crossing of the bacterial cell wall, required for the action of the AGAs. The accumulation of the AGAs in the cell is against a concentration gradient, so it is predicted to be an energy-requiring process. However, this process has only been studied for two AGAs, streptomycin and gentamycin (Hancock 1981; Hancock et al. 1991). The mechanism of uptake of AGAs by the bacterial cell remains elusive, but several hypotheses have been suggested. It is thought that positively charged AGAs enter the cell by electrostatically interacting with the negatively charged lipopolysaccharides distributed in the outer bacterial membrane, a process denominated as self-promoted uptake (Taber et al. 1987; Hancock et al. 1991). These hypotheses comprise changes in membrane composition, overexpression of efflux pumps, and changes in membrane permeability. Mechanisms of resistance to this crossing involve modification of lipopolysaccharides of the outer membrane or downregulation of porins. A less negative outer membrane will have decreased affinity for aminoglycosides, and this occurs, most commonly, by incorporation of positively charged 4-amino-4-deoxy-L-arabinose in the lipopolysaccharide layer (Macfarlane et al. 2000; Kwon and Lu 2006; Fernandez et al. 2010), but incorporation of phosphoethanolamine was also reported (Nowicki et al. 2015). The presence of diacyl phosphatidylinositol dimannoside in mycobacteria is also proposed to decrease the membrane fluidity and, consequently, decrease its permeability (Bansal-Mutalik and Nikaido 2014).

Concerning porins, some mycobacteria have approximately 50-fold less MspA-like porins than other Gram-negative bacteria, which may account for a low permeability of the mycobacterial outer membrane (Jarlier 1994; Niederweis 2003). Mutations that lead to a lower number of porins may be part of a resistance strategy, but more studies are needed to support this.

In a different perspective, an alternative resistance mechanism is the removal of AGAs from the interior of bacteria by efflux pumps. The AcrAD-TolC type is a multidrug transporter and the main AGA efflux pump present in several

Gram-negative bacteria, extensively reviewed in Li et al. (2015). A well-studied example is the role of MexXY-OprM efflux pump in the resistance of *P. aeruginosa* to AGAs (Morita et al. 2012). In addition, other AGA efflux pumps were identified in Gram-negative (VcmB in *Vibrio cholerae*) (Begum et al. 2005), Gram-positive (LmrA in *Lactobacillus lactis*) (Poelarends et al. 2002), and mycobacteria (Rv1258c in *M. tuberculosis*) (Balganesh et al. 2012). In a recent work, overexpression of MexXY in *P. aeruginosa* was associated with reduced aminoglycoside susceptibility and development of chronic lung infections in patients with multidrug-resistant cystic fibrosis (Singh et al. 2017). In Gram-negative bacteria, aminoglycosides are expelled mainly by an intrinsic 3-component AcrAD-TolC-type efflux pump (classified in the resistance-nodulation-division (RND) family of efflux pumps), which comprises the drug–proton membrane antiporter AcrD, the membrane fusion protein AcrA, and the outer membrane component TolC (Nikaido 2011). Although this is the main type of transporter, AGAs may be a substrate for efflux pumps of different families (extensively reviewed in Li et al. (2015)) The crystal structure of AcrD, shown to capture aminoglycosides from both the periplasm and the cytoplasm (Aires and Nikaido 2005), was reported (unpublished, PDB accession code: 4r86), however, in the absence of an aminoglycoside substrate. A cryo-electron microscopy structure of the AcrAB-TolC multidrug efflux pump from *E. coli*, in resting and drug transport states, has been recently reported, increasing the knowledge on the assembly and operation of the AcrAB-TolC pump (Wang et al. 2017). Comparison with the known structure of the RND multidrug transporter AcrB (e.g. PDB accession code: 5eno (Sjuts et al. 2016)) revealed that large periplasmic loops of AcrD seem to play a role in AcrD's selectivity for aminoglycosides (Elkins and Nikaido 2003). In a recent work (Ramaswamy et al. 2017), apart from the flexible protein loops, selectivity differences may be attributed to a higher lipophilicity of AcrB. Furthermore, as described by Nakashima and colleagues (2011) for the 3D structures of AcrB bound to the high-molecular-mass drugs rifampicin and erythromycin, these transporters enclose two substrate binding pockets: a proximal multisite binding pocket (accommodating high-molecular-mass drugs) and a phenylalanine cluster region (distal pocket, accommodating low-molecular-mass drugs), separated by a loop. According to the authors, the remarkably broad substrate recognition of AcrB is due to the existence of these two distinct high-volume multisite pockets. A mutagenesis study demonstrated that the MexY transporter from *P. aeruginosa* is capable of binding aminoglycosides (Lau et al. 2014), but comparison with a homology model for MexY produced from the 3D structures of AcrB and MexB (PDB accession code: 3w9j) explains that although similar, MexY cannot be inhibited by pyridopyrimidine derivatives. This is due to the presence of a tryptophan residue in MexY, instead of the phenylalanine in AcrB (Nakashima et al. 2013). Structural characterization of AcrD and MexY or many other efflux pumps known to bind aminoglycosides is needed. Current research aims at the design of aminoglycosides capable of evading efflux pumps and efflux pump inhibitors (Opperman and Nguyen 2015; Venter et al. 2015).

1.3 Development of New AGAs: The Potential of Glycomics

Outstanding problems of AGAs are related with their efficacy due to increasing antibiotic-resistant bacteria and aminoglycoside inherent toxicity. The rational design for novel AGAs will have to consider (i) the ability of the drug to bind strongly and specifically to the bacterial therapeutic target and (ii) the lack or weak binding to resistance- and toxicity-causing enzymes. The available structural data involving aminoglycoside-bound rRNA molecules and details of the resistance mechanisms with insights obtained from X-ray structures of AMEs has made AGAs' rational design possible. This topic has been extensively reviewed (Thamban Chandrika and Garneau-Tsodikova 2018). A relevant example is plazomicin (ACHN-490, Achaogen Inc.), a semisynthetic *N*-acylated AGA of the 4,6-disubstituted deoxystreptamine (4,6-DOS) type (Aggen et al. 2010), which is in clinical development as a new antibacterial agent (Table 1.1). The evidence shows that ACHN-490 targets the bacterial ribosome with high binding affinity and could be resistant to most of aminoglycoside resistance-causing enzymes (Endimiani et al. 2009; Aggen et al. 2010; Armstrong and Miller 2010; Labby and Garneau-Tsodikova 2013; López-Diaz et al. 2016).

In recent decades, the study of carbohydrates (or glycans) and their derivatives has emerged as a challenging area providing new avenues of research at the interface of chemistry and biology. The advances in glycomic approaches have helped to elucidate biological functions of carbohydrates. Development of methods for chemical or chemoenzymatic synthesis of carbohydrates (Boltje et al. 2009; Muthana et al. 2009) and for their analysis by NMR (Kato and Peters 2017) or mass spectrometry (Kailemia et al. 2014) has contributed to these advances, as well as the development of high-throughput methods for interaction studies, such as carbohydrate microarrays (Rillahan and Paulson 2011; Palma et al. 2014). The convergence of these methodologies is leading glycomics at the forefront of research to design glycan mimetics, offering the opportunity to develop drugs, including aminoglycosides targeting diverse structures of RNA (Thamban Chandrika and Garneau-Tsodikova 2018 and references therein). In the subsections below, carbohydrate chemistry and microarray technology are briefly highlighted as powerful approaches toward development of new AGAs and analysis of aminoglycoside target interactions.

1.3.1 Exploitation of Carbohydrate Chemistry to Study Structure–Activity Relationship of Aminoglycoside Derivatives

As introduced in Section 1.1, aminoglycosides are small molecule drugs that contain carbohydrates as part of their core structure and possess several amino and hydroxyl functionalities (Figure 1.1). The chemical modification using carbohydrate chemistry has been applied to find the SAR and to design new molecules (Li and Tom Chang 2006; Guo et al. 2012; Matsushita et al. 2015; Thamban Chandrika and Garneau-Tsodikova 2018). The evidence from the mechanisms of molecular action and bacterial drug resistance showed that (i) the effectiveness of antibiotics is related to the amount of charge and the number of modifiable sites (e.g. AGAs with a higher number of amino groups can bind more strongly to RNA); (ii) ring I is the main target of inactivating bacterial enzymes, and ring III seems to be less sensitive to structural modifications; and (iii) the substitution of C-3′ or C-4′ hydroxyl groups can protect the antibiotics from APH and ANT enzymes, keeping the AGA activity.

Thus, the chemical strategies to address resistance mechanisms can include modification of the AGA core and insertion of substituents at different positions of AGA scaffolds, protecting the enzymatically modified sites and changing the hydrophobicity of the molecule; generation of new AGAs by glycodiversification and chemoenzymatic reactions; and synthesis of inhibitors for the modifying enzymes. The preparation of the AGA derivatives can be carried out using the AGA natural scaffold or by synthesis of AGA analogues. Here, we will highlight some examples; a more extensive review on these can be found in Guo et al. (2012) and Thamban Chandrika and Garneau-Tsodikova (2018). One example was the synthesis of kanamycin derivatives modified at C4′ position of ring I of the neamine moiety with different substituents to protect against the action of ANT(4′), APH(4′), or APH(3′) (Yan et al. 2011). Some of these derivatives exhibited antibiotic activity not only against nonresistant bacteria but also drug-resistant bacteria that express AMEs (e.g. those containing a nitrogen atom at the end and three carbon–carbon bonds between the end nitrogen atom and the carbonyl group). A second example was the synthesis of amide-linked aminoglycoside-CoA bi-substrates as inhibitors of AMEs (Gao et al. 2008). Most of these bi-substrates were competitive inhibitors of AAC(6′)-Ii in the nanomolar range. The crystal structure of the complexes revealed that they are good mimics of the enzymatic reaction intermediates. In a recent study, glycodiversification of paromomycin was carried out, leading to synthesis of a series of derivatives modified at the 4′-*O*-glycoside with different carbohydrates (Matsushita et al. 2015). The target selectivity to inhibit protein synthesis (i.e. inhibition of the bacterial ribosome over eukaryotic mitochondrial and cytosolic ribosomes with reduced toxicity) was greater for the equatorial than for the axial pyranosides and greater for the D-pentopyranosides than for the L-pentopyranosides and D-hexopyranosides. As an example, the 4′-*O*-β-D-xylopyranosyl paromomycin showed antibacterial ribosomal activity comparable with that of paromomycin, but was significantly more selective with reduced affinity for the cytosolic ribosome and drug-induced ototoxicity. Importantly, the 4′-*O*-glycosylated paromomycin derivatives were not susceptible to modification by ANT(4′,4″) AME.

These approaches highlighted that single synergistic modifications are promising strategies to be used in new generations of AGAs to reduce toxicity and overcome resistance mechanisms.

1.3.2 Aminoglycoside Microarrays to Screen Interactions of Antibiotics with RNAs and Proteins

The study of protein–carbohydrate molecular interactions has been revolutionized by the advent of the high-throughput carbohydrate microarray technology (Rillahan and Paulson 2011; Palma et al. 2014). Carbohydrate microarrays are composed of diverse carbohydrate probe libraries printed onto a microarray solid surface as spatially oriented microspots. This miniaturization feature of the microarrays enables the screening of a wide range of interactions and generation of a large amount of information on different recognition systems. In addition, the multivalent display of arrayed probes enables the microarray to detect with high-sensitivity carbohydrate–protein interactions.

Early on the advent of carbohydrate microarray technology, Seeberger and colleagues (Disney and Seeberger 2004; Disney et al. 2004) adapted the chemistry of preparing carbohydrate microarrays to covalently immobilize unmodified aminoglycosides, through their amino groups, onto functionalized surfaces, such as *N*-hydroxysuccinimide (NHS)-activated glass slides, through an amide bond, or aldehyde-coated glass slides, through a carbon–nitrogen double bond (Figure 1.4a). In this proof-of-concept study, the authors proved the potential of the microarrays derived from 4,5- and 4,6-linked 2-DOS derivatives to screen interactions with mimics of aminoglycoside binding sites in the ribosome (bacterial and human rRNA A-sites) and with model proteins to study aminoglycoside toxicity (DNA polymerase and phospholipase C). These studies opened new avenues for the rapid screening and identification of new compounds that bind RNA tightly and exhibit decreased affinity toward resistance- and toxicity-causing proteins (Barrett et al. 2008; Disney et al. 2008).

Although attractive, a major limitation of these microarray strategies was the nonspecific way the aminoglycosides were immobilized onto the surface, which could lead to erroneous interpretation of the binding affinities (Disney and Seeberger 2004). New libraries immobilized through functional groups not predicted to interact with the target were an important development (Disney and Childs-Disney 2007; Barrett et al. 2008; Disney 2012). For example, Disney and colleagues (Barrett et al. 2008) modified aminoglycosides with an azido group at the hydroxyl group, enabling immobilization onto alkyne-displaying

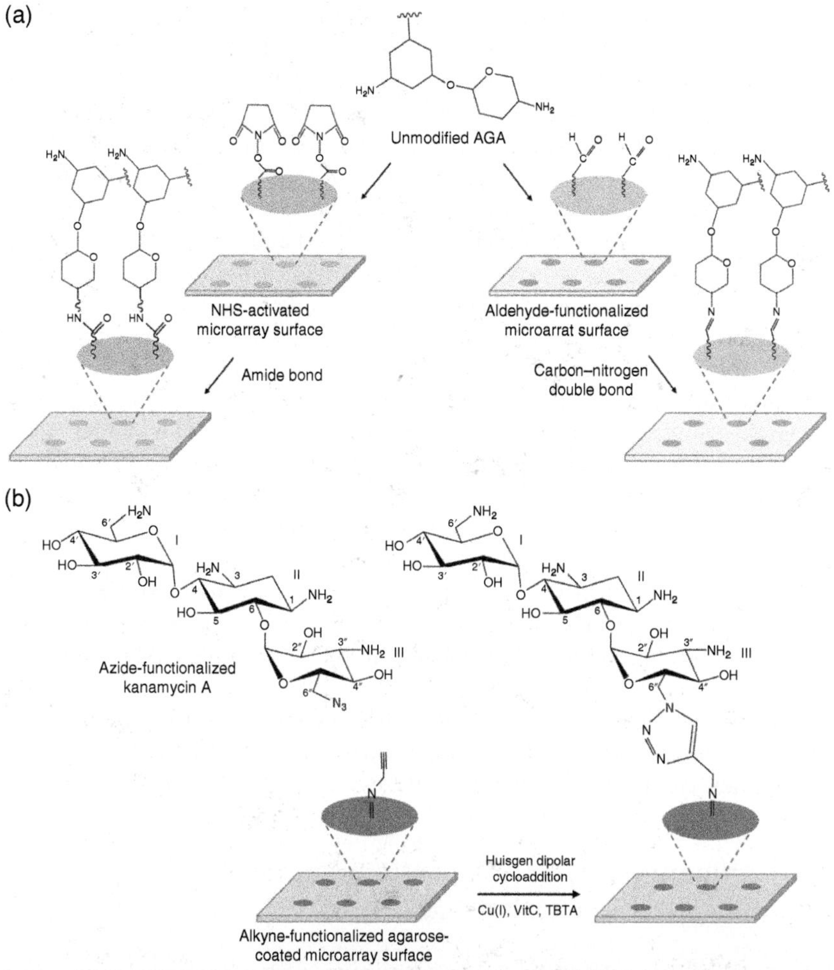

Figure 1.4 Examples of chemical approaches to prepare covalent aminoglycoside microarrays for studies of RNA and protein target interactions. (a) Immobilization of unmodified aminoglycosides through an amide bond onto *N*-hydroxysuccinimide (NHS)-activated glass slides or through a carbon–nitrogen double bond onto aldehyde-functionalized glass slides; a general depiction of an AGA is shown. (b) Immobilization of azide-functionalized kanamycin A at the C-6′ hydroxyl group onto an alkyne-functionalized agarose-coated microarray surface using the Huisgen dipolar cycloaddition reaction.

agarose-coated surfaces by a Huisgen dipolar cycloaddition reaction (Figure 1.4b). The same group also modified aminoglycosides with an alkyne group and constructed microarrays onto azido-displaying surfaces (Disney and Childs-Disney 2007). These strategies were used to develop a high-throughput and sensitive microarray-based method to study the modification of AGAs by acetyltransferases (Barrett et al. 2008; Tsitovich et al. 2010) and directly monitoring antibiotic resistance (Disney and Childs-Disney 2007).

Not only the binding to the target but also the specificity, defined as the ability to discriminate between different RNA sequences, is critical for the development of new therapeutics. Several studies determined the specificity of rRNA interaction with different aminoglycosides. An advantage of the microarray approach is that the specificities can be estimated in a high-throughput manner by comparing relative intensities of each aminoglycoside binding to different RNA sequences. Importantly, the covalent aminoglycoside microarrays depicted in Figure 1.4b provided the basis for development, by Disney and colleagues, of a microarray-based two-dimensional combinatorial screening (2DCS) method for interactions of AGAs with RNA (Childs-Disney et al. 2007; Disney et al. 2008). In this method, ligand and RNA motifs are screened in parallel to identify selective interactions that can be used to target RNA. Later, this method was combined with a statistical analysis scoring method, designated structure–activity relationship through sequencing (StARTS) (Velagapudi and Disney 2013). This method scores RNA motif–ligand partners selected by 2DCS to map RNA/ligand interactions. The integration of the microarray-based 2DCS and computational StARTS methods led to the development of the Inforna technology, which has been applied in the identification of small molecules that bind to precursor miRNAs (pre-miRNAs) (Velagapudi et al. 2014).

With the development of microarray technology and the combination of chemoenzymatic synthesis, computational chemistry, and SAR-derived insight, new AGAs are anticipated to address effective antibacterial drugs.

1.4 Influence of the Human Microbiome in Aminoglycoside Resistance

In the quest for novel approaches to AGA usage, and to overcome factors that diminish the effectiveness of aminoglycoside therapy, the human gut microbiome has gained importance as it constitutes a reservoir of resistance genes and has a major potential to modulate the action of antibiotics in the human organism (Francino 2016; Langdon et al. 2016).

The human microbiota is a large microbial community primarily found on the skin and in gastrointestinal (GI), respiratory, and urogenital tracts that corresponds to 50% of the human body cells (Human Microbiome Project Consortium et al. 2012; Sender et al. 2016). Thus the human microbiome may be defined as the collection of all microbiota strains, genomes, metabolites, and host's by-products (Marchesi and Ravel 2015). This community carries a gene pool 150 times larger than the human's, including unique genes that expand our own metabolic abilities (Qin et al. 2010). The complex molecular crosstalk with the human cells is essential to maintain the nutritional balance, stimulate gut cell maturation and proliferation, protect against enteric or opportunistic pathogens, and modulate the host immune system.

It is well established that changes in our diet alter the composition and metabolic output of the human gut microbiome (Sonnenburg and Bäckhed 2016), and it is not surprising that broad-spectrum antibiotic uptake also induces collateral damages in this community, altering the gut ecosystem (Dethlefsen and Relman 2011). Either diet- or antibiotic-induced microbiome perturbations can result in transient or permanent changes, spanning from days up to years. This imbalance in a normal healthy microbiome composition, termed dysbiosis, has been linked to many pathologies including inflammatory bowel disease (IBD), diabetes, obesity, colon cancer, allergies, and infectious diseases (Honda and Littman 2016; Thaiss et al. 2016).

The widespread antibiotic (mis)use is creating a heavy selective pressure in the gut microbiome, forcing the propagation of super-resistant strains. In addition, the correlation between antibiotics consumption and the growing onset of several pathologies from nosocomial resistant infections to chronic and autoimmune conditions, not only in adults but also in infants of increasingly low ages, is raising concern (Francino 2016; Langdon et al. 2016).

Due to the decreased usage of AGAs, their impact on human microbiome has not been as explored as for other antibiotic classes (Rafii et al. 2008; Langdon et al. 2016) (information on AGAs is summarized in Table 1.1). However, the renewed interest in the AGAs and their known role in antibiotic resistance makes it imperative to assess their specific impact and raise awareness to the need of more AGA-targeted studies. Here, we revise the main features of what is known for the general impact of antibiotics, establishing comparisons and particularizing for the AGAs whenever possible. We give an overview of the effect of altered gut microbiome function and composition and of antibiotic-induced alterations on host health and homeostasis and in the accumulation and spread of resistances, which is of major relevance concerning the usage of AGAs. Alternative strategies to modulate human microbiome responses to antibiotics are also discussed.

1.4.1 The Effect of Antibiotic-Induced Alterations

In the crowded gut environment, commensals, opportunists, and pathogens continuously struggle and compete for resources and attachment sites (Human Microbiome Project Consortium et al. 2012). In return for food and shelter, bacteria provide many advantages for the host by producing important metabolites and modulating the immune system. Alternative glycan sources used as microbial substrates will result in the production of short-chain fatty acids (SCFAs) such as butyrate, propionate, or acetate (Hamer et al. 2008). Human nutrition and glycan uptake are factors shaping the microbiota (Sonnenburg and Bäckhed 2016). In the same way, antibiotic uptake can disturb the microbiome function and composition, resulting in an irreversible loss of taxonomic and metabolic diversity (Dethlefsen and Relman 2011). Owing to their broad bactericidal effect, different antibiotics, particularly broad-spectrum antibiotics, will have different implications for the human microbiome by eliminating alternative bacterial strains (Rafii et al. 2008; Langdon et al. 2016). The extent of the damage caused by antibiotics is also influenced by the pharmacokinetics of the drug, time of the treatment, and the administered dose. Repeated antibiotic uptake or prolonged treatments will also weaken the gut ecosystem and render the microbiome more resilient to restore the initial composition and taxonomic diversity (Dethlefsen and Relman 2011). This dysbiosis will alter gut colonization resistance and metabolite production, impacting not only on the microbial community but also on the host.

With the emerging field of human microbiome, many diseases are being linked to an imbalance in normal microbiota composition, due to antibiotic exposure that will compromise host homeostasis and tolerance, deregulate the metabolism, and increase the susceptibility to infections. In this subsection, we give some examples and try to highlight studies on the impact of AGAs (Table 1.1). Extended information concerning these can be found in the following reviews (Francino 2016; Sonnenburg and Bäckhed 2016; Kim et al. 2017).

1.4.1.1 Immunological Diseases: Compromised Host Homeostasis and Tolerance

IBD, a group of inflammatory disorders that include Crohn's and necrotizing enterocolitis (NEC), have a higher incidence rate in children that have been exposed to broad-spectrum antibiotics before five years of age (i.e. gentamycin and ampicillin) (Hildebrand et al. 2008; Greenwood et al. 2014) (Table 1.1). During antibiotic-provoked dysbiosis, the levels of SCFAs drop, promoting GI inflammation. Crohn's disease is associated with diversity loss and specific microbial changes, namely, increased abundance of *Enterobacteriaceae* and reduction of *Faecalibacterium prausnitzii*, while NEC in newborns is linked to a low abundance of *Bifidobacterium* spp.

The uptake of broad-spectrum antibiotics in early life, a critical period for immune system development and maturation, has also been implicated in the development of atopic diseases. In the case of asthma, a lower abundance of a specific bacterial cohort – *Faecalibacterium, Lachnospira, Veillonella*, and *Rothia* genera – was observed after treatment with the glycopeptide antibiotic vancomycin, but not with AGA streptomycin (Russell et al. 2012; Arrieta et al. 2015). As a result of microbiome depletion, SCFA low levels caused an increase in immunoglobulin E (IgE) and invariant natural killer T (iNKT)-expressing cells and a decrease in regulatory T cell (Treg) levels, promoting allergic inflammation and systemic intolerance to a large spectrum of environmental antigens.

Equally important, the risk of developing type 1 diabetes, an insulin-dependent autoimmune disease, has been associated with dysbiosis due to repeated use of broad-spectrum antibiotics (i.e. streptomycin, polymyxin E and ampicillin, or vancomycin) (Boursi et al. 2015; Candon et al. 2015) (Table 1.1).

1.4.1.2 Metabolic Diseases: Deregulated Metabolism

The onset of many metabolic diseases such as obesity has been linked to an early antibiotic exposure in mice and man (Ajslev et al. 2011; Cho et al. 2012). Additionally, antibiotic use seems to be a risk factor for advancing from obesity to a metabolic syndrome (cohort of cardiovascular disease, fatty liver disease, steatohepatitis, and type 2 diabetes) (Emanuela et al. 2012).

In a project led by Kroemer and Zitvogel, researchers discovered that antibiotics inhibited the clinical benefit of a novel immunotherapy (immune checkpoint inhibitors [ICI]) in patients with advanced cancer (Routy et al. 2018). In fact, the administration of ampicillin, colistin and streptomycin in mice and beta-lactam inhibitors, fluoroquinolones, or macrolides in humans suppressed ICI effect by reducing *Akkermansia muciniphila* abundance in the microbiome (Routy et al. 2018) (Table 1.1). This is in concordance with the recently recognized role of the microbiome in tumorigenesis and cancer progression (Roy and Trinchieri 2017).

1.4.1.3 Infectious Diseases: Increased Susceptibility to Pathogens

The lack of microbial competition in the gut lumen as a result of antibiotic-provoked dysbiosis can lead to overgrowth and colonization by opportunists or pathogens such as *Clostridium difficile, S. aureus*, or *Klebsiella pneumonia* (Song et al. 2008; Buffie et al. 2012). These pathogenic bacteria are usually the cause of antibiotic-associated diarrhea (AAD), a major concern in nosocomial infections. *C. difficile*, a spore-forming bacterium, is the leading healthcare-acquired infection, causing a recurrent chronic infection and often lethal pseudomembranous colitis (Buffie et al. 2012). Likewise, susceptibility to *Salmonella enterica* serovar Typhimurium increased in a mouse model after streptomycin and vancomycin use (Sekirov et al. 2008) (Table 1.1). After antibiotic treatment, other pathogenic and opportunistic bacteria, such as resistant *Enterococcus* strains, can disrupt the gut mucosa, leading to bloodstream infections or sepsis, particularly in immunocompromised patients and children (Ubeda et al. 2010; Mai et al. 2013).

Additionally, the lack of microbiome signaling (microbial-associated molecular patterns [MAMPS]) following broad-spectrum antibiotics prescription (including gentamycin) can also reduce the innate and adaptive immune responses to microorganisms, making the host more vulnerable to other infections, e.g. *Candida albicans* (Shankar et al. 2015) (Table 1.1).

1.4.2 A Reservoir of Antibiotic Resistance

The antibiotic uptake impacts the microbiome by promoting the selection of resistant organisms and by increasing the mutation rate and exchange of resistance genes. The human gut microbiome is a major reservoir of antibiotic resistance (Marshall et al. 2009; Sommer et al. 2009; Clemente et al. 2015; Moore et al. 2015). In this promiscuous environment, commensals, opportunists, and pathogens are in proximity, favoring the horizontal transfer of antibiotic resistance genes present in plasmids and mobile genetic elements (e.g. transposons and insertion sequences) (Smillie et al. 2011; Brito et al. 2016).

Among the different antibiotic classes used in therapeutics, AGAs do not seem to be the major cause of bacterial depletion. Treatment with streptomycin even altered the gut microbiome of a cystic fibrosis mouse model improving airway hyperresponsiveness (Bazett et al. 2016). In fact, many gut bacteria are naturally resistant to AGAs, e.g. highly abundant *Bacteroidetes*. Yet, AGAs have a central role in the emergence and dissemination of many mechanisms of antibiotic resistance. As mentioned in Section 1.2, bacterial resistance to AGAs is mediated by AMEs and 16S RMTases. Antibiotic resistance genes encoding these enzymes populate the human microbiome and the resistome – i.e. the array of mobile genomic elements encoding antibiotic resistance mechanisms (Schaik 2015). For example, in clinically relevant Gram-positive *Staphylococcus* and *Enterococcus* spp., the most frequently identified gene encodes the bifunctional enzyme AAC(6′)-Ie/APH(2″)-Ia that confers resistance to gentamycin, kanamycin, and tobramycin (Smith et al. 2014). In addition, over fifty different variants of the gene codifying acetyltransferases AAC(6′)-Ib are found in mobile genetic elements, conferring broad-spectrum aminoglycoside resistance in many Gram-negative pathogens. A peculiar case is the variant encoding the AAC(6′)-Ib-cr that targets both AGAs and beta-lactams (Ramirez et al. 2013). Recently, the chromosomally encoded enzymes 16S RMTase EfmM and acetyltransferase AAC(6′)-Ii, intrinsically present in actinomycetes and Gram-negative bacteria, were found to confer resistance to kanamycin and tobramycin in *Enterococcus* spp. (Galimand et al. 2011). Other genes encoding relevant AMEs (Figure 1.3, Section 1.2.1) were found in *S. aureus* (APH(2″)-Ia) (Caldwell et al. 2016), *Acinetobacter haemolyticus* (AAC(6′)-Ig) (Stogios et al. 2017), and *K. pneumoniae* (ANT(2″)-Ia) (Cox et al. 2015).

Over the last decades, the almost constant and undue antibiotic uptake through pharmaceuticals or food and feed industries has created a selective pressure not only in the human microbiome but also in animals and soil-dwelling strains. Mechanisms of resistance and mutation are passing between different zootrophic levels and back to man through the food chain/soil/water/etc. Even newborns and isolated human populations that have never been administrated with antibiotics present high levels of resistance genes (Clemente et al. 2015; Moore et al. 2015).

The ArmA and RmtB are the main 16S RMTases conferring resistance to AGAs in bacteria, and genes encoding these enzymes are alarmingly spreading from human to animals around the world (Wachino and Arakawa 2012). If co-located with a carbapenemase gene on the same mobile genetic element, ArmA can even cause pandrug resistance (Milan et al. 2016). Also of concern is the identification of resistance RMTases in foodborne bacteria, as food may become a relevant route of resistance transmission (Granier et al. 2011). In addition, the ArmA and RmtC methyltransferases were reported to cause resistance in *Enterobacteriaceae* strains, even to newly developed plazomicin AGA (Zhanel et al. 2012).

1.4.3 Strategies to Modulate the Human Microbiome

Understanding the complex interactions occurring in the gut and how we can manipulate these to benefit human health is one of the key challenges of our time. Personalized medicine solutions with the combined use of probiotics and prebiotics can be key to circumvent antibiotic resistance, including AGAs, and the last resort for many human diseases.

Probiotics have been defined as "live microorganisms that, when administered in adequate amounts, confer a health benefit on the host" (Hill et al. 2014). The use of probiotics or probiotic bacteria to restore the microbiota composition and avoid disease after antibiotic treatment is a well-established approach (Liévin-Le Moal and Servin 2014). For example, *Lactobacillus* spp. have a protective effect against AAD (*L. reuteri*) (Cimperman et al. 2011), while *Bifidobacterium* strains reduce the risk of infections in infancy (Taipale et al. 2016). *Escherichia coli* Nissle 1917 (Mutaflor) is also a known probiotic to fight antibiotic-resistant pathogens (Wassenaar 2016). The development of genetically engineered probiotic lactic acid bacteria to target and compete with specific pathogens or to produce antimicrobial peptides is a promising approach that can drastically reduce the spread of resistance mechanisms (Hwang et al. 2014). However, caution must be taken as probiotic cultures have the potential to carry antibiotic resistance genes themselves, including aminoglycoside resistance genes (Zheng et al. 2017).

Another way to restore and enhance the human gut microbiome is by controlling or modifying the carbohydrate uptake – via controlled diet or supplementation with prebiotics. Prebiotics are substrates that are "selectively utilized by host microorganisms, conferring a health benefit" (Gibson et al. 2017). These compounds are usually nondigestible carbohydrates or fibers, fermented by colonic bacteria to produce beneficial SCFAs (Hamer et al. 2008). High fiber diets and inulin supplementation stimulate the growth of *Bifidobacterium* and reduce the risk of obesity, although the exact molecular mechanism is not clear (Weitkunat et al. 2017; Schroeder et al. 2018). Furthermore, prebiotics can be utilized to alter the side effects of antibiotics. In a recent study, Johnson and colleagues demonstrated that supplementation with pectin and inulin could ameliorate the effect of gentamycin and ampicillin uptake in an *in vitro batch* culture of human gut samples (Johnson et al. 2015). Knowing the substrate specificity of each microbiome strain and how the glycans are recognized and shared by the community is central to design novel prebiotics. This will open endless possibilities to recover from dysbiosis and fight infections by increasing the growth of specific competitors or simply by limiting the growth of targeted pathogens (nutrient depletion).

Combined therapeutics with prebiotics and fecal microbiota therapy (FMT) have shown some promising results in treating or ameliorating *C. difficile* chronic infections and IBD (Colman and Rubin 2014; Moayyedi et al. 2015). Despite the intrinsic resistance to AGAs, *C. difficile* is a reservoir of mobile elements and genetic mutations, hence a menace to increasing antibiotic resistance (Johanesen et al. 2015). The FMT is a crude form of probiotic administration enabling delivery of unknown strains that are thought to suppress *C. difficile* infection by competing for the same niche. The study of FMT can give rise to tailored microbial preparations of defined composition to treat any dysbiosis scenario.

Other approaches to modulate the human gut microbiome include the production of anti-quorum sensing drugs to disrupt bacterial communication and consequently virulence in *P. aeruginosa* (Starkey et al. 2014); antitoxin or small molecule inhibitors of type II–IV secretion systems, the major virulence systems in *Bacillus anthracis, Shigella flexneri,* or *Brucella* spp. (Felise et al. 2008; Smith et al. 2012); and anti-biofilm compounds to block biofilm formation, a mechanism of resistance to AGAs (Poole 2011).

1.5 Conclusions and Outlook

In this chapter, an overview on AGAs was made highlighting important aspects related to their historical clinical use as effective antibacterial drugs and their structural motifs and the action of AGA-modifying enzymes and other mechanisms causing bacterial resistance. In recent decades, there has been a renewed interest from the scientific community and the pharmaceutical industry on this well-studied broad-spectrum class of antibiotics. It is also worth mentioning that the study of AGAs as ribosome-targeting antibiotics has been crucial to the elucidation of the bacterial ribosome and to understand at molecular level the fidelity and efficiency of the translation process. The knowledge gained from the structural studies, concerning their mode of action and on resistance mechanisms applied by bacteria, has hinted into new strategies to fine-tune the action of AGAs and their pharmacological profiles. In a recent review by Thamban Chandrika and Garneau-Tsodikova (2018), the authors provide a comprehensive analysis on strategies (synthetic and chemoenzymatic) developed by different groups for generation of novel aminoglycosides and their derivatives. The development of plazomicin by Achaogen as a potential selective, nontoxic, and bacterial resistant antimicrobial drug is an example of these efforts. It is anticipated that the rational design based on SAR studies and also on analysis of the affinity and specificity of AGA–target interactions will culminate in the generation of new effective "third-generation" AGAs. Here, developments made on carbohydrate chemistry and high-throughput microarray technologies offer promising approaches to be used in new generations of AGAs to reduce toxicity and for their effectiveness against rapidly emerging resistant bacterial strains.

This chapter focused on the use of aminoglycosides as broad-spectrum antibiotics against pathogenic bacteria. But the knowledge on these molecules has recently expanded to other applications, including as antifungal, antiprotozoal, and genetic regulating compounds due to the ability of binding different RNA targets (Fosso et al. 2014). Aminoglycosides are also used for growth promotion in animal and crop husbandry, and despite the strict regulation in the EU and the United States for some aminoglycosides, these molecules are uncontrollably used in developing countries (Mitema 2010). Therefore, the use of aminoglycosides offers a highly important challenge for different applications. In this context, the optimization of the use of these and prevention of factors that diminish their therapy effectiveness, such as toxicity and bacterial resistance, should be done by taking into consideration the human gut microbiome. The microbiome constitutes a reservoir of resistance genes (the resistome) and has a major potential to modulate the action of aminoglycosides in the human organism. In the future, multidisciplinary approaches and combined micro- and nanotechnologies (Biteen et al. 2016), such as in genomics, metabolomics, and glycomics, should be applied to study the microbiome as an ecological niche and identify the different interactions occurring herein – bacteria–bacteria, bacteria–nutrients, and bacteria–host. This knowledge can be applied to construct predictive models to assess the deleterious effect of each antibiotic and help simulate novel or optimized drugs.

Acknowledgments

The authors acknowledge the Foundation for Science and Technology of the Ministry of Science, Technology and Higher Education (FCT-MCTES) of Portugal through financed projects PTDC/QUI-QUI/112537/2009, PTDC/BIA-MIC/5947/2014, and PTDC/BBB-BEP/0869/2014 and grants PD/BD/105727/2014 to VGC and IF IF/00033/2012 to ASP. The Unidade de Ciências Biomoleculares Aplicadas (UCIBIO) is financed by national funds from FCT-MCTES (UID/Multi/04378/2013) and co-financed by the European Regional Development Fund under the PT2020 Partnership Agreement (POCI-01-0145-FEDER-007728).

References

Aggen, J.B., Armstrong, E.S., Goldblum, A.A. et al. (2010). Synthesis and spectrum of the neoglycoside ACHN-490. *Antimicrobial Agents and Chemotherapy* 54 (11): 4636–4642.

Aires, J.R. and Nikaido, H. (2005). Aminoglycosides are captured from both periplasm and cytoplasm by the AcrD multidrug efflux transporter of *Escherichia coli*. *Journal of Bacteriology* 187 (6): 1923–1929.

Ajslev, T.A., Andersen, C.S., Gamborg, M. et al. (2011). Childhood overweight after establishment of the gut microbiota: the role of delivery mode, pre-pregnancy weight and early administration of antibiotics. *International Journal of Obesity* 35 (4): 522–529.

Antunes, L.C.M., Han, J., Ferreira, R.B.R. et al. (2011). Effect of antibiotic treatment on the intestinal metabolome. *Antimicrobial Agents and Chemotherapy* 55 (4): 1494–1503.

Armstrong, E.S. and Miller, G.H. (2010). Combating evolution with intelligent design: the neoglycoside ACHN-490. *Current Opinion in Microbiology* 13 (5): 565–573.

Arrieta, M.-C., Stiemsma, L.T., Dimitriu, P.A. et al. (2015). Early infancy microbial and metabolic alterations affect risk of childhood asthma. *Science Translational Medicine* 7 (307): –307ra152.

Balganesh, M., Dinesh, N., Sharma, S. et al. (2012). Efflux pumps of *Mycobacterium tuberculosis* Play a significant role in antituberculosis activity of potential drug candidates. *Antimicrobial Agents and Chemotherapy* 56 (5): 2643–2651.

Bansal-Mutalik, R. and Nikaido, H. (2014). Mycobacterial outer membrane is a lipid bilayer and the inner membrane is unusually rich in diacyl phosphatidylinositol dimannosides. *Proceedings of the National Academy of Sciences* 111 (13): 4958–4963.

Barrett, O.J., Pushechnikov, A., Wu, M. et al. (2008). Studying aminoglycoside modification by the acetyltransferase class of resistance-causing enzymes via microarray. *Carbohydrate Research* 343 (17): 2924–2931.

Bazett, M., Bergeron, M.-E., and Haston, C.K. (2016). Streptomycin treatment alters the intestinal microbiome, pulmonary T cell profile and airway hyperresponsiveness in a cystic fibrosis mouse model. *Scientific Reports* 6 (1): 19189.

Becker, B. and Cooper, M.A. (2013). Aminoglycoside antibiotics in the 21st century. *ACS Chemical Biology* 8 (1): 105–115.

Begum, A., Rahman, M.M., Ogawa, W. et al. (2005). Gene cloning and characterization of four MATE Family multidrug efflux pumps from *Vibrio cholerae* non-O1. *Microbiology and Immunology* 49 (11): 949–957.

Biteen, J.S., Blainey, P.C., Cardon, Z.G. et al. (2016). Tools for the microbiome: nano and beyond. *ACS Nano* 10 (1): 6–37.

Boltje, T.J., Buskas, T., and Boons, G.J. (2009). Opportunities and challenges in synthetic oligosaccharide and glycoconjugate research. *Nature Chemistry* 1 (8): 611–622.

Borovinskaya, M.A., Shoji, S., Fredrick, K. et al. (2008). Structural basis for hygromycin B inhibition of protein biosynthesis. *RNA* 14 (8): 1590–1599.

Boursi, B., Mamtani, R., Haynes, K. et al. (2015). The effect of past antibiotic exposure on diabetes risk. *European Journal of Endocrinology* 172 (6): 639–648.

Brito, I.L., Yilmaz, S., Huang, K. et al. (2016). Mobile genes in the human microbiome are structured from global to individual scales. *Nature* 535 (7612): 435–439.

Buelow, E., Bello González, T.D.J., Fuentes, S. et al. (2017). Comparative gut microbiota and resistome profiling of intensive care patients receiving selective digestive tract decontamination and healthy subjects. *Microbiome* 5 (1): 88.

Buffie, C.G., Jarchum, I., Equinda, M. et al. (2012). Profound alterations of intestinal microbiota following a single dose of clindamycin results in sustained susceptibility to *Clostridium difficile*-induced colitis. *Infection and Immunity* 80 (1): 62–73.

Caldwell, S.J., Huang, Y., and Berghuis, A.M. (2016). Antibiotic binding drives catalytic activation of aminoglycoside kinase APH(2″)-Ia. *Structure* 24 (6): 935–945.

Candon, S., Perez-Arroyo, A., Marquet, C. et al. (2015). Antibiotics in early life alter the gut microbiome and increase disease incidence in a spontaneous mouse model of autoimmune insulin-dependent diabetes. *PLoS ONE* 10 (5): 1–16.

Carter, A.P., Clemons, W.M., Brodersen, D.E. et al. (2000). Functional insights from the structure of the 30S ribosomal subunit and its interactions with antibiotics. *Nature* 407 (6802): 340–348.

Childs-Disney, J.L., Wu, M., Pushechnikov, A. et al. (2007). A small molecule microarray platform to select RNA internal loop: ligand interactions. *ACS Chemical Biology* 2 (11): 745–754.

Cho, I., Yamanishi, S., Cox, L. et al. (2012). Antibiotics in early life alter the murine colonic microbiome and adiposity. *Nature* 488 (7413): 621–626.

Cimperman, L., Bayless, G., Best, K. et al. (2011). A randomized, double-blind, placebo-controlled pilot study of *Lactobacillus reuteri* ATCC 55730 for the prevention of antibiotic-associated diarrhea in hospitalized adults. *Journal of Clinical Gastroenterology* 45 (9): 785–789.

Clemente, J.C., Pehrsson, E.C., Blaser, M.J. et al. (2015). The microbiome of uncontacted Amerindians. *Science Advances* 1 (3): e1500183.

Colman, R.J. and Rubin, D.T. (2014). Fecal microbiota transplantation as therapy for inflammatory bowel disease: a systematic review and meta-analysis. *Journal of Crohn's and Colitis* 8 (12): 1569–1581.

Cox, G., Stogios, P.J., Savchenko, A. et al. (2015). Structural and molecular basis for resistance to aminoglycoside antibiotics by the adenylyltransferase ANT(2″)-Ia. *mBio* 6 (1): e02180-14.

Davies, J. (2007). In the beginning there was streptomycin. In: *Aminoglycoside Antibiotics: From Chemical Biology to Drug Discovery,* (ed. D.P. Arya), 1–13. Hoboken, NJ: Wiley.

Davis, B.D., Chen, L., and Tai, P.C. (1986). Misread protein creates membrane channels: an essential step in the bactericidal action of aminoglycosides. *Proceedings of the National Academy of Sciences of the United States of America* 83 (16): 6164–6168.

Dethlefsen, L. and Relman, D.A. (2011). Incomplete recovery and individualized responses of the human distal gut microbiota to repeated antibiotic perturbation. *Proceedings of the National Academy of Sciences* 108 (S1): 4554–4561.

Disney, M.D., 2012. Studying modification of aminoglycoside antibiotics by resistance-causing enzymes via microarray. In *Carbohydrate Microarrays. Methods in Molecular Biology*, Chevolot Y. (Ed), 808(8), pp.303–320. New York; Heidelberg: Humana Press.

Disney, M.D. and Childs-Disney, J.L. (2007). Using selection to identify and chemical microarray to study the RNA internal loops recognized by 6′-N-acylated kanamycin A. *ChemBioChem* 8 (6): 649–656.

Disney, M.D., Labuda, L.P., Paul, D.J. et al. (2008). Two-dimensional combinatorial screening identifies specific aminoglycoside–RNA internal loop partners. *Journal of the American Chemical Society* 130 (33): 11185–11194.

Disney, M.D., Magnet, S., Blanchard, J.S. et al. (2004). Aminoglycoside microarrays to study antibiotic resistance. *Angewandte Chemie International Edition* 43 (12): 1591–1594.

Disney, M.D. and Seeberger, P.H. (2004). Aminoglycoside microarrays to explore interactions of antibiotics 14th RNAs and proteins. *Chemistry* 10 (13): 3308–3314.

Elkins, C.A. and Nikaido, H. (2003). Chimeric analysis of AcrA function reveals the importance of its C-terminal domain in its interaction with the AcrB multidrug efflux pump. *Journal of Bacteriology* 185 (18): 5349–5356.

Emanuela, F., Grazia, M., Marco, D.R. et al. (2012). Inflammation as a link between obesity and metabolic syndrome. *Journal of Nutrition and Metabolism* 2012: 476380.

Endimiani, A., Hujer, K.M., Hujer, A.M. et al. (2009). ACHN-490, a neoglycoside with potent in vitro activity against multidrug-resistant *Klebsiella pneumoniae* isolates. *Antimicrobial Agents and Chemotherapy* 53 (10): 4504–4507.

Felise, H.B., Nguyen, H.V., Pfuetzner, R.A. et al. (2008). An inhibitor of Gram-negative bacterial virulence protein secretion. *Cell Host and Microbe* 4 (4): 325–336.

Fernandez, L., Gooderham, W.J., Bains, M. et al. (2010). Adaptive resistance to the "last hope" antibiotics polymyxin B and colistin in *Pseudomonas aeruginosa* is mediated by the novel two-component regulatory system ParR-ParS. *Antimicrobial Agents and Chemotherapy* 54 (8): 3372–3382.

Fosso, M.Y., Li, Y., and Garneau-Tsodikova, S. (2014). New trends in the use of aminoglycosides. *MedChemCommun* 5 (8): 1075–1091.

Francino, M.P. (2016). Antibiotics and the human gut microbiome: dysbioses and accumulation of resistances. *Frontiers in Microbiology* 6 (1): 1543.

Galimand, M., Schmitt, E., Panvert, M. et al. (2011). Intrinsic resistance to aminoglycosides in *Enterococcus faecium* is conferred by the 16S rRNA m5C1404-specific methyltransferase EfmM. *RNA* 17 (2): 251–262.

Gao, F., Yan, X., Zahr, O. et al. (2008). Synthesis and use of sulfonamide-, sulfoxide-, or sulfone-containing aminoglycoside-CoA bisubstrates as mechanistic probes for aminoglycoside N-6′-acetyltransferase. *Bioorganic and Medicinal Chemistry Letters* 18 (20): 5518–5522.

Garneau-Tsodikova, S. and Labby, K.J. (2016). Mechanisms of resistance to aminoglycoside antibiotics: overview and perspectives. *MedChemComm* 7 (1): 11–27.

Georghiou, S.B., Magana, M., Garfein, R.S. et al. (2012). Evaluation of genetic mutations associated with *Mycobacterium tuberculosis* resistance to amikacin, kanamycin and capreomycin: a systematic review. *PLoS ONE* 7 (3): e33275.

Gibson, G.R., Hutkins, R., Sanders, M.E. et al. (2017). Expert consensus document: The International Scientific Association for Probiotics and Prebiotics (ISAPP) consensus statement on the definition and scope of prebiotics. *Nature Reviews Gastroenterology and Hepatology* 14 (8): 491–502.

Granier, S.A., Hidalgo, L., San Millan, A. et al. (2011). ArmA methyltransferase in a monophasic *Salmonella enterica* isolate from food. *Antimicrobial Agents and Chemotherapy* 55 (11): 5262–5266.

Greenwood, C., Morrow, A.L., Lagomarcino, A.J. et al. (2014). Early empiric antibiotic use in preterm infants is associated with lower bacterial diversity and higher relative abundance of enterobacter. *The Journal of Pediatrics* 165 (1): 23–29.

Guo, L., Wan, Y., Wang, X. et al. (2012). Development of aminoglycoside antibiotics by carbohydrate chemistry. *Mini-Reviews in Medicinal Chemistry* 12 (14): 1533–1541.

Hamer, H.M., Jonkers, D., Venema, K. et al. (2008). Review article: the role of butyrate on colonic function. *Alimentary Pharmacology and Therapeutics* 27 (2): 104–119.

Han, Q., Zhao, Q., Fish, S. et al. (2005). Molecular recognition by glycoside pseudo base pairs and triples in an apramycin-RNA complex. *Angewandte Chemie International Edition* 44 (18): 2694–2700.

Hanberger, H., Edlund, C., Furebring, M. et al. (2013). Rational use of aminoglycosides: review and recommendations by the Swedish Reference Group for Antibiotics (SRGA). *Scandinavian Journal of Infectious Diseases* 45 (3): 161–175.

Hancock, R.E. (1981). Aminoglycoside uptake and mode of action-with special reference to streptomycin and gentamycin. *The Journal of Antimicrobial Chemotherapy* 8 (4): 249–276.

Hancock, R.E.W., Farmer, S.W., Li, Z.S. et al. (1991). Interaction of aminoglycosides with the outer membranes and purified lipopolysaccharide and OmpF porin of *Escherichia coli*. *Antimicrobial Agents and Chemotherapy* 35 (7): 1309–1314.

Heinsen, F.A., Knecht, H., Neulinger, S.C. et al. (2015). Dynamic changes of the luminal and mucosa-associated gut microbiota during and after antibiotic therapy with paromomycin. *Gut Microbes* 6 (4): 243–254.

Hildebrand, H., Malmborg, P., Askling, J. et al. (2008). Early-life exposures associated with antibiotic use and risk of subsequent Crohn's disease. *Scandinavian Journal of Gastroenterology* 43 (8): 961–966.

Hill, C., Guarner, F., Reid, G. et al. (2014). The International Scientific Association for Probiotics and Prebiotics consensus statement on the scope and appropriate use of the term probiotic. *Nature Reviews Gastroenterology and Hepatology* 11 (8): 506–514.

Hobbie, S.N., Akshay, S., Kalapala, S.K. et al. (2008). Genetic analysis of interactions with eukaryotic rRNA identify the mitoribosome as target in aminoglycoside ototoxicity. *Proceedings of the National Academy of Sciences* 105 (52): 20888–20893.

Honda, K. and Littman, D.R. (2016). The microbiota in adaptive immune homeostasis and disease. *Nature* 535 (7610): 75–84.

Human Microbiome Project Consortium, T., Huttenhower, C., Gevers, D. et al. (2012). Structure, function and diversity of the healthy human microbiome. *Nature* 486 (7402): 207–214.

Hwang, I.Y., Tan, M.H., Koh, E. et al. (2014). Reprogramming microbes to be pathogen-seeking killers. *ACS Synthetic Biology* 3 (4): 228–237.

Jarlier, V. (1994). Mycobacterial cell wall: structure and role in natural resistance to antibiotics. *FEMS Microbiology Letters* 123 (1–2): 11–18.

Johanesen, P.A., Mackin, K.E., Hutton, M.L. et al. (2015). Disruption of the gut microbiome: *Clostridium difficile* infection and the threat of antibiotic resistance. *Genes* 6 (4): 1347–1360.

Johnson, L.P., Walton, G.E., Psichas, A. et al. (2015). Prebiotics modulate the effects of antibiotics on gut microbial diversity and functioning in vitro. *Nutrients* 7 (6): 4480–4497.

Kailemia, M.J., Ruhaak, L.R., Lebrilla, C.B. et al. (2014). Oligosaccharide analysis by mass spectrometry: a review of recent developments. *Analytical Chemistry* 86 (1): 196–212.

Kato, K. and Peters, T. (eds.) (2017). *NMR in Glycoscience and Glycotechnology*. Cambridge: Royal Society of Chemistry.

Kim, S., Covington, A., and Pamer, E.G. (2017). The intestinal microbiota: antibiotics, colonization resistance, and enteric pathogens. *Immunological Reviews* 279 (1): 90–105.

Kindrachuk, K.N., Fernández, L., Bains, M. et al. (2011). Involvement of an ATP-dependent protease, PA0779/AsrA, in inducing heat shock in response to tobramycin in *Pseudomonas aeruginosa. Antimicrobial Agents and Chemotherapy* 55 (5): 1874–1882.

Krahn, T., Gilmour, C., Tilak, J. et al. (2012). Determinants of intrinsic aminoglycoside resistance in *Pseudomonas aeruginosa. Antimicrobial Agents and Chemotherapy* 56 (11): 5591–5602.

Kwon, D.H. and Lu, C.-D. (2006). Polyamines induce resistance to cationic peptide, aminoglycoside, and quinolone antibiotics in *Pseudomonas aeruginosa* PAO1. *Antimicrobial Agents and Chemotherapy* 50 (5): 1615–1622.

Labby, K.J. and Garneau-Tsodikova, S. (2013). Strategies to overcome the action of aminoglycoside-modifying enzymes for treating resistant bacterial infections. *Future Medicinal Chemistry* 5 (11): 1285–1309.

Langdon, A., Crook, N., and Dantas, G. (2016). The effects of antibiotics on the microbiome throughout development and alternative approaches for therapeutic modulation. *Genome Medicine* 8 (1): 39.

Lau, C.H.-F., Hughes, D., and Poole, K. (2014). MexY-promoted aminoglycoside resistance in *Pseudomonas aeruginosa*: involvement of a putative proximal binding pocket in aminoglycoside recognition. *mBio* 5 (2): e01068-14.

Li, J. and Tom Chang, C.-W. (2006). Recent developments in the synthesis of novel aminoglycoside antibiotics. *Anti-Infective Agents in Medicinal Chemistry* 5 (3): 255–271.

Li, X.-Z., Plésiat, P., and Nikaido, H. (2015). The challenge of efflux-mediated antibiotic resistance in Gram-negative bacteria. *Clinical Microbiology Reviews* 28 (2): 337–418.

Lichtman, J.S., Ferreyra, J.A., Ng, K.M. et al. (2016). Host-microbiota interactions in the pathogenesis of antibiotic-associated diseases. *Cell Reports* 14 (5): 1049–1061.

Liévin-Le Moal, V. and Servin, A.L. (2014). Anti-infective activities of Lactobacillus strains in the human intestinal microbiota: from probiotics to gastrointestinal anti-infectious biotherapeutic agents. *Clinical Microbiology Reviews* 27 (2): 167–199.

López-Diaz, M.D.C., Culebras, E., Rodríguez-Avial, I. et al. (2016). Plazomicin activity against 346 extended-spectrum-β-lactamase/AmpC-producing *Escherichia coli* urinary isolates, related to aminoglycoside-modifying enzymes characterized. *Antimicrobial Agents and Chemotherapy* 61 (2): AAC.02454-16.

Macfarlane, E.L.A., Kwasnicka, A., and Hancock, R.E.W. (2000). Role of *Pseudomonas aeruginosa* PhoP-PhoQ in resistance to antimicrobial cationic peptides and aminoglycosides. *Microbiology* 146 (10): 2543–2554.

Magiorakos, A.-P., Srinivasan, A., Carey, R.B. et al. (2012). Multidrug-resistant, extensively drug-resistant and pandrug-resistant bacteria: an international expert proposal for interim standard definitions for acquired resistance. *Clinical Microbiology and Infection* 18 (3): 268–281.

Mai, V., Torrazza, R.M., Ukhanova, M. et al. (2013). Distortions in development of intestinal microbiota associated with late onset sepsis in preterm infants. *PLoS ONE* 8 (1): e52876.

Marchesi, J.R. and Ravel, J. (2015). The vocabulary of microbiome research: a proposal. *Microbiome* 3: 31.

Marshall, B.M., Ochieng, D.J., and Levy, S.B. (2009). Commensals: underappreciated reservoir of antibiotic resistance. *Microbe Magazine* 4 (5): 231–238.

Matsushita, T., Chen, W., Juskeviciene, R. et al. (2015). Influence of 4′-O-glycoside constitution and configuration on ribosomal selectivity of paromomycin. *Journal of the American Chemical Society* 137 (24): 7706–7717.

McCoy, L.S., Xie, Y., and Tor, Y. (2011). Antibiotics that target protein synthesis. *Wiley Interdisciplinary Reviews: RNA* 2 (2): 209–232.

Milan, A., Furlanis, L., Cian, F. et al. (2016). Epidemic dissemination of a carbapenem-resistant *Acinetobacter baumannii* clone carrying armA two years after its first isolation in an Italian hospital. *Microbial Drug Resistance* 22 (8): 668–674.

Mitema, E.S. (2010). The role of unregulated sale and dispensing of antimicrobial agents on the development of antimicrobial resistance in developing countries. In: *Antimicrobial Resistance in Developing Countries* (ed. A. Sosa, D. Byarugaba, C. Amábile-Cuevas, et al.), 403–411. Cham: Springer Nature, Switzerland AG.

Moayyedi, P., Surette, M.G., Kim, P.T. et al. (2015). Fecal microbiota transplantation induces remission in patients with active ulcerative colitis in a randomized controlled trial. *Gastroenterology* 149 (1): 102–109.

Moore, A.M., Ahmadi, S., Patel, S. et al. (2015). Gut resistome development in healthy twin pairs in the first year of life. *Microbiome* 3: 27.

Morita, Y., Tomida, J., and Kawamura, Y. (2012). MexXY multidrug efflux system of *Pseudomonas aeruginosa*. *Frontiers in Microbiology* 3: 408.

Muthana, S., Cao, H., and Chen, X. (2009). Recent progress in chemical and chemoenzymatic synthesis of carbohydrates. *Current Opinion in Chemical Biology* 13 (5–6): 573–581.

Nakashima, R., Sakurai, K., Yamasaki, S. et al. (2011). Structures of the multidrug exporter AcrB reveal a proximal multisite drug-binding pocket. *Nature* 480 (7378): 565–569.

Nakashima, R., Sakurai, K., Yamasaki, S. et al. (2013). Structural basis for the inhibition of bacterial multidrug exporters. *Nature* 500 (7460): 102–106.

Nessar, R., Reyrat, J.M., Murray, A. et al. (2011). Genetic analysis of new 16S rRNA mutations conferring aminoglycoside resistance in *Mycobacterium abscessus*. *Journal of Antimicrobial Chemotherapy* 66 (8): 1719–1724.

Niederweis, M. (2003). Mycobacterial porins: new channel proteins in unique outer membranes. *Molecular Microbiology* 49 (5): 1167–1177.

Nikaido, H. (2011). Structure and mechanism of RND-type multidrug efflux pumps. *Advances in Enzymology and Related Areas of Molecular Biology* 77 (11): 1–60.

Nowicki, E.M., O'Brien, J.P., Brodbelt, J.S. et al. (2015). Extracellular zinc induces phosphoethanolamine addition to *Pseudomonas aeruginosa* lipid A via the ColRS two-component system. *Molecular Microbiology* 97 (1): 166–178.
Opperman, T.J. and Nguyen, S.T. (2015). Recent advances toward a molecular mechanism of efflux pump inhibition. *Frontiers in Microbiology* 6: 421.
Palma, A.S., Feizi, T., Childs, R.A. et al. (2014). The neoglycolipid (NGL)-based oligosaccharide microarray system poised to decipher the meta-glycome. *Current Opinion in Chemical Biology* 18: 87–94.
Perry, J.A., Westman, E.L., and Wright, G.D. (2014). The antibiotic resistome: what's new? *Current Opinion in Microbiology* 21: 45–50.
Pettersen, E.F., Goddard, T.D., Huang, C.C. et al. (2004). UCSF Chimera: a visualization system for exploratory research and analysis. *Journal of Computational Chemistry* 25 (13): 1605–1612.
Piepersberg, W., Aboshanab, K.M., Schmidt-Beißner, H. et al. (2007). The biochemistry and genetics of aminoglycoside producers. In: *Aminoglycoside Antibiotics: From Chemical Biology to Drug Discovery* (ed. D.P. Arya), 15–118. Hoboken, NJ: Wiley.
Poelarends, G.J., Mazurkiewicz, P., and Konings, W.N. (2002). Multidrug transporters and antibiotic resistance in *Lactococcus lactis*. *Biochimica et Biophysica Acta* 1555 (1–3): 1–7.
Poole, K. (2011). *Pseudomonas aeruginosa*: resistance to the max. *Frontiers in Microbiology* 2: 65.
Qin, J., Li, R., Raes, J. et al. (2010). A human gut microbial gene catalogue established by metagenomic sequencing. *Nature* 464 (7285): 59–65.
Rafii, F., Sutherland, J.B., and Cerniglia, C.E. (2008). Effects of treatment with antimicrobial agents on the human colonic microflora. *Therapeutics and Clinical Risk Management* 4 (6): 1343–1357.
Ramaswamy, V.K., Vargiu, A.V., Malloci, G. et al. (2017). Molecular rationale behind the differential substrate specificity of bacterial RND multi-drug transporters. *Scientific Reports* 7: 8075.
Ramirez, M.S., Nikolaidis, N., and Tolmasky, M.E. (2013). Rise and dissemination of aminoglycoside resistance: the aac(6′)-Ib paradigm. *Frontiers in Microbiology* 4: 121.
Ramirez, M.S. and Tolmasky, M.E. (2010). Aminoglycoside modifying enzymes. *Drug Resistance Updates* 13 (6): 151–171.
Rillahan, C.D. and Paulson, J.C. (2011). Glycan microarrays for decoding the glycome. *Annual Review of Biochemistry* 80 (1): 797–823.
Routy, B., Le Chatelier, E., Derosa, L. et al. (2018). Gut microbiome influences efficacy of PD-1–based immunotherapy against epithelial tumors. *Science* 359 (6371): 91–97.
Roy, S. and Trinchieri, G. (2017). Microbiota: a key orchestrator of cancer therapy. *Nature Reviews Cancer* 17 (5): 271–285.

Russell, S.L., Gold, M.J., Hartmann, M. et al. (2012). Early life antibiotic-driven changes in microbiota enhance susceptibility to allergic asthma. *EMBO Reports* 13 (5): 440–447.

Sadovskaya, I., Vinogradov, E., Li, J. et al. (2010). High-level antibiotic resistance in *Pseudomonas aeruginosa* biofilm: the ndvB gene is involved in the production of highly glycerol-phosphorylated beta-(1->3)-glucans, which bind aminoglycosides. *Glycobiology* 20 (7): 895–904.

Schroeder, B.O., Birchenough, G.M.H., Ståhlman, M. et al. (2018). Bifidobacteria or fiber protects against diet-induced microbiota-mediated colonic mucus deterioration. *Cell Host and Microbe* 23 (1): –27-40.e7.

Sekirov, I., Tam, N.M., Jogova, M. et al. (2008). Antibiotic-induced perturbations of the intestinal microbiota alter host susceptibility to enteric infection. *Infection and Immunity* 76 (10): 4726–4736.

Sender, R., Fuchs, S., and Milo, R. (2016). Are we really vastly outnumbered? Revisiting the ratio of bacterial to host cells in humans. *Cell* 164 (3): 337–340.

Shankar, J., Solis, N.V., Mounaud, S. et al. (2015). Using Bayesian modelling to investigate factors governing antibiotic-induced *Candida albicans* colonization of the GI tract. *Scientific Reports* 5: 8131.

Singh, M., Yau, Y.C.W., Wang, S. et al. (2017). MexXY efflux pump overexpression and aminoglycoside resistance in cystic fibrosis isolates of *Pseudomonas aeruginosa* from chronic infections. *Canadian Journal of Microbiology* 63 (12): 929–938.

Sjuts, H., Vargiu, A.V., Kwasny, S.M. et al. (2016). Molecular basis for inhibition of AcrB multidrug efflux pump by novel and powerful pyranopyridine derivatives. *Proceedings of the National Academy of Sciences* 113 (13): 3509–3514.

Smillie, C.S., Smith, M.B., Friedman, J. et al. (2011). Ecology drives a global network of gene exchange connecting the human microbiome. *Nature* 480 (7376): 241–244.

Smith, C.A., Toth, M., Weiss, T.M. et al. (2014). Structure of the bifunctional aminoglycoside-resistance enzyme AAC(6′)-Ie-APH(2″)-Ia revealed by crystallographic and small-angle X-ray scattering analysis. *Acta Crystallographica Section D Biological Crystallography* 70 (10): 2754–2764.

Smith, M.A., Coinon, M., Paschos, A. et al. (2012). Identification of the binding site of Brucella VirB8 interaction inhibitors. *Chemistry and Biology* 19 (8): 1041–1048.

Sommer, M.O.A., Dantas, G., and Church, G.M. (2009). Functional characterization of the antibiotic resistance reservoir in the human microflora. *Science* 325 (5944): 1128–1131.

Song, H.J., Shim, K.N., Jung, S.A. et al. (2008). Antibiotic-associated diarrhea: candidate organisms other than *Clostridium difficile*. *Korean Journal of Internal Medicine* 23 (1): 9–15.

Sonnenburg, J.L. and Bäckhed, F. (2016). Diet–microbiota interactions as moderators of human metabolism. *Nature* 535 (7610): 56–64.

Starkey, M., Lepine, F., Maura, D. et al. (2014). Identification of anti-virulence compounds that disrupt quorum-sensing regulated acute and persistent pathogenicity. *PLoS Pathogens* 10 (8): e1004321.

Stogios, P.J., Kuhn, M.L., Evdokimova, E. et al. (2017). Structural and biochemical characterization of *Acinetobacter* spp. aminoglycoside acetyltransferases highlights functional and evolutionary variation among antibiotic resistance enzymes. *ACS Infectious Diseases* 3 (2): 132–143.

Taber, H.W., Mueller, J.P., Miller, P.F. et al. (1987). Bacterial uptake of aminoglycoside antibiotics. *Microbiological reviews* 51 (4): 439–457.

Taipale, T.J., Pienihäkkinen, K., Isolauri, E. et al. (2016). Bifidobacterium animalis subsp. lactis BB-12 in reducing the risk of infections in early childhood. *Pediatric Research* 79 (1): 65–69.

Thaiss, C.A., Zmora, N., Levy, M. et al. (2016). The microbiome and innate immunity. *Nature* 535 (7610): 65–74.

Thamban Chandrika, N. and Garneau-Tsodikova, S. (2018). Comprehensive review of chemical strategies for the preparation of new aminoglycosides and their biological activities. *Chemical Society Reviews* 47 (4): 1189–1249.

Tsitovich, P.B., Pushechnikov, A., French, J.M. et al. (2010). A chemoenzymatic route to diversify aminolgycosides enables a microarray-based method to probe acetyltransferase activity. *ChemBioChem* 11 (12): 1656–1660.

Ubeda, C., Taur, Y., Jenq, R.R. et al. (2010). Vancomycin-resistant Enterococcus domination of intestinal microbiota is enabled by antibiotic treatment in mice and precedes bloodstream invasion in humans. *Journal of Clinical Investigation* 120 (12): 4332–4341.

Valle, M., Zavialov, A., Sengupta, J. et al. (2003). Locking and unlocking of ribosomal motions. *Cell* 114 (1): 123–134.

Van Schaik, W. (2015). The human gut resistome. *Royal Society Publisher* 370: 20140087.

Velagapudi, S.P. and Disney, M.D. (2013). Defining RNA motif–aminoglycoside interactions via two-dimensional combinatorial screening and structure–activity relationships through sequencing. *Bioorganic & Medicinal Chemistry* 21 (20): 6132–6138.

Velagapudi, S.P., Gallo, S.M., and Disney, M.D. (2014). Sequence-based design of bioactive small molecules that target precursor microRNAs. *Nature Chemical Biology* 10 (4): 291–297.

Venter, H., Mowla, R., Ohene-Agyei, T. et al. (2015). RND-type drug efflux pumps from Gram-negative bacteria: molecular mechanism and inhibition. *Frontiers in Microbiology* 6: 377.

Wachino, J. and Arakawa, Y. (2012). Exogenously acquired 16S rRNA methyltransferases found in aminoglycoside-resistant pathogenic Gram-negative bacteria: an update. *Drug Resistance Updates* 15 (3): 133–148.

Wang, L., Pulk, A., Wasserman, M.R. et al. (2012). Allosteric control of the ribosome by small-molecule antibiotics. *Nature Structural and Molecular Biology* 19 (9): 957–963.

Wang, Z., Fan, G., Hryc, C.F. et al. (2017). An allosteric transport mechanism for the AcrAB-TolC multidrug efflux pump. *eLife* 6: e24905.

Wassenaar, T.M. (2016). Insights from 100 years of research with probiotic *E. coli*. *European Journal of Microbiology and Immunology* 6 (3): 147–161.

Weitkunat, K., Stuhlmann, C., Postel, A. et al. (2017). Short-chain fatty acids and inulin, but not guar gum, prevent diet-induced obesity and insulin resistance through differential mechanisms in mice. *Scientific Reports* 7 (1): 6109.

Yan, R.-B., Yuan, M., Wu, Y. et al. (2011). Rational design and synthesis of potent aminoglycoside antibiotics against resistant bacterial strains. *Bioorganic and Medicinal Chemistry* 19 (1): 30–40.

Zhanel, G.G., Lawson, C.D., Zelenitsky, S. et al. (2012). Comparison of the next-generation aminoglycoside plazomicin to gentamycin, tobramycin and amikacin. *Expert Review of Anti-infective Therapy* 10 (4): 459–473.

Zhao, Y., Wu, J., Li, J.V. et al. (2013). Gut microbiota composition modifies fecal metabolic profiles in mice. *Journal of Proteome Research* 12 (6): 2987–2999.

Zheng, M., Zhang, R., Tian, X. et al. (2017). Assessing the risk of probiotic dietary supplements in the context of antibiotic resistance. *Frontiers in Microbiology* 8: 908.

2

Mechanisms of Action and of Resistance to Quinolones

José L. Martínez

Departamento de Biotecnología Microbiana, Centro Nacional de Biotecnología, Consejo Superior de Investigaciones Científicas, CSIC, Madrid, Spain

2.1 Introduction

The first rationally developed antibiotics had synthetic origin (Domagk 1935; Horlein 1936). However, the explosive development of new antibiotics came from the search of natural compounds produced by environmental microorganisms (Waksman and Woodruff 1940). Soon after the introduction of antibiotics for therapy, resistant variants of the microorganisms began to be selected under the selective pressure of the treatment. Two were (and are still now) the genetic causes of this resistance, mutation and acquisition of resistance determinants by means of horizontal gene transfer (HGT). Antibiotic-producing microorganisms require to encode in their genome resistance determinants in order to avoid the activity of the antimicrobials that they produce. Consequently, it was proposed and later on widely accepted that resistance genes acquired through HGT by human pathogens were originated in the antibiotic-producing organisms (Benveniste and Davies 1973; Davies 1994; Webb and Davies 1993). In line with these thoughts and since quinolones are purely synthetic drugs, not present in nature before their synthesis, it was thought that quinolone resistance genes would not exist in nature, and consequently the only mechanisms leading to quinolone resistance should be mutations in the genes encoding their topoisomerase targets (Crumplin and Odell 1987) or the transporters of these antimicrobials (Piddock 1999). Nowadays we know that this viewpoint was oversimplistic and different genes, some of which are plasmid encoded, are involved in the acquisition of quinolone resistance by bacterial pathogens. Along with the current chapter, we will review updated information on the mechanisms and the genetic basis of quinolone resistance among virulent microorganisms.

Antibiotic Drug Resistance, First Edition. Edited by José-Luis Capelo-Martínez and Gilberto Igrejas.

2.2 Mechanism of Action of Quinolones

Quinolones are a group of synthetic antibiotics obtained during chloroquine synthesis, which were discovered more than five decades ago (Lesher et al. 1962). The first member of the family, nalidixic acid, was soon introduced into clinics, but just for treating urinary infections by Gram-negative microorganisms (Deitz et al. 1963). In the next decade there was a constant release into clinics of novel quinolones such as the oxolinic acid. However the range of infections for which these antimicrobial compounds were useful remained to be quite limited. In the decade of the 1980s, a new generation of quinolones that contain a fluorine and novel ring moieties (fluoroquinolones) was developed (Ball 2000). This second generation of quinolones presented better pharmacological properties as well as a better entrance into Gram-positive microorganisms. Consequently, they were considered broad-spectrum antibiotics. Among them, ciprofloxacin still remains as the antimicrobial of choice not only for treating a variety of infections, mainly those by Gram-negative organisms, but also for treating Gram-positive infections. A newer generation of quinolones with improved pharmacological properties includes widely used antimicrobials such as levofloxacin (Anderson and Perry 2008).

The targets of the quinolones are the bacterial topoisomerases DNA gyrase and topoisomerase IV (Khodursky and Cozzarelli 1998; Shen et al. 1989). Both enzymes comprise two different subunits that form a tetramer and play essential roles in most processes of the DNA activity, including replication, transcription, recombination, and repair. While gyrase is the main responsible for modulating DNA supercoiling and removing the torsional stress associated with DNA replication and transcription, the role of topoisomerase IV in these processes is less relevant, being mainly involved in removing knots that accumulate in bacterial DNA as a result of different processes as well as in decatenating daughter chromosomes after replication (Champoux 2001; Levine et al. 1998). To perform their functions, bacterial topoisomerases generate double-strand breaks in the bacterial DNA. In such a way, while topoisomerases ara essential for microbial life, their uncontrolled activity may cause fragmentation of the bacterial genome. Quinolones act by increasing the concentration of DNA–enzyme cleavage complexes and inhibit DNA ligation, hence increasing DNA breaks and stalling the replication fork progression (Anderson and Osheroff 2001). Thus, these antimicrobials have been dubbed as "topoisomerase poisons" because they convert these two enzymes into intracellular toxins (Kreuzer and Cozzarelli 1979).

It is worth mentioning that due to this mechanism of action that leads to DNA damage, quinolones induce SOS response and increase recombination and mutagenesis (Lopez et al. 2007; Lopez and Blazquez 2009; Phillips et al. 1987; Ysern et al. 1990). Therefore quinolones have been considered as promoters of genetic variation (Blazquez et al. 2012), including those processes leading to antibiotic resistance.

2.3 Mutations in the Genes Encoding the Targets of Quinolones

Although, as above stated, different quinolone resistance genes have been described, mutations in the genes encoding the targets of these antimicrobials still remain as the most prevalent cause for the acquisition of high-level resistance to quinolones (Jacoby 2005). Notably the main target of quinolones is different in Gram-negative (gyrase) and in Gram-positive (topoisomerase IV) organisms (Drlica and Zhao 1997; Ferrero et al. 1994). In agreement with this situation, mutations in the genes encoding the two subunits of the bacterial gyrase, *gyrA* and *gyrB* (mainly in *gyrA*), are the most relevant in Gram-negative resistant bacteria, whereas Gram-positive microorganisms most frequently present mutations in the genes *parC* and *parE* (mainly in *parC*), encoding the subunits of the bacterial topoisomerase IV. The analysis of quinolone-resistant mutants from different organisms have shown that there are hot spots within *gyrA* and *parC*, dubbed as quinolone resistance-determining region (QRDR), where most mutations accumulate (Piddock 1999). The amino acids more relevant for acquiring resistance, based on the *Escherichia coli* sequences of the topoisomerases, are serine 83 and aspartate 87 for GyrA and serine 79 and aspartate 83 for ParC. These mutants present a reduced quinolone–enzyme binding affinity, without reducing the cleavage activity of the topoisomerase in the absence of the antibiotic. While these mutants can present fitness costs on occasions (Redgrave et al. 2014), there are some examples showing that, at least for some mutations, a substantial fitness cost is not observed (Balsalobre and de la Campa 2008), or the cost associated with resistance depends on the genomic context of the strain that has acquired quinolone resistance (Luo et al. 2005).

In addition, it has been shown that even in the case of mutations in the topoisomerases producing a relevant fitness cost, compensatory mutations that alleviate such cost can be selected. Notably, some of these compensatory mutations occur in other elements as regulators of efflux pumps or a different topoisomerase (or topoisomerase subunit), which are of relevance for the acquisition of quinolone resistance (Andersson and Hughes 2010). Consequently, in some of these cases, compensation of fitness costs is associated with an increased level of resistance to quinolones. This is the situation of *E. coli*, an organism in which a combination of different mutations leading to high-level quinolone resistance is not an infrequent event. In this bacterium, a *gyrA* mutant presenting the S83L and D87N double substitution displays a ciprofloxacin minimal inhibitory concentration (MIC) of 0.38 $\mu g\,ml^{-1}$, with a 3% reduction in fitness as compared with the wild-type strain. The acquisition of a S80I mutation in ParC increases MIC to 32 $\mu g\,ml^{-1}$, and the fitness of this strain presenting mutations in both topoisomerases is 1% higher than the wild-type strain (Marcusson et al. 2009). These results indicate that the accumulation of quinolone resistance mutations does not necessarily produce a higher burden to the resistant microorganism.

Rather, increase in resistance may correlate with a higher fitness of the mutant strain. Whether or not this type of compensatory mutations can be selected in the absence of selection remains to be fully elucidated. However this situation could be a good example of the potential selection of antibiotic-resistant mutants even in the absence of antibiotics.

2.4 Multidrug Efflux Pumps and Quinolone Resistance

The discovery of multidrug resistance (MDR) efflux pumps and the analysis of the substrates they can extrude challenged the ideas that mutations in the genes encoding the targets of these antimicrobials would be the unique cause of quinolone resistance. Indeed bacterial MDR efflux pumps can extrude quinolone. Even more, despite their synthetic origin, quinolones are among the most usual substrates of these pumps (Hooper 1999).

MDR efflux pumps are encoded in the core genomes of all bacterial species (Alonso et al. 1999; Li and Nikaido 2009; Nikaido and Takatsuka 2009; Piddock 2006a; Saier et al. 1998), and they contribute to intrinsic and acquired resistance to different antibiotics, including quinolones (Fernández 2012; Garcia-Leon et al. 2014a; Hernando-Amado et al. 2016; Li et al. 2015; Martinez et al. 2009; Piddock 2006a; Poole 2007; Vila and Martínez 2008). The ubiquitous presence of these elements in all living cells together with their high conservation suggests that they are ancient elements in bacterial genomes, which evolved long before the wide use of antibiotics for treating infections, and that they likely present functions with relevance for the bacterial physiology besides antibiotic resistance (Alvarez-Ortega et al. 2013; Blanco et al. 2016; Fetar et al. 2011; García-León et al. 2014b; Martinez et al. 2009; Piddock 2006b; Poole 2012).

Until now five different families of MDR efflux pumps have been described (Hernando-Amado et al. 2016), and four of these five families of MDR systems – the ATP-binding cassette (ABC) family, the major facilitator superfamily (MFS), the resistance-nodulation-division (RND) family, and the multidrug and toxic compound extrusion (MATE) family – may participate in quinolone resistance (Poole 2000a, b). Although in a few cases such as NorC in *Staphylococcus aureus* (Truong-Bolduc et al. 2006) or Rv1634 in *Mycobacterium tuberculosis* (De Rossi et al. 2002) quinolones seem to be the unique (or at least the predominant) antibiotics extruded by these resistance determinants, efflux pumps usually present little selectivity in the sense that one single efflux pump can extrude a variety of different substrates (Martinez et al. 2009).

The expression of MDR efflux pumps is tightly controlled by local and global regulators; they are usually expressed at low level under regular growing conditions in the laboratory (Grkovic et al. 2002). The basal level of expression varies depending on the efflux pump. In this regard, some efflux pumps present an expression level enough for contributing to intrinsic resistance to quinolones (Li et al. 1994; Vila and Martinez 2008), whereas for others, presenting lower level of expression, such contribution is minimal if any. In any case one important consequence of these studies is that inhibitors of efflux pumps may improve the activity of quinolones (Renau et al. 1999).

Expression of efflux pumps can be triggered in the presence of an effector or under some specific growing conditions (García-León et al. 2014b; Hernandez et al. 2011; Rosenberg et al. 2003). In addition, mutants presenting high derepressed levels of expression of efflux pumps can be selected in the presence of antibiotics, both *in vitro* and *in vivo*. These mutants can present reduced susceptibility to quinolones and to other antibiotics (acquired resistance) (Alonso and Martinez 2001; Cohen et al. 1989; Jalal et al. 2000; Ziha-Zarifi et al. 1999), although in several occasions (with some exceptions such as *Stenotrophomonas maltophilia*; see below) the observed MICs are below the breakpoint levels used for defining clinical resistance (Martinez et al. 2015) and high-level resistance is achieved just in combination with other mutations (Marcusson et al. 2009).

In occasions, high-level quinolone resistance can be achieved upon the simultaneous expression of some different MDR efflux pumps (Yang et al. 2003). However, in most studied cases, the highest level of resistance to quinolones is achieved when mutations in the target genes and increased efflux occur simultaneously (Llanes et al. 2006).

As above stated, MDR efflux pumps present a wide range of substrates, which include antibiotics belonging to different structural families (Paulsen 2003). This means that the overexpression of one single efflux pump increases the MICs for a variety of antibiotics. Consequently, efflux pumps overexpressing mutants can be selected by the selective pressure of any of the substrates of the pump. In other words, selective pressure by one antibiotic can also select for quinolone resistance (cross-selection) if both antimicrobials are substrates of the efflux pump (Cohen et al. 1989).

2.5 Transferable Quinolone Resistance

The above described mechanisms are based on the selection of mutants. For them, resistance can spread just by clonal expansion. Nevertheless, the possibility of transferrable mechanisms of resistance to quinolones was firstly proposed in the basis of *in vitro* analysis (Gomez-Gomez and Blazquez 1997; Martinez et al. 1998), and plasmids carrying quinolone resistance genes (dubbed *qnr*) were described soon afterwards (Martinez-Martinez et al. 1998). Qnr is a member of the pentapeptide repeat protein (PRP) family (Vetting et al. 2006). It binds the bacterial topoisomerases (DNA gyrase and topoisomerase IV) protecting them from the activity of quinolones. Five *qnr* families, namely, *qnrA*, *qnrB*, *qnrS*, *qnrC*, and *qnrD* (Cavaco et al. 2009; Hata et al. 2005; Jacoby et al. 2006; Strahilevitz et al. 2007; Tran and Jacoby 2002; Wang et al. 2009), have been so far described in plasmids disseminated among bacterial pathogens. Other members of the PRP family besides Qnr, such as MfpA that contributes to the intrinsic resistance of *Mycobacterium* to quinolones (Montero et al. 2001), might be relevant for resistance to these antimicrobials.

It is important to be noticed that *qnr* genes have been found to be present in the chromosomes of environmental bacteria, mainly in water-dwelling microorganisms (Arsene and Leclercq 2007; Rodriguez-Martinez et al. 2008; Sanchez et al. 2008). Indeed, it has been clearly demonstrated that *Shewanella algae* is the origin of plasmid-encoded *qnrA* genes (Poirel et al. 2005b), and it has been proposed, although the evidence is not so clear, that *Vibrionaceae* might be a reservoir for other plasmid-encoded *qnr* genes (Cattoir et al. 2007; Poirel et al. 2005a).

Quinolones have been (and are still in some countries) largely used in fish farms without major concerns about the associated risks for human health until recently (Baquero et al. 2008; Cabello 2006). The reasoning behind was that mutation-driven resistance will select resistant mutants in bacteria infecting (or colonizing) fishes as well as in the inhabitants of water and sludge; none of them will be of relevance as a human pathogen. Since antibiotic resistance due to mutation is not usually transferrable (see below), the clonal expansion of these resistance microorganism will not impact on the potential acquisition of resistance by human pathogens. Nevertheless, once the possibility of the acquisition of quinolone resistance through HGT has been demonstrated, the possibility that the intense quinolone selective pressure in aquaculture (Baquero et al. 2008; Cabello 2006; Martinez 2008, 2009) is likely favoring the selection and spread of their resistance elements must be taken into consideration. A very first step in this process can be the transfer of a *qnr* gene from its original host to a mobile genetic element present in an environmental microorganism (Martinez 2008). This could be the case of the *qnrS2* gene, found on the same broad-host-range IncU-type plasmid from two different *Aeromonas* isolates from two non-connected geographical locations (Cattoir et al. 2008; Picao et al. 2008).

It is worth mentioning that although *qnr* genes have been found in a large range of plasmids (Strahilevitz et al. 2009), the modules involved in their transmission present some common characteristics. *qnrA* and *qnrB* are usually located in *sul1*-type integrons, which usually harbor other antibiotic resistance genes, including ß-lactamases and aminoglycoside-inactivating enzymes (Wang et al. 2003) and are associated with IS*CR1* (Garnier et al. 2006; Nordmann and Poirel 2005). *qnrS* genes are not harbored by integrons; however they are frequently associated with the TEM-1 ß-lactamase-containing Tn*3* transposon (Hata et al. 2005; Strahilevitz et al. 2009). Finally, the plasmids themselves might contain other resistance determinants, being particularly relevant the association of *qnr* genes with genes coding for extended-spectrum ß-lactamases and AmpC ß-lactamases (Strahilevitz et al. 2009). These co-resistance associations, in addition to the presence of other resistance genes in *qnr*-encoding plasmids, have likely favored the dissemination of *qnr* genes even when quinolones are not used for therapy.

Although less prevalent at the moment, other transferrable mechanisms of quinolone resistance besides Qnr proteins have been described. Among them, two quinolone efflux pumps, QepA (Yamane et al. 2007) and OqxAB (Hansen et al. 2004, 2007), have been found to be encoded in different plasmids. In addition to quinolones, QepA extrudes a narrow range of substrates, including erythromycin, ethidium bromide, and acriflavine. OqxAB (almost exclusively present in organisms causing animal infections) however presents a wider substrate range that includes tetracycline, chloramphenicol, benzalkonium chloride, and triclosan (Hansen et al. 2007).

A quinolone-inactivating enzyme (ciprofloxacin and norfloxacin), encoded by the *aac(6')-Ib-cr* gene, has been described more than 10 years ago. This enzyme evolved toward quinolone resistance from an original aminoglycoside acetyltransferase through the acquisition of two amino acid changes (Trp102Arg and Asp179Tyr) (Robicsek et al. 2006), which allow the inactivation of quinolones at the cost of reducing the efficacy of the enzyme against aminoglycosides (Robicsek et al. 2006). This modification strongly suggests that the new evolved enzyme has been selected under strong quinolone selective pressure, likely during antimicrobial treatment. Since association of *aac(6')-Ib-cr* with genes encoding the ß-lactamase CTX-M-15 or extended-spectrum ß-lactamases has been reported (Coque et al. 2008; Pitout et al. 2008), co-selection of this determinant by other antibiotics besides quinolones is a suitable possibility.

It has been stated that the risk for the dissemination of resistance is lower in the case of mutations than in the case of HGT-acquired genes. For the first, spread is achieved through clonal expansion, whereas for the second both clonal expansion and gene transfer are both at work. While this is the most common rule, there are some exceptions, and one of them consists in antibiotic resistance mutations in genes encoding bacterial topoisomerases. Indeed, it was described that *parC* and *gyrA* mutations, and the associated phenotype of resistance to quinolones, can be transferred by transformation in *Streptococcus pneumoniae* clinical isolates (Ferrandiz et al. 2000). Further, mutation-acquired resistance to quinolones can be transferred from the viridans group streptococci to *S. pneumoniae* (Balsalobre et al. 2003), indicating that commensal bacteria might contribute by means of transferring target mutations to the development of quinolone resistance, at least in the case of *S. pneumoniae*.

2.6 *Stenotrophomonas maltophilia* and Its Uncommon Mechanisms of Resistance to Quinolones

As above stated, the selection of mutations in genes encoding the bacterial topoisomerases still remains as the main mechanism of resistance to quinolones. Nevertheless the Gram-negative opportunistic pathogen *S. maltophilia* is an exception to such rule. Studies with *in vitro* obtained quinolone-resistant mutants as well as with clinical isolates have shown that quinolone resistance is not associated with mutations in bacterial topoisomerases (Garcia-Leon et al. 2014a; Ribera et al. 2002; Valdezate et al. 2002). Some recent works have described the presence of amino acid changes in the topoisomerases of clinical *S. maltophilia* isolates, but the observed changes are outside the QRDRs and seem to be likely allelic forms of these genes more than mutations involved in quinolone resistance (Cha et al. 2016; Jia et al. 2015).

The analysis of quinolone-resistant mutants and clinical isolates of *S. maltophilia* has shown that the MDR efflux pump SmeDEF is a major contributor to the intrinsic resistance to quinolones of this microorganism when expressed at its basal wild-type level (Zhang et al. 2001). Overexpression of this efflux pump increase MIC levels of quinolones above the breakpoints defining resistance in this microorganism (Alonso and Martinez 2000; Sanchez et al. 2002; Sanchez and Moreno 2005), and mutants overexpressing SmeDEF are frequently found among clinical *S. maltophilia* isolates (Alonso and Martinez 2001; Cho et al. 2012; Gould and Avison 2006; Liaw et al. 2010; Sanchez et al. 2004). While overexpression of this efflux pump produces a fitness cost (Alonso et al. 2004), it might be possible that this cost might be lower than those associated with mutations in genes encoding the bacterial topoisomerases. This differential fitness might be the reason for the lack of topoisomerase mutants among the quinolone-resistant mutants of this bacterial species. Notably, even when the SmeDEF efflux pump is removed, mutations in topoisomerases are not selected in presence of quinolones. The main mechanism of resistance in this case becomes to be the overexpression of another efflux pump, SmeVWX (Garcia-Leon et al. 2014a). Overexpression of this efflux pump has been reported for clinical *S. maltophilia* isolates, indicating that these types of mutants are selected *in vivo* (Garcia-Leon 2015).

In addition to efflux pumps capable of efficiently extruding quinolones, *S. maltophilia* presents in its genome (Sanchez et al. 2008; Shimizu et al. 2008) a *qnr* determinant (Sm*qnr*) that contributes to intrinsic resistance to these antimicrobials (García-León et al. 2012; Sanchez and Martinez 2010; Sanchez and Martinez 2015b). Nevertheless, Sm*qnr* mutants have not been described neither upon *in vitro* selection by quinolones nor in clinical isolates, which cast doubts on the role of this determinant in acquired resistance to quinolones of *S. maltophilia*. More recently, a novel mechanism of resistance based on the increased expression of the heat shock response upon the inactivation of

RNase G has been described in *S. maltophilia* (Bernardini et al. 2015). However, although it has been proposed that the responses to stress are relevant players in the development of antibiotic resistance among bacterial pathogens (Poole 2012), the clinical significance of this mechanism of resistance remains to be fully established.

One important aspect concerning quinolone resistance in *S. maltophilia* is the way it can impact resistance to other drugs. The mutations in the genes encoding bacterial topoisomerases, which are regularly found in quinolone-resistant bacteria, do not alter the susceptibility to other antibiotics of such mutants. However, the increased expression of efflux pumps can simultaneously alter the susceptibility to several different antibiotics. *Stenotrophomonas maltophilia* is an organism with a characteristic low level of susceptibility to antibiotics (Sanchez et al. 2009). The combination trimethoprim/sulfamethoxazole is sometimes the last resort of antimicrobial therapy of infections caused by this pathogen. It has been described that the efflux pump SmeDEF, in addition to being a main player in the acquisition of resistance to quinolones, is also involved in trimethoprim/sulfamethoxazole resistance (Sanchez and Martinez 2015a).

Acknowledgments

Work in the laboratory is supported by was supported by Instituto de Salud Carlos III (grant RD16/0016/0011)—cofinanced by the European Development Regional Fund "A Way to Achieve Europe," by grant S2017/BMD-3691 InGEMICS-CM, funded by Comunidad de Madrid (Spain) and European Structural and Investment Funds and by the Spanish Ministry of Economy and Competitivity (BIO2017-83128-R).

References

Alonso, A. and Martinez, J.L. (2000). Cloning and characterization of SmeDEF, a novel multidrug efflux pump from *Stenotrophomonas maltophilia*. *Antimicrob Agents Chemother* 44 (11): 3079–3086.

Alonso, A. and Martinez, J.L. (2001). Expression of multidrug efflux pump SmeDEF by clinical isolates of *Stenotrophomonas maltophilia*. *Antimicrob Agents Chemother* 45 (6): 1879–1881.

Alonso, A., Morales, G., Escalante, R. et al. (2004). Overexpression of the multidrug efflux pump SmeDEF impairs *Stenotrophomonas maltophilia* physiology. *J Antimicrob Chemother* 53 (3): 432–434.

Alonso, A., Rojo, F., and Martinez, J.L. (1999). Environmental and clinical isolates of *Pseudomonas aeruginosa* show pathogenic and biodegradative properties irrespective of their origin. *Environ Microbiol* 1 (5): 421–430.

Alvarez-Ortega, C., Olivares, J., and Martínez, J.L. (2013). RND multidrug efflux pumps: what are they good for? *Front Microbiol* 4: 7.

Anderson, V.E. and Osheroff, N. (2001). Type II topoisomerases as targets for quinolone antibacterials: turning Dr. Jekyll into Mr. Hyde. *Curr Pharm Des* 7 (5): 337–353.

Anderson, V.R. and Perry, C.M. (2008). Levofloxacin: a review of its use as a high-dose, short-course treatment for bacterial infection. *Drugs* 68 (4): 535–565.

Andersson, D.I. and Hughes, D. (2010). Antibiotic resistance and its cost: is it possible to reverse resistance? *Nat Rev Microbiol* 8 (4): 260–271.

Arsene, S. and Leclercq, R. (2007). Role of a *qnr*-like gene in the intrinsic resistance of *Enterococcus faecalis* to fluoroquinolones. *Antimicrob Agents Chemother* 51 (9): 3254–3258.

Ball, P. (2000). Quinolone generations: natural history or natural selection? *J Antimicrob Chemother* 46 (Suppl T1): 17–24.

Balsalobre, L. and de la Campa, A.G. (2008). Fitness of *Streptococcus pneumoniae* fluoroquinolone-resistant strains with topoisomerase IV recombinant genes. *Antimicrob Agents Chemother* 52 (3): 822–830.

Balsalobre, L., Ferrandiz, M.J., Linares, J. et al. (2003). Viridans group streptococci are donors in horizontal transfer of topoisomerase IV genes to *Streptococcus pneumoniae*. *Antimicrob Agents Chemother* 47 (7): 2072–2081.

Baquero, F., Martinez, J.L., and Canton, R. (2008). Antibiotics and antibiotic resistance in water environments. *Curr Opin Biotechnol* 19 (3): 260–265.

Benveniste, R. and Davies, J. (1973). Aminoglycoside antibiotic-inactivating enzymes in actinomycetes similar to those present in clinical isolates of antibiotic-resistant bacteria. *Proc Natl Acad Sci U S A* 70 (8): 2276–2280.

Bernardini, A., Corona, F., Dias, R. et al. (2015). The inactivation of RNase G reduces the *Stenotrophomonas maltophilia* susceptibility to quinolones by triggering the heat shock response. *Front Microbiol* 6: 1068.

Blanco, P., Hernando-Amado, S., Reales-Calderon, J.A. et al. (2016). Bacterial multidrug efflux pumps: much more than antibiotic resistance determinants. *Microorganisms* 4: 14.

Blazquez, J., Couce, A., Rodriguez-Beltran, J., and Rodriguez-Rojas, A. (2012). Antimicrobials as promoters of genetic variation. *Curr Opin Microbiol* 15 (5): 561–569.

Cabello, F.C. (2006). Heavy use of prophylactic antibiotics in aquaculture: a growing problem for human and animal health and for the environment. *Environ Microbiol* 8 (7): 1137–1144.

Cattoir, V., Poirel, L., Aubert, C. et al. (2008). Unexpected occurrence of plasmid-mediated quinolone resistance determinants in environmental *Aeromonas* spp. *Emerg Infect Dis* 14 (2): 231–237.

Cattoir, V., Poirel, L., Mazel, D. et al. (2007). *Vibrio splendidus* as the source of plasmid-mediated QnrS-like quinolone resistance determinants. *Antimicrob Agents Chemother* 51 (7): 2650–2651.

Cavaco, L.M., Hasman, H., Xia, S., and Aarestrup, F.M. (2009). qnrD, a novel gene conferring transferable quinolone resistance in *Salmonella enterica* serovar Kentucky and Bovismorbificans strains of human origin. *Antimicrob Agents Chemother* 53 (2): 603–608.

Cha, M.K., Kang, C.I., Kim, S.H. et al. (2016). Emergence of fluoroquinolone-resistant *Stenotrophomonas maltophilia* in blood isolates causing bacteremia: molecular epidemiology and microbiologic characteristics. *Diagn Microbiol Infect Dis* 85 (2): 210–212.

Champoux, J.J. (2001). DNA topoisomerases: structure, function, and mechanism. *Annu Rev Biochem* 70: 369–413.

Cho, H.H., Sung, J.Y., Kwon, K.C., and Koo, S.H. (2012). Expression of Sme efflux pumps and multilocus sequence typing in clinical isolates of *Stenotrophomonas maltophilia*. *Ann Lab Med* 32 (1): 38–43.

Cohen, S.P., McMurry, L.M., Hooper, D.C. et al. (1989). Cross-resistance to fluoroquinolones in multiple-antibiotic-resistant (Mar) *Escherichia coli* selected by tetracycline or chloramphenicol: decreased drug accumulation associated with membrane changes in addition to OmpF reduction. *Antimicrob Agents Chemother* 33 (8): 1318–1325.

Coque, T.M., Novais, A., Carattoli, A. et al. (2008). Dissemination of clonally related *Escherichia coli* strains expressing extended-spectrum beta-lactamase CTX-M-15. *Emerg Infect Dis* 14 (2): 195–200.

Crumplin, G.C. and Odell, M. (1987). Development of resistance to ofloxacin. *Drugs* 34 (Suppl 1): 1–8.

Davies, J. (1994). Inactivation of antibiotics and the dissemination of resistance genes. *Science* 264 (5157): 375–382.

De Rossi, E., Arrigo, P., Bellinzoni, M. et al. (2002). The multidrug transporters belonging to major facilitator superfamily in *Mycobacterium tuberculosis*. *Mol Med* 8 (11): 714–724.

Deitz, W.H., Bailey, J.H., and Froelich, E.J. (1963). In vitro antibacterial properties of nalidixic acid, a new drug active against gram-negative organisms. *Antimicrob Agents Chemother (Bethesda)* 161: 583–587.

Domagk, G. (1935). Ein beitrag zur chemotherapie der bakteriellen infektionen. *DMW-Dtsch Med Wochenschr* 61 (07): 250–253.

Drlica, K. and Zhao, X. (1997). DNA gyrase, topoisomerase IV, and the 4-quinolones. *Microbiol Mol Biol Rev* 61 (3): 377–392.

Fernández, L. and Hancock, R.E. (2012). Adaptive and mutational resistance: role of porins and efflux pumps in drug resistance. *Clin Microbiol Rev* 25: 661–681.

Ferrandiz, M.J., Fenoll, A., Linares, J., and De La Campa, A.G. (2000). Horizontal transfer of *parC* and *gyrA* in fluoroquinolone-resistant clinical isolates of *Streptococcus pneumoniae*. *Antimicrob Agents Chemother* 44 (4): 840–847.

Ferrero, L., Cameron, B., Manse, B. et al. (1994). Cloning and primary structure of *Staphylococcus aureus* DNA topoisomerase IV: a primary target of fluoroquinolones. *Mol Microbiol* 13 (4): 641–653.

Fetar, H., Gilmour, C., Klinoski, R. et al. (2011). *mexEF-oprN* multidrug efflux operon of *Pseudomonas aeruginosa*: regulation by the MexT activator in response to nitrosative stress and chloramphenicol. *Antimicrob Agents Chemother* 55: 508–514.

Garcia-Leon, G., Ruiz de Alegria Puig, C., Garcia de la Fuente, C. et al. (2015). High-level quinolone resistance is associated with the overexpression of *smeVWX* in *Stenotrophomonas maltophilia* clinical isolates. *Clin Microbiol Infect* 21 (5): 464–467.

Garcia-Leon, G., Salgado, F., Oliveros, J.C. et al. (2014a). Interplay between intrinsic and acquired resistance to quinolones in *Stenotrophomonas maltophilia*. *Environ Microbiol* 16: 1282–1296.

García-León, G., Hernández, A., Hernando-Amado, S. et al. (2014b). A function of SmeDEF, the major quinolone resistance determinant of *Stenotrophomonas maltophilia*, is the colonization of plant roots. *Appl Environ Microbiol* 80: 4559–4565.

García-León, G., Sánchez, M.B., and Martínez, J.L. (2012). The inactivation of intrinsic antibiotic resistance determinants widens the mutant selection window for quinolones in *Stenotrophomonas maltophilia*. *Antimicrob Agents Chemother* 56: 6397–6399.

Garnier, F., Raked, N., Gassama, A. et al. (2006). Genetic environment of quinolone resistance gene qnrB2 in a complex sul1-type integron in the newly described *Salmonella enterica* serovar Keurmassar. *Antimicrob Agents Chemother* 50 (9): 3200–3202.

Gomez-Gomez, J.M., Blazquez, J., Espinosa De Los Monteros, L.E. et al. (1997). In vitro plasmid-encoded resistance to quinolones. *FEMS Microbiol Lett* 154 (2): 271–276.

Gould, V.C. and Avison, M.B. (2006). SmeDEF-mediated antimicrobial drug resistance in *Stenotrophomonas maltophilia* clinical isolates having defined phylogenetic relationships. *J Antimicrob Chemother* 57 (6): 1070–1076.

Grkovic, S., Brown, M.H., and Skurray, R.A. (2002). Regulation of bacterial drug export systems. *Microbiol Mol Biol Rev* 66 (4): 671–701.

Hansen, L.H., Jensen, L.B., Sorensen, H.I., and Sorensen, S.J. (2007). Substrate specificity of the OqxAB multidrug resistance pump in *Escherichia coli* and selected enteric bacteria. *J Antimicrob Chemother* 60 (1): 145–147.

Hansen, L.H., Johannesen, E., Burmolle, M. et al. (2004). Plasmid-encoded multidrug efflux pump conferring resistance to olaquindox in *Escherichia coli*. *Antimicrob Agents Chemother* 48 (9): 3332–3337.

Hata, M., Suzuki, M., Matsumoto, M. et al. (2005). Cloning of a novel gene for quinolone resistance from a transferable plasmid in *Shigella flexneri* 2b. *Antimicrob Agents Chemother* 49 (2): 801–803.

Hernandez, A., Ruiz, F.M., Romero, A., and Martinez, J.L. (2011). The binding of triclosan to SmeT, the repressor of the multidrug efflux pump SmeDEF, induces antibiotic resistance in *Stenotrophomonas maltophilia*. *PLoS Pathog* 7 (6): e1002103.

Hernando-Amado, S., Blanco, P., Alcalde-Rico, M. et al. (2016). Multidrug efflux pumps as main players in intrinsic and acquired resistance to antimicrobials. *Drug Resist Updat* 28: 13–27.

Hooper, D.C. (1999). Mechanisms of fluoroquinolone resistance. *Drug Resist Updat* 2 (1): 38–55.

Horlein, H. (1936). The chemotherapy of infectious diseases caused by protozoa and bacteria: (section of tropical diseases and parasitology). *Proc R Soc Med* 29 (4): 313–324.

Jacoby, G.A. (2005). Mechanisms of resistance to quinolones. *Clin Infect Dis* 41 (Suppl 2): S120–S126.

Jacoby, G.A., Walsh, K.E., Mills, D.M. et al. (2006). qnrB, another plasmid-mediated gene for quinolone resistance. *Antimicrob Agents Chemother* 50 (4): 1178–1182.

Jalal, S., Ciofu, O., Hoiby, N. et al. (2000). Molecular mechanisms of fluoroquinolone resistance in *Pseudomonas aeruginosa* isolates from cystic fibrosis patients. *Antimicrob Agents Chemother* 44 (3): 710–712.

Jia, W., Wang, J., Xu, H., and Li, G. (2015). Resistance of *Stenotrophomonas maltophilia* to fluoroquinolones: prevalence in a university hospital and possible mechanisms. *Int J Environ Res Public Health* 12 (5): 5177–5195.

Khodursky, A.B. and Cozzarelli, N.R. (1998). The mechanism of inhibition of topoisomerase IV by quinolone antibacterials. *J Biol Chem* 273 (42): 27668–27677.

Kreuzer, K.N. and Cozzarelli, N.R. (1979). *Escherichia coli* mutants thermosensitive for deoxyribonucleic acid gyrase subunit A: effects on deoxyribonucleic acid replication, transcription, and bacteriophage growth. *J Bacteriol* 140 (2): 424–435.

Lesher, G.Y., Froelich, E.J., Gruett, M.D. et al. (1962). 1,8-Naphthyridine derivatives. A new class of chemotherapeutic agents. *J Med Pharm Chem* 91: 1063–1065.

Levine, C., Hiasa, H., and Marians, K.J. (1998). DNA gyrase and topoisomerase IV: biochemical activities, physiological roles during chromosome replication, and drug sensitivities. *Biochim Biophys Acta* 1400 (1-3): 29–43.

Li, X.Z., Livermore, D.M., and Nikaido, H. (1994). Role of efflux pump(s) in intrinsic resistance of *Pseudomonas aeruginosa*: resistance to tetracycline, chloramphenicol, and norfloxacin. *Antimicrob Agents Chemother* 38 (8): 1732–1741.

Li, X.Z. and Nikaido, H. (2009). Efflux-mediated drug resistance in bacteria: an update. *Drugs* 69 (12): 1555–1623.

Li, X.-Z., Plésiat, P., and Nikaido, H. (2015). The challenge of efflux-mediated antibiotic resistance in Gram-negative bacteria. *Clin Microbiol Rev* 28: 337–418.

Liaw, S.J., Lee, Y.L., and Hsueh, P.R. (2010). Multidrug resistance in clinical isolates of *Stenotrophomonas maltophilia*: roles of integrons, efflux pumps, phosphoglucomutase (SpgM), and melanin and biofilm formation. *Int J Antimicrob Agents* 35 (2): 126–130.

Llanes, C., Neuwirth, C., El Garch, F. et al. (2006). Genetic analysis of a multiresistant strain of *Pseudomonas aeruginosa* producing PER-1 beta-lactamase. *Clin Microbiol Infect* 12 (3): 270–278.

Lopez, E. and Blazquez, J. (2009). Effect of subinhibitory concentrations of antibiotics on intrachromosomal homologous recombination in *Escherichia coli*. *Antimicrob Agents Chemother* 53 (8): 3411–3415.

Lopez, E., Elez, M., Matic, I., and Blazquez, J. (2007). Antibiotic-mediated recombination: ciprofloxacin stimulates SOS-independent recombination of divergent sequences in *Escherichia coli*. *Mol Microbiol* 64 (1): 83–93.

Luo, N., Pereira, S., Sahin, O. et al. (2005). Enhanced in vivo fitness of fluoroquinolone-resistant *Campylobacter jejuni* in the absence of antibiotic selection pressure. *Proc Natl Acad Sci U S A* 102 (3): 541–546.

Marcusson, L.L., Frimodt-Moller, N., and Hughes, D. (2009). Interplay in the selection of fluoroquinolone resistance and bacterial fitness. *PLoS Pathog* 5 (8): e1000541.

Martinez, J.L. (2008). Antibiotics and antibiotic resistance genes in natural environments. *Science* 321 (5887): 365–367.

Martinez, J.L. (2009). The role of natural environments in the evolution of resistance traits in pathogenic bacteria. *Proc Biol Sci* 276 (1667): 2521–2530.

Martinez, J.L., Alonso, A., Gomez-Gomez, J.M., and Baquero, F. (1998). Quinolone resistance by mutations in chromosomal gyrase genes. Just the tip of the iceberg? *J Antimicrob Chemother* 42 (6): 683–688.

Martinez, J.L., Coque, T.M., and Baquero, F. (2015). What is a resistance gene? Ranking risk in resistomes. *Nat Rev Microbiol* 13 (2): 116–123.

Martinez, J.L., Sanchez, M.B., Martinez-Solano, L. et al. (2009). Functional role of bacterial multidrug efflux pumps in microbial natural ecosystems. *FEMS Microbiol Rev* 33 (2): 430–449.

Martinez-Martinez, L., Pascual, A., and Jacoby, G.A. (1998). Quinolone resistance from a transferable plasmid. *Lancet* 351 (9105): 797–799.

Montero, C., Mateu, G., Rodriguez, R., and Takiff, H. (2001). Intrinsic resistance of *Mycobacterium smegmatis* to fluoroquinolones may be influenced by new pentapeptide protein MfpA. *Antimicrob Agents Chemother* 45 (12): 3387–3392.

Nikaido, H. and Takatsuka, Y. (2009). Mechanisms of RND multidrug efflux pumps. *Biochim Biophys Acta* 1794 (5): 769–781.

Nordmann, P. and Poirel, L. (2005). Emergence of plasmid-mediated resistance to quinolones in Enterobacteriaceae. *J Antimicrob Chemother* 56 (3): 463–469.

Paulsen, I.T. (2003). Multidrug efflux pumps and resistance: regulation and evolution. *Curr Opin Microbiol* 6 (5): 446–451.

Phillips, I., Culebras, E., Moreno, F., and Baquero, F. (1987). Induction of the SOS response by new 4-quinolones. *J Antimicrob Chemother* 20 (5): 631–638.

Picao, R.C., Poirel, L., Demarta, A. et al. (2008). Plasmid-mediated quinolone resistance in *Aeromonas allosaccharophila* recovered from a Swiss lake. *J Antimicrob Chemother* 62 (5): 948–950.

Piddock, L.J. (1999). Mechanisms of fluoroquinolone resistance: an update 1994-1998. *Drugs* 58: 11–18.

Piddock, L.J. (2006a). Clinically relevant chromosomally encoded multidrug resistance efflux pumps in bacteria. *Clin Microbiol Rev* 19 (2): 382–402.

Piddock, L.J. (2006b). Multidrug-resistance efflux pumps: not just for resistance. *Nat Rev Microbiol* 4 (8): 629–636.

Pitout, J.D., Wei, Y., Church, D.L., and Gregson, D.B. (2008). Surveillance for plasmid-mediated quinolone resistance determinants in *Enterobacteriaceae* within the Calgary Health Region, Canada: the emergence of *aac(6')-Ib-cr*. *J Antimicrob Chemother* 61 (5): 999–1002.

Poirel, L., Liard, A., Rodriguez-Martinez, J.M., and Nordmann, P. (2005a). Vibrionaceae as a possible source of Qnr-like quinolone resistance determinants. *J Antimicrob Chemother* 56 (6): 1118–1121.

Poirel, L., Rodriguez-Martinez, J.M., Mammeri, H. et al. (2005b). Origin of plasmid-mediated quinolone resistance determinant QnrA. *Antimicrob Agents Chemother* 49 (8): 3523–3525.

Poole, K. (2000a). Efflux-mediated resistance to fluoroquinolones in gram-negative bacteria. *Antimicrob Agents Chemother* 44 (9): 2233–2241.

Poole, K. (2000b). Efflux-mediated resistance to fluoroquinolones in gram-positive bacteria and the mycobacteria. *Antimicrob Agents Chemother* 44 (10): 2595–2599.

Poole, K. (2007). Efflux pumps as antimicrobial resistance mechanisms. *Ann Med* 39 (3): 162–176.

Poole, K. (2012). Stress responses as determinants of antimicrobial resistance in Gram-negative bacteria. *Trends Microbiol* 20: 227–234.

Redgrave, L.S., Sutton, S.B., Webber, M.A., and Piddock, L.J. (2014). Fluoroquinolone resistance: mechanisms, impact on bacteria, and role in evolutionary success. *Trends Microbiol* 22 (8): 438–445.

Renau, T.E., Leger, R., Flamme, E.M. et al. (1999). Inhibitors of efflux pumps in *Pseudomonas aeruginosa* potentiate the activity of the fluoroquinolone antibacterial levofloxacin. *J Med Chem* 42 (24): 4928–4931.

Ribera, A., Domenech-Sanchez, A., Ruiz, J. et al. (2002). Mutations in *gyrA* and *parC* QRDRs are not relevant for quinolone resistance in epidemiological unrelated *Stenotrophomonas maltophilia* clinical isolates. *Microb Drug Resist* 8 (4): 245–251.

Robicsek, A., Strahilevitz, J., Jacoby, G.A. et al. (2006). Fluoroquinolone-modifying enzyme: a new adaptation of a common aminoglycoside acetyltransferase. *Nat Med* 12 (1): 83–88.

Rodriguez-Martinez, J.M., Velasco, C., Briales, A. et al. (2008). Qnr-like pentapeptide repeat proteins in Gram-positive bacteria. *J Antimicrob Chemother* 61 (6): 1240–1243.

Rosenberg, E.Y., Bertenthal, D., Nilles, M.L. et al. (2003). Bile salts and fatty acids induce the expression of *Escherichia coli* AcrAB multidrug efflux pump through their interaction with Rob regulatory protein. *Mol Microbiol* 48 (6): 1609–1619.

Saier, M.H. Jr., Paulsen, I.T., Sliwinski, M.K. et al. (1998). Evolutionary origins of multidrug and drug-specific efflux pumps in bacteria. *FASEB J* 12 (3): 265–274.

Sanchez, M.B., Hernandez, A., and Martinez, J.L. (2009). *Stenotrophomonas maltophilia* drug resistance. *Future Microbiol* 4 (6): 655–660.

Sanchez, M.B., Hernandez, A., Rodriguez-Martinez, J.M. et al. (2008). Predictive analysis of transmissible quinolone resistance indicates *Stenotrophomonas maltophilia* as a potential source of a novel family of Qnr determinants. *BMC Microbiol* 8: 148.

Sanchez, M.B. and Martinez, J.L. (2010). SmQnr contributes to intrinsic resistance to quinolones in *Stenotrophomonas maltophilia*. *Antimicrob Agents Chemother* 54 (1): 580–581.

Sanchez, M.B. and Martinez, J.L. (2015a). The efflux pump SmeDEF contributes to trimethoprim-sulfamethoxazole resistance in *Stenotrophomonas maltophilia*. *Antimicrob Agents Chemother* 59 (7): 4347–4348.

Sanchez, M.B. and Martinez, J.L. (2015b). Regulation of Smqnr expression by SmqnrR is strain-specific in *Stenotrophomonas maltophilia*. *J Antimicrob Chemother* 70 (10): 2913–2914.

Sanchez, P., Alonso, A., and Martinez, J.L. (2002). Cloning and characterization of SmeT, a repressor of the *Stenotrophomonas maltophilia* multidrug efflux pump SmeDEF. *Antimicrob Agents Chemother* 46 (11): 3386–3393.

Sanchez, P., Alonso, A., and Martinez, J.L. (2004). Regulatory regions of *smeDEF* in *Stenotrophomonas maltophilia* strains expressing different amounts of the multidrug efflux pump SmeDEF. *Antimicrob Agents Chemother* 48 (6): 2274–2276.

Sanchez, P., Moreno, E., and Martinez, J.L. (2005). The biocide triclosan selects *Stenotrophomonas maltophilia* mutants that overproduce the SmeDEF multidrug efflux pump. *Antimicrob Agents Chemother* 49: 781–782.

Shen, L.L., Kohlbrenner, W.E., Weigl, D., and Baranowski, J. (1989). Mechanism of quinolone inhibition of DNA gyrase. Appearance of unique norfloxacin binding sites in enzyme-DNA complexes. *J Biol Chem* 264 (5): 2973–2978.

Shimizu, K., Kikuchi, K., Sasaki, T. et al. (2008). Sm*qnr*, a new chromosome-carried quinolone resistance gene in *Stenotrophomonas maltophilia*. *Antimicrob Agents Chemother* 52 (10): 3823–3825.

Strahilevitz, J., Engelstein, D., Adler, A. et al. (2007). Changes in qnr prevalence and fluoroquinolone resistance in clinical isolates of *Klebsiella pneumoniae* and

Enterobacter spp. collected from 1990 to 2005. *Antimicrob Agents Chemother* 51 (8): 3001–3003.

Strahilevitz, J., Jacoby, G.A., Hooper, D.C., and Robicsek, A. (2009). Plasmid-mediated quinolone resistance: a multifaceted threat. *Clin Microbiol Rev* 22 (4): 664–689.

Tran, J.H. and Jacoby, G.A. (2002). Mechanism of plasmid-mediated quinolone resistance. *Proc Natl Acad Sci U S A* 99 (8): 5638–5642.

Truong-Bolduc, Q.C., Strahilevitz, J., and Hooper, D.C. (2006). NorC, a new efflux pump regulated by MgrA of *Staphylococcus aureus*. *Antimicrob Agents Chemother* 50 (3): 1104–1107.

Valdezate, S., Vindel, A., Echeita, A. et al. (2002). Topoisomerase II and IV quinolone resistance-determining regions in *Stenotrophomonas maltophilia* clinical isolates with different levels of quinolone susceptibility. *Antimicrob Agents Chemother* 46 (3): 665–671.

Vetting, M.W., Hegde, S.S., Fajardo, J.E. et al. (2006). Pentapeptide repeat proteins. *Biochemistry* 45 (1): 1–10.

Vila, J. and Martinez, J.L. (2008). Clinical impact of the over-expression of efflux pump in nonfermentative gram-negative bacilli, development of efflux pump inhibitors. *Curr Drug Targets* 9 (9): 797–807.

Waksman, S.A. and Woodruff, H.B. (1940). The soil as a source of microorganisms antagonistic to disease-producing bacteria. *J Bacteriol* 40 (4): 581–600.

Wang, M., Guo, Q., Xu, X. et al. (2009). New plasmid-mediated quinolone resistance gene, qnrC, found in a clinical isolate of *Proteus mirabilis*. *Antimicrob Agents Chemother* 53: 1892.

Wang, M., Tran, J.H., Jacoby, G.A. et al. (2003). Plasmid-mediated quinolone resistance in clinical isolates of *Escherichia coli* from Shanghai, China. *Antimicrob Agents Chemother* 47 (7): 2242–2248.

Webb, V. and Davies, J. (1993). Antibiotic preparations contain DNA: a source of drug resistance genes? *Antimicrob Agents Chemother* 37 (11): 2379–2384.

Yamane, K., Wachino, J., Suzuki, S. et al. (2007). New plasmid-mediated fluoroquinolone efflux pump, QepA, found in an *Escherichia coli* clinical isolate. *Antimicrob Agents Chemother* 51 (9): 3354–3360.

Yang, S., Clayton, S.R., and Zechiedrich, E.L. (2003). Relative contributions of the AcrAB, MdfA and NorE efflux pumps to quinolone resistance in *Escherichia coli*. *J Antimicrob Chemother* 51 (3): 545–556.

Ysern, P., Clerch, B., Castano, M. et al. (1990). Induction of SOS genes in *Escherichia coli* and mutagenesis in *Salmonella typhimurium* by fluoroquinolones. *Mutagenesis* 5 (1): 63–66.

Zhang, L., Li, X.Z., and Poole, K. (2001). SmeDEF multidrug efflux pump contributes to intrinsic multidrug resistance in *Stenotrophomonas maltophilia*. *Antimicrob Agents Chemother* 45 (12): 3497–3503.

Ziha-Zarifi, I., Llanes, C., Kohler, T. et al. (1999). In vivo emergence of multidrug-resistant mutants of *Pseudomonas aeruginosa* overexpressing the active efflux system MexA-MexB-OprM. *Antimicrob Agents Chemother* 43 (2): 287–291.

3

Beta-Lactams

Luz Balsalobre[1], Ana Blanco[1], and Teresa Alarcón[1,2]

[1] *Department of Microbiology, Hospital Universitario La Princesa, Instituto de Investigación Sanitaria Princesa, Madrid, Spain*
[2] *Department of Preventive Medicine, Public Health and Microbiology, Medical School, Autonomous University of Madrid, Madrid, Spain*

3.1 Introduction

Beta-lactam antibiotics are a large group of antibiotics that share in common a basic structure called beta-lactam ring. The diverse molecular structures of these antibiotics allow them to be classified into several subgroups with unique characteristics. Beta-lactam antibiotics are the most popular antibacterial agents used for treating bacterial infections due to their bactericidal activity and low toxicity, except for patients suffering from allergies.

The antibacterial activity of the first beta-lactam antibiotic, penicillin G, was described by Fleming in 1928 (Fleming 2001), which was widely used to treat wound infections during World War II and in the 1940s (Bush and Bradford 2016).

However, some years later, penicillin-resistant strains of *Staphylococcus aureus* emerged. Those strains expressed a beta-lactamase able to hydrolyze the beta-lactam ring (Kirby 1944). Methicillin, a beta-lactamase-insensitive semisynthetic penicillin, was introduced afterward, but methicillin-resistant *S. aureus* (MRSA) was identified soon after the introduction of methicillin into clinical practice (Saroglou et al. 1980).

The most important mechanism of resistance is the production of beta-lactamases, which catalyze the hydrolysis of the beta-lactam ring and are frequently encoded on plasmids that are able to disseminate among strains of the same species or genus or even among non-related bacteria. The production of low-affinity penicillin-binding proteins (PBPs) is the second most frequent mechanism of resistance in Gram-positive and Gram-negative bacteria such as MRSA or *Streptococcus pneumoniae*, where resistance is acquired by

Antibiotic Drug Resistance, First Edition. Edited by José-Luis Capelo-Martínez and Gilberto Igrejas.

transformation. And finally a decrease in the production of the outer membrane proteins or bacterial efflux system also contributes to lower accumulation of antibiotics inside the cell, reducing the efficacy of the antibiotics (Zervosen et al. 2012).

The emergence of resistance to beta-lactams led to the development and discovery of new groups of drugs. Development of new drugs attempts to address the problem of resistance, but resistance to this new group of drugs occurred once again, making the fight against resistant bacteria a long one.

Due to lack of new antibiotics to treat infections produced by multiresistant bacteria, the Infectious Diseases Society of America (IDSA) launched the "10 × '20 Initiative" in 2010 (Boucher et al. 2013), a campaign calling for the development and approval of 10 antibiotics by 2020. Beta-lactamases producing bacteria, especially carpabenemases, are responsible for many difficult-to-treat infections worldwide, and the development of new antibiotics to treat these infections is a priority (Boucher et al. 2013).

3.2 Chemical Structure

The beta-lactam ring is an integral part of the chemical structure of several antibiotic families of beta-lactams. It is a heterocyclic ring, formed by cycling of an amide group, and therefore, composed of three atoms of carbon and one of nitrogen (Gomez et al. 2015). The presence (and structure) or absence of a secondary ring allows beta-lactams to be classified based on their core ring structures (Figure 3.1).

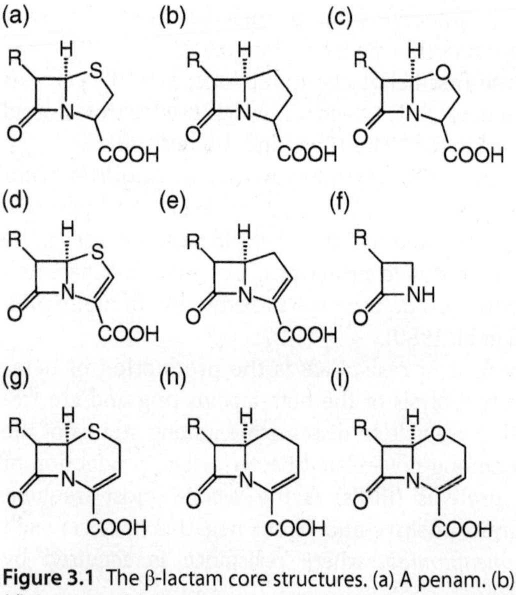

Figure 3.1 The β-lactam core structures. (a) A penam. (b) A carbapenem. (c) An oxapenam. (d) A penam. (e) A carbapenem. (f) A monobactam. (g) A cephem. (h) A carbacephem. (i) An oxacephem.

Penam, oxapenam, carbapenem, monobactam, and cephem are beta-lactam subclasses of which the clinically relevant beta-lactam antibiotics are derived. The radicals joined to the core rings define the different antibiotics and also their antimicrobial activity, pharmacokinetic and toxicity.

The high reactivity of beta-lactam ring is responsible for the inherent allergenicity to beta-lactam antibiotics. It is estimated that six to eight percent of the population is allergic to these drugs. Hydroxyl or sulfhydryl groups found in certain proteins react with the ring system, creating a covalent penicillin-protein conjugate that can induce an allergic response (Bousquet et al. 2008).

3.3 Classification and Spectrum of Activity

According to their beta-lactam core ring structure and chemical structure, beta-lactam antibiotics are classified into five important groups with a particular spectrum of activity.

3.3.1 Penicillins

Penicillins are included in the large group of penams. They possess a bicyclic core structure, 6-aminopenicillanic acid (6-APA), which is formed by condensation of L-cysteine and D-valine (Long et al. 2005). Due to the highly reactive nature of the beta-lactam ring, penicillins are susceptible to degradation under certain conditions such as acidity. This is because they must be protected from gastric acid pH when used by oral administration (Fernandes and Prudêncio 2013). The different penicillins are defined by the side chain in the 6-amino group (Figure 3.2).

Penicillins can be classified as natural (benzylpenicillin or penicillin G and phenoxymethylpenicillin or penicillin V) and semisynthetic penicillins, which in turn can be grouped into penicillinase-resistant (cloxacillin, dicloxacillin, nafcillin, oxacillin, temocillin) or extended-spectrum penicillins (including amoxicillin, ampicillin, and mecillinam that are aminopenicillins, ticarcillin that is a carboxypenicillin, and piperacillin that is a ureidopenicillin) (Fernandes and Prudêncio 2013).

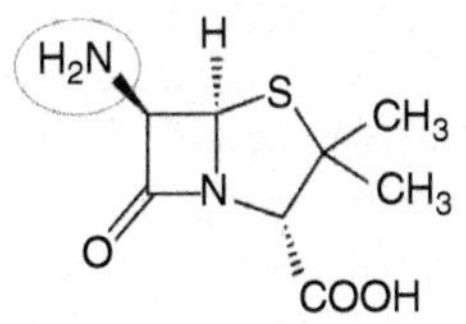

Figure 3.2 Molecule of 6-aminopenicillanic acid. Circle: amino group in position 6.

3.3.1.1 Natural Penicillins

These penicillins are produced from a number of species of *Penicillium* spp. Penicillin G was the first antimicrobial agent commercialized and is still active against different bacteria.

Natural penicillins (penicillin G and penicillin V) have activity against (Miller 2002; Bush and Bradford 2016):

- Non-beta-lactamase-producing Gram-positive cocci, including *viridans* streptococci, group A streptococci, *S. pneumoniae*, enterococci, *S. aureus*, coagulase-negative staphylococci, and anaerobic streptococci (*Peptostreptococcus, Peptococcus* spp.).
- Gram-positive bacilli such as *Listeria monocytogenes, Erysipelothrix rhusiopathiae, and Bacillus anthracis.*
- Gram-negative cocci such as *Neisseria meningitidis* and non-penicillinase producing *Neisseria gonorrhoeae* strains.
- *Treponema pallidum* and many other spirochetes.
- Anaerobic organisms such as *Clostridium* spp. (excluding *Clostridium difficile*), *Actinomyces* spp., and non-beta-lactamase-producing Gram-negative rods.

3.3.1.2 Semisynthetic Penicillins

Semisynthetic penicillins are prepared by using 6-APA as starting block and adding side chain to the 6-amino group.

3.3.1.2.1 Penicillinase-Resistant Penicillins

The addition of an isoxazolyl side chain to the 6-amino group of the penicillin compound protects the beta-lactam ring from hydrolysis by penicillinases produced by staphylococci. However, the agents in this group, also known as antistaphylococcal penicillins, showed lower activity against streptococci than natural penicillins and a null activity against enterococci. Anaerobic activity ranges from minimal to none and Gram-negative activity is virtually nonexistent (Miller 2002; Bush and Bradford 2016). Methicillin, oxacillin, and cloxacillin are included in this group.

3.3.1.2.2 Aminopenicillins

Aminopenicillins (ampicillin and amoxicillin) were developed by adding an amino group to benzylpenicillin. This modification improved coverage against Gram-negative organisms, whereas the spectrum of activity against Gram-positive organisms remained similar to that of the natural penicillins.

These agents showed activity against streptococci and had slightly greater activity against *Enterococcus* spp. and *L. monocytogenes* than the natural penicillins. Among Gram-negative bacilli, these drugs showed activity against *Haemophilus influenzae, Escherichia coli, Proteus mirabilis, Salmonella* spp., and *Shigella* spp. However, the side chain added did not inhibit hydrolysis by staphylococcal penicillinases or Gram-negative beta-lactamases (Bush and Bradford 2016).

3.3.1.2.3 Antipseudomonal Penicillins: Carboxypenicillins and Ureidopenicillins

In order to increase Gram-negative coverage and particularly against *Pseudomonas aeruginosa,* a carboxyl group was added to the penicillin structure to produce carboxypenicillins (ticarcillin). This modification enhanced Gram-negative coverage by improving its bacterial penetration through the cell wall. Their spectrum of activity includes *Enterobacter, Providencia, Morganella,* indole-positive *Proteus,* and *P. aeruginosa.*

To produce piperacillin, an ureido group plus a piperazine side chain was added. The Gram-negative coverage of piperacillin includes that of the carboxypenicillins plus coverage against *Klebsiella, Serratia,* and *Enterococcus* and increased anaerobic coverage (Miller 2002; Bush and Bradford 2016).

3.3.2 Cephalosporins

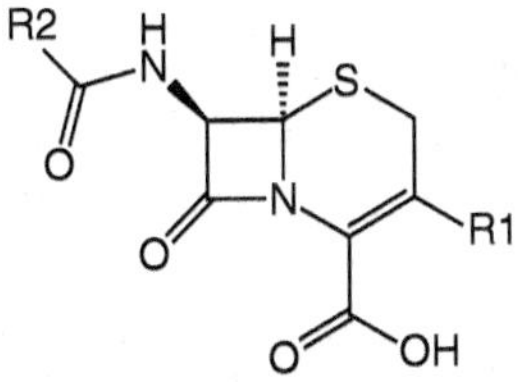

Figure 3.3 7-Aminocephalosporanic acid, core structure of cephalosporins.

Cephalosporins are derivatives of the fermentation products of the fungus *Acremonium chrysogenum.* Its cephem core structure is called 7-aminocephalosporanic acid (7-ACA) (Figure 3.3). The chemical structure in radicals R1 and R2 define the different cephalosporin molecules as well as their antibacterial activities and pharmacokinetic properties (Garau et al. 1997; Fernandes and Prudêncio 2013).

Cephalosporins are more resistant against hydrolysis by beta-lactamases than penicillins; therefore, they showed a broader spectrum of activity and they are classified into five generations according to their antibacterial activity (Fernandes and Prudêncio 2013). In general, the first- and second-generation cephalosporins have good activity against Gram-positive microorganisms, and the third- and fourth-generation are more active against Gram-negative pathogens. However, *Listeria, Pasteurella* spp., enterococci, and all methicillin-resistant staphylococci are cephalosporin resistant (with the exception of fifth-generation cephalosporins).

3.3.2.1 First Generation

The following cephalosporins are included: cephalothin, cephapirin, cefazolin, cephalexin, cephradine, and cefadroxil. They are very active against Gram-positive cocci, except for enterococci and MRSA, moderately active against some Gram-negative rods (*E. coli, P. mirabilis,* and *Klebsiella*), and poorly active against *Moraxella catarrhalis, H. influenzae,* and *Neisseria* spp. Anaerobic bacteria are often sensitive, excluding *Bacteroides fragilis.* None of the drugs in this group penetrate the central nervous system (CNS), and they are not first choice drugs for any infection.

3.3.2.2 Second Generation

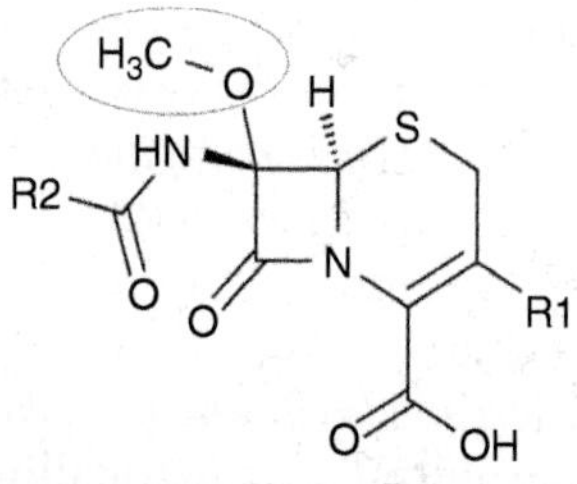

Figure 3.4 Core structure of cephamycins. Circle: methoxy group in position 7.

The following cephalosporins are included within this group: cefamandole, cefuroxime, cefonicid, and ceforanide. It is a heterogeneous group with activity against organisms covered by first-generation drugs, with a slight loss of activity against staphylococci and an extended coverage against Gram-negative bacilli, excluding *P. aeruginosa*. Cephamycins (cefoxitin, cefmetazole, cefminox, and cefotetan) are not cephalosporins since they have a methoxy group in position 7 (Figure 3.4), but they are included in this group for their clinical utility; some of them such as cefoxitin and cefotetan are also active against *B. fragilis*.

3.3.2.3 Third Generation

Cephalosporins included in this group are cefotaxime, ceftizoxime, ceftriaxone, ceftazidime, cefoperazone, cefixime, ceftibuten, and cefdinir. They have decreased activity against Gram-positive cocci but remarkable activity against Gram-negative rods due to their enhanced beta-lactamase stability. Among Gram-positive cocci they show a marked activity against streptococci but a limited activity against staphylococci. Among Gram-negative pathogens, they are very active against *E. coli*, *Klebsiella* spp., *Proteus*, *Providencia*, *Citrobacter*, and *Serratia*.

Ceftazidime has an α-carbon dimethylacetic acid rather than the more common methoxyamino group, which is responsible for its enhanced potency against *Pseudomonas* spp. Cefoperazone also has activity against *P. aeruginosa*. Several drugs in this group can reach the CNS in sufficient concentrations to treat meningitis caused by Gram-negative rods.

3.3.2.4 Fourth Generation

Fourth-generation cephalosporins, such as cefpirome and cefepime, exhibit a greater antibacterial spectrum than third-generation cephalosporins, including *P. aeruginosa* and many *Enterobacteriaceae*, because they are active against inducible and derepressed AmpC beta-lactamase-producing species (such as *Enterobacter* and *Citrobacter* species). In addition, they are more active than third-generation cephalosporins against Gram-positive cocci, including methicillin-susceptible *S. aureus* (Garau et al. 1997; Bush and Bradford 2016). However, they have minimal anaerobic coverage.

3.3.2.5 Fifth Generation

The following cephalosporins are included within this group: ceftobiprole, ceftaroline, and ceftolozane.

Ceftaroline, the active metabolite of the prodrug ceftaroline fosamil, was developed to specifically target resistant strains of bacteria, such as MRSA, and it binds with high affinity to PBP2a of staphylococci, which confers methicillin resistance, and also binds to all six PBPs in *S. pneumoniae.* Due to this capability, it shows a potent activity against MRSA and *S. pneumoniae* penicillin-resistant strains. In addition, ceftaroline has activity against enterococci, hetero-resistant vancomycin-intermediate *S. aureus* (hVISA), vancomycin-resistant *S. aureus* (VRSA), and many Gram-negative pathogens excluding extended-spectrum beta-lactamase (ESBL)-producing or AmpC-overexpressing strains. Among anaerobes, ceftaroline is active against *Propionibacterium* spp. and *Actinomyces* spp. but inactive against *B. fragilis* and *Prevotella* spp. (Bassetti et al. 2013; Shirley et al. 2013).

Ceftobiprole medocaril, another fifth-generation cephalosporin, is a prodrug that is almost completely metabolized to the active drug, ceftobiprole, following intravenous administration. It shows a broad spectrum of activity against Gram-positive and Gram-negative pathogens, similar to that of ceftaroline. Among Gram-positive pathogens, it shows activity against MRSA due to its high affinity for PBP2a, vancomycin-resistant *S. aureus*, penicillin-resistant pneumococci, and *Enterococcus faecalis.* Among Gram-negative pathogens, this drug exhibits activity against *H. influenzae* and *M. catarrhalis*, irrespective of beta-lactamase production; *P. aeruginosa*; *Acinetobacter* spp.; and non-ESBL-producing *Enterobacteriaceae.* For anaerobic bacteria, ceftobiprole is generally active against *Clostridium* spp. and *Fusobacterium* spp., but inactive against *Bacteroides* spp., *Prevotella* spp., and *Veillonella* spp. (Bassetti et al. 2013; Farrell et al. 2014).

Ceftolozane is a fifth-generation cephalosporin that differs from other cephalosporins due to its increased activity against some AmpC beta-lactamase producers including *P. aeruginosa.*

3.3.3 Monobactams

Monobactams are monocyclic beta-lactams active against Gram-negative bacilli including *Pseudomonas* spp. However, they have no activity against Gram-positive bacteria or anaerobes. Aztreonam is the only clinically available antibiotic of this group (Figure 3.5). Its activity is similar but slightly inferior to ceftazidime but can be

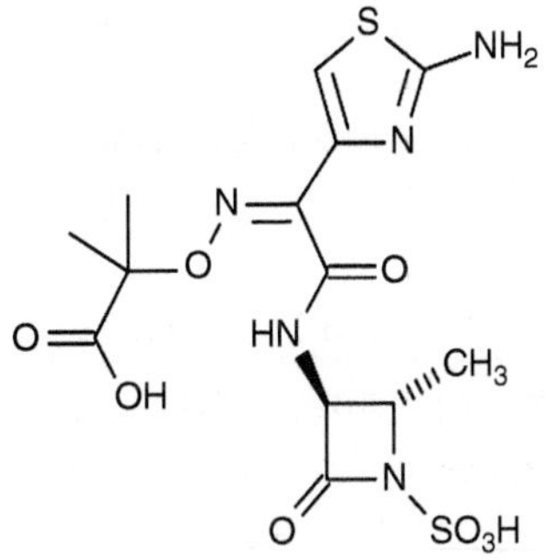

Figure 3.5 Aztreonam molecule.

used in patients with type 2 penicillin or cephalosporin hypersensitivity. There exists a nebulized formulation useful for cystic fibrosis patients colonized by Gram-negative rods, including *P. aeruginosa* (Gomez et al. 2015).

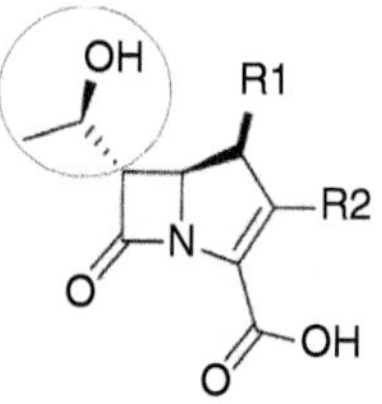

Figure 3.6 Carbapenem core structure. Circle: 6-*trans*-hydroxyethyl group.

3.3.4 Carbapenems

Carbapenems are the widest spectrum antibiotics available among beta-lactam agents. They differ from other beta-lactams in that they have a carbon atom instead of a sulfur or oxygen atom in the bicyclic nucleus and a hydroxyethyl side chain in *trans* configuration at position 6 (Figure 3.6), which confers stability against most beta-lactamases.

The first carbapenem known was thienamycin, produced by the Gram-positive bacteria *Streptomyces cattleya* (Birnbaum et al. 1985). Imipenem was subsequently obtained by chemical modification. In total, there are four carbapenems that are widely commercialized (doripenem, ertapenem, imipenem, and meropenem) and other two (biapenem and tebipenem) available only in Japan (Bush and Bradford 2016).

In general, their spectrum of activity extends to the majority of Gram-positive and Gram-negative pathogens, including both aerobes and anaerobes, due to their efficient bacterial penetration, stability against hydrolysis by most beta-lactamases, and high affinity for multiple PBPs. However, this class of beta-lactams shows intrinsic inactivity against methicillin-resistant staphylococci, *Enterococcus faecium*, and some non-fermenting rods, such as *Stenotrophomonas maltophilia* and *Burkholderia cepacia*.

Imipenem and doripenem are potent antibiotics against Gram-positive bacteria, whereas meropenem and ertapenem are slightly more effective against Gram-negative organisms. However, ertapenem has a more limited spectrum, because it is not as active as imipenem or meropenem against *P. aeruginosa* and other non-fermenting rods (Papp-Wallace et al. 2011).

3.3.5 Beta-Lactam Associated with Beta-Lactamase Inhibitors

Beta-lactamase inhibitors in clinical medicine are introduced in 1970 and are a good approach to combat beta-lactam resistance. They are used in combination with a beta-lactam and are able to restore the activity of beta-lactam.

They can be classified in two groups (Table 3.1):

- Beta-lactam inhibitors. Clavulanic acid, tazobactam, and sulbactam are structurally beta-lactams and act as inactivators or "suicide inhibitors" of class A beta-lactamases.

Table 3.1 Combination of beta-lactamase inhibitor with a beta-lactam antibiotic.

Inhibitor type	β-Lactamase inhibitor	β-Lactam antibiotic
Beta-lactam	Clavulanic acid	Amoxicillin
		Ticarcillin
	Sulbactam	Ampicillin
	Tazobactam	Piperacillin
		Ceftolozane
Non-beta-lactam	Avibactam	Ceftazidime

- Non-beta-lactam inhibitors. Avibactam is a non-beta-lactam inhibitor. Its structure contains a bridged diazabicyclooctane (DBO) and is a tight-binding, covalent reversible inhibitor for most enzymes. It is used in combination with extended-spectrum cephalosporins (ceftazidime) and is under development for use with ceftaroline and aztreonam. Some other non-beta-lactam inhibitors under development include the DBOs RG6080 and relebactam (MK-7655) in combination with imipenem and the boronic acid RPX7009 in combination with meropenem (Bush and Bradford 2016).

The beta-lactamase inhibitors, which share structural similarity with penicillin, exert their activity by binding to beta-lactamases at their active site, decreasing the quantity of enzyme available for hydrolysis of antimicrobial beta-lactam. This class of beta-lactamase inhibitors is able to inhibit most Ambler class A beta-lactamases (excluding carbapenemases such as KPC) but not those from Ambler class B, C, or D (Drawz and Bonomo 2010; Zasowski et al. 2015). Thus, they confer activity against beta-lactamase-producing organisms, such as methicillin-susceptible staphylococci and some Gram-negative organisms including *H. influenzae*, *Moraxella* spp., and virtually all anaerobes (Drawz and Bonomo 2010).

Two new combinations of cephalosporin plus beta-lactamase inhibitor were approved recently by the Food and Drugs Administration (FDA) (Goodlet et al. 2016) as well as by the EMEA.

Ceftazidime plus avibactam is a combination of third-generation cephalosporin with a new non-beta-lactamase inhibitor. This combination protects ceftazidime from Ambler class A enzymes (including KPC carbapenemases), Ambler class C enzymes (AmpC cephalosporinases), and some class D enzymes (including OXA-type carbapenemases) (Hidalgo et al. 2016). Therefore, through the addition of avibactam, ceftazidime's activity is expanded to many ceftazidime-resistant and carbapenem-resistant *Enterobacteriaceae* and

P. aeruginosa strains. However, ceftazidime-avibactam does not possess any appreciable activity against the Ambler class B metallo-beta-lactamases (Zasowski et al. 2015).

Ceftolozane-tazobactam is a new combination of a well-established beta-lactamase inhibitor with a new fifth-generation cephalosporin, ceftolozane, which differs from other cephalosporins due to its increased activity against some AmpC beta-lactamase producers including *P. aeruginosa*. Both combinations have activity against some anaerobic bacteria, including *Fusobacterium* spp. and *Propionibacterium* spp. However, activity against *Bacteroides* spp. is less predictable and they are not active against *Clostridium* spp. (Snydman et al. 2014).

3.4 Mechanism of Action

Beta-lactams are bactericidal agents that kill susceptible bacteria by inhibiting the synthesis of the cell wall in both Gram-negative and Gram-positive bacteria.

Peptidoglycan, the major constituent of the bacterial cell wall, is a heteropolymer made of linear glycan strands of alternating β-1,4-linked *N*-acetylglucosamine and *N*-acetylmuramic acid residues (Figure 3.7). The carboxyl group of each *N*-acetylmuramic acid residue is substituted by a pentapeptide subunit, which contains alternating L- and D-amino acids and one dibasic amino acid, which is often meso-diaminopimelic acid (m-DAP) in most Gram-negative bacteria and some Gram-positive bacteria, such as some *Bacillus* species, or L-lysine in most Gram-positive bacteria.

The three-dimensional network structure of peptidoglycan is made by cross-linking between the peptide subunit of one chain and that of a neighboring chain (Vollmer and Bertsche 2008). In contrast to the uniform structure of the glycan, the peptide moiety shows considerable variations from one organism to another. The most common pentapeptide found is L-$Ala_{(1)}$-D-$Glu_{(2)}$-(m-DAP or L-Lys)$_{(3)}$-D-$Ala_{(4)}$-D-$Ala_{(5)}$.The peptide cross-linking is formed by the action of an enzyme that links D-$Ala_{(4)}$ from one peptide chain to the free amino group of m-DAP or L-$Lys_{(3)}$ on another peptide chain.

Cross-linking of the peptidoglycan is made either directly or through a short peptide bridge as in many Gram-positive organisms. For instance, in *S. aureus* the cross-linking is made through five glycines (van Heijenoort and Gutmann 2000). The final stage of peptidoglycan synthesis occurs outside the cell when PBPs catalyze the polymerization of the glycan strand (transglycosylation) and the cross-linking between glycan chains (transpeptidation).

PBPs are often divided into:

- High-molecular-mass (HMM) PBPs. HMM PBPs can be further classified as class A or class B PBPs, according to their functional domains.

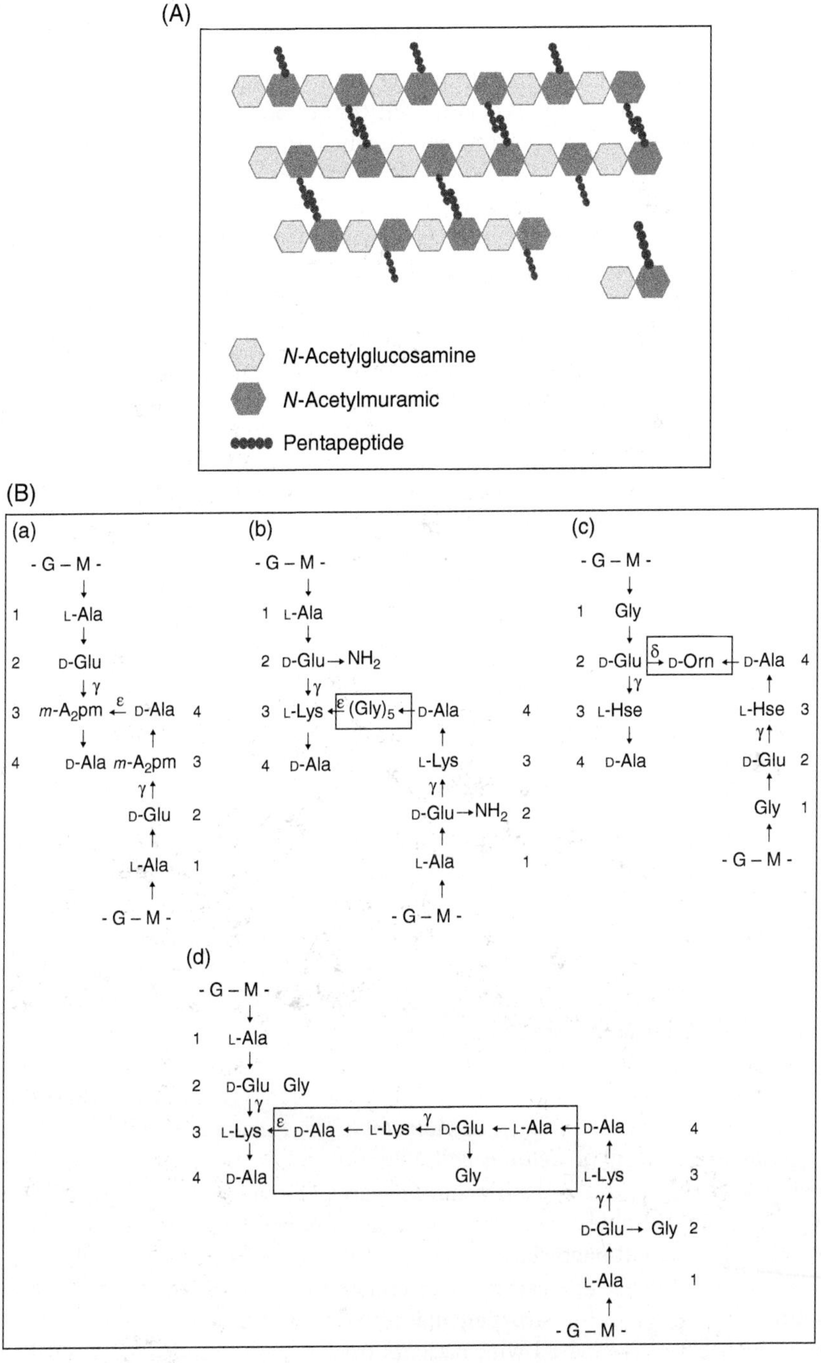

Figure 3.7 **A**: Peptidoglycan structure. **B**: Examples of cross-linkage and interpeptide bridge. (a) *Escherichia coli* (direct 3-4 cross-link). (b) *Staphylococcus aureus* (3-4 cross-link with a pentaglycine bridge). (c) *Corynebacterium poinsettiae* (2-4 cross-link with a D-ornithine bridge). (d) *Micrococcus luteus* (3-4 cross-link with a bridge consisting of a peptide stem). G, *N*-acetylglucosamine; M, *N*-acetylmuramic acid.

Class A PBPs are bifunctional, having both transglycosylase and transpeptidase activities, whereas class B PBPs have only transpeptidase activity. HMM PBPs are essential to cell survival and are the actual target of beta-lactams.

- Low-molecular-mass (LMM) PBPs. LMM PBPs are usually peptidases dispensable for bacterial survival, and thus, they represent minor target of beta-lactam antibiotics (Sauvage and Terrak 2016).

Beta-lactams are able to interact with PBPs because of the stereochemical similarity of their beta-lactam structure with the D-alanine-D-alanine residues that participate in the synthesis of the peptidoglycan. As a result, the blockage of the transpeptidation reaction compromises the structural integrity of the bacterial cell wall leading to cell lysis and death.

Modification of the PBPs structure, either by mutations or by formation of mosaic structures after recombination of heterologous genes, results in PBPs with low affinities for beta-lactams and explains the acquired resistance to these antibiotics, in particular that of Gram-positive organisms (Zapun et al. 2008).

3.5 Activity of Beta-Lactams Against Multiresistant Bacteria

The World Health Organization (WHO) recently published a list of the 12 families of bacteria for which new antibiotics are needed (WHO 2017). From the 12 families, 9 of them corresponds to Gram-negative bacteria, which highlights the particular threat of this group. The list is divided into three categories according to the urgency of need for new antibiotics. In Table 3.2, the beta-lactam antibiotics that could be used for each type of bacteria according to the level of the WHO priority are shown.

Priority 1: CRITICAL. They include multidrug-resistant bacteria producing infection in hospitals, nursing homes, and patients that requires the use ventilators or blood catheters, such as *Acinetobacter*, *Pseudomonas*, and various *Enterobacteriaceae* (*Klebsiella*, *E. coli*, *Serratia*, and *Proteus*) that have become resistant to most antibiotics, including carbapenems and third-generation cephalosporins. Some of the new beta-lactams have activity against these multidrug-resistant bacteria, such as ceftazidime-avibactam or ceftolozane-tazobactam. Moreover, carbapenem-containing combinations are good options for treatment for carbapenemase-producing bacteria as lower mortality was observed compared with patients treated with monotherapy (Daikos et al. 2014).

Table 3.2 Bacteria and resistance considered for each level of priority as well as the beta-lactam antibiotics that could be used for each type of bacteria.

Priority	Bacteria	Resistance	Beta-lactams that could be active
Priority 1: CRITICAL	*Acinetobacter*	Carbapenem-R	Carbapenems[b] Sulbactam
	Pseudomonas	Carbapenem-R	Carbapenems[b] Ceftolozane-tazobactam Ceftazidime-avibactam
	Enterobacteriaceae[a]	Carbapenem-R, third-generation cephalosporin-R	Carbapenem[b] Ceftazidime-avibactam Ceftolozane-tazobactam
Priority 2: HIGH	*Enterococcus faecium*	Vancomycin-R	None
	Staphylococcus aureus	Methicillin-R Vancomycin-I-R	Ceftaroline Ceftobiprole
	Helicobacter pylori	Clarithromycin-R	Amoxicillin
	Campylobacter spp.	Fluoroquinolone-R	Amoxicillin-clavulanic acid Imipenem
	Salmonellae	Fluoroquinolone-R	Ampicillin or amoxicillin, third-gen cephalosporin
	Neisseria gonorrhoeae	Third-generation cephalosporin-R fluoroquinolone-R	None
Priority 3: MEDIUM	*Streptococcus pneumoniae*	Penicillin-non-S	Cefotaxime, ceftaroline, cefditoren, carbapenem
	Haemophilus influenzae	Ampicillin-R	Amoxicillin/clavulanic acid, cephalosporins, and others
	Shigella spp.	Fluoroquinolone-R	Ampicillin, third-generation cephalosporin

I, intermediate; R, resistant; S, susceptible.
[a] *Enterobacteriaceae*: *K. pneumonia*, *E. coli*, *Enterobacter* spp., *Serratia* spp., *Proteus* spp., *Providencia* spp., and *Morganella* spp.
[b] It is possible to use the carbapenem with the lower MIC, combined with other active antibiotic.

Priority 2: HIGH. They are bacteria that may produce nosocomial or outpatient infection (vancomycin-resistant *E. faecium*, MRSA, vancomycin-intermediate or vancomycin-resistant *S. aureus)*, but others produce infection only in outpatients (clarithromycin-resistant *Helicobacter pylori*, fluoroquinolone-resistant *Campylobacter* spp., fluoroquinolone-resistant *Salmonella*, cephalosporin-resistant or fluoroquinolone-resistant *N. gonorrhoeae*).

Priority 3: MEDIUM. They are bacteria that produce community-acquired infection although patients may require hospital admission: penicillin-non-susceptible *S. pneumoniae*, ampicillin-resistant *H. influenzae*, and fluoroquinolone-resistant *Shigella* spp.

3.6 Conclusions

Since the clinical introduction of penicillin in the 1940s, beta-lactam antibiotics have remained the most popular drugs for treating bacterial infections. Moreover, after more than 78 years of clinical use, these antibiotics are still widely used due to their great activity, broad spectrum, and safety profile. However, as a consequence of their widespread use, bacterial resistance to this family of drugs is growing up day by day in both community and hospital settings. Emergence and dissemination of beta-lactam resistance have renewed interest in the development of novel beta-lactam antibiotics or beta-lactamase inhibitors, and some of them are currently available for multiresistant bacteria treatment.

References

Bassetti, M., Merelli, M., Temperoni, C., and Astilean, A. (2013). New antibiotics for bad bugs: where are we? *Ann Clin Microbiol Antimicrob* 12: 22.

Birnbaum, J., Kahan, F.M., Kropp, H., and MacDonald, J.S. (1985). Carbapenems, a new class of beta-lactam antibiotics. Discovery and development of imipenem/cilastatin. *Am J Med* 78: 3–21.

Boucher, H.W., Talbot, G.H., Benjamin, D.K. Jr. et al. (2013). 10 x '20 Progress: development of new drugs active against gram-negative bacilli: an update from the Infectious Diseases Society of America. *Clin Infect Dis* 56: 1685–1694.

Bousquet, P.J., Pipet, A., Bousquet-Rouanet, L., and Demoly, P. (2008). Oral challenges are needed in the diagnosis of beta-lactam hypersensitivity. *Clin Exp Allergy* 38: 185–190.

Bush, K. and Bradford, P.A. (2016). beta-Lactams and beta-lactamase inhibitors: an overview. *Cold Spring Harb Perspect Med* 6: pii: a025247.

Daikos, G.L., Tsaousi, S., Tzouvelekis, L.S. et al. (2014). Carbapenemase-producing *Klebsiella pneumoniae* bloodstream infections: lowering mortality by antibiotic combination schemes and the role of carbapenems. *Antimicrob Agents Chemother* 58: 2322–2328.

Drawz, S.M. and Bonomo, R.A. (2010). Three decades of beta-lactamase inhibitors. *Clin Microbiol Rev* 23: 160–201.

Farrell, D.J., Flamm, R.K., Sader, H.S., and Jones, R.N. (2014). Ceftobiprole activity against over 60,000 clinical bacterial pathogens isolated in Europe, Turkey, and Israel from 2005 to 2010. *Antimicrob Agents Chemother* 58: 3882–3888.

Fernandes, R., Amador, P., and Prudêncio, C. (2013). BetaLactams: chemical structure, mode of action and mechanisms of resistance. *Rev Med Microbiol* 24: 11.

Fleming, A. (2001). On the antibacterial action of cultures of a penicillium, with special reference to their use in the isolation of *B. influenzae*. 1929. *Bull World Health Organ* 79: 780–790.

Garau, J., Wilson, W.W., Wood, M., and Carlet, J. (1997). Fourth-generation cephalosporins: a review of in vitro activity, pharmacokinetics, pharmacodynamics and clinical utility. *Clin Microbiol Infect* 3: s87–s101.

Gomez, J., Garcia-Vazquez, E., and Hernandez-Torres, A. (2015). Betalactams in clinical practice. *Rev Esp Quimioter* 28: 1–9.

Goodlet, K.J., Nicolau, D.P., and Nailor, M.D. (2016). Ceftolozane/tazobactam and ceftazidime/avibactam for the treatment of complicated intra-abdominal infections. *Ther Clin Risk Manag* 12: 1811–1826.

Hidalgo, J.A., Vinluan, C.M., and Antony, N. (2016). Ceftazidime/avibactam: a novel cephalosporin/nonbeta-lactam beta-lactamase inhibitor for the treatment of complicated urinary tract infections and complicated intra-abdominal infections. *Drug Des Devel Ther* 10: 2379–2386.

Kirby, W.M. (1944). Extraction of a highly potent penicillin inactivator from penicillin resistant staphylococci. *Science* 99: 452–453.

Long, A.J., Clifton, I.J., Roach, P.L. et al. (2005). Structural studies on the reaction of isopenicillin N synthase with the truncated substrate analogues delta-(L-alpha-aminoadipoyl)-L-cysteinyl-glycine and delta-(L-alpha-aminoadipoyl)-L-cysteinyl-D-alanine. *Biochemistry* 44: 6619–6628.

Miller, E.L. (2002). The penicillins: a review and update. *J Midwifery Womens Health* 47: 426–434.

Papp-Wallace, K.M., Endimiani, A., Taracila, M.A., and Bonomo, R.A. (2011). Carbapenems: past, present, and future. *Antimicrob Agents Chemother* 55: 4943–4960.

Saroglou, G., Cromer, M., and Bisno, A.L. (1980). Methicillin-resistant *Staphylococcus aureus*: interstate spread of nosocomial infections with emergence of gentamicin-methicillin resistant strains. *Infect Control* 1: 81–89.

Sauvage, E. and Terrak, M. (2016). Glycosyltransferases and transpeptidases/ penicillin-binding proteins: valuable targets for new antibacterials. *Antibiotics (Basel)* 5: pii: E12.

Shirley, D.A., Heil, E.L., and Johnson, J.K. (2013). Ceftaroline fosamil: a brief clinical review. *Infect Dis Ther* 2: 95–110.

Snydman, D.R., McDermott, L.A., and Jacobus, N.V. (2014). Activity of ceftolozane-tazobactam against a broad spectrum of recent clinical anaerobic isolates. *Antimicrob Agents Chemother* 58: 1218–1223.

Van Heijenoort, J. and Gutmann, L. (2000). Correlation between the structure of the bacterial peptidoglycan monomer unit, the specificity of transpeptidation, and susceptibility to beta-lactams. *Proc Natl Acad Sci U S A* 97: 5028–5030.

Vollmer, W. and Bertsche, U. (2008). Murein (peptidoglycan) structure, architecture and biosynthesis in *Escherichia coli*. *Biochim Biophys Acta* 1778: 1714–1734.

WHO (2017). Global priority list of antibiotic-resistant bacteria to guide research, discovery, and development of new antibiotics. http://www.who.int/medicines/publications/WHO-PPL-Short_Summary_25Feb-ET_NM_WHO.pdf?ua=1 (accessed 8 March 2017).

Zapun, A., Contreras-Martel, C., and Vernet, T. (2008). Penicillin-binding proteins and beta-lactam resistance. *FEMS Microbiol Rev* 32: 361–385.

Zasowski, E.J., Rybak, J.M., and Rybak, M.J. (2015). The beta-lactams strike back: ceftazidime-avibactam. *Pharmacotherapy* 35: 755–770.

Zervosen, A., Sauvage, E., Frere, J.M. et al. (2012). Development of new drugs for an old target: the penicillin binding proteins. *Molecules* 17: 12478–12505.

4

Glycopeptide Antibiotics

Mechanism of Action and Recent Developments

Paramita Sarkar[1] *and Jayanta Haldar*[1,2]

[1] *Antimicrobial Research Laboratory, New Chemistry Unit, Jawaharlal Nehru Centre for Advanced Scientific Research, Bengaluru, Karnataka, India*
[2] *School of Advanced Materials, Jawaharlal Nehru Centre for Advanced Scientific Research, Bengaluru, Karnataka, India*

4.1 Introduction

Glycopeptide antibiotics are an important class of antibiotics that inhibit bacterial cell wall biosynthesis. Vancomycin (Figure 4.1a) was the first member of this class of antibiotics to be discovered by Eli Lilly and approved for clinical use by the Food and Drug Administration (FDA) in 1958 (Butler et al. 2014). The discovery of vancomycin came at a time when resistance to most widely used antibiotics, β-lactam, had emerged. It has been used as the antibiotic of last resort for treatment of multidrug-resistant Gram-positive bacterial infections since then. It is still vital for the treatment of Gram-positive bacterial infections caused by methicillin-resistant *Staphylococcus aureus* (MRSA) and *Clostridium difficile*. Later in 1988, teicoplanin (Figure 4.1a), which was discovered in the Lepetit Research Center (Milan, Italy), was approved for use in Europe for treating similar Gram-positive infections (Butler et al. 2014). The glycopeptides were isolated from the soil bacteria *Actinomycetales*. Numerous other glycopeptides have been discovered since then, which never made it to the clinic. A year after the discovery of vancomycin, ristocetin was isolated. Although it seemed to be a promising antibacterial agent, it was discontinued from clinical use because it caused platelet aggregation in patients missing a platelet factor in platelet-type von Willebrand disease (Jenkins et al. 1974; Meyer et al. 1974). Glycopeptides such as actaplanin and avoparcin were shown to promote growth in farm animals. The use of avoparcin was banned due to

Antibiotic Drug Resistance, First Edition. Edited by José-Luis Capelo-Martínez and Gilberto Igrejas.

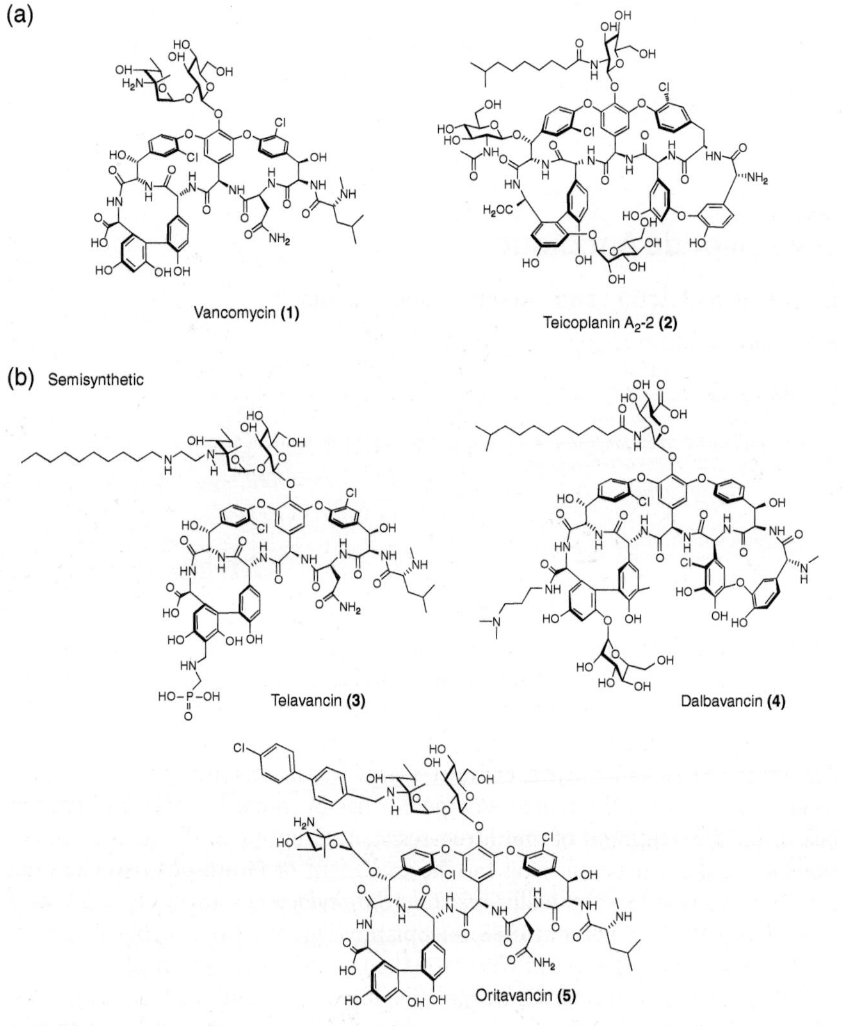

Figure 4.1 (a) Natural and (b) semisynthetic glycopeptide antibiotics approved for use in humans.

its association with the transfer of vancomycin resistance to *Enterococcus faecium* (Bager et al. 1997). Later, a newer generation of semisynthetic derivatives of glycopeptide antibiotics, telavancin, dalbavancin, and oritavancin (Figure 4.1b), was approved for clinical use (Table 4.1). However, these are yet to be approved for the treatment of vancomycin-resistant infections. The clinically relevant glycopeptides and their mechanisms of action have been described in detail herein.

Table 4.1 Clinical use of glycopeptide antibiotics.

Antibiotic		Date of approval	Treatment of
Natural glycopeptides	Vancomycin	1958	Pseudomembranous colitis, staphylococcal enterocolitis, endocarditis, bone and joint infection caused by MRSA and enterococci
	Teicoplanin	1988	Complicated skin and soft tissue infections, bone and joint infections, pneumonia, complicated urinary tract infections, endocarditis, bacteremia caused by MRSA and enterococci
Semisynthetic glycopeptides	Telavancin	2009	Complicated skin and skin structure infections (cSSSi), hospital-acquired and ventilator-associated pneumonia caused by *S. aureus*, enterococci, and streptococci
	Dalbavancin	2014	Acute skin and skin structure infections caused by methicillin-susceptible *S. aureus* (MSSA) and methicillin-resistant *S. aureus* (MRSA), streptococci, and vancomycin-sensitive *Enterococcus faecalis*
	Oritavancin	2014	Acute skin and tissue infections caused by MRSA, MSSA, streptococci, and vancomycin-susceptible *E. faecalis*

4.2 Naturally Occurring Glycopeptide Antibiotics

This class of antibiotics consists of glycosylated tricyclic or tetracyclic heptapeptides. These antibiotics have been divided into five distinct structural subclasses (I–V) based on the substitution and the residues at positions 1 and 3 of the heptapeptide (Nicolaou et al. 1999). The first three groups differ in the amino acids at the N-terminal of the peptide core. Group I, the vancomycin-type group, possesses aliphatic amino acids at positions 1 and 3. Avoparcin, an example of group II, possesses two aromatic amino acids at the first and third position. Ristocetin and actaplanin are examples of group III, which possesses an ether linkage between the amino acids at positions 1 and 3. Teicoplanin is an example of the fourth group, which consists of a lipophilic moiety attached to the vancosamine sugar. In group V, exemplified by complestatin, a tryptophan is linked to the central amino acid.

The first-generation glycopeptides vancomycin (**1**) and teicoplanin (**2**) are naturally occurring antibiotics approved for use in humans, and their structures are closely related. Initially, vancomycin was limited due to its toxicity. However, as purification techniques improved, the purity of vancomycin increased and led to their extensive use for treatment of multidrug-resistant infections. Teicoplanin differs from vancomycin in the presence of an acyl chain on the vancosamine sugar, an additional glycosylation and an ether linkage between 4-hydroxyphenylglycine and 3,5-dihydroxyphenylglycine. Teicoplanin is a complex consisting of five similar compounds that differ in the acyl residue. Vancomycin has a short half-life of about 4–11 hours for intravenous administration and hence requires multiple dosing (Butler et al. 2014). Teicoplanin exhibits a better half-life of approximately 30 hours for intravenous or intramuscular administration (Butler et al. 2014). Overall, these two glycopeptides bear similar toxicity and activity profiles.

4.3 Mechanism of Action of Glycopeptide Antibiotics

It was found that the treatment of bacteria with vancomycin led to the accumulation of precursors of cell wall such as UDP-MurNAc-pentapeptide in the cytoplasm. This indicated that these glycopeptide antibiotics inhibit the later stage of cell wall biosynthesis. The structure of the glycopeptide backbone is such that it forms a cleft that exactly fits amino acid sequences that have an L-D-D configuration (Reynolds 1989). The glycopeptide antibiotics non-covalently bind to the C-terminal D-Ala-D-Ala of the murein precursor, lipid II and immature peptidoglycan, through five H-bonds (Figure 4.2a). The strong binding between the antibiotic and the target peptide is a cooperative effect of hydrogen bonding between the carboxylate of the ligand and three amide NH and two amide–amide hydrogen bonds between the ligand and antibiotic and hydrophobic interactions between the methyl groups of alanine and the hydrophobic moieties of the antibiotic (Williams and Bardsley 1999). Upon binding, the glycopeptide antibiotics shield the D-Ala-D-Ala from the action of the transpeptidase enzyme. Further, the glycopeptide-D-Ala-D-Ala complex sterically hinders the positioning of the transglycosylase enzyme to cleave the muramyl-phosphate bond. Both transglycosylation and transpeptidation during cell wall biosynthesis are inhibited, thereby preventing further cross-linking. This leads to a weakening of the peptidoglycan, leaving the bacteria susceptible to lysis due to changes in osmotic pressure. Most glycopeptides with the exception of teicoplanin are known to dimerize in solution, and this enhances target binding through a cooperative effect (Mackay et al. 1994). The importance of this cooperative effect is demonstrated in glycopeptides like eremomycin, which exhibit significant antibacterial activity in spite of having moderate binding affinity to peptidoglycan analogues. The dimer of chloroeremomycin involves four amide–amide hydrogen bonds and a hydrogen bond between ammonium ion of the epivancosamine moiety on residue 6 and a carbonyl on the other molecule, thereby forming six hydrogen bonds on the interface of the dimer (Butler et al. 2014). In the case of vancomycin, which

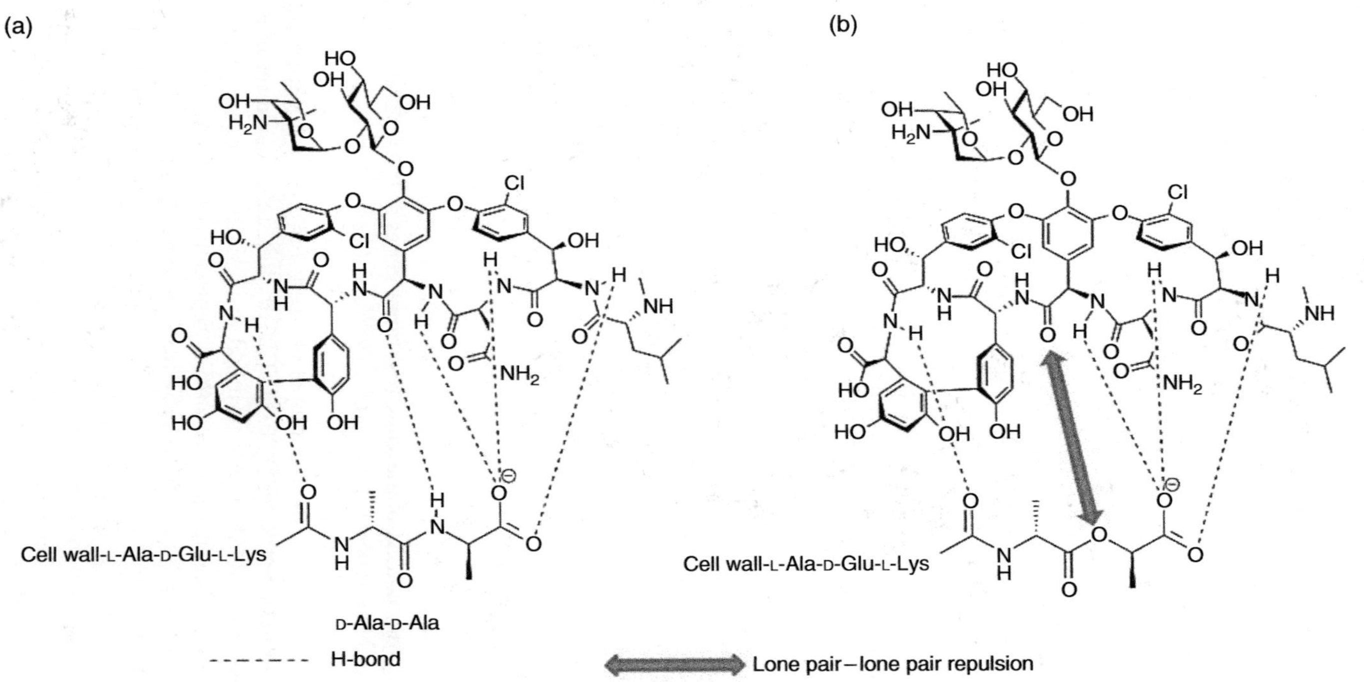

Figure 4.2 (a) Vancomycin binds to the D-Ala-D-Ala terminal through five H-bonds. (b) Lone–lone pair repulsion and loss of one H-bond leads to 1000-fold loss in binding affinity to the mutated target peptide D-Ala-D-Lac.

lacks the epivancosamine moiety on residue 6, the dimer consists of four hydrogen bonds at the interface. The antibiotic teicoplanin possesses a lipophilic moiety that imparts membrane-anchoring properties to it. The second-generation glycopeptide antibiotics (telavancin, dalbavancin, and oritavancin) exhibit enhanced antibacterial activity by virtue of the presence of lipophilic groups that facilitate localization of antibiotic on the membrane surfaces and also perturb the integrity of the bacterial membrane (Zhanel et al. 2010).

4.4 Resistance to Glycopeptides

It was found early on that bacteria found it difficult to develop resistance to vancomycin. The minimum inhibitory concentration (MIC) of vancomycin was found to show a meager eight-fold increase after 25 serial passages against *S. aureus* (McGuire et al. 1955). Resistance development to vancomycin by alteration of the target of glycopeptides (the D-Ala-D-Ala terminus of lipid II and/or the immature peptidoglycan) is difficult as the process involves simultaneous modifications of multiple enzymes in the pathway to peptidoglycan synthesis. Resistance to vancomycin was observed in enterococci 30 years after its approval (Murray 2000). The incidence of vancomycin-resistant enterococci (VRE) in hospitalized patients with enterococcal infections in the United States had increased to 30%. In 2001, hetero-resistant vancomycin-intermediate *S. aureus* (VISA) and the first vancomycin-resistant *S. aureus* (VRSA) were reported (Hiramatsu 2001; Hiramatsu et al. 1997). Resistance to vancomycin in enterococci (VRE) resulted from the transfer of resistance genes from other glycopeptide-resistant bacteria such as those resulting from the overuse of avoparcin as animal growth promoter (Bager et al. 1997). Courvalin and Walsh groups elucidated the mechanism of vancomycin resistance in enterococci in the 1990s (Arthur and Courvalin 1993; Bugg et al. 1991; Evers et al. 1996). Subsequent work on glycopeptide resistance in producer organisms has revealed that they consist of the same resistance genes as the resistant enterococcal strains. Nine gene clusters – *vanA, vanB, vanC, vanD, vanE, vanG, vanL, vanM*, and *vanN* – that confer resistance to vancomycin in enterococci in various ways have been identified (Binda et al. 2014; Boyd et al. 2008; Lebreton et al. 2011; McKessar et al. 2000; Xu et al. 2010). Resistance could result from the modification of D-Ala-D-Ala terminal to D-Ala-D-Ser as in the case of the *vanC, vanE, vanG, vanL*, and *vanN* or replacement of D-Ala-D-Ala with D-Ala-D-Lac in the case of *vanA, vanB, vanD*, and *vanM* genes. The replacement of the D-Ala-D-Ala terminal of the cell wall precursor pentapeptide with D-Ala-D-Lac (Figure 4.2b) leads to a 1000-fold loss of binding affinity and therefore resistance in these phenotypes (McComas et al. 2003). Resistance by mutation to D-Ala-D-Ser results in a sevenfold decrease in binding affinity (McKessar et al. 2000). The VanA and VanB resistance phenotypes are the most common and differ in their susceptibility to vancomycin and teicoplanin. Although the VanA phenotype shows reduced susceptibility to both teicoplanin and vancomycin, the VanB phenotype retains sensitivity to teicoplanin. The genes *vanRSHAX* encode for an alternate biosynthetic pathway that produces the mutated cell wall precursors (Walsh et al. 1996). In VISA the thickening of the bacterial cell wall leads to an increase in the number of binding sites for vancomycin binding and hence resistance (Hiramatsu 2001). Genotypic analysis of VRSA indicated that the resistance genes were acquired from VRE (Weigel et al. 2003). In some strains of VRSA, resistance was found to be due to both thickening of cell wall and mutation of the pentapeptide terminal (Hiramatsu 2001).

4.5 Second-Generation Glycopeptides

The second-generation glycopeptides are semisynthetic lipoglycopeptide antibiotics with enhanced antibacterial activity over vancomycin in both vancomycin-sensitive and vancomycin-resistant strains. There are three antibiotics that belong to this generation, namely, telavancin, dalbavancin, and oritavancin (Figure 4.1b). These glycopeptides possess a hydrophobic side chain that serves as a membrane anchor. The anchoring to the bacterial membrane increases localization in the membrane region and enhances the binding affinity to the membrane-associated lipid II. They strongly inhibit both transglycosylation and transpeptidation. This leads to enhanced antibacterial activity. All the three antibiotics have been briefly discussed below.

4.5.1 Telavancin

Telavancin (**3**) is a semisynthetic derivative of vancomycin consisting of a lipophilic decylaminoethyl moiety conjugated to the vancosamine sugar and a (phosphomethyl)aminomethyl moiety at the *para* position of the aromatic ring of the C-terminal dihydroxyphenylglycine residue (Charneski et al. 2009). The presence of the lipophilic moiety provides membrane interaction properties. This leads to permeabilization and depolarization of the bacterial cell membrane and a dual mechanism of action (Higgins et al. 2005). This membrane activity leads to improved bactericidal activity in comparison with the first-generation of glycopeptides against MRSA and methicillin-susceptible *S. aureus* (MSSA) as well as activity vancomycin-resistant strains such as VISA, VRSA, and VRE (Karlowsky et al. 2015). The hydrophilic (phosphomethyl)aminomethyl moiety enhances tissue distribution and clearance, hence reducing nephrotoxicity (Leadbetter et al. 2004). Telavancin has a half-life of approximately 8 hours in humans and is hence administered intravenously once daily for the required time duration of treatment. It was approved by the US FDA for the treatment of complicated skin and skin structure infections (cSSSi) in 2009 and for treatment of hospital-acquired and ventilator-associated pneumonia caused by *S. aureus* later in 2013 (Table 4.1).

4.5.2 Dalbavancin

Dalbavancin (**4**) is a semisynthetic derivative of the glycopeptide A40926 belonging to the teicoplanin family, which has been modified through the amidation of the C-terminal carboxyl group with a *N,N*-dimethylpropylamine group (Anderson and Keating 2009; Candiani et al. 1999). The terminally branched dodecyl fatty acid chain of dalbavancin connected through an amide linkage with the glucosamine moiety helps in membrane anchoring. Dalbavancin binds to D-Ala-D-Ala with higher affinity than its parent compound, by virtue of its ability to form dimers and membrane-anchoring property (Malabarba et al. 1995). This results in better potency than vancomycin and teicoplanin against MRSA and susceptible enterococci. The MIC of dalbavancin was approximately 0.1 $\mu g\,ml^{-1}$ against VRE (VanB phenotype), although it exhibited poor activity against VRE (VanA phenotype) (Candiani et al. 1999). It was approved by the FDA in 2014 for the treatment of acute skin and skin structure infections caused by MSSA, MRSA, streptococci, and vancomycin-sensitive *Enterococcus faecalis*. The half-life ranges from 149 to 250 hours in humans and requires a once-weekly intravenous injection (Zhanel et al. 2010).

4.5.3 Oritavancin

Oritavancin (**5**) is a derivative of chloroeremomycin with an acyl substitution and a lipophilic 4-chlorobiphenyl group attached to the amino group of the epivancosamine moiety (Bouza and Burillo 2010). The biphenyl moiety imparts membrane interaction properties leading to its permeabilization and promotes dimer formation prior to binding to the target peptides, resulting in enhanced binding affinity. Oritavancin has a significant inhibitory effect on transpeptidation in addition to inhibition of transglycosylation (Patti et al. 2009). Its activity against VRE results from its ability to bind the pentaglycyl bridging segment in the peptidoglycan. Its multiple modes of action result in enhanced antibacterial activity against MRSA, VRSA, and VRE (Arhin et al. 2009). This antibiotic exhibits strong bactericidal properties at concentrations where vancomycin had a static effect. It was approved by the FDA in 2014 for the treatment of acute skin and tissue infections caused by MRSA, MSSA, streptococci, and vancomycin-susceptible *E. faecalis*. It has a terminal half-life of approximately 393 hours, which enables treatment with just a single dose administered intravenously (Saravolatz and Stein 2015).

4.6 Strategies to Overcome Resistance to Glycopeptides

With the emergence of resistance to the first-generation glycopeptides, the need for newer antibiotics to treat vancomycin-resistant bacteria grew. Although the second generation of glycopeptides exhibited efficacy against vancomycin-resistant bacteria, they are yet to be approved for treating vancomycin-resistant infections. The presence of several functional groups including amino, carboxyl, and hydroxyl groups and the chloro-substituents of the triarylbiether backbone made structural modifications to the natural glycopeptides a feasible option (Sarkar et al. 2017). Although new glycopeptides have been approved, research toward developing newer glycopeptides to overcome resistance continues. The strategies to overcome glycopeptide resistance involve modifications to enhance binding affinity to mutated termini of pentapeptide and introduction of additional membrane interaction properties. Numerous literature reviews have covered the modifications made to glycopeptides in detail (Ashford and Bew 2012; Sarkar et al. 2017). The various developments toward overcoming glycopeptide resistance have been described briefly in this section.

4.6.1 Modifications That Enhance the Binding Affinity to Target Pentapeptide

The approaches to enhance binding affinity to the target pentapeptide involve the modification of the heptapeptide backbone, attachment of moieties that can form H-bonds with the target, and the development of homomeric multivalent analogues.

4.6.1.1 Peptide Backbone Modification

In order to improve the binding affinity to the target peptide, the carboxamide of residue 3 (L-asparagine), the substitutions at residue 1 (L-leucine), and the carboxamide of residue 4 (D-hydroxyphenylglycine) have been functionalized by various groups. It was found that the repulsive lone pair repulsions between vancomycin and the mutated target peptides greatly contributed to the 1000-fold loss in binding affinity while the H-bond loss contributed to a lesser extent (James et al. 2012; McComas et al. 2003). It was hypothesized that designs that focus on eliminating the destabilizing lone pair interaction rather than the reintroduction of the lost H-bond will be more effective. The Boger group thus developed a vancomycin aglycon analogue that contains a methylene group instead of the residue 4 amide carbonyl ([$\Psi[CH_2NH]Tpg^4$]vancomycin aglycon) (Crowley and Boger 2006). This analogue demonstrated a 35-fold reduction in affinity for D-Ala-D-Ala and a 40-fold increase in affinity for D-Ala-D-Lac. Other derivatives in which the carbonyl at residue 4 of vancomycin aglycon was replaced by sulfur and amidine group, respectively, were synthesized

(Xie et al. 2011, 2012). Upon replacement with sulfur, the binding affinity to D-Ala-D-Ala and D-Ala-D-Lac was reduced due to the larger size of sulfur. The [Ψ[C(=NH)NH]Tpg^4]vancomycin aglycon (**6**), with an amidine in place of the carbonyl at residue 4 of vancomycin aglycon, binds to both the unaltered peptidoglycan terminal D-Ala-D-Ala and the mutated ligand D-Ala-D-Lac due to its ability to serve as a hydrogen bond donor or acceptor (Xie et al. 2011). This amidine derivative of vancomycin aglycon displayed an MIC of less than 0.5 $\mu g\,ml^{-1}$ against sensitive and resistant bacteria. Later, Boger et al. developed appended a (4-chlorobiphenyl)methyl moiety to the vancosamine moiety of [Ψ[C(=NH)NH]Tpg^4]vancomycin (**7**) to incorporate both enhanced binding affinity and favorable hydrophobicity within the same molecule (Okano et al. 2014). This resulted in high activity against vancomycin-resistant bacteria with MICs in the range of 0.005–0.06 $\mu g\,ml^{-1}$.

4.6.1.2 Attachment of H-Bond-Forming Moieties

It was observed that a water molecule bridged the carboxylic group of vancomycin and the ligand in the crystal of vancomycin–ligand complex (Nitanai et al. 2009). This indicated that modification of the C-terminus with moieties that can form hydrogen bonds with the target peptide could enhance binding affinity. Various cyclic and acyclic sugars such as maltose, lactobionic acid, gluconic acid, and cellobiose were conjugated to vancomycin to increase the binding affinity to the mutated target peptide (D-Ala-D-Lac) (Yarlagadda et al. 2015a). The binding affinity of these sugar–vancomycin conjugates was comparable to that of vancomycin. The association constant (K_a) of vancomycin was found to be $1.1 \times 10^5\ M^{-1}$ and $5 \times 10^2\ M^{-1}$ to D-Ala-D-Ala and mutated terminal, D-Ala-D-Lac, respectively. The derivative in which lactobionic acid was conjugated to the carboxylic acid group of vancomycin (**8**, Figure 4.3) exhibited approximately 150-fold ($K_a = 8.8 \times 10^4\ M^{-1}$) higher affinity for *N,N′*-diacetyl-Lys-D-Ala-D-Lac as compared with vancomycin. An improved antibacterial activity leads to an MIC of 36 μM against VRE (VanA phenotype), as opposed to 750 μM against vancomycin. To further enhance the activity against vancomycin-resistant strains, alkyl chains (octyl to dodecyl) were conjugated to the amino group of vancosamine of compound to yield the lipophilic vancomycin–sugar conjugates. This was expected to impart additional membrane-anchoring properties. The lead molecule (**9**) bearing decyl alkyl chain at the vancosamine moiety and lactobionic acid at the carboxyl group of vancomycin exhibited more than 1000-fold (MIC = 0.7 μM) and 250-fold (MIC = 1 μM) better activity against VanA and VanB strains of VRE, respectively, as compared with vancomycin. The incorporation of the lipophilic moiety into the glycopeptide-lactobionic acid scaffold imparted membrane interaction properties, resulting in improved activity against VRE (Yarlagadda et al. 2015a). Further, the pharmacokinetic and pharmacodynamic properties of this derivative were better than that of vancomycin (Yarlagadda et al. 2015b).

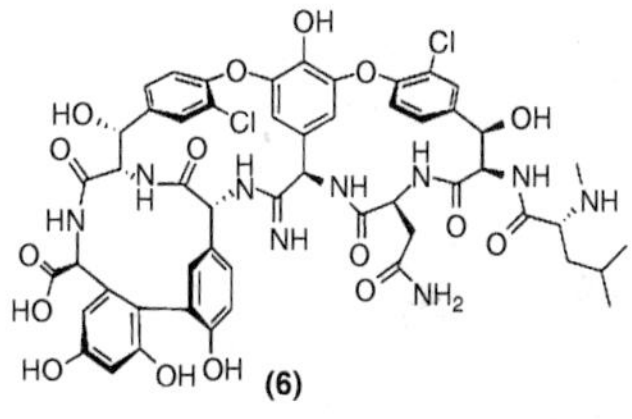

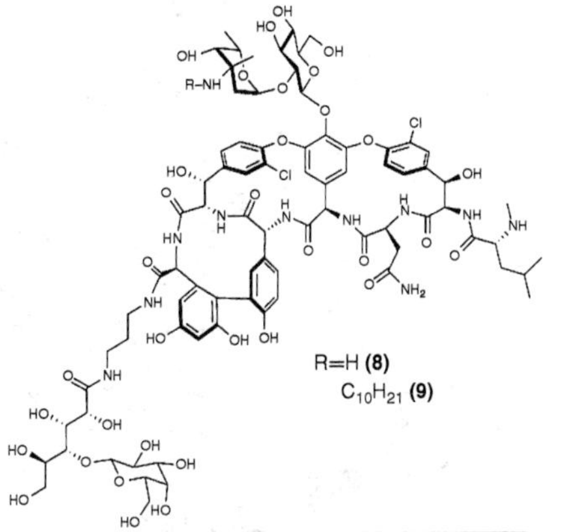

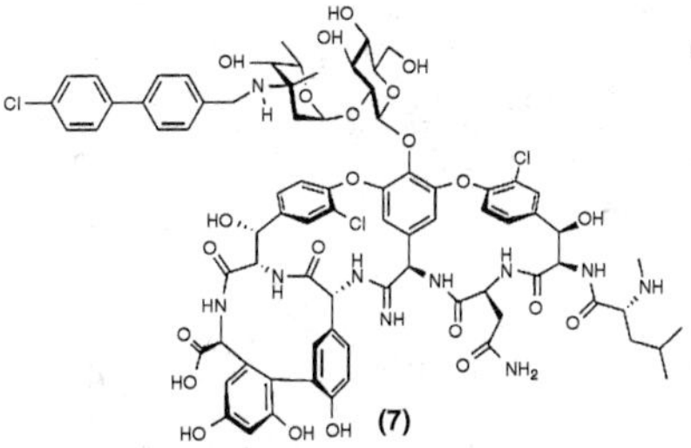

Compound	Association constant K_a (M^{-1})	
	Susceptible	Resistant
Vancomycin	1.1×10^5	5.0×10^2
6	7.3×10^4	6.9×10^4
8	2.1×10^5	8.8×10^4

Susceptible (model ligand): *N,N'*-diacetyl-Lys-D-Ala-D-Ala
Resistant (model ligand): *N,N'*-diacetyl-Lys-D-Ala-D-Lac

Figure 4.3 Glycopeptide derivatives with enhanced binding affinity to mutated target peptide through H-bonding.

4.6.1.3 Development of Homomeric Multivalent Analogues

The non-covalent dimers formed by glycopeptides have a cooperative effect and result in enhanced antibacterial efficacy (Mackay et al. 1994). The development of multivalent analogues has been considered to be a useful strategy to increase binding efficiency to the target peptide (Figure 4.4). Bis(vancomycin)carboxamides with variable linkers were developed by Griffin's group in 1996 (**10a–c**, Figure 4.4) (Sundram et al. 1996). These derivatives exhibited improved activity against VRE compared with monomeric vancomycin. The derivative **10b** showed the highest activity with approximately 60-fold increase (MIC = 11 μM) against VRE as compared with vancomycin. Later in 1998, Whitesides' group developed trivalent vancomycin derivatives (**11**) that showed an enhanced binding affinity compared with monomeric vancomycin by virtue of polyvalency (Rao et al. 1998). Arimoto et al. synthesized a multivalent polymer of vancomycin (**12**), which exhibited an approximately 60-fold increase in antibacterial activity against VRE compared with vancomycin (MIC = 31 μg ml^{-1} against the VRE VanA phenotype and 2 μg ml^{-1} against the VRE VanB phenotype) (Arimoto et al. 1999). Later in 2000, Nicolaou's group selected vancomycin dimers through target-accelerated combinatorial synthesis (Nicolaou et al. 2000). The best activity against VRE (MIC = 1–2 μg ml^{-1}) was observed for dimer (**13**) with an olefinic

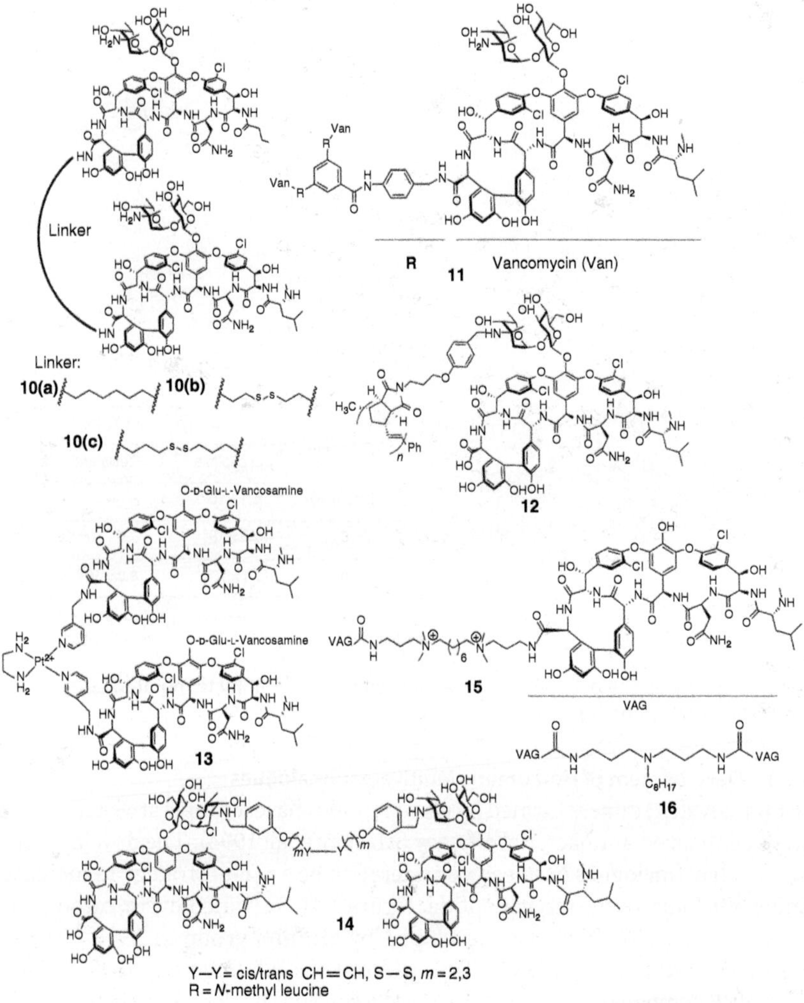

Figure 4.4 Multivalent analogues to increase activity through enhanced binding affinity.

linker and a *N*-methyl leucine substitution. Dimers with flexible organic linkers showed reduced avidity of multivalent binding due to a loss in conformational entropy. The loss of this entropy could be reduced by utilizing metal complexes with a specific geometry and structural rigidity, thereby increasing the binding affinity. With this concept, Bing Xu et al. developed dimers of vancomycin (**14**), in which two vancomycin molecules were complexed with $[Pt(en)_2(H_2O)_2]$ (Xing et al. 2003). A ~720-fold increase (MIC = $0.8\,\mu g\,ml^{-1}$) in activity as compared with vancomycin was observed. Vancomycin aglycon dimers with linkers varying in lipophilicity and charge were developed by Yarlagadda et al. (2015d). The vancomycin dimer that contained two positively charged centers in the octylene linker (**15**) exhibited a 15-fold (MIC = 48 μM) and 130-fold (MIC = 0.1 μM) increase in activity over vancomycin against VRE and VISA, respectively. In order to impart membrane-anchoring properties to vancomycin aglycon dimers, a pendant octyl lipophilic moiety was attached to the linker (**16**) (Yarlagadda et al. 2015d). This molecule showed a 300-fold higher activity than vancomycin against VRE (MIC = 2.5 μM).

4.6.2 Incorporation of Lipophilicity

Teicoplanin possesses a lipophilic chain that helps anchor the drug to the cell membrane. Thus, it was hypothesized that the membrane interaction properties could be imparted by the attachment of lipophilic moieties to vancomycin or other glycopeptides.

Thorson et al. incorporated a lipophilic 6-azido glucose onto the vancomycin aglycon (**17**, Figure 4.5) (Fu et al. 2005). This compound was found to be more

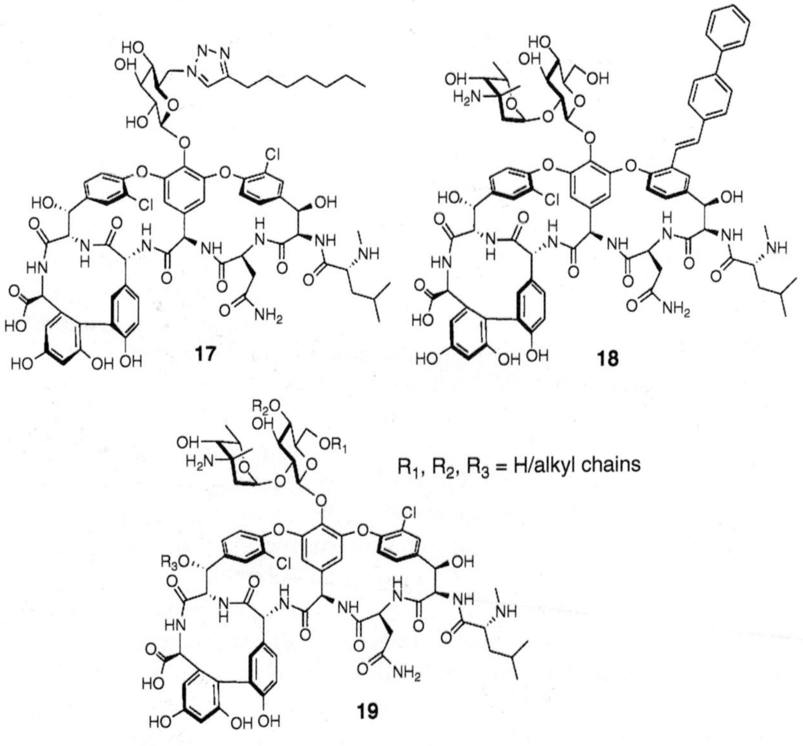

Figure 4.5 Vancomycin derivatives with lipophilic moieties to enhance activity.

effective than vancomycin against the VanB phenotype of VRE (MIC = 1 μg ml^{-1}). Arimoto et al. (**18**) reported derivatives in which the chloro group of amino acid residue 2 was replaced with a lipophilic group. This derivative exhibited enhanced antibacterial activity (MIC = 0.5 μg ml^{-1}) against the VanB phenotype of VRE although it remained inactive against the VanA phenotype of VRE (Nakama et al. 2010). However, the additional introduction of such a group at the amino acid residue 6 led to a reduction in activity even against vancomycin-susceptible strains. In 2015, Miller et al. developed three regioselective peptide catalysts that specifically substitute the aliphatic hydroxyls on vancomycin, generating three lipidated vancomycin analogues (**19**) (Yoganathan and Miller 2015). The compounds exhibited good activity against both VanA and VanB phenotypic VRE with MIC = 0.25 μg ml^{-1}.

4.6.3 Incorporation of Lipophilic Cationic Moieties to Impart Membrane Disruption Properties

The conjugation of cationic lipophilic moieties with vancomycin has been shown to impart membrane disruption properties through enhanced interactions with the negatively charged bacterial membrane (Yarlagadda et al. 2014, 2015c). Cationic lipophilic derivatives with varying alkyl chain lengths were attached to the carboxylic acid group of vancomycin. The compound with an octyl chain (**20a**, Figure 4.6) appended to vancomycin showed a ~32-fold (MIC ~ 0.4 μM), 320-fold (MIC ~ 0.3 μM), and 60-fold (MIC ~ 12.5 μM) increase in activity against VISA, VRSA, and VRE VanA phenotypes (Yarlagadda et al. 2014). Activity against vancomycin-resistant bacteria was found to increase with increase in hydrophobicity. The derivative with the tetradecyl chain was found to be the most potent, with an MIC of 0.7 μM (>1000-fold more active than the parent drug, vancomycin). The efficacy of these derivatives was attributed to their ability to permeabilize and depolarize the bacterial membrane. Based on the activity and toxicity profiles, **20a** was found to be the optimum compound. This derivative was also found to reduce the intracellular bacterial load of MRSA (Yarlagadda et al. 2016b). In order to further enhance antibacterial efficacy, derivatives with enhanced binding affinity to the pentapeptide terminal and membrane disruption properties were developed. To achieve this, a cationic lipophilic moiety was conjugated to the vancomycin–sugar conjugate (**8**), resulting in a lipophilic cationic vancomycin–sugar conjugate (**21**) (Yarlagadda et al. 2015c). This dual approach resulted in an 8000-fold (MIC = 0.09 μM) increase in *in vitro* activity as compared with vancomycin and showed promising *in vivo* activity against VRE. Both (**20a**) and (**21**) exhibited increased cell wall biosynthesis inhibition compared with vancomycin, in addition to membrane disruption properties. These compounds showed no propensity to induce resistance in bacteria (MRSA) even after multiple serial passages although vancomycin showed an increase in MIC after the seventh passage.

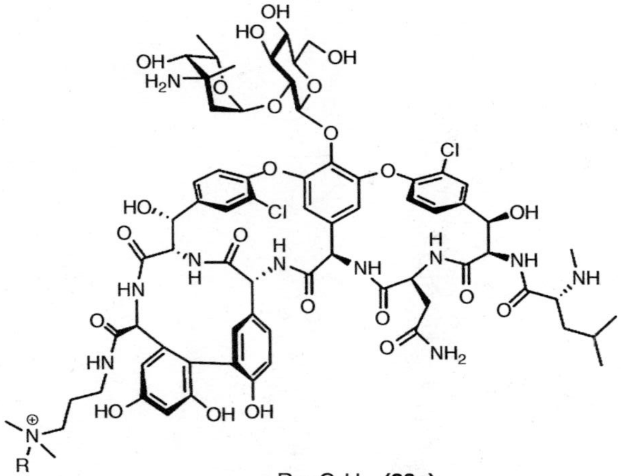

R = C_8H_{17} **(20a)**
$C_{14}H_{29}$ **(20b)**

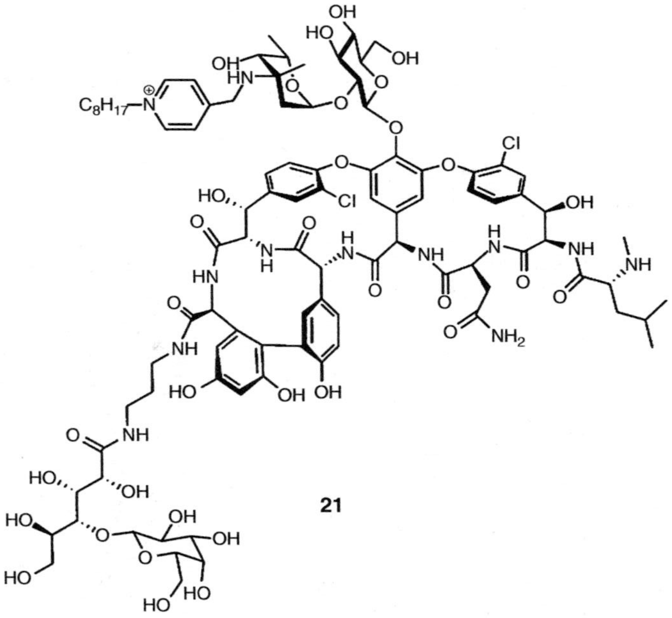

21

Figure 4.6 Cationic lipophilic vancomycin derivatives with membrane disruption properties.

4.6.4 Incorporation of Metal Chelating Moiety to Vancomycin to Impart New Mechanism of Action

The pyrophosphate of the bacterial cell wall is necessary for the transfer of cell wall precursors from the cytoplasm to the outer side. In a report, vancomycin was conjugated to a Zn^{2+} binding ligand (Figure 4.7) to overcome vancomycin resistance (Yarlagadda et al. 2016c). The dipicolyl amine moiety is known to capture the divalent zinc ion with high selectivity. Thus, this dipicolyl–vancomycin conjugate (**22**) forms a complex with the pyrophosphate groups of cell wall lipids in addition to binding to the cell wall precursor peptides. A 375-fold (MIC = 2 μM) higher *in vitro* activity than vancomycin against VRE and enhanced cell wall biosynthesis inhibition was achieved by this design. This compound was found to reduce the bacterial load in VRE kidney infection model by 5 log CFU ml^{-1} as compared with the untreated control. Further, the compound did not induce resistance in MRSA.

4.7 Glycopeptides Under Clinical Trials

There are currently two glycopeptide derivatives under clinical trials (Figure 4.8). These are vancomycin–cephalosporin hybrids developed by Theravance Biopharma Inc. – cefilavancin (**23**) and TD-1607 (**24**) are currently

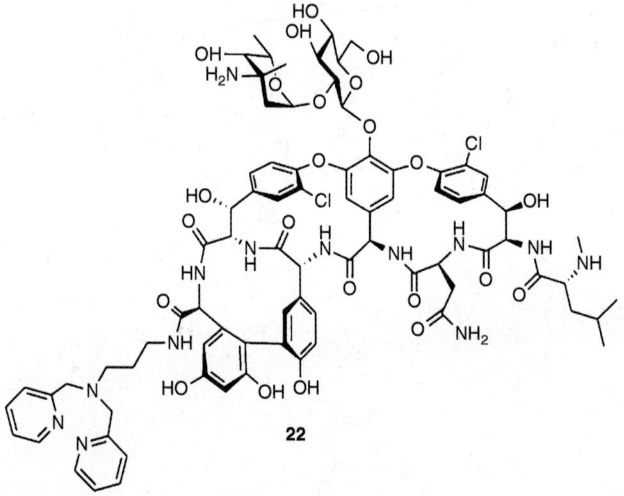

22

Figure 4.7 Vancomycin–dipicolyl derivative that interacts with the pyrophosphates of bacterial membrane.

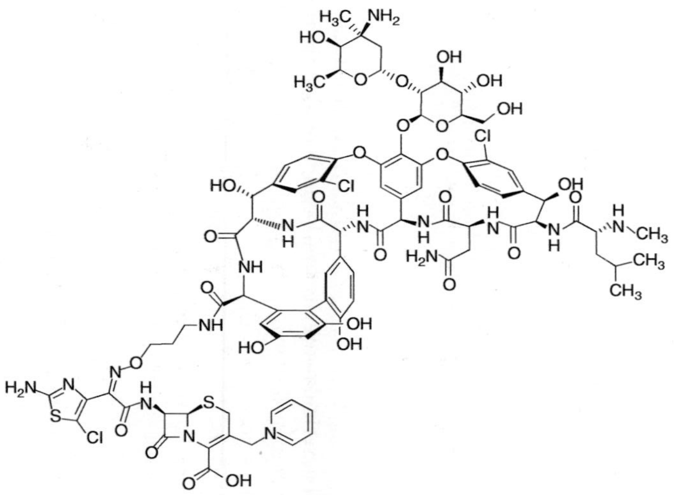

Cefilavancin **(23)**

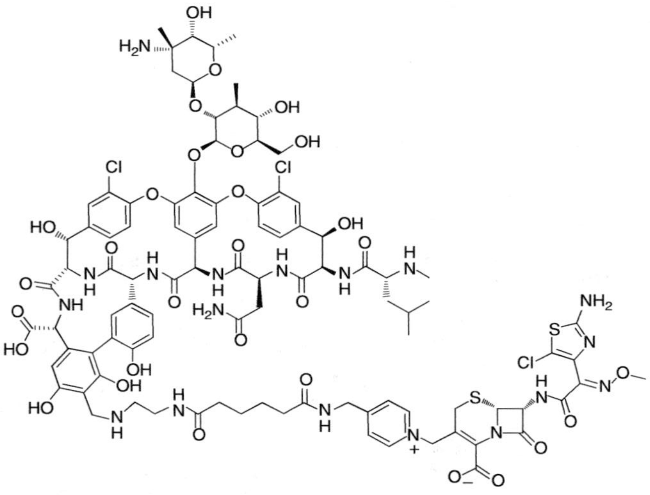

TD-1607 **(24)**

Figure 4.8 Glycopeptide-based new molecules under clinical trials.

in phase III and phase I of clinical trials, respectively (Long et al. 2008; Sader et al. 2014 (https://www.theravance.com/programs/infectious-disease)). It has been observed that when antibiotics bind multiple targets, resistance development would be less feasible. Thus, a multipronged approach results in enhanced target affinity and lower propensity of resistance development. Since the targets of β-lactams and glycopeptides are in close proximity, it was hypothesized that incorporating the two into a single molecule would result in enhanced bactericidal activity due to multiple target binding. The results confirmed rapid antibacterial activity. These derivatives were however not effective against vancomycin-resistant strains.

4.8 Glycopeptide Antibiotics: The Challenges

Glycopeptide antibiotics, specifically vancomycin, have been important for the treatment of multidrug-resistant Gram-positive infections. However, there are some aspects that need to be addressed with respect to the use of these antibiotics. In this respect, the ability of the glycopeptides and their derivatives to overcome the intrinsic resistance in Gram-negative bacteria, treat biofilm-associated infections, and treat intracellular infections remains underexplored.

The outer membrane of Gram-negative bacteria serves as a permeability barrier for larger molecules, hence resulting in the intrinsic resistance toward glycopeptide antibiotics. In order to overcome the resistance to vancomycin, few strategies have been developed. In 2013, Collins' group found that on using silver, Gram-negative bacteria could be sensitized to various antibiotics including vancomycin (Morones-Ramirez et al. 2013). In another instance, multivalent Au–vancomycin conjugates were developed and found to exhibit activity against *Escherichia coli* (Gu et al. 2003). When vancomycin was loaded into fusogenic liposomes, the antibacterial activity against clinical isolates of *E. coli* and *A. baumannii* was obtained (Nicolosi et al. 2010). Our group developed a lipophilic cationic vancomycin derivative bearing a tetradecyl chain (**20b**) that permeabilizes the outer and inner membranes of Gram-negative bacteria, thereby resulting in bactericidal activity (Yarlagadda et al. 2016a). However, a lot more needs to be achieved to extend the applicability of glycopeptide antibiotics toward Gram-negative bacteria.

Vancomycin has been used for the treatment of infections caused by *C. difficile*. The drug is not selective over the beneficial gut bacteria. The efficacy of the glycopeptides against spores of *C. difficile* needs to be improved. The inherent resistance of biofilm-associated infections and dormant bacteria in bacterial infections is a growing health concern. Telavancin and oritavancin have been shown to be active against slowly dividing cells and dormant bacteria and also to disrupt biofilms of *S. aureus in vitro* (Belley et al. 2009; LaPlante and Mermel 2009). Glycopeptides are known to aggravate sepsis. Since sepsis claims numerous lives in hospital settings, glycopeptides could be developed to address this issue as well. There are literature reports on the activity of glycopeptides against influenza virus, hepatitis virus, and Ebola virus, which imply the possibility of extending the application of these antibiotics to other diseases (Maieron and Kerschner 2012; Wang et al. 2016).

There are still numerous challenges that need to be faced and overcome in the area of antibiotics research. As bacteria continue to evolve, so should we in our quest for the "magic bullet" such that we can remain a step ahead in this race.

References

Anderson, V.R. and Keating, G.M. (2009). Dalbavancin. *Drugs* 68 (5): 639–648.

Arhin, F.F., Draghi, D.C., Pillar, C.M. et al. (2009). Comparative in vitro activity profile of oritavancin against recent gram-positive clinical isolates. *Antimicrob. Agents Chemother.* 53 (11): 4762–4771.

Arimoto, H., Nishimura, K., Hayakawa, I. et al. (1999). Multi-valent polymer of vancomycin: enhanced antibacterial activity against VRE. *Chem. Commun.* (15): 1361–1362.

Arthur, M. and Courvalin, P. (1993). Genetics and mechanisms of glycopeptide resistance in enterococci. *Antimicrob. Agents Chemother.* 37 (8): 1563–1571.

Ashford, P.A. and Bew, S.P. (2012). Recent advances in the synthesis of new glycopeptide antibiotics. *Chem. Soc. Rev.* 41 (3): 957–978.

Bager, F., Madsen, M., Christensen, J., and Aarestrup, F.M. (1997). Avoparcin used as a growth promoter is associated with the occurrence of vancomycin-resistant Enterococcus faecium on Danish poultry and pig farms. *Prev. Vet. Med.* 31 (1–2): 95–112.

Belley, A., Neesham-Grenon, E., McKay, G. et al. (2009). Oritavancin kills stationary-phase and biofilm *Staphylococcus aureus* cells in vitro. *Antimicrob. Agents Chemother.* 53 (3): 918–925.

Binda, E., Marinelli, F., and Marcone, G.L. (2014). Old and new glycopeptide antibiotics: action and resistance. *Antibiotics* 3 (4): 572–594.

Bouza, E. and Burillo, A. (2010). Oritavancin: a novel lipoglycopeptide active against Gram-positive pathogens including multiresistant strains. *Int. J. Antimicrob. Agents* 36 (5): 401–407.

Boyd, D.A., Willey, B.M., Fawcett, D. et al. (2008). Molecular characterization of *Enterococcus faecalis* N06-0364 with low-level vancomycin resistance harboring a novel D-Ala-D-Ser gene cluster, vanL. *Antimicrob. Agents Chemother.* 52 (7): 2667–2672.

Bugg, T.D., Wright, G.D., Dutka-Malen, S. et al. (1991). Molecular basis for vancomycin resistance in *Enterococcus faecium* BM4147: biosynthesis of a

depsipeptide peptidoglycan precursor by vancomycin resistance proteins VanH and VanA. *Biochemistry* 30 (43): 10408–10415.
Butler, M.S., Hansford, K.A., Blaskovich, M.A. et al. (2014). Glycopeptide antibiotics: back to the future. *J. Antibiot.* 67 (9): 631–644.
Candiani, G., Abbondi, M., Borgonovi, M. et al. (1999). In-vitro and in-vivo antibacterial activity of BI 397, a new semi-synthetic glycopeptide antibiotic. *J. Antimicrob. Chemother.* 44 (2): 179–192.
Charneski, L., Patel, P.N., and Sym, D. (2009). Telavancin: a novel lipoglycopeptide antibiotic. *Ann. Pharmacother.* 43 (5): 928–938.
Crowley, B.M. and Boger, D.L. (2006). Total synthesis and evaluation of [Ψ [CH2NH] Tpg4] vancomycin aglycon: reengineering vancomycin for dual D-Ala-D-Ala and D-Ala-D-Lac binding. *J. Am. Chem. Soc.* 128 (9): 2885–2892.
Evers, S., Quintiliani, R. Jr., and Courvalin, P. (1996). Genetics of glycopeptide resistance in enterococci. *Microb. Drug Resist.* 2 (2): 219–223.
Fu, X., Albermann, C., Zhang, C., and Thorson, J.S. (2005). Diversifying vancomycin via chemoenzymatic strategies. *Org. Lett.* 7 (8): 1513–1515.
Gu, H.W., Ho, P.L., Tong, E. et al. (2003). Presenting vancomycin on nanoparticles to enhance antimicrobial activities. *Nano Lett.* 3 (9): 1261–1263.
Higgins, D.L., Chang, R., Debabov, D.V. et al. (2005). Telavancin, a multifunctional lipoglycopeptide, disrupts both cell wall synthesis and cell membrane integrity in methicillin-resistant Staphylococcus aureus. *Antimicrob. Agents Chemother.* 49 (3): 1127–1134.
Hiramatsu, K. (2001). Vancomycin-resistant *Staphylococcus aureus*: a new model of antibiotic resistance. *Lancet Infect. Dis.* 1 (3): 147–155.
Hiramatsu, K., Hanaki, H., Ino, T. et al. (1997). Methicillin-resistant *Staphylococcus aureus* clinical strain with reduced vancomycin susceptibility. *J. Antimicrob. Chemother.* 40 (1): 135–136.
James, R.C., Pierce, J.G., Okano, A. et al. (2012). Redesign of glycopeptide antibiotics: back to the future. *ACS Chem. Biol.* 7 (5): 797–804.
Jenkins, C.S., Meyer, D., Dreyfus, M.D., and Larrieu, M.J. (1974). Willebrand factor and ristocetin. I. Mechanism of rustocetin-induced platelet aggregation. *Br. J. Haematol.* 28 (4): 561–578.
Karlowsky, J.A., Nichol, K., and Zhanel, G.G. (2015). Telavancin: mechanisms of action, in vitro activity, and mechanisms of resistance. *Clin. Infect. Dis.* 61 (Suppl 2): S58–S68.
LaPlante, K.L. and Mermel, L.A. (2009). In vitro activities of telavancin and vancomycin against biofilm-producing *Staphylococcus aureus, S. epidermidis,* and *Enterococcus faecalis* strains. *Antimicrob. Agents Chemother.* 53 (7): 3166–3169.
Leadbetter, M.R., Adams, S.M., Bazzini, B. et al. (2004). Hydrophobic vancomycin derivatives with improved ADME properties: discovery of telavancin (TD-6424). *J. Antibiot.* 57 (5): 326–336.

Lebreton, F., Depardieu, F., Bourdon, N. et al. (2011). D-Ala-D-Ser VanN-type transferable vancomycin resistance in *Enterococcus faecium*. *Antimicrob. Agents Chemother.* 55 (10): 4606–4612.

Long, D.D., Aggen, J.B., Chinn, J. et al. (2008). Exploring the positional attachment of glycopeptide/beta-lactam heterodimers. *J. Antibiot.* 61 (10): 603–614.

Mackay, J.P., Gerhard, U., Beauregard, D.A. et al. (1994). Glycopeptide antibiotic activity and the possible role of dimerization: a model for biological signaling. *J. Am. Chem. Soc.* 116 (11): 4581–4590.

Maieron, A. and Kerschner, H. (2012). Teicoplanin therapy leading to a significant decrease in viral load in a patient with chronic hepatitis C. *J.Antimicrob. Chemother.* 67 (10): 2537–2538.

Malabarba, A., Ciabatti, R., Scotti, R. et al. (1995). New semisynthetic glycopeptides MDL 63,246 and MDL 63,042, and other amide derivatives of antibiotic A-40,926 active against highly glycopeptide-resistant VanA enterococci. *J. Antibiot.* 48 (8): 869–883.

McComas, C.C., Crowley, B.M., and Boger, D.L. (2003). Partitioning the loss in vancomycin binding affinity for D-Ala-D-Lac into lost H-bond and repulsive lone pair contributions. *J. Am. Chem. Soc.* 125 (31): 9314–9315.

McGuire, J.M., Wolfe, R.N., and Ziegler, D.W. (1955). Vancomycin, a new antibiotic. II. In vitro antibacterial studies. *Antibiot. Annu.* 3: 612–618.

McKessar, S.J., Berry, A.M., Bell, J.M. et al. (2000). Genetic characterization of vanG, a novel vancomycin resistance locus of *Enterococcus faecalis*. *Antimicrob. Agents Chemother.* 44 (11): 3224–3228.

Meyer, D., Jenkins, C.S., Dreyfus, M.D. et al. (1974). Willebrand factor and ristocetin. II. Relationship between Willebrand factor, Willebrand antigen and factor VIII activity. *Br. J. Haematol.* 28 (4): 579–599.

Morones-Ramirez, J.R., Winkler, J.A., Spina, C.S., and Collins, J.J. (2013). Silver enhances antibiotic activity against gram-negative bacteria. *Sci. Transl. Med.* 5 (190): 190ra81.

Murray, B.E. (2000). Vancomycin-resistant enterococcal infections. *N. Engl. J. Med.* 342 (10): 710–721.

Nakama, Y., Yoshida, O., Yoda, M. et al. (2010). Discovery of a novel series of semisynthetic vancomycin derivatives effective against vancomycin-resistant bacteria. *J. Med. Chem.* 53 (6): 2528–2533.

Nicolaou, K.C., Boddy, C.N., Brase, S., and Winssinger, N. (1999). Chemistry, biology, and medicine of the glycopeptide antibiotics. *Angew. Chem. Int. Ed.* 38 (15): 2096–2152.

Nicolaou, K.C., Hughes, R., Cho, S.Y. et al. (2000). Target-accelerated combinatorial synthesis and discovery of highly potent antibiotics effective against vancomycin-resistant bacteria. *Angew. Chem. Int. Ed.* 39 (21): 3823–3828.

Nicolosi, D., Scalia, M., Nicolosi, V.M., and Pignatello, R. (2010). Encapsulation in fusogenic liposomes broadens the spectrum of action of vancomycin against Gram-negative bacteria. *Int. J. Antimicrob. Agents* 35 (6): 553–558.

Nitanai, Y., Kikuchi, T., Kakoi, K. et al. (2009). Crystal structures of the complexes between vancomycin and cell-wall precursor analogs. *J. Mol. Biol.* 385 (5): 1422–1432.

Okano, A., Nakayama, A., Schammel, A.W., and Boger, D.L. (2014). Total synthesis of [Ψ [C (=NH) NH] Tpg4] vancomycin and its (4-chlorobiphenyl) methyl derivative: impact of peripheral modifications on vancomycin analogues redesigned for dual D-Ala-D-Ala and D-Ala-D-Lac binding. *J. Am. Chem. Soc.* 136 (39): 13522–13525.

Patti, G.J., Kim, S.J., Yu, T.Y. et al. (2009). Vancomycin and oritavancin have different modes of action in *Enterococcus faecium*. *J. Mol. Biol.* 392 (5): 1178–1191.

Rao, J., Lahiri, J., Isaacs, L. et al. (1998). A trivalent system from vancomycin.D-ala-D-Ala with higher affinity than avidin.biotin. *Science* 280 (5364): 708–711.

Reynolds, P.E. (1989). Structure, biochemistry and mechanism of action of glycopeptide antibiotics. *Eur. J. Clin. Microbiol. Infect. Dis.* 8 (11): 943–950.

Sader, H.S., Rhomberg, P.R., Farrell, D.J. et al. (2014). Antimicrobial activity of TD-1607 tested against contemporary (2010–2012) methicillin-resistant Staphylococcus aureus (MRSA) strains. *Proceedings of the 54th Interscience Conference on Antimicrobial Agents and Chemotherapy (ICAAC) 2014, Poster F-970*, Washington, DC (5–9 September 2014).

Saravolatz, L.D. and Stein, G.E. (2015). Oritavancin: a long-half-life lipoglycopeptide. *Clin. Infect. Dis.* 61 (4): 627–632.

Sarkar, P., Yarlagadda, V., Ghosh, C., and Haldar, J. (2017). A review on cell wall synthesis inhibitors with an emphasis on glycopeptide antibiotics. *MedChemComm.* 8 (3): 516–533.

Sundram, U.N., Griffin, J.H., and Nicas, T.I. (1996). Novel vancomycin dimers with activity against vancomycin-resistant enterococci. *J. Am. Chem. Soc.* 118 (51): 13107–13108.

Walsh, C.T., Fisher, S.L., Park, I.S. et al. (1996). Bacterial resistance to vancomycin: five genes and one missing hydrogen bond tell the story. *Chem. Biol.* 3 (1): 21–28.

Wang, Y., Cui, R., Li, G. et al. (2016). Teicoplanin inhibits Ebola pseudovirus infection in cell culture. *Antiviral Res.* 125: 1–7.

Weigel, L.M., Clewell, D.B., Gill, S.R. et al. (2003). Genetic analysis of a high-level vancomycin-resistant isolate of *Staphylococcus aureus*. *Science* 302 (5650): 1569–1571.

Williams, D.H. and Bardsley, B. (1999). The vancomycin group of antibiotics and the fight against resistant bacteria. *Angew. Chem. Int. Ed. Engl.* 38 (9): 1172–1193.

Xie, J., Okano, A., Pierce, J.G. et al. (2012). Total synthesis of [Ψ [C(=S)NH]Tpg^4] vancomycin aglycon, [Ψ[C(=NH)NH]Tpg^4]vancomycin aglycon, and related key compounds: reengineering vancomycin for dual D-Ala-D-Ala and D-Ala-D-Lac binding. *J.Am. Chem. Soc.* 134 (2): 1284–1297.

Xie, J., Pierce, J.G., James, R.C. et al. (2011). A redesigned vancomycin engineered for dual D-Ala-D-ala And D-Ala-D-Lac binding exhibits potent antimicrobial

activity against vancomycin-resistant bacteria. *J.Am. Chem. Soc.* 133 (35): 13946–13949.
Xing, B., Yu, C.W., Ho, P.L. et al. (2003). Multivalent antibiotics via metal complexes: potent divalent vancomycins against vancomycin-resistant enterococci. *J. Med. Chem.* 46 (23): 4904–4909.
Xu, X., Lin, D., Yan, G. et al. (2010). vanM, a new glycopeptide resistance gene cluster found in *Enterococcus faecium. Antimicrob. Agents Chemother.* 54 (11): 4643–4647.
Yarlagadda, V., Akkapeddi, P., Manjunath, G.B., and Haldar, J. (2014). Membrane active vancomycin analogues: a strategy to combat bacterial resistance. *J. Med. Chem.* 57 (11): 4558–4568.
Yarlagadda, V., Konai, M.M., Manjunath, G.B. et al. (2015a). Tackling vancomycin-resistant bacteria with 'lipophilic-vancomycin-carbohydrate conjugates. *J. Antibiot.* 68 (5): 302–312.
Yarlagadda, V., Konai, M.M., Paramanandham, K. et al. (2015b). In vivo efficacy and pharmacological properties of a novel glycopeptide (YV4465) against vancomycin-intermediate *Staphylococcus aureus. Int. J. Antimicrob. Agents* 46 (4): 446–450.
Yarlagadda, V., Manjunath, G.B., Sarkar, P. et al. (2016a). Glycopeptide antibiotic to overcome the intrinsic resistance of Gram-negative bacteria. *ACS Infect. Dis.* 2 (2): 132–139.
Yarlagadda, V., Samaddar, S., and Haldar, J. (2016b). Intracellular activity of a membrane-active glycopeptide antibiotic against meticillin-resistant *Staphylococcus aureus* infection. *J. Glob. Antimicrob. Resist.* 5: 71–74.
Yarlagadda, V., Samaddar, S., Paramanandham, K. et al. (2015c). Membrane disruption and enhanced inhibition of cell-wall biosynthesis: a synergistic approach to tackle vancomycin-resistant bacteria. *Angew. Chem. Int. Ed.* 54 (46): 13644–13649.
Yarlagadda, V., Sarkar, P., Manjunath, G.B., and Haldar, J. (2015d). Lipophilic vancomycin aglycon dimer with high activity against vancomycin-resistant bacteria. *Bioorg. Med. Chem. Lett.* 25 (23): 5477–5480.
Yarlagadda, V., Sarkar, P., Samaddar, S., and Haldar, J. (2016c). A vancomycin derivative with a pyrophosphate binding group: a strategy to combat vancomycin-resistant bacteria. *Angew. Chem. Int. Ed. Engl.* 15 (27): 7836–7840.
Yoganathan, S. and Miller, S.J. (2015). Structure diversification of vancomycin through peptide-catalyzed, site-selective lipidation: a catalysis-based approach to combat glycopeptide-resistant pathogens. *J. Med. Chem.* 58 (5): 2367–2377.
Zhanel, G.G., Calic, D., Schweizer, F. et al. (2010). New lipoglycopeptides: a comparative review of dalbavancin, oritavancin and telavancin. *Drugs* 70 (7): 859–886.

5

Current Macrolide Antibiotics and Their Mechanisms of Action

S. Lohsen and D.S. Stephens

School of Medicine, Emory University, Atlanta, GA, USA

5.1 Introduction

Macrolide antibiotics are characterized by a large macrocyclic lactone ring to which one or more deoxy sugars (usually cladinose and desosamine) are attached. More than 2000 macrolides have been isolated from a variety of natural sources as diverse as actinobacteria, fungi, plants, insects, and vertebrates (Ōmura 2002). Nearly all of the macrolide antibiotics used today are chemical derivatives of naturally occurring macrolides, especially erythromycin that is derived from the actinobacteria now known as *Saccharopolyspora erythraea* (Labeda 1987). Clinically relevant macrolide antibiotics fall into the category of 14-, 15-, or 16-membered lactone ring macrolides, while those with other antimicrobial and immunosuppressive activities may have more varied ring structures. The prototypical erythromycin has a 14-membered lactone ring as seen in Figure 5.1.

Erythromycin, the first macrolide antibiotic, was discovered in 1952 (McGuire et al. 1952). Erythromycin is a mixture of related compounds, erythromycins A, B, C, and D, with erythromycin A the primary active compound. Most modern macrolides are second-generation macrolides and are derived from erythromycin A, including clarithromycin, dirithromycin, roxithromycin, and azithromycin, among others. Other macrolides are available for use in humans outside of the United States (spiramycin, josamycin, midecamycin, and miocamycin) or are used exclusively in animals (including tylosin, ivermectin, and kitasamycin, among many others).

By modifying the structure of erythromycin A, pharmacokinetics and tolerance were improved, resulting in the second-generation macrolides. Derivatives of second-generation macrolides are characterized as third-generation

Antibiotic Drug Resistance, First Edition. Edited by José-Luis Capelo-Martínez and Gilberto Igrejas.

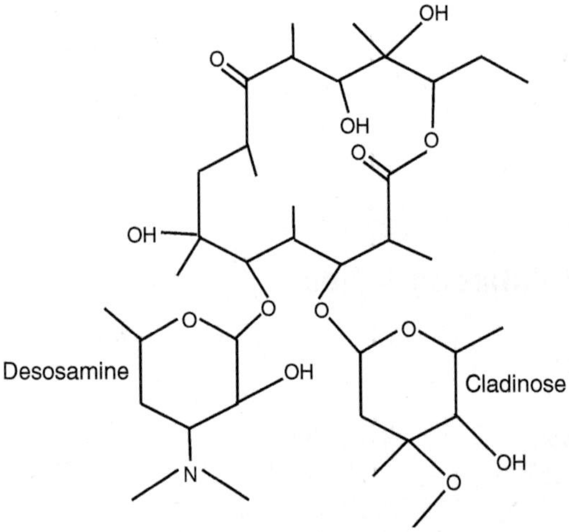

Figure 5.1 Erythromycin structure.

macrolides, which include the group designated as ketolides. Total synthesis and semisynthesis are keys in the production of new clinically relevant macrolide compounds, with aims to both improve tolerance of current macrolides and to combat antibiotic resistance against earlier-generation macrolides.

The general mode of action of macrolide antibiotics is through inhibiting protein assembly by reversible binding to the large 50S subunit of the bacterial ribosome, specifically to the 23s rRNA. Erythromycin works to prevent both ribosomal assembly and the extension of the peptide chain by blocking the exit tunnel of the ribosome (Champney and Tober 2000). Macrolides are often grouped with lincosaminide and streptogramin B classes of antibiotics despite their differing structures because they have similar mechanisms of action (Kirst 2010). Depending on the sensitivity of the bacteria and the particular macrolide, macrolides can either be bacteriostatic (most common) or bactericidal (Zuckerman 2000).

Due to widespread use, many different pathogenic bacteria have acquired mechanisms to resist macrolide antibiotics. These mechanisms include target modification of the ribosome and efflux or modification of the macrolides themselves. To circumvent these mechanisms of macrolide resistance, novel macrolide compounds are continually being created and evaluated in an attempt to stay "ahead of the curve," as macrolide antibiotics have been classified by the World Health Organization as critically important antimicrobials for human medicine. An in-depth examination of the state of macrolide research as of 2000 included descriptions of many of the naturally derived macrolides as well as semisynthetic macrolides (Ōmura 2002). This chapter focuses attention on in-use or in the pipeline macrolides used in humans and examines their structure, mechanism of action, spectrum of use, and future challenges.

5.2 Structure of Macrolides

Erythromycin has a macrocyclic 14-membered lactone ring with cladinose (a neutral sugar) and desosamine (an amino sugar) moieties attached. Much work has been done to modify erythromycin A to improve its stability, as erythromycin A is highly acid labile. Modifications to erythromycin A have led to a variety of novel clinically relevant macrolide compounds generated by semisynthesis, including clarithromycin, azithromycin, telithromycin, and other ketolides. A more complete synthesis approach has also been taken using a macrolide ring and a variety of building blocks/diversifiable elements, which can be synthesized in thousands of different combinations before testing. This novel approach has thus far yielded at least one novel macrolide, solithromycin, in which recently completed phase III clinical trials as of early 2017 (Seiple et al. 2016).

From the erythromycin A molecule, substituting a methoxy group for the C_6 hydroxyl group of the erythronolide A ring creates clarithromycin, which is a more acid-stable macrolide. A C-9 oxazine derivative of erythromycylamine (amination of C-9 ketone of erythromycin) is dirithromycin. A methyl-substituted nitrogen in place of the carbonyl group at position 9 of erythromycin A creates the azalide azithromycin.

Third-generation macrolides, the ketolides, are created when the cladinose moiety is removed from the erythronolide A ring and replaced with a keto group. This modification makes the macrolide less acid labile and increases its antibacterial activity. The ketolide telithromycin has two additional modifications, the first of which creates a carbamate ring at the $C_{11\text{-}12}$ positions and an attachment to this ring of an imidazo-pyridyl group. A second modification, as in clarithromycin, is the replacement of the C_6 hydroxyl group with a methoxy group.

Many among the newest generation of macrolides created and undergoing clinical testing to combat macrolide resistance are derived from or similar to the ketolide antibiotics and include solithromycin, cethromycin, and nafithromycin. Solithromycin is similar in structure to telithromycin but has a fluorine group at C_2 and no pyridine group on the alkyl-aryl side chain and is classified as a fluoroketolide. Nafithromycin (WCK 4873) is an 11,12-substituted lactone ketolide derivative (Vo 2004). Ketolides with 6-*O*-substitutions have shown more activity than telithromycin against macrolide-resistant *Staphylococcus aureus, Streptococcus pyogenes, Streptococcus pneumoniae,* and *Haemophilus influenzae*. An example is cethromycin, which has the typical keto substitution in addition to a quinolylallyl side chain at the C_6 position of the lactone ring (Mansour et al. 2013; Or et al. 2000). The ketolides telithromycin and cethromycin show activity against *S. pneumoniae* strains resistant to macrolides mediated either by efflux or ribosomal modification (Jorgensen et al. 2004).

Unique among macrolide antibiotics, fidaxomicin has an 18-membered ring (Venugopal and Johnson 2012). Non-antibiotic macrolides have quite varied structures and fall into two general classes: antifungal polyene macrolides with an amphipathic structure containing active lipophilic polyene chains and immunosuppressive/immunomodulatory macrolides such as the 23-membered macrolide lactone tacrolimus (Hirokazu Tanaka et al. 1987; Mesa-Arango et al. 2012; Tevyashova et al. 2013).

Many additional modifications of existing macrolides have been created but have not yet been tested clinically. Many of these compounds have shown effectiveness against previously macrolide-resistant bacterial strains such as *S. pneumoniae, S. pyogenes, S. aureus,* and *H. influenzae,* among others, *in vitro,* and some are detailed below.

Some modifications of erythromycin have resulted in alkylides and carbamolides. The alkylides are derivatives of 3-*O*-alkyl-6-*O*-methylerythromycin and possess a 3-*O*-arylalkyl group instead of the 3-*O*-cladinose (Liang et al. 2012). 3-*O*-Modifications of erythromycin with carbamoyl groups combined with 6-*O*-methyl and the fusion across the carbons at ring positions 11 and 12 (similar to ketolides) with a carbamate (to form an oxycarbonylimino) create the carbamolide macrolide group (Magee et al. 2013).

Modifications of clarithromycin have resulted in acylides. The acylides have an added acyl group at the 3-*O* position of clarithromycin, which are effective *in vitro* against both methylase- and efflux-based erythromycin-resistant *S. pneumoniae* and *H. influenzae* (Tanikawa et al. 2001, 2003). The acylide backbone improves activity against efflux and inducible macrolide resistance (Pavlovic et al. 2017). Independent of the acylide modifications, fusion of a five-membered lactone ring to the 11- and 12-positions of the clarithromycin carbon ring results in products that have similar antibacterial properties to telithromycin (Hunziker et al. 2004).

Modifications of azithromycin result in several compounds, including 3-*O*-declandinosylazithromycin derivatives that have increased antibacterial activity against resistant *S. pneumoniae* and *S. pyogenes* strains (Yan et al. 2017). Additional active antibacterial compounds have been synthesized through modification of azithromycin at position four with arylalkyl side chains, such as the addition of an arylcarbamate moiety, yielding compounds that show promising activity against macrolide-resistant *S. pneumoniae* strains (Ma et al. 2009). Similarly, introducing a novel arylalkylcarbamoyl side chain at position 4 results in improved activity against erythromycin-resistant *S. pneumoniae* (Wang et al. 2017). Macrolones have been engineered by adding ligand groups to conventional macrolides, particularly being successful with the addition of a quinolone moiety onto an azithromycin scaffold (Cipcic Paljetak et al. 2016). Characterization of macrolones has shown that they too show promising activity against macrolide-resistant strains of *S. pneumoniae, S. pyogenes, S. aureus,* and *H. influenzae* (Cipcic Paljetak et al. 2016).

Similar to the substitutions seen in the ketolide cethromycin, arylbutynyl substitutions at the C_6 position also show improved activity against erythromycin-resistant *S. aureus*, *S. pneumoniae*, and *S. pyogenes* relative to telithromycin (Keyes et al. 2003). Further compounds with higher levels of activity against methicillin-resistant *Staphylococcus epidermidis* and *S. pneumoniae* (compared with cethromycin) have been synthesized, which principally contain a variously substituted phenyl carbamate propyl chain to the 11-N ketolide backbone (Zheng et al. 2016). Novel macrolides with a 9-oxime ketolide structure have had mixed results in *in vivo* assays and led to further modification of 9-oxime-contianing macrolides (Agouridas et al. 1998). The successful 2-fluorination of ketolides such as solithromycin works in conjunction with aminopyridyl or carbamoylpyridyl groups to contribute to potency in the context of bactericidal 9-oxime erythromycins (Tian et al. 2017).

Further novel configurations of macrolides include bicyclolide oximes that have the 11,12-carbamate ring present in telithromycin with an additional 3,6-ring structure, which have been termed C-9 alkenylidine bridged macrolides (Poce et al. 2009; Tang et al. 2008). Bicyclolides including modithromycin and the oxime bicyclolide EDP-788 showed initial promising activity against macrolide-resistant *Mycobacterium avium* and *Neisseria gonorrhoeae* (Bermudez et al. 2007; Jacobsson et al. 2015).

5.3 Macrolide Mechanisms of Action

The primary mechanism of action of macrolide antibiotics is through binding to the bacterial ribosome to inhibit protein assembly and then translation. This is achieved through binding to the 50S subunit of the ribosome, specifically the 23s rRNA, which causes steric hindrance of peptide movement (Schlunzen et al. 2001). These interactions generally are bacteriostatic in nature, but for clarithromycin and azithromycin acting on *S. pyogenes*, *S. pneumoniae*, and *H. influenzae*, these interactions are bactericidal (Zuckerman 2000). Macrolides such as azithromycin accumulate to high levels in the primary space where bacterial infections occur, particularly the interstitial fluid of soft tissues (Kobuchi et al. 2016)

Macrolides and lincosamides such as clindamycin bind primarily to the peptidyl transferase ring of the 23s rRNA domain V and block the exit tunnel for nascent peptides (Schlunzen et al. 2001). The desosamine sugar and the lactone ring of the macrolide are responsible for mediating the hydrogen bond interactions with the peptidyl transferase cavity. In particular, the 2′ OH of the desosamine sugar interacts with a crucial nucleotide (A2058 of *Escherichia coli*) of the ribosome that is a target of resistance-mediating methylases and ribosomal mutations that mediate macrolide resistance (Weisblum 1995). This nucleotide and others that are involved with forming interactions with macrolides are all located within domain V of the 23s rRNA (Schlunzen et al. 2001). Domain II of the 23s rRNA may also play a role in stabilizing binding of macrolides with a carbamate ring at $C_{11\text{-}12}$ as seen with telithromycin (Douthwaite et al. 2000). A new interaction with the nascent peptide exit channel is mediated with telithromycin by the alkyl-aryl arm interacting with residue A752 of domain II of 23s rRNA (Dunkle et al. 2010). Replacing the 3-*O*-cladinose with aryl side chains also creates a new interaction with the nascent peptide exit tunnel as seen in the case with carbamolides that are 3-*O*-carbamoyl erythromycin A-derived analogues (Magee et al. 2013). 14- and 15-membered macrolides allow 6–8 peptides to assemble in the ribosome, whereas the ketolide telithromycin allows 9–10 amino acid resides to accumulate in the ribosome, pointing to more space between the peptidyl transferase center and the ketolide (Tenson et al. 2003).

In addition to 23s rRNA binding, some macrolides (including the ketolide subclass) have been shown to have a secondary mechanisms of function. These macrolides, particularly clarithromycin, inhibit the formation of the 50S subunit (and clarithromycin along with several ketolide antibiotics also inhibits 30S ribosomal subunit formation) (Champney and Tober 2000; Champney et al. 1998). Different macrolides show preference for this mode of action versus steric translation inhibition. For example, in *H. influenzae*, inhibition of 30S ribosomal subunit formation is seen with roxithromycin, flurithromycin, and ketolide antibiotics (Champney and Tober 2003; Mabe et al. 2004). Despite findings that mutations in ribosomal proteins L4 and L22 proteins involved in the peptide tunnel formation lead to macrolide resistance, no direct interactions between ribosomal proteins and macrolides have been identified (Canu et al. 2002; Nissen et al. 2000; Schlunzen et al. 2001).

Some experimental macrolides have been shown to have broad bactericidal activity, speculated to be due to the expression of a fragment of nascent peptide, the accumulation of which is believed to be fatal (Tian et al. 2017). The newer generations of macrolides including telithromycin and solithromycin allow for a higher level of translation to persist, which is dependent on a protein's N-terminal sequence, the structure of the nascent peptide, and the structure of the nascent peptide exit tunnel (Kannan et al. 2012). 9-oxime ketolides have additional interactions with the 23s rRNA at the ribosomal exit tunnel to create slightly different steric interference and antibacterial activity (Han et al. 2015). 16-Membered macrolides in particular (spiramycin and tylosin) increase the rates of stop codon readthrough, resulting in the general loss of translational accuracy (Kannan and Mankin 2011; Thompson et al. 2004). Rokitamycin has shown bactericidal activity *in vitro* but is rapidly converted to bacteriostatic metabolites *in vivo*, and as such, it is unclear what pharmacokinetic properties at the site of infection will allow mediation of the bactericidal or bacteriostatic activities (Hardy et al. 1988).

One macrolide in which the mode of activity is completely bactericidal is fidaxomicin. Its unique 18-membered ring structure among the macrolide antibiotics gives it a unique mode of function that is not yet completely understood, but it is known to bind to and inhibit RNA polymerase (Venugopal and Johnson 2012).

Other macrolide activities include gastrointestinal motor stimulating, anti-inflammatory, immunomodulatory, and anticancer. Additionally, at least one report of antiviral activity has been recorded, specifically against rhinovirus for the oleandomycin derivative Mac5, though the mechanism through which this occurred was unclear (Porter et al. 2016).

Gastrointestinal motor-stimulating activities (motilide) have been reported for 14-membered macrolides, and this activity appears not to be related to the antibacterial quality of the compounds (Omura et al. 1985). Non-antibiotic "motilides" have been synthesized, which are more potent than erythromycin when it comes to facilitating gastric emptying (Cowles et al. 2000). The activity of motilides is most likely due to the beta-turn conformation of the macrolide ring mimicking a functional portion of motilin, a peptide that causes duodenal contractions (Steinmetz et al. 2002). These motilides are potentially useful in the context of anorexia-cachexia associated with delayed gastric transit (Asakawa et al. 2003).

Low doses of antibacterial macrolides including erythromycin, clarithromycin, azithromycin, and roxithromycin have been used in some countries to treat the inflammatory airway disease diffuse panbronchiolitis (Kadota et al. 1993). Erythromycin and several of its derivatives (including some with only weak antibacterial activity) suppress interleukin-8 release, which could potentially lead to reductions in inflammation, specifically in bronchial epithelial cells (Sunazuka et al. 1999). Pointing to an additional role for reducing inflammation, erythromycin and some derivatives inhibit leukocyte chemotaxis (Oohori et al. 2000), while clarithromycin (as well as josamycin and midecamycin) reduces interleukin-2 production in T cells (Morikawa et al. 1994). Several intermediate metabolites of erythromycin increase monocyte differentiation, implying that the effect seen by erythromycin is as a result of its metabolism and is unrelated to the antibacterial activity of erythromycin (Sunazuka et al. 2003). The immunomodulatory effect of azithromycin is used clinically where its activity on airway epithelial cells to promote function and reduce mucus secretion is used in the management of many chronic lung diseases (Sivapalasingam and Neal 2015).

The macrolide rapamycin (sirolimus) has immunosuppressive functions and is also used as an anticancer drug. Rapamycin can be paired with a chemosensitizer such as piperine in a nanoparticle formulation to overcome efflux-based drug resistance in the context of breast cancer treatment (Katiyar et al. 2016). Other macrolide interactions with the immune system involve the promotion of macrophage recruitment and increase C–C motif chemokine ligand

2 production from macrophages (Iwanaga et al. 2015). Similarly, erythromycin (and roxithromycin) stimulates neutrophil migration *in vitro* (Anderson 1989).

Study of the immunomudulatory activities of macrolides has led to the exploration of non-antibacterial erythromycin derivatives. Less potent macrolides (GS-459755 [2′-desoxy-9-(*S*)-erythromycylamine] and GSA-560660 [azithromycin-based 2′-desoxy molecules]) are under development to reduce inflammation via inhibition of the NLRP3 inflammasome in chronic lung diseases while at the same time not affecting increases in rates of bacterial resistance (Hodge et al. 2017). Other non-antibacterial macrolides are well known for their use in modification of the immune system. Tacrolimus is a macrolide immunosuppressant/immunomodulator that is a calcineurin inhibitor and therefore suppresses T-cell activity (Kitahara and Kawai 2007).

5.4 Clinical Use of Macrolides

In the United States, the Food Drug Administration has approved five macrolide antibiotics: erythromycin, clarithromycin, azithromycin, dirithromycin, and telithromycin. Several non-antibiotic macrolides have also been approved for clinical use and include the immunosuppressives rapamycin and tacrolimus, as well as the antifungals amphotericin B and nystatin A1. Orphan drug status has been given to cethromycin for some prophylactic treatments. Macrolides have been classified as critically important to human medicine by the Federal Drug Administration (Powers 1998). Globally, a broader variety of macrolides have been approved for use and include the following: spiramycin, rokitamycin, midecamycin, kitasamycin/leucomycin, josamycin, trichomycin (antifungal and antiprotozoal), and pimaricin/natamycin (antifungal).

Macrolides have varied bioavailability. Erythromycin is highly acid labile and has the lowest bioavailability, which varies greatly with the formulation (Ginsburg 1986). As previously noted, due to its similarity to motilin, erythromycin is associated with gastrointestinal motor-stimulating activity and accompanying side effects (Omura et al. 1985). Other macrolides are more acid stable and as a result have both increased bioavailability and reduced gastrointestinal side effects – in the case of clarithromycin and telithromycin, 52–55 and 57% bioavailability, respectively (Chu et al. 1992; Douthwaite et al. 2000). Upon oral administration, clarithromycin is converted into a 14-hydroxy metabolite, which accounts for the majority of clarithromycin's antimicrobial activity (Chu et al. 1992). Clarithromycin and other 14-membered macrolides are oxidized by hepatic enzymes in the cytochrome P450 system by the CYP3A subclass of enzymes (Rodrigues et al. 1997). They inhibit CYP3A; however, azithromycin does not, nor do the ketolides (Chavan et al. 2016).

Macrolides are absorbed well into tissues, as seen in the case of telithromycin, where significantly higher concentrations are observed in the target tissues of the respiratory tract, which is similar to previous studies of other macrolides (Shain and Amsden 2002). Excellent tissue penetration is also seen for clarithromycin and azithromycin, where tissue concentrations are much higher than serum levels of macrolide, and azithromycin has a prolonged tissue half-life (Zuckerman 2000).

Contraindications for macrolides are varied, and some are restricted to the 15-membered macrolide azithromycin. Azithromycin has been shown to increase rates of cardiac arrhythmias in patients with coexisting risk factors (Albert and Schuller 2014). Telithromycin's use is restricted to treatment of mild to moderate community-acquired pneumonia due to its hepatotoxicity (Zuckerman et al. 2009). The imidazopyridyl side chain of telithromycin is thought to be associated with adverse health events including hepatic failure and exacerbation of myasthenia gravis (Fernandes et al. 2017).

Macrolides are often used to treat community-acquired pneumonia and other respiratory tract infections (both upper and lower) and are effective against both Gram-positive and Gram-negative bacteria. Generally, the 14-membered macrolides have similar activities to the prototypical erythromycin, while the 15-membered macrolide azithromycin shows less activity against Gram-positive bacteria, but higher activity against Gram-negative bacteria, and 16-membered macrolides are generally less active than erythromycin (Hardy et al. 1988). Ketolides have been shown to have better activity against many Gram-positive organisms than the other macrolides (Champney and Tober 2003). Macrolides such as dirithromycin are approved for acute exacerbations of chronic bronchitis and exacerbations of asthma and chronic obstructive pulmonary disease due to rhinoviruses and uncontrolled inflammatory pathways (FDA 1997; Porter et al. 2016). Roxithromycin has a similar profile to other macrolides for treatment of respiratory tract infections (Hayashi and Kawashima 2012).

Combination therapies that include macrolides have shown success in the treatment of pulmonary disease caused by nontuberculous mycobacteria, and azithromycin and clarithromycin are considered cornerstones of treatment for *M. avium* complex lung disease (van Ingen et al. 2013; Jeong et al. 2016). Macrolides are also currently recommended for treatment of pertussis (Hardy et al. 2016).

Macrolides are generally considered a first-line alternative to those allergic to beta-lactams. They are the first-line recommended treatment for *Legionella pneumophila* (Descours et al. 2017). Macrolides are the most effective treatment of *Helicobacter pylori* infection including duodenal ulcer disease in a standard sequential treatment along with a proton pump inhibitor and amoxicillin (Branquinho et al. 2017; Sivapalasingam and Neal 2015). Macrolides can be used with a potassium-competitive acid blocker to improve the efficacy of a clarithromycin-containing seven-day triple therapy of *H. pylori*. Even with clarithromycin-resistant bacteria, the combination results in robust clearance (Noda et al. 2016).

In babies under six months of age, macrolides are indicated only for pertussis and *Chlamydia trachomatis*-induced pneumonia or eye infection (Subcommittee of the Expert Committee on the Selection and Use of Essential Medicines 2008). In older children, macrolides are used as first-line drugs for pertussis, *C. trachomatis*, Legionnaires' disease, *Campylobacter jejuni*, *M. avium* complex, *H. pylori*, and as an alternative for penicillin-allergic individuals in acute otitis media, syphilis, community-acquired pneumonia, streptococcal sore throat, acute bacterial sinusitis, and acute otitis media (Subcommittee of the Expert Committee on the Selection and Use of Essential Medicines 2008).

Macrolides such as erythromycin, dirithromycin, azithromycin, and roxithromycin can also be used in the treatment of skin and soft tissue infections (Counter et al. 1991; FDA 1997, Hayashi and Kawashima 2012; Sivapalasingam and Neal 2015). Oral and topical macrolides such as dirithromycin and roxithromycin are used in the treatment of acne (Hayashi and Kawashima 2012; Nakase et al. 2016).

Macrolides are also used in the treatment of some sexually transmitted infections. Azithromycin and clarithromycin are effective in treating infections caused by C. *trachomatis*, and azithromycin has the benefit of higher adherence to treatment, as only a single dose is required (Geisler et al. 2016; Zuckerman et al. 2009). Azithromycin is also used to treat urethritis and cervicitis more generally, due to *C. trachomatis* or *N. gonorrhoeae*, and along with clarithromycin is used in the treatment of pelvic inflammatory disease (Sivapalasingam and Neal 2015). Historically azithromycin has been used to treat syphilis and, in some trials, shows higher efficacy than penicillin G benzathine treatment (Molini et al. 2016; Zuckerman et al. 2009). Erythromycin has been used in the treatment of chancroid, nongonococcal urethritis, and granuloma inguinale (Sivapalasingam and Neal 2015).

There are also a number of more atypical antibacterial treatments beyond those listed above for which macrolides are used. Azithromycin has activity against Gram-negative enteric organisms such as *E. coli*, *Salmonella* species, *Yersinia enterocolitica*, and *Shigella* species, while clarithromycin and telithromycin do not (Zuckerman et al. 2009). In vulnerable populations in which antimicrobial treatment is needed for *Campylobacter* infections, macrolides are critically important (Bolinger and Kathariou 2017). Trials of prophylactic azithromycin administration in Gambia showed a decrease in *S. aureus*, group B streptococcus, and *S. pneumoniae* colonization in mothers and newborns and may help lower death by bacterial sepsis rates (Roca et al. 2016). Cethromycin (Restanza) was given orphan drug status by the Federal Drug Administration in 2007 for the prophylactic treatment of inhalation anthrax exposure, tularemia due to *Francisella tularensis*, and plague due to *Yersinia pestis* (Mansour et al. 2013). Fidaxomicin, the 18-membered ring macrolide, has minimal systemic absorption and is used in the treatment of *Clostridium difficile* (Venugopal and Johnson 2012).

Apart from their antibacterial uses, macrolides can also be used to treat some parasitic and fungal infections. Azithromycin is active against *Plasmodium falciparum* and *Plasmodium vivax* and can be an effective antimalarial treatment, especially when used in conjunction with other antimalarial drugs in the context of pregnant women with malaria, as well as being used in a liposomal preparation topically for the treatment of cutaneous leishmaniasis (Nakornchai and Konthiang 2006; Phong et al. 2016; Rajabi et al. 2016). Roxithromycin is used in the treatment of amoeboid infections caused by *Acanthamoeba castellanii* (Mattana et al. 2004). Polyene antimycotics are a subclass of macrolides, which include amphotericin B and nystatin A1, which are used clinically for their antifungal activity. This activity is mediated through binding to ergosterol in fungal cell membranes, causing ion leakage of potassium and sodium (Mesa-Arango et al. 2012).

Completely apart from their antimicrobial activity, macrolides such as tacrolimus, pimecrolimus, and rapamycin/sirolimus additionally can be used as immunosuppressants in the context of organ transplants and immune disorders. Tacrolimus therapy shows efficacy in rheumatoid arthritis and can also be used to treat cyclosporine A-resistant or cyclosporine A-dependent minimal change disease as well as vernal keratoconjunctivitis (Chatterjee and Agrawal 2016; Kitahara and Kawai 2007; Xu et al. 2017). Rapamycin blocks the mTOR signaling pathway and can be used alone or in conjunction with tacrolimus to provide steroid-free immunosuppression (Asante-Korang et al. 2017).

5.5 Next-Generation Macrolides and Future Use

Continuing to address the bacterial evolution of antibiotic resistance necessitates the development of novel antibiotics. The development of novel macrolides, however, in recent years has encountered a number of stumbling blocks. Concerns over the ketolides, specifically with respect to telithromycin and its associated hepatotoxicity, led many companies to halt development of ketolides (Fernandes et al. 2017; Georgopapadakou 2014). While having gained orphan drug status for some prophylactic use, cethromycin failed to demonstrate an ability to accumulate to clinically useful concentrations during human trials and was not approved by the Federal Drug Administration for use in treatment of mild to moderate community-acquired pneumonia (FDA 2009). Other macrolide candidates discontinued during clinical trials in recent years include modithromycin (EDP-420) and another bridged bicyclic macrolide, EDP-788 (Bermudez et al. 2007; Jacobsson et al. 2015).

Another novel macrolide, solithromycin, showed activity against traditional macrolide-resistant *S. pneumoniae* strains, and trials of efficacy relative to the fluoroquinolones typically used for community-acquired pneumonia were encouraging (Farrell et al. 2010, 2015; Figueira et al. 2016; Sivapalasingam and Neal 2015; Viasus et al. 2017). As a result, solithromycin recently underwent phase III clinical trials for use in the treatment of moderate to moderately severe community-acquired bacterial pneumonia. However, the risk of hepatotoxicity was not fully characterized; thus solithromycin's new drug application was not approvable as filed, and further testing in a larger population is required to exclude drug-induced liver injury (Buege et al. 2017).

Nafithromycin (WCK 4873) is the only remaining macrolide currently undergoing clinical trials in the United States. It demonstrates a broad range of potent activity against community-acquired bacterial pneumonia pathogens *in vitro* (Farrell et al. 2016). This, along with its ability to accumulate in the epithelial lining fluid and alveolar macrophages, points to its potential usefulness in the treatment of lower respiratory tract infections (Rodvold et al. 2017). Unlike other macrolides and ketolides, nafithromycin shows no inhibition of CYP activity levels and therefore is unlikely to have significant drug–drug interaction issues (Chavan et al. 2016). Nafithromycin is currently undergoing phase II clinical trials for the treatment of community-acquired bacterial pneumonia in adults and was awarded Qualified Infectious Disease Product status in 2015 by the US Federal Drug Administration (Farrell et al. 2016).

In order to more specifically design modifications of current macrolide antibiotics, molecular dynamics simulation and structure modeling have become useful tools. By using modeling along with nuclear magnetic resonance, crystallography, or other high-resolution characterization of antibiotic–ribosomal interactions, modifications to existing macrolides such as azithromycin are predicted to result in increased binding affinity to their ribosomal targets (Pavlova et al. 2017). By gaining a greater understanding of the structure of macrolide-inactivating enzymes such as macrolide phosphotransferases (types I and II) encoded by the *mph*(A) and *mph*(B) genes, the design of next-generation macrolides can avoid inactivation by such phosphotransferases, particularly by the incorporation of additional substitutions at positions C_9–C_{14} of the macrolide ring (Fong et al. 2017).

The future of established classes of macrolides may not lie with simply modifying ring structures and side chains of existing macrolide backbones, but it may be in combination therapies with other compounds to potentiate the activity of existing antibiotics. This has been seen in the case of enhanced susceptibility of *S. aureus* strains with a variety of functionally distinct antibiotics, including those in lipopeptide, lincosamide, and beta-lactam classes, of which the lincosamides are functionally similar to the macrolide mechanism of function (Quave et al. 2012). Community-acquired pneumonia treatment in mice shows that even bacterial strains that are resistant to macrolides are effectively treated by a combination therapy with ceftriaxone (a beta-lactam) and azithromycin (Yoshioka et al. 2016).

Augmenting macrolide treatment with blue light therapy and nanosilver has been shown to increase the effectiveness of treatment of previously macrolide-resistant strains with a triple treatment of blue light, nanosilver, and azithromycin against methicillin-resistant *S. aureus* (Akram et al. 2016). Additionally, efficacy of macrolide antifungals such as amphotericin B can be boosted by conjugation to silver nanoparticles (Ahmad et al. 2016).

Further ways to augment activity of macrolides include loading into liposomes and combining therapy with other antimicrobials. The cationic antimicrobial peptide AMP DP7 conjugated with cholesterol has been modified into azithromycin antibiotic-loaded liposomes. In a mouse model of methicillin-resistant *S. aureus* infection, previously subeffective doses have enhanced activity with these liposomes, indicating this could be an aid in boosting antibiotic effectiveness (Liu et al. 2016).

Beyond engineering novel macrolides and augmenting activities of existing macrolides, a third approach to expand the repertoire of macrolide antibiotics lies in identifying further naturally occurring macrolides. Extremophilic fungi grown in fungal cocultures to encourage production of novel secondary metabolites have produced a new class of 16-membered macrolides called berkeleylactones (Stierle et al. 2017). Further similar fungal cocultures could encourage the crosstalk necessary to facilitate formation of novel macrolide metabolites that could also have antibacterial activity.

While macrolide antibiotics currently play an important role in the treatment of a variety of infections, from respiratory to sexually transmitted diseases, the perpetual evolution of antibiotic resistance in bacteria poses a very appreciated risk to the future usefulness of these and other classes of antibiotics. The development of novel macrolides in recent years has encountered challenges, and the immediate future of macrolides appears to lie in the development of combination and augmentative therapies. However, this class of molecules continues to be foundational in human drug development.

References

Agouridas, C., Denis, A., Auger, J.M. et al. (1998). Synthesis and antibacterial activity of ketolides (6-O-methyl-3-oxoerythromycin derivatives): a new class of antibacterials highly potent against macrolide-resistant and -susceptible respiratory pathogens. *J Med Chem* 41: 4080–4100.

Ahmad, A., Wei, Y., Syed, F. et al. (2016). Amphotericin B-conjugated biogenic silver nanoparticles as an innovative strategy for fungal infections. *Microb Pathog* 99: 271–281.

Akram, F.E., El-Tayeb, T., Abou-Aisha, K., and EL-AZIZI, M. (2016). A combination of silver nanoparticles and visible blue light enhances the antibacterial efficacy of ineffective antibiotics against methicillin-resistant *Staphylococcus aureus* (MRSA). *Ann Clin Microbiol Antimicrob* 15: 48.

Albert, R.K. and Schuller, J.L. (2014). Macrolide antibiotics and the risk of cardiac arrhythmias. *Am J Respir Crit Care Med* 189: 1173–1180.

Anderson, R. (1989). Erythromycin and roxithromycin potentiate human neutrophil locomotion in vitro by inhibition of leukoattractant-activated superoxide generation and autooxidation. *J Infect Dis* 159: 966–973.

Asakawa, A., Akio, I., Ohinata, K. et al. (2003). EM574, a motilide, has an orexigenic activity with affinity for growth-hormone secretagogue receptor. *J Gastroenterol Hepatol* 18: 881–882.

Asante-Korang, A., Carapellucci, J., Krasnopero, D. et al. (2017). Conversion from calcineurin inhibitors to mTOR inhibitors as primary immunosuppressive drugs in pediatric heart transplantation. *Clin Transplant* 31: e13054.

Bermudez, L.E., Motamedi, N., Chee, C. et al. (2007). EDP-420, a bicyclolide (bridged bicyclic macrolide), is active against *Mycobacterium avium*. *Antimicrob Agents Chemother* 51: 1666–1670.

Bolinger, H. and Kathariou, S. (2017). The current state of macrolide resistance in Campylobacter spp.: trends and impacts of resistance mechanisms. *Appl Environ Microbiol* 83: e00416-17.

Branquinho, D., Almeida, N., Gregorio, C. et al. (2017). Levofloxacin or Clarithromycin-based quadruple regimens: what is the best alternative as first-line treatment for *Helicobacter pylori* eradication in a country with high resistance rates for both antibiotics? *BMC Gastroenterol* 17: 31.

Buege, M.J., Brown, J.E., and Aitken, S.L. (2017). Solithromycin: a novel ketolide antibiotic. *Am J Health Syst Pharm* 74 (12): 875–887.

Canu, A., Malbruny, B., Coquemont, M. et al. (2002). Diversity of ribosomal mutations conferring resistance to macrolides, clindamycin, streptogramin, and telithromycin in *Streptococcus pneumoniae*. *Antimicrob Agents Chemother* 46: 125–131.

Champney, W.S. and Tober, C.L. (2000). Specific inhibition of 50S ribosomal subunit formation in *Staphylococcus aureus* cells by 16-membered macrolide, lincosamide, and streptogramin B antibiotics. *Curr Microbiol* 41: 126–135.

Champney, W.S. and Tober, C.L. (2003). Preferential inhibition of protein synthesis by ketolide antibiotics in *Haemophilus influenzae* cells. *Curr Microbiol* 46: 103–108.

Champney, W.S., Tober, C.L., and Burdine, R. (1998). A comparison of the inhibition of translation and 50S ribosomal subunit formation in *Staphylococcus aureus* cells by nine different macrolide antibiotics. *Curr Microbiol* 37: 412–417.

Chatterjee, S. and Agrawal, D. (2016). Tacrolimus in corticosteroid-refractory vernal keratoconjunctivitis. *Cornea* 35: 1444–1448.

Chavan, R., Zope, V., Yeole, R., and Patel, M. (2016). WCK 4873 (nafithromycin): assessment of in vitro human CYP inhibitory potential of a novel lactone-ketolide. *Open Forum Infect Dis* 3 (Suppl 1): S515.

Chu, S.Y., Deaton, R., and Cavanaugh, J. (1992). Absolute bioavailability of clarithromycin after oral administration in humans. *Antimicrob Agents Chemother* 36: 1147–1150.

Cipcic Paljetak, H., Verbanac, D., Padovan, J. et al. (2016). Macrolones: novel class of macrolide antibiotics active against key resistant respiratory pathogens in vitro and in vivo. *Antimicrob Agents Chemother* 60 (9): 5337–5348.

Counter, F.T., Ensminger, P.W., Preston, D.A. et al. (1991). Synthesis and antimicrobial evaluation of dirithromycin (AS-E 136; LY237216), a new macrolide antibiotic derived from erythromycin. *Antimicrob Agents Chemother* 35: 1116–1126.

Cowles, V.E., Nellans, H.N., Seifert, T.R. et al. (2000). Effect of novel motilide ABT-229 versus erythromycin and cisapride on gastric emptying in dogs. *J Pharmacol Exp Ther* 293: 1106–1111.

Descours, G., Ginevra, C., Jacotin, N. et al. (2017). Ribosomal mutations conferring macrolide resistance in Legionella pneumophila. *Antimicrob Agents Chemother* 61 (3): e02188-16-e02188-16.

Douthwaite, S., Hansen, L.H., and Mauvais, P. (2000). Macrolide-ketolide inhibition of MLS-resistant ribosomes is improved by alternative drug interaction with domain II of 23S rRNA. *Mol Microbiol* 36: 183–193.

Dunkle, J.A., Xiong, L., Mankin, A.S., and Cate, J.H. (2010). Structures of the *Escherichia coli* ribosome with antibiotics bound near the peptidyl transferase center explain spectra of drug action. *Proc Natl Acad Sci U S A* 107: 17152–17157.

Farrell, D.J., Mendes, R.E., and Jones, R.N. (2015). Antimicrobial activity of solithromycin against serotyped macrolide-resistant *Streptococcus pneumoniae* isolates collected from U.S. medical centers in 2012. *Antimicrob Agents Chemother* 59: 2432–2434.

Farrell, D.J., Sader, H.S., Castanheira, M. et al. (2010). Antimicrobial characterisation of CEM-101 activity against respiratory tract pathogens, including multidrug-resistant pneumococcal serogroup 19A isolates. *Int J Antimicrob Agents* 35: 537–543.

Farrell, D.J., Sader, H.S.; Rhomberg, P.R. et al. (2016). In vitro activity of lactone ketolide WCK 4873 when tested against contemporary community-acquired bacterial pneumonia pathogens from a global surveillance program. ASM Microbe 2016, Boston, MA.

FDA (1997). NDA 50-678/SE1-003 Approval Letter. Center for Drug Evaluation and Research, 19 December 1997.

FDA (2009). Cethromycin. Briefing Document for the Anti-Infective Drugs Advisory Committee, 2 June 2009. NDA 22-398. http://www.fda.gov/downloads/AdvisoryCommittees/CommitteesMeetingMaterials/Drugs/

Anti-InfectiveDrugsAdvisoryCommittee/UCM166989.pdf (Accessed 7 March 2009).

Fernandes, P., Martens, E., and Pereira, D. (2017). Nature nurtures the design of new semi-synthetic macrolide antibiotics. *J Antibiot (Tokyo)* 70 (5): 527–533.

Figueira, M., Fernandes, P., and Pelton, S.I. (2016). Efficacy of a fourth generation macrolide, solithromycin (CEM-101), for experimental otitis media (EOM) caused by non-typeable *Haemophilus influenzae* (NTHi) and *Streptococcus pneumoniae* (SP). *Antimicrob Agents Chemother* 60 (9): 5533–5538.

Fong, D.H., Burk, D.L., Blanchet, J. et al. (2017). Structural basis for kinase-mediated macrolide antibiotic resistance. *Structure* 25 (5): 750–761.

Geisler, W.M., Perry, R.C., and Kerndt, P.R. (2016). Azithromycin versus doxycycline for chlamydia. *N Engl J Med* 374: 1787.

Georgopapadakou, N.H. (2014). The wobbly status of ketolides: where do we stand? *Expert Opin Investig Drugs* 23: 1313–1319.

Ginsburg, C.M. (1986). Pharmacology of erythromycin in infants and children. *Pediatr Infect Dis* 5: 124–129.

Han, X., Lv, W., Guo, S.Y. et al. (2015). Synthesis and structure-activity relationships of novel 9-oxime acylides with improved bactericidal activity. *Bioorg Med Chem* 23: 6437–6453.

Hardy, D.J., Hensey, D.M., Beyer, J.M. et al. (1988). Comparative in vitro activities of new 14-, 15-, and 16-membered macrolides. *Antimicrob Agents Chemother* 32: 1710–1719.

Hardy, D.J., Vicino, D., and Fernandes, P. (2016). In vitro activity of solithromycin against Bordetella pertussis: an emerging respiratory pathogen. *Antimicrob Agents Chemother* 60 (12): 7043–7045.

Hayashi, N. and Kawashima, M. (2012). Multicenter randomized controlled trial on combination therapy with 0.1% adapalene gel and oral antibiotics for acne vulgaris: comparison of the efficacy of adapalene gel alone and in combination with oral faropenem. *J Dermatol* 39: 511–515.

Hodge, S.J., Tran, H., Hamon, R. et al. (2017). Non-antibiotic macrolides restore airway macrophage phagocytic function with potential anti-inflammatory effects in chronic lung diseases. *Am J Physiol Lung Cell Mol Physiol* 312 (5): L678–L687.

Hunziker, D., Wyss, P.C., Angehrn, P. et al. (2004). Novel ketolide antibiotics with a fused five-membered lactone ring: synthesis, physicochemical and antimicrobial properties. *Bioorg Med Chem* 12: 3503–3519.

Iwanaga, N., Nakamura, S., Oshima, K. et al. (2015). Macrolides promote CCL2-mediated macrophage recruitment and clearance of nasopharyngeal pneumococcal colonization in mice. *J Infect Dis* 212: 1150–1159.

Jacobsson, S., Golparian, D., Phan, L.T. et al. (2015). In vitro activities of the novel bicyclolides modithromycin (EDP-420, EP-013420, S-013420) and EDP-322 against MDR clinical *Neisseria gonorrhoeae* isolates and international reference strains. *J Antimicrob Chemother* 70: 173–177.

Jeong, B.H., Jeon, K., Park, H.Y. et al. (2016). Peak plasma concentration of azithromycin and treatment responses in *Mycobacterium avium* complex lung disease. *Antimicrob Agents Chemother* 60: 6076–6083.

Jorgensen, J.H., Crawford, S.A., McElmeel, M.L., and Whitney, C.G. (2004). Activities of cethromycin and telithromycin against recent North American isolates of *Streptococcus pneumoniae*. *Antimicrob Agents Chemother* 48: 605–607.

Kadota, J., Sakito, O., Kohno, S. et al. (1993). A mechanism of erythromycin treatment in patients with diffuse panbronchiolitis. *Am Rev Respir Dis* 147: 153–159.

Kannan, K. and Mankin, A.S. (2011). Macrolide antibiotics in the ribosome exit tunnel: species-specific binding and action. *Ann N Y Acad Sci* 1241: 33–47.

Kannan, K., Vazquez-Laslop, N., and Mankin, A.S. (2012). Selective protein synthesis by ribosomes with a drug-obstructed exit tunnel. *Cell* 151: 508–520.

Katiyar, S.S., Muntimadugu, E., Rafeeqi, T.A. et al. (2016). Co-delivery of rapamycin- and piperine-loaded polymeric nanoparticles for breast cancer treatment. *Drug Deliv* 23: 2608–2616.

Keyes, R.F., Carter, J.J., Englund, E.E. et al. (2003). Synthesis and antibacterial activity of 6-O-arylbutynyl ketolides with improved activity against some key erythromycin-resistant pathogens. *J Med Chem* 46: 1795–1798.

Kirst, H.A. (2010). New macrolide, lincosaminide and streptogramin B antibiotics. *Expert Opin Ther Pat* 20: 1343–1357.

Kitahara, K. and Kawai, S. (2007). Cyclosporine and tacrolimus for the treatment of rheumatoid arthritis. *Curr Opin Rheumatol* 19: 238–245.

Kobuchi, S., Aoki, M., Inoue, C. et al. (2016). Transport of azithromycin into extravascular space in rats. *Antimicrob Agents Chemother* 60 (11): 6823–6827.

Labeda, D.P. (1987). Transfer of the type strain of *Streptomyces erythraeus* (Waksman 1923) Waksman and Henrici 1948 to the genus Saccharopolyspora Lacey and Goodfellow 1975 as *Saccharopolyspora erythraea* sp. no., and designation of a Neotype Strain for *Streptomyces erythraeus*. *International Journal of Systematic Bacteriology* 37: 19–22.

Liang, J.H., Li, X.L., Wang, H. et al. (2012). Structure-activity relationships of novel alkylides: 3-O-arylalkyl clarithromycin derivatives with improved antibacterial activities. *Eur J Med Chem* 49: 289–303.

Liu, X., Li, Z., Wang, X. et al. (2016). Novel antimicrobial peptide-modified azithromycin-loaded liposomes against methicillin-resistant *Staphylococcus aureus*. *Int J Nanomedicine* 11: 6781–6794.

Ma, S., Jiao, B., Liu, Z. et al. (2009). Synthesis and antibacterial activity of 4″,11-di-O-arylalkylcarbamoyl azithromycin derivatives. *Bioorg Med Chem Lett* 19: 1698–1701.

Mabe, S., Eller, J., and Champney, W.S. (2004). Structure-activity relationships for three macrolide antibiotics in *Haemophilus influenzae*. *Curr Microbiol* 49: 248–254.

Magee, T.V., Han, S., McCurdy, S.P. et al. (2013). Novel 3-O-carbamoyl erythromycin A derivatives (carbamolides) with activity against resistant staphylococcal and streptococcal isolates. *Bioorg Med Chem Lett* 23: 1727–1731.

Mansour, H., Chahine, E.B., Karaoui, L.R., and El-Lababidi, R.M. (2013). Cethromycin: a new ketolide antibiotic. *Ann Pharmacother* 47: 368–379.

Mattana, A., Biancu, G., Alberti, L. et al. (2004). In vitro evaluation of the effectiveness of the macrolide rokitamycin and chlorpromazine against *Acanthamoeba castellanii*. *Antimicrob Agents Chemother* 48: 4520–4527.

McGuire, J.M., Bunch, R.L., Anderson, R.C. et al. (1952). Ilotycin, a new antibiotic. *Schweiz Med Wochenschr* 82: 1064–1065.

Mesa-Arango, A.C., Scorzoni, L., and Zaragoza, O. (2012). It only takes one to do many jobs: amphotericin B as antifungal and immunomodulatory drug. *Front Microbiol* 3: 286.

Molini, B.J., Tantalo, L.C., Sahi, S.K. et al. (2016). Macrolide resistance in *Treponema pallidum* correlates with 23S rDNA mutations in recently isolated clinical strains. *Sex Transm Dis* 43: 579–583.

Morikawa, K., Oseko, F., Morikawa, S., and Iwamoto, K. (1994). Immunomodulatory effects of three macrolides, midecamycin acetate, josamycin, and clarithromycin, on human T-lymphocyte function in vitro. *Antimicrob Agents Chemother* 38: 2643–2647.

Nakase, K., Nakaminami, H., Takenaka, Y. et al. (2016). A novel 23S rRNA mutation in *Propionibacterium acnes* confers resistance to 14-membered macrolides. *J Glob Antimicrob Resist* 6: 160–161.

Nakornchai, S. and Konthiang, P. (2006). Activity of azithromycin or erythromycin in combination with antimalarial drugs against multidrug-resistant Plasmodium falciparum in vitro. *Acta Trop* 100: 185–191.

Nissen, P., Hansen, J., Ban, N. et al. (2000). The structural basis of ribosome activity in peptide bond synthesis. *Science* 289: 920–930.

Noda, H., Noguchi, S., Yoshimine, T. et al. (2016). A novel potassium-competitive acid blocker improves the efficacy of clarithromycin-containing 7-day triple therapy against *Helicobacter pylori*. *J Gastrointestin Liver Dis* 25: 283–238.

Ōmura, S. (2002). *Macrolide Antibiotics: Chemistry, Biology, and Practice*. Amsterdam, Boston: Academic Press.

Omura, S., Tsuzuki, K., Sunazuka, T. et al. (1985). Gastrointestinal motor-stimulating activity of macrolide antibiotics and the structure-activity relationship. *J Antibiot (Tokyo)* 38: 1631–1632.

Oohori, M., Otoguro, K., Sunazuka, T. et al. (2000). Effect of 14-membered ring macrolide compounds on rat leucocyte chemotaxis and the structure-activity relationships. *J Antibiot (Tokyo)* 53: 1219–1222.

Or, Y.S., Clark, R.F., Wang, S. et al. (2000). Design, synthesis, and antimicrobial activity of 6-O-substituted ketolides active against resistant respiratory tract pathogens. *J Med Chem* 43: 1045–1049.

Pavlova, A., Parks, J.M., Oyelere, A.K., and Gumbart, J.C. (2017). Toward the rational design of macrolide antibiotics to combat resistance. *Chem Biol Drug Des* 90 (5): 641–652.

Pavlovic, D., Kimmins, S., and Mutak, S. (2017). Synthesis of novel 15-membered 8a-azahomoerythromycin A acylides: consequences of structural modification at the C-3 and C-6 position on antibacterial activity. *Eur J Med Chem* 125: 210–224.

Phong, N.C., Quang, H.H., Thanh, N.X. et al. (2016). In vivo efficacy and tolerability of artesunate-azithromycin for the treatment of falciparum malaria in Vietnam. *Am J Trop Med Hyg* 95: 164–167.

Poce, G., Cesare Porretta, G., and Biava, M. (2009). C-9 Alkenylidine bridged macrolides: WO2008061189. Enanta Pharmaceuticals, Inc. *Expert Opin Ther Pat* 19: 901–906.

Porter, J.D., Watson, J., Roberts, L.R. et al. (2016). Identification of novel macrolides with antibacterial, anti-inflammatory and type I and III IFN-augmenting activity in airway epithelium. *J Antimicrob Chemother* 71 (10): 2767–2781.

Powers, J. H. 1998. Importance of Macrolides in Human Medicine. *In*: RESEARCH, C. F. D. E. A. (ed.). https://slideplayer.com/slide/5661947 (accessed 23 April 2019).

Quave, C.L., Estevez-Carmona, M., Compadre, C.M. et al. (2012). Ellagic acid derivatives from *Rubus ulmifolius* inhibit *Staphylococcus aureus* biofilm formation and improve response to antibiotics. *PLoS One* 7: e28737.

Rajabi, O., Layegh, P., Hashemzadeh, S., and Khoddami, M. (2016). Topical liposomal azithromycin in the treatment of acute cutaneous leishmaniasis. *Dermatol Ther* 29: 358–363.

Roca, A., Oluwalana, C., Bojang, A. et al. (2016). Oral azithromycin given during labour decreases bacterial carriage in the mothers and their offspring: a double-blind randomized trial. *Clin Microbiol Infect* 22: 565.e1-9.

Rodrigues, A.D., Roberts, E.M., Mulford, D.J. et al. (1997). Oxidative metabolism of clarithromycin in the presence of human liver microsomes. Major role for the cytochrome P4503A (CYP3A) subfamily. *Drug Metab Dispos* 25: 623–630.

Rodvold, K.A., Gotfried, M.H., Chugh, R. et al. (2017). Comparison of plasma and intrapulmonary concentrations of nafithromycin (WCK 4873) in healthy adult subjects. *Antimicrob Agents Chemother* 61: e01096-17.

Schlunzen, F., Zarivach, R., Harms, J. et al. (2001). Structural basis for the interaction of antibiotics with the peptidyl transferase centre in eubacteria. *Nature* 413: 814–821.

Seiple, I.B., Zhang, Z., Jakubec, P. et al. (2016). A platform for the discovery of new macrolide antibiotics. *Nature* 533: 338–345.

Shain, C.S. and Amsden, G.W. (2002). Telithromycin: the first of the ketolides. *Ann Pharmacother* 36: 452–464.

Sivapalasingam, S.S. and Neal, H. (2015). Macrolides, clindamycin, and ketolides. In: *Mandell, Douglas, and Bennett's Principles and Practice of Infectious*

Diseases, Updated Edition, 8e (ed. J.E. Bennett, R. Dolin and M.J. Blaser). Philadelphia, PA: Saunders.

Steinmetz, W.E., Shapiro, B.L., and Roberts, J.J. (2002). The structure of erythromycin enol ether as a model for its activity as a motilide. *J Med Chem* 45: 4899–4902.

Stierle, A.A., Stierle, D.B., Decato, D. et al. (2017). The berkeleylactones, antibiotic macrolides from fungal coculture. *J Nat Prod* 80: 1150–1160.

Subcommittee of the Expert Committee on the Selection and Use of Essential Medicines (2008). Macrolides in children. Second Meeting of the Subcommittee of the Expert Committee on the Selection and Use of Essential Medicines, 2008, Geneva, Switzerland.

Sunazuka, T., Takizawa, H., Desaki, M. et al. (1999). Effects of erythromycin and its derivatives on interleukin-8 release by human bronchial epithelial cell line BEAS-2B cells. *J Antibiot (Tokyo)* 52: 71–74.

Sunazuka, T., Yoshida, K., Oohori, M. et al. (2003). Effect of 14-membered macrolide compounds on monocyte to macrophage differentiation. *J Antibiot (Tokyo)* 56: 721–724.

Tanaka, H., Kuroda, A., Marusawa, H. et al. (1987). Structure of FK506, a novel immunosuppressant isolated from Streptomyces. *J Am Chem Soc* 109: 5031–5033.

Tang, D., Gai, Y., Polemeropoulos, A. et al. (2008). Design, synthesis, and antibacterial activities of novel 3,6-bicyclolide oximes: length optimization and zero carbon linker oximes. *Bioorg Med Chem Lett* 18: 5078–5082.

Tanikawa, T., Asaka, T., Kashimura, M. et al. (2001). Synthesis and antibacterial activity of acylides (3-O-acyl-erythromycin derivatives): a novel class of macrolide antibiotics. *J Med Chem* 44: 4027–4030.

Tanikawa, T., Asaka, T., Kashimura, M. et al. (2003). Synthesis and antibacterial activity of a novel series of acylides: 3-O-(3-pyridyl)acetylerythromycin A derivatives. *J Med Chem* 46: 2706–2715.

Tenson, T., Lovmar, M., and Ehrenberg, M. (2003). The mechanism of action of macrolides, lincosamides and streptogramin B reveals the nascent peptide exit path in the ribosome. *J Mol Biol* 330: 1005–1014.

Tevyashova, A.N., Olsufyeva, E.N., Solovieva, S.E. et al. (2013). Structure-antifungal activity relationships of polyene antibiotics of the amphotericin B group. *Antimicrob Agents Chemother* 57: 3815–3822.

Thompson, J., Pratt, C.A., and Dahlberg, A.E. (2004). Effects of a number of classes of 50S inhibitors on stop codon readthrough during protein synthesis. *Antimicrob Agents Chemother* 48: 4889–4891.

Tian, J.C., Han, X., Lv, W. et al. (2017). Design, synthesis and structure-bactericidal activity relationships of novel 9-oxime ketolides and reductive epimers of acylides. *Bioorg Med Chem Lett* 27 (7): 1513–1524.

Van Ingen, J., Ferro, B.E., Hoefsloot, W. et al. (2013). Drug treatment of pulmonary nontuberculous mycobacterial disease in HIV-negative patients: the evidence. *Expert Rev Anti Infect Ther* 11: 1065–1077.

Venugopal, A.A. and Johnson, S. (2012). Fidaxomicin: a novel macrocyclic antibiotic approved for treatment of *Clostridium difficile* infection. *Clin Infect Dis* 54: 568–574.

Viasus, D., Ramos, O., Ramos, L. et al. (2017). Solithromycin for the treatment of community-acquired bacterial pneumonia. *Expert Rev Respir Med* 11: 5–12.

Vo, N.H., Phan, L.T., Hou, Y. et al. (2004). Novel 11,12-substituted lactone ketolide derivatives having antibacterial activity. 10/223144. US Patent US20040038915A1.

Wang, Y., Cong, C., Chern Chai, W. et al. (2017). Synthesis and antibacterial activity of novel 4″-O-(1-aralkyl-1,2,3-triazol-4-methyl-carbamoyl) azithromycin analogs. *Bioorg Med Chem Lett* 27 (16): 3872–3877.

Weisblum, B. (1995). Erythromycin resistance by ribosome modification. *Antimicrob Agents Chemother* 39: 577–585.

Xu, D., Gao, X., Bian, R. et al. (2017). Tacrolimus improves proteinuria remission in adults with cyclosporine A-resistant or -dependent minimal change disease. *Nephrology (Carlton)* 22: 251–256.

Yan, M., Ma, R., Jia, L. et al. (2017). Synthesis and antibacterial activity of novel 3-O-descladinosylazithromycin derivatives. *Eur J Med Chem* 127: 874–884.

Yoshioka, D., Kajiwara, C., Ishii, Y. et al. (2016). The efficacy of beta-lactam plus macrolide combination therapy in a mouse model of lethal *Pneumococcal pneumonia. Antimicrob Agents Chemother.* https://doi.org/10.1128/AAC.01024-16.

Zheng, Z., Du, D., Cao, L. et al. (2016). Synthesis and antibacterial activity of novel 11-[3-[(arylcarbamoyl)oxy]propylamino]-11-deoxy-6-O-methyl-3-oxoerythromycin A 11-N,12-O-cyclic carbamate derivatives. *J Antibiot (Tokyo)* 69: 811–817.

Zuckerman, J.M. (2000). The newer macrolides: azithromycin and clarithromycin. *Infect Dis Clin North Am* 14: 449–462, x.

Zuckerman, J.M., Qamar, F., and Bono, B.R. (2009). Macrolides, ketolides, and glycylcyclines: azithromycin, clarithromycin, telithromycin, tigecycline. *Infect Dis Clin North Am* 23: 997–1026, ix-x.

Part II

Mechanism of Antibiotic Resistance

6

Impact of Key and Secondary Drug Resistance Mutations on Structure and Activity of β-Lactamases

Egorov Alexey, Ulyashova Mariya, and Rubtsova Maya

Department of Chemistry, M.V. Lomonosov Moscow State University, Moscow, Russia

6.1 Introduction

The resistance of bacteria to various classes of antibiotics is determined by the activity of a large number of bacterial enzymes. They either represent the targets for antibiotics or modify antibiotics and their targets. The family of genes responsible for the resistance has been called "the resistome" (Wright 2007), and the family of enzymes involved in the implementation of resistance has been called "the enzystome" (Egorov et al. 2018). β-Lactam antibiotics (penicillins, cephalosporins, carbapenems, monobactams) are the most widely used in medicine and agriculture; they account for more than 60% of all used antibiotics. The rise of antibiotic resistance of the causative agents of infectious diseases in humans, animals, and the environment has become a threat and is responsible for serious problems in clinical microbiology (WHO 2018; Cassini et al. 2019). The problem became significantly complicated by the emergence of multidrug-resistant (MDR) and pandrug-resistant (PDR) Gram-negative bacteria (Davies and Davies 2010; Theuretzbacher 2017). β-Lactam antibiotics irreversibly inhibit penicillin-binding proteins (PBPs) participating in the synthesis of the cell wall, and this leads to the death of bacterial pathogens. In Gram-positive bacteria, PBPs mutated, which reduced their ability to bind β-lactams and caused the resistance. Gram-negative bacteria evolved by other means, synthesizing a new class of enzymes, β-lactamases, which catalytically hydrolyze β-lactams. This protects the bacterial cells by thousands of times more efficiently, despite the emergence of new β-lactams (Bonomo 2017). The high rate of spread of resistant pathogenic strains is explained by the localization of the β-lactamase genes mainly on the mobile genetic elements.

Antibiotic Drug Resistance, First Edition. Edited by José-Luis Capelo-Martínez and Gilberto Igrejas.

The evolution of β-lactamases takes more than two billion years; some of their genes have chromosomal localization. Active use of penicillins and then cephalosporins in clinical practice triggered an exponential increase of new types of β-lactamases and their mutant forms; more than 2700 enzymes have been described so far (Bonomo 2017). The study of β-lactamases using modern methods of analysis allowed us to obtain data on the structure, stability, molecular dynamics, and mechanism of action of these enzymes (Knox et al. 2018; Wiedorn et al. 2018).

β-Lactamases represent the superfamily of genetically and functionally different enzymes, which catalyze the hydrolysis of the β-lactam ring of antibiotics (Bush 2018). Based on the homology of the primary protein sequences, all β-lactamases are divided into four molecular classes A, B, C, and D (Hall and Barlow 2005). The enzymes of molecular classes A, C, and D are serine hydrolases, while the enzymes of molecular class B are metalloenzymes and contain one or two zinc atoms. Based on differences in catalytic efficiency with respect to different groups of β-lactams, enzymes of each molecular class are divided into types, each of which represents a separate family consisting of many mutant forms. The most common are serine β-lactamases of molecular class A; their changes in the properties occur due to point substitutions in acquired resistance genes and their combinations (Bush 2018). Among them, extended-spectrum β-lactamases (ESBLs) represent the greatest threat in clinical practice due to their broad substrate specificity and ability to destroy third-generation cephalosporins, the most commonly used β-lactams.

Due to structural diversity leading to a change in functional properties, β-lactamases are the most actively studied enzymes. The number of publications in MEDLINE (https://www.ncbi.nlm.nih.gov/pubmed) is about 30 000 (Bush 2018). The chapter presents data on the prevalence of single mutations and their combinations in TEM-type β-lactamases of molecular class A, isolated from resistant clinical strains, and on the effects of these mutations on the structural and catalytic features of these enzymes. All these are of interest in microbiology, biotechnology, and pharmacy.

6.2 Structure of the Protein Globule of TEM-Type β-Lactamases: Catalytic and Mutated Residues

A convenient model for studying the effects of mutations and their combinations on biosynthesis, structure, stability, and catalytic properties is the TEM-type β-lactamase family (Pimenta et al. 2014; Palzkill 2018), consisting of the parent enzyme TEM-1 and more than 200 mutant forms (http://www.lahey.org/studies). A feature of this family is high mutational variability: each mutant

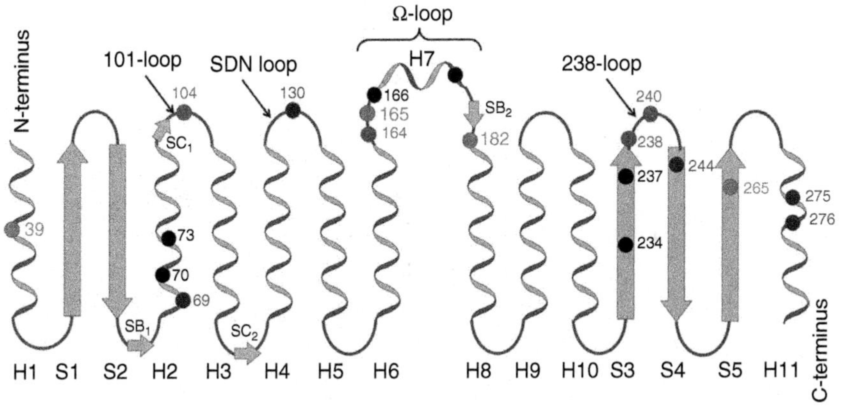

Figure 6.1 The secondary structure of β-lactamase TEM-1 (PDB ID 1ERO (Ness et al. 2000)) and the location of catalytic residues and residues whose mutations are the key (ESBL; IRT), and the most common residues whose mutations are secondary.

contains from one to seven single substitutions, and the number of mutating amino acids, found in β-lactamases isolated from clinical strains, incorporates 32% of the residues in the primary sequence.

Protein globule of TEM-type β-lactamases is formed by a single chain of 288 amino acid residues. The secondary structure consists of 11 α-helixes, five β-strands, and the loops of irregular structure (Figure 6.1). Structural fold of these β-lactamases is sandwich-like and consists of three domains: the first includes eight α-helices (H2–H9), the second in the middle five-stranded (S1–S5) β-sheet, and the third three α-helices (H1, H10, and H11) (Pimenta et al. 2014) (Figure 6.2a). The domains are structured by a network of ionic and hydrogen bonds. By SCOP classification it relates to β-lactamase/D-Ala carboxypeptidase family (SCOP 56602).

The structural feature of TEM-type β-lactamases is a compact core scaffold including the H2 α-helix with catalytic S70 and the S3 β-strand located close to each other (Figure 6.2b). The rest of the protein includes regular and mobile elements of the secondary structure (loops). Although the structure of the protein globule is generally quite rigid, its mobility is maintained by the movement of the loops (Fisette et al. 2010). The Ω-loop (residues 164–179), located in the lower part at the entrance to the active site of the enzyme, is an important conservative structural element of all serine β-lactamases (Fink 1985). According to the molecular dynamics data, the root mean square deviation (RMSD) of atoms in β-lactamase TEM-1 at 300 K is 1.3–1.4 Å; in the mutant form with substitutions of residues 238 and 240, it increases by 0.5 Å, with the highest mobility observed in the Ω-loop (Shcherbinin et al. 2017).

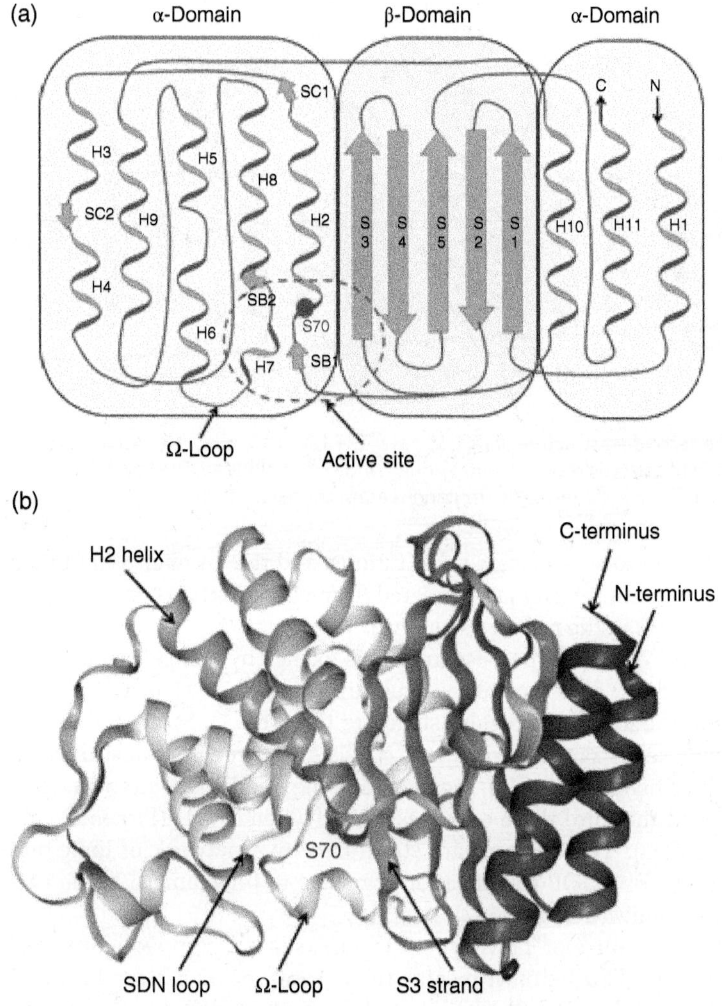

Figure 6.2 Schematic representation of the structural fold of β-lactamase TEM-1 consisting of three domains (a). The dotted line indicates the active site with catalytic S70. The presented scheme corresponds to the location of domains in the tertiary structure of β-lactamase TEM-1 (PDB ID 1ERO (Ness et al. 2000)) (b). The locations of the catalytic S70, H2 α-helix, S3 β-strand, and the Ω-loop are highlighted. *(See insert for color representation of the figure.)*

6.2.1 Catalytic Site of β-Lactamase TEM-1

The β-lactamase active site is located between H2 α-helix and S3 β-strand. Catalytically important residues are located in the conservative regions of these α-helix (S70, K73) and β-strand (K234, A237), as well as on the elements of irregular structure (SDN loop [S130, N132], Ω-loop [G166, N170]) (Figures 6.1 and 6.2).

Figure 6.3 shows the scheme of enzymatic hydrolysis of β-lactam antibiotics, which includes three main processes: binding (i), acylation (ii), and deacylation (iii). These stages can be described by a set of kinetic parameters (K_M and k_{cat}), which are a function of the kinetic constants of the elementary stages of binding (k_1, k_{-1}), acylation (k_2), and deacylation (k_3) (Raquet et al. 1994). At the first stage, the binding of an antibiotic molecule occurs in the active site of the enzyme. The main process is the deprotonation of the catalytic S70 with the participation of the Ω-loop residues E166 and N170, which form a network of hydrogen bonds with the water molecule (Stec et al. 2005). The alternative proton transfer pathway includes the participation of residues K73 and S130 (Massova and Kollman 2002). As a result of the nucleophilic attack of deprotonated S70 on the carbonyl group of the antibiotic, a high-energy acyl intermediate is formed. The charged side groups of residues K234 and N132, which form ionic bonds with the charged groups of the antibiotic, are involved in its stabilization, as well as residue A237, which stabilizes the negative charge on the carbonyl carbon atom.

In the second stage, the transition of the intermediate to the low-energy covalent acyl-enzyme complex occurs, wherein the nitrogen atom is protonated and the amide bond of the β-lactam ring of the antibiotic is cleaved.

In the process of deacylation (stage 3), the water molecule coordinated by E166 and N170 attacks the carbonyl bound to the serine oxygen, and this bond is hydrolyzed. As a result, an enzyme molecule and a hydrolyzed antibiotic molecule are released.

Amino acid residues S70, K73, K234, E166, and N170 involved in the catalytic cycle of antibiotic hydrolysis are conservative and do not mutate, and residues S130 and A237 are rarely mutating.

6.2.2 Mutations Causing Phenotypes of TEM-Type β-Lactamases

Based on differences in substrate specificity due to amino acid mutations, TEM-type β-lactamases are divided into four phenotypes: 2b, which are β-lactamases hydrolyzing penicillins and first-generation cephalosporins; 2be (ESBL), which are β-lactamases hydrolyzing penicillins, first- to fourth-generation cephalosporins, and monobactams; 2br (inhibitor-resistant type [IRT]), which are β-lactamases resistant to inhibitors of β-lactam structure (clavulanic acid, sulbactam, tazobactam); and 2ber (complex mutation type [CMT]), which are inhibitor-resistant ESBLs, mixed type (Bush and Jacoby 2010). The mutations are divided into several types according to their influence on the catalytic properties of β-lactamases: the key mutations, which change the phenotype, and secondary mutations, which affect mainly the stability and folding of the protein globule. Secondary mutations do not change the phenotype; some of them exhibit a compensatory suppressor role (Zimmerman et al. 2017).

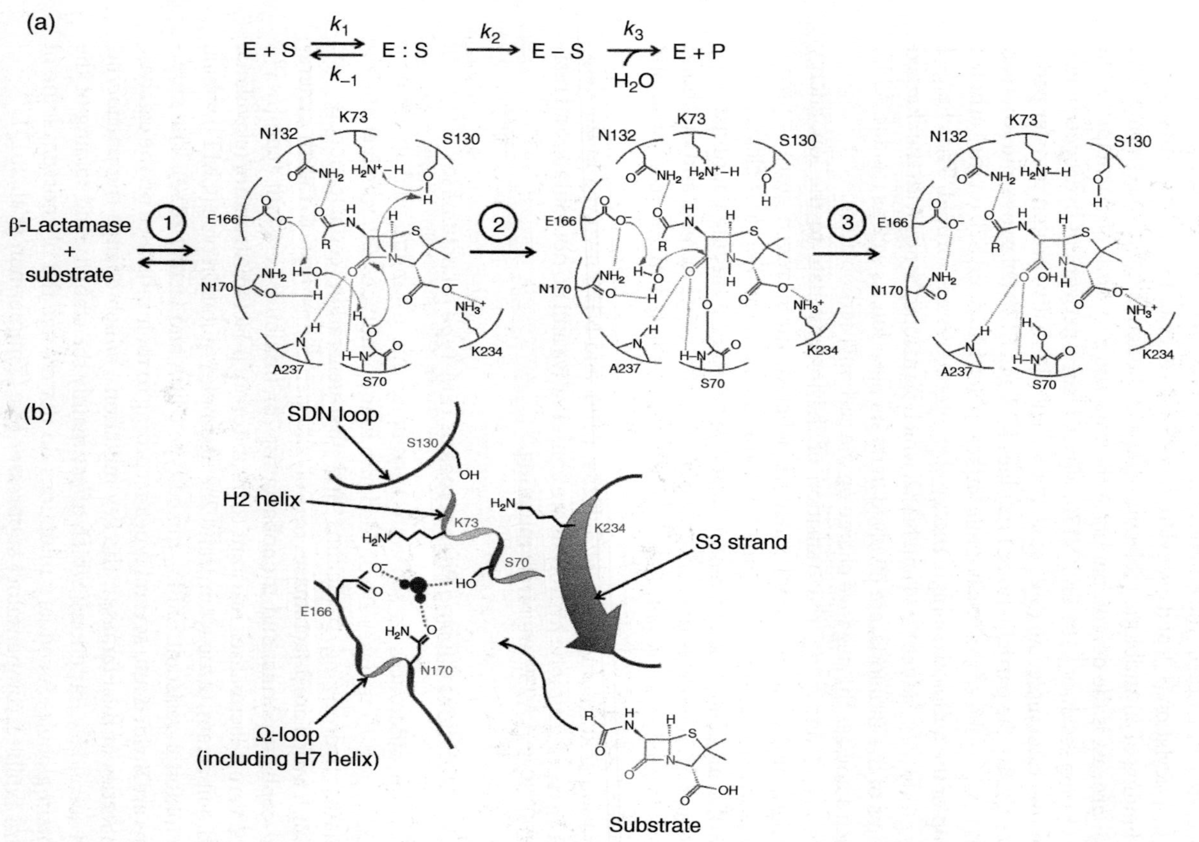

Figure 6.3 Scheme of β-lactam antibiotic hydrolysis by β-lactamase TEM-1 (a). The structure of the active site of β-lactamase TEM-1 with the location of catalytically important residues (b).

Key mutations alter the activity and substrate specificity of enzymes; they constitute no more than 10% of the total number of mutating residues. Enzymes with mutations of residues E104, R164, A237, G238, and E240 correspond to the phenotype 2be; enzymes with mutations of residues M69, S130, R244, R275, and N276 correspond to the phenotype 2br; enzymes with a combination of mutations of these groups of residues correspond to the phenotype 2ber.

All residues, which mutations are the key, are located near the catalytic site in the tertiary structure of the protein. Their location in the secondary structure of the protein is shown in Figure 6.1. The residues, whose mutations lead to the expansion of the substrate specificity (2be), are located in functionally significant loops (E104, R164, E240, and G238) and on S3 β-strand (A237). The residues, whose mutations affect the resistance to β-lactam inhibitors (2br), are located on conservative elements with a regular structure: on H2 and H11 α-helices (M69, R275, N276) and on S4 β-strand (R244). The residue S130 is located in the SDN loop. Key mutations lead, as a rule, to destabilization of the protein globule.

Most of the residues whose mutations are attributed to the secondary ones are located in loops far from the active site (Abriata et al. 2012). The role of most secondary mutations remains unclear. For some of them, it has been established that they can compensate the destabilizing effects of the key mutations (see Section 6.4).

6.3 Effect of the Key Mutations on Activity of TEM-Type β-Lactamases

To date, 84 TEM-type β-lactamases isolated from clinical bacterial strains have an ESBL phenotype (2be) and contain from one to four key mutations. The analysis of the distribution of combinations of these mutations in TEM-type β-lactamases was carried out relying on information from http://www.lahey.org/studies as of 1 November 2018. Figure 6.4 shows the classification of mutated residues of TEM-type β-lactamases into the key and secondary and their localization in the secondary structure and the frequency of the key mutations occurred in single and double combinations, as well as in multiple combinations with the other key and secondary mutations. Figure 6.5 presents the data on the frequency of occurrence of the key single mutations and their combinations in the enzymes with different phenotypes. To analyze the effect of combinations of mutations on the catalytic properties of β-lactamases, a parameter of catalytic efficiency (k_{cat}/K_M) was used as it is more consistent and less dependent on experimental conditions (Palzkill 2018) (Figure 6.6).

(a)

	104 101-loop	164 Ω-loop	237 S3 strand	238 238-loop	240 238-loop	69 H2 helix	130 SDN loop	244 S4 β-strand	275 H11 helix	276 H11 helix
104 101-loop	6	11	–	15	–	–	–	–	–	–
164 Ω-loop	10	12	–	–	10	1	–	–	–	–
237 S3 strand	4	7	–	–	–	–	–	–	–	–
238 238-loop	6	3	1	6	9	–	–	–	–	–
240 238-loop	5	10	7	1	1	–	–	–	–	–
69 H2 helix	2	5	–	1	1	8	–	1	5	6
130 SDN loop	1	–	–	1	–	–	2	–	–	–
244 S4 β-strand	1	1	1	–	1	–	–	12	–	–
275 H11 helix	–	–	–	1	1	–	–	–	3	–
276 H11 helix	1	4	–	1	1	5	–	–	–	1

(b)

	104 101-loop	164 Ω-loop	237 S3 β-strand	238 238-loop	240 238-loop	69 H2 helix	130 SDN-loop	244 S4 β-strand	275 H11 helix	276 H11 helix
39 H1 helix	19	15	6	12	10	1	2	4	–	–
165 Ω-loop	–	1	–	–	–	7	–	–	2	3
182 H8 helix	17	9	2	10	5	2	1	–	1	–
265 S5 β-strand	4	8	1	9	8	–	–	2	1	–

Figure 6.4 Distribution of combinations of mutations in TEM-type β-lactamases isolated from clinical strains. For each residue, the location in the secondary structure of the protein is indicated. (a) Amount of the enzymes with combinations of the key mutations: single (diagonal cells), double (cells located above the diagonal), and multiple (cells located below the diagonal). (b) Amount of the enzymes with combinations of the key and secondary mutations.

6.3.1 Single Key Mutations in TEM-Type ESBLs (2be)

TEM-type ESBLs with a single key mutation account for about 30% ($n = 25$) of all β-lactamases of this phenotype (Figure 6.4). The residue R164 is the most frequently mutable ($n = 12$), it is located in the Ω-loop, and the salt bridge between R164 and D179 defines its configuration. Substitution of arginine leads to the disappearance of this bond, which increases the mobility of the loop and changes the position of the catalytically important residues E166 and N170 (Knox 1995). As a result, catalytic efficiency for the hydrolysis of ceftazidime enhances significantly (by 4 orders of magnitude) (Figure 6.6). In contrast, catalytic efficiency toward penicillins, which are substrates with small

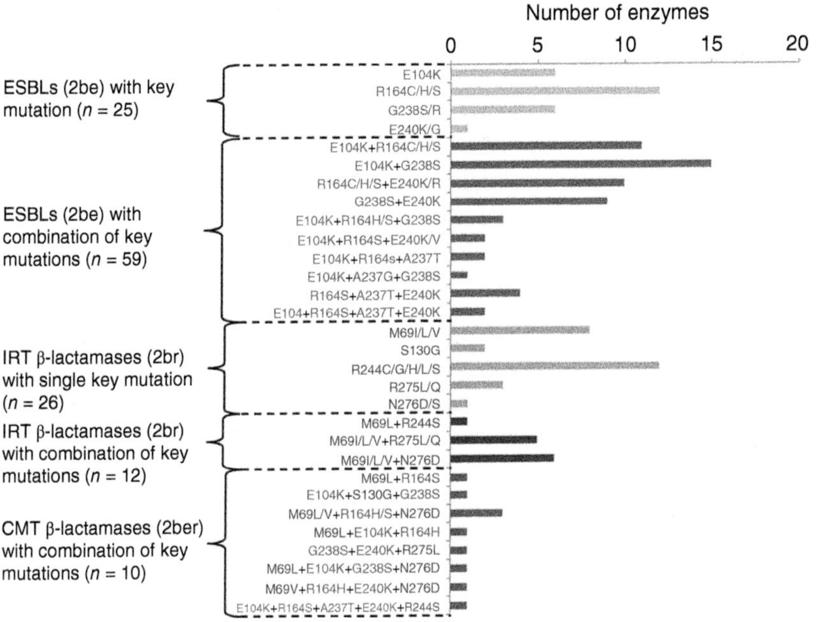

Figure 6.5 Single key mutations and their combinations in TEM-type β-lactamases with different phenotypes: 2be, extended-spectrum β-lactamases (ESBLs) hydrolyzing penicillins, cephalosporins, and monobactams; 2br, inhibitor-resistant β-lactamases (IRT); 2ber, inhibitor-resistant ESBLs (CMT).

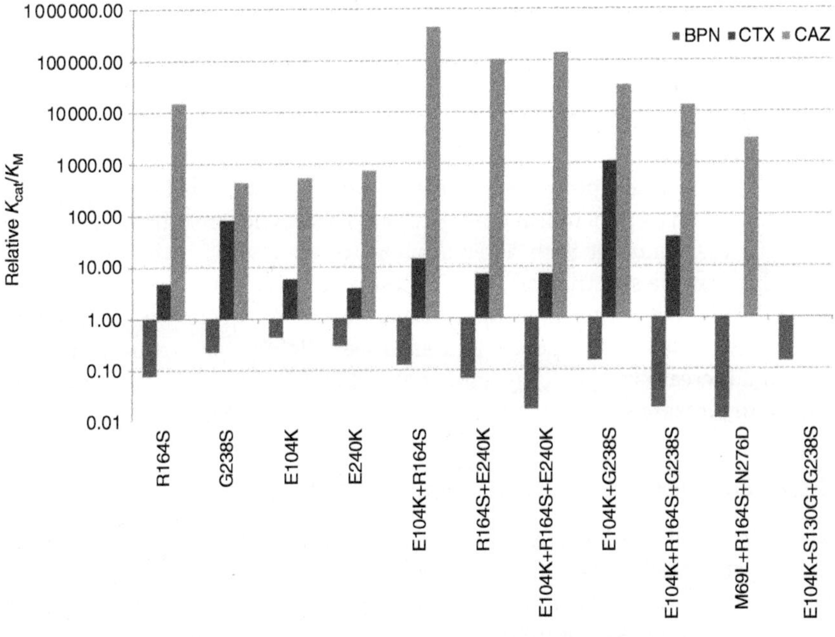

Figure 6.6 Relative catalytic efficiency (k_{cat}/K_M) of TEM ESBLs and CMT β-lactamases compared with that of TEM-1 against benzylpenicillin (BNP), cefotaxime (CTX), and ceftazidime (CAZ) (Sowek et al. 1991; Sirot et al. 1997; Neuwirth et al. 2001; Perilli et al. 2007; Robin et al. 2007; Palzkill 2018). A value larger than one indicates that the mutant is more active than TEM-1. *(See insert for color representation of the figure.)*

side chains, is 10 times lower. R164 has the highest variability of substitutions, most often changing with serine or histidine and more rarely with cysteine (Figure 6.5). These mutants differ in the efficiency of hydrolysis of various cephalosporins.

The residue G238 is replaced by serine only in six enzymes. It is located in the loop 238–243 near S3 β-strand, and the substitution leads to significant displacement of the loop, which results in the widening of the active site and promotes an accommodation of third-generation cephalosporins with larger oxyimino side chains (Orencia et al. 2001). G238S mutant is active toward both cefotaxime and ceftazidime; the value of k_{cat}/K_M increases by 2–3 orders of magnitude (Figure 6.6). Catalytic efficiency toward penicillins becomes 5–10-fold less.

The mutation E104K was found as a single substitution in six β-lactamases. The residue 104 is located in the 101-loop, near the catalytically important H2 α-helix, on which catalytic S70 is located. As a result of the replacement, the residue 104 changes its orientation, and the volume of the enzyme active site expands. When the oxyimino cephalosporins bind, a new ionic bond is formed between the negatively charged carboxylic group of antibiotic and positive charge of K104. This leads to an increase in k_{cat}/K_M values for ceftazidime by 2–3 orders of magnitude (Palzkill 2018). Effect of E104K on hydrolysis of penicillins and first-generation cephalosporins is slightly negative.

The substitution E240K is rarely found as a single mutation ($n = 1$), but it is often found in combinations with other mutations (Figures 6.4 and 6.5). The residue is located in the loop 238–242, near the catalytically significant S3 β-strand. The effect of this substitution consists in the formation of an additional electrostatic bond between K240 and side chain of the cephalosporins (Knox 1995).

The residue A237 precedes the key residues G238 and E240 on the S3 β-strand, which forms a wall of the active site. There is no structural information available on a mutant with a replacement of alanine with threonine. It is assumed that the substitution A237T improves the interaction of the residue side chain with a carbonyl group of the antibiotic molecule, which improves coordination of cephalosporins and enhances slightly cefotaxime hydrolysis (1.3-fold increase in k_{cat}/K_M) (Cantu et al. 1997). In addition, A237G has been shown to increase resistance to aztreonam (Cantu et al. 1996).

6.3.2 Combinations of Key Mutations in TEM-Type ESBLs (2be)

More than half of ESBLs ($n = 59$) contain combinations of two or more key mutations related to this phenotype (Figure 6.4). It should be noted that only certain combinations of mutations occur in β-lactamases isolated from clinical strains (104/164, 104/238, 164/240, and 238/240). Combinations of mutations of residues 164/238 and 104/240 occur only in triple mutants (104/164/238 and 104/164/240).

Combination of key mutations in double mutants results generally in strong additive effects on catalysis, leading to higher hydrolysis efficiency (Figure 6.6). For example, the addition of E104K to R164S results in a 30-fold increase in catalytic efficiency for ceftazidime compared with the R164S enzyme. The addition of E104K to the G238S enzyme results in a 75-fold increase in the efficiency for the same antibiotic compared with the G238S enzyme. Apparently, mutations of residues 104 and 240 located in the loops near the catalytically important α-helix and β-sheet contribute to additional changes in the active site conformation and improve the orientation of cephalosporins with bulky substituents due to additional ionic interactions (Page 2008). The loop between the S3 and S4 β-strands in TEM-52 (G238S/E104K) was shown to move by 2.8 Å with respect to its position in TEM-1, and the side chain of E240 was shifted from the active site (Orencia et al. 2001). In this respect, it has been suggested that E240K and E104K might act as stabilizing mutations that can compensate for the destabilizing effect of R164 and G238 substitutions (Raquet et al. 1995).

Combinations of R164S/G238S and E104K/E240K not found together in double mutants may show negative effects on the hydrolysis in triple mutants. For example, catalytic efficiency toward cefotaxime drops markedly (by 30-fold) in triple mutant TEM-134 (E104K/R164S/G238) compared with double mutant TEM-3 (E104K/G238S) (Perilli et al. 2007). While the residues are not in direct contact (more than 10 Å apart), it was found that both G238S and R164S introduce conformational changes in the Ω-loop, resulting in nonoptimal conformations (Dellus-Gur et al. 2015).

6.3.3 Key Mutations in IRT TEM-Type β-Lactamases (2br)

The TEM-type β-lactamases with the IRT phenotype (2br), which are characterized by resistance to the inhibitors with the β-lactam ring (clavulanic acid, sulbactam, and tazobactam), involve 38 enzymes. These inhibitors compete with β-lactams for the binding in the active site of β-lactamase. Unlike antibiotics, the acyl-enzyme complex with inhibitors is stable and characterized by a very low deacylation rate defined by the value of k_3 (Figure 6.3). TEM-type β-lactamases that have substitutions of residues M69, S130, R244, R275, and N276 are resistant to these inhibitors. In general, the total number of β-lactamases with mutations of these residues is substantially less than enzymes with mutations related to the ESBL phenotype. Apparently, this is due to the significantly smaller amount of inhibitors used in clinical practice as compared with antibiotics.

6.3.4 Single Key Mutations in IRT TEM-Type β-Lactamases (2br)

Single key mutations in TEM-type β-lactamases isolated from clinical samples most often occur in residues R244 (substitutions C/G/H/L/S, $n = 12$) and M69 (substitutions I/L/V, $n = 8$) (Figures 6.4 and 6.5).

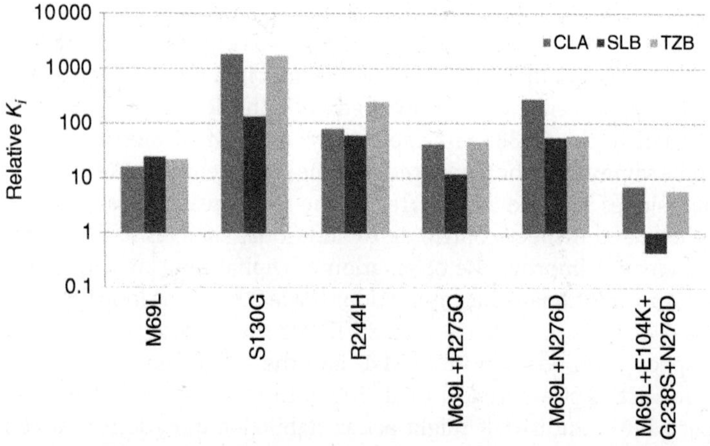

Figure 6.7 Relative Ki of IRT and CMT TEM-type β-lactamases compared with K_i of TEM-1 for inhibition by clavulanic acid (CLA), sulbactam (SLB), and tazobactam (TZB) (Sirot et al. 1997; Bermudes et al. 1999; Chaïbi et al. 1999). A value larger than one indicates that a complex of mutant enzyme with the inhibitor is less stable than TEM-1 with this inhibitor. *(See insert for color representation of the figure.)*

The residue R244 is located on S4 β-strand and may serve as a counterion for the carboxyl group of β-lactams or inhibitors, improving their orientation in the active site of β-lactamase (Salverda et al. 2010). Significant variability of R244 substitutions was found by a number of amino acids with shorter side chains (serine, cysteine, histidine, leucine, or glycine). The first three substitutions were found more frequently. The mutations may result in electrostatic disruption and disturb the hydrogen bond network involving residues R244, S130, and others (Wang et al. 2002a). The substitutions of R244 are characterized by increased K_i values for β-lactam inhibitors that are higher by 2 orders of magnitude compared with K_i of TEM-1 (K_i is an equilibrium dissociation constant showing the stability of enzyme-inhibitor complex) (Figure 6.7).

The residue M69 is located on the H2 α-helix in the proximity with catalytic S70, and its side chain lies behind H3 α-helix, forming a wall of the oxyanion pocket. It has an important role in the structural position of both S70 and S130. By means of steric interactions it also affects the positions of A237 and G238. Mutations of this residue lead to distortion of S70, which results in weakening of the hydrogen bond network between S70 and S130. The latter changes its conformation, causing diminished inhibition (Wang et al. 2002a). The substitution M69V/I affects mostly k_{cat} value, while the substitution M69L has an effect on the binding of β-lactams and inhibitors. All substitutions result in increased values of K_i by more than 10 times (Chaïbi et al. 1999) (Figure 6.7).

The residue S130 is located on a conservative SDN loop connecting H4 and H5 α-helices, and it is part of a complex hydrogen bonding network that contributes to the stable conformation of the active site and the deprotonation of S70 (Pimenta et al. 2014). The mutation S130G is quite rare ($n = 2$), but it is characterized by the highest decrease in sensitivity to all three inhibitors, and the K_i values are enhanced by 2–3 orders of magnitude compared with K_i of TEM-1 (Thomas et al. 2005) (Figure 6.7).

The substitutions R275L/Q and N276D are rarely found in β-lactamases isolated from clinical samples as single mutations ($n = 3$ and 1, correspondingly) (Figure 6.4). These residues are located on H11 α-helix rather far from the active site. It explains non-marked effects on inhibition; at the same time substitutions R275Q and N276D were revealed to be involved in the stabilization of the enzyme (Osuna et al. 2002).

Increased interaction between D276 and R244 due to forming a salt bridge between the positive charge of arginine and the negatively charged carboxyl of aspartate (Brown et al. 2010) neutralizes positive electrostatic potential of R244 guanidinium group and results in the decreased affinity of the mutant for β-lactams and inhibitors (Swarén et al. 1999; Brown et al. 2010).

6.3.5 Combinations of Key Mutations in IRT TEM-Type β-Lactamases (2br)

Combinations of the key IRT mutations are less common than the key ESBL mutations. They are found only in 12 TEM-type β-lactamases and include mainly combinations of M69/R275 and M69/N276 substitutions (Figure 6.4). In general, an additive effect of combinations of mutations is observed: in double mutants, K_i is one order higher compared to the effect of single M69 substitutions (Figure 6.7). This is revealed mainly toward the resistance to clavulanic acid.

6.3.6 Combinations of Key ESBL and IRT Mutations in CMT TEM-Type β-Lactamases (2ber)

β-Lactamases with a confirmed CMT phenotype (2ber) involve 10 enzymes, which combine the key mutations of two types (ESBL and IRT) (Figure 6.4). The combination of M69L(I), R164S(H), and N276D is found in three enzymes, and the remaining combinations are found only in single β-lactamases (Figure 6.5). The analysis of changes in the catalytic properties of these mutants shows that ESBL and IRT mutations have a negative effect on each other: the efficiency of hydrolysis of various β-lactams and resistance to inhibitors decrease in all CMT enzymes in comparison with the corresponding ESBL and IRT enzymes, and the magnitude of this effect depends on combinations of mutations (Figures 6.6 and 6.7).

6.4 Effect of Secondary Mutations on the Stability of TEM-Type β-Lactamases

About 80% of substitutions in TEM-type β-lactamases are secondary, and for most of them the functional role has not been established. The prevalence of these substitutions varies greatly: mutations of some residues (Q39, M182, T265) are common (Figure 6.4), but most of the mutations are rare, occurring in 1–2 enzymes. As a rule, if these mutations are single, they do not affect significantly either the catalytic properties or the substrate specificity of the enzyme. The feature of these mutations is that they are most often combined with the key ESBL mutations (Figure 6.5).

The residue W165 is located at the beginning of the Ω-loop, which forms the entrance to the active site of the enzyme, and it precedes the residue E166, which is involved in coordinating the water molecule during deacylation (Figure 6.3). It is the only one of the four tryptophan residues susceptible to mutations. In TEM-type β-lactamases, the residue W165 is substituted by several amino acids (arginine, cysteine, glycine, and leucine) (Guthrie et al. 2011). The mutant W165R exhibits a slight decrease in the inhibitory effect of clavulanic acid. Molecular modeling suggests that the side chain of R165 is able to form an additional salt bond with the E168 of Ω-loop (Chaïbi et al. 1999).

One of the most studied secondary mutations is the M182T substitution found in 29 TEM-type β-lactamases. It is frequently found in combination with substitutions that have a deleterious effect on thermal stability of the protein (e.g. E104K). The residue M182 is located in the primary sequence at the beginning of the H8 α-helix after the Ω-loop. It has a contact with a β-strand (62–65) leading to residue M69 behind the oxyanion pocket and forms an additional hydrogen bond with A185 (Wang et al. 2002b). It is the first mutation in TEM-type β-lactamases, which was shown to increase the thermodynamic stability of the enzyme by 6.5 °C. Thus, it was called as a global suppressor that compensates for decreasing protein stability caused by the key mutations (Brown et al. 2010; Zimmerman et al. 2017).

Other secondary mutations located far from the active site were actively investigated in relation to their effects on the thermal stability of the protein globule. For T265M mutation, similar properties were first predicted (Salverda et al. 2010). Analysis of its location in the protein globule showed that the hydroxyl group of threonine forms a hydrogen bond with R43. When replacing with M265, the hydrogen bond disappears, and the distance between S70 and M265 increases. The long side chain of methionine fills the hydrophobic region inside the protein globule, located next to it. The hydrophobic interaction of aliphatic residues leads to compaction of the globule, which lowers the free energy of the protein and compensates for the disappearance of the hydrogen bond. Compaction of the protein globule may lead to an increase in the thermal stability of the enzyme. Later it was confirmed experimentally that the mutant is characterized by a T_m value higher by 1.6 °C (Brown et al. 2010). The effect of the V84I mutation on increasing the kinetic stability of β-lactamase TEM-171 has also been demonstrated recently (Grigorenko et al. 2018a). This mutation is rarely found in β-lactamases from clinical strains.

The Q39K substitution has long been thought to be an example of a neutral secondary mutation in TEM-type β-lactamases. It is located on the surface of a protein globule on H1 α-helix and is quite frequently observed in different β-lactamase phenotypes both as a single mutation and in combination with the key substitutions as well. It was shown to increase slightly the minimal inhibitory concentration (MIC) for cephaloridine, ceftazidime, and aztreonam (Blazquez et al. 1995). It was recently shown that a combination of Q39K with the key substitutions related to ESBL (E104K, R164S) and IRT (M69V) phenotypes results in decreased K_M and k_{cat} values, and a thermal stability of mutants with Q39K substitution was also reduced (Grigorenko et al. 2018b). In this case, the weakening of the protein structure occurs due to the exposure of hydrophilic side chains to the solvent. Thus, depending on the location of the mutations, the type of amino acid substitution, and combination with other mutations, the mechanisms of their influence can be oppositely directed.

6.5 Conclusions

The resistance of pathogens to β-lactam antibiotics is a unique mechanism of bacterial biological protection against foreign toxins due to production of enzymes – β-lactamases. The peculiarity of serine β-lactamases of molecular class A is their universal three-domain structure. Each domain has its own conservative fold, consisting of α-helices or β-strands connected by flexible loops. The domains are structured by a network of ionic and hydrogen bonds, and the active site is located in a compact core scaffold. TEM-type β-lactamases isolated from clinical strains are characterized by high mutability. Substitutions of amino acid residues occur mainly in the loops. Key mutations provide a broad hydrolytic specificity of β-lactamases to different groups of β-lactam antibiotics. They contribute to an increase in active site size and improve its availability for antibiotics with bulky substituents like the third- and fourth-generation cephalosporins. On the other hand, these changes lead to a decrease in the stability of the mutant enzyme. Secondary mutations occur, as a rule, in peripheral regions of the protein globule and may regulate the protein structure. That is why some of them (for example, M182T mutation) with increasing β-lactamase stability are called "the global suppressors." Combinations of different mutations increase the plasticity of protein structure and keep the activity–stability compromise.

The key mutations occur only in 3–5% of the total number of amino acids in the primary sequence. However, a large number of their possible combinations with each other and with secondary mutations lead to a diversity of enzymes with different phenotypes (ESBL, IRT, and CMT). This results in the ineffectiveness of β-lactams and the crisis in the development and production of their new modifications. The mechanism of genetic regulation of mutation selection, the role of various environmental factors, and the antibiotics themselves in the formation of resistance remain unclear.

An alternative way to combat resistance is to use β-lactamase inhibitors in combinations with antibiotics. Such combinations were applied successfully in clinical practice; however, the key mutations led to the appearance of enzymes with an IRT phenotype and a loss of the inhibitory ability of these substances. The total number of such mutations is small, but they are quite common. For the development of this promising way of combating the resistance, it is necessary to search for new mechanisms of inhibition and new targets, for example, sites of allosteric inhibition. Such sites may be located in the hinge regions and the loops, and new inhibitors may have the structures based on peptides. Flexible fragments of β-lactamases, in particular the Ω-loop, containing the catalytic residue E166 and playing an important role in deacylation may represent a great potential for that matter.

Microbiomes of the soil, plants, animals, and humans can become a new source of antibacterial molecules. The use of modern molecular methods for the determination of mutations in enzymes and the sensitivity of bacteria to antibiotics and inhibitors in clinical practice allows the targeted selection and use of antibacterial drugs, which should restrict the development and spread of the resistance.

The use of new methods for studying the structure of β-lactamases and their complexes with antibiotics and inhibitors gives hope for overcoming the resistance of pathogenic bacteria to both existing and new antibiotics.

This work was supported by the Russian Science Foundation (Project 15-14-00014-C).

Abbreviations

CMT	complex mutant type
ESBL	extended-spectrum β-lactamase
IRT	inhibitor-resistant type
MDR	multidrug resistant
MIC	minimal inhibitory concentration
PBP	penicillin-binding protein
PDR	pan-drug resistant
RMSD	root mean square deviation

References

Abriata, L.A., Merijn, M.L., and Tomatis, P.E. (2012). Sequence-function-stability relationships in proteins from datasets of functionally annotated variants: the case of TEM β-lactamases. *FEBS Lett.* 586 (19): 3330–3335.

Bermudes, H., Jude, F., Chaibi, E.B. et al. (1999). Molecular characterization of TEM-59 (IRT-17), a novel inhibitor-resistant TEM-derived β-lactamase in a clinical isolate of *Klebsiella oxytoca*. *Antimicrob. Agents Chemother.* 43 (7): 1657–1661.

Blazquez, J., Morosini, M., Negri, M. et al. (1995). Single amino acid replacements at positions altered in naturally occurring extended-spectrum TEM β-lactamases. *Antimicrob. Agents Chemother.* 39 (1): 145–149.

Bonomo, R.A. (2017). β-Lactamases: a focus on current challenges. *Cold Spring Harb. Perspect. Med.* 7 (1): 1–16.

Brown, N.G., Pennington, J.M., Huang, W. et al. (2010). Multiple global suppressors of protein stability defects facilitate the evolution of extended-spectrum TEM β-lactamases. *J. Mol. Biol.* 404 (5): 832–846.

Bush, K. (2018). Past and present perspectives on β-lactamases. *Antimicrob. Agents Chemother.* 62 (10): 1–20.

Bush, K. and Jacoby, G.A. (2010). Updated functional classification of β-lactamases. *Antimicrob. Agents Chemother.* 54 (3): 969–976.

Cantu, C., Huang, W., and Palzkill, T. (1996). Selection and characterization of amino acid substitutions at residues 237-240 of TEM-1 β-lactamase with altered substrate specificity for aztreonam and ceftazidime. *J. Biol. Chem.* 271 (37): 22538–22545.

Cantu, C., Huang, W., and Palzkill, T. (1997). Cephalosporin substrate specificity determinants of TEM-1 β-lactamase. *J. Biol. Chem.* 272 (46): 29144–29150.

Cassini, A., Högberg, L.D., Plachouras, D. et al. (2019). Attributable deaths and disability-adjusted life-years caused by infections with antibiotic-resistant bacteria in the EU and the European Economic Area in 2015: a population-level modelling analysis. *Lancet Infect Dis.* 19: 56–66.

Chaïbi, E.B., Sirot, D., Paul, G., and Labia, R. (1999). Inhibitor-resistant TEM β-lactamases: phenotypic, genetic and biochemical characteristics. *J. Antimicrob. Chemother.* 43 (4): 447–458.

Davies, J. and Davies, D. (2010). Origins and evolution of antibiotic resistance. *Microbiol. Mol. Biol. Rev.* 74 (3): 417–433.

Dellus-Gur, E., Elias, M., Caselli, E. et al. (2015). Negative epistasis and evolvability in TEM-1 β-lactamase – the thin line between an enzyme's conformational freedom and disorder. *J. Mol. Biol.* 427 (14): 2396–2409.

Egorov, A.M., Ulyashova, M.M., and Rubtsova, M.Ю. (2018). Bacterial enzymes and antibiotic resistance. *Acta Nat.* 10 (4): 33–48.

Fink, A.L. (1985). The molecular basis of β-lactamase catalysis and inhibition. *Pharmaceutical Research*. An Official Journal of the American Association of Pharmaceutical Scientists 2 (2): 55–61.

Fisette, O., Morin, S., Savard, P.-Y. et al. (2010). TEM-1 backbone dynamics – insights from combined molecular dynamics and nuclear magnetic resonance. *Biophys. J.* 98: 637–645.

Grigorenko, V., Uporov, I., Rubtsova, M. et al. (2018b). Mutual influence of secondary and key drug-resistance mutations on catalytic properties and thermal stability of TEM-type β-lactamases. *FEBS Open. Bio.* 8 (1): 117–129.

Grigorenko, V.G., Rubtsova, M.Y., Uporov, I.V. et al. (2018a). Bacterial TEM-type serine β-lactamases: structure and analysis of mutations. Biochemistry (Moscow). *Suppl. Ser. B Biomed. Chem.* 12 (2): 87–95.

Guthrie, V.B., Allen, J., Camps, M., and Karchin, R. (2011). Network models of TEM β-lactamase mutations coevolving under antibiotic selection show modular structure and anticipate evolutionary trajectories. *PLoS Comput. Biol.* 7 (9): e1002184.

Hall, B.G. and Barlow, M. (2005). Revised Ambler classification of β-lactamases. *J. Antimicrob. Chemother.* 55 (6): 1050–1051.

Knox, J.R. (1995). Extended-spectrum and inhibitor-resistant TEM-type β-lactamases: mutations, specificity, and three-dimensional structure. *Antimicrob. Agents Chemother.* 39 (12): 2593–2601.

Knox, R., Lento, C., and Wilson, D.J. (2018). Mapping conformational dynamics to individual steps in the TEM-1 β-lactamase catalytic mechanism. *J. Mol. Biol.* 430 (18): 3311–3322.

Massova, I. and Kollman, P.A. (2002). pKa, MM, and QM studies of mechanisms of β-lactamases and penicillin-binding proteins: acylation step. *J. Comput. Chem.* 23 (16): 1559–1576.

Ness, S., Martin, R., Kindler, A.M. et al. (2000). Structure-based design guides the improved efficacy of deacylation transition state analogue inhibitors of TEM-1 β-lactamase. *Biochemistry* 39: 5312–5321.

Neuwirth, C., Madec, S., Siebor, E. et al. (2001). TEM-89 β-lactamase produced by a *Proteus mirabilis* clinical isolate: new complex mutant (CMT 3) with mutations in both TEM-59 (IRT-17) and TEM-3. *Antimicrob. Agents Chemother.* 45 (12): 3591–3594.

Orencia, M.C., Yoon, J.S., Ness, J.E. et al. (2001). Predicting the emergence of antibiotic resistance by directed evolution and structural analysis. *Nat. Struct. Biol.* 8 (3): 238–242.

Osuna, J., Pérez-Blancas, A., and Soberón, X. (2002). Improving a circularly permuted TEM-1 β-lactamase by directed evolution. *Protein Eng. Des Sel.* 15 (6): 463–470.

Page, M.G.P. (2008). Extended-spectrum β-lactamases: structure and kinetic mechanism. *Clin. Microbiol. Infect.* 14 (Suppl. 1): 63–74.

Palzkill, T. (2018). Structural and mechanistic basis for extended-spectrum drug-resistance mutations in altering the specificity of TEM, CTX-M, and KPC β-lactamases. *Front. Mol. Biosci.* 5: 1–19.

Perilli, M., Celenza, G., Fiore, M. et al. (2007). Biochemical analysis of TEM-134, a new TEM-type extended-spectrum β-lactamase variant produced in a *Citrobacter koseri* clinical isolate from an Italian hospital. *J. Antimicrob. Chemother.* 60 (4): 877–880.

Pimenta, A.C., Fernandes, R., and Moreira, I.S. (2014). Evolution of drug resistance: insight on TEM β-lactamases structure and activity and β-lactam antibiotics. *Mini Rev. Med. Chem.* 14 (2): 111–122.

Raquet, X., Lamotte-Brasseur, J., Fonzé, E. et al. (1994). TEM β-lactamase mutants hydrolysing third-generation cephalosporins: a kinetic and molecular modelling analysis. *J. Mol. Biol.* 244 (5): 625–639.

Raquet, X., Vanhove, M., Goussard, S. et al. (1995). Stability of TEM β-lactamase mutants hydrolyzing third generation cephalosporins. *Proteins* 23: 63–72.

Robin, F., Delmas, J., Brebion, A. et al. (2007). TEM-158 (CMT-9), a new member of the CMT-type extended-spectrum β-lactamases. *Antimicrob. Agents Chemother.* 51 (11): 4181–4183.

Salverda, M.L.M., de Visser, J.A.G.M., and Barlow, M. (2010). Natural evolution of TEM-1 β-lactamase: experimental reconstruction and clinical relevance. *FEMS Microbiol. Rev.* 34 (6): 1015–1036.

Shcherbinin, D.S., Rubtsova, M.Y., Grigorenko, V.G. et al. (2017). The study of the role of mutations M182T and Q39K in the TEM-72 β-lactamase structure by the molecular dynamics method. Biochemistry (Moscow). *Suppl. Ser. B Biomed. Chem.* 11 (2): 120–127.

Sirot, D., Recule, C., Chaibi, E.B. et al. (1997). A complex mutant of TEM-1 β-lactamase with mutations encountered in both IRT-4 and extended-spectrum TEM-15, produced by an *Escherichia coli* clinical isolate. *Antimicrob. Agents Chemother.* 41 (6): 1322–1325.

Sowek, J.A., Singer, S.B., Ohringer, S. et al. (1991). Substitution of lysine at position 104 or 240 of TEM-1(pTZ18R) β-lactamase enhances the effect of serine-164 substitution on hydrolysis or affinity for cephalosporins and the monobactam aztreonam. *Biochemistry* 30: 3179–3188.

Stec, B., Holtz, K.M., Wojciechowski, C.L., and Kantrowitz, E.R. (2005). Structure of the wild-type TEM-1 β-lactamase at 1.55 Å and the mutant enzyme Ser70Ala at 2.1 Å suggest the mode of noncovalent catalysis for the mutant enzyme. *Acta Crystallogr. Sect. D Biol. Crystallogr.* 61 (8): 1072–1079.

Swarén, P., Golemi, D., Cabantous, S. et al. (1999). X-ray structure of the Asn276Asp variant of the *Escherichia coli* TEM-1 β-lactamase: direct observation of electrostatic modulation in resistance to inactivation by clavulanic acid. *Biochemistry* 38 (30): 9570–9576.

Theuretzbacher, U. (2017). Global antimicrobial resistance in Gram-negative pathogens and clinical need. *Curr. Opin. Microbiol.* 39: 106–112.

Thomas, V.L., Golemi-Kotra, D., Kim, C. et al. (2005). Structural consequences of the inhibitor-resistant Ser130Gly substitution in TEM β-lactamase. *Biochemistry* 44 (26): 9330–9338.

Wang, X., Minasov, G., and Shoichet, B.K. (2002a). The structural bases of antibiotic resistance in the clinically derived mutant β-lactamases TEM-30, TEM-32, and TEM-34. *J. Biol. Chem.* 277 (35): 32149–32156.

Wang, X., Minasov, G., and Shoichet, B.K. (2002b). Evolution of an antibiotic resistance enzyme constrained by stability and activity trade-offs. *J. Mol. Biol.* 320 (1): 85–95.

WHO (2018). *WHO Report on Surveillance of Antibiotic Consumption: 2016–2018 Early Implementation*. Geneva: World Health Organization. https://www.who.int/medicines/areas/rational_use/oms-amr-amc-report-2016-2018/en (accessed 23 April 2019).

Wiedorn, M.O., Oberthür, D., Bean, R. et al. (2018). Megahertz serial crystallography. *Nat. Commun.* 9 (1): 4025.

Wright, G.D. (2007). The antibiotic resistome: the nexus of chemical and genetic diversity. *Nat. Rev. Microbiol.* 5 (3): 175–186.

Zimmerman, M.I., Hart, K.M., Sibbald, C.A. et al. (2017). Prediction of new stabilizing mutations based on mechanistic insights from Markov state models. *ACS Cent. Sci.* 3 (12): 1311–1321.

7

Acquired Resistance from Gene Transfer

Elisabeth Grohmann[1], Verena Kohler[2,3], and Ankita Vaishampayan[1]

[1] *Life Sciences and Technology, Beuth University of Applied Sciences Berlin, Berlin, Germany*
[2] *Institute of Molecular Biosciences, University of Graz, Graz, Austria*
[3] *Department of Molecular Biosciences, The Wenner-Gren Institute, Stockholm University, Stockholm, Sweden*

7.1 Introduction

Antibiotic drugs emerged as unambiguously the most powerful medical tool to combat infectious diseases. They substantially improved human health and significantly increased lifespan. Antibiotic drugs can be divided into several classes due to their structural features and their target site, e.g. β-lactams, aminoglycosides, glycopeptides, tetracyclines, macrolides, lincosamides, streptogramins, sulfonamides, quinolones, and carbapenems (Davies and Davies 2010; Sultan et al. 2018). Antibiotic drugs can render bacteria harmless by interfering with cell wall synthesis, protein synthesis, or the nucleic acid machinery, by affecting metabolic pathways, or by disintegrating bacterial membrane structures (Kohanski et al. 2010) (see Figure 7.1). Several issues, including antibiotic overuse and wrong prescription, fueled the increasing appearance of bacterial strains no longer responding to conventional antibiotic treatment strategies. While antibiotics by themselves do not provoke resistance, frequent exposure and high doses of antibiotic drugs exert selection pressure on bacteria, triggering several resistance strategies. Antibiotic resistances (ABRs) in bacteria have emerged as a global health threat. Several human pathogens associated with epidemics have evolved into multidrug-resistant (MDR) forms. This then leads to a massive reduction of therapeutic options (Davies and Davies 2010; Karam et al. 2016; Aslam et al. 2018; Yelin and Kishony 2018). One alarming development is that bacteria become increasingly resistant to different antibiotic classes via acquisition of resistance genes originating from the same and/or different bacterial species. Dissemination of resistance determinants mostly occurs by horizontal gene transfer (HGT).

Antibiotic Drug Resistance, First Edition. Edited by José-Luis Capelo-Martínez and Gilberto Igrejas.

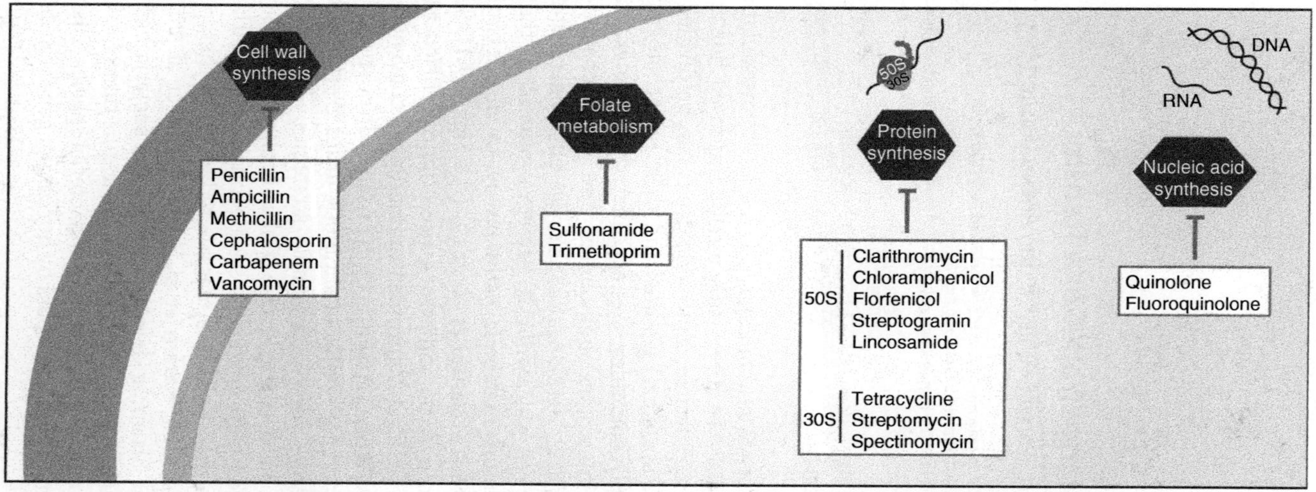

Figure 7.1 Common classes of antibiotic drugs and their targets. Antibiotics relevant for this chapter and their respective targets within a bacterial cell are illustrated.

This chapter will concentrate on conjugative transfer as the most important means of HGT, including several examples from leading human pathogens, and will discuss their resistance and spreading mechanisms.

7.2 Horizonal Gene Transfer: A Brief Overview

HGT is the non-genealogical transfer of genetic material between frequently nonrelated organisms without the need of cell division (Goldenfeld and Woese 2007). HGT can be divided into three different mechanisms as illustrated in Figure 7.2. Transformation involves uptake of free extracellular DNA originating from different organisms from the environment.

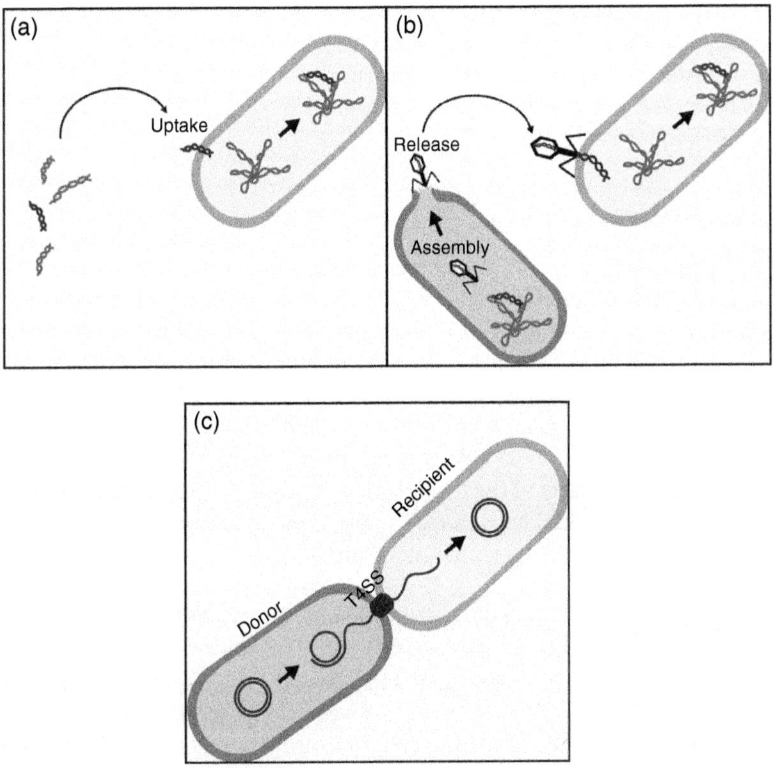

Figure 7.2 Mechanisms of horizontal gene transfer. Horizontal gene transfer can be divided into three main mechanisms: (a) transformation as the uptake of free DNA from the environment, (b) transduction as transfer of genetic material via bacteriophages, and (c) conjugation, a mechanism requiring a type IV secretion system (T4SS) for the directed shuttling of a mobile genetic element.

It requires natural competence of the recipient cell. Transduction is bacteriophage-facilitated transfer of genetic material, and conjugation requires direct cell-to-cell contact and a sophisticated protein machinery accomplishing DNA transfer from donor to recipient cell (Perry and Wright 2013; Sultan et al. 2018). To make sure that the incorporated genetic material is inherited via cell division, distinct stabilization mechanisms are required. This can be accomplished either by replication of the new DNA as a plasmid or via integration into a replicon (the host chromosome), which is the case in successful transformation or transduction events. Evolutionary success and/or gain of fitness by the foreign genetic material include advantageous features allowing better adaptation of the recipient to environmental changes or survival under poor and otherwise lethal conditions (Thomas and Nielsen 2005; Stokes and Gillings 2011; Bellanger et al. 2014). HGT is a major process in moving and rearranging bacterial genetic material and up to one-fourth of distinct bacterial genomes are estimated to originate from it (Ochman et al. 2000; Stokes and Gillings 2011; Soucy et al. 2015).

Gene transfer processes are amplified under selection pressure, which includes the presence of antimicrobial substances in clinical and environmental settings (Baquero et al. 2013).

7.2.1 Transformation

Transformation was the first mechanism of genetic exchange discovered (Griffith 1966; Stewart and Carlson 1986). It is the uptake and incorporation of extracellular genetic material from the surrounding environment (Chen et al. 2005). Transformation is an active mechanism, where free DNA is taken up into the cytoplasm of a bacterial cell (Daubin and Szöllősi 2016). Naturally competent bacteria such as *Acinetobacter* spp., *Bacillus subtilis, Streptococcus pneumoniae,* and *Haemophilus influenzae* take up free DNA, degrade it partially, and incorporate parts of the foreign genetic material into their genomes. These naturally transformable bacteria usually express several genes for DNA uptake and processing systems (Juhas et al. 2009).

7.2.2 Transduction

Transduction is a bacteriophage-mediated transfer of genetic material and was first described in the 1950s (Zinder and Lederberg 1952). Bacteriophages are viruses that are specialized to infect bacteria and can be classified into two distinct types. Lytic phages can infect and reproduce only when destroying the host cell. In contrast, lysogenic phages are capable to integrate into the host genome and are thus part of the bacterial chromosome. The so-called prophage remains in the host's DNA until the lytic cycle is induced under certain environmental conditions. This then leads to excision of the prophage from the host's chromosome (Canchaya et al. 2003; Sørensen et al. 2005; Kelly et al. 2009). Bacteriophages assemble DNA stretches of the host in their capsid and insert this genetic material into new host bacteria. Here, the foreign bacterial DNA can be integrated into the genome and can then be stably inherited (Frost et al. 2005; Sultan et al. 2018). Bacteriophages that carry ABR genes have been rarely isolated in clinical and/or environmental settings (Davies and Davies 2010).

7.2.3 Conjugation

Conjugative transfer is the most common mechanism of HGT for dissemination of ABRs (Davies and Davies 2010). Conjugation needs close physical contact between bacterial cells and requires the formation of a pore, where the substrate (the DNA molecule) can be passed through (Thomas and Nielsen 2005; Perry and Wright 2013). Conjugative transfer processes occur, among others, in the intestines of humans and animals, and diverse ABR genes have been isolated from the human gut microbiome. This so-called gut resistome represents a resistance reservoir that is available to diverse bacterial species, among them human pathogens (Shoemaker et al. 2001; Sommer et al. 2009). Conjugation has been described between unrelated bacterial species over large taxonomic distances (Tamminen et al. 2012). Bacteria can spread genes conveying ABRs via mobile genetic elements (MGE), including plasmids and integrative elements of distinct subclasses (Sultan et al. 2018).

7.3 Conjugative Transfer Mechanisms

Conjugative transfer is a one-way mechanism, where DNA is transported from the donor to the recipient cell via a sophisticated machinery (Guglielmini et al. 2011). MGEs, including conjugative plasmids and integrative conjugative elements (ICEs), harbor all genetic information for conjugative transfer. Two different types of conjugative transfer are known in both plasmids and ICEs (Bañuelos-Vazquez et al. 2017). The most common type is the transport of single-stranded plasmid DNA, found in both Gram-negative (G−) and Gram-positive (G+) species. The second type involves transport of double-stranded DNA. This mechanism has so far only been described in MGEs of G+ pluricellular actinomycetes, e.g. *Streptomyces* (recent work summarizes this process extensively (Reuther et al. 2006; Thoma and Muth 2015; Thoma and Muth 2016; Grohmann et al. 2017b; Pettis 2018)).

Conjugative systems are part of the multifaceted type IV secretion systems (T4SSs). These systems are usually encoded by multiple genes that are organized in a single operon. Type F and Type P T4SSs, also called Type IVA, resemble the prototype VirB/D4 system from *Agrobacterium tumefaciens,* where most mechanistic and functional insights have come from (Christie et al. 2017). Type I secretion systems, also named Type IVB systems, are similar to the much bigger Dot/Icm system and can be found in *Legionella pneumophila* and *Coxiella burnetii*, two intracellular pathogens (Juhas et al. 2009).

Conjugative processes require a relaxosome, a coupling protein and a T4SS. The relaxase binds and nicks the circular, double-stranded plasmid DNA at the origin of transfer (*oriT*). By interacting with the coupling protein, the DNA–relaxosome complex is guided to the channel, the T4SS, that facilitates the actual translocation. T4SSs are large protein complexes spanning the entire cell envelope, consisting of several mating pair formation proteins, at least one ATPase and factors facilitating the contact to recipients. The contact is formed either via conjugative pili in G– bacteria or via surface adhesins in G+ systems. In parallel, replication processes ensure that both old and new hosts have a double-stranded version of the MGE (Guglielmini et al. 2011; Grohmann et al. 2017a).

7.3.1 Conjugative Transfer of Plasmids

Plasmids are autonomously replicating genetic elements that can vary greatly in size. These elements can be subgrouped according to their replication and partitioning systems into so-called incompatibility (Inc) groups. This is defined as the inability of two plasmids from the same Inc group to stably coexist in one bacterial cell. Plasmids usually harbor nonessential genetic information, which might become important under certain circumstances, e.g. in the presence of antibiotic selection pressure (Bañuelos-Vazquez et al. 2017). Plasmids can be self-transmissible (or conjugative) and thus encode all necessary proteins for mobilization and subsequent transfer processes (Figure 7.3). Non-self-transmissible but mobilizable plasmids do not harbor the necessary genetic information but can be disseminated if factors required for transfer are provided *in trans*. Interestingly, mobilizable elements outnumber self-conjugative MGEs; thus conjugative systems seem to be extensively used *in trans* (Guglielmini et al. 2011). Transfer and dissemination of plasmids in natural environments greatly rely on biofilms (Kelly et al. 2009). Frequency of HGT in general was shown to be higher in bacterial biofilms. A biofilm is a community of bacteria surrounded by a self-produced matrix made of extracellular polymeric substances, e.g. exopolysaccharides, proteins, and nucleic acids (Madsen et al. 2012).

Plasmids and the associated spread of encoded genes have significantly contributed to the rise of antibiotic-resistant bacteria (Sultan et al. 2018). Even though conjugation is a conserved mechanism, the exact sequence of events and the involved factors deviate slightly between different Inc groups (de La Cruz et al. 2010).

ABR genes were identified on plasmids of the IncP, IncN, and IncQ families. A common feature of these subgroups is their broad host range (Perry and Wright 2013). Plasmids of the IncP-1 family disseminate very efficiently via

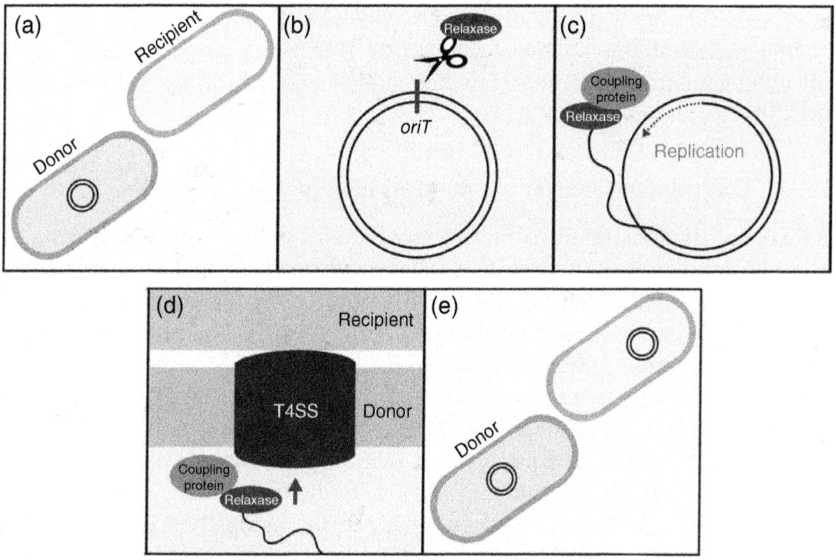

Figure 7.3 Transfer mechanism of conjugative plasmids. Upon a signal from the recipient and/or the environment (a), the relaxase nicks the conjugative plasmid at the origin of transfer (*oriT*) in the donor cell (b). While the coupling protein interacts with the relaxase bound to the nicked single-stranded plasmid DNA, replication processes take place (c). By interaction of the coupling protein with the type IV secretion system (T4SS) at the donor's cell envelope, the single-stranded plasmid DNA is shuttled to the recipient cell (d). After conjugative processes and plasmid replication have taken place, both cells harbor a double-stranded version of the conjugative plasmid (e).

conjugation and can exist in virtually all G- bacteria. IncP-1 plasmids frequently harbor resistance genes against heavy metals, resulting in a co-selection with ABR genes. It was assumed that the propagation of MDR in soil, water, and wastewater treatment plants might be predominantly caused by IncP-1 plasmids (Perry and Wright 2013; Popowska and Krawczyk-Balska 2013).

IncA/C-family plasmids are common among enterobacteria and were also found in *Vibrio cholerae.* These plasmids are widely distributed among food-borne pathogens. These groups of plasmids carry multidrug resistance gene cassettes and are usually self-transmissible (Lindsey et al. 2009; Poole et al. 2009; Bañuelos-Vazquez et al. 2017). Plasmids of the IncH11-family are essential carriers of ABR in *Salmonella typhi* and, as early as 1998, IncH11 plasmids were isolated from MDR *S. typhi* worldwide (Ugboko and De 2014).

Plasmids belonging to the Inc18 family play a primordial role in several G+ pathogens. This family of plasmids encodes several ABRs, e.g. conferring resistance to vancomycin, chloramphenicol, and the macrolide–lincosamide–streptogramin group of antibiotics. Inc18 plasmids show an exceptionally broad host range and are widely distributed in clinical settings. These plasmids have been demonstrated to be responsible for transfer of vancomycin resistance via the *vanA* gene from enterococci to methicillin-resistant *Staphylococcus aureus* (MRSA) (Kohler et al. 2018).

7.3.2 Conjugative Transfer of Integrative Conjugative Elements

ICEs were demonstrated to be the most abundant conjugative elements in prokaryotes (Guglielmini et al. 2011). Different from plasmids, ICEs must integrate into the host's genome for stable inheritance. ICEs can harbor ABR genes and have a broad host range (Perry and Wright 2013). The exact mechanism of ICE conjugation is not completely elucidated; it is assumed that transfer processes resemble those of single-stranded plasmid transport via T4SS. Nevertheless, two additional steps are required, excision and reintegration (Figure 7.4). Thus, ICEs carry genes resembling factors of lysogenic phages (Wozniak and Waldor 2010). ICEs show a modular structure, and genes that exert the same/similar functions are usually clustered together. These elements frequently carry three modules, the integration/excision (maintenance) module, the conjugative (dissemination) module, and the regulation module.

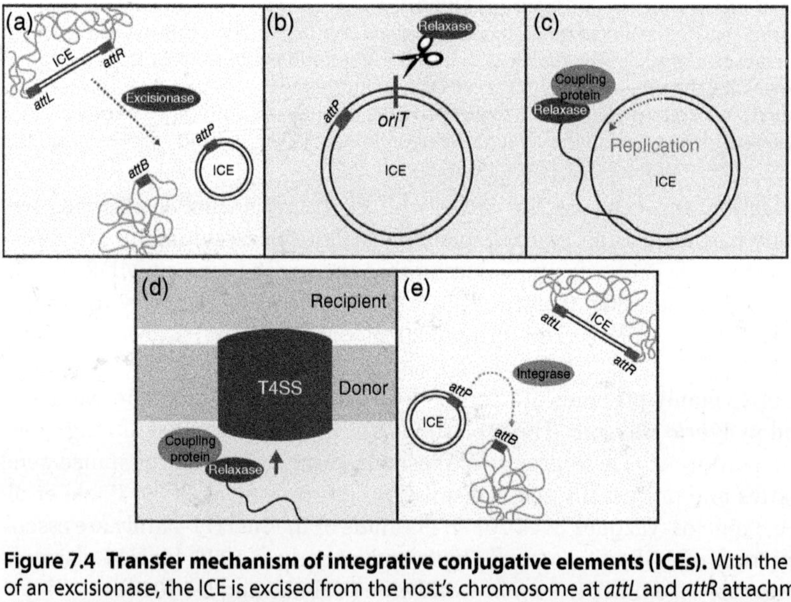

Figure 7.4 Transfer mechanism of integrative conjugative elements (ICEs). With the aid of an excisionase, the ICE is excised from the host's chromosome at *attL* and *attR* attachment sites, creating *attP* and *attB* on the respective DNA molecule (a). Nicking of plasmid DNA by the relaxase (b), interaction with the coupling protein (c), and shuttling of single-stranded DNA via the type IV secretion system (d) follow the same mechanisms as described for conjugative plasmids (Figure 7.3). After successful conjugation and replication, the circularized ICE is integrated into the host's chromosome (at *attB*), catalyzed by an integrase.

The ICE integration/excision module encodes a recombinase, frequently a tyrosine recombinase. This enzyme assists in the reaction between *attP*, a sequence on the recombination module of the ICE, and *attB* on the host chromosome, forming the attachment sites *attL* and *attR* that flank the element after successful integration (Burrus and Waldor 2004). In most cases of tyrosine recombinase-mediated reactions, an excisionase is needed for excision. Site-specific recombination between *attR* and *attL* forms again an *attP* site on the circularized ICE and an *attB* site on the host's chromosome. ICEs can exhibit high to weak integration specificity, depending on the respective integration/excision modules. ICEs harboring a tyrosine recombinase usually integrate into several specific sites, not only at the 3′ end of tRNAs but also at both ends of housekeeping genes (Burrus and Waldor 2004; Johnson and Grossman 2015).

SXT is a 100 kb ICE that was initially isolated from *V. cholerae* in India and codes for multiple ABR genes. SXT and related MGEs have become alarmingly widespread in Asian and African *V. cholerae* isolates. SXT is similar to the IncJ element R391 and both integrate into the same chromosomal site, the *prfC* gene, with the aid of a tyrosine recombinase (Hochhut and Waldor 1999; Hochhut et al. 2001). The mechanism of integration and excision was described to be similar to that in lambdoid phages. Both the integrase and the excisionase are needed for successful SXT excision (Burrus and Waldor 2003). These two genes are in a convergent orientation and further most probably not co-regulated. The proteins forming the conjugative machinery have significant homology with transfer (Tra) proteins from pCAR1 from *Pseudomonas resinovorans* and Rts1 from *Proteus vulgaris* (Murata et al. 2002; Maeda et al. 2003; Burrus et al. 2006). The SXT *tra*-genes are located in four clusters spanning more than 25 kb. These *tra*-genes are conserved among SXT-related ICEs. The R391 *tra*-genes reveal more than 94% identity to their counterparts in SXT (Beaber et al. 2002; Böltner and Osborn 2004). Both SXT and R391 harbor exclusion systems that prevent redundant transmission of these ICEs (Burrus et al. 2006).

Tn*916* is regarded as prototype of the Tn*916*-Tn*1545* family of ICEs and is one of the most studied conjugative transposons in G+ bacteria. Tn*916* was originally isolated from *Enterococcus faecalis* (Santoro et al. 2014). Conjugative processes of Tn*916* are again described to be similar to those of conjugative plasmids. Following conjugative transfer of the nicked single-stranded DNA and replication, Tn*916* integrates into A + T rich sequences. Thus, this MGE reveals a low specificity for integration, which results in frequent intracellular transposition (Jaworski and Clewell 1995; Roberts and Mullany 2009). In contrast, Tn*5276* from *Lactococcus lactis* reveals a higher integration specificity (Rauch and de Vos 1992). Tn*916* and related MGEs are regulated by tetracycline. Tn*916* encodes a tetracycline resistance gene and under exposure to this antibiotic drug, conjugative transfer increased 19-fold in *B. subtilis* (Showsh and Andrews 1992).

7.3.3 Conjugative Transfer of Other Integrative Elements

Loss of conjugative and/or integrative functions in ICEs can generate genomic islands (Gillings et al. 2017). These MGEs are frequently referred to as integrative mobilizable elements (IMEs). IMEs were also described in *Clostridia* as clostridial mobilizable transposons (Adams et al. 2002). These transposons, including Tn*4451* from *Clostridium perfringens* and Tn*4453* from *Clostridium difficile*, can be mobilized by another conjugative element.

Genomic islands in general reveal a substantially different G + C content and an alternative codon usage. Further, they encode mobility genes and frequently harbor other genetic information, e.g. for virulence (on pathogenicity islands). It is assumed that these elements were previously disseminated by HGT, since these contain the remains of mobility-enabling elements (integrase, transposase, flanking direct repeats, etc.) (Hacker and Carniel 2001; Bellanger et al. 2014).

The genomic island 1 from salmonella (SGI1) is a 48 kb genomic element with an ABR gene cluster, conferring resistance to ampicillin, chloramphenicol, florfenicol, streptomycin, spectinomycin, sulfonamides, and tetracycline. SGI1 has been found on several *Salmonella enterica* serovars. The encoded resistance gene cluster, a class I integron, leads to the common multidrug resistance phenotype of *S. enterica* Typhimurium DT104. SGI1 integrates site-specifically and the encoded integrase, similar to the lambdoid integrase family, is needed for excision from the host chromosome. This integrase is encoded as the first gene of SGI1, which overlaps with a putative excisionase gene transcribed from the opposite direction. SGI1 is non-self-transmissible, but mobilizable, belonging to the group of IMEs. Conjugative transfer requires the aid of a helper plasmid, e.g. of IncA/C plasmid R55. Several conjugative IncA/C MDR plasmids can mobilize SGI1 from *S. enterica* to *Escherichia coli* (Doublet et al. 2005; Douard et al. 2010). Interestingly, SGI1 shares similar features with SXT, an ICE from *V. cholerae* (Doublet et al. 2005).

Integrons are genetic elements that can capture gene cassettes lacking promoters using site-specific recombination (Rowe-Magnus and Mazel 2001). These elements consist of integrase genes and an integration site for gene cassettes. Integrons have been identified on various MGEs, including plasmids, transposons, and even chromosomes (Fluit and Schmitz 2004). Integrons can supply promoters for the expression of gene cassettes. These elements are flanked by inverted repeats and can move between/within replicons (Rowe-Magnus and Mazel 2001). The main role in bacterial adaptation lies on their capability to express, rearrange, and capture diverse gene cassettes. Class I integrons have gathered more than 130 resistance genes and can be found in the majority of G− pathogens and in several G+ bacteria as well (Partridge et al. 2009; Stokes and Gillings 2011; Ugboko and De 2014). These captured gene cassettes confer resistance to most antibiotic classes used in modern treatment strategy plans.

7.4 Antibiotic Resistances and Their Transfer

In 2017, the World Health Organization (WHO) published a list of bacterial pathogens for which alternative drugs or treatments need to be urgently developed. This list is categorized into three groups: (a) critical priority pathogens, which include carbapenem-resistant *Acinetobacter baumannii* and *Pseudomonas aeruginosa* and carbapenem-resistant and third-generation cephalosporin-resistant *Enterobacteriaceae*; (b) high priority pathogens, which consist of vancomycin-resistant enterococci (VRE), vancomycin-resistant *S. aureus* (VRSA), methicillin-resistant *S. aureus* (MRSA), clarithromycin-resistant *Helicobacter pylori*, fluoroquinolone-resistant *Campylobacter* spp. and *Salmonella* spp., and cephalosporin-resistant *Neisseria gonorrhoeae*; and (c) medium priority, with penicillin non-susceptible *S. pneumoniae*, ampicillin-resistant *H. influenzae*, and fluoroquinolone-resistant *Shigella* spp. (Tacconelli et al. 2017; WHO 2017).

Resistance to the β-lactam group of antibiotics is the major concern with the listed pathogens, in addition to fluoroquinolones, clarithromycin, and to the last resort antibiotic vancomycin. In this part of the chapter, we will discuss these ABRs, dissemination of these resistances, and their transfer mechanisms.

7.4.1 Dissemination of Carbapenem Resistance Among Bacterial Pathogens

Carbapenem, methicillin, ampicillin, penicillin, and cephalosporin belong to the broad-spectrum β-lactam group of antibiotics (Berglund et al. 2017). Production of β-lactamases in bacteria leads to resistance against β-lactam antibiotics (Bonnet 2004).

Carbapenem resistance is widely observed in *A. baumannii*, *P. aeruginosa*, and bacteria belonging to the *Enterobacteriaceae* and is a critical problem in healthcare due to challenges in treating the infections caused by these pathogens (WHO 2017). Carbapenem resistance is brought by several β-lactamase genes, such as bla_{IMP} encoding imipenem resistance, bla_{VIM} (Verona integron-encoded metallo-β-lactamases), bla_{OXA} encoding oxacillin resistance, and bla_{NDM} encoding New Delhi metallo-β-lactamase (Pagano et al. 2016; Berglund et al. 2017). These genes, conferring carbapenem resistance, are often found on MGEs, like plasmids, transposons, and integrons (Pagano et al. 2016; Berglund et al. 2017). MGEs have the ability to transfer diverse types of resistances like ABR and resistance to toxic heavy metals, such as silver, mercury, and cadmium, among diverse bacterial species and even genera (Broaders et al. 2013). Acquired resistance is usually mediated by HGT (van Hoek et al. 2011). MGEs including plasmids, phages, transposons, genomic islands, integrons, and gene cassettes (Brown-Jaque et al. 2015) can serve as a reservoir of ABR genes and are major players in the dissemination of ABR among microorganisms in different environments (von Wintersdorff et al. 2016; Ma et al. 2017).

Numerous ABR genes are carried on MGEs. For example, RepAci6 plasmids in *A. baumannii* carry the bla_{OXA-23} gene on Tn*2006*, and pNDM-BJ01-related plasmids carry the bla_{NDM-1} gene on Tn*125* (Partridge et al. 2018). Tn*2006* is currently considered to be the most common MGE involved in carbapenem-resistant *A. baumannii*, with a great potential to disseminate the resistance among *A. baumannii* isolates (Lee et al. 2012; Pagano et al. 2016). Transferable plasmids, such as IncP-2-like plasmids from carbapenem-resistant *P. aeruginosa* isolates, mostly harbor cassette-borne carbapenemase genes like bla_{IMP-9}, bla_{SIM-2}, and bla_{VIM-2}, which encode the clinically most significant metallo-β-lactamases, found in class 1 integrons and are inserted into transposons of the Tn*21* family (Moosavian and Rahimzadeh 2015; Partridge et al. 2018). In *Enterobacteriaceae*, the bla_{OXA-48} gene conferring resistance to oxacillin is located on Tn*1999*, on the IncL/M type plasmid pOXA-48a. This plasmid has been identified in several species and has a broad host range among *Enterobacteriaceae* with interspecies transfer frequency of 3.3×10^{-5} transconjugants per recipient (Evans and Amyes 2014), confirming the resistance spread through HGT.

HGT can be promoted or induced by certain factors, as shown by Stecher et al. (2012). They demonstrated that pathogen-induced gut inflammation promoted HGT between pathogenic and commensal *Enterobacteriaceae* in the gut. In their study in a mouse colitis model, enteropathy caused by *Salmonella* triggered parallel blooms of the pathogen and resident commensal *E. coli*, which enhanced conjugative transfer of the colicin plasmid p2 from *S. enterica* serovar *Typhimurium* to *E. coli*. Under normal conditions, commensal bacteria were able to block HGT by inhibiting contact-dependent conjugation between *Enterobacteriaceae*. These blooms may result in rearrangement of plasmid-encoded genes between pathogens and commensals, stimulating the dissemination of genes involved in fitness, virulence, and ABR (Stecher et al. 2012). In another study highlighting intergeneric HGT, Domingues et al. (2012) evaluated the extent to which natural transformation can mediate transfer of MGEs between bacterial species by exposing *Acinetobacter baylyi* to purified DNA or cell lysates from integron-carrying strains of *A. baumannii*, *P. aeruginosa*, *S. enterica* serovar *Typhimurium*, *Enterobacter cloacae*, *E. coli*, and *Citrobacter freundii*. After 24 hours of exposure, not only ABR traits but also whole integrons and transposons had been transferred to *A. baylyi*. DNA integration did not occur solely due to genetic relatedness or due to integrase functions, but many mechanisms facilitated the stable incorporation of DNA in the recipient genome; homologous and heterologous recombination and Tn*21*-like and IS*26*-like transpositions resulted in successful integration of integron containing DNA in the recipient genome (Domingues et al. 2012).

7.4.2 Dissemination of Cephalosporin Resistance Among Bacterial Pathogens

In addition to carbapenem resistance, the WHO has also listed resistance to third-generation cephalosporins in *Enterobacteriaceae* and in *N. gonorrhoeae* as a serious problem (WHO 2017). Cephalosporin resistance occurs due to production of extended-spectrum β-lactamases (ESBLs), especially the ones belonging to the CTX-M family, and these are also a major cause of increased resistance rates to third-generation cephalosporins in *Enterobacteriaceae* (Ruppé et al. 2015). The $bla_{\text{CTX-M}}$ genes originated in *Kluyvera ascorbata* and *Kluyvera georgiana* and were mobilized by integration into plasmids (Bonnet 2004). Elements like IS*Ecp1*and Orf513 (located in a class 1 integron, encoding a putative site-specific recombinase likely involved in acquiring genes from the chromosome) may be involved in the transfer of the $bla_{\text{CTX-M}}$ genes and cephalosporinase genes such as the bla_{CMY}- type (Bonnet 2004; Poirel et al. 2005). The CTX-M encoding genes are often located on plasmids of different sizes ranging from 7 kb (pIP843) to 160 kb (pMG267) (Pai et al. 2001; Cao et al. 2002; Bonnet 2004). *In vitro,* these plasmids are transmissible by conjugation with transfer frequencies from 10^{-7} to 10^{-2} transconjugants per donor cell; these plasmids also harbor genes encoding resistance to other antibiotics like aminoglycosides, chloramphenicol, sulfonamides, trimethoprim, and tetracycline (Bonnet 2004). The spread of cephalosporin resistance is plasmid-mediated in animals and humans as well, as was observed by de Been and coworkers (2014). The authors studied the epidemiology of ESBL-producing *E. coli* isolates (from the Netherlands between 2006 and 2011) from human clinical infections, chickens raised on production farms, and chicken retail meat, using whole genome sequencing. Based on the whole genome sequencing data, the cephalosporin resistance genes, $bla_{CTX\text{-}M\text{-}1}$ and $bla_{CMY\text{-}2}$ were found on plasmids of the Inc1 and IncK group, respectively, and it was observed that the resistance spread was plasmid mediated (de Been et al. 2014).

7.4.3 Dissemination of Methicillin Resistance Among Bacterial Pathogens

Another crucial β-lactam ABR is methicillin resistance. Methicillin resistance in *S. aureus* is well known and MRSA is listed as high priority pathogen (WHO 2017). The gene *mecA,* which confers resistance to methicillin, is carried on the MGE, Staphylococcal Cassette Chromosome mec (SCCmec), which can be transferred via HGT (Jamrozy et al. 2017). The HGT mechanism most likely involved in the spread of resistance in *S. aureus* is transduction because conjugative plasmids and transposons have been identified only in a small proportion of isolates. Also, there is little evidence of transformation in *S. aureus,* since *S. aureus* strains are difficult to transform as they possess type I and IV restriction modification barriers, which identify and restrict foreign DNA (Lindsay et al. 2012; Jones et al. 2015).

7.4.4 Dissemination of Vancomycin Resistance Among Bacterial Pathogens

Along with methicillin resistance, vancomycin resistance in staphylococci is a threat to healthcare. Vancomycin resistance is also observed in enterococci and is a serious problem as vancomycin serves as a drug of last resort. Vancomycin resistance is encoded in the *van* gene cluster consisting of *vanA*, *vanB*, *vanH*, *vanR*, *vanS*, *vanW*, *vanX*, *vanY*, and *vanZ*, typically named after the ligases they encode (Kristich et al. 2014). Genes conferring resistance to vancomycin are carried on MGEs; *vanA* is carried on Tn*1546* and *vanB* can be chromosomally or plasmid-encoded and is carried on Tn*1549*-like transposons or on Tn*5382* (Kristich et al. 2014). Vancomycin resistance can be transferred by conjugation. The vancomycin resistance observed in *S. aureus* was likely acquired from VRE by transfer of Inc18 plasmids comprising insertions of Tn*1546*, encoding vancomycin resistance (Zhu et al. 2008; Malachowa and DeLeo 2010).

7.4.5 Dissemination of Fluoroquinolone Resistance Among Bacterial Pathogens

Among the high priority resistances listed by WHO are also clarithromyin resistance and fluoroquinolone resistance (WHO 2017). Clarithromycin resistance observed in *H. pylori* occurs due to point mutations of the 23S rRNA gene. Such ABRs, which are mediated by chromosomal mutations, are transmitted vertically among bacteria instead by HGT (Mégraud 2013).

Fluoroquinolones are mainly prescribed to treat gastrointestinal and urinary tract infections (Sanchez and de Melo 2018). Among the pathogens listed by WHO, *Campylobacter* spp., *Salmonella* spp., and *Shigella* spp., exhibit high levels of fluoroquinolone resistance (WHO 2017). Quinolone resistance can be chromosomally encoded or through acquisition of plasmid-mediated quinolone resistance genes like *qnrA*, *qnrB*, *qnrS*, and *aac(6′)-lb-cr* (Patel et al. 2011; Sanchez and de Melo 2018). Quinolone resistance can be transferred to other bacteria by conjugation or transformation (Mannion et al. 2018). For example, plasmid DNA from *Shigella flexneri* isolates was transformed and transported by conjugative transfer into an antibiotic susceptible *E. coli* strain. The *E. coli* strain acquired resistance to quinolones and β-lactams in addition (Mannion et al. 2018). *qnr* harboring plasmids often also carry resistance genes for other antibiotics, which can also be transferred by conjugation as discovered by Hata et al. (2005). The conjugative plasmid pAH0376 (~47 kb) in *S. flexneri*

2b isolates, which caused a food poisoning outbreak in Japan in 2003, was observed to harbor the TEM-1 β-lactamase gene and the *qnrS* gene (Hata et al. 2005; Mannion et al. 2018).

7.4.6 Dissemination of Penicillin and Ampicillin Resistance Among Bacterial Pathogens

Penicillin and ampicillin resistance are listed as medium priority (WHO 2017). The mosaic penicillin-binding protein (PBP) genes are observed in streptococci and are considered to be the result of gene transfer from penicillin-resistant species to various *Streptococcus* species, which might also be the reason behind reduced susceptibility to penicillin observed in *S. pneumoniae* (Dowson et al. 1990; Sibold et al. 1994; von Wintersdorff et al. 2016).

Recently, ampicillin resistance in *H. influenzae* has also become a problem in treating infections. One of the mechanisms of ampicillin resistance in *H. influenzae* is through production of β-lactamases. *H. influenzae* carries the β-lactamase genes on large ICEs or on small plasmids ranging from 4000 to 6000 bp with a structural backbone consisting of bla_{TEM} β-lactamase and *rep,* a replicase-encoding gene. bla_{TEM} is the dominant β-lactamase gene in *H. influenzae.* β-lactam resistance in *H. influenzae* can be spread through clonal expansion of resistant clones and by horizontal plasmid transfer (Fleury et al. 2014). Fleury and coworkers observed that small plasmids can transfer *in vivo* between *H. influenzae* strains and plasmid pN223 can be transferred even intergenerically, to *E. coli,* rendering it resistant to ampicillin (Fleury et al. 2014).

7.5 Nanotubes Involved in Acquisition of Antibiotic Resistances

Apart from the classical mechanisms of HGT, recently some less known mechanisms of exchange of material between bacteria have been discovered. For example, bacteria such as *S. aureus* and *E. coli* use 1 μm long and 30–130 nm wide tubular extrusions called nanotubes to directly exchange cytoplasmic molecules like enzymes, which makes the recipient bacteria resistant to antibiotics. In addition, nanotubes also enable transfer of genetic material that is carried on elements like non-conjugative plasmids likely leading to hereditary ABR in bacteria (Dubey and Ben-Yehuda 2011; Jansen and Aktipis 2014). The spread of ABR reduces the efficiency of antibiotics and restricts their use as an infection treatment option. Hence, it is imperative to monitor and understand ABR patterns and their mechanism of dissemination to determine effective solutions to combat the problem of ABR and its spread (Goulas et al. 2018).

7.6 Conclusions and Outlook

Monitoring programs to assess the abundance and spread of ABRs in the environment have been initiated and clearly demonstrate that not only has the problem been tackled in the clinical setting, but also the transmission of ABR from the clinic to the environment and vice versa has to be a matter of concern. Goulas et al. performed a comprehensive literature search to point out potential solutions to get the contamination of the environment with ABRs and their spread in the environment better under control. Those include optimization of antibiotic treatments; alternative treatments; better hygiene conditions; solutions to improve the treatment of human and animal wastes, e.g. better wastewater treatment, composting, and disinfection; and environmental management strategies such as better regulation of sludge application, protection of drinking water catchment areas, soil management, and bioremediation (Goulas et al. 2018).

A good start to reduce the generation of antibiotic-resistant pathogens in the clinic would be focusing on alternative treatments of bacterial infections such as bacteriophage therapy as recently demonstrated by Llanos-Chea and coworkers in a human intestinal organoid-derived infection model for a resistant strain of the enteric pathogen *S. flexneri* (Llanos-Chea et al. 2018). In addition, strategies to reduce the transmission of ABR in the clinic should include the application of conjugation inhibitors, small molecules that inhibit the activity of key T4SS proteins involved in the transmission of resistance plasmids as proposed by Casu et al. (2017) and Sharifahmadian and Baron (2017).

Abbreviations

ABR	antibiotic resistance
ESBLs	extended-spectrum β-lactamases
HGT	horizontal gene transfer
ICE	integrative conjugative element
IME	integrative mobilizable element
Inc	incompatibility
MDR	multidrug resistant
MGE	mobile genetic element
MRSA	methicillin-resistant *S. aureus*
oriT	origin of transfer
PBP	penicillin-binding protein
SGI1	*Salmonella* genomic island 1
tra	transfer

T4SS	type IV secretion system
VRE	vancomycin-resistant enterococci
VRSA	vancomycin-resistant *S. aureus*
WHO	World Health Organization

References

Adams, V., Lyras, D., Farrow, K.A., and Rood, J.I. (2002). The clostridial mobilisable transposons. *Cellular and Molecular Life Sciences* 59 (12): 2033–2043. https://doi.org/10.1007/s000180200003.

Aslam, B., Wang, W., Arshad, M.I. et al. (2018). Antibiotic resistance: a rundown of a global crisis. *Infection and Drug Resistance* 11: 1645–1658. https://doi.org/10.2147/IDR.S173867.

Bañuelos-Vazquez, L.A., Torres Tejerizo, G., and Brom, S. (2017). Regulation of conjugative transfer of plasmids and integrative conjugative elements. *Plasmid* https://doi.org/10.1016/j.plasmid.2017.04.002.

Baquero, F., Tedim, A.P., and Coque, T.M. (2013). Antibiotic resistance shaping multi-level population biology of bacteria. *Frontiers in Microbiology* https://doi.org/10.3389/fmicb.2013.00015.

Beaber, J.W., Hochhut, B., and Waldor, M.K. (2002). Genomic and functional analyses of SXT, an integrating antibiotic resistance gene transfer element derived from *Vibrio cholerae*. *Journal of Bacteriology* 184 (15): 4259–4269. https://doi.org/10.1128/JB.184.15.4259-4269.2002.

Bellanger, X., Payot, S., Leblond-Bourget, N., and Guédon, G. (2014). Conjugative and mobilizable genomic islands in bacteria: evolution and diversity. *FEMS Microbiology Reviews* https://doi.org/10.1111/1574-6976.12058.

Berglund, F., Marathe, N.P., Österlund, T. et al. (2017). Identification of 76 novel B1 metallo-β-lactamases through large-scale screening of genomic and metagenomic data. *Microbiome* 1–13. https://doi.org/10.1186/s40168-017-0353-8.

Böltner, D. and Osborn, A.M. (2004). Structural comparison of the integrative and conjugative elements R391, pMERPH, R997, and SXT. *Plasmid* 51 (1): 12–23. https://doi.org/10.1016/j.plasmid.2003.10.003.

Bonnet, R. (2004). Growing group of extended-spectrum-lactamases: the CTX-M enzymes. *Antimicrobial Agents and Chemotherapy* 48 (1): 1–14. https://doi.org/10.1128/AAC.48.1.1.

Broaders, E., Gahan, C.G.M., Marchesi, J.R. et al. (2013). Mobile genetic elements of the human gastrointestinal tract. *Gut Microbes* 4 (4): 271–280. https://doi.org/10.4161/gmic.24627.

Brown-Jaque, M., Calero-Cáceres, W., and Muniesa, M. (2015). Mobile elements. *Plasmid* 79: 1–7. https://doi.org/10.1016/j.plasmid.2015.01.001.

Burrus, V., Marrero, J., and Waldor, M.K. (2006). The current ICE age: biology and evolution of SXT-related integrating conjugative elements. *Plasmid* https://doi.org/10.1016/j.plasmid.2006.01.001.

Burrus, V. and Waldor, M.K. (2003). Control of SXT integration and excision. *Journal of Bacteriology* 185 (17): 5045–5054. https://doi.org/10.1128/JB.185.17.5045-5054.2003.

Burrus, V. and Waldor, M.K. (2004). Shaping bacterial genomes with integrative and conjugative elements. *Research in Microbiology* 155 (5): 376–386. https://doi.org/10.1016/j.resmic.2004.01.012.

Canchaya, C., Fournous, G., Chibani-Chennoufi, S. et al. (2003). Phage as agents of lateral gene transfer. *Current Opinion in Microbiology* https://doi.org/10.1016/S1369-5274(03)00086-9.

Cao, V., Lambert, T., and Courvalin, P. (2002). ColE1-Like Plasmid pIP843 of *Klebsiella pneumoniae* encoding. *Antimicrobial Agents and Chemotherapy* 46 (5): 1212–1217. https://doi.org/10.1128/AAC.46.5.1212.

Casu, B., Arya, T., Bessette, B., and Baron, C. (2017). Fragment-based screening identifies novel targets for inhibitors of conjugative transfer of antimicrobial resistance by plasmid pKM101. *Scientific Reports* 7 (1): 14907. https://doi.org/10.1038/s41598-017-14953-1.

Chen, I., Christie, P.J., and Dubnau, D. (2005). The ins and outs of DNA transfer in bacteria. *Science* 310 (5753): 1456–1460.

Christie, P.J., Valero, L.G., and Buchrieser, C. (2017). Biological diversity and evolution of type IV secretion systems. *Current Topics in Microbiology and Immunology* 413: 1–30. https://doi.org/10.1007/978-3-319-75241-9_1.

Daubin, V. and Szöllősi, G.J. (2016). Horizontal gene transfer and the history of life. *Cold Spring Harbor Perspectives in Biology* 8 (4): https://doi.org/10.1101/cshperspect.a018036.

Davies, J. and Davies, D. (2010). Origins and evolution of antibiotic resistance. *Microbiology and Molecular Biology Reviews* 74 (3): 417–433. https://doi.org/10.1128/MMBR.00016-10.

De Been, M., Lanza, V.F., de Toro, M. et al. (2014). Dissemination of cephalosporin resistance genes between *Escherichia coli* strains from farm animals and humans by specific plasmid lineages. *PLoS Genetics* 10 (12): https://doi.org/10.1371/journal.pgen.1004776.

De La Cruz, F., Frost, L.S., Meyer, R.J., and Zechner, E.L. (2010). Conjugative DNA metabolism in Gram-negative bacteria. *FEMS Microbiology Reviews* https://doi.org/10.1111/j.1574-6976.2009.00195.x.

Domingues, S., Harms, K., Fricke, F.W. et al. (2012). Natural transformation facilitates transfer of transposons, integrons and gene cassettes between bacterial species. *PLoS Pathogens* 8 (8): https://doi.org/10.1371/journal.ppat.1002837.

Douard, G., Praud, K., Cloeckaert, A., and Doublet, B. (2010). The *Salmonella* Genomic Island 1 is specifically mobilized in trans by the IncA/C multidrug resistance plasmid family. *PLoS One* 5 (12): e15302. https://doi.org/10.1371/journal.pone.0015302.

Doublet, B., Boyd, D., Mulvey, M.R., and Cloeckaert, A. (2005). The *Salmonella* genomic island 1 is an integrative mobilizable element. *Molecular Microbiology* 55 (6): 1911–1924. https://doi.org/10.1111/j.1365-2958.2005.04520.x.

Dowson, C.G., Hutchison, A., Woodfordt, N. et al. (1990). Penicillin-resistant viridans streptococci have obtained altered penicillin-binding protein genes from penicillin-resistant strains of *Streptococcus pneumoniae*. *Proceedings of the National Academy of Sciences* 87 (15): 5858–5862.

Dubey, G.P. and Ben-Yehuda, S. (2011). Intercellular nanotubes mediate bacterial communication. *Cell* 144 (4): 590–600. https://doi.org/10.1016/j.cell.2011.01.015.

Evans, B.A. and Amyes, S.G.B. (2014). OXA-Lactamases. *Clinical Microbiology Reviews* 27 (2): 241–263.

Fleury, C., Resman, F., Rau, J., and Riesbeck, K. (2014). Prevalence, distribution and transfer of small β-lactamase-containing plasmids in Swedish *Haemophilus influenzae*. *Journal of Antimicrobial Chemotherapy* 69 (5): 1238–1242. https://doi.org/10.1093/jac/dkt511.

Fluit, A.C. and Schmitz, F.J. (2004). Resistance integrons and super-integrons. *Clinical Microbiology and Infection* https://doi.org/10.1111/j.1198-743X.2004.00858.x.

Frost, L.S., Leplae, R., Summers, A.O., and Toussaint, A. (2005). Mobile genetic elements: the agents of open source evolution. *Nature Reviews Microbiology* https://doi.org/10.1038/nrmicro1235.

Gillings, M.R., Paulsen, I.T., and Tetu, S.G. (2017). Genomics and the evolution of antibiotic resistance. *Annals of the New York Academy of Sciences* 1388 (1): 92–107. https://doi.org/10.1111/nyas.13268.

Goldenfeld, N. and Woese, C. (2007). Biology's next revolution. *Nature* 445 (7126): 369–369. https://doi.org/10.1038/445369a.

Goulas, A., Livoreil, B., Grall, N. et al. (2018). What are the effective solutions to control the dissemination of antibiotic resistance in the environment? A systematic review protocol. *Environmental Evidence* 7 (3): 1–9. https://doi.org/10.1186/s13750-018-0118-2.

Griffith, F. (1966). The significance of pneumococcal types. *Journal of Hygiene* 64 (2): 129–175. https://doi.org/10.1017/S0022172400040420.

Grohmann, E., Christie, P.J., Waksman, G., and Backert, S. (2017a). Type IV secretion in Gram-negative and Gram-positive bacteria. *Molecular Microbiology* 107 (4): 455–471. https://doi.org/10.1111/mmi.13896.

Grohmann, E., Keller, W., and Muth, G. (2017b). Mechanisms of conjugative transfer and Type IV secretion-mediated effector transport in Gram-positive bacteria. *Current Topics in Microbiology and Immunology* 413: 115–141. https://doi.org/10.1007/978-3-319-75241-9_5.

Guglielmini, J., Quintais, L., Garcillán-Barcia, M.P. et al. (2011). The repertoire of ICE in prokaryotes underscores the unity, diversity, and ubiquity of conjugation. *PLoS Genetics* 7 (8): https://doi.org/10.1371/journal.pgen.1002222.

Hacker, J. and Carniel, E. (2001). Ecological fitness, genomic islands and bacterial pathogenicity. *EMBO Reports* 2 (5): 376–381. https://doi.org/10.1093/embo-reports/kve097.

Hata, M., Suzuki, M., Matsumoto, M. et al. (2005). Cloning of a novel gene for quinolone resistance from a transferable plasmid in *Shigella flexneri* 2b. *Antimicrobial Agents and Chemotherapy* 49 (2): 801–803. https://doi.org/10.1128/AAC.49.2.801.

Hochhut, B., Beaber, J.W., Woodgate, R., and Waldor, M.K. (2001). Formation of chromosomal tandem arrays of the SXT element and R391, two conjugative chromosomally integrating elements that share an attachment site. *Journal of Bacteriology* 183 (4): 1124–1132. https://doi.org/10.1128/JB.183.4.1124-1132.2001.

Hochhut, B. and Waldor, M.K. (1999). Site-specific integration of the conjugal *Vibrio cholerae* SXT element into *PrfC*. *Molecular Microbiology* 32 (1): 99–110. https://doi.org/10.1046/j.1365-2958.1999.01330.x.

van Hoek, A.H., Mevius, D.J., Guerra, B. et al. (2011). Acquired antibiotic resistance genes: an overview. *Frontiers in Microbiology* 2 (203): https://doi.org/10.3389/fmicb.2011.00203.

Jamrozy, D., Coll, F., Mather, A.E. et al. (2017). Evolution of mobile genetic element composition in an epidemic methicillin- resistant *Staphylococcus aureus*: temporal changes correlated with frequent loss and gain events. *BMC Genomics* 18 (1): 1–12. https://doi.org/10.1186/s12864-017-4065-z.

Jansen, G. and Aktipis, C.A. (2014). Resistance is mobile: the accelerating evolution of mobile genetic elements encoding resistance. *Journal of Evolutionary Medicine* 2: 3–5. https://doi.org/10.4303/jem/235873.

Jaworski, D.D. and Clewell, D.B. (1995). A functional origin of transfer (*oriT*) on the conjugative transposon *Tn*916. *Journal of Bacteriology* 177 (22): 6644–6651. https://doi.org/10.1128/jb.177.22.6644-6651.1995.

Johnson, C.M. and Grossman, A.D. (2015). Integrative and conjugative elements (ICEs): what they do and how they work. *Annual Review of Genetics* 49 (1): 577–601. https://doi.org/10.1146/annurev-genet-112414-055018.

Jones, M.J., Donegan, N.P., Mikheyeva, I.V., and Cheung, A.L. (2015). Improving transformation of *Staphylococcus aureus* belonging to the CC1, CC5 and CC8 clonal complexes. *PLoS One* 10 (3): 1–14. https://doi.org/10.1371/journal.pone.0119487.

Juhas, M., Van Der Meer, J.R., Gaillard, M. et al. (2009). Genomic islands: tools of bacterial horizontal gene transfer and evolution. *FEMS Microbiology Reviews* https://doi.org/10.1111/j.1574-6976.2008.00136.x.

Karam, G., Chastre, J., Wilcox, M.H., and Vincent, J.L. (2016). Antibiotic strategies in the era of multidrug resistance. *Critical Care* https://doi.org/10.1186/s13054-016-1320-7.

Kelly, B.G., Vespermann, A., and Bolton, D.J. (2009). The role of horizontal gene transfer in the evolution of selected foodborne bacterial pathogens. *Food and Chemical Toxicology* https://doi.org/10.1016/j.fct.2008.02.006.

Kohanski, M.A., Dwyer, D.J., and Collins, J.J. (2010). How antibiotics kill bacteria: from targets to networks. *Nature Reviews Microbiology* https://doi.org/10.1038/nrmicro2333.

Kohler, V., Vaishampayan, A., and Grohmann, E. (2018). Broad-host-range Inc18 plasmids: occurrence, spread and transfer mechanisms. *Plasmid* https://doi.org/10.1016/J.PLASMID.2018.06.001.

Kristich, C.J., Rice, L.B., and Arias, C.A. (2014). Enterococcal infection-treatment and antibiotic resistance. In: *Enterococci: From Commensals to Leading Causes of Drug Resistant Infection* (ed. M.S. Gilmore, D.B. Clewell, Y. Ike, et al.). Boston: Massachusetts Eye and Ear Infirmary.

Lee, H.-Y., Chang, R.-C., Su, L.-H. et al. (2012). Wide spread of Tn*2006* in an AbaR4-type resistance island among carbapenem-resistant *Acinetobacter baumannii* clinical isolates in Taiwan. *International Journal of Antimicrobial Agents* 40 (2): 163–167. https://doi.org/10.1016/j.ijantimicag.2012.04.018.

Lindsay, J.A., Knight, G.M., Budd, E.L., and McCarthy, A.J. (2012). Shuffling of mobile genetic elements (MGEs) in successful healthcare-associated MRSA (HA-MRSA). *Mobile Genetic Elements* 2 (5): 1–5. https://doi.org/10.4161/mge.22085.

Lindsey, R.L., Fedorka-Cray, P.J., Frye, J.G., and Meinersmann, R.J. (2009). Inc A/C plasmids are prevalent in multidrug-resistant *Salmonella enterica* isolates. *Applied and Environmental Microbiology* 75 (7): 1908–1915. https://doi.org/10.1128/AEM.02228-08.

Llanos-Chea, A., Citorik, R.J., Nickerson, K.P. et al. (2018). Bacteriophage therapy testing against *Shigella flexneri* in a novel human intestinal organoid-derived infection model. *Journal of Pediatric Gastroenterology and Nutrition* https://doi.org/10.1097/MPG.0000000000002203.

Ma, L., Li, A.-d., Yin, X.-l., and Zhang, T. (2017). The prevalence of integrons as the carrier of antibiotic resistance genes in natural and man-made environments. *Environmental Science and Technology* 51 (10): 5721–5728. https://doi.org/10.1021/acs.est.6b05887.

Madsen, J.S., Burmølle, M., Hansen, L.H., and Sørensen, S.J. (2012). The interconnection between biofilm formation and horizontal gene transfer. *FEMS Immunology and Medical Microbiology* https://doi.org/10.1111/j.1574-695X.2012.00960.x.

Maeda, K., Nojiri, H., Shintani, M. et al. (2003). Complete nucleotide sequence of carbazole/dioxin-degrading Plasmid pCAR1 in *Pseudomonas resinovorans* strain CA10 indicates its mosaicity and the presence of large catabolic transposon Tn*4676*. *Journal of Molecular Biology* 326 (1): 21–33. https://doi.org/10.1016/S0022-2836(02)01400-6.

Malachowa, N. and DeLeo, F.,.R. (2010). Mobile genetic elements of *Staphylococcus aureus*. *Cellular and Molecular Life Sciences* 67 (18): 3057–3071. https://doi.org/10.1007/s00018-010-0389-4.

Mannion, A.J., Martin, H.R., Shen, Z. et al. (2018). Plasmid-mediated quinolone resistance in *Shigella flexneri* isolated from macaques. *Frontiers in Microbiology* 9 (311): https://doi.org/10.3389/fmicb.2018.00311.

Mégraud, F. (2013). Current recommendations for *Helicobacter pylori* therapies in a world of evolving resistance. *Gut Microbes* 4 (6): 541–548. https://doi.org/10.4161/gmic.25930.

Moosavian, M. and Rahimzadeh, M. (2015). Molecular detection of metallo-β-lactamase genes bla_{IMP-1}, bla_{VIM-2}, and bla_{SPM-1} in imipenem resistant *Pseudomonas aeruginosa* isolated from clinical specimens in teaching hospitals of Ahvaz, Iran. *Iranian Journal of Microbiology* 7 (1): 2–6.

Murata, T., Ohnishi, M., Ara, T. et al. (2002). Complete nucleotide sequence of Plasmid Rts1: implications for evolution of large plasmid genomes. *Journal of Bacteriology* 184 (12): 3194–3202. https://doi.org/10.1128/JB.184.12.3194-3202.2002.

Ochman, H., Lawrence, J.G., and Groisman, E.A. (2000). Lateral gene transfer and the nature of bacterial evolution. *Nature* 405 (18): 299–304. https://doi.org/10.1021/jf202185m.

Pagano, M., Francisco, A., and Luis, A. (2016). Mobile genetic elements related to carbapenem resistance in *Acinetobacter baumannii*. *Brazilian Journal of Microbiology* 47 (4): 785–792. https://doi.org/10.1016/j.bjm.2016.06.005.

Pai, H., Choi, E.-H., Lee, H.-J. et al. (2001). Identification of CTX-M-14 extended-spectrum β-lactamase in clinical isolates of *Shigella sonnei, Escherichia coli*, and *Klebsiella pneumoniae* in Korea. *Journal of Clinical Microbiology* 39 (10): 3747–3749. https://doi.org/10.1128/JCM.39.10.3747.

Partridge, S.R., Kwong, S.M., Firth, N., and Jensen, S.O. (2018). Mobile genetic elements associated with antimicrobial resistance. *Clinical Microbiology Reviews* 31 (4): 1–61.

Partridge, S.R., Tsafnat, G., Coiera, E., and Iredell, J.R. (2009). Gene cassettes and cassette arrays in mobile resistance integrons. *FEMS Microbiology Reviews* https://doi.org/10.1111/j.1574-6976.2009.00175.x.

Patel, A.L., Chaudhry, U., Sachdev, D. et al. (2011). An insight into the drug resistance profile & mechanism of drug resistance in *Neisseria gonorrhoeae*. *Indian Journal of Medical Research* 134 (4): 419–431.

Perry, J.A. and Wright, G.D. (2013). The antibiotic resistance 'Mobilome': searching for the link between environment and clinic. *Frontiers in Microbiology* 4: 138. https://doi.org/10.3389/fmicb.2013.00138.

Pettis, G.S. (2018). Spreading the news about the novel conjugation mechanism in *Streptomyces* bacteria. *Environmental Microbiology Reports* https://doi.org/10.1111/1758-2229.12659.

Poirel, L., Decousser, J.-W., and Nordmann, P. (2005). ISEcp1B -mediated transposition of bla_{CTX-M} in *Escherichia coli*. *Antimicrobial Agents and Chemotherapy* 49 (1): 1–4. https://doi.org/10.1128/AAC.49.1.447.

Poole, T.L., Edrington, T.S., Brichta-Harhay, D.M. et al. (2009). Conjugative transferability of the A/C plasmids from *Salmonella enterica* isolates that possess or lack bla_{CMY} in the A/C plasmid backbone. *Foodborne Pathogens and Disease* 6 (10): 1185–1194. https://doi.org/10.1089/fpd.2009.0316.

Popowska, M. and Krawczyk-Balska, A. (2013). Broad-host-range IncP-1 plasmids and their resistance potential. *Frontiers in Microbiology*. https://doi.org/10.3389/fmicb.2013.00044.

Rauch, P.J.G. and De Vos, W.M. (1992). Characterization of the novel nisin-sucrose conjugative transposon Tn*5276* and its insertion in *Lactococcus lactis*. *Journal of Bacteriology* 174 (4): 1280–1287. https://doi.org/10.1128/jb.174.4.1280-1287.1992.

Reuther, J., Gekeler, C., Tiffert, Y. et al. (2006). Unique conjugation mechanism in mycelial *Streptomycetes*: a DNA-binding ATPase translocates unprocessed plasmid DNA at the hyphal tip. *Molecular Microbiology* 61 (2): 436–446. https://doi.org/10.1111/j.1365-2958.2006.05258.x.

Roberts, A.P. and Mullany, P. (2009). A modular master on the move: the Tn*916* *f*amily of mobile genetic elements. *Trends in Microbiology* https://doi.org/10.1016/j.tim.2009.03.002.

Rowe-Magnus, D.A. and Mazel, D. (2001). Integrons: natural tools for bacterial genome evolution. *Current Opinion in Microbiology* https://doi.org/10.1016/S1369-5274(00)00252-6.

Ruppé, É., Woerther, P.L., and Barbier, F. (2015). Mechanisms of antimicrobial resistance in Gram-negative bacilli. *Annals of Intensive Care* 5 (1): https://doi.org/10.1186/s13613-015-0061-0.

Sanchez, D.G. and de Melo, F.M. (2018). Detection of different β-lactamases encoding genes, including Bla_{NDM}, and plasmid-mediated quinolone resistance genes in different water sources from Brazil. *Environmental Monitoring and Assessment* 190 (407): https://doi.org/10.1007/s10661-018-6801-5.

Santoro, F., Vianna, M.E., and Roberts, A.P. (2014). Variation on a theme; an overview of the Tn*916*/Tn*1545* family of mobile genetic elements in the oral and nasopharyngeal streptococci. *Frontiers in Microbiology* 5: 535. https://doi.org/10.3389/fmicb.2014.00535.

Sharifahmadian, M. and Baron, C. (2017). Type IV secretion in *Agrobacterium tumefaciens* and development of specific inhibitors. In: *Type IV Secretion in Gram-Negative and Gram-Positive Bacteria. Current Topics in Microbiology and Immunology*, vol. 413 (ed. S. Backert and E. Grohmann), 169–186. Cham: Springer.

Shoemaker, N.B., Vlamakis, H., Hayes, K., and Salyers, A.A. (2001). Evidence for extensive resistance gene transfer among *Bacteroides* spp. and among *Bacteroides* and other genera in the human colon. *Applied and Environmental Microbiology* 67 (2): 561–568. https://doi.org/10.1128/AEM.67.2.561-568.2001.

Showsh, S.A. and Andrews, R.E. (1992). Tetracycline enhances Tn*916*-mediated conjugal transfer. *Plasmid* 28 (3): 213–224. https://doi.org/10.1016/0147-619X(92)90053-D.

Sibold, C., Henrichsen, J., König, A. et al. (1994). Mosaic *pbpX* genes of major clones of penicillin-resistant *Streptococcus pneumoniae* have evolved from *pbp*X genes of a penicillin-sensitive *Streptococcus oralis*. *Molecular Microbiology* 12 (6): 1013–1023.

Sommer, M.O.A., Dantas, G., and Church, G.M. (2009). Functional characterization of the antibiotic resistance reservoir in the human microflora. *Science* 325 (5944): 1128–1131. https://doi.org/10.1126/science.1176950.

Sørensen, S.J., Bailey, M., Hansen, L.H. et al. (2005). Studying plasmid horizontal transfer in situ: a critical review. *Nature Reviews Microbiology* https://doi.org/10.1038/nrmicro1232.

Soucy, S.M., Huang, J., and Gogarten, J.P. (2015). Horizontal gene transfer: building the web of life. *Nature Reviews Genetics* https://doi.org/10.1038/nrg3962.

Stecher, B., Denzler, R., Maier, L. et al. (2012). Gut inflammation can boost horizontal gene transfer between pathogenic and commensal *Enterobacteriaceae*. *Proceedings of the National Academy of Sciences* 109 (4): https://doi.org/10.1073/pnas.1113246109.

Stewart, G.J. and Carlson, C.A. (1986). The biology of natural transformation. *Annual Review of Microbiology* 40 (1): 211–231. https://doi.org/10.1146/annurev.mi.40.100186.001235.

Stokes, H.W. and Gillings, M.R. (2011). Gene flow, mobile genetic elements and the recruitment of antibiotic resistance genes into Gram-negative pathogens. *FEMS Microbiology Reviews* 35: 790–819. https://doi.org/10.1111/j.1574-6976.2011.00273.x.

Sultan, I., Rahman, S., Jan, A.T. et al. (2018). Antibiotics, resistome and resistance mechanisms: a bacterial perspective. *Frontiers in Microbiology* 9: 2066. https://doi.org/10.3389/fmicb.2018.02066.

Tacconelli, E., Carrara, E., Savoldi, A. et al. (2017). Articles discovery, research, and development of new antibiotics: the WHO priority list of antibiotic-resistant bacteria and tuberculosis. *The Lancet Infectious Diseases* https://doi.org/10.1016/S1473-3099(17)30753-3.

Tamminen, M., Virta, M., Fani, R., and Fondi, M. (2012). Large-scale analysis of plasmid relationships through gene-sharing networks. *Molecular Biology and Evolution* 29 (4): 1225–1240. https://doi.org/10.1093/molbev/msr292.

Thoma, L. and Muth, G. (2015). The conjugative DNA-transfer apparatus of *Streptomyces. International Journal of Medical Microbiology* https://doi.org/10.1016/j.ijmm.2014.12.020.

Thoma, L. and Muth, G. (2016). Conjugative DNA-transfer in *Streptomyces*, a mycelial organism. *Plasmid* https://doi.org/10.1016/j.plasmid.2016.09.004.

Thomas, C.M. and Nielsen, K.M. (2005). Mechanisms of, and barriers to, horizontal gene transfer between bacteria. *Nature Reviews Microbiology* https://doi.org/10.1038/nrmicro1234.

Ugboko, H. and De, N. (2014). Review article mechanisms of antibiotic resistance in *Salmonella typhi. International Journal of Current Microbiology and Applied Sciences* 3 (12): 461–476.

Von Wintersdorff, C.J.H., Penders, J., van Niekerk, J.M. et al. (2016). Dissemination of antimicrobial resistance in microbial ecosystems through horizontal gene transfer. *Frontiers in Microbiology* 7 (173): 1–10. https://doi.org/10.3389/fmicb.2016.00173.

WHO (2017). Meet WHO's dirty dozen: the 12 bacteria for which new drugs are most urgently needed. *Science* https://doi.org/10.1126/science.aal0829.

Wozniak, R.A.F. and Waldor, M.K. (2010). Integrative and conjugative elements: mosaic mobile genetic elements enabling dynamic lateral gene flow. *Nature Reviews Microbiology* https://doi.org/10.1038/nrmicro2382.

Yelin, I. and Kishony, R. (2018). Antibiotic resistance. *Cell* https://doi.org/10.1016/j.cell.2018.02.018.

Zhu, W., Lark, N.C., McDougal, L.K. et al. (2008). Vancomycin resistant *Staphylococcus aureus* isolates associated with Inc18-like *vanA* plasmids in Michigan. *Antimicrobial Agents and Chemotherapy* 52 (2): 452–457. https://doi.org/10.1128/AAC.00908-07.

Zinder, N.D. and Lederberg, J. (1952). Genetic exchange in *Salmonella. Journal of Bacteriology* 64 (5): 679–699.

8

Antimicrobial Efflux Pumps

Manuel F. Varela

Eastern New Mexico University, Portales, NM, USA

8.1 Bacterial Antimicrobial Efflux Pumps

8.1.1 Active Drug Efflux Systems

Bacterial microorganisms that are causative agents of infectious disease in human and animal populations exploit two chief antimicrobial efflux systems that serve to extrude inhibitory antimicrobial solutes from their internal cellular milieu, effectively compromising the efficacy of clinical chemotherapy (Floyd et al. 2013; Kumar and Varela 2013). In general, active drug efflux proteins fall into two transporter types, known as primary and secondary active transport (Levy 2002). Both types of drug efflux systems consist of integral membrane transporter proteins that bind to antibacterial solutes and actively catalyze their translocation through the bacterial membrane in an outward direction. These active drug transporters serve the bacteria to reduce the intracellular concentrations of internally located inhibitory agents from the bacterial cell (Pu et al. 2017). In many cases, active drug efflux systems confer bacterial resistance to multiple antimicrobial drugs, producing a serious public health concern on a worldwide basis (Levy 2001).

8.1.1.1 Primary Active Drug Transporters

The first active drug efflux system is frequently referred to as primary active transport, and these systems take advantage of the biological energy stored in the form of adenosine 5′-triphosphate (ATP) and its associated hydrolysis to transport antimicrobial agents across the membrane to the extracellular location (Davidson and Maloney 2007). One well-studied primary drug efflux class consists of the highly related ATP-binding cassette (ABC) superfamily of transporters (Kathawala et al. 2015). In terms of overall structure, the ABC

Antibiotic Drug Resistance, First Edition. Edited by José-Luis Capelo-Martínez and Gilberto Igrejas.

transporters harbor at least two or more sets each of 6-helical hydrophobic membrane spanning domains and two highly conserved intracellularly located hydrophilic nucleotide-binding domains (Nikaido and Hall 1998). Other elements of the ABC drug transporters include an ATP signature structural motif, conserved Walker A and Walker B sequence and structural motifs, and loop structures, denoted as Q, D, and H loops (Altenberg 2004).

8.1.1.2 Bacterial ABC Drug Transporters

Several key bacterial pathogens possess ABC drug transporters. The *Staphylococcus aureus* bacteria, serious clinical pathogens in their own right, harbor several known ABC transporters. One of these, Sav1866, confers resistance to vinblastine, tetraphenylphosphonium, Hoechst 33342, and ethidium bromide and has had its crystal structure determined to a 3 Å resolution (Dawson and Locher 2006). Another drug efflux system of the ABC transporter type, Isa(E), bestows resistance to lincosamides, streptogramin A, and the pleuromutilins in both methicillin-susceptible and methicillin-resistant strains of *S. aureus* (Wendlandt et al. 2013). The cholera-causing *Vibrio cholerae* bacteria express the ABC efflux pump called VcaM, which transports doxorubicin, daunomycin, and fluoroquinolones, such as norfloxacin and ciprofloxacin, conferring resistance to these agents (Huda et al. 2003b; Lu et al. 2018). In *Escherichia coli*, one example of an intensively studied ABC transporter is the MacB pump, which transports primarily members of the macrolide class of antimicrobial agents (Kobayashi et al. 2001). Interestingly, MacB associates with an accessory secondary active transporter, called MacA, a peripheral membrane associated protein of the membrane fusion superfamily, and TolC, an outer membrane protein, all of which come together to confer bacterial drug resistance (Kobayashi et al. 2001; Lin et al. 2009). A similar tripartite ABC drug transporter system also resides in key *Salmonella enterica* bacteria, although this particular system is less well understood (Bogomolnaya et al. 2013). In the respiratory pathogen *Streptococcus pneumoniae*, an ABC drug exporter system called PatAB was discovered to confer resistance against fluoroquinolone agents (Baylay et al. 2015). Another tripartite ABC drug efflux system, FuaABC, was found in the pathogen *Stenotrophomonas maltophilia*, and its transport system was demonstrated to provide resistance against fusaric acid (Hu et al. 2012). The causative agent of tuberculosis, *Mycobacterium tuberculosis*, was found to contain an ABC transporter, called Rv1218c, which confers resistance to a variety of antimicrobials, like the pyridones, pyrroles, biaryl-piperazines, and bisanilino-pyrimidines (Balganesh et al. 2010; Wang et al. 2013). These ABC transporters and others will undoubtedly continue to be important model systems for investigations of solute transport involving antimicrobial agents not only in bacterial microorganisms but also in conferring resistance to anticancer chemotherapeutics in human medicine (Li et al. 2016).

8.1.2 Secondary Active Drug Transporters

The second active drug efflux system is commonly called secondary active transport. These secondary active drug transporters utilize the biological energy inherent within ion motive forces to transport inhibitory agents across the membrane to reduce their cellular concentrations, providing resistance multiple drugs. Such secondary transporters have been referred to as antiporters and exchangers. Currently, several secondary active drug transporter families have been established. These include the major facilitator superfamily (MFS) (Saier et al. 1999), the resistance-nodulation-cell division (RND) superfamily (Ruggerone et al. 2013), the multidrug and toxic compound extrusion (MATE) superfamily (Kuroda and Tsuchiya 2009), the small multidrug resistance (SMR) superfamily (Chung and Saier. 2001), and the recently described proteobacterial antimicrobial compound efflux (PACE) family (Hassan et al. 2018). Important key members within each of these secondary active efflux pump families are considered below.

8.1.2.1 Bacterial MFS Drug Efflux Pumps

The MFS of solute transporters is currently composed of tens of thousands of members from across all living taxa (Pao et al. 1998). In general, these MFS transport systems harbor 12 or 14 transmembrane domains and consist of both passive and secondary active transporters with diverse substrates and highly conserved structures and sequence motifs (Law et al. 2008; Ranaweera et al. 2015; Kumar et al. 2016b; Kakarla et al. 2017b). Because many of the MFS proteins function to extrude antimicrobial agents from microbial pathogens (Saidijam et al. 2006), these drug and multidrug efflux pumps are considered good targets for modulation (Kumar et al. 2013, 2016a).

The first MFS drug efflux pump to be discovered occurred in the laboratory of Levy in which they had found active bacterial efflux of the tetracyclines from *E. coli* (Levy 2002). Another drug efflux pump from *E. coli* called MdfA (previously known as CmlA) confers multiple drug efflux and has been intensively studied, especially at the structural level (Nagarathinam et al. 2017; Yardeni et al. 2017). Together, with the structural features of MdfA, along with the crystal structure for the *E. coli* YajR, which was also solved to high resolution (Jiang et al. 2013), these known transporter structures have served to help demonstrate the functional importance of highly conserved sequence motifs that are found in many members of the MFS (Kakarla et al. 2017b). A related efflux pump, CraA, from *Acinetobacter baumannii* confers resistance to the chloramphenicols in this bacterium (Roca et al. 2009). In *S. aureus*, NorA, QacA, and LmrS were all demonstrated at the physiological level to be effective in conferring multidrug efflux of a variety of structurally unrelated antimicrobial agents (Brown and Skurray 2001; Floyd et al. 2010; Schindler and Kaatz 2016). In particular, NorA has been shown to be an excellent target for modulation by efflux pump inhibitors (Zhang and Ma 2010). In our laboratory, we demonstrated that the natural spice compound cumin effectively inhibited drug transport activity from LmrS (Kakarla et al. 2017a). Because these MFS drug efflux mechanisms share homologous sequences and structures, it is predicted that they share similar mechanisms of transport across the membrane. Thus, transporters in this large superfamily will become important as investigators seek to elucidate these mechanistic commonalities for modulation of clinical drug and multidrug resistances.

8.1.2.2 Bacterial RND Multidrug Efflux Pumps

The RND antimicrobial efflux pump systems represent another large superfamily of transporters (Nikaido 2018). Many of these RND efflux pumps are driven by a proton motive force as their prime source of energy (Nikaido and Takatsuka 2009). Several key RND transporters assemble into a tripartite complex, consisting of an integral cytoplasmic membrane component, a periplasm-associated fusion protein, and an outer membrane channel protein (Misra and Bavro 2009).

Two of the most extensively studied RND drug efflux pumps are the AcrB multidrug efflux pump from *E. coli* (Nikaido and Zgurskaya 2001; Pos 2009) and the MexB multidrug efflux pump from *Pseudomonas aeruginosa* (Poole 2001; Blair and Piddock 2009). Both of these RND drug transporters have been structurally crystallized to high resolutions and extensively studied at the mechanistic level (Murakami et al. 2002; Sennhauser et al. 2009). The AcrB pump is part of an overall multimeric complex, AcrAB-TolC, which is composed also of the periplasmic fusion protein, AcrA, and the outer membrane associated channel, called TolC, all residing within the Gram-negative cell wall of the *E. coli* bacterium (Blair and Piddock 2009). Likewise, the MexAB-OprM efflux system is structurally organized in a similar fashion (Poole 2001; Puzari and Chetia 2017). Together, these two RND antimicrobial transporter systems constitute suitable targets for efflux inhibition in bacteria (Wang et al. 2016). Other RND efflux pumps include the MtrCDE system from *Neisseria gonorrhoeae* specific for structurally distinctive hydrophobic chemicals (Rouquette et al. 1999; Johnson and Shafer 2015) and the CusCBA system from *E. coli* conferring heavy metal resistance (Long et al. 2012). The crystal structures of both MtrD and CusA have been determined at about 3.5 Å resolution (Long et al. 2010; Bolla et al. 2014). A related RND zinc-specific pump called ZneA from *Cupriavidus metallidurans* has been crystallized, and its structure was revealed to 3 Å resolution (Pak et al. 2013). Since the RND transporters are found in both prokaryotic and all eukaryotic organisms, it will be interesting to gain a finer understanding of the tripartite assemblages in organisms without Gram-negative walls.

8.1.2.3 Bacterial MATE Drug Pumps

The transporters belonging to the MATE superfamily in bacteria provide multidrug resistance functions (Kuroda and Tsuchiya 2009). The members within the MATE superfamily have been subcategorized into three smaller subfamilies (Hvorup et al. 2003). Family 1 primarily harbors MATE drug efflux systems inherent in bacterial microorganisms (Hvorup et al. 2003). Family 2 members are composed of MATE transporter systems found mainly in eukaryotic organisms (Hvorup et al. 2003). Members of MATE family 3 are found chiefly in the eubacteria (bacteria) and the archaebacteria (archaea), which are all prokaryotic organisms (Hvorup et al. 2003). In general, the MATE transporters are energized by ion motive forces and work via an antiport system in an ion–substrate exchange manner (Huda et al. 2003b; Otsuka et al. 2005; Steed et al. 2013; Song et al. 2014). The primary amino acid sequences of the MATE transporters vary in length between 400 and 1000 residues (Chen et al. 2002; Huda et al. 2003a). The secondary structures consist of 12 transmembrane α-helices that traverse the membrane in a zigzag manner (He et al. 2010; Zhang et al. 2012).

The tertiary structures for several key bacterial MATE transporters have been elucidated (Lu et al. 2013a, b; Tanaka et al. 2013; Symersky et al. 2015). The first of these MATE transporter crystal structures to be elucidated is NorM from the bacterium *V. cholerae,* also denoted as NorM-VC (He et al. 2010). The NorM structure was found to devoid of bound substrate and in an open configuration facing the periplasmic side of the cytoplasmic membrane (He et al. 2010). Shortly after the elucidation of the NorM-VC structure, other MATE transporter structures were reported, for NorM-NG from *N. gonorrhoeae* (Lu 2016), DinF from *Bacillus halodurans* (Lu et al. 2013a), and PfMATE from *Pyrococcus furiosus* (Tanaka et al. 2013). These MATE transporters shared similarities in structures with a pseudo-twofold axis of symmetry involving two helical bundles composed of helices 1 through 6 and helices 1 through 12 (Lu 2016). Studies of highly conserved amino acid residues within the MATE transporters have served as a basis for the development of a mechanistic model involving an ion-driven solute transport system across in the membrane through the transporters (Kuroda and Tsuchiya 2009; Nies et al. 2016). Future work will be interesting to determine to what extent other related MATE transporters, or even non-MATE transporters, harbor the current ion-coupled substrate transport models, as well. In any case, members of the MATE transporter superfamily will also certainly constitute clinically appropriate bacterial and cancer targets for putative modulation.

8.1.2.4 SMR Superfamily of Drug Efflux Pumps

The SMR family of antimicrobial transporters belongs to the overall larger drug/metabolite transporter (DMT) superfamily (Paulsen et al. 1996; Jack et al. 2001). As a whole, members of the SMR family confer bacterial resistance to antibiotic and antiseptic agents, especially those agents that fall under the quaternary ammonium compound (QAC) class of antimicrobial agents (Paulsen et al. 1996). The SMR pumps are also secondary active transporters, being driven by ion motive forces, such as that provided by protons (Schuldiner 2009; Dastvan et al. 2016). Additionally, transporters belonging to the SMR are small in terms of their primary structural features, ranging between 100 and 200 amino acid residues in their polypeptide chains (Grinius and Goldberg 1994). Consequently, the relatively short polypeptides possessed by transporters of the SMR family leave room for only approximately four transmembrane segments to be present per polypeptide, requiring that these monomer polypeptides assemble into larger multimers in order to form complete structural configurations that accommodate solute transport across the membrane (Paulsen et al. 1996; Nara et al. 2007; Bay et al. 2008). Topological studies have indicated that the SMR transporters consist of either homooligomers or heterooligomers, depending on the nature of the particular transport system of interest (Nara et al. 2007; Kolbusz et al. 2010; Stockbridge et al. 2013). Interestingly, certain members of the SMR family have been implicated to assemble into a multipartite complex involving OmpW, an outer membrane protein in *E. coli* (Beketskaia et al. 2014). Extensive effort has been directed upon the SMR efflux pumps from a structure–function perspective (Wong et al. 2014; Padariya et al. 2015; Qazi et al. 2015), substrate-binding site analyses (Bay and Turner 2009; Wong et al. 2014; Banigan et al. 2015; Padariya et al. 2015; Dastvan et al. 2016), mechanistic studies of substrate translocation (Schuldiner 2009; Wong et al. 2014), effects of membrane lipid composition on transport activities (Bay and Turner 2013; Mors et al. 2013), and in terms of transport modulation (Ovchinnikov et al. 2018). One of the best studied transporters of the SMR family is EmrE, from *E. coli* (Schuldiner 2009; Padariya et al. 2015; Qazi et al. 2015; Dastvan et al. 2016; Ovchinnikov et al. 2018). As such, EmrE is an excellent model system for efflux inhibitor studies (Nasie et al. 2012; Dutta et al. 2014; Ovchinnikov et al. 2018).

8.1.2.5 PACE Family of Drug Transporters

Transporters of the PACE family represent a recent development in solute transport biology (Hassan et al. 2015, 2018). A key member of the PACE family to emerge, AceI, had originated from *A. baumannii*, which has been demonstrated to confer multidrug resistance in host cells (Hassan et al. 2013). Homologues of AceI and belonging to the PACE family have been located in a variety of bacterial species (Hassan et al. 2013, 2015). Interestingly, it was reported that chlorhexidine, a known substrate for AceI, enhanced the expression of the *aceI* gene and that exposure to the inducer provided increased levels of resistance to chlorhexidine in *Acinetobacter baylyi* host cells (Fuangthong et al. 2011). It is anticipated that as newer investigations are reported for members of the PACE family of drug transporters, these transporter systems will become increasingly important both at the biomedical science research and at clinical medicine levels.

References

Altenberg, G.A. (2004). Structure of multidrug-resistance proteins of the ATP-binding cassette (ABC) superfamily. *Curr. Med. Chem. Anticancer Agents* 4 (1): 53–62.

Balganesh, M., Kuruppath, S., Marcel, N. et al. (2010). Rv1218c, an ABC transporter of *Mycobacterium tuberculosis* with implications in drug discovery. *Antimicrob. Agents Chemother.* 54 (12): 5167–5172.

Banigan, J.R., Gayen, A., Cho, M.K., and Traaseth, N.J. (2015). A structured loop modulates coupling between the substrate-binding and dimerization domains in the multidrug resistance transporter EmrE. *J. Biol. Chem.* 290 (2): 805–814.

Bay, D.C., Rommens, K.L., and Turner, R.J. (2008). Small multidrug resistance proteins: a multidrug transporter family that continues to grow. *Biochim. Biophys. Acta* 1778 (9): 1814–1838.

Bay, D.C. and Turner, R.J. (2009). Diversity and evolution of the small multidrug resistance protein family. *BMC Evol. Biol.* 9: 140.

Bay, D.C. and Turner, R.J. (2013). Membrane composition influences the topology bias of bacterial integral membrane proteins. *Biochim. Biophys. Acta* 1828 (2): 260–270.

Baylay, A.J., Ivens, A., and Piddock, L.J. (2015). A novel gene amplification causes upregulation of the PatAB ABC transporter and fluoroquinolone resistance in *Streptococcus pneumoniae*. *Antimicrob. Agents Chemother.* 59 (6): 3098–3108.

Beketskaia, M.S., Bay, D.C., and Turner, R.J. (2014). Outer membrane protein OmpW participates with small multidrug resistance protein member EmrE in quaternary cationic compound efflux. *J. Bacteriol.* 196 (10): 1908–1914.

Blair, J.M. and Piddock, L.J. (2009). Structure, function and inhibition of RND efflux pumps in Gram-negative bacteria: an update. *Curr. Opin. Microbiol.* 12 (5): 512–519.

Bogomolnaya, L.M., Andrews, K.D., Talamantes, M. et al. (2013). The ABC-type efflux pump MacAB protects *Salmonella enterica* serovar typhimurium from oxidative stress. *MBio* 4 (6): e00630-13.

Bolla, J.R., Su, C.C., Do, S.V. et al. (2014). Crystal structure of the *Neisseria gonorrhoeae* MtrD inner membrane multidrug efflux pump. *PLoS One* 9 (6): e97903.

Brown, M.H. and Skurray, R.A. (2001). Staphylococcal multidrug efflux protein QacA. *J. Mol. Microbiol. Biotechnol.* 3 (2): 163–170.

Chen, J., Morita, Y., Huda, M.N. et al. (2002). VmrA, a member of a novel class of Na(+)-coupled multidrug efflux pumps from *Vibrio parahaemolyticus*. *J. Bacteriol.* 184 (2): 572–576.

Chung, Y.J. and Saier, M.H. Jr. (2001). SMR-type multidrug resistance pumps. *Curr. Opin. Drug Discov. Devel.* 4 (2): 237–245.

Dastvan, R., Fischer, A.W., Mishra, S. et al. (2016). Protonation-dependent conformational dynamics of the multidrug transporter EmrE. *Proc. Natl. Acad. Sci. U. S. A.* 113 (5): 1220–1225.

Davidson, A.L. and Maloney, P.C. (2007). ABC transporters: how small machines do a big job. *Trends Microbiol.* 15 (10): 448–455.

Dawson, R.J. and Locher, K.P. (2006). Structure of a bacterial multidrug ABC transporter. *Nature* 443 (7108): 180–185.

Dutta, S., Morrison, E.A., and Henzler-Wildman, K.A. (2014). Blocking dynamics of the SMR transporter EmrE impairs efflux activity. *Biophys. J.* 107 (3): 613–620.

Floyd, J.L., Smith, K.P., Kumar, S.H. et al. (2010). LmrS is a multidrug efflux pump of the major facilitator superfamily from *Staphylococcus aureus*. *Antimicrob. Agents Chemother.* 54 (12): 5406–5412.

Floyd, J.T., Kumar, S., Mukherjee, M.M. et al. (2013). A review of the molecular mechanisms of drug efflux in pathogenic bacteria: a structure-function perspective. In: *Recent Research Developments in Membrane Biology* (ed. P. Shankar), 15–66. Kerala, India: Research Signpost.

Fuangthong, M., Julotok, M., Chintana, W. et al. (2011). Exposure of *Acinetobacter baylyi* ADP1 to the biocide chlorhexidine leads to acquired resistance to the biocide itself and to oxidants. *J. Antimicrob. Chemother.* 66 (2): 319–322.

Grinius, L.L. and Goldberg, E.B. (1994). Bacterial multidrug resistance is due to a single membrane protein which functions as a drug pump. *J. Biol. Chem.* 269 (47): 29998–30004.

Hassan, K.A., Jackson, S.M., Penesyan, A. et al. (2013). Transcriptomic and biochemical analyses identify a family of chlorhexidine efflux proteins. *Proc. Natl. Acad. Sci. U. S. A.* 110 (50): 20254–20259.

Hassan, K.A., Liu, Q., Elbourne, L.D.H. et al. (2018). Pacing across the membrane: the novel PACE family of efflux pumps is widespread in Gram-negative pathogens. *Res. Microbiol.* 169 (7–8): 450–454.

Hassan, K.A., Liu, Q., Henderson, P.J., and Paulsen, I.T. (2015). Homologs of the *Acinetobacter baumannii* AceI transporter represent a new family of bacterial multidrug efflux systems. *MBio* 6 (1): e01982-14.

He, X., Szewczyk, P., Karyakin, A. et al. (2010). Structure of a cation-bound multidrug and toxic compound extrusion transporter. *Nature* 467 (7318): 991–994.

Hu, R.M., Liao, S.T., Huang, C.C., Huang, Y.W., and Yang, T.C. (2012). An inducible fusaric acid tripartite efflux pump contributes to the fusaric acid resistance in *Stenotrophomonas maltophilia* PloS one 7: e51053.

Huda, M.N., Chen, J., Morita, Y. et al. (2003a). Gene cloning and characterization of VcrM, a Na^+-coupled multidrug efflux pump, from *Vibrio cholerae* non-O1. *Microbiol. Immunol.* 47 (6): 419–427.

Huda, N., Lee, E.W., Chen, J. et al. (2003b). Molecular cloning and characterization of an ABC multidrug efflux pump, VcaM, in non-O1 *Vibrio cholerae*. *Antimicrob. Agents Chemother.* 47 (8): 2413–2417.

Hvorup, R.N., Winnen, B., Chang, A.B. et al. (2003). The multidrug/oligosaccharidyl-lipid/polysaccharide (MOP) exporter superfamily. *Eur. J. Biochem./FEBS* 270 (5): 799–813.

Jack, D.L., Yang, N.M., and Saier, M.H. Jr. (2001). The drug/metabolite transporter superfamily. *Eur. J. Biochem./FEBS* 268 (13): 3620–3639.

Jiang, D., Zhao, Y., Wang, X. et al. (2013). Structure of the YajR transporter suggests a transport mechanism based on the conserved motif A. *Proc. Natl. Acad. Sci. U. S. A.* 110 (36): 14664–14669.

Johnson, P.J. and Shafer, W.M. (2015). The transcriptional repressor, MtrR, of the *mtrCDE* efflux pump operon of *Neisseria gonorrhoeae* can also serve as an activator of "off target" gene (*glnE*) expression. *Antibiotics (Basel)* 4 (2): 188–197.

Kakarla, P., Floyd, J., Mukherjee, M. et al. (2017a). Inhibition of the multidrug efflux pump LmrS from *Staphylococcus aureus* by cumin spice *Cuminum cyminum*. *Arch. Microbiol.* 199 (3): 465–474.

Kakarla, P., Ranjana, K.C., Shrestha, U. et al. (2017b). Functional roles of highly conserved amino acid sequence motifs A and C in solute transporters of the major facilitator superfamily. In: *Drug Resistance in Bacteria, Fungi, Malaria, and Cancer* (ed. G. Arora, A. Sajid and V.C. Kalia), 111–140. Cham: Springer International Publishing.

Kathawala, R.J., Gupta, P., Ashby, C.R. Jr., and Chen, Z.S. (2015). The modulation of ABC transporter-mediated multidrug resistance in cancer: a review of the past decade. *Drug Resist. Updat.* 18: 1–17.

Kobayashi, N., Nishino, K., and Yamaguchi, A. (2001). Novel macrolide-specific ABC-type efflux transporter in *Escherichia coli*. *J. Bacteriol.* 183 (19): 5639–5644.

Kolbusz, M.A., ter Horst, R., Slotboom, D.J., and Lolkema, J.S. (2010). Orientation of small multidrug resistance transporter subunits in the membrane: correlation with the positive-inside rule. *J. Mol. Biol.* 402 (1): 127–138.

Kumar, S., He, G., Kakarla, P. et al. (2016a). Bacterial multidrug efflux pumps of the major facilitator superfamily as targets for modulation. *Infect. Disord. Drug Targets* 16 (1): 28–43.

Kumar, S., Mukherjee, M.M., and Varela, M.F. (2013). Modulation of bacterial multidrug resistance efflux pumps of the major facilitator superfamily. *Int J Bacteriol.* 2013: 1–15.

Kumar, S., Ranjana, K.C., Sanford, L.M. et al. (2016b). Structural and functional roles of two evolutionarily conserved amino acid sequence motifs within solute transporters of the major facilitator superfamily. *Trends Cell Mol. Biol.* 11: 41–53.

Kumar, S. and Varela, M.F. (2013). Molecular mechanisms of bacterial resistance to antimicrobial agents. In: *Microbial Pathogens and Strategies for Combating Them: Science, Technology and Education*, vol. 14 (ed. A. Méndez-Vilas), 522–534. Badajoz, Spain: Formatex Research Center.

Kuroda, T. and Tsuchiya, T. (2009). Multidrug efflux transporters in the MATE family. *Biochim. Biophys. Acta* 1794 (5): 763–768.

Law, C.J., Maloney, P.C., and Wang, D.N. (2008). Ins and outs of major facilitator superfamily antiporters. *Annu. Rev. Microbiol.* 62: 289–305.

Levy, S.B. (2001). Antibiotic resistance: consequences of inaction. *Clin. Infect. Dis.* 33 (Suppl 3): S124–S129.

Levy, S.B. (2002). Active efflux, a common mechanism for biocide and antibiotic resistance. *J. Appl. Microbiol.* 92 (Suppl): 65S–71S.

Li, W., Zhang, H., Assaraf, Y.G. et al. (2016). Overcoming ABC transporter-mediated multidrug resistance: molecular mechanisms and novel therapeutic drug strategies. *Drug Resist. Updat.* 27: 14–29.

Lin, H.T., Bavro, V.N., Barrera, N.P. et al. (2009). MacB ABC transporter is a dimer whose ATPase activity and macrolide-binding capacity are regulated by the membrane fusion protein MacA. *J. Biol. Chem.* 284 (2): 1145–1154.

Long, F., Su, C.C., Lei, H.T. et al. (2012). Structure and mechanism of the tripartite CusCBA heavy-metal efflux complex. *Philos. Trans. R. Soc. Lond. Ser. B Biol. Sci.* 367 (1592): 1047–1058.

Long, F., Su, C.C., Zimmermann, M.T. et al. (2010). Crystal structures of the CusA efflux pump suggest methionine-mediated metal transport. *Nature* 467 (7314): 484–488.

Lu, M. (2016). Structures of multidrug and toxic compound extrusion transporters and their mechanistic implications. *Channels (Austin)* 10 (2): 88–100.

Lu, M., Radchenko, M., Symersky, J. et al. (2013a). Structural insights into H^+-coupled multidrug extrusion by a MATE transporter. *Nat. Struct. Mol. Biol.* 20 (11): 1310–1317.

Lu, M., Symersky, J., Radchenko, M. et al. (2013b). Structures of a Na^+-coupled, substrate-bound MATE multidrug transporter. *Proc. Natl. Acad. Sci. U. S. A.* 110 (6): 2099–2104.

Lu, W.J., Lin, H.J., Janganan, T.K. et al. (2018). ATP-binding cassette transporter VcaM from *Vibrio cholerae* is dependent on the outer membrane factor family for its function. *Int. J. Mol. Sci.* 19 (4): 1–15.

Misra, R. and Bavro, V.N. (2009). Assembly and transport mechanism of tripartite drug efflux systems. *Biochim. Biophys. Acta* 1794 (5): 817–825.

Mors, K., Hellmich, U.A., Basting, D. et al. (2013). A lipid-dependent link between activity and oligomerization state of the *M. tuberculosis* SMR protein TBsmr. *Biochim. Biophys. Acta* 1828 (2): 561–567.

Murakami, S., Nakashima, R., Yamashita, E., and Yamaguchi, A. (2002). Crystal structure of bacterial multidrug efflux transporter AcrB. *Nature* 419 (6907): 587–593.

Nagarathinam, K., Jaenecke, F., Nakada-Nakura, Y. et al. (2017). The multidrug-resistance transporter MdfA from *Escherichia coli*: crystallization and X-ray diffraction analysis. *Acta Crystallogr. F Struct. Biol. Commun.* 73 (Pt 7): 423–430.

Nara, T., Kouyama, T., Kurata, Y. et al. (2007). Anti-parallel membrane topology of a homo-dimeric multidrug transporter, EmrE. *J. Biochem.* 142 (5): 621–625.

Nasie, I., Steiner-Mordoch, S., and Schuldiner, S. (2012). New substrates on the block: clinically relevant resistances for EmrE and homologues. *J. Bacteriol.* 194 (24): 6766–6770.

Nies, A.T., Damme, K., Kruck, S. et al. (2016). Structure and function of multidrug and toxin extrusion proteins (MATEs) and their relevance to drug therapy and personalized medicine. *Arch. Toxicol.* 90 (7): 1555–1584.

Nikaido, H. (2018). RND transporters in the living world. *Res. Microbiol.* 169 (7–8): 363–371.

Nikaido, H. and Hall, J.A. (1998). Overview of bacterial ABC transporters. *Methods Enzymol.* 292: 3–20.

Nikaido, H. and Takatsuka, Y. (2009). Mechanisms of RND multidrug efflux pumps. *Biochim. Biophys. Acta* 1794 (5): 769–781.

Nikaido, H. and Zgurskaya, H.I. (2001). AcrAB and related multidrug efflux pumps of *Escherichia coli*. *J. Mol. Microbiol. Biotechnol.* 3 (2): 215–218.

Otsuka, M., Yasuda, M., Morita, Y. et al. (2005). Identification of essential amino acid residues of the NorM Na^+/multidrug antiporter in *Vibrio parahaemolyticus*. *J. Bacteriol.* 187 (5): 1552–1558.

Ovchinnikov, V., Stone, T.A., Deber, C.M., and Karplus, M. (2018). Structure of the EmrE multidrug transporter and its use for inhibitor peptide design. *Proc. Natl. Acad. Sci. U. S. A.* 115 (34): E7932–E7941.

Padariya, M., Kalathiya, U., and Baginski, M. (2015). Structural and dynamic changes adopted by EmrE, multidrug transporter protein – studies by molecular dynamics simulation. *Biochim. Biophys. Acta* 1848 (10 Pt A): 2065–2074.

Pak, J.E., Ekende, E.N., Kifle, E.G. et al. (2013). Structures of intermediate transport states of ZneA, a Zn(II)/proton antiporter. *Proc. Natl. Acad. Sci. U. S. A.* 110 (46): 18484–18489.

Pao, S.S., Paulsen, I.T., and Saier, M.H. Jr. (1998). Major facilitator superfamily. *Microbiol. Mol. Biol. Rev.* 62 (1): 1–34.

Paulsen, I.T., Skurray, R.A., Tam, R. et al. (1996). The SMR family: a novel family of multidrug efflux proteins involved with the efflux of lipophilic drugs. *Mol. Microbiol.* 19 (6): 1167–1175.

Poole, K. (2001). Multidrug efflux pumps and antimicrobial resistance in *Pseudomonas aeruginosa* and related organisms. *J. Mol. Microbiol. Biotechnol.* 3 (2): 255–264.

Pos, K.M. (2009). Drug transport mechanism of the AcrB efflux pump. *Biochim. Biophys. Acta* 1794 (5): 782–793.

Pu, Y., Ke, Y., and Bai, F. (2017). Active efflux in dormant bacterial cells – new insights into antibiotic persistence. *Drug Resist. Updat.* 30: 7–14.

Puzari, M. and Chetia, P. (2017). RND efflux pump mediated antibiotic resistance in Gram-negative bacteria *Escherichia coli* and *Pseudomonas aeruginosa*: a major issue worldwide. *World J. Microbiol. Biotechnol.* 33 (2): 24.

Qazi, S.J.S., Chew, R., Bay, D.C., and Turner, R.J. (2015). Structural and functional comparison of hexahistidine tagged and untagged forms of small multidrug resistance protein, EmrE. *Biochem. Biophys. Rep.* 1: 22–32.

Ranaweera, I., Shrestha, U., Ranjana, K.C. et al. (2015). Structural comparison of bacterial multidrug efflux pumps of the major facilitator superfamily. *Trends Cell Mol. Biol.* 10: 131–140.

Roca, I., Marti, S., Espinal, P. et al. (2009). CraA, a major facilitator superfamily efflux pump associated with chloramphenicol resistance in *Acinetobacter baumannii*. *Antimicrob. Agents Chemother.* 53 (9): 4013–4014.

Rouquette, C., Harmon, J.B., and Shafer, W.M. (1999). Induction of the *mtrCDE*-encoded efflux pump system of *Neisseria gonorrhoeae* requires MtrA, an AraC-like protein. *Mol. Microbiol.* 33 (3): 651–658.

Ruggerone, P., Murakami, S., Pos, K.M., and Vargiu, A.V. (2013). RND efflux pumps: structural information translated into function and inhibition mechanisms. *Curr. Top. Med. Chem.* 13 (24): 3079–3100.

Saidijam, M., Benedetti, G., Ren, Q. et al. (2006). Microbial drug efflux proteins of the major facilitator superfamily. *Curr. Drug Targets* 7 (7): 793–811.

Saier, M.H. Jr., Beatty, J.T., Goffeau, A. et al. (1999). The major facilitator superfamily. *J. Mol. Microbiol. Biotechnol.* 1 (2): 257–279.

Schindler, B.D. and Kaatz, G.W. (2016). Multidrug efflux pumps of Gram-positive bacteria. *Drug Resist. Updat.* 27: 1–13.

Schuldiner, S. (2009). EmrE, a model for studying evolution and mechanism of ion-coupled transporters. *Biochim. Biophys. Acta* 1794 (5): 748–762.

Sennhauser, G., Bukowska, M.A., Briand, C., and Grutter, M.G. (2009). Crystal structure of the multidrug exporter MexB from *Pseudomonas aeruginosa*. *J. Mol. Biol.* 389 (1): 134–145.

Song, J., Ji, C., and Zhang, J.Z. (2014). Insights on Na^+ binding and conformational dynamics in multidrug and toxic compound extrusion transporter NorM. *Proteins* 82 (2): 240–249.

Steed, P.R., Stein, R.A., Mishra, S. et al. (2013). Na^+-substrate coupling in the multidrug antiporter norm probed with a spin-labeled substrate. *Biochemistry* 52 (34): 5790–5799.

Stockbridge, R.B., Robertson, J.L., Kolmakova-Partensky, L., and Miller, C. (2013). A family of fluoride-specific ion channels with dual-topology architecture. *eLife* 2: e01084.

Symersky, J., Guo, Y., Wang, J., and Lu, M. (2015). Crystallographic study of a MATE transporter presents a difficult case in structure determination with low-resolution, anisotropic data and crystal twinning. *Acta Crystallogr. D Biol. Crystallogr.* 71 (Pt 11): 2287–2296.

Tanaka, Y., Hipolito, C.J., Maturana, A.D. et al. (2013). Structural basis for the drug extrusion mechanism by a MATE multidrug transporter. *Nature* 496 (7444): 247–251.

Wang, K., Pei, H., Huang, B. et al. (2013). The expression of ABC efflux pump, Rv1217c-Rv1218c, and its association with multidrug resistance of *Mycobacterium tuberculosis* in China. *Curr. Microbiol.* 66 (3): 222–226.

Wang, Y., Venter, H., and Ma, S. (2016). Efflux pump inhibitors: a novel approach to combat efflux-mediated drug resistance in bacteria. *Curr. Drug Targets* 17 (6): 702–719.

Wendlandt, S., Lozano, C., Kadlec, K. et al. (2013). The enterococcal ABC transporter gene *lsa(E)* confers combined resistance to lincosamides, pleuromutilins and streptogramin A antibiotics in methicillin-susceptible and methicillin-resistant *Staphylococcus aureus*. *J. Antimicrob. Chemother.* 68 (2): 473–475.

Wong, K., Ma, J., Rothnie, A. et al. (2014). Towards understanding promiscuity in multidrug efflux pumps. *Trends Biochem. Sci.* 39 (1): 8–16.

Yardeni, E.H., Zomot, E., and Bibi, E. (2017). The fascinating but mysterious mechanistic aspects of multidrug transport by MdfA from *Escherichia coli*. *Res. Microbiol.* 169: 455–460.

Zhang, L. and Ma, S. (2010). Efflux pump inhibitors: a strategy to combat P-glycoprotein and the NorA multidrug resistance pump. *ChemMedChem* 5 (6): 811–822.

Zhang, X., He, X., Baker, J. et al. (2012). Twelve transmembrane helices form the functional core of mammalian MATE1 (multidrug and toxin extruder 1) protein. *J. Biol. Chem.* 287 (33): 27971–27982.

9

Bacterial Persistence in Biofilms and Antibiotics

Mechanisms Involved

Anne Jolivet-Gougeon[1,2] *and Martine Bonnaure-Mallet*[1,2]

[1] *Univ Rennes, INSERM, INRA, CHU Rennes, Institut NUMECAN (Nutrition Metabolisms and Cancer), Rennes, France*
[2] *Teaching Hospital of Rennes, Rennes, France*

9.1 Introduction

Biofilms are more often observed in aqueous media or with exposure to moisture. They can develop on various surfaces, both in humans and on industrial surfaces or medical devices. A biofilm is formed when microorganisms organize themselves in community within an exopolysaccharide (EPS) matrix, adhering to inert or living surfaces (Ferreira et al. 2011). Components of the biofilm matrix consist of polysaccharides, DNA, and proteins. Many examples can be cited to illustrate the involvement of biofilms in human clinical practice (e.g. skin, pacemakers, catheters, intrauterine catheters, naso-laryngeal tubes, stents, alloplastic materials, hydrocephalus shunts, artificial hearts, prosthesis, implants, artificial devices, contact lenses, dental plaque, and cystic fibrosis) (Bauer et al. 2002; Coman et al. 2008; Blango and Mulvey 2010; Goerke and Wolz 2010; Lopez-Causape et al. 2015; Antony and Farran 2016) or the industrial environment (e.g. work surfaces, pipelines). When a biofilm is made, its destruction is necessary to eliminate the risk of infection or persistent contamination, but antibiotics are not always effective in eradicating bacteria organized in a dense network, especially when they are fixed on metal or plastic material. Biofilms are alive and evolve over time, becoming enriched with other bacterial species that cooperate to counteract the action of the antibiotic(s). They can also disperse and colonize other places with bacteria detached from the mature biofilm. Biofilms are up to 1000-fold more antibiotic resistant than planktonic cultures (Mulcahy et al. 2008; Chin et al. 2017). Nucleo et al. (2010) tested the formation of *Proteus mirabilis* biofilms and showed that the β-lactamase-positive strains (multidrug resistant, extended spectrum, β-lactamase producing) were significantly better in biofilm

Antibiotic Drug Resistance, First Edition. Edited by José-Luis Capelo-Martínez and Gilberto Igrejas.

formation than negative strains, regardless of growth medium. Quantification of biofilms has been performed using crystal violet assay in carbapenemase-producing *Acinetobacter baumannii* isolates harboring the blaOXA-23 gene, and biofilm-associated protein BAP was always detected in biofilm-producing strains, suggesting that biofilm formation ability is responsible for their persistence and colonization (Sung et al. 2016). Increased bacterial persistence is also observed in biofilms and provides cross-tolerance to different clinically important antibiotics (Harms et al. 2016; Michiels et al. 2016). Several molecular mechanisms can explain this failure of antibiotics to overcome these biofilms, but many methods are currently available to cope with them.

The purpose of this review is to define the reasons for antibiotic failure in biofilms and the usual and innovative means to overcome biofilm resistance in biofilms.

9.2 Reasons for Failure of Antibiotics in Biofilms

Inside the biofilm, a high density of bacteria promotes genetic transfer, enhances the selection of resistant strains, and increases the frequency of mutation (Rodriguez-Rojas et al. 2012). A subpopulation of bacteria can enter a starved state (persisters), with modifications to physiology and metabolic rate, leading to more cells resistant to antibiotics (Mah 2012; Carvalho et al. 2018). This may be a significant contributor to recurrent infections. Kwan et al. (2013) succeeded in inducing the formation of persisters by pretreating *Escherichia coli* with rifampin to stop transcription, tetracycline to stop translation, and carbonyl cyanide m-chlorophenylhydrazone to stop ATP synthesis. There may be several explanations for the failure of antibiotics in biofilms, including a defect of penetration inside the biofilm, increased resistance of bacteria, or better adaptation of bacterial cells to their environment (Stewart 2002; Hoiby et al. 2010a, b; Jolivet-Gougeon and Bonnaure-Mallet 2014; Li et al. 2017). Kaldalu et al. (2016) also argued that mechanisms involved in persistence are more complex than supposed and a function of bacterial species, strain, growth conditions, and antibiotics is used in the experiments. Phenotypic and genetic heterogeneity within biofilms, with particular emphasis on persistence and antimicrobial tolerance, were discussed in a recent review (Sadiq et al. 2017).

9.2.1 Failure of Antibiotics to Penetrate Biofilm: Active Antibiotics on the Biofilm

Changes in biofilm architecture resulted in the accumulation of metabolites (e.g. indole for *E. coli*), which control the competition dynamics between bacterial species and influence antibiotic susceptibilities (Bhattacharjee et al. 2017). The resistance of some strains within the biofilm has been widely demonstrated (Greene et al. 2016), but antibiotic activity depends on biofilm maturity and bacterial strain (Bauer et al. 2013). The measures minimal biofilm inhibition concentration (MBIC) and minimal biofilm eradication concentration (MBEC) have been determined by an *in vitro* biofilm model to give an indication of the minimal inhibitory concentration (MIC) closer to real conditions in biofilms (Reiter et al. 2012). Stewart (1996) investigated antibiotic penetration into microbial biofilm and showed the theoretic inability of beta-lactams to penetrate inside the biofilm. Overproduction of chromosomally encoded AmpC cephalosporinase by gene derepression is considered the main mechanism of resistance of *Pseudomonas aeruginosa* isolates in sputum to beta-lactam antibiotics. Giwercman et al. (1992) demonstrated that strong β-lactamase inducers, such as carbapenems, allow the production of β-lactamase through all of the bacterial layers, whereas poorer inducers, such as ceftazidime, influence just the superficial layers of the biofilm, probably due to inactivation of the antibiotic by the β-lactamase. Hill et al. (2005) proposed using, in cystic fibrosis patients, colistin (tested at concentrations suitable for nebulization) either alone or in combination with tobramycin ($10\,\mu g\,ml^{-1}$), followed by meropenem combined with tobramycin or ciprofloxacin. The addition of aztreonam also improved the efficacy of ceftazidime for treatment of the biofilm, probably because aztreonam acts as a β-lactamase inhibitor (Giwercman et al. 1992).

Dynamic biofilms stained with Live/Dead probes have also been observed by confocal microscopy. Bauer et al. (2013) demonstrated that rifampin, tigecycline, and moxifloxacin are effective against mature methicillin-resistant *Staphylococcus aureus* ATCC 33591 biofilms, whereas oxacillin has demonstrated activity against methicillin-sensitive *S. aureus* ATCC 25923. Delafloxacin and daptomycin were the most potent, and fusidic acid, vancomycin, and linezolid the less potent overall. Rosales-Reyes et al. (2016) demonstrated susceptibility to polymyxin B of multidrug-resistant *A. baumannii* with a biofilm phenotype.

However, all of these results are to be considered with caution. Torres et al. (2017) used microdialysis to evaluate ciprofloxacin penetration into the lungs of healthy rats infected with *P. aeruginosa* in biofilm and showed that bacterial biofilm infection reduced the ciprofloxacin free interstitial lung concentrations and increased plasma exposure, suggesting that plasma concentrations alone are not a good surrogate for lung concentrations.

9.2.2 Outer Membrane Vesicles (OMVs)

Bacterial outer membrane vesicles are spheres of lipids released from the outer membranes (OM) of Gram-negative bacteria that are involved in cell-to-cell communication and biofilm formation (Wang et al. 2015a). Ciofu et al. (2000) have shown that the source of β-lactamase in biofilm may also be the membrane vesicles containing β-lactamase liberated by resistant *P. aeruginosa* bacteria. He et al. (2017) studied the association between biofilm formation and membrane vesicle secretion of methicillin-resistant *S. aureus* under conditions of subinhibitory concentrations of vancomycin. Vancomycin treatment led to modification of the contents of the vesicles, which were enriched in proteins, facilitating the formation of a biofilm.

9.2.3 Horizontal Transfer of Encoding β-Lactamase Genes

A significant increase in the concentration of antibiotic resistance genes has been observed in biofilms collected downstream of wastewater treatment plant discharge points, favoring the increase and spread of antibiotic resistance among streambed biofilms (Proia et al. 2016). The health relevance of biofilms is widely demonstrated by the increase in antibiotic-resistant bacteria hosted in biofilms in hospitals and the interaction of these bacteria with immune system cells (Obst et al. 2006; Dupin et al. 2015). Some bacterial strains belonging to serotypes associated with human infections are able to persist within the processing plants, forming biofilms (Osman et al. 2016). In addition, commonly used prophylactic antibiotic treatments, such as cefazolin, have been unable to eradicate bacterial biofilms and prevent colonization on model implant surfaces (Dastgheyb et al. 2015), questioning the use of these antibiotics for prophylaxis.

Biofilms are also facilitators of genetic material transfer and plasmid stability (Marti et al. 2017; Ridenhour et al. 2017). Mo et al. (2017) demonstrated dissemination of $bla_{CMY\text{-}2}$-carrying plasmids in European broiler production, the intergenus transfer of $bla_{CMY\text{-}2}$-carrying plasmids from *E. coli* to environmental bacteria in the food-processing chain, and their capacity for conjugative transfer between different poultry-associated bacterial genera. Li et al. (2001) also showed that biofilm-grown *Streptococcus mutans* cells were transformed at a rate 10- to 600-fold higher than planktonic cells and that dead bacteria in the biofilms could act as donors of a chromosomally encoded erythromycin resistance determinant under control of the active pheromone ComC.

However, Yoon et al. (2015) showed that natural transformation and plasmid transfer are diminished in recipients overproducing the multidrug efflux pump AdeABC in biofilm conditions following alteration of the membrane.

Finally, co-transfer of genes encoding antibiotic resistance and facilitating biofilm formation and/or persistence has been observed in large plasmidic pathogenicity islands (Van Schaik et al. 2010).

9.2.4 Influence of Subinhibitory Concentrations of Antibiotics on Biofilm

At sub-MIC concentrations, some macrolide antibiotics (e.g. azithromycin) (Hoffmann et al. 2007; Starner et al. 2008), beta-lactam antibiotics (e.g. ceftazidime) (Otani et al. 2018), and fluoroquinolone antibiotics (e.g. ciprofloxacin)

(Skindersoe et al. 2008) have been demonstrated to inhibit quorum sensing in *P. aeruginosa*. The timing of DNA repair following fluoroquinolone treatment has also been reported to be important for persistent survival (Mok and Brynildsen 2018).

Sub-MIC concentrations of β-lactam antibiotics induce increased alginate synthesis in *P. aeruginosa* biofilms (e.g. imipenem) (Bagge et al. 2004) or slime production (e.g. cefamandole or vancomycin) (Dunne 1990). Nucleo et al. (2009) argued that exposure of *A. baumannii* to subinhibitory concentrations of imipenem results in biofilm stimulation and increased production of iron uptake proteins, especially in glucose-based medium. Otani et al. (2018) demonstrated that ceftazidime reduces the gene expression of *lecA*, *lecB*, *pel*, and *psl*, which are involved in adhesion and polysaccharide matrix synthesis of *P. aeruginosa*, suggesting that sub-MICs of ceftazidime affect biofilm formation at different levels. Subinhibitory concentrations of tigecycline and oxacillin exhibited significant biofilm-inducing activity in *Staphylococcus epidermidis*, with tigecycline exerting its effects on *embp* expression through SarA (Weiser et al. 2016).

Some bacterial virulence factors, including biofilm development, are controlled by a major regulatory locus called staphylococcal accessory regulator A (*sarA*), which is a negative regulator of the extracellular matrix-binding protein (Embp). Biofilm formation of the parental *S. aureus* strain was demonstrated to be enhanced in an experimental model of endocarditis, but mutation in the *sarA* gene decreased the ability to form a biofilm. Sub-MICs of vancomycin significantly increased *sarA* expression, and the authors suggested that *sarA* mutant strains form significantly thinner and/or less well-structured biofilms *in vivo* compared with their respective parental strains. This latter phenotype may conceivably allow greater penetration of vancomycin into such defective biofilms (Abdelhady et al. 2014). In *S. aureus*, *sarA* mutants also exhibit significant reductions in oxacillin resistance compared with their parental counterparts and oxacillin persistence in an experimental endocarditis model *in vivo* (Li et al. 2016a).

Hoffman et al. (2005) showed that subinhibitory concentrations of aminoglycoside antibiotics induce biofilm formation in *P. aeruginosa* and *E. coli* via induction of *arr*, a gene encoding an inner-membrane phosphodiesterase whose substrate is cyclic di-guanosine monophosphate (c-di-GMP), which is involved in cell surface adhesion. A common feature of all rugose small colony variants of *S. aureus* is increased levels of the intracellular signaling molecule c-di-GMP (Starkey et al. 2009). The new bis-(3′-5′)-cyclic dimeric GMP (c-di-GMP), called YfiBNR, was identified by Xu et al. (2016) in *P. aeruginosa* with the same properties as a regulator of biofilm formation.

Interrelations between activation of the toxin-antitoxin (TA) system by nutrient starvation, persister formation, and antibiotic resistance (or tolerance) is summarized in Figure 9.1.

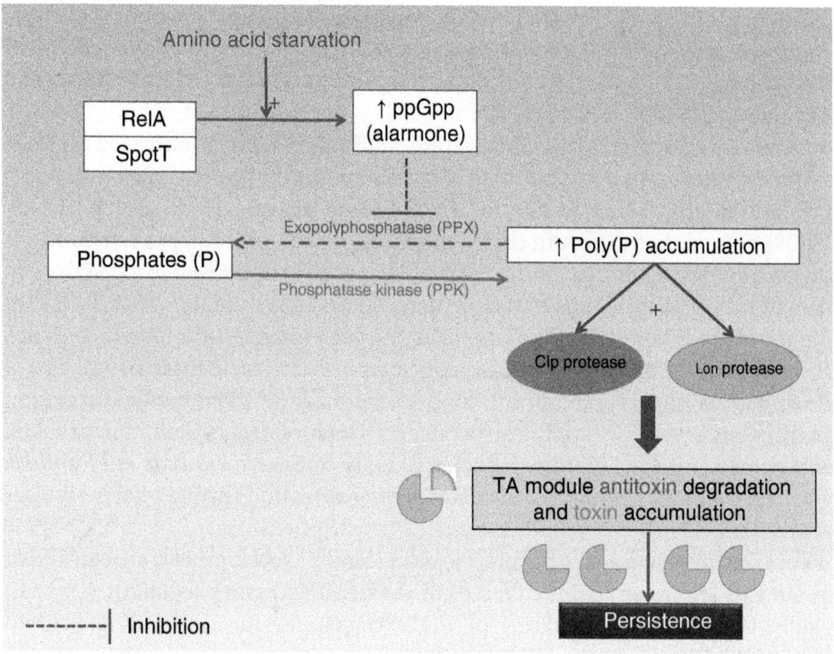

Figure 9.1 Interrelation between activation of the toxin–antitoxin (TA) system by nutrient starvation, persister formation, and antibiotic resistance (or tolerance). Examples of toxins in TA modules: couple cell division (CcdB), ParE, pneumococcal epsilon-zeta (pezT), *Escherichia* zeta toxin (ezeTA), death on cure (Doc), HokB, MqsR (motility quorum-sensing regulator), MosT (TA module, within the integrative and conjugative element SXT), MazF, RelE, VapC, HipA.

9.2.5 Small Colony Variants (SCVs), Persistence (Persisters), and Toxin–Antitoxin (TA) Systems

9.2.5.1 Small Colony Variants (SCVs)

SCVs are characterized by impaired growth, with downregulation of genes involved in metabolism and virulence, whereas genes important for persistence and biofilm formation are upregulated (Mirani et al. 2015). Rugose SCVs of *S. aureus* have increased expression of the *pel* and *psl* polysaccharide gene clusters, decreased expression of motility functions, and a defect in growth with some amino acid and tricarboxylic acid cycle intermediates as sole carbon sources (Starkey et al. 2009). An extensive inflammatory response was also stimulated in *S. aureus*, causing significant damage to the surrounding host tissue, promoting phagocytic evasion, and stimulating neutrophil reactive oxygen species (ROS) production (Pestrak et al. 2018). Rugose SCVs also elicited a reduced chemokine response from polarized airway epithelium cells compared with wild-type strains (Starkey et al. 2009). Bacterial aggregation and

TLR-mediated pro-inflammatory cytokine production contribute to the immune response of rugose SCVs of *S. aureus* (Pestrak et al. 2018). Furthermore, SCVs are resistant to various antibiotics, including aminoglycosides, trimethoprim-sulfamethoxazole, fluoroquinolones, and fusidic acid, and even antiseptics, such as triclosan (Garcia et al. 2013; Kahl 2014). Xia et al. (2017) described a deletion of the *yig*P locus in *E. coli* involved in the formation of SCVs. They investigated the antibiotic resistance profile of the *E. coli* SCV and found increased erythromycin, kanamycin, and D-cycloserine resistance but sensitivity to ampicillin, polymyxin, chloramphenicol, tetracycline, rifampin, and nalidixic acid. Wang et al. (2015b) sequenced the whole genome of *Pseudomonas chlororaphis* with an SCV phenotype, showing several mutations, especially in *yfi*R (cyclic-di-GMP production) and *fusA* (elongation factor). Genetic analysis revealed that the *yfi*R locus plays a major role in controlling SCV phenotypes, including colony size, growth, motility, and biofilm formation. DnpA, a putative de-*N*-acetylase of the PIG-L superfamily, is part of the conserved LPS and required for antibiotic tolerance in *P. aeruginosa*. Mutation in DnpA leads to fluoroquinolone tolerance, and its overexpression in the wild-type strain is related to the development of a persistent phenotype (Liebens et al. 2014, 2016). Moreover, a point mutation in *fusA* contributed to kanamycin resistance. Bui et al. (2015) also sequenced an SCV strain of *S. aureus* and found mutations in genes coding MgrA, a global regulator, and RsbU, a phosphoserine phosphatase within the regulatory pathway of sigma factor SigB. MgrA positively regulated genes involved in antibiotic resistance, such as the quinolone resistance efflux pump (*norABC*) and *tetK38*, which encodes tetracycline resistance proteins (Truong-Bolduc et al. 2005). Trotonda et al. (2008) determined that MgrA represses biofilm formation, and biofilm formation by *mgrA* mutants is not in association with the *sigB* and *ica* loci, but in association with the expression of surface proteins (*srtA*, sortase) and the presence of extracellular DNA. Bui et al. (2015) induced *S. aureus* WCH-SK2 into a stable SCV cell type using methylglyoxal. No reversion was observed after subculturing, and cells possessed a metabolic and surface profile that was different from that of previously described SCVs or biofilm cells (protein extracellular matrix Ebh and extracellular DNA, but not polysaccharide). However, several other phenotypic and genetic changes induced by antibiotics have been observed in bacteria with the SCV phenotype (Lim et al. 2016).

9.2.5.2 Persisters

Persisters are in a dormant metabolic state, even while remaining genetically identical to the actively growing cells, and their regulation is complex (Fasani and Savageau 2013; Feng et al. 2014; Hayes and Kedzierska 2014; Kedzierska and Hayes 2016; Cabral et al. 2018). They are a subpopulation of cells surviving antibiotic treatment but, in contrast to resistant bacteria, cannot grow in the presence of antibiotics. They have been described in several bacterial species, including intracellular bacteria such as *Mycobacteria* (Basaraba and Ojha 2017; Singh et al. 2017) or *Borrelia* (Berndtson 2013). Their tolerance arises from physiological processes rather than genetic mutations (Allison et al. 2011a). Metabolites entering upper glycolysis (glucose, mannitol, and fructose) and pyruvate induce a rapid gentamicin killing of persisters in biofilm via the generation of a proton motive force, which facilitates aminoglycoside uptake (Allison et al. 2011b).

Several molecules (Figure 9.2) have been implicated in the formation of persisters in the viable but non-cultivable state: RelA, ppGpp, PPK, Lon and Clp proteases, level of adenosine triphosphate (ATP), and TA systems (Ayrapetyan et al. 2015; Cui et al. 2016). The stringent response protein RelA synthesizes ppGpp (alarmones) during amino acid starvation, which can inhibit exopolyphosphatases, inducing the accumulation of PolyP. PolyP activates Lon or Clp proteases able to degrade antitoxins of TA modules, freeing toxins that induce persistence. Ortiz-Severin et al. (2015) demonstrated that

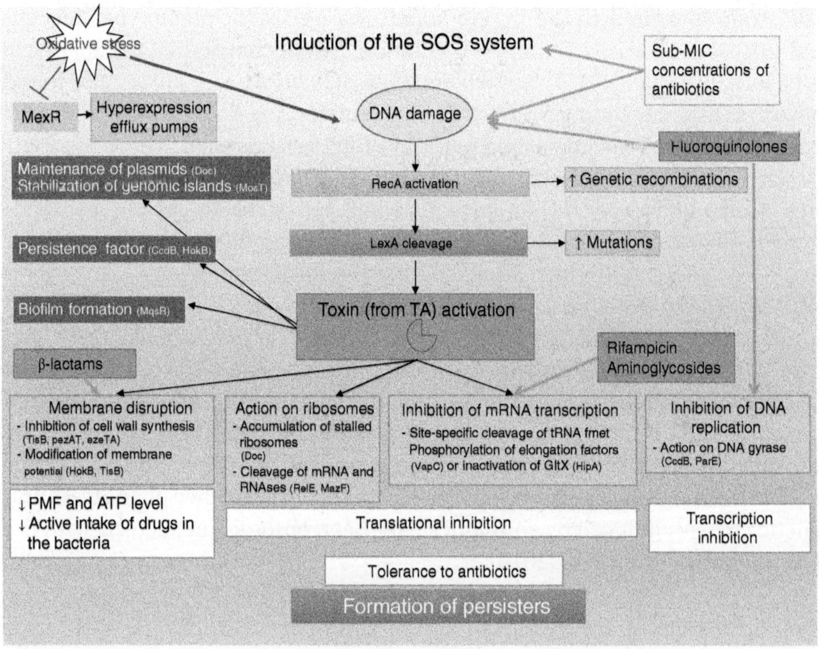

Figure 9.2 Example of persister formation under starvation in a toxin–antitoxin (TA) system. SpoT and RelA, the two ppGpp synthetases of *Escherichia coli*, are induced upon amino acid starvation. The resulting increase in the amount of this alarmone inhibits the exopolyphosphatase (PPX) and leads to the accumulation of polyphosphates. The high level of polyphosphates, combined with Lon or Clp proteases, stimulates the cleavage of antitoxin, which increases the amount of active toxin.

polyphosphate kinase (ppk) mutants are deficient in motility, quorum sensing, biofilm formation, and virulence but also showed phenotypic changes related to susceptibility toward antibiotics, including ciprofloxacin, chloramphenicol, and rifampicin. Other studies demonstrated the induction of a VBNC state or its persistence in biofilm-embedded *S. aureus* in the presence of vancomycin or quinupristin/dalfopristin (Pasquaroli et al. 2013), as well as the role of daptomycin in the induction and persistence of the viable but non-cultivable state of *S. aureus* biofilms (Pasquaroli et al. 2014). Mechler et al. (2015) described a *S. aureus* mutant with an adaptive point mutation in the putative inorganic phosphate (Pi) transporter gene *pitA*. This mutant enhanced tolerance toward daptomycin and was accompanied by elevated intracellular concentrations of Pi and polyphosphate. The Lon protease of *P. aeruginosa*, which is involved in biofilm formation, swimming, swarming, and twitching motility, can be induced by subinhibitory concentrations of aminoglycosides (Marr et al. 2007). The cellular ATP level is predictive of bactericidal antibiotic efficacy, and persisted formation in *S. aureus* is associated with ATP depletion (Conlon et al. 2016). Various other molecules have been described as being involved in the formation of persisters (De Groote et al. 2009). Screening the library for defective mutants in persistence or tolerance to rifampicin has revealed genes involved in purine biosynthesis (adenylosuccinate lyase *purB* and phosphoribosylaminoimidazole synthetase *purM*). The *purB* and *purM* mutants exhibit defective persistence to various antibiotics, low pH, and heat stress compared with the parental strain *S. aureus* USA300 (Yee et al. 2015). Studies by Amato and Brynildsen (Amato et al. 2013; Amato and Brynildsen 2015) tend to prove that formation of ampicillin persisters requires RelA, ClpA, SsrA, and SmpB and that ppGpp, DksA, SsrA, and SmpB participate in ofloxacin and ampicillin persister formation.

9.2.5.3 Toxin–Antitoxin (TA) Modules

TA modules in bacteria are one possible cause of persistence, as they are associated with cellular dormancy and involved in biofilm formation, with persistent multidrug resistance of many human pathogens (Maeda et al. 2017) and maintenance of multiresistance (Yang and Walsh 2017). Transcription studies of persisters have detected the overexpression of TA modules of several different toxins possibly involved in multidrug tolerance, such as RelA, MazF, HipA, and YgiU. Several genes that act as global regulators have been identified, such as DksA, SsrS-YgfA, DnaKJ, HupAB, IhfAB (Hansen et al. 2008; Singh et al. 2012), SarA, and SigB (Li et al. 2016a).

A two-gene operon, HipBA, is one of many chromosomally encoded TA modules in *E. coli* (Feng et al. 2014). Zhao et al. (2013) studied the function of the *hip*BA TA system in biofilm formation by *E. coli* and showed that inactivation of protein kinase HipA reduced the level of extracellular DNA present in biofilm formation. The antitoxin HipB forms a complex with HipA and holds it in the nucleoid. HipA is then no longer able to phosphorylate glutamyl-tRNA synthetase, preventing initiation of the forthcoming stringent response (Wen et al. 2016). Overexpression of toxins that inhibit essential functions, such as translation, may then contribute to persister formation and tolerance to some antibiotics. Maeda et al. (2017) also described a new TA system, YjjJ, which is encoded by a single gene and is a homologue of the toxin HipA. YjjJ seems to have different cellular targets than HipA, but HipB, the cognate antitoxin of HipA, also acts as an antitoxin for YjjJ.

The RelE and MazF toxins cause dormancy by cleaving mRNA (Lewis 2012). During amino acid starvation, RelE cleaves mRNA in the ribosomal A-site, inhibiting protein translation. RelE is structurally similar to bacterial RNases that employ general acid–base catalysis to facilitate RNA cleavage (Dunican et al. 2015).

The TisB toxin forms a membrane pore, leading to a decrease in the proton motive force and ATP consistent with its role in forming dormant cells (Lewis 2012). Dörr et al. (2010) showed the induction of persisted formation by ciprofloxacin and tolerance to multiple antibiotics by cells producing TisB toxin. Ciprofloxacin kills cells primarily by converting its target proteins, which are DNA topoisomerases, into DNA endonucleases. A decrease in ATP will prevent topoisomerases from damaging the DNA, indicating a link between fluoroquinolones and TisB induction.

The induction of Doc toxin of the Phd-Doc TA system mimics the effects of treatment with the aminoglycoside hygromycin B; both interact with 30S ribosomal subunits, stabilize polysomes, and significantly increase mRNA half-life. The antibiotic also competes with ribosome-bound Doc, whereas hygromycin B-resistant mutants suppress Doc toxicity, suggesting that the Doc-binding site includes that of aminoglycosides (Liu et al. 2008).

As in *Mycobacteria*, multiple *vapBC* modules have been described in *E. coli*, contributing to enhanced survival within the host. VapCs are endoribonucleases that can inhibit translation by site-specific cleavage of initiator tRNA or the universally conserved region in 23S rRNA (Winther et al. 2016), and we can easily imagine that they interfere with macrolides or aminoglycosides, which have sites of action on the ribosome.

Van Acker et al. (2014) investigated whether TA modules contribute to persistence toward antibiotics in *Burkholderia cenocepacia*, a well-known resistant pathogen that colonizes cystic fibrosis patients. The overexpression of toxins results in growth inhibition, often increasing the number of surviving persisters, especially in untreated sessile cells. Nine toxin-encoding genes are upregulated after treatment with tobramycin, but none after treatment with ciprofloxacin.

TA systems are constantly being discovered in genetic analysis and bioinformatics searches (Budde et al. 2007; Harrison et al. 2009; Chan et al. 2012; Gil et al. 2015; Jaiswal et al. 2016; Thakur et al. 2018). Some are associated with small RNA regulators (Kim and Wood 2010). A recent review of the literature described the different classes of TA modules, their mechanisms of action, and their role in antibiotic resistance (Yang and Walsh 2017).

9.2.6 Quorum Sensing: Bacterial Metabolites

Quorum sensing is the ability of bacteria to detect and respond to cell population density via gene regulation. Within eight hours of infection in thermally injured mice, *P. aeruginosa* forms biofilms on specific host tissues independent of quorum sensing, indicating the importance of the state of bacterial growth in signal induction (Schaber et al. 2007). Autoinducers are signaling molecules produced in response to changes in cell population density and regulate a wide variety of physiological activities, including antibiotic production and biofilm formation. In otitis, *Haemophilus influenzae* can promote *Moraxella catarrhalis* persistence within polymicrobial biofilms via interspecies quorum signaling autoinducer AI-2 (Armbruster et al. 2010), and *Streptococcus pneumoniae* increases *in vivo* colonization by *M. catarrhalis* in a quorum signal-dependent manner (Perez et al. 2014). The most described quorum-sensing regulator, LuxS, is required for maximal biofilm formation in many bacterial species, including strict anaerobes (Đapa et al. 2013).

Butt et al. (2016) showed that deletion of kynurenine formamidase (KynB), which is involved in the metabolism of tryptophan to anthranilate and then 2-alkyl-4-quinolone, in *Burkholderia pseudomallei* K96243 results in increased biofilm formation and increased tolerance to ciprofloxacin. Addition of exogenous anthranilic acid restores the biofilm phenotype, but not the persister phenotype. Deletion of other 2-alkyl-4-quinolone-encoding genes, such as *pqsA* in *P. aeruginosa*, also increases ciprofloxacin tolerance (Haussler and Becker 2008). The addition of pyocyanin, paraquat, or 3-(oxododecanoyl)-L-homo-serine lactone significantly increased persister numbers of logarithmic phase *P. aeruginosa* (Moker et al. 2010), whereas indole increased persistence in *E. coli* (Vega et al. 2012).

9.2.7 Extracellular DNA

Mulcahy et al. (2008) showed that extracellular DNA in the biofilm matrix contributes to cation gradients, genomic DNA release, and inducible antibiotic resistance. Extracellular DNA can chelate cations that stabilize lipopolysaccharide (LPS) and the outer membrane (OM), inducing cell lysis with the release of cytoplasmic contents and genomic DNA. These authors demonstrated that subinhibitory concentrations of DNA created a cation-limited environment, resulting in induction of the PhoPQ- and PmrAB-regulated cationic antimicrobial peptide resistance operon PA3552–PA3559 in *P. aeruginosa*. Overexpression of PA3552–PA3559 resulted in increased resistance to cationic antimicrobial peptides and aminoglycosides but had no effect on β-lactam and fluoroquinolone resistance. Doroshenko et al. (2014) demonstrated an increase in the extracellular DNA concentration and a protective effect in *S. epidermidis* biofilms preexposed to sub-MIC vancomycin.

9.2.8 Nutrient Limitation

Nutrient limitation can induce persistence. Greene et al. (2016) described a relationship between antibiotic resistance and desiccation in biofilms. In addition, glucose deprivation has been shown to increase the formation of persisters and increase biofilm tolerance to fluoroquinolone and β-lactam treatment (Amato et al. 2013). Stationary phase *E. coli* cells are more susceptible to ciprofloxacin by supplementing oxygen and carbon sources (Gutierrez et al. 2017), and *E. coli* grown in minimal glucose media expresses higher levels of the efflux pump-encoding gene *acrB* (Bailey et al. 2006). Amino acid deprivation is a prerequisite for bacterial tolerance, and global nutrient limitation induces bacteria with a higher tolerance for β-lactams, fluoroquinolones, and aminoglycosides (Fung et al. 2010).

9.2.9 SOS Inducers (Antibiotics and Others)

Oxygen limitation increases antibiotic tolerance, robust biofilms, and alginate biosynthesis, which contribute to the persistence of *P. aeruginosa* (Schobert and Tielen 2010).

Nitric oxide (NO) is an endogenous signaling molecule and therapeutic agent that triggers biofilm dispersal. Allan et al. (2016) used quantitative proteomic analysis (isobaric tag for relative and absolute quantitation) and identified 13 proteins involved in bacterial metabolism or translation that are differentially expressed following low-concentration NO treatment, indicating that NO modulates pneumococcal metabolism. Low-concentration NO also enhances pneumococcal killing when combined with amoxicillin-clavulanic acid.

Bile is a bactericidal agent and can lead to the generation of ROS in *S. Typhi*. The quorum-sensing systems can regulate the level of superoxide dismutase and catalase enzymes produced in response to oxidative stress. In the presence of ciprofloxacin and ampicillin, *S. typhi* forms persister cells, which increase greater than threefold in the presence of bile (Walawalkar et al. 2016). The presence of cis-2-unsaturated fatty acids produced by *B. cenocepacia* and *Stenotrophomonas maltophilia* in sputum from patients with cystic fibrosis leads to altered biofilm formation, increased resistance to antibiotics, and persistence of *P. aeruginosa* (Twomey et al. 2012).

9.2.10 Hypermutator Phenotype

Kovacs et al. (2013) failed to find a relationship between the hypermutator phenotype in *Enterobacteriaceae* isolated from urinary tract infection and their ability to initiate a biofilm. Proteomic analysis of one hypermutator multidrug-resistant *A. baumannii* strain revealed 31 differentially expressed proteins, including three proteins involved in biofilm suppression and the oxidative stress response (Karami-Zarandi et al. 2017). These authors hypothesized that it may be the result of adaptation of the hypermutator strain. Due to mutations, likely random mutations, one can imagine that the impact of this phenotype varies according to strains and situations.

9.2.11 Multidrug Efflux Pumps

Planktonic bacteria and bacteria in biofilm exhibit differences in cell permeability, including efflux pump activity, which could be related to the differences observed in the susceptibility to antibiotics (Berlanga et al. 2017). Overproduction of the resistance-nodulation-cell division (RND)-type efflux pumps AdeABC and AdeIJK was demonstrated to alter bacterial membrane composition, resulting in decreased biofilm formation but not motility (Yoon et al. 2015). Similar results were described by Vieira et al. (2017) for the CmeABC MDR pump in *Campylobacter jejuni* with *Acanthamoeba polyphaga* by using a modified gentamicin protection assay. Mutants of *tolC*, a gene encoding a part of the AcrAB-TolC system in *E. coli*, exhibit decreased adhesion and biofilm formation, but expression of AcrAB only protects *E. coli* from forming biofilm against low concentrations of ciprofloxacin (Soto 2013). Pamp et al. (2008) also showed that *P. aeruginosa* mutants defective in mexAB-oprM-mediated antimicrobial efflux are not able to develop a tolerant subpopulation in biofilms. Buffet-Bataillon et al. (2016) showed that quaternary ammonium compounds (QACs) are able to induce overexpression of efflux pumps, which could lead to a stress response, facilitating mutation in the quinolone resistance determining region and biofilm formation with an increased risk of the transfer of mobile genetic elements carrying fluoroquinolone or QAC resistance determinants. Wang et al. (2017) investigated the interactions of *S. mutans* with the quaternary ammonium monomer dimethylaminohexadecyl methacrylate and showed that it could induce persister formation in biofilms, but the authors did not describe the molecular mechanisms that explain these results.

Another explanation for the implication of efflux pumps in the formation of biofilm could be increased extrusion or intrusion of quorum-sensing molecules (Soto 2013).

9.3 Usual and Innovative Means to Overcome Biofilm Resistance in Biofilms

Faced with the difficulty of treating infections caused by bacteria in biofilms, many strategies have been attempted to tackle biofilms (Kostakioti et al. 2013; Lee et al. 2016; Ribeiro et al. 2016; Wood 2016, 2017; Taha et al. 2018).

9.3.1 Antibiotics (Bacteriocins) Natural and Synthetic Molecules: Phages

Some antibiotics are able to counteract biofilm development, such as aztreonam, by blocking quorum sensing and increasing sensitivity to hydrogen peroxide and the complement system. Aztreonam can also inhibit alginate production and polymerization of *P. aeruginosa* alginate by its incomplete precipitation and high levels of readily dialyzable uronic acids (Hoffmann et al. 2007). The development of antibiotic adjuvants, such as meridianin D, that increase the effectiveness of currently available antibiotics is a promising alternative approach (Huggins et al. 2018). Topical, inhaled, combined, and sequential antibiotic treatments have been tested in various situations, with variable results according to the study (Ciofu et al. 2017). In particular, Gallo et al. (2017) tested meropenem on *Acinetobacter calcoaceticus–Acinetobacter baumannii* persisters and claimed that the meropenem concentration did not influence persistent fractions, even when far above the MIC. Pamp et al. (2008) observed that biofilm *P. aeruginosa* cells exhibit low metabolic activity and are effectively killed by colistin. New technologies, such as the use of nanoparticles (Qayyum and Khan 2016; Kulshrestha et al. 2017; Zaidi et al. 2017) or quantum dots (QDs) (Li et al. 2016b), allow better penetration of antibiotics within the biofilm (Ficai et al. 2018). QDs are colloidal semiconductor nanocrystals that emit photoluminescence in proportion to the dot size. Their association with biofilm components can be utilized to determine the impact of surface chemistry on QD mobility and distribution in bacterial biofilms, which can be evaluated by epifluorescent or confocal microscopy (Morrow et al. 2010).

Bacteriocins are a family of peptides or proteins naturally synthesized by certain bacteria; they are not antibiotics but have antibiotic properties, causing the formation of pores in the bacterial membrane. Bacteriocins play an important role in the competition between bacterial strains, and their production is stimulated by quorum sensing. Oliveira et al. (2015) showed that antibiotics can act as stress inducers in a bacterial biofilm environment. At high density, cells can better control the environment with secreted products that favor their genotype over others, such as *P. aeruginosa* secreting pyocin, which acts as a narrow-spectrum antibiotic but may also stimulate biofilm formation by increasing cell attachment. Many bacteriocins have been described after the first discovery of colicin and nisin (Mathur et al. 2018; Yi et al. 2018), including derivatives (Seal et al. 2018).

Many natural or synthetic molecules have been tested, alone or in combination with antibiotics, for their anti-biofilm properties, such as skyllamycins (Navarro et al. 2014). Laser scanning confocal microscopy has revealed that aloe-emodin treatment inhibits extracellular protein production in *S. aureus*, and the Congo red assay showed that it also reduces the accumulation of polysaccharide intercellular adhesin on the cell surface (Xiang et al. 2017). Many other molecules extracted from essential oils have been tested for their antibiofilm properties (Miladi et al. 2017; Oh et al. 2017; Artini et al. 2018) and been reported to inhibit bacteria by damaging the cell membrane, altering the lipid profile, and inhibiting ATPases, cell division, membrane porins, motility, and biofilm formation, but the molecular mechanisms involved remain unknown.

Lytic bacteriophages are viruses that are able to infect specific bacterial species and can be used in combination with antibiotics to potentiate their action for the treatment of biofilm infections (Chaudhry et al. 2017). Alves et al. (2016) tested two biofilm models (static and dynamic) with a phage cocktail to assess the ability to reduce and disperse *P. aeruginosa* biofilm biomass. In the static model, after four hours of contact with the phage suspension, more than 95% of the biofilm biomass was eliminated. In the flow biofilm model, the biofilm was dispersed after 48 hours. Kumaran et al. (2018) investigated the ability of phage treatment to enhance the activity of antibiotics cefazolin, vancomycin, dicloxacillin, tetracycline, and linezolid against biofilm-forming *S. aureus*. They showed a significant reduction of up to 3 log colony forming unit (CFU) per milliliter when the phage treatment preceded antibiotics and a more important effect with vancomycin and cefazolin, particularly at lower antibiotic concentrations.

9.3.2 Efflux Pump Inhibitors

Efflux pump inhibitors are compounds that meet the following criteria (Soto 2013): enhanced activity of the pump substrates, no activity in efflux pump mutants, increased accumulation and decreased extrusion of efflux pump substrates, no action on the proton gradient across the cytoplasmic membrane, and inhibition of this efflux activity.

9.3.3 Anti-Persisters: Quorum-Sensing Inhibitors

Many anti-persister strategies have been attempted, as prolonged treatment with aminoglycosides that cause mistranslation, leading to misfolded peptides, can sterilize a stationary culture of *P. aeruginosa*, a pathogen responsible for chronic, highly tolerant infection of cystic fibrosis patients. One of the best bactericidal agents is rifampin, an inhibitor of RNA polymerase, and Keren et al. (2012) suggested that it "kills" by preventing persister resuscitation. The anticancer drug cisplatin [*cis*-diamminodichloroplatinum(II)], which mainly forms intra-strand DNA cross-links, can eradicate *E. coli* K-12 persister cells through a growth-independent mechanism (Chowdhury et al. 2016). A combination of tobramycin and fumarate has been tested successfully (Koeva et al. 2017) as an antibacterial potentiator for eliminating recurrent *P. aeruginosa* infections in cystic fibrosis patients through the eradication of bacterial persisters. Lysine decarboxylase converts lysine to cadaverine, which is able to enhance the tolerance of *P. aeruginosa* to carbenicillin and ticarcillin by reducing persister formation (Manuel et al. 2010). Lysine-ε-aminotransferase inhibitors have also been tested against *Mycobacterium tuberculosis* (Reshma et al. 2017). For a review of other anti-persister molecules, see Defraine et al. (2018).

Quorum-sensing inhibitors are therapeutics that attack bacterial virulence rather than kill bacteria. They were recently developed, and cell density-dependent gene regulation in bacteria was proposed as target (Otto 2004; Rasmussen et al. 2005; Yadav et al. 2012; Pan et al. 2013a, b; Ma et al. 2018; Virmani et al. 2018). Inhibition of the *P. aeruginosa* quorum-sensing regulator MvfR (PqsR) using a benzamide-benzimidazole compound interferes with biofilm formation and potentiates biofilm sensitivity to antibiotics (Maura and Rahme 2017). Rasamiravaka et al. (2015) described a new bioactive compound (oleanolic aldehyde coumarate) extracted from *Dalbergia trichocarpa* bark that is able to disrupt *P. aeruginosa* PAO1 *las* and *rhl* quorum-sensing mechanisms. It could also improve the bactericidal activity of tobramycin against biofilm-encapsulated PAO1 cells. Domenech and García (2017) successfully tested antioxidants that are widely used in the clinical setting, specifically *N*-acetyl-l-cysteine and cysteamine, against mixed biofilms of non-encapsulated *S. pneumoniae* and nontypeable *H. influenzae.*

9.3.4 Enzymes

Alipour et al. (2009) showed that administration of DNase and alginate lyase enhances the activity of tobramycin in biofilms by dissolving the biofilm matrix.

9.3.5 Electrical Methods

Niepa et al. (2012) tested low-level electrochemical currents on *P. aeruginosa* using stainless steel with tobramycin and demonstrated a synergistic effect. Canty et al. (2017) used *in vitro* electrostimulation by cathodic voltage-controlled electrical stimulation (CVCES) of commercially pure titanium incubated in cultures of methicillin-resistant *S. aureus* and *A. baumannii* as a method for preventing bacterial attachment. Nodzo et al. (2016) evaluated the antimicrobial effects of combining CVCES with prolonged vancomycin therapy against methicillin-resistant *S. aureus* to clear infection on a shoulder implant with established biofilm, concluding that this combination may be beneficial in treating biofilm-related implant infections.

9.3.6 Photodynamic Therapy

Photodynamic therapy has been applied in some studies (Giannelli et al. 2017; Briggs et al. 2018) to help treat infections. Activation of a photosensitizer leads to an acute stress response, with changes in calcium and lipid metabolism and the production of cytokines and stress proteins, such as protein kinases, which are activated, and transcription factors (Castano et al. 2005).

9.4 Conclusion

At high bacterial density, antibiotics can select some pathogenic genotypes over others, such as a *P. aeruginosa* strain secreting pyocin (Oliveira et al. 2015) or a highly antibiotic-resistant strain. We must caution against this escalation of ecological competition, which may result in an unexpected outcome, more abundant pathogen. Many therapies are currently being tested to deal with the increase in antibiotic resistance, which is largely favored in biofilm conditions. The future will tell us which approaches are most effective, especially in the long term.

Acknowledgments

We thank Adina Pascu for helping us format the manuscript. We send our thanks to *Fondation des Gueules cassées; sourire quand même* for their financial support for the translation of this publication.

Conflict of Interest

None.

References

Abdelhady, W., Bayer, A.S., Seidl, K. et al. (2014). Impact of vancomycin on *sarA*-mediated biofilm formation: role in persistent endovascular infections due to methicillin-resistant *Staphylococcus aureus*. *J. Infect. Dis.* 209: 1231–1240.

Alipour, M., Suntres, Z.E., and Omri, A. (2009). Importance of DNase and alginate lyase for enhancing free and liposome encapsulated aminoglycoside activity against *Pseudomonas aeruginosa*. *J. Antimicrob. Chemother.* 64: 317–325.

Allan, R.N., Morgan, S., Brito-Mutunayagam, S. et al. (2016). Low concentrations of nitric oxide modulate *Streptococcus pneumoniae* biofilm metabolism and antibiotic tolerance. *Antimicrob. Agents Chemother.* 60: 2456–2466.

Allison, K.R., Brynildsen, M.P., and Collins, J.J. (2011a). Heterogeneous bacterial persisters and engineering approaches to eliminate them. *Curr. Opin. Microbiol.* 14: 593–598.

Allison, K.R., Brynildsen, M.P., and Collins, J.J. (2011b). Metabolite-enabled eradication of bacterial persisters by aminoglycosides. *Nature* 473: 216–220.

Alves, D.R., Perez-Esteban, P., Kot, W. et al. (2016). A novel bacteriophage cocktail reduces and disperses *Pseudomonas aeruginosa* biofilms under static and flow conditions. *Microb. Biotechnol.* 9: 61–74.

Amato, S.M. and Brynildsen, M.P. (2015). Persister heterogeneity arising from a single metabolic stress. *Curr. Biol.* 25: 2090–2098.

Amato, S.M., Orman, M.A., and Brynildsen, M.P. (2013). Metabolic control of persister formation in *Escherichia coli*. *Mol. Cell* 50: 475–487.

Antony, S. and Farran, Y. (2016). Prosthetic joint and orthopedic device related infections: the role of biofilm in the pathogenesis and treatment. *Infect. Disord. Drug Targets* 16: 22–27.

Armbruster, C.E., Hong, W., Pang, B. et al. (2010). Indirect pathogenicity of *Haemophilus influenzae* and *Moraxella catarrhalis* in polymicrobial otitis media occurs via interspecies quorum signaling. *MBio* 1 (3): pii: e00102-10.

Artini, M., Patsilinakos, A., Papa, R. et al. (2018). Antimicrobial and antibiofilm activity and machine learning classification analysis of essential oils from different Mediterranean plants against *Pseudomonas aeruginosa*. *Molecules* 23 (2): pii: E482.

Ayrapetyan, M., Williams, T.C., and Oliver, J.D. (2015). Bridging the gap between viable but non-culturable and antibiotic persistent bacteria. *Trends Microbiol.* 23: 7–13.

Bagge, N., Schuster, M., Hentzer, M. et al. (2004). *Pseudomonas aeruginosa* biofilms exposed to imipenem exhibit changes in global gene expression and beta-lactamase and alginate production. *Antimicrob. Agents Chemother.* 48: 1175–1187.

Bailey, A.M., Webber, M.A., and Piddock, L.J. (2006). Medium plays a role in determining expression of acrB, marA, and soxS in *Escherichia coli*. *Antimicrob. Agents Chemother.* 50: 1071–1074.

Basaraba, R.J. and Ojha, A.K. (2017). Mycobacterial biofilms: revisiting tuberculosis bacilli in extracellular necrotizing lesions. *Microbiol. Spectr.* 5 (3): https://doi.org/10.1128/microbiolspec.TBTB2-0024-2016.

Bauer, J., Siala, W., Tulkens, P.M., and Van Bambeke, F. (2013). A combined pharmacodynamic quantitative and qualitative model reveals the potent activity of daptomycin and delafloxacin against *Staphylococcus aureus* biofilms. *Antimicrob. Agents Chemother.* 57: 2726–2737.

Bauer, T.T., Torres, A., Ferrer, R. et al. (2002). Biofilm formation in endotracheal tubes: association between pneumonia and the persistence of pathogens. *Monaldi Arch. Chest Dis.* 57: 84–87.

Berlanga, M., Gomez-Perez, L., and Guerrero, R. (2017). Biofilm formation and antibiotic susceptibility in dispersed cells versus planktonic cells from clinical, industry and environmental origins. *Antonie Van Leeuwenhoek* 110: 1691–1704.

Berndtson, K. (2013). Review of evidence for immune evasion and persistent infection in Lyme disease. *Int. J. Gen. Med.* 6: 291–306.

Bhattacharjee, A., Khan, M., Kleiman, M., and Hochbaum, A.I. (2017). Effects of growth surface topography on bacterial signaling in coculture biofilms. *ACS Appl. Mater. Interfaces* 9: 18531–18539.

Blango, M.G. and Mulvey, M.A. (2010). Persistence of uropathogenic *Escherichia coli* in the face of multiple antibiotics. *Antimicrob. Agents Chemother.* 54: 1855–1863.

Briggs, T., Blunn, G., Hislop, S. et al. (2018). Antimicrobial photodynamic therapy-a promising treatment for prosthetic joint infections. *Lasers Med. Sci.* 33: 523–532.

Budde, P.P., Davis, B.M., Yuan, J., and Waldor, M.K. (2007). Characterization of a higBA toxin-antitoxin locus in *Vibrio cholerae*. *J. Bacteriol.* 189: 491–500.

Buffet-Bataillon, S., Tattevin, P., Maillard, J.Y. et al. (2016). Efflux pump induction by quaternary ammonium compounds and fluoroquinolone resistance in bacteria. *Future Microbiol.* 11: 81–92.

Bui, L.M., Hoffmann, P., Turnidge, J.D. et al. (2015). Prolonged growth of a clinical *Staphylococcus aureus* strain selects for a stable small-colony-variant cell type. *Infect. Immun.* 83: 470–481.

Butt, A., Halliday, N., Williams, P. et al. (2016). *Burkholderia pseudomallei* kynB plays a role in AQ production, biofilm formation, bacterial swarming and persistence. *Res. Microbiol.* 167: 159–167.

Cabral, D.J., Wurster, J.I., and Belenky, P. (2018). Antibiotic persistence as a metabolic adaptation: stress, metabolism, the host, and new directions. *Pharmaceuticals (Basel)* 11 (1): pii: E14.

Canty, M., Luke-Marshall, N., Campagnari, A., and Ehrensberger, M. (2017). Cathodic voltage-controlled electrical stimulation of titanium for prevention of methicillin-resistant *Staphylococcus aureus* and *Acinetobacter baumannii* biofilm infections. *Acta Biomater.* 48: 451–460.

Carvalho, G., Balestrino, D., Forestier, C., and Mathias, J.D. (2018). How do environment-dependent switching rates between susceptible and persister cells affect the dynamics of biofilms faced with antibiotics? *NPJ Biofilms Microbiomes* 4: 6.

Castano, A.P., Demidova, T.N., and Hamblin, M.R. (2005). Mechanisms in photodynamic therapy: part two-cellular signaling, cell metabolism and modes of cell death. *Photodiagn. Photodyn. Ther.* 2: 1–23.

Chan, W.T., Moreno-Cordoba, I., Yeo, C.C., and Espinosa, M. (2012). Toxin-antitoxin genes of the Gram-positive pathogen *Streptococcus pneumoniae*: so few and yet so many. *Microbiol. Mol. Biol. Rev.* 76: 773–791.

Chaudhry, W.N., Concepcion-Acevedo, J., Park, T. et al. (2017). Synergy and order effects of antibiotics and phages in killing *Pseudomonas aeruginosa* biofilms. *PLoS One* 12: e0168615.

Chin, K.C.J., Taylor, T.D., Hebrard, M. et al. (2017). Transcriptomic study of *Salmonella enterica* subspecies enterica serovar Typhi biofilm. *BMC Genomics* 18: 836.

Chowdhury, N., Wood, T.L., Martinez-Vazquez, M. et al. (2016). DNA-crosslinker cisplatin eradicates bacterial persister cells. *Biotechnol. Bioeng.* 113: 1984–1992.

Ciofu, O., Beveridge, T.J., Kadurugamuwa, J. et al. (2000). Chromosomal beta-lactamase is packaged into membrane vesicles and secreted from *Pseudomonas aeruginosa*. *J. Antimicrob. Chemother.* 45: 9–13.

Ciofu, O., Rojo-Molinero, E., Macia, M.D., and Oliver, A. (2017). Antibiotic treatment of biofilm infections. *APMIS* 125: 304–319.

Coman, G., Mardare, N., Copacianu, B., and Covic, M. (2008). Persistent and recurrent skin infection with small-colony variant methicillin-resistant *Staphylococcus aureus*. *Rev. Med. Chir. Soc. Med. Nat. Iasi.* 112: 104–107.

Conlon, B.P., Rowe, S.E., Gandt, A.B. et al. (2016). Persister formation in *Staphylococcus aureus* is associated with ATP depletion. *Nat. Microbiol.* 1: 16051.

Cui, P., Xu, T., Zhang, W.H., and Zhang, Y. (2016). Molecular mechanisms of bacterial persistence and phenotypic antibiotic resistance. *Yi Chuan* 38: 859–871.

Đapa, T., Leuzzi, R., Ng, Y.K. et al. (2013). Multiple factors modulate biofilm formation by the anaerobic pathogen *Clostridium difficile*. *J. Bacteriol.* 195: 545–555.

Dastgheyb, S.S., Hammoud, S., Ketonis, C. et al. (2015). Staphylococcal persistence due to biofilm formation in synovial fluid containing prophylactic cefazolin. *Antimicrob. Agents Chemother.* 59: 2122–2128.

De Groote, V.N., Verstraeten, N., Fauvart, M. et al. (2009). Novel persistence genes in *Pseudomonas aeruginosa* identified by high-throughput screening. *FEMS Microbiol. Lett.* 297: 73–79.

Defraine, V., Fauvart, M., and Michiels, J. (2018). Fighting bacterial persistence: current and emerging anti-persister strategies and therapeutics. *Drug Resist. Updat.* 38: 12–26.

Domenech, M. and García, E. (2017). N-Acetyl-L-cysteine and cysteamine as new strategies against mixed biofilms of nonencapsulated *Streptococcus pneumoniae* and nontypeable *Haemophilus influenzae*. *Antimicrob. Agents Chemother.* 61 (2): pii: e01992-16.

Doroshenko, N., Tseng, B.S., Howlin, R.P. et al. (2014). Extracellular DNA impedes the transport of vancomycin in *Staphylococcus epidermidis* biofilms preexposed to subinhibitory concentrations of vancomycin. *Antimicrob. Agents Chemother.* 58: 7273–7282.

Dörr, T., Vulic, M., and Lewis, K. (2010). Ciprofloxacin causes persister formation by inducing the TisB toxin in *Escherichia coli*. *PLoS Biol.* 8: e1000317.

Dunican, B.F., Hiller, D.A., and Strobel, S.A. (2015). Transition state charge stabilization and acid-base catalysis of mRNA cleavage by the endoribonuclease RelE. *Biochemistry* 54: 7048–7057.

Dunne, W.M. Jr. (1990). Effects of subinhibitory concentrations of vancomycin or cefamandole on biofilm production by coagulase-negative Staphylococci. *Antimicrob. Agents Chemother.* 34: 390–393.

Dupin, C., Tamanai-Shacoori, Z., Ehrmann, E. et al. (2015). Oral Gram-negative anaerobic bacilli as a reservoir of beta-lactam resistance genes facilitating infections with multiresistant bacteria. *Int. J. Antimicrob. Agents* 45: 99–105.

Fasani, R.A. and Savageau, M.A. (2013). Molecular mechanisms of multiple toxin-antitoxin systems are coordinated to govern the persister phenotype. *Proc. Natl. Acad. Sci. U.S.A.* 110: E2528–E2537.

Feng, J., Kessler, D.A., Ben-Jacob, E., and Levine, H. (2014). Growth feedback as a basis for persister bistability. *Proc. Natl. Acad. Sci. U.S.A.* 111: 544–549.

Ferreira, A.S., Silva, I.N., Oliveira, V.H. et al. (2011). Insights into the role of extracellular polysaccharides in Burkholderia adaptation to different environments. *Front. Cell Infect. Microbiol.* 1: 16.

Ficai, D., Grumezescu, V., Fufă, O.M. et al. (2018). Antibiofilm coatings based on PLGA and nanostructured cefepime-functionalized magnetite. *Nanomaterials (Basel)* 8 (9): pii: E633.

Fung, D.K., Chan, E.W., Chin, M.L., and Chan, R.C. (2010). Delineation of a bacterial starvation stress response network which can mediate antibiotic tolerance development. *Antimicrob. Agents Chemother.* 54: 1082–1093.

Gallo, S.W., Donamore, B.K., Pagnussatti, V.E. et al. (2017). Effects of meropenem exposure in persister cells of *Acinetobacter calcoaceticus-baumannii*. *Future Microbiol.* 12: 131–140.

Garcia, L.G., Lemaire, S., Kahl, B.C. et al. (2013). Antibiotic activity against small-colony variants of *Staphylococcus aureus*: review of *in vitro*, animal and clinical data. *J. Antimicrob. Chemother.* 68: 1455–1464.

Giannelli, M., Landini, G., Materassi, F. et al. (2017). Effects of photodynamic laser and violet-blue led irradiation on *Staphylococcus aureus* biofilm and *Escherichia coli* lipopolysaccharide attached to moderately rough titanium surface: *in vitro* study. *Lasers Med. Sci.* 32: 857–864.

Gil, F., Pizarro-Guajardo, M., Alvarez, R. et al. (2015). *Clostridium difficile* recurrent infection: possible implication of TA systems. *Future Microbiol.* 10: 1649–1657.

Giwercman, B., Meyer, C., Lambert, P.A. et al. (1992). High-level beta-lactamase activity in sputum samples from cystic fibrosis patients during antipseudomonal treatment. *Antimicrob. Agents Chemother.* 36: 71–76.

Goerke, C. and Wolz, C. (2010). Adaptation of *Staphylococcus aureus* to the cystic fibrosis lung. *Int. J. Med. Microbiol.* 300: 520–525.

Greene, C., Vadlamudi, G., Newton, D. et al. (2016). The influence of biofilm formation and multidrug resistance on environmental survival of clinical and environmental isolates of *Acinetobacter baumannii*. *Am. J. Infect. Control* 44: e65–e71.

Gutierrez, A., Jain, S., Bhargava, P. et al. (2017). Understanding and sensitizing density-dependent persistence to quinolone antibiotics. *Mol. Cell* 68: 1147.e3–1154.e3.

Hansen, S., Lewis, K., and Vulić, M. (2008). Role of global regulators and nucleotide metabolism in antibiotic tolerance in *Escherichia coli*. *Antimicrob. Agents Chemother.* 52: 2718–2726.

Harms, A., Maisonneuve, E., and Gerdes, K. (2016). Mechanisms of bacterial persistence during stress and antibiotic exposure. *Science* 354 (6318): pii: aaf4268.

Harrison, J.J., Wade, W.D., Akierman, S. et al. (2009). The chromosomal toxin gene yafQ is a determinant of multidrug tolerance for *Escherichia coli* growing in a biofilm. *Antimicrob. Agents Chemother.* 53: 2253–2258.

Haussler, S. and Becker, T. (2008). The pseudomonas quinolone signal (PQS) balances life and death in *Pseudomonas aeruginosa* populations. *PLoS Pathog.* 4: e1000166.

Hayes, F. and Kedzierska, B. (2014). Regulating toxin-antitoxin expression: controlled detonation of intracellular molecular timebombs. *Toxins (Basel)* 6: 337–358.

He, X., Yuan, F., Lu, F. et al. (2017). Vancomycin-induced biofilm formation by methicillin-resistant *Staphylococcus aureus* is associated with the secretion of membrane vesicles. *Microb. Pathog.* 110: 225–231.

Hill, D., Rose, B., Pajkos, A. et al. (2005). Antibiotic susceptabilities of *Pseudomonas aeruginosa* isolates derived from patients with cystic fibrosis under aerobic, anaerobic, and biofilm conditions. *J. Clin. Microbiol.* 43: 5085–5090.

Hoffman, L.R., D'argenio, D.A., Maccoss, M.J. et al. (2005). Aminoglycoside antibiotics induce bacterial biofilm formation. *Nature* 436: 1171–1175.

Hoffmann, N., Lee, B., Hentzer, M. et al. (2007). Azithromycin blocks quorum sensing and alginate polymer formation and increases the sensitivity to serum and stationary-growth-phase killing of *Pseudomonas aeruginosa* and attenuates chronic *P. aeruginosa lung* infection in Cftr(–/–) mice. *Antimicrob. Agents Chemother.* 51: 3677–3687.

Hoiby, N., Bjarnsholt, T., Givskov, M. et al. (2010a). Antibiotic resistance of bacterial biofilms. *Int. J. Antimicrob. Agents* 35: 322–332.

Hoiby, N., Ciofu, O., and Bjarnsholt, T. (2010b). *Pseudomonas aeruginosa* biofilms in cystic fibrosis. *Future Microbiol.* 5: 1663–1674.

Huggins, W.M., Barker, W.T., Baker, J.T. et al. (2018). Meridianin D analogues display antibiofilm activity against MRSA and increase colistin efficacy in Gram-negative bacteria. *ACS Med. Chem. Lett.* 9: 702–707.

Jaiswal, S., Paul, P., Padhi, C. et al. (2016). The Hha-TomB toxin-antitoxin system shows conditional toxicity and promotes persister cell formation by inhibiting apoptosis-like death in *S. Typhimurium*. *Sci Rep* 6: 38204.

Jolivet-Gougeon, A. and Bonnaure-Mallet, M. (2014). Biofilms as a mechanism of bacterial resistance. *Drug Discov. Today Technol.* 11: 49–56.

Kahl, B.C. (2014). Small colony variants (SCVs) of *Staphylococcus aureus* – a bacterial survival strategy. *Infect Genet. Evol.* 21: 515–522.

Kaldalu, N., Hauryliuk, V., and Tenson, T. (2016). Persisters-as elusive as ever. *Appl. Microbiol. Biotechnol.* 100: 6545–6553.

Karami-Zarandi, M., Douraghi, M., Vaziri, B. et al. (2017). Variable spontaneous mutation rate in clinical strains of multidrug-resistant *Acinetobacter baumannii* and differentially expressed proteins in a hypermutator strain. *Mutat. Res.* 800-802: 37–45.

Kędzierska, B. and Hayes, F. (2016). Emerging roles of toxin-antitoxin modules in bacterial pathogenesis. *Molecules* 21 (6): pii: E790.

Keren, I., Mulcahy, L.R., and Lewis, K. (2012). Persister eradication: lessons from the world of natural products. *Methods Enzymol.* 517: 387–406.

Kim, Y. and Wood, T.K. (2010). Toxins Hha and CspD and small RNA regulator Hfq are involved in persister cell formation through MqsR in *Escherichia coli. Biochem. Biophys. Res. Commun.* 391: 209–213.

Koeva, M., Gutu, A.D., Hebert, W. et al. (2017). An antipersister strategy for treatment of chronic *Pseudomonas aeruginosa* infections. *Antimicrob. Agents Chemother.* 61 (12): pii: e00987-17.

Kostakioti, M., Hadjifrangiskou, M., and Hultgren, S.J. (2013). Bacterial biofilms: development, dispersal, and therapeutic strategies in the dawn of the postantibiotic era. *Cold Spring Harb. Perspect. Med.* 3: a010306.

Kovacs, B., Le Gall-David, S., Vincent, P. et al. (2013). Is biofilm formation related to the hypermutator phenotype in clinical Enterobacteriaceae isolates? *FEMS Microbiol. Lett.* 347: 116–122.

Kulshrestha, S., Qayyum, S., and Khan, A.U. (2017). Antibiofilm efficacy of green synthesized graphene oxide-silver nanocomposite using *Lagerstroemia speciosa* floral extract: a comparative study on inhibition of Gram-positive and Gram-negative biofilms. *Microb. Pathog.* 103: 167–177.

Kumaran, D., Taha, M., Yi, Q. et al. (2018). Does treatment order matter? Investigating the ability of bacteriophage to augment antibiotic activity against *Staphylococcus aureus* biofilms. *Front. Microbiol.* 9: 127.

Kwan, B.W., Valenta, J.A., Benedik, M.J., and Wood, T.K. (2013). Arrested protein synthesis increases persister-like cell formation. *Antimicrob. Agents Chemother.* 57: 1468–1473.

Lee, J.H., Kim, Y.G., Yong Ryu, S., and Lee, J. (2016). Calcium-chelating alizarin and other anthraquinones inhibit biofilm formation and the hemolytic activity of *Staphylococcus aureus. Sci. Rep.* 6: 19267.

Lewis, K. (2012). Persister cells: molecular mechanisms related to antibiotic tolerance. *Handb. Exp. Pharmacol.* 211: 121–133.

Li, J., Xie, S., Ahmed, S. et al. (2017). Antimicrobial activity and resistance: influencing factors. *Front. Pharmacol.* 8: 364.

Li, L., Cheung, A., Bayer, A.S. et al. (2016a). The global regulon sarA regulates beta-lactam antibiotic resistance in methicillin-resistant *Staphylococcus aureus in vitro* and in endovascular infections. *J. Infect. Dis.* 214: 1421–1429.

Li, Y.H., Lau, P.C., Lee, J.H. et al. (2001). Natural genetic transformation of *Streptococcus mutans* growing in biofilms. *J. Bacteriol.* 183: 897–908.

Li, Y.J., Harroun, S.G., Su, Y.C. et al. (2016b). Synthesis of self-assembled spermidine-carbon quantum dots effective against multidrug-resistant bacteria. *Adv. Healthc. Mater.* 5: 2545–2554.

Liebens, V., Defraine, V., Van Der Leyden, A. et al. (2014). A putative de-N-acetylase of the PIG-L superfamily affects fluoroquinolone tolerance in *Pseudomonas aeruginosa. Pathog. Dis.* 71: 39–54.

Liebens, V., Frangipani, E., Van Der Leyden, A. et al. (2016). Membrane localization and topology of the DnpA protein control fluoroquinolone tolerance in *Pseudomonas aeruginosa. FEMS Microbiol. Lett.* 363: fnw184.

Lim, W.S., Phang, K.K., Tan, A.H. et al. (2016). Small colony variants and single nucleotide variations in Pf1 region of PB1 phage-resistant *Pseudomonas aeruginosa. Front. Microbiol.* 7: 282.

Liu, M., Zhang, Y., Inouye, M., and Woychik, N.A. (2008). Bacterial addiction module toxin Doc inhibits translation elongation through its association with the 30S ribosomal subunit. *Proc. Natl. Acad. Sci. U.S.A.* 105: 5885–5890.

Lopez-Causape, C., Rojo-Molinero, E., Macia, M.D., and Oliver, A. (2015). The problems of antibiotic resistance in cystic fibrosis and solutions. *Expert. Rev. Respir. Med.* 9: 73–88.

Ma, Z.P., Song, Y., Cai, Z.H. et al. (2018). Anti-quorum sensing activities of selected coral symbiotic bacterial extracts from the South China Sea. *Front. Cell Infect. Microbiol.* 8: 144.

Maeda, Y., Lin, C.Y., Ishida, Y. et al. (2017). Characterization of YjjJ toxin of *Escherichia coli. FEMS Microbiol. Lett.* 364 (11): https://doi.org/10.1093/femsle/fnx086.

Mah, T.F. (2012). Biofilm-specific antibiotic resistance. *Future Microbiol.* 7: 1061–1072.

Manuel, J., Zhanel, G.G., and De Kievit, T. (2010). Cadaverine suppresses persistence to carboxypenicillins in *Pseudomonas aeruginosa* PAO1. *Antimicrob. Agents Chemother.* 54: 5173–5179.

Marr, A.K., Overhage, J., Bains, M., and Hancock, R.E. (2007). The Lon protease of *Pseudomonas aeruginosa* is induced by aminoglycosides and is involved in biofilm formation and motility. *Microbiology* 153: 474–482.

Marti, R., Schmid, M., Kulli, S. et al. (2017). Biofilm formation potential of heat-resistant *Escherichia coli* dairy isolates and the complete genome of multidrug-resistant, heat-resistant strain FAM21845. *Appl. Environ. Microbiol.* 83 (15): pii: e00628-17.

Mathur, H., Field, D., Rea, M.C. et al. (2018). Fighting biofilms with lantibiotics and other groups of bacteriocins. *NPJ Biofilms Microbiomes* 4: 9.

Maura, D. and Rahme, L.G. (2017). Pharmacological inhibition of the *Pseudomonas aeruginosa* MvfR quorum-sensing system interferes with biofilm formation and potentiates antibiotic-mediated biofilm disruption. *Antimicrob. Agents Chemother.* 61 (12): pii: e01362-17.

Mechler, L., Herbig, A., Paprotka, K. et al. (2015). A novel point mutation promotes growth phase-dependent daptomycin tolerance in *Staphylococcus aureus*. *Antimicrob. Agents Chemother.* 59: 5366–5376.

Michiels, J.E., Van Den Bergh, B., Verstraeten, N. et al. (2016). *In vitro* emergence of high persistence upon periodic aminoglycoside challenge in the ESKAPE pathogens. *Antimicrob. Agents Chemother.* 60: 4630–4637.

Miladi, H., Zmantar, T., Kouidhi, B. et al. (2017). Synergistic effect of eugenol, carvacrol, thymol, p-cymene and gamma-terpinene on inhibition of drug resistance and biofilm formation of oral bacteria. *Microb. Pathog.* 112: 156–163.

Mirani, Z.A., Aziz, M., and Khan, S.I. (2015). Small colony variants have a major role in stability and persistence of *Staphylococcus aureus* biofilms. *J. Antibiot. (Tokyo)* 68: 98–105.

Mo, S.S., Sunde, M., Ilag, H.K. et al. (2017). Transfer potential of plasmids conferring extended-spectrum-cephalosporin resistance in *Escherichia coli* from poultry. *Appl. Environ. Microbiol.* 83 (12): pii: e00654-17.

Mok, W.W.K. and Brynildsen, M.P. (2018). Timing of DNA damage responses impacts persistence to fluoroquinolones. *Proc. Natl. Acad. Sci. U.S.A.* 115: E6301–E6309.

Moker, N., Dean, C.R., and Tao, J. (2010). *Pseudomonas aeruginosa* increases formation of multidrug-tolerant persister cells in response to quorum-sensing signaling molecules. *J. Bacteriol.* 192: 1946–1955.

Morrow, J.B., Arango, C.P., and Holbrook, R.D. (2010). Association of quantum dot nanoparticles with Pseudomonas aeruginosa biofilm. *J. Environ. Qual.* 39: 1934–1941.

Mulcahy, H., Charron-Mazenod, L., and Lewenza, S. (2008). Extracellular DNA chelates cations and induces antibiotic resistance in *Pseudomonas aeruginosa* biofilms. *PLoS Pathog.* 4: e1000213.

Navarro, G., Cheng, A.T., Peach, K.C. et al. (2014). Image-based 384-well high-throughput screening method for the discovery of skyllamycins A to C as biofilm inhibitors and inducers of biofilm detachment in *Pseudomonas aeruginosa*. *Antimicrob. Agents Chemother.* 58: 1092–1099.

Niepa, T.H., Gilbert, J.L., and Ren, D. (2012). Controlling *Pseudomonas aeruginosa* persister cells by weak electrochemical currents and synergistic effects with tobramycin. *Biomaterials* 33: 7356–7365.

Nodzo, S.R., Tobias, M., Ahn, R. et al. (2016). Cathodic voltage-controlled electrical stimulation plus prolonged Vancomycin reduce bacterial burden of a titanium implant-associated infection in a rodent model. *Clin. Orthop. Relat. Res.* 474: 1668–1675.

Nucleo, E., Fugazza, G., Migliavacca, R. et al. (2010). Differences in biofilm formation and aggregative adherence between beta-lactam susceptible and beta-lactamases producing *P. mirabilis* clinical isolates. *New Microbiol.* 33: 37–45.

Nucleo, E., Steffanoni, L., Fugazza, G. et al. (2009). Growth in glucose-based medium and exposure to subinhibitory concentrations of imipenem induce biofilm formation in a multidrug-resistant clinical isolate of *Acinetobacter baumannii*. *BMC Microbiol.* 9: 270.

Obst, U., Schwartz, T., and Volkmann, H. (2006). Antibiotic resistant pathogenic bacteria and their resistance genes in bacterial biofilms. *Int. J. Artif. Organs* 29: 387–394.

Oh, S.Y., Yun, W., Lee, J.H. et al. (2017). Effects of essential oil (blended and single essential oils) on anti-biofilm formation of *Salmonella* and *Escherichia coli*. *J. Anim. Sci. Technol.* 59: 4.

Oliveira, N.M., Martinez-Garcia, E., Xavier, J. et al. (2015). Biofilm formation as a response to ecological competition. *PLoS Biol* 13: e1002191.

Ortiz-Severin, J., Varas, M., Bravo-Toncio, C. et al. (2015). Multiple antibiotic susceptibility of polyphosphate kinase mutants (ppk1 and ppk2) from *Pseudomonas aeruginosa* PAO1 as revealed by global phenotypic analysis. *Biol. Res.* 48: 22.

Osman, K.M., Samir, A., Abo-Shama, U.H. et al. (2016). Determination of virulence and antibiotic resistance pattern of biofilm producing Listeria species isolated from retail raw milk. *BMC Microbiol.* 16: 263.

Otani, S., Hiramatsu, K., Hashinaga, K. et al. (2018). Sub-minimum inhibitory concentrations of ceftazidime inhibit *Pseudomonas aeruginosa* biofilm formation. *J. Infect. Chemother.* 24: 428–433.

Otto, M. (2004). Quorum-sensing control in Staphylococci: a target for antimicrobial drug therapy? *FEMS Microbiol. Lett.* 241: 135–141.

Pamp, S.J., Gjermansen, M., Johansen, H.K., and Tolker-Nielsen, T. (2008). Tolerance to the antimicrobial peptide colistin in *Pseudomonas aeruginosa* biofilms is linked to metabolically active cells, and depends on the pmr and mexAB-oprM genes. *Mol. Microbiol.* 68: 223–240.

Pan, J., Song, F., and Ren, D. (2013a). Controlling persister cells of *Pseudomonas aeruginosa* PDO300 by (Z)-4-bromo-5-(bromomethylene)-3-methylfuran-2(5H)-one. *Bioorg. Med. Chem. Lett.* 23: 4648–4651.

Pan, J., Xie, X., Tian, W. et al. (2013b). (Z)-4-bromo-5-(bromomethylene)-3-methylfuran-2(5H)-one sensitizes *Escherichia coli* persister cells to antibiotics. *Appl. Microbiol. Biotechnol.* 97: 9145–9154.

Pasquaroli, S., Citterio, B., Cesare, A.D. et al. (2014). Role of daptomycin in the induction and persistence of the viable but non-culturable state of *Staphylococcus aureus* biofilms. *Pathogens* 3: 759–768.

Pasquaroli, S., Zandri, G., Vignaroli, C. et al. (2013). Antibiotic pressure can induce the viable but non-culturable state in *Staphylococcus aureus* growing in biofilms. *J. Antimicrob. Chemother.* 68: 1812–1817.

Perez, A.C., Pang, B., King, L.B. et al. (2014). Residence of *Streptococcus pneumoniae* and *Moraxella catarrhalis* within polymicrobial biofilm promotes antibiotic resistance and bacterial persistence *in vivo*. *Pathog. Dis.* 70: 280–288.

Pestrak, M.J., Chaney, S.B., Eggleston, H.C. et al. (2018). *Pseudomonas aeruginosa* rugose small-colony variants evade host clearance, are hyper-inflammatory, and persist in multiple host environments. *PLoS Pathog.* 14: e1006842.

Proia, L., Von Schiller, D., Sanchez-Melsio, A. et al. (2016). Occurrence and persistence of antibiotic resistance genes in river biofilms after wastewater inputs in small rivers. *Environ. Pollut.* 210: 121–128.

Qayyum, S. and Khan, A.U. (2016). Biofabrication of broad range antibacterial and antibiofilm silver nanoparticles. *IET Nanobiotechnol.* 10: 349–357.

Rasamiravaka, T., Vandeputte, O.M., Pottier, L. et al. (2015). *Pseudomonas aeruginosa* biofilm formation and persistence, along with the production of quorum sensing-dependent virulence factors, are disrupted by a triterpenoid coumarate ester isolated from *Dalbergia trichocarpa*, a tropical legume. *PLoS One* 10: e0132791.

Rasmussen, T.B., Bjarnsholt, T., Skindersoe, M.E. et al. (2005). Screening for quorum-sensing inhibitors (QSI) by use of a novel genetic system, the QSI selector. *J. Bacteriol.* 187: 1799–1814.

Reiter, K.C., Villa, B., Da Silva Paim, T.G. et al. (2012). Enhancement of antistaphylococcal activities of six antimicrobials against sasG-negative methicillin-susceptible *Staphylococcus aureus:* an *in vitro* biofilm model. *Diagn. Microbiol. Infect. Dis.* 74: 101–105.

Reshma, R.S., Jeankumar, V.U., Kapoor, N. et al. (2017). Mycobacterium tuberculosis lysine-varepsilon-aminotransferase a potential target in dormancy: Benzothiazole based inhibitors. *Bioorg. Med. Chem.* 25: 2761–2771.

Ribeiro, S.M., Felicio, M.R., Boas, E.V. et al. (2016). New frontiers for anti-biofilm drug development. *Pharmacol. Ther.* 160: 133–144.

Ridenhour, B.J., Metzger, G.A., France, M. et al. (2017). Persistence of antibiotic resistance plasmids in bacterial biofilms. *Evol. Appl.* 10: 640–647.

Rodriguez-Rojas, A., Oliver, A., and Blazquez, J. (2012). Intrinsic and environmental mutagenesis drive diversification and persistence of *Pseudomonas aeruginosa* in chronic lung infections. *J. Infect. Dis.* 205: 121–127.

Rosales-Reyes, R., Alcantar-Curiel, M.D., Jarillo-Quijada, M.D. et al. (2016). Biofilm formation and susceptibility to polymyxin B by a highly prevalent clone of multidrug-resistant *Acinetobacter baumannii* from a Mexican Tertiary Care Hospital. *Chemotherapy* 61: 8–14.

Sadiq, F.A., Flint, S., Li, Y. et al. (2017). Phenotypic and genetic heterogeneity within biofilms with particular emphasis on persistence and antimicrobial tolerance. *Future Microbiol.* 12: 1087–1107.

Schaber, J.A., Triffo, W.J., Suh, S.J. et al. (2007). *Pseudomonas aeruginosa* forms biofilms in acute infection independent of cell-to-cell signaling. *Infect. Immun.* 75: 3715–3721.

Schobert, M. and Tielen, P. (2010). Contribution of oxygen-limiting conditions to persistent infection of *Pseudomonas aeruginosa*. *Future Microbiol.* 5: 603–621.

Seal, B.S., Drider, D., Oakley, B.B. et al. (2018). Microbial-derived products as potential new antimicrobials. *Vet. Res.* 49: 66.

Singh, K.S., Sharma, R., Keshari, D. et al. (2017). Down-regulation of malate synthase in *Mycobacterium tuberculosis* H37Ra leads to reduced stress tolerance, persistence and survival in macrophages. *Tuberculosis (Edinb)* 106: 73–81.

Singh, V.K., Syring, M., Singh, A. et al. (2012). An insight into the significance of the DnaK heat shock system in *Staphylococcus aureus*. *Int. J. Med. Microbiol.* 302: 242–252.

Skindersoe, M.E., Alhede, M., Phipps, R. et al. (2008). Effects of antibiotics on quorum sensing in *Pseudomonas aeruginosa*. *Antimicrob. Agents Chemother.* 52: 3648–3663.

Soto, S.M. (2013). Role of efflux pumps in the antibiotic resistance of bacteria embedded in a biofilm. *Virulence* 4: 223–229.

Starkey, M., Hickman, J.H., Ma, L. et al. (2009). *Pseudomonas aeruginosa* rugose small-colony variants have adaptations that likely promote persistence in the cystic fibrosis lung. *J. Bacteriol.* 191: 3492–3503.

Starner, T.D., Shrout, J.D., Parsek, M.R. et al. (2008). Subinhibitory concentrations of azithromycin decrease nontypeable *Haemophilus influenzae* biofilm formation and diminish established biofilms. *Antimicrob. Agents Chemother.* 52: 137–145.

Stewart, P.S. (1996). Theoretical aspects of antibiotic diffusion into microbial biofilms. *Antimicrob. Agents Chemother.* 40: 2517–2522.

Stewart, P.S. (2002). Mechanisms of antibiotic resistance in bacterial biofilms. *Int. J. Med. Microbiol.* 292: 107–113.

Sung, J.Y., Koo, S.H., Kim, S., and Kwon, G.C. (2016). Persistence of multidrug-resistant *Acinetobacter baumannii* isolates harboring blaOXA-23 and bap for 5 years. *J. Microbiol. Biotechnol.* 26: 1481–1489.

Taha, M., Abdelbary, H., Ross, F.P., and Carli, A.V. (2018). New innovations in the treatment of PJI and biofilms-clinical and preclinical topics. *Curr. Rev. Musculoskelet Med.* 11 (3): 380–388.

Thakur, Z., Saini, V., Arya, P. et al. (2018). Computational insights into promoter architecture of toxin-antitoxin systems of *Mycobacterium tuberculosis*. *Gene* 641: 161–171.

Torres, B. G. S., Helfer, V. E., Bernardes, P. M., Macedo, A. J., Nielsen, E. I., Friberg, L. E. & Dalla Costa, T. 2017. Population pharmacokinetic modeling as a tool to characterize the decrease in ciprofloxacin free interstitial levels caused by *Pseudomonas aeruginosa* biofilm lung infection in Wistar rats. *Antimicrob. Agents Chemother.*, 61, pii: e02553-16.

Trotonda, M.P., Tamber, S., Memmi, G., and Cheung, A.L. (2008). MgrA represses biofilm formation in *Staphylococcus aureus*. *Infect. Immun.* 76: 5645–5654.

Truong-Bolduc, Q.C., Dunman, P.M., Strahilevitz, J. et al. (2005). MgrA is a multiple regulator of two new efflux pumps in *Staphylococcus aureus*. *J. Bacteriol.* 187: 2395–2405.

Twomey, K.B., O'connell, O.J., Mccarthy, Y. et al. (2012). Bacterial cis-2-unsaturated fatty acids found in the cystic fibrosis airway modulate virulence and persistence of *Pseudomonas aeruginosa*. *ISME J.* 6: 939–950.

Van Acker, H., Sass, A., Dhondt, I. et al. (2014). Involvement of toxin-antitoxin modules in *Burkholderia cenocepacia* biofilm persistence. *Pathog. Dis.* 71: 326–335.

Van Schaik, W., Top, J., Riley, D.R. et al. (2010). Pyrosequencing-based comparative genome analysis of the nosocomial pathogen *Enterococcus faecium* and identification of a large transferable pathogenicity island. *BMC Genomics* 11: 239.

Vega, N.M., Allison, K.R., Khalil, A.S., and Collins, J.J. (2012). Signaling-mediated bacterial persister formation. *Nat. Chem. Biol.* 8: 431–433.

Vieira, A., Ramesh, A., Seddon, A.M., and Karlyshev, A.V. (2017). CmeABC multidrug efflux pump contributes to antibiotic resistance and promotes *Campylobacter jejuni* survival and multiplication in *Acanthamoeba polyphaga*. *Appl. Environ. Microbiol.* 83 (22): pii: e01600-17.

Virmani, R., Hasija, Y., and Singh, Y. (2018). Effect of homocysteine on biofilm formation by Mycobacteria. *Indian J. Microbiol.* 58: 287–293.

Walawalkar, Y.D., Vaidya, Y., and Nayak, V. (2016). Response of *Salmonella typhi* to bile-generated oxidative stress: implication of quorum sensing and persister cell populations. *Pathog. Dis.* 74 (8): pii: ftw090.

Wang, D., Dorosky, R.J., Han, C.S. et al. (2015b). Adaptation genomics of a small-colony variant in a *Pseudomonas chlororaphis* 30-84 biofilm. *Appl. Environ. Microbiol.* 81: 890–899.

Wang, S., Zhou, C., Ren, B. et al. (2017). Formation of persisters in *Streptococcus mutans* biofilms induced by antibacterial dental monomer. *J. Mater. Sci. Mater. Med.* 28: 178.

Wang, W., Chanda, W., and Zhong, M. (2015a). The relationship between biofilm and outer membrane vesicles: a novel therapy overview. *FEMS Microbiol. Lett.* 362: fnv117.

Weiser, J., Henke, H.A., Hector, N. et al. (2016). Sub-inhibitory tigecycline concentrations induce extracellular matrix binding protein Embp dependent *Staphylococcus epidermidis* biofilm formation and immune evasion. *Int. J. Med. Microbiol.* 306: 471–478.

Wen, Y., Sobott, F., and Devreese, B. (2016). ATP and autophosphorylation driven conformational changes of HipA kinase revealed by ion mobility and crosslinking mass spectrometry. *Anal. Bioanal. Chem.* 408: 5925–5933.

Winther, K., Tree, J.J., Tollervey, D., and Gerdes, K. (2016). VapCs of *Mycobacterium tuberculosis* cleave RNAs essential for translation. *Nucleic Acids Res.* 44: 9860–9871.

Wood, T.K. (2016). Combatting bacterial persister cells. *Biotechnol. Bioeng.* 113: 476–483.

Wood, T.K. (2017). Strategies for combating persister cell and biofilm infections. *Microb. Biotechnol.* 10: 1054–1056.

Xia, H., Tang, Q., Song, J. et al. (2017). A yigP mutant strain is a small colony variant of *E. coli* and shows pleiotropic antibiotic resistance. *Can. J. Microbiol.* 63: 961–969.

Xiang, H., Cao, F., Ming, D. et al. (2017). Aloe-emodin inhibits *Staphylococcus aureus* biofilms and extracellular protein production at the initial adhesion stage of biofilm development. *Appl. Microbiol. Biotechnol.* 101: 6671–6681.

Xu, M., Yang, X., Yang, X.A. et al. (2016). Structural insights into the regulatory mechanism of the *Pseudomonas aeruginosa* YfiBNR system. *Protein Cell* 7: 403–416.

Yadav, M.K., Chae, S.W., and Song, J.J. (2012). Effect of 5-azacytidine on in vitro biofilm formation of *Streptococcus pneumoniae. Microb. Pathog.* 53: 219–226.

Yang, Q.E. and Walsh, T.R. (2017). Toxin-antitoxin systems and their role in disseminating and maintaining antimicrobial resistance. *FEMS Microbiol. Rev.* 41: 343–353.

Yee, R., Cui, P., Shi, W. et al. (2015). Genetic screen reveals the role of purine metabolism in *Staphylococcus aureus* persistence to rifampicin. *Antibiotics (Basel)* 4: 627–642.

Yi, L., Luo, L., and Lu, X. (2018). Efficient exploitation of multiple novel bacteriocins by combination of complete genome and peptidome. *Front. Microbiol.* 9: 1567.

Yoon, E.J., Chabane, Y.N., Goussard, S. et al. (2015). Contribution of resistance-nodulation-cell division efflux systems to antibiotic resistance and biofilm formation in *Acinetobacter baumannii. MBio* 6 (2): pii: e00309-15.

Zaidi, S., Misba, L., and Khan, A.U. (2017). Nano-therapeutics: a revolution in infection control in post antibiotic era. *Nanomedicine* 13: 2281–2301.

Zhao, J., Wang, Q., Li, M. et al. (2013). *Escherichia coli* toxin gene *hipA* affects biofilm formation and DNA release. *Microbiology* 159: 633–640.

Part III

Socio-Economical Perspectives and Impact of AR

10

Sources of Antibiotic Resistance

Zoonotic, Human, Environment

Ivone Vaz-Moreira[1,2], Catarina Ferreira[1*], Olga C. Nunes[2], and Célia M. Manaia[1]*

[1] Universidade Católica Portuguesa, CBQF - Centro de Biotecnologia e Química Fina – Laboratório Associado, Escola Superior de Biotecnologia, Porto, Portugal
[2] LEPABE – Laboratory for Process Engineering, Environment, Biotechnology and Energy, Faculdade de Engenharia, Universidade do Porto, Porto, Portugal
**These authors contributed equally to this work.*

10.1 The Antibiotic Era

The industrial revolution nurtured the need to combat infectious diseases and encouraged the efforts of scientists and pharmaceutical companies to find drugs with antimicrobial activity capable of overcoming infection, the main cause of mortality until the middle of the twentieth century (Cohen 2000; Aminov 2010). Therefore, antibiotherapy processes that were successful in the control of humans and animals infectious diseases became one of the most important revolutions of the modern human and veterinary medicine. The recognition that penicillin, a natural fungal metabolite, was able to control the development of pathogenic bacteria set a mark in mankind's history and started what can be called the antibiotic era (Aminov 2010). The search for antibiotics able to control pathogenic bacteria led to the recognition that the production of molecules with antibacterial activity is widespread in the microbial world, being produced by both fungi and bacteria. The effectiveness of these compounds explains why most of the antibiotics used nowadays in human and veterinary medicine are natural (biosynthetic) or semisynthetic derivatives of these natural products (Butler and Buss 2006).

Regardless of being biosynthetic, semisynthetic, or fully synthetic, the therapeutic action of antibiotics is due to their capacity to interfere with key structural components and/or functions of the bacterial cell, which are not present in the host's cells. Most of the antibiotic classes with clinical relevance

Antibiotic Drug Resistance, First Edition. Edited by José-Luis Capelo-Martínez and Gilberto Igrejas.

interfere with three major types of target in the bacterial cell – the cell wall synthesis, proteins synthesis, and DNA access, mainly during replication. Hence, although secondary toxic effects may occur, antibiotics selectively target bacteria, whose cells are destroyed or inhibited to divide, while no harm is anticipated in the host's cells.

Unfortunately, the enthusiasm put on antibiotics as therapeutic agents would not last. Resistance, meaning the ability to survive and proliferate in the presence of antibiotics at concentrations used for therapeutic purposes, is found for all antibiotics, sooner or later after their commercialization (Alanis 2005). When bacteria acquired the capability of recurrently grow in the presence of a clinically used dose of antibiotic, they cause the failure of the antibiotic as therapeutic agent and are, hence, named as antibiotic-resistant bacteria (ARB). The mechanisms used by ARB can be summarized as the ability to (i) alter the antibiotic molecule (degradation, transformation), (ii) control the antibiotic intracellular concentration (efflux, cell impermeabilization), or (iii) modify the cellular antibiotic target (Blair et al. 2015).

10.2 Intrinsic and Acquired Antibiotic Resistance

Some bacteria harbor ancestral traits that confer intrinsic resistance to one or more classes of antibiotics. Intrinsic resistance can result from specific cell properties, such as the absence of cell wall or the presence of an external membrane, or some specific chromosomal genes. In these cases, the genes encoding for such properties are part of the core genome of a given species or genus (EUCAST; Davies and Davies, 2010). Consequently, all members of a given taxonomic group (species, genus, family) share the same resistance phenotype. In contrast, when the genes encoding for antibiotic resistance (ARGs) are part of the accessory genome of a bacterial strain, which includes genetic information that was acquired, antibiotic resistance is only observed in some representatives of a given species (EUCAST). Acquired antibiotic resistance may result from gene mutation or genetic recombination (Martinez and Baquero 2000; Zhang et al. 2009a; Davies and Davies 2010). Gene mutations occur randomly in the genome, often potentiated by mutagenic agents, and when they represent an evolutionary advantage to the cell, they may become dominant through dissemination by vertical transmission (from one generation to the next). Genetic recombination, frequently referred to as horizontal gene transfer (HGT), is believed to be common among bacteria, representing one of the major driving forces for bacterial evolution (Ochman et al. 2000; Davies and Davies 2010; Wiedenbeck and Cohan 2011). The HGT may occur by (i) conjugation among bacteria, which involves the transfer of genetic material from a donor to a recipient cell, requiring that both share the same space, but not necessarily the same species; (ii) transformation, consisting on the uptake of naked DNA released by dead cells; and (iii) transduction, mediated by bacteriophages (Andersson and Hughes 2010).

10.3 The Natural Antibiotic Resistome

Antimicrobial biosynthesis and the associated evasion mechanisms are common in bacteria that coexist in a given microbial habitat (D'Costa et al. 2007). Antibiotic resistance is a natural property of bacteria, hypothetically favored in nature to cope with microbial community members that naturally produce antibiotic residues. Hence, resistance mechanisms may be regarded as survival traits in bacteria thriving in natural communities, being observed in bacteria that were never exposed to antibiotics of anthropogenic origin (Allen et al. 2010). The natural antibiotic resistome, i.e. the whole set of genes that contribute to cope with the presence of antibiotics, encodes possibly a wide panoply of functions that can span from microbial cell defense, inhibition of competitors growth, to biochemical signals, modulators of metabolic activity, or even natural substrates (Davies et al. 2006; Dantas et al. 2008; Martinez 2009). Remarkably, the soil natural resistome has been one of the most studied, with genetic determinants of resistance evidencing either high or low resemblance with the resistance determinants observed nowadays in clinically relevant bacteria (Riesenfeld et al. 2004; D'Costa et al. 2006; D'Costa et al. 2011; Forsberg et al. 2012). The diversity of the natural resistome is also evidenced by the fact that it is probably spread over the whole domain *Bacteria*, with frequent descriptions of occurrence in members of the phyla *Actinobacteria*, *Proteobacteria*, or *Bacteroidetes* (Riesenfeld et al. 2004; D'Costa et al. 2006; D'Costa et al. 2011; Forsberg et al. 2012).

10.4 The Contaminant Resistome

Due to the widespread dissemination in the environment, ARB and ARG are nowadays considered environmental contaminants (Pruden et al. 2006; Berendonk et al. 2015). Contaminant ARB and ARG are emitted by a wide diversity of sources, mainly human and animal excreta. Hence, domestic and animal farm effluents as well as animal manure are the major sources for ARB and ARG, which continuously enrich what has been called the contaminant resistome (Vaz-Moreira et al. 2014; Manaia 2017).

The human impact on the dissemination of antibiotic resistance is demonstrated by the observation of a direct correlation between human intervention and abundance and diversity of ARB and ARG. For example, studies with wildlife species, such as small mammals, gulls and birds of the prey, iguanas, or permafrost soils, show the wide dissemination of antibiotic resistance and, eventually, the effects of the continuous exposure of the biota to antibiotics and some pollutants (Thaller et al. 2010; D'Costa et al. 2011; Vredenburg et al. 2014; Furness et al. 2017). The risks that new resistance determinants jump from the natural resistome to clinically relevant bacteria, jeopardizing the efficiency of antibiotics still regarded as valuable therapeutic tools, exist and have been deeply discussed (Wright 2010; Perry and Wright 2013; Martinez et al. 2015; Manaia 2017). However, it is the continuous and wide dissemination of the contaminant antibiotic resistome that represents nowadays a major threat for human health. Since the acquisition of ARGs represents an additional cost for the bacterial cell, which may be justified only if it brings a survival advantage, it is believed that it is stimulated by external selective pressures (Goh et al. 2002; Allen et al. 2010; Andersson and Hughes 2014). Among the different selective pressures that are identified (e.g. metals or biocides), antibiotics are those considered the most important selectors for resistance acquisition and maintenance. Indeed, it is impossible to dissociate the evolution of the contaminant resistome from that of antibiotic usage, to which the next section is dedicated.

10.5 Evolution of Antibiotics Usage

Since the 1940s the global production and consumption of antibiotics has increased not only due to the rise of the human population but also due to prosperity (CDDEP 2015). It is estimated that during the first decade of the twenty-first century occurred an overall increase of 36% on antibiotic use for human consumption. This percentage corresponds to an average increase from 5×10^{10} to 7×10^{10} standard units, i.e. the number of doses sold, in 71 countries with different incomes (Van Boeckel et al. 2014). In general, the consumption of antibiotics per capita is higher in high-income countries than in middle-income countries. However, in middle-income countries the increase in human antibiotic consumption has been higher than in high-income countries, where, in general, the consumption stabilized or decreased (Van Boeckel et al. 2014). Brazil, Russia, India, China, and South Africa account for 76% increase in antibiotic sales for human consumption (Van Boeckel et al. 2014).

The absence of adequate measures to prevent infectious diseases and the inappropriate antibiotic consumption has probably significantly contributed to the excessive antibiotic use. Indeed, it has been argued that in some low-income countries, antibiotic consumption increase is mainly a compensation of the lack of efficient programs of vaccination and sanitation (Laxminarayan et al. 2016). In turn, inadequate antibiotics prescription is associated with the uncertainty in diagnosis (e.g. frequently to treat upper respiratory tract infections caused by viruses), motivating the prescription of an unnecessary broad-spectrum antibiotic, with an incorrect dosage or duration (Starrels et al. 2009; Om et al. 2016). Notoriously, these situations are not necessarily related with the country development index. In hospitals, the excessive use of broad-spectrum antibiotics is leading to dangerous and almost difficult to treat infections. However, the lack of or delayed access to antibiotics still kills more people than resistant infections (CDDEP 2015).

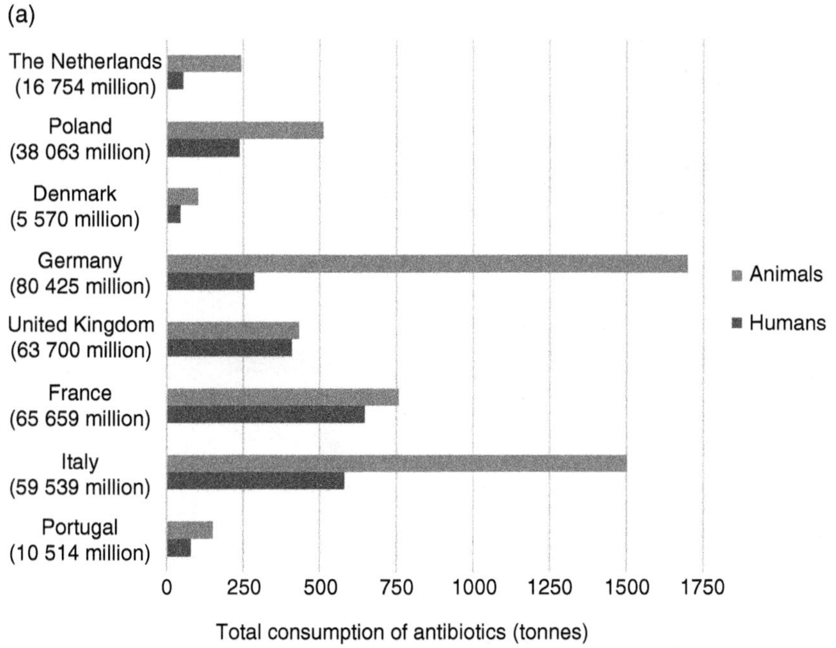

Figure 10.1 Consumption of antibiotics in humans and food-producing animals (a) by selected European countries in 2012 and (b) by class of antibiotics for each European country. Data of consumption in tonnes was recovered from ECDC, EFSA, and EMA report (ECDC/EFSA/EMA 2015). Data from human consumption includes the antibiotics used in the community and hospitals. Data from food-producing animal's consumption includes the use in pigs, cattle, and broilers. The countries' population size data are estimates from the World Bank for the year 2012 (http://data.worldbank.org/indicator/SP.POP.TOTL). *(See insert for color representation of the figure.)*

The simultaneous increase of the per capita income and population growth have been driving pressure for availability of animal protein, and consequently, the need to use antibiotics to optimize intensive animal/aquaculture farming. Antibiotics and other antimicrobials have been extensively used in livestock animals and aquaculture not only to treat diseases but mainly to improve growth or to prevent infections (FAO 2016; Liu et al. 2017). As a consequence, the utilization of antibiotics in animal farming (e.g. poultry, swine, and cattle) is higher than human consumption (Figure 10.1a), constituting up to 70% of the annual consumption of antibiotics in each country (FDA 2015). Although with a smaller impact, antibiotics are used also to control diseases and pests in household pets and agriculture crops, respectively (Lloyd 2007; Prescott 2008; CDDEP 2015). The utilization of antimicrobials in intensive farming, sometimes without the supervision of veterinary professionals, raises a major concern for human health since many compounds used in food animals

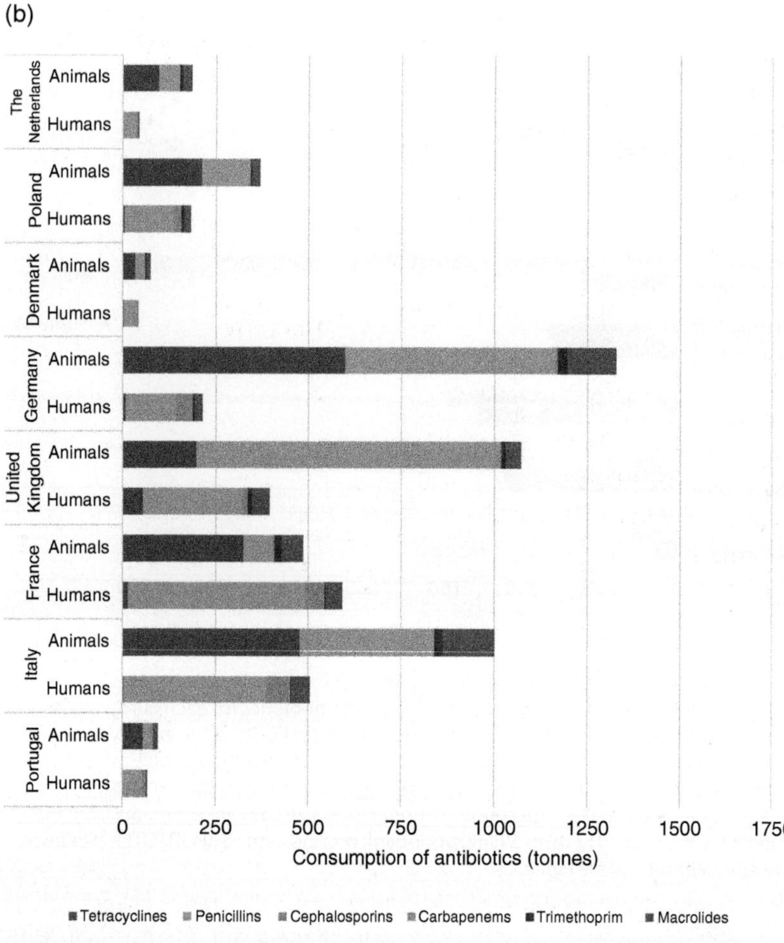

Figure 10.1 (Continued)

belong to the same antimicrobial classes as those used in humans (Figure 10.1b) (FAO 2016).

Although the worldwide data on the antimicrobial sales for animal production is limited, Van Boeckel and collaborators (2015) conducted the first study assessing the antimicrobial consumption in food-producing animals around the world. The authors estimated a total consumption of more than 63 000 tonnes of antibiotics in livestock in 2010 and predicted an increase to more than 105 000 tons by 2030. Among the countries with the highest antibiotics consumption in 2010 were China (23%), United States of America (13%), Brazil (9%), India (3%), and Germany (3%). The predictions by Van Boeckel et al. (2015) for food animal production in 2030 identify China (30%), United

States of America (10%), Brazil (8%), India (4%), and Mexico (2%) as the five major consumers.

10.6 Antibiotic Resistance Evolution

It is assumed that the presence of antimicrobials in a given ecological niche will lead to the successive elimination of susceptible bacteria with the simultaneous Darwinian selection of ARB (Founou et al. 2016). Hence, any host, human or animal, consuming antibiotics or environmental compartment contaminated with antibiotics, are potential reservoirs of ARB and ARG. In parallel with selection, the stress imposed by the presence of antibiotic residues may favor the acquisition of ARG by susceptible bacteria, making antibiotic residues important evolution drivers (Andersson and Hughes 2011). Although antibiotic resistance acquisition is common in nature, these processes are facilitated by the massive and intensive use of antibiotics (Sørensen et al. 2005). A high number of reports and studies produced over the last decades show an increase not only of resistance prevalence but also of the diversity and distribution of ARG (EARS-Net; ESAC; NARMS; Knapp et al. 2010). The exposure of the bacterial communities to antibiotics has been suggested as a major driver for the emergence and spread of resistance. It is believed that mechanisms such as the disturbance of the microbial communities, interference with gene expression, or biofilm formation can trigger or act in combination with the horizontal transfer of ARG (Martinez 2008; Andersson and Hughes 2012, 2014; You and Silbergeld 2014). In addition, it has been argued that the levels of resistance observed at any time and place are the result not only of the recent conditions but also of the global history of antibiotic resistance acquisition (O'Brien 2002). This can be explained based on the fact that when the carriage of resistance determinants does not impose a fitness cost, they will not be lost by their host, even in the absence of selective pressures (Andersson and Hughes 2010, 2011).

10.7 Stressors for Antibiotic Resistance

Besides the antibiotic residues or metabolites thereof, some other chemical compounds have been described to contribute to ARB and ARG enrichment due to processes of co- or cross-resistance. Co-resistance to antibiotics and other chemicals occurs when the genes specifying the resistance phenotypes are located together in the same genetic element. In turn, cross-resistance arises from a given mechanism that confers resistance to two or more antimicrobials from different classes. The best example of co-resistance is the genetic linkage of ARG and metal translocation genes (e.g. Hg, Cu, Cd, Zn) (Seiler and Berendonk 2012; Pal et al. 2015). Examples of cross-resistance mechanisms are the broad substrate spectrum efflux pumps or neutralizing enzymes, reduced cell envelop permeability, or alteration of the target site, with the consequent invulnerability to antimicrobial agents belonging to different classes (Seiler and Berendonk 2012; Wales and Davies 2015).

The potential of metals as antibiotic resistance stressors is relevant, given their widespread distribution. Metal supplementation is used in animal and aquaculture feed and is found in the composition of organic and inorganic fertilizers, in pesticides, and in anti-fouling products. In addition, some metals are also used as biocides, controlling diseases, mainly in the pig and poultry sectors (Seiler and Berendonk 2012; Wales and Davies 2015). Consequently, they are found not only in agricultural soils and sediments but also in aquaculture and in domestic and industrial wastewater, through which they reach the wastewater treatment plants (WWTPs) (Nicholson et al. 2003; Gall et al. 2015). Besides metals, many other compounds are used as biocides. These antimicrobial compounds have a wide application, being used as preservatives of pharmaceuticals, cosmetics or feed/food products, or disinfectants or antiseptics in diverse human activities (agricultural, industrial, and healthcare settings and at community level). Like metals, biocides residues are found in different ecological niches, as domiciliary, healthcare, or aquatic environments (Bloomfield 2002; Maillard 2005; Antizar-Ladislao 2008; Meyer and Cookson 2010; Chen et al. 2014), where they are pointed out as favoring the emergence of antibiotic resistance through co-selection. Such conclusions are based on the fact that some bacterial strains have low susceptibility (i.e. high MICs) to simultaneously metals and/or biocides and antibiotics (Wales and Davies 2015).

Co-resistance of antibiotics and metals, such as Hg, Cd, Cu, and Zn, has been described (Seiler and Berendonk 2012; Pal et al. 2015). Mercury (Hg) has been detected in fish feed and is frequently found in sewage and activated sludge (Olson et al. 1991; Choi and Cech 1998). Genetic linkage of multiple antibiotic resistance (sulfonamides, tetracyclines, beta-lactams, streptomycin, and florfenicol) and Hg was found in an *Aeromonas salmonicida* subsp. *salmonicida* strain isolated from salmon from aquaculture facilities (Seiler and Berendonk 2012). Co-resistance of macrolide/aminoglycoside antibiotics and Cu, Cd, and Zn used as animal growth promoters and frequently found in different environments has also been described (Seiler and Berendonk 2012; Pal et al. 2015). Hence, even at very low concentrations, metals have the potential to select bacteria with multidrug-resistant plasmids, contributing to the emergence, maintenance, and transmission of antibiotic-resistant bacteria (Gullberg et al. 2011). Co-resistance of Hg and quaternary ammonium compounds (QAC) (Pal et al. 2015) as well as QAC and sulfonamides has been described (Wales and Davies 2015). Cross-resistance of fluoroquinolone or tetracycline and phenolic biocides is also reported in *Escherichia* and *Salmonella* strains (Wales and Davies 2015).

Besides the reduced fitness costs explained above, other mechanisms are believed to contribute to antibiotic resistance persistence even in the absence of selective pressures. The toxin–antitoxin system has been suggested as an important factor for the stabilization of ARGs and metal or biocide resistance genes that are co-transferred by conjugative plasmids. It is supposed that these plasmids would be more persistent when vertically transferred from one generation to the next, even in the absence of selective pressure, since the toxin–antitoxin systems stabilize plasmids in their hosts by killing daughter cells that do not inherit the plasmid (Pal et al. 2015). The toxin–antitoxin system is responsible for the production of a toxin and of an antidote for that toxin (antitoxin). If the plasmid is absent in a daughter cell, the unstable antitoxin is degraded and the stable toxic protein kills the new cell; this is known as post-segregation killing (Van Melderen and De Bast 2009).

Besides exposure to antimicrobials, many other adaptive stress responses may influence the susceptibility to antibiotics since they may impact many of the same components and processes that are targeted by antimicrobials. Some non-antimicrobial agents have been described as potential selectors for antibiotic resistance. Some examples are feed or food preservation agents, such as phenazopyridine or sepiolite, or sub-lethal concentrations of salt and acidic pH as well as herbicides, such as glyphosate, widely used in agriculture and gardening (McGowan et al. 2006; Amabile-Cuevas and Arredondo-Garcia 2013; Rodriguez-Beltran et al. 2013; Kurenbach et al. 2015). In addition, nutrient starvation, oxidative stress, thermal shock, and cell envelope damage are among the factors compromising the cell growth by stimulating protective changes in cell physiology or in lifestyle (biofilm formation) or inducing mutations (Poole 2012).

In summary, a wide range of conditions, including those promoted by micropollutants, prevailing in the different ecological niches (animal and human hosts, feed and food products, anthropogenic impacted environments) may contribute to the overall increase and dissemination of antibiotic resistance.

10.8 Paths of Antibiotic Resistance Dissemination

The human and animal intestinal tracts favor the occurrence of antibiotic resistance selection and/or HGT (Shoemaker et al. 2001; Sommer et al. 2010), as these are considered important reservoirs of ARB and ARG (Salyers et al. 2004) as well as livestock and aquaculture as other reservoirs in the environment (Landers et al. 2012; CDC 2013; Woolhouse et al. 2015). All these sources have the potential to supply ARB and ARG to the environment, where they can accumulate and spread through aquatic systems and soils (Figure 10.2). The use of animal manure or compost to amend agriculture soils, a practice that is in line with the nowadays so appreciated organic farming, is another potential

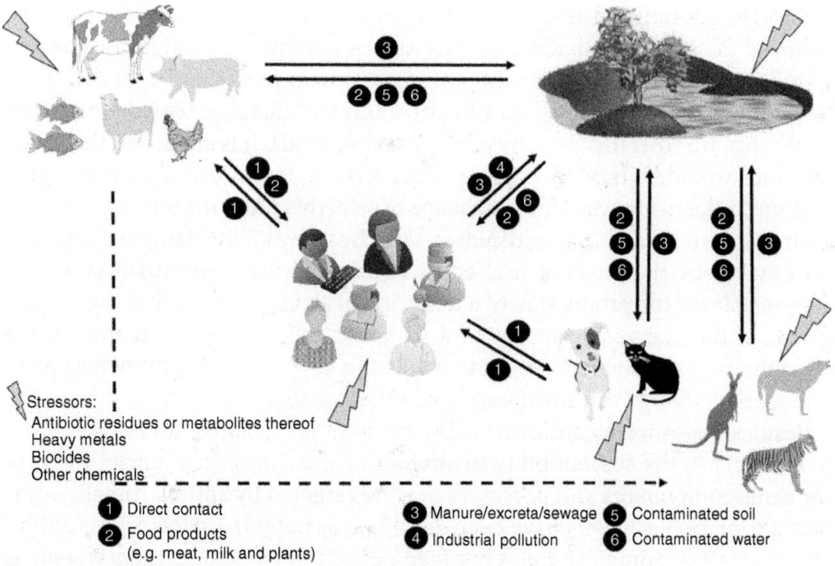

Figure 10.2 Antibiotic resistance dissemination pathways.

source of contamination of soil, surface, and groundwater with ARB and ARG from animal origin. The point source contamination will diffuse through processes such as water runoff or percolation, invading the aquatic systems (Figure 10.2). While these diffuse paths of ARB dissemination may be associated with the transmission from the environment to humans, some point source contamination can also be recognized. For example, humans with a high degree of exposure to reservoirs of resistance, such as healthcare facilities, animal farms, and abattoirs, as well as food handlers are major receptors and potential disseminators of ARB and ARG into the community and healthcare settings (Marshall and Levy 2011). Also humans living in contact with livestock and pets are important links in the antibiotic resistance dissemination network (Guardabassi et al. 2004; Lloyd 2007; Smith et al. 2009).

Domestic wastewater is another relevant source of ARB and ARG, which, in world regions with adequate sanitation network, are collected in sewage systems and treated at urban wastewater treatment plants (UWTP). Because most of these plants use conventional treatments designed mainly to reduce the organic and microbial loads of wastewater, part of the ARB and ARG of the incoming sewage is discharged into the natural water body receptors that include rivers, lakes, and the sea in coastal areas (Michael et al. 2013; Rizzo et al. 2013; Czekalski et al. 2014). Studies conducted in different world regions show that every minute a well-functioning domestic UWTP can release more than 10^9 antibiotic-resistant enteric bacteria and more than 10^{14} copies of antibiotic resistance genes (Vaz-Moreira et al. 2014; Manaia et al. 2016). This is a very high microbial load that will result in the contamination of rivers or lakes, widely demonstrated in the literature (Novais et al. 2005; LaPara et al. 2011; Tacão et al. 2014), as well as their sediments (LaPara et al. 2011; Czekalski et al. 2014) or soils (Jones-Dias et al. 2016a). The conditions prevailing in the biological reactors of the UWTPs (high load of readily metabolizable nutrients and of bacteria together with micropollutants) may not favor the elimination of ARB or even enhance the selection of ARB and/or the occurrence of HGT of ARG. Hence, the implementation of tertiary treatments including disinfection is desirable (Manaia et al. 2016). These processes may reduce the microbial load of the treated wastewater (up to 4 log units) to lower values than those with conventional treatments (up to 2 log units) (Michael et al. 2013; Sousa et al. 2017). Nevertheless, disinfected treated wastewater still contains ARB and ARG, regardless the method employed (e.g. UV 254 nm, chlorination, ozonation) (LaPara et al. 2011; Moreira et al. 2016; Sousa et al. 2017). Moreover, some authors have suggested that disinfection may increase the prevalence of ARB and ARG in the treated wastewater (Chen and Zhang 2013; Alexander et al. 2016; Hu et al. 2016).

The reduction of the organic load of the raw wastewater is accompanied with the increase of the biomass of the degrading microorganisms. The spent activated sludge is further stabilized, digested, or eventually used for composting (Kelessidis and Stasinakis 2012). ARB and ARG surviving these processes may reach soils (farms and gardens) through the direct application of activated sludge, digested sludge, or compost as soil amendments (Ma et al. 2011; Zhang et al. 2011). Wastewater reuse is another key issue regarding antibiotic resistance dissemination. The human population growth and prosperity combined with climate changes are increasingly pressing for reuse of treated wastewater (Becerra-Castro et al. 2015). The consequent contamination of soils and of the human food chain is a major concern (Campos et al. 2013; Jones-Dias et al. 2016b).

ARB and ARG have been detected in produce and food products (e.g. vegetables, meat, poultry, milk, eggs), which can function as vehicles of ARG transfer to humans, mainly if consumed raw (Aarestrup et al. 2000; van den Bogaard and Stobberingh 2000; Sunde and Norstrom 2006; Bezanson et al. 2008; Mena et al. 2008; Leverstein-van Hall et al. 2011; Marti et al. 2013). Indeed, ARGs that encode resistance against antibiotics only used in animals (e.g. nourseothricin, apramycin) were detected in commensal bacteria of both human and animal origin, in zoonotic pathogens like *Salmonella* spp., and in strictly human pathogens, like *Shigella* spp. (van den Bogaard and Stobberingh 2000).

Soils and water contamination may also affect different forms of wildlife such as the Iberian lynx or seagulls that fly over our cities (Sousa et al. 2014; Vredenburg et al. 2014). In these cases, the migratory routes of these wild animals can represent important forms of dispersion of resistance between distant zones. The widespread environmental contamination with ARB is also detected indoor mainly in urban areas, as have been shown in studies conducted on bus lines (Mendes et al. 2015) or with domestic animals that cohabit with their owners (Leite-Martins et al. 2014). In fact, from what is known nowadays, one can wonder which places are antibiotic-resistance-free. This contamination originates essentially from animals and humans. Whether a part of this contamination returns to humans is an issue yet to be clarified (Manaia 2017).

10.9 Antibiotic Resistance in Humans and Animals

Prophylaxis and metaphylaxis, i.e. the use of therapeutic or sub-therapeutic doses of antibiotics in healthy animals to prevent infectious diseases, to both healthy and infected animals are relevant contributions for the contaminant resistome. Although some antibiotics are restricted to the animal use (e.g. tylosin, apramycin, avoparcin), others are used in both food animals and the treatment of humans' infections (e.g. gentamicin, tetracyclines, penicillins, and sulfonamides) (Kemper 2008; Landers et al. 2012). In addition, although some antibiotics are restricted to veterinary use, they belong to the same general classes of those used in humans; thus, even if they are not the same exact compound, their mode of action is the same or similar (Phillips et al. 2004). The major fraction of antibiotic consumption is associated with animal production, not only because more individual doses are used but also because the therapeutic doses, which are proportional to the body weight, are higher for some animals than for humans. The influence of the antibiotic use on the enrichment of ARB in animals has been demonstrated. For example, in Denmark it was observed a reduction, from 80 to 3%, in the levels of vancomycin-resistant enterococci in poultry after the banning of avoparcin as growth promoter (Singer et al. 2003). Although the association of avoparcin, and other glycopeptides, with the prevalence of vancomycin resistance in livestock is one of the most studied, other associations have been demonstrated for other classes of antimicrobials, such as virginiamycin use and quinupristin/dalfopristin resistance; tylosin use and erythromycin resistance; avilamycin use and avilamycin resistance (Wegener 2003; Bengtsson and Wierup 2006; Cogliani et al. 2011). Some examples of co-selection were also documented, as the genetic linkage of the genes *van*A and the macrolide resistance gene *erm*B in glycopeptide-resistant enterococci from porcine origin. Thus, the ban of avoparcin was not enough to decrease the prevalence of glycopeptide-resistant enterococci, as this decrease was observed only after the ban of macrolide growth promoters (Aarestrup 2000; Boerlin et al. 2001).

Recently the World Health Organization published a list of antibiotic-resistant "priority pathogens," including 12 groups of bacteria that pose the greatest threat to human health (WHO 2017). The most critical groups given their potential to develop multidrug-resistant phenotypes, include *Acinetobacter baumannii*, *Pseudomonas aeruginosa*, and different *Enterobacteriaceae* genera (including *Klebsiella pneumoniae*, *Enterobacter* spp., *Serratia* spp., *Proteus* spp., *Providencia* spp., and *Morganella* spp.). In the same list, and considered to be of high or medium risk, are *Enterococcus faecium*, *Streptococcus pneumoniae*, *Staphylococcus aureus*, *Helicobacter pylori*, *Campylobacter*, *Haemophilus influenzae*, *Salmonella* spp., *Shigella* spp., and *Neisseria gonorrhoeae*. Some of these bacteria are also important commensals or pathogens in animals. Indeed, animals can be vehicles of *A. baumannii*, *Enterobacteriaceae* (e.g. *K. pneumoniae*), *Enterococcus*, *S. aureus*, *H. pylori*, *Campylobacter*, *Salmonella* spp., and *Shigella* spp. (Lahuerta et al. 2011; Abdel-Raouf et al. 2014; Cantas and Suer 2014; EFSA and ECDC 2015; Damborg et al. 2016). Among these, *Salmonella* and *Campylobacter* are considered as the most frequent foodborne bacterial agents in Europe (EFSA and ECDC 2015), while *A. baumannii* have been more associated with companion animals (Damborg et al. 2016; Ewers et al. 2017; Lupo et al. 2017). In general, farm animals are recognized as important antimicrobial resistance reservoirs (Bywater et al. 2004; EFSA and ECDC 2017), including resistance to some recent drugs as carbapenems (Fischer et al. 2013), that are detected both in animals and humans (Gupta et al. 2003). Nevertheless, there are few studies establishing the direction of the movement of resistance between human and livestock populations (Woolhouse et al. 2015). Indeed, the demonstration of transmission of ARB from animals to humans may be not straightforward.

In 1976, Levy et al. (1976), maybe in one of the first reports, published a study demonstrating the dissemination of bacteria from animals to humans, observing the same tetracycline resistant *Escherichia coli* strain in chickens and in the workers of the farm. Later studies, based on the whole genome sequencing (WGS), showed that the lineage CC97 of methicillin-resistant *S. aureus* (MRSA) was observed to enter the human population from a livestock source in more than one instance over the past 100 years (Spoor et al. 2013). In contrast, for the lineage ST5, globally disseminated in poultry, it was concluded that it was transmitted from humans (Lowder et al. 2009). However, other authors quantifying the occurrence of cross-species transmission for the MRSA strain CC398 concluded that the transmissions from livestock-to-human were more frequent than from human-to-livestock over the evolutionary history of the strain (Ward et al. 2014). Similarly, Sørensen et al. (2001) studying the transmission of glycopeptide-resistant *E. faecium*, confirmed the risk associated with the consumption of meat products contaminated with resistant bacteria, showing that these bacteria ingested via chicken or pork meat lasted in human stool for up to 14 days.

Homologous ARGs identified in human and farm animal isolates provided evidences for the possible cross transfer of ARB between animals and humans. For example, the gene responsible for methicillin resistance (*mec*A) in *S. aureus* found in animal isolates (e.g. cattle, pigs, chickens) was also detected in isolates from farmers (Lee 2003). Also in farms where apramycin was used as growth promoter, the gene *aac(3)-IV*, encoding resistance to aminoglycosides, was detected in *E. coli* isolates of swine, poultry, and farm workers (Chaslus-Dancla et al. 1991; Zhang et al. 2009b; Li et al. 2015).

Cephalosporins are extensively used for production of food animals, and beta-lactam resistance has been used in the literature to illustrate antibiotic resistance dissemination between both potential reservoirs. For example, there are some evidences of possible cross-contamination of bacteria carrying enzymes conferring resistance against this type of antibiotic, such as the beta-lactamase CTX-M-14, across humans, pets, and poultry in Asian countries (Ewers et al. 2012). In addition, the CTX-M-15 subtype, observed to be spread by different *Enterobacteriaceae*, including a *K. pneumoniae* clone in humans, is also found in pets and horses (Ewers et al. 2014). Contrary to cephalosporins, carbapenems (a class of last resort antibiotics) are not used in livestock. However, carbapenemase-encoding genes (e.g. the bla_{NDM}, bla_{IMP}, bla_{VIM}, and bla_{KPC} genes) were already detected in isolates of livestock animals (Mollenkopf et al. 2017). Woodford et al. (2014) demonstrated that the ARB from swine and poultry animals carrying the bla_{VIM} carbapenemase were also found in other farm animals (e.g. insects and rodents) via the transmission carried out through manure. Yet, studies on evidencing the transmission of carbapenem resistance from humans to animals are scarce (Mollenkopf et al. 2017). Similarly, for the beta-lactamase, NDM-1, which emerged recently as a major clinical threat that was spread via intercontinental dissemination, was never found as an indication of animal–human transmission (Kumarasamy et al. 2010; Nordmann 2011; Cabanes et al. 2012). Though, the gene bla_{NDM-1} was already found in *E. coli* isolates of companion animals, suggesting the possibility of interspecies transmission (Shaheen et al. 2013).

Colistin is considered a last resort antibiotic, namely, for NDM-1-producing bacteria. Originally described as being a chromosomally encoded resistance type, recently colistin resistance became transferable through the gene *mcr-1*, identified in human and animal *E. coli* isolates from China (Liu et al. 2016; Schwarz and Johnson 2016). The higher prevalence of *mcr-1*-carrying bacteria in animals over humans suggests that this resistance gene has animal origin and has spread to humans (Liu et al. 2016). Supporting this hypothesis is the extended use of polymyxins as animal growth promoters, for prophylaxis and therapy, which may have selected for plasmid-mediated colistin resistance (Al-Tawfiq et al. 2017). Notoriously, recent reports highlighted the continuous spread of plasmid-mediated *mcr-1* gene to imported food, to urban rivers, and to humans (Arcilla et al. 2016; Al-Tawfiq et al. 2017; do Monte et al. 2017). Another MCR subtype, the *mcr-2* gene, has been identified only in livestock *E. coli* isolates, reinforcing the animal origin for plasmid-mediated resistance to colistin (Xavier et al. 2016).

In a screen for the presence of 260 ARGs in different habitats, including wastewater, soil, human, and livestock samples, Li et al. (2015) observed that the most abundant and commonly distributed ARGs were the ones associated with antibiotics extensively used in human and veterinary medicine. These included the growth promotors (e.g. aminoglycoside, bacitracin, β-lactam, chloramphenicol, macrolide–lincosamide–streptogramin, quinolone, sulfonamide, and tetracycline). Ninety-nine out of 260 ARGs shared between human and young livestock feces, it was possible to observe that some were more abundant in humans than in animals (e.g. bacitracin resistance genes), while the opposite was observed for others (e.g. macrolide–lincosamide–streptogramin, tetracycline, sulfonamide, aminoglycoside, and multidrug resistance genes) (Li et al. 2015). The association between humans and domestic wastewater was also reported by Li et al. (2015), who found that 68 ARGs (out of 260) encoding resistance to tetracycline, beta-lactam, and aminoglycosides detected in human feces samples and domestic raw wastewater showed high similarity. While it is clear that most of the ARGs detected in humans can reach the wastewater treatment plants and even persist after treatment, there are evidences that the dissemination in the environment may favor some, probably harbored by stable mobile genetic elements and/or ubiquitous bacteria (Czekalski et al. 2014; Munck et al. 2015). Some ARGs, despite the low general prevalence, are good examples of high persistence or dissemination potential. For example, the $bla_{\text{KPC-2}}$ gene carried by carbapenem-resistant bacteria has been detected in the rivers receiving wastewater treatment plants discharges (Picão et al. 2013; Yang et al. 2017). In India, New Delhi, the $bla_{\text{NDM-1}}$ gene was detected in drinking water bacteria (*Achromobacter* spp., *Kingella denitrificans*, and *P. aeruginosa*), after the observation of some cases of human contamination with *Enterobacteriaceae* and *A. baumannii* isolates carrying this gene (Walsh et al. 2011). In summary, although some evidences are available, the impact of animal and environmental reservoirs onto the human health is still an issue deserving thorough research (Phillips et al. 2004; Martinez 2008; Woolhouse et al. 2015).

10.10 Final Considerations

This overview reinforces the need for an urgent implementation of the One Health perspective, in which the health and well-being of the citizens, the safety of the food chain, and the protection of the environment are comprehensively surveilled and controlled. Along with other contaminants and human-health threats, antibiotic resistance is a key issue in the One Health priorities for the next decades.

References

Aarestrup, F.M. (2000). Characterization of glycopeptide-resistant *Enterococcus faecium* (GRE) from broilers and pigs in Denmark: genetic evidence that persistence of GRE in pig herds is associated with coselection by resistance to macrolides. *J Clin Microbiol* 38 (7): 2774–2777.

Aarestrup, F.M., Kruse, H., Tast, E. et al. (2000). Associations between the use of antimicrobial agents for growth promotion and the occurrence of resistance among *Enterococcus faecium* from broilers and pigs in Denmark, Finland, and Norway. *Microb Drug Resist Mech Epidemiol Dis* 6 (1): 63–70.

Abdel-Raouf, M., Abdel-Gleel, Y., and Enab, A. (2014). Study on the role of pet animals for *Helicobacter pylori* transmission. *J Am Sci* 10: 20–28.

Alanis, A.J. (2005). Resistance to antibiotics: are we in the post-antibiotic era? *Arch Med Res* 36 (6): 697–705.

Alexander, J., Knopp, G., Dotsch, A. et al. (2016). Ozone treatment of conditioned wastewater selects antibiotic resistance genes, opportunistic bacteria, and induce strong population shifts. *Sci Total Environ* 559: 103–112.

Allen, H.K., Donato, J., Wang, H.H. et al. (2010). Call of the wild: antibiotic resistance genes in natural environments. *Nat Rev Microbiol* 8 (4): 251–259.

Al-Tawfiq, J.A., Laxminarayan, R., and Mendelson, M. (2017). How should we respond to the emergence of plasmid-mediated colistin resistance in humans and animals? *Int J Infect Dis* 54: 77–84.

Amabile-Cuevas, C.F. and Arredondo-Garcia, J.L. (2013). Nitrofurantoin, phenazopyridine, and the superoxide-response regulon soxRS of *Escherichia coli*. *J Infect Chemother* 19 (6): 1135–1140.

Aminov, R.I. (2010). A brief history of the antibiotic era: lessons learned and challenges for the future. *Front Microbiol* 1: 134.

Andersson, D.I. and Hughes, D. (2010). Antibiotic resistance and its cost: is it possible to reverse resistance? *Nat Rev Microbiol* 8 (4): 260–271.

Andersson, D.I. and Hughes, D. (2011). Persistence of antibiotic resistance in bacterial populations. *FEMS Microbiol Rev* 35 (5): 901–911.

Andersson, D.I. and Hughes, D. (2012). Evolution of antibiotic resistance at non-lethal drug concentrations. *Drug Resist Updat* 15 (3): 162–172.

Andersson, D.I. and Hughes, D. (2014). Microbiological effects of sublethal levels of antibiotics. *Nat Rev Microbiol* 12 (7): 465–478.

Antizar-Ladislao, B. (2008). Environmental levels, toxicity and human exposure to tributyltin (TBT)-contaminated marine environment. A review. *Environ Int* 34 (2): 292–308.

Arcilla, M.S., van Hattem, J.M., Matamoros, S. et al. (2016). Dissemination of the mcr-1 colistin resistance gene. *Lancet Infect Dis* 16 (2): 147–149.

Becerra-Castro, C., Lopes, A.R., Vaz-Moreira, I. et al. (2015). Wastewater reuse in irrigation: a microbiological perspective on implications in soil fertility and human and environmental health. *Environ Int* 75: 117–135.

Bengtsson, B. and Wierup, M. (2006). Antimicrobial resistance in Scandinavia after a ban of antimicrobial growth promoters. *Anim Biotechnol* 17 (2): 147–156.

Berendonk, T.U., Manaia, C.M., Merlin, C. et al. (2015). Tackling antibiotic resistance: the environmental framework. *Nat Rev Microbiol* 13 (5): 310–317.

Bezanson, G.S., MacInnis, R., Potter, G., and Hughes, T. (2008). Presence and potential for horizontal transfer of antibiotic resistance in oxidase-positive bacteria populating raw salad vegetables. *Int J Food Microbiol* 127 (1–2): 37–42.

Blair, J.M., Webber, M.A., Baylay, A.J. et al. (2015). Molecular mechanisms of antibiotic resistance. *Nat Rev Microbiol* 13 (1): 42–51.

Bloomfield, S.F. (2002). Significance of biocide usage and antimicrobial resistance in domiciliary environments. *J Appl Microbiol* 92: 144s–157s.

Boerlin, P., Wissing, A., Aarestrup, F.M. et al. (2001). Antimicrobial growth promoter ban and resistance to macrolides and vancomycin in enterococci from pigs. *J Clin Microbiol* 39 (11): 4193–4195.

van den Bogaard, A.E. and Stobberingh, E.E. (2000). Epidemiology of resistance to antibiotics: links between animals and humans. *Int J Antimicrob Agents* 14 (4): 327–335.

Butler, M.S. and Buss, A.D. (2006). Natural products: the future scaffolds for novel antibiotics? *Biochem Pharmacol* 71 (7): 919–929.

Bywater, R., Deluyker, H., Deroover, E. et al. (2004). A European survey of antimicrobial susceptibility among zoonotic and commensal bacteria isolated from food-producing animals. *J Antimicrob Chemother* 54 (4): 744–754.

Cabanes, F., Lemant, J., Picot, S. et al. (2012). Emergence of *Klebsiella pneumoniae* and *Salmonella* metallo-beta-lactamase (NDM-1) producers on Reunion Island. *J Clin Microbiol* 50 (11): 3812–3812.

Campos, J., Mourao, J., Pestana, N. et al. (2013). Microbiological quality of ready-to-eat salads: an underestimated vehicle of bacteria and clinically relevant antibiotic resistance genes. *Int J Food Microbiol* 166 (3): 464–470.

Cantas, L. and Suer, K. (2014). Review: the important bacterial zoonoses in "one health" concept. *Front Public Health* 2: 144.

CDC (2013). *Antibiotic Resistance Threats in the United States*. Atlanta, GA: Centres for Disease Control and Prevention.

CDDEP (2015). *The State of the World's Antibiotics, 2015*. Washington, DC: CDDEP.

Chaslus-Dancla, E., Pohl, P., Meurisse, M. et al. (1991). High genetic homology between plasmids of human and animal origins conferring resistance to the aminoglycosides gentamicin and apramycin. *Antimicrob Agents Chemother* 35 (3): 590–593.

Chen, H. and Zhang, M. (2013). Effects of advanced treatment systems on the removal of antibiotic resistance genes in wastewater treatment plants from Hangzhou. *China Environ Sci Technol* 47 (15): 8157–8163.

Chen, Z.F., Ying, G.G., Liu, Y.S. et al. (2014). Triclosan as a surrogate for household biocides: an investigation into biocides in aquatic environments of a highly urbanized region. *Water Res* 58: 269–279.

Choi, M.H. and Cech, J.J. (1998). Unexpectedly high mercury level in pelleted commercial fish feed. *Environ Toxicol Chem* 17 (10): 1979–1981.

Cogliani, C., Goossens, H., and Greko, C. (2011). Restricting antimicrobial use in food animals: lessons from Europe. *Microbe* 6 (6): 274.

Cohen, M.L. (2000). Changing patterns of infectious disease. *Nature* 406 (6797): 762–767.

Czekalski, N., Gascon Diez, E., and Burgmann, H. (2014). Wastewater as a point source of antibiotic-resistance genes in the sediment of a freshwater lake. *ISME J* 8 (7): 1381–1390.

Damborg, P., Broens, E.M., Chomel, B.B. et al. (2016). Bacterial zoonoses transmitted by household pets: state-of-the-art and future perspectives for targeted research and policy actions. *J Comp Pathol* 155 (1): S27–S40.

Dantas, G., Sommer, M.O., Oluwasegun, R.D., and Church, G.M. (2008). Bacteria subsisting on antibiotics. *Science* 320 (5872): 100–103.

Davies, J. and Davies, D. (2010). Origins and evolution of antibiotic resistance. *Microbiol Mol Biol Rev* 74 (3): 417–433.

Davies, J., Spiegelman, G.B., and Yim, G. (2006). The world of subinhibitory antibiotic concentrations. *Curr Opin Microbiol* 9 (5): 445–453.

D'Costa, V.M., Griffiths, E., and Wright, G.D. (2007). Expanding the soil antibiotic resistome: exploring environmental diversity. *Curr Opin Microbiol* 10 (5): 481–489.

D'Costa, V.M., King, C.E., Kalan, L. et al. (2011). Antibiotic resistance is ancient. *Nature* 477 (7365): 457–461.

D'Costa, V.M., McGrann, K.M., Hughes, D.W., and Wright, G.D. (2006). Sampling the antibiotic resistome. *Science* 311 (5759): 374–377.

do Monte, D.F., Mem, A., Fernandes, M.R. et al. (2017). Chicken meat as reservoir of colistin-resistant *Escherichia coli* carrying mcr-1 genes in South America. *Antimicrob Agents Chemother* 61: e02718-16-e02718-16.

ECDC/EFSA/EMA (2015). ECDC/EFSA/EMA first joint report on the integrated analysis of the consumption of antimicrobial agents and occurrence of antimicrobial resistance in bacteria from humans and food-producing animals. *EFSA J* 13 (1): 114.

EFSA, ECDC (2015). *The European Union Summary Report on Trends and Sources of Zoonoses, Zoonotic Agents and Food-Borne Outbreaks in 2014*. Europen Food Safety Authority.

EFSA, ECDC (2017). The European Union summary report on antimicrobial resistance in zoonotic and indicator bacteria from humans, animals and food in 2015. *EFSA J* 15 (2): 4694.

Ewers, C., Bethe, A., Semmler, T. et al. (2012). Extended-spectrum β-lactamase-producing and AmpC-producing *Escherichia coli* from livestock and companion animals, and their putative impact on public health: a global perspective. *Clin Microbiol Infect* 18 (7): 646–655.

Ewers, C., Klotz, P., Leidner, U. et al. (2017). OXA-23 and ISAba1–OXA-66 class D β-lactamases in *Acinetobacter baumannii* isolates from companion animals. *Int J Antimicrob Agents* 49 (1): 37–44.

Ewers, C., Stamm, I., Pfeifer, Y. et al. (2014). Clonal spread of highly successful ST15-CTX-M-15 *Klebsiella pneumoniae* in companion animals and horses. *J Antimicrob Chemother* 69 (10): 2676–2680.

FAO (2016). *Drivers, Dynamics and Epidemiology of Antimicrobial Resistance in Animal Production.* Rome: Food and Agriculture Organization of the United Nations.

FDA (2015). *Summary Report on Antimicrobials Sold or Distributed for Use in Food-Producing Animals.* FDA.

Fischer, J., Rodriguez, I., Schmoger, S. et al. (2013). *Salmonella enterica* subsp. enterica producing VIM-1 carbapenemase isolated from livestock farms. *J Antimicrob Chemother* 68 (2): 478–480.

Forsberg, K.J., Reyes, A., Wang, B. et al. (2012). The shared antibiotic resistome of soil bacteria and human pathogens. *Science* 337 (6098): 1107–1111.

Founou, L.L., Founou, R.C., and Essack, S.Y. (2016). Antibiotic resistance in the food chain: a developing country-perspective. *Front Microbiol* 7: 1881.

Furness, L.E., Campbell, A., Zhang, L. et al. (2017). Wild small mammals as sentinels for the environmental transmission of antimicrobial resistance. *Environ Res* 154: 28–34.

Gall, J.E., Boyd, R.S., and Rajakaruna, N. (2015). Transfer of heavy metals through terrestrial food webs: a review. *Environ Monit Assess* 187 (4): 201.

Goh, E.B., Yim, G., Tsui, W. et al. (2002). Transcriptional modulation of bacterial gene expression by subinhibitory concentrations of antibiotics. *Proc Natl Acad Sci U S A* 99 (26): 17025–17030.

Guardabassi, L., Schwarz, S., and Lloyd, D.H. (2004). Pet animals as reservoirs of antimicrobial-resistant bacteria. *J Antimicrob Chemother* 54 (2): 321–332.

Gullberg, E., Cao, S., Berg, O.G. et al. (2011). Selection of resistant bacteria at very low antibiotic concentrations. *PLoS Pathog* 7 (7): e1002158.

Gupta, A., Fontana, J., Crowe, C. et al. (2003). Emergence of multidrug-resistant *Salmonella enterica* serotype Newport infections resistant to expanded-spectrum cephalosporins in the United States. *J Infect Dis* 188 (11): 1707–1716.

Hu, Q., Zhang, X.X., Jia, S. et al. (2016). Metagenomic insights into ultraviolet disinfection effects on antibiotic resistome in biologically treated wastewater. *Water Res* 101: 309–317.

Jones-Dias, D., Manageiro, V., and Canica, M. (2016a). Influence of agricultural practice on mobile bla genes: IncI1-bearing CTX-M, SHV, CMY and TEM in *Escherichia coli* from intensive farming soils. *Environ Microbiol* 18 (1): 260–272.

Jones-Dias, D., Manageiro, V., Ferreira, E. et al. (2016b). Architecture of Class 1, 2, and 3 integrons from gram negative bacteria recovered among fruits and vegetables. *Front Microbiol* 7: 1400.

Kelessidis, A. and Stasinakis, A.S. (2012). Comparative study of the methods used for treatment and final disposal of sewage sludge in European countries. *Waste Manag* 32 (6): 1186–1195.

Kemper, N. (2008). Veterinary antibiotics in the aquatic and terrestrial environment. *Ecol Indic* 8 (1): 1–13.

Knapp, C.W., Dolfing, J., Ehlert, P.A., and Graham, D.W. (2010). Evidence of increasing antibiotic resistance gene abundances in archived soils since 1940. *Environ Sci Technol* 44 (2): 580–587.

Kumarasamy, K.K., Toleman, M.A., Walsh, T.R. et al. (2010). Emergence of a new antibiotic resistance mechanism in India, Pakistan, and the UK: a molecular, biological, and epidemiological study. *Lancet Infect Dis* 10 (9): 597–602.

Kurenbach, B., Marjoshi, D., Amabile-Cuevas, C.F. et al. (2015). Sublethal exposure to commercial formulations of the herbicides dicamba, 2,4-dichlorophenoxyacetic acid, and glyphosate cause changes in antibiotic susceptibility in *Escherichia coli* and *Salmonella enterica* serovar Typhimurium. *MBio* 6 (2): pii: e00009-15.

Lahuerta, A., Westrell, T., Takkinen, J. et al. (2011). Zoonoses in the European Union: origin, distribution and dynamics: the EFSA-ECDC summary report 2009. *Eurosurveillance (Online Edition)* 16 (13): 5–8.

Landers, T.F., Cohen, B., Wittum, T.E., and Larson, E.L. (2012). A review of antibiotic use in food animals: perspective, policy, and potential. *Public Health Rep* 127 (1): 4–22.

LaPara, T.M., Burch, T.R., McNamara, P.J. et al. (2011). Tertiary-treated municipal wastewater is a significant point source of antibiotic resistance genes into Duluth-Superior Harbor. *Environ Sci Technol* 45 (22): 9543–9549.

Laxminarayan, R., Sridhar, D., Blaser, M. et al. (2016). Achieving global targets for antimicrobial resistance. *Science* 353 (6302): 874–875.

Lee, J.H. (2003). Methicillin (oxacillin)-resistant *Staphylococcus aureus* strains isolated from major food animals and their potential transmission to humans. *Appl Environ Microbiol* 69 (11): 6489–6494.

Leite-Martins, L., Meireles, D., Bessa, L.J. et al. (2014). Spread of multidrug-resistant *Enterococcus faecalis* within the household setting. *Microb Drug Resist* 20 (5): 501–507.

Leverstein-van Hall, M.A., Dierikx, C.M., Stuart, J.C. et al. (2011). Dutch patients, retail chicken meat and poultry share the same ESBL genes, plasmids and strains. *Clin Microbiol Infect* 17 (6): 873–880.

Levy, S.B., Fitzgerald, G.B., and Macone, A.B. (1976). Spread of antibiotic-resistant plasmids from chicken to chicken and from chicken to man. *Nature* 260 (5546): 40–42.

Li, B., Yang, Y., Ma, L. et al. (2015). Metagenomic and network analysis reveal wide distribution and co-occurrence of environmental antibiotic resistance genes. *ISME J* 9 (11): 2490–2502.

Liu, X., Steele, J.C., and Meng, X.-Z. (2017). Usage, residue, and human health risk of antibiotics in Chinese aquaculture: A review. *Environ Pollut* 223: 161–169.

Liu, Y.Y., Wang, Y., Walsh, T.R. et al. (2016). Emergence of plasmid-mediated colistin resistance mechanism MCR-1 in animals and human beings in China: a microbiological and molecular biological study. *Lancet Infect Dis* 16 (2): 161–168.

Lloyd, D.H. (2007). Reservoirs of antimicrobial resistance in pet animals. *Clin Infect Dis* 45 (Suppl 2): S148–S152.

Lowder, B.V., Guinane, C.M., Ben Zakour, N.L. et al. (2009). Recent human-to-poultry host jump, adaptation, and pandemic spread of Staphylococcus aureus. *Proc Natl Acad Sci U S A* 106 (46): 19545–19550.

Lupo, A., Châtre, P., Ponsin, C. et al. (2017). Clonal spread of *Acinetobacter baumannii* sequence type 25 carrying blaOXA-23 in companion animals in France. *Antimicrob Agents Chemother* 61 (1): e01881-16.

Ma, Y.J., Wilson, C.A., Novak, J.T. et al. (2011). Effect of various sludge digestion conditions on sulfonamide, macrolide, and tetracycline resistance genes and class I integrons. *Environ Sci Technol* 45 (18): 7855–7861.

Maillard, J.Y. (2005). Antimicrobial biocides in the healthcare environment: efficacy, usage, policies, and perceived problems. *Ther Clin Risk Manag* 1 (4): 307–320.

Manaia, C.M. (2017). Assessing the risk of antibiotic resistance transmission from the environment to humans: non-direct proportionality between abundance and risk. *Trends Microbiol* 25 (3): 173–181.

Manaia, C.M., Macedo, G., Fatta-Kassinos, D., and Nunes, O.C. (2016). Antibiotic resistance in urban aquatic environments: can it be controlled? *Appl Microbiol Biotechnol* 100 (4): 1543–1557.

Marshall, B.M. and Levy, S.B. (2011). Food animals and antimicrobials: impacts on human health. *Clin Microbiol Rev* 24 (4): 718–733.

Marti, R., Scott, A., Tien, Y.C. et al. (2013). Impact of manure fertilization on the abundance of antibiotic-resistant bacteria and frequency of detection of antibiotic resistance genes in soil and on vegetables at harvest. *Appl Environ Microbiol* 79 (18): 5701–5709.

Martinez, J.L. (2008). Antibiotics and antibiotic resistance genes in natural environments. *Science* 321 (5887): 365–367.

Martinez, J.L. (2009). Environmental pollution by antibiotics and by antibiotic resistance determinants. *Environ Pollut* 157 (11): 2893–2902.

Martinez, J.L. and Baquero, F. (2000). Mutation frequencies and antibiotic resistance. *Antimicrob Agents Chemother* 44 (7): 1771–1777.

Martinez, J.L., Coque, T.M., and Baquero, F. (2015). What is a resistance gene? Ranking risk in resistomes. *Nat Rev Microbiol* 13 (2): 116–123.

McGowan, L.L., Jackson, C.R., Barrett, J.B. et al. (2006). Prevalence and antimicrobial resistance of enterococci isolated from retail fruits, vegetables, and meats. *J Food Prot* 69 (12): 2976–2982.

Mena, C., Rodrigues, D., Silva, J. et al. (2008). Occurrence, identification, and characterization of Campylobacter species isolated from portuguese poultry samples collected from retail establishments. *Poult Sci* 87 (1): 187–190.

Mendes, A., Martins da Costa, P., Rego, D. et al. (2015). Contamination of public transports by *Staphylococcus aureus* and its carriage by biomedical students: point-prevalence, related risk factors and molecular characterization of methicillin-resistant strains. *Public Health* 129 (8): 1125–1131.

Meyer, B. and Cookson, B. (2010). Does microbial resistance or adaptation to biocides create a hazard in infection prevention and control? *J Hosp Infect* 76 (3): 200–205.

Michael, I., Rizzo, L., McArdell, C.S. et al. (2013). Urban wastewater treatment plants as hotspots for the release of antibiotics in the environment: a review. *Water Res* 47 (3): 957–995.

Mollenkopf, D.F., Stull, J.W., Mathys, D.A. et al. (2017). Carbapenemase-producing *Enterobacteriaceae* recovered from the environment of a swine farrow-to-finish operation in the United States. *Antimicrob Agents Chemother* 61 (2): e01298-16-e01298-16.

Moreira, N.F.F., Sousa, J.M., Macedo, G. et al. (2016). Photocatalytic ozonation of urban wastewater and surface water using immobilized TiO2 with LEDs: micropollutants, antibiotic resistance genes and estrogenic activity. *Water Res* 94: 10–22.

Munck, C., Albertsen, M., Telke, A. et al. (2015). Limited dissemination of the wastewater treatment plant core resistome. *Nat Commun* 6: 8452.

Nicholson, F.A., Smith, S.R., Alloway, B.J. et al. (2003). An inventory of heavy metals inputs to agricultural soils in England and Wales. *Sci Total Environ* 311 (1–3): 205–219.

Nordmann, P., Naas, T., and Poirel, L. (2011). Global spread of Carbapenemase-producing *Enterobacteriaceae. Emerg Infect Dis J* 17 (10): 1791–1798.

Novais, C., Coque, T.M., Ferreira, H. et al. (2005). Environmental contamination with vancomycin-resistant enterococci from hospital sewage in Portugal. *Appl Environ Microbiol* 71 (6): 3364–3368.

O'Brien, T.F. (2002). Emergence, spread, and environmental effect of antimicrobial resistance: how use of an antimicrobial anywhere can increase resistance to any antimicrobial anywhere else. *Clin Infect Dis* 34 (Suppl 3): S78–S84.

Ochman, H., Lawrence, J.G., and Groisman, E.A. (2000). Lateral gene transfer and the nature of bacterial innovation. *Nature* 405 (6784): 299–304.

Olson, B.H., Cayless, S.M., Ford, S., and Lester, J.N. (1991). Toxic element contamination and the occurrence of mercury-resistant bacteria in Hg-contaminated soil, sediments, and sludges. *Arch Environ Contam Toxicol* 20 (2): 226–233.

Om, C., Daily, F., Vlieghe, E. et al. (2016). "If it's a broad spectrum, it can shoot better": inappropriate antibiotic prescribing in Cambodia. *Antimicrob Resist Infect Control* 5: 58.

Pal, C., Bengtsson-Palme, J., Kristiansson, E., and Larsson, D.G. (2015). Co-occurrence of resistance genes to antibiotics, biocides and metals reveals novel insights into their co-selection potential. *BMC Genomics* 16: 964.

Perry, J.A. and Wright, G.D. (2013). The antibiotic resistance "mobilome": searching for the link between environment and clinic. *Front Microbiol* 4: 138.

Phillips, I., Casewell, M., Cox, T. et al. (2004). Does the use of antibiotics in food animals pose a risk to human health? A critical review of published data. *J Antimicrob Chemother* 53 (1): 28–52.

Picão, R.C., Cardoso, J.P., Campana, E.H. et al. (2013). The route of antimicrobial resistance from the hospital effluent to the environment: focus on the occurrence of KPC-producing *Aeromonas* spp. and *Enterobacteriaceae* in sewage. *Diagn Microbiol Infect Dis* 76 (1): 80–85.

Poole, K. (2012). Stress responses as determinants of antimicrobial resistance in Gram-negative bacteria. *Trends Microbiol* 20 (5): 227–234.

Prescott, J.F. (2008). Antimicrobial use in food and companion animals. *Anim Health Res Rev* 9 (2): 127–133.

Pruden, A., Pei, R., Storteboom, H., and Carlson, K.H. (2006). Antibiotic resistance genes as emerging contaminants: studies in northern Colorado. *Environ Sci Technol* 40 (23): 7445–7450.

Riesenfeld, C.S., Goodman, R.M., and Handelsman, J. (2004). Uncultured soil bacteria are a reservoir of new antibiotic resistance genes. *Environ Microbiol* 6 (9): 981–989.

Rizzo, L., Manaia, C., Merlin, C. et al. (2013). Urban wastewater treatment plants as hotspots for antibiotic resistant bacteria and genes spread into the environment: a review. *Sci Total Environ* 447: 345–360.

Rodriguez-Beltran, J., Rodriguez-Rojas, A., Yubero, E., and Blazquez, J. (2013). The animal food supplement sepiolite promotes a direct horizontal transfer of antibiotic resistance plasmids between bacterial species. *Antimicrob Agents Chemother* 57 (6): 2651–2653.

Salyers, A.A., Gupta, A., and Wang, Y. (2004). Human intestinal bacteria as reservoirs for antibiotic resistance genes. *Trends Microbiol* 12 (9): 412–416.

Schwarz, S. and Johnson, A.P. (2016). Transferable resistance to colistin: a new but old threat. *J Antimicrob Chemother* 71 (8): 2066–2070.

Seiler, C. and Berendonk, T.U. (2012). Heavy metal driven co-selection of antibiotic resistance in soil and water bodies impacted by agriculture and aquaculture. *Front Microbiol* 3: 399.

Shaheen, B.W., Nayak, R., and Boothe, D.M. (2013). Emergence of a New Delhi metallo-β-lactamase (NDM-1)-encoding gene in clinical *Escherichia coli* isolates recovered from companion animals in the United States. *Antimicrob Agents Chemother* 57 (6): 2902–2903.

Shoemaker, N.B., Vlamakis, H., Hayes, K., and Salyers, A.A. (2001). Evidence for extensive resistance gene transfer among *Bacteroides* spp. and among *Bacteroides* and other genera in the human colon. *Appl Environ Microbiol* 67 (2): 561–568.

Singer, R.S., Finch, R., Wegener, H.C. et al. (2003). Antibiotic resistance: the interplay between antibiotic use in animals and human beings. *Lancet Infect Dis* 3 (1): 47–51.

Smith, T.C., Male, M.J., Harper, A.L. et al. (2009). Methicillin-resistant *Staphylococcus aureus* (MRSA) strain ST398 is present in midwestern U.S. swine and swine workers. *PLoS One* 4 (1): e4258.

Sommer, M.O., Church, G.M., and Dantas, G. (2010). The human microbiome harbors a diverse reservoir of antibiotic resistance genes. *Virulence* 1 (4): 299–303.

Sørensen, S.J., Bailey, M., Hansen, L.H. et al. (2005). Studying plasmid horizontal transfer in situ: a critical review. *Nat Rev Microbiol* 3 (9): 700–710.

Sørensen, T.L., Blom, M., Monnet, D.L. et al. (2001). Transient intestinal carriage after ingestion of antibiotic-resistant *Enterococcus faecium* from chicken and pork. *N Engl J Med* 345 (16): 1161–1166.

Sousa, J.M., Macedo, G., Pedrosa, M. et al. (2017). Ozonation and UV254nm radiation for the removal of microorganisms and antibiotic resistance genes from urban wastewater. *J Hazard Mater* 323 (Pt A): 434–441.

Sousa, M., Goncalves, A., Silva, N. et al. (2014). Acquired antibiotic resistance among wild animals: the case of Iberian Lynx (Lynx pardinus). *Vet Q* 34 (2): 105–112.

Spoor, L.E., McAdam, P.R., Weinert, L.A. et al. (2013). Livestock origin for a human pandemic clone of community-associated methicillin-resistant *Staphylococcus aureus*. *MBio* 4 (4): e00356-13.

Starrels, J.L., Barg, F.K., and Metlay, J.P. (2009). Patterns and determinants of inappropriate antibiotic use in injection drug users. *J Gen Intern Med* 24 (2): 263–269.

Sunde, M. and Norstrom, M. (2006). The prevalence of, associations between and conjugal transfer of antibiotic resistance genes in *Escherichia coli* isolated from Norwegian meat and meat products. *J Antimicrob Chemother* 58 (4): 741–747.

Tacão, M., Moura, A., Correia, A., and Henriques, I. (2014). Co-resistance to different classes of antibiotics among ESBL-producers from aquatic systems. *Water Res* 48: 100–107.

Thaller, M.C., Migliore, L., Marquez, C. et al. (2010). Tracking acquired antibiotic resistance in commensal bacteria of Galapagos land iguanas: no man, no resistance. *PLoS One* 5 (2): e8989.

Van Boeckel, T.P., Brower, C., Gilbert, M. et al. (2015). Global trends in antimicrobial use in food animals. *Proc Natl Acad Sci U S A* 112 (18): 5649–5654.

Van Boeckel, T.P., Gandra, S., Ashok, A. et al. (2014). Global antibiotic consumption 2000 to 2010: an analysis of national pharmaceutical sales data. *Lancet Infect Dis* 14 (8): 742–750.

Van Melderen, L. and De Bast, M.S. (2009). Bacterial toxin–antitoxin systems: more than selfish entities? *PLoS Genet* 5 (3): e1000437.

Vaz-Moreira, I., Nunes, O.C., and Manaia, C.M. (2014). Bacterial diversity and antibiotic resistance in water habitats: searching the links with the human microbiome. *FEMS Microbiol Rev* 38 (4): 761–778.

Vredenburg, J., Varela, A.R., Hasan, B. et al. (2014). Quinolone-resistant *Escherichia coli* isolated from birds of prey in Portugal are genetically distinct from those isolated from water environments and gulls in Portugal, Spain and Sweden. *Environ Microbiol* 16 (4): 995–1004.

Wales, A.D. and Davies, R.H. (2015). Co-selection of resistance to antibiotics, biocides and heavy metals, and its relevance to foodborne pathogens. *Antibiotics (Basel)* 4 (4): 567–604.

Walsh, T.R., Weeks, J., Livermore, D.M., and Toleman, M.A. (2011). Dissemination of NDM-1 positive bacteria in the New Delhi environment and its implications for human health: an environmental point prevalence study. *Lancet Infect Dis* 11 (5): 355–362.

Ward, M., Gibbons, C., McAdam, P. et al. (2014). Time-scaled evolutionary analysis of the transmission and antibiotic resistance dynamics of *Staphylococcus aureus* clonal complex 398. *Appl Environ Microbiol* 80 (23): 7275–7282.

Wegener, H.C. (2003). Antibiotics in animal feed and their role in resistance development. *Curr Opin Microbiol* 6 (5): 439–445.

WHO (2017). *Global Priority List of Antibiotic-Resistant Bacteria to Guide Research, Dicovery, and Development of New Antibiotics.* WHO.

Wiedenbeck, J. and Cohan, F.M. (2011). Origins of bacterial diversity through horizontal genetic transfer and adaptation to new ecological niches. *FEMS Microbiol Rev* 35 (5): 957–976.

Woodford, N., Wareham, D.W., Guerra, B., and Teale, C. (2014). Carbapenemase-producing *Enterobacteriaceae* and non-*Enterobacteriaceae* from animals and the environment: an emerging public health risk of our own making? *J Antimicrob Chemother* 69 (2): 287–291.

Woolhouse, M., Ward, M., van Bunnik, B., and Farrar, J. (2015). Antimicrobial resistance in humans, livestock and the wider environment. *Philos Trans R Soc Lond B Biol Sci* 370 (1670): 20140083.

Wright, G.D. (2010). Antibiotic resistance in the environment: a link to the clinic? *Curr Opin Microbiol* 13 (5): 589–594.

Xavier, B.B., Lammens, C., Ruhal, R. et al. (2016). Identification of a novel plasmid-mediated colistin-resistance gene, mcr-2, in *Escherichia coli*, Belgium, June 2016. *Eurosurveillance* 21 (27): 30280.

Yang, F., Huang, L., Li, L. et al. (2017). Discharge of KPC-2 genes from the WWTPs contributed to their enriched abundance in the receiving river. *Sci Total Environ* 581–582: 136–143.

You, Y. and Silbergeld, E.K. (2014). Learning from agriculture: understanding low-dose antimicrobials as drivers of resistome expansion. *Front Microbiol* 5: 284.

Zhang, T., Zhang, X.X., and Ye, L. (2011). Plasmid metagenome reveals high levels of antibiotic resistance genes and mobile genetic elements in activated sludge. *PLoS One* 6 (10): e26041.

Zhang, X.X., Zhang, T., and Fang, H.H. (2009a). Antibiotic resistance genes in water environment. *Appl Microbiol Biotechnol* 82 (3): 397–414.

Zhang, X.Y., Ding, L.J., and Fan, M.Z. (2009b). Resistance patterns and detection of aac (3)-IV gene in apramycin-resistant *Escherichia coli* isolated from farm animals and farm workers in northeastern of China. *Res Vet Sci* 87 (3): 449–454.

11

Antibiotic Resistance

Immunity-Acquired Resistance: Evolution of Antimicrobial Resistance Among Extended-Spectrum β-Lactamases and Carbapenemases in *Klebsiella pneumoniae* and *Escherichia coli*

Isabel Carvalho[1–5], *Nuno Silva*[6], *João Carrola*[7], *Vanessa Silva*[1–5], *Carol Currie*[6], *Gilberto Igrejas*[2,3], *and Patrícia Poeta*[1,4–5]

[1] *Department of Veterinary Sciences, University of Trás-os-Montes and Alto Douro, Vila Real, Portugal*
[2] *Department of Genetics and Biotechnology, University of Trás-os-Montes and Alto Douro, Vila Real, Portugal*
[3] *Functional Genomics and Proteomics Unit, University of Trás-os-Montes and Alto Douro, Vila Real, Portugal*
[4] *Laboratory Associated for Green Chemistry (LAQV-REQUIMTE), New University of Lisbon, Monte da Caparica, Portugal;*
[5] *MicroART- Antibiotic Resistance Team, University of Trás-os-Montes and Alto Douro, Vila real, Portugal*
[6] *Moredun Research Institute Pentlands Science Park, Penicuik, Scotland UK*
[7] *Centre for the Research and Technology of Agro-Environmental and Biological Sciences (CITAB), University of Trás-os-Montes and Alto Douro, Vila Real, Portugal*

11.1 Overview of Antibiotic Resistance as a Worldwide Health Problem

In 1928, Alexander Fleming discovered that the penicillin allows to treat different infectious diseases such as meningitis, tuberculosis, and pneumonia (Tan and Tatsumura 2015). Antimicrobials were first used to treat serious infections, in the 1940s (Chopra and Roberts 2001). Since the 1980s, antibiotics have been considered as one of the wonder discoveries of the century, and they were widely used in prophylaxis or treatment of bacterial infections, saving millions of lives and transforming modern medicine (Centers for Disease control and Prevention (2013); Spellberg 2010). Today, these drugs are extensively used not only in human medicine and in veterinary practices but also in animal production (Poeta et al. 2006).

The gastrointestinal tract (GI tract) has an important role in human nutrition and health by promoting nutrient supply, preventing pathogen colonization, and maintaining normal immunity (Silva et al. 2011). The GI tract is continuously in contact with commensal bacteria that are composed of more than 500 different species (Silva et al. 2012). According to Ventola (2015), the rapid emergence of resistant bacteria is occurring worldwide, endangering the efficacy of antibiotics. Human and animal populations taking antibiotics have

Antibiotic Drug Resistance, First Edition. Edited by José-Luis Capelo-Martínez and Gilberto Igrejas.

become reservoirs of these resistant organisms, including endangered unintentional hosts (Day et al. 2016; Iredell et al. 2016).

In more detail, it is said that the main sources of antibiotic contamination come from medicine practices related with urban and hospital effluents (Freire et al. 2016), pharmaceutical production facilities (Rummukainen et al. 2013; Tang et al. 2016), and wastewater from animal production farms (Van Boeckel et al. 2015). All these lead to a substantial increase of antimicrobial accumulation and their metabolites in the environment (Ulstad et al. 2016). The spread still continues, daily, all around the world, but it is more alarming in developing countries (Sosa et al. 2010; Reardon 2014).

Over the years, new classes of antibiotics were developed, such as erythromycin, tetracycline, and chloramphenicol, in the 1940s with the aim of preventing or treating infectious diseases (Kong 2010). But more recently, these have emerged as new pollutants in the environment, mainly because some metabolites are persistent and localized to groundwater and can even cause problems even at low concentrations (Manzetti and Ghisi 2014).

According to CDC (2015), antibiotics are among the most commonly prescribed drugs used in human medicine. However, up to 50% of the time, antibiotics are not optimally prescribed, with incorrect dosing, improper antibiotic selection, or inappropriate treatment duration, or even wrong prescription in case of nonbacterial problems (Ventola 2015).

This constant use of numerous antimicrobial compounds in different activities enables them to reach easily the environment (Li et al. 2014). Consequently, these drugs suffer transformations that induce the production of active compounds with unknown environmental consequences and the rise of new drug-resistant bacteria (DuPont and Steffen 2016; Manzetti and Ghisi 2014). Furthermore, bacteria have become adaptable and more competent in facing environmental stress (Chen et al. 2016; Planta 2007). So, there are different kinds of resistance mechanisms to counter these drugs (Figure 11.1), classified into four main categories: (i) modification or protection of the target, (ii) enzymatic modification or inhibition of the antibiotics, (iii) active expulsion of the antibiotics (efflux pump), and (iv) alteration of membrane permeability (Spellberg et al. 2011; Tenover 2006).

The increment of drug resistance outpaces the pharmaceutical development of new drugs and, according to Centers for Disease control and Prevention (2013), "the more antibiotics are used, the more quickly bacteria develop resistance."

The worldwide antimicrobial use is one of the main concerns because it turns this problem as a global menace, but we have to consider, as mentioned previously, the endless inappropriate use, overuse, and negligent disposal (Potter et al. 2016; Righi et al. 2016).

Today, in the twenty-first century, the antimicrobial resistance (AMR) threat continue to be a huge problem for the treatments of many pathogenic microorganisms with expected rise of resistant bacteria and the increase of

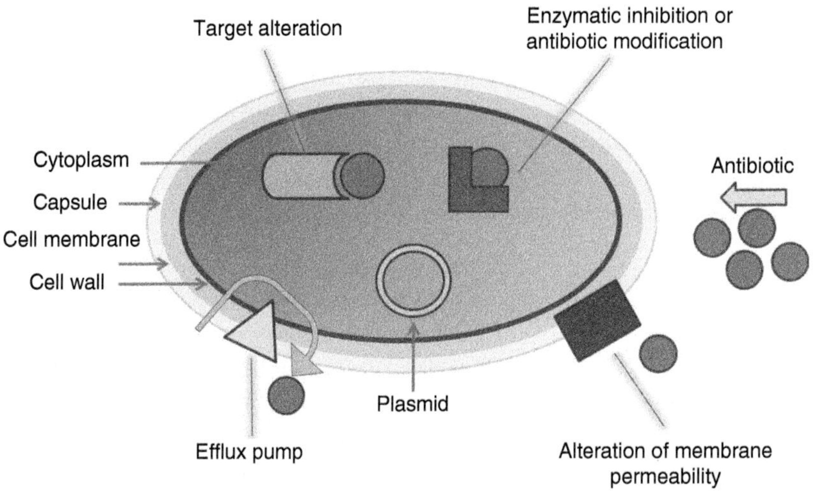

Figure 11.1 Main mechanisms of antibiotic resistance: modification of the target, enzymatic modification or inhibition of the antibiotic, active antibiotic expulsion (efflux pump), and change in membrane permeability.

multidrug resistance rate, usually called as *superbugs* or *superbacteria* (Simoneit et al. 2015; Viegas et al. 2015). This problem is greater in developing countries, due to lack of regulation or inefficient control of weak regulation (Chen et al. 2015).

The lack of new antimicrobial drugs in the market is an actual main concern, but the absence of new possibly efficient antibiotic drug to be released in the next years means that we will have a tricky task for future antibiotic supply with long-term effects (Blair et al. 2015; Höjgård 2012; Sharma and Towse 2011). According to Carvalho et al. (2017), microbial resistance to antibiotics is becoming possible: the bacteria are evolving and are now affecting unintentional hosts, such as domestic animals in urban areas.

Methicillin-resistant *Staphylococcus aureus* (MRSA), extended-spectrum β-lactamases (ESBLs), and carbapenem-resistant *Enterobacteriaceae* (CRE) are the most important antimicrobial-resistant bacteria with huge clinical impact on patient outcomes, public health, and healthcare (Hamilton and Wenlock 2016).

11.2 Objectives

Escherichia coli and *Klebsiella pneumoniae* are the main agents implicated in severe sepsis, including urinary infections, among other pathologies (Iredell et al. 2016). Modulation of the phenotype by host bacteria explains why tracking and control of CRE has been particularly problematic in *Enterobacteriaceae* (Belluco et al. 2016).

The main goal of this review is to approach AMR in bacteria as a worldwide health problem, namely, in two main medically important Gram-negative bacteria, *E. coli* and *K. pneumoniae*. We also focus on the causes of AMR, widespread dissemination of AMR genes that can affect animals and humans, main causes of that resistances, and consequences and future strategies to alleviate the threat.

11.3 Causes of Antimicrobial Resistance

As we mentioned before, the main factors for the development of AMR are use, overuse, and misuse of antimicrobials (Bengtsson-Palme and Larsson 2015; Höjgård 2012). However, lengthy hospitalization, invasive procedures, and admission to the intensive care unit (ICU) are other important factors of colonization and infection with MDR organisms (Cristina et al. 2016). Centers for Disease control and Prevention (2013) highlighted the antibiotic use in a population as a primary driver of the AMR, associated with the development of resistant bacteria.

The abuse and misuse of drug compounds (such as antibiotics, antifungals, and antivirals) can also lead to this global health concern and compromised prevention and treatment of an ever-increasing range of infections caused by bacteria, fungi, and viruses (WHO 2016). According to Sosa et al. (2010), the antibacterial drug rate is higher when compared with antifungal or antiviral agents. However, it is also important to consider other factors like human migration (He et al. 2016), the failure to complete antibiotic treatment (Brauner et al. 2016), incorrect dose treatment (Cornejo-Juárez et al. 2015), or wrong selection of the antimicrobial (Lynch et al. 2013).

It is also important to consider the production and distribution of counterfeit medication, including fake antibiotics, in general presenting low and variable concentration of the active compounds that can contribute to the development of AMR (Khuluza et al. 2016; Venhuis et al. 2016). In the case of developing countries, antimicrobials can often be purchased without prescription. According to WHO (2016), "poverty and inadequate access to drugs continue to be a major force in the development of resistance." In the same way, particular MDR bacteria are common because of poor sanitary conditions, in contrast with the developed world (Okeke et al. 2005; Sosa et al. 2010). The inadequate access to effective drugs, medication sharing, self-prescription, and manufacture of antimicrobials of questionable quality represent some factors that contribute to the growth of MDR organisms (Planta 2007).

Another situation is the frequent and widespread use of antimicrobials in agricultural procedures. Animals and humans are compromised because resistant bacteria can transfer potentially resistant genes to humans by environmental exposure and food contamination (Ahmed et al. 2010; Guerrero-Ramos et al. 2016). Nowadays, antibiotics are often used not only for treatments but also for prophylactic purposes, mainly in food animals such as chickens (Saenz et al. 2001) or pigs (Li et al. 2014). A review on antibiotic consumption in the United States showed that 80% of the antimicrobial drugs are widely used in different animal species to prevent infections and also obtain a higher-quality product, as supplements in livestock (Ventola 2015).

The exchange of resistance determinants between bacteria can be majored by the horizontal gene transfer (HGT) via mobile genetic elements such as plasmids, transposons, or integrons (Lupo et al. 2012). Quick mutations cause new drug resistance rates in the microorganism population because of their shorter generation times and larger population sizes (Blair et al. 2015; Woodford et al. 2006). Therefore, MDR is common in pathogens with short life cycles, like viruses, bacteria, and protozoa (Medicine 2013).

11.4 *Enterobacteriaceae*: General Characterization

The GI tract of humans and animals are inhabited predominately by Gram-negative and Gram-positive anaerobes (Centers for Disease control and Prevention 2013). Bacteria within the family *Enterobacteriaceae* includes common *K. pneumoniae, E. coli, Salmonella enterica,* and the rare *Proteus mirabilis, Raoultella planticola, and Citrobacter freundii.* All of these strains are considered important human pathogens in nosocomial and community settings (Hamilton and Wenlock 2016; Ruppé et al. 2015). The work of Burow et al. (2014) reported that commensal bacteria have an important role in the spread of AMR because they can be reservoirs of antibiotic resistance genes and drug-resistant opportunistic pathogens. A recent study done in Bangladesh by Das et al. (2016) showed that *E. coli* (69%) and *Klebsiella* spp. (15%) were the most commonly isolated pathogens present in the urinary tract from children hospitalized with diarrhea.

11.4.1 Escherichia coli

Different commensal bacteria may be reservoirs for MDR determinants. *Escherichia coli* is a rod of *Enterobacteriaceae* family and it is also an inhabitant of intestinal microbiota in warm-blooded animals. *Escherichia coli* strains ferment lactose and has an important role in studies about microbiological safety of food products (Van Boeckel et al. 2015), water (Bulycheva et al. 2014), and environment (Kuhnert et al. 2000). Moreover, this bacterium is a causative agent of diarrhea, neonatal meningitis, pneumonia, and extraintestinal infections (Das et al. 2016; In et al. 2016).

In the European context, Portugal has the seventh prevalence of resistance to fluoroquinolones in *E. coli*, with the highest rate among the 29 countries of Europe in 2014 (Jones-Dias et al. 2016). The antibiotic administration has decreased by 6% in the last seven years but, in the last 12 months, a third of Europeans have taken antibiotics (Comission 2014). There are also reported 120 million cases of community-acquired urinary tract infections (UTI) and 868 mil sepsis-associated mortalities due to *E. coli* worldwide per year (Wu et al. 2013). *Escherichia coli* strains might act as reservoirs of antibacterial-resistant genes that could be a main public health concern (Frye and Jackson 2013).

Thus, a key indicator of environmental contamination caused by the growing of antimicrobial-resistant bacteria can be different in animal species like companion animals (Seni et al. 2016), donkeys (Carvalho et al. 2017), swine (Herskin et al. 2016), equines (Moura et al. 2013), wild animals like Iberian lynx (Gonçalves et al. 2014), and birds (Radhouani et al. 2012).

For example, a research done by Huiting (2015) with broilers from farms in Morocco showed higher rates of AMR for colistin (3% resistant), gentamicin (25% resistant), florfenicol (51% resistant), trimethoprim–sulfamethoxazole (68% resistant), enrofloxacin (87% resistant), amoxicillin (75% resistant), and doxycycline (100% resistant). The data is similar with Spain and the United Kingdom (Shaikh et al. 2015). One possible explanation, according to Cohen et al. (2007), is the very low rate of healthcare in different shops in Morocco.

Another study with equine fecal *E. coli* isolates from two livery stables in North West England showed high tetracycline-resistant rate (Ahmed et al. 2010).

11.4.2 *Klebsiella pneumoniae*

Klebsiella pneumoniae is a saprophyte in humans and other mammals. This type of enterobacteria colonizes the GI tract, skin, and nasopharynx, but it can be also found in different environmental niches, such as water, soil, vegetables, fruits, cereals, and fecal samples (Cristina et al. 2016).

MDR *K. pneumoniae* is one of the global causes of nosocomial and life-threatening infections and it is responsible for roughly 15% of Gram-negative infections in hospital ICUs (Onori et al. 2015). It is an important causative agent responsible for pneumonia and UTIs but can also cause liver abscess and intra-abdominal infections, particularly in hospitalized immunosuppressed patients (Lederman and Crum 2005; Tzouvelekis et al. 2012).

In the early 1970s, high levels of *K. pneumoniae* were detected in patient's nasopharynges and it became an important cause of nosocomial infections. At the end of the twentieth century and beginning of the twenty-first, due to migration and travel of persons all around the world, these strains had increased speed and widened areas of colonization. They also become a "collector" of successive addition of genetic elements encoding resistance to aminoglycosides and extended-spectrum lactams (Bengtsson-Palme et al. 2015; Karanika et al. 2016).

In recent years, *K. pneumoniae* has assumed greater magnitude. According to ECDC (2017), the rate of resistance to third-generation cephalosporins in this strain increased from 16.5 to 40.9%, between 2007 and 2014.

11.5 Current Antibiotic Resistance Threats

The emergence of MDR clones containing novel resistance genes, specially ESBLs and CREs, will reduce the therapeutic options and efficacy of treatments (Lynch et al. 2013).

11.5.1 Carbapenem-Resistant *Enterobacteriaceae*

CRE including *Klebsiella* species and *E. coli* has been an increasing concern in the last two decades (Morrill et al. 2015; Moucheraud et al. 2015). In recent years, these strains have developed different mechanisms to combat the carbapenem action, which is one of the classes of antibiotics with broader spectrum.

According to a recent study, Reardon (2014) observed that 95% of adults in India and Pakistan have bacteria resistance to β-lactam antibiotics, including carbapenems. This bacteria group has become resistant to "all or nearly all" available commercial antibiotics, presenting health problems mainly after liver transplant and bloodstream infections (Freire et al. 2016; Righi et al. 2016).

Consequently, bacteria have become resistant to antibiotics, making simple infections lethal by ineffective treatment. This is a global-scale problem, as in Figure 11.1 illustrating carbapenem resistance in *Enterobacteriaceae* in 38 European countries and epidemiological stages, for *K. pneumoniae.*

According to the Clinical Laboratory Standards Institute (CLSI 2015), *Enterobacteriaceae* are carbapenem-resistant if they have minimum inhibitory concentrations (MICs) of $\geq 2 \mu g\,ml^{-1}$ against ertapenem and $\geq 4 \mu g\,ml^{-1}$ against doripenem, meropenem, or imipenem (Potter et al. 2016).

In the European Union (EU) (plus Norway and Iceland), 5–12% of hospital patients acquire an infection during their stay. Furthermore, on average 400 000 people per year present a resistant strain and, of whom, 25 000 die (WHO 2016). Approximately 600 deaths per year in the United States result from infections caused by the two most common types of CRE, CRKP and carbapenem-resistant *E. coli* (CREC) (Ventola 2015). A study done by Freire et al. (2016) showed that 17.6% of the patients are positive for CRE and 16.2% developed CRE infection. However, reported infections caused by CRE are scarce, as well as the treatment failure related to these pathogenic organisms (van Duin et al. 2013).

Resistance in *K. pneumoniae* is widespread in all regions of the world (WHO 2016). In 1996, the first strain of CRKP was isolated and associated with bla_{KPC}

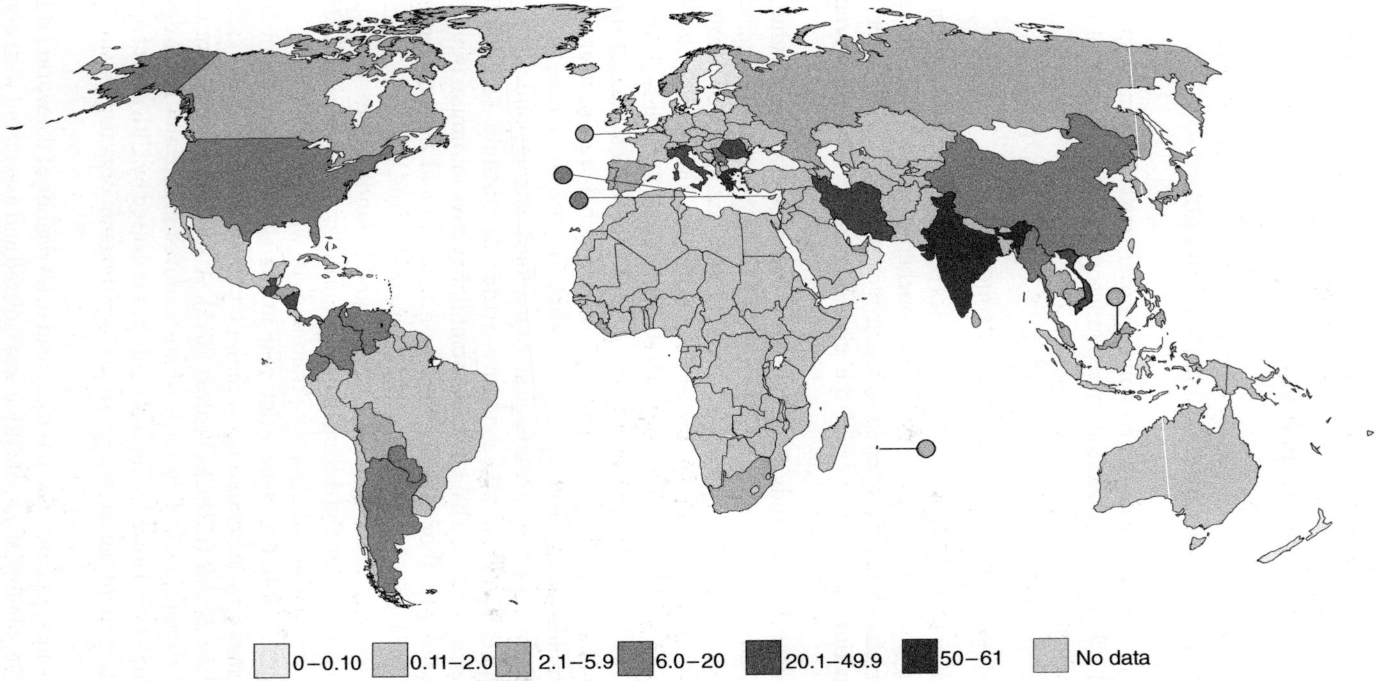

Figure 11.2 Wide distribution of carbapenem-resistant *Klebsiella pneumoniae*, by country. The frequency of prescription antibiotics varies greatly from state to state, suggesting areas where improvements in antibiotic prescription (fewer unnecessary prescriptions) would be most helpful. *Source*: Figure courtesy of the Center for Disease Dynamics, Economics & Policy (2015). Used with permission via Creative Commons license. *(See insert for color representation of the figure.)*

gene. Nowadays, this resistance is frequent in industrialized countries (Cristina et al. 2016) and also in developing countries like Southern Asia (Das et al. 2016). Additional carbapenemases have now been reported as *bla*$_{NDM}$, *bla*$_{OXA-48}$, *bla*$_{VIM}$, and *bla*$_{IMP-1}$ (Cristina et al. 2016; Onori et al. 2015). The resistance can be sourced from different ways as we explained before, including the production of β-lactamases *amp*C type and *K. pneumoniae* carbapenemase (KPC) type (Abdallah et al. 2015; Ruppé et al. 2015).

According to WHO (2016), carbapenem antibiotics are ineffective in more than half of people treated for *K. pneumoniae* infections in some European countries. Human mortality rates caused by CRKP are high in case of infections, 26–44%, and very high related with bacteremia, 70% (Amit et al. 2015; Hoxha et al. 2016). According to Cristina et al. (2016), the infection rates have been dramatically increasing worldwide over the past 10 years.

Furthermore, 187 cases of KPC infections were detected in Brazil in 2010 and 18 deaths were registered (Dos Santos et al. 2015). In a single Italian hospital, Onori et al. (2015) analyzed the diversity of *K. pneumoniae* strains and observed 16 CRKP isolates that were added to a database of 319 genomes. A study done in a university hospital in Milan (Italy) by Ridolfo et al. (2016) showed that CRKP was isolated during hospital stays in 46 of the patients (73%). In the same way, Poulou et al. (2012) reported 73 CRKP infections associated with ICU (43.8%) and medical wards (41.1%) in a university hospital in Greece. Most of the cases (90.4%) were from general medical and surgery wards, and the remaining 9.6% were from the ICU. Furthermore, a study done by Laurent et al. (2008) showed that 9 in 30 patients in a Belgium hospital developed an infection caused by KPC.

Recent studies reported that the infections caused by MDR *E. coli* will increase and it is expected to cause three million deaths each year by 2050 (Potter et al. 2016; Tang et al. 2016). Infections caused by MDR *K. pneumoniae* represent a global public health concern and a major therapeutic challenge for present and future generations.

11.5.2 Extended-Spectrum β-Lactamase

The prevalence of ESBL-producing bacteria is increasing in humans, animals, and their surrounding environment and is of global concern (Day et al. 2016; Seni et al. 2016). But it is important to note that the number of cases reported in ESBL are higher than CRE, particularly in animals.

The spread of resistance to many β-lactam antibiotics has been associated with one important mechanism called HGT, potentially carrying AMR genes of β-lactamases on plasmids (Lynch et al. 2013; Shaikh et al. 2015). ESBL-producing bacteria are now commonly isolated both in industrialized and in developing countries (Figure 11.3).

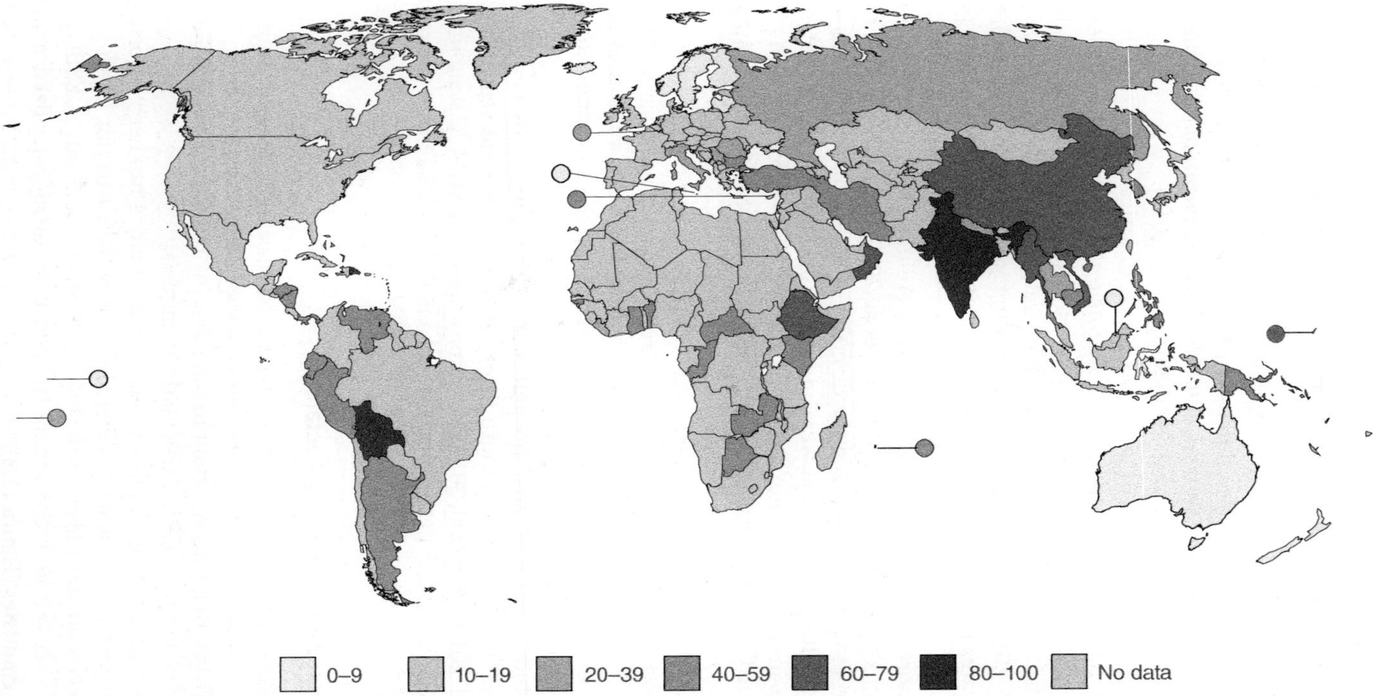

Figure 11.3 Wide distribution of extended-spectrum β-lactamase-producing *Escherichia coli*, by country (resistance to one or more of the following drugs were used: ceftazidime, ceftriaxone and cefotaxime). *Source*: Adapted from https://cddep.org/wp-content/uploads/2017/06/swa_edits_9.16.pdf. *(See insert for color representation of the figure.)*

In this way, the dissemination into the environment and the transfer between other bacteria have been observed (Caniça et al. 2015). These genes can be disseminated in the environment and be transferred between strains of the same or different bacterial species, such as *Enterobacteriaceae* (DuPont and Steffen 2016; Su et al. 2008).

Consequently, the use of common antimicrobials renders the therapy inefficient against resistant bacteria (by positive selection for more resistant bacteria in the body for each generation), not only in patients with weakened immune system (van Duin et al. 2013) but also in healthy people (Founou et al. 2016; Ulstad et al. 2016).

WHO (2016) considers that travel to countries with high prevalence or to reservoirs of resistant bacteria in animal production, such as poultry farms, is one of the factors that contributed to the resistance to these antimicrobial agents. It is also caused by the production of ESBLs in the case of *Enterobacteriaceae* (Valentin et al. 2014). Furthermore, Ulstad et al. (2016) observed that 9.9% of the 284 volunteers in Norway were ciprofloxacin non-susceptible isolates. According to the same authors, 4.9% of ESBL and 3.2% of *Klebsiella* spp. isolated from healthy people were detected. Consequently, they are invariably associated with high treatment failure rates.

Wu et al. (2013) reported that human-to-human contact and animal-derived foods are the main factors for the dissemination of ESBL-producing bacteria among humans. The authors added the importance of minimizing this transmission to control the spread of ESBL-positive *E. coli* in humans. According to Overdevest et al. (2011), ESBL was present in 45 human fecal samples (4.9%) and 23 blood culture isolates (74.2%) from hospitalized patients in the Netherlands. Additionally, recent studies show the presence of ESBL-producing *E. coli* strains isolated from children with UTI (Al Atya et al. 2016; Clemente et al. 2015).

In another study realized in Germany, Valentin et al. (2014) analyzed fecal samples from humans, livestock populations (poultry, swine, cattle), and companion animals (dogs and horses) and verified that approximately 70% of the animal isolates and 50% of the human isolates contained ESBL genes. According to Overdevest et al. (2011), the use of antimicrobial drugs in food production animals in the Netherlands is among the highest in Europe. These authors verified a high prevalence of ESBL genes in chicken meat (79.8%). Poeta et al. (2010) reported that 22 of the 52 *E. coli* strains isolated from rabbits were producing ESBL. According to Carvalho et al. (2015), the presence of β-lactamases was not detected in Miranda donkeys (North of Portugal). However, ESBLs were reported in other studies performed in Portugal (Clemente et al. 2015; Costa et al. 2009) and European countries, such as equines in Germany (Schmiedel et al. 2014) and Iberian lynx in Portugal (Gonçalves et al. 2014).

Hernández et al. (2012) verified that human-associated *Enterobacteriaceae* were not common in the sampled penguins from Antarctic because none of these animals were positive for ESBL-producing bacteria. Hammerum et al. (2014) verified that ESBL-producing *E. coli* were detected in 19 of the 195 farmers who were taking care of Danish pigs. One possible explanation is the transfer of ESBL-producing *E. coli* or plasmids between pigs and farmers.

As mentioned earlier, the increase of ESBL-producing bacteria in humans, animals, and their surrounding environments is of global concern (Seni et al. 2016). *Escherichia coli* pathogenic strains may cause many different infections in food animals, affecting the skin, urinary tract, central nervous system, and cardiovascular system (Frye and Jackson 2013). Colibacillosis is particularly preoccupant because this disease has significant morbidity and mortality in swine and poultry (Bustamante et al. 2003; Silva et al. 2011).

11.6 Consequences and Future Strategies to Brace the Antibiotic Backbone

Antimicrobials have revolutionized the first clinical treatment of infections, in the 1940–1950, caused by bacterial agents, promoting a decrease in the number of pathogens or preventing the proliferation of pathogenic strains (Cantas et al. 2013; Poeta 2006).

AMR presents different problems and concerns for humans. The reduction of the efficacy of the treatments related to the increase of the duration of hospitalization (longer treatments and more invasive), causing higher costs per treatment, and ultimately the complete inefficacity of the drug treatments cause the mortality of patients, particularly for immunocompromised patients, e.g. of cancer therapy or after transplantation (Freire et al. 2016). It can be potentially lethal for epidemics (which can be spread globally), will be more frequent, and will reach to more people, increasing the morbidity and mortality and creating a huge impact in the human society.

The nosocomial infections and other bacterial diseases are a major cause of death in the developing world (Okeke et al. 2005; Sosa et al. 2010). The same authors add that poverty is "a major force driving the development of antimicrobial resistance." Moreover, antimicrobials can often be purchased without prescription. Furthermore, some drugs are counterfeit, while others do not have any active ingredient or a correct dose (Centers for Disease control and Prevention 2013; Hajjou et al. 2015).

Recent studies show that antibiotics affect the normal microbiome, causing different kinds of effects in human health, particularly obesity (Grupper and Nicolau 2017; Poulsen et al. 2017). These authors add that the use of antibiotics by children less than two years old has been linked to an increased risk of early childhood obesity.

Another main consequence is that no new effective antibiotics has been developed and introduced in the pharmaceutical market in the last years to cure bacterial infections (Reardon 2014), principally to treat that resistant superbugs.

The global main consequences are the increasing healthcare costs, prolonged hospital stays with a high probability of acquire nosocomial infections, treatment failures, and a significant number of deaths associated with increased suffering.

In addition, AMR has also huge economic implications in the EU, estimated to cause an economic loss of more than €1.5 billion each year (Jones-Dias et al. 2016), and a huge social impact. It is expected that 400 000 people will die because of MDR over the next 35 years (WHO 2016).

Coordinated actions are required in order to reduce the spread of AMR. There are some procedures to minimize this problem:

- Decrease the unnecessary antibiotic treatments and minimize overprescription Marinho et al. (2016).
- Use antibacterial drugs only for bacterial infections (Cantas et al. 2013).
- Do correctly the prescription and therapy of the antibiotic defined by the doctor (Michael et al. 2014).
- Implement effective containment of infections (Cristina et al. 2016).
- Make a prudent use or avoid these drugs in prophylaxis in agriculture and animal production (Frye and Jackson 2013; Shaikh et al. 2015).
- Change the public perception about common campaign related to the idea that humans need to be completely and permanently "clean" at a microscopic level (Tang et al. 2016).

11.7 Concluding Remarks and Future Perspectives

Due to antibiotic overuse, over prescription, non-prescription purchase, hoarding, commercial pressures, agriculture applications, and the failure of control measures to prevent the spread of resistant bacteria, MDR bacteria producing ESBLs and CREs are increasing problems in human and veterinary medicine. Furthermore, new resistance mechanisms are emerging and new developed antimicrobials are required.

This will lead to high economic and social costs, alongside increasing human suffering; thus worldwide coordinated actions must be implemented to minimize transmission between human and animals and reduce the spread and negative clinical effects.

These actions should be implemented at a global scale, and intense campaigns for the correct use of antibiotics should be done, in order to change the global perception of antimicrobial usage among common people, doctors, pharmaceuticals, farmers, etc. The task is huge, and not easy to resolve, but it is too important to not take it seriously.

Acknowledgments

The authors gratefully acknowledge the financial support of "Fundação para a Ciência e Tecnologia" (FCT – Portugal), through the reference SFRH/BD/133266/2017 (Medicina Clínica e Ciências da Saúde).

References

Abdallah, H.M., Reuland, E.A., Wintermans, B.B. et al. (2015). Extended-Spectrum β-Lactamases and/or Carbapenemases-Producing *Enterobacteriaceae* Isolated from retail chicken meat in Zagazig, Egypt. *PLoS One* 10 (8): e0136052.

Ahmed, M.O., Clegg, P.D., Williams, N.J. et al. (2010). Antimicrobial resistance in equine faecal *Escherichia coli* isolates from North West England. *Ann Clin Microbiol Antimicrob* 9: 12.

Al Atya, A.K., Drider-Hadiouche, K., Vachee, A., and Drider, D. (2016). Potentialization of β-lactams with colistin: in case of extended spectrum β-lactamase producing *Escherichia coli* strains isolated from children with urinary infections. *Res Microbiol* 167 (3): 215–221.

Amit, S., Mishali, H., Kotlovsky, T. et al. (2015). Bloodstream infections among carriers of carbapenem-resistant *Klebsiella pneumoniae*: etiology, incidence and predictors. *Clin Microbiol Infect* 21 (1): 30–34.

Belluco, S., Barco, L., Roccato, A., and Ricci, A. (2016). *Escherichia coli* and *Enterobacteriaceae* counts on poultry carcasses along the slaughterline: a systematic review and meta-analysis. *Food Control* 60: 269–280.

Bengtsson-Palme, J., Angelin, M., Huss, M. et al. (2015). The Human Gut Microbiome as a Transporter of Antibiotic Resistance Genes between Continents. *Antimicrob Agents Chemother* 59 (10): 6551–6560.

Bengtsson-Palme, J. and Larsson, D.G.J. (2015). Antibiotic resistance genes in the environment: prioritizing risks. *Nat Rev Microbiol* 13 (6): 396.

Blair, J.M., Webber, M.A., Baylay, A.J. et al. (2015). Molecular mechanisms of antibiotic resistance. *Nat Rev Microbiol* 13 (1): 42–51.

Brauner, A., Fridman, O., Gefen, O., and Balaban, N.Q. (2016). Distinguishing between resistance, tolerance and persistence to antibiotic treatment. *Nat Rev Microbiol* 14 (5): 320–330.

Bulycheva, E.V., Korotkova, E.I., Voronova, O.A. et al. (2014). Fluorescence analysis of *E. coli* bacteria in water. *Proc Chem* 10: 179–183.

Burow, E., Simoneit, C., Tenhagen, B.A., and Käsbohrer, A. (2014). Oral antimicrobials increase antimicrobial resistance in porcine *E. coli*: a systematic review. *Prev Vet Med* 113 (4): 364–375.

Bustamante, W., Alpízar, A., Hernández, S. et al. (2003). Predominance of *vanA* Genotype among Vancomycin-Resistant *Enterococcus* Isolates from Poultry and Swine in Costa Rica. *Appl Environ Microbiol* 69 (12): 7414–7419.

Caniça, M., Manageiro, V., Jones-Dias, D. et al. (2015). Current perspectives on the dynamics of antibiotic resistance in different reservoirs. *Res Microbiol* 166 (7): 594–600.

Cantas, L., Shah, S.Q., Cavaco, L.M. et al. (2013). A brief multi-disciplinary review on antimicrobial resistance in medicine and its linkage to the global environmental microbiota. *Front Microbiol* 14 (4): 96.

Carvalho, I., Del Campo, R., Sousa, M. et al. (2017). Antimicrobial resistant *Escherichia coli* and *Enterococcus* spp. isolated from Miranda Donkey (*Equus asinus*): an old problem from a new source with a different approach. *J Med Microbiol* 66 (2): 191–202.

Carvalho, I., Sousa, M., Silva, N. et al. (2015). Resistência a antibióticos em *Escherichia coli* e *Enterococcus* spp. isolados de amostras fecais da raça Asinina de Miranda (*Equus asinus*). *Livro de Resumos das VI Jornadas de Segurança Alimentar e Saúde Pública*, 27. Vila Real: Universidade de Trás-os-Montes e Alto Douro.

CDC (2015). About antimicrobial resistance. *National Center for Emerging and Zoonotic.*

Center for Disease Dynamics, Economics & Policy (2015). State of the World's Antibiotics. CDDEP: Washington, D.C.

Centers for Disease Control and Prevention (2013). *Antibiotic resistance threats in the United States*. 1–113.

Chen, Q., An, X., Li, H. et al. (2016). Long-term field application of sewage sludge increases the abundance of antibiotic resistance genes in soil. *Environ Int* 92–93: 1–10.

Chen, X., Wen, Y., Li, L. et al. (2015). The stability, and efficacy against penicillin-resistant *Enterococcus faecium*, of the plectasin peptide efficiently produced by *Escherichia coli*. *J Microbiol Biotechnol* 25 (7): 1007–1014.

Chopra, I. and Roberts, M. (2001). Tetracycline antibiotics: mode of action, applications, molecular biology, and epidemiology of bacterial resistance. *Microbiol Mol Biol Rev* 65: 232–260.

Clemente, L., Manageiro, V., Jones-Dias, D. et al. (2015). Antimicrobial susceptibility and oxymino-β-lactam resistance mechanisms in *Salmonella enterica* and *Escherichia coli* isolates from different animal sources. *Res Microbiol* 166 (7): 574–583.

CLSI (2015). *Performance Standards for Antimicrobial Susceptibility Testing*. Wayne, PA: Clinical and Laboratory Standards Institute.

Cohen, N., Ennaji, H., Bouchrif, B. et al. (2007). Comparative study of microbiological quality of raw poultry meat at various seasons and for different slaughtering processes in Casablanca (Morocco). *J Appl Poultry Res* 16 (4): 502–508.

Cornejo-Juárez, P., Vilar-Compte, D., Pérez-Jiménez, C. et al. (2015). The impact of hospital-acquired infections with multidrug-resistant bacteria in an oncology intensive care unit. *Int J Infect Dis* 31: 31–34.

Costa, D., Vinue, L., Poeta, P. et al. (2009). Prevalence of extended-spectrum beta-lactamase-producing *Escherichia coli* isolates in faecal samples of broilers. *Vet Microbiol* 138 (3–4): 339–344.

Cristina, M.L., Sartini, M., Ottria, G. et al. (2016). Epidemiology and biomolecular characterization of carbapenem-resistant *Klebsiella pneumoniae* in an Italian hospital. *J Prev Med Hyg* 57 (3): E149–E156.

Das, R., Ahmed, T., Saha, H. et al. (2016). Clinical risk factors, bacterial aetiology, and outcome of urinary tract infection in children hospitalized with diarrhoea in Bangladesh. *Epidemiol Infect* 145 (5): 1018–1024.

Day, M.J., Rodriguez, I., van Essen-Zandbergen, A. et al. (2016). Diversity of STs, plasmids and ESBL genes among *Escherichia coli* from humans, animals and food in Germany, the Netherlands and the UK. *J Antimicrob Chemother* 71 (5): 1178–1182.

Dos Santos, K.M., Vieira, A.D., Salles, H.O. et al. (2015). Safety, beneficial and technological properties of *Enterococcus faecium* isolated from Brazilian cheeses. *Braz J Microbiol* 46 (1): 237–249.

DuPont, H.L. and Steffen, R.M. (2016). Use of antimicrobial agents for treatment and prevention of travellers' diarrhoea in the face of enhanced risk of transient fecal carriage of multi-drug resistant Enterobacteriaceae: setting the stage for consensus recommendations. *J Travel Med* 23 (6): taw054.

European Commission (2014). *Antimicrobial Resistance*. Brussels. https://ec.europa.eu/health/amr/antimicrobial-resistance_en. This is the pdf https://ec.europa.eu/health/amr/sites/amr/files/amr_factsheet_en.pdf

European Centre for Disease Prevention and Control (2017). Antimicrobial resistance surveillance in Europe 2015. Annual Report of the European Antimicrobial Resistance Surveillance Network (EARS-Net). Stockholm.

Founou, L.L., Founou, R.C., and Essack, S.Y. (2016). Antibiotic resistance in the food chain: a developing country-perspective. *Front Microbiol* 7: 1881.

Freire, M.P., Oshiro, I.C., Pierrotti, L.C. et al. (2016). Carbapenem-resistant *Enterobacteriaceae* acquired before liver transplantation: impact on recipient outcomes. *Transplantation*.

Frye, J.G. and Jackson, C.R. (2013). Genetic mechanisms of antimicrobial resistance identified in *Salmonella enterica*, *Escherichia coli*, and *Enteroccocus* spp. isolated from U.S. food animals. *Front Microbiol* 4: 22.

Gonçalves, A., Poeta, P., Monteiro, R. et al. (2014). Comparative proteomics of an extended spectrum beta-lactamase producing *Escherichia coli* strain from the Iberian wolf. *J Proteomics* 104: 80–93.

Grupper, M. and Nicolau, D.P. (2017). Obesity and skin and soft tissue infections: how to optimize antimicrobial usage for prevention and treatment? *Curr Opin Infect Dis* 30 (2): 180–191.

Guerrero-Ramos, E., Cordero, J., Molina-Gonzalez, D. et al. (2016). Antimicrobial resistance and virulence genes in *enterococci* from wild game meat in Spain. *Food Microbiol* 53: 156–164.

Hajjou, M., Krech, L., Lane-Barlow, C. et al. (2015). Monitoring the quality of medicines: results from Africa, Asia, and South America. *Am J Trop Med Hyg* 92 (6 Suppl): 68–74.

Hamilton, W.L., and Wenlock, R. (2016). Antimicrobial resistance: a major threat to public health. *Cambridge Medicine* doi: 10.7244/cmj.2016.01.001.

Hammerum, A.M., Larsen, J., Andersen, V.D. et al. (2014). Characterization of extended-spectrum beta-lactamase (ESBL)-producing *Escherichia coli* obtained from Danish pigs, pig farmers and their families from farms with high or no consumption of third- or fourth-generation cephalosporins. *J Antimicrob Chemother* 69 (10): 2650–2657.

He, L.Y., Ying, G.G., Liu, Y.S. et al. (2016). Discharge of swine wastes risks water quality and food safety: antibiotics and antibiotic resistance genes from swine sources to the receiving environments. *Environ Int* 92–93: 210–219.

Hernández, J., Stedt, J., Bonnedahl, J. et al. (2012). Human-associated extended-spectrum β-lactamase in the Antarctic. *Appl Environ Microbiol* 78 (6): 2056–2058.

Herskin, M.S., Jensen, H.E., Jespersen, A. et al. (2016). Impact of the amount of straw provided to pigs kept in intensive production conditions on the occurrence and severity of gastric ulceration at slaughter. *Res Vet Sci* 104: 200–206.

Höjgård, S. (2012). Antibiotic resistance – why is the problem so difficult to solve? *Infect Ecol Epidemiol.* 2: 10.3402/iee.v2i0.18165. doi:10.3402/iee.v2i0.18165.

Hoxha, A., Kärki, T., Giambi, C. et al. (2016). Attributable mortality of carbapenem-resistant in *Klebsiella pneumoniae* infections in a prospective matched cohort study in Italy. *J Hosp Infect* 92 (1): 61–66.

Huiting, J. (2015). Antimicrobial resistance of Escherichia coli in broilers with colibacillosis in Morocco. Faculty of Veterinary Medicine Theses.

In, J., Foulke-Abel, J., Zachos, N.C. et al. (2016). Enterohemorrhagic reduce mucus and intermicrovillar bridges in human stem cell-derived colonoids. *Cell Mol Gastroenterol Hepatol* 2 (1): 48–63.

Institute of Medicine (2013). *Countering the Problem of Falsified and Substandard Drugs*. Washington, DC: The National Academies Press.

Iredell, J., Brown, J., and Tagg, K. (2016). Antibiotic resistance in *Enterobacteriaceae*: mechanisms and clinical implications. *BMJ* 352: h6420.

Jones-Dias, D., Manageiro, V., Graça, R. et al. (2016). *qnr*S1- and *aac*(6′)-Ib-cr-producing *Escherichia coli* among isolates from animals of different sources: susceptibility and genomic characterization. *Front Microbiol* 7: 671.

Karanika, S., Karantanos, T., Arvanitis, M. et al. (2016). Fecal colonization with extended-spectrum beta-lactamase-producing *Enterobacteriaceae* and risk factors among healthy individuals: a systematic review and metaanalysis. *Clin Infect Dis* 63 (3): 310–318.

Khuluza, F., Kigera, S., Jahnke, R.W., and Heide, L. (2016). Use of thin-layer chromatography to detect counterfeit sulfadoxine/pyrimethamine tablets with the wrong active ingredient in Malawi. *Malar J* 15: 215.

Kong, K.F., L. Schneper, and Mathee, K. (2010). Beta-lactam antibiotics: from antibiosis to resistance and bacteriology. *Apmis* 118 (1): 1–36.

Kuhnert, P., Boerlin, P., and Frey, J. (2000). Target genes for virulence assessment of *Escherichia coli* isolates from water, food and the environment. *FEMS Microbiol Rev* 24 (1): 107–117.

Laurent, C., Rodriguez-Villalobos, H., Rost, F. et al. (2008). Intensive care unit outbreak of extended-spectrum beta-lactamase-producing *Klebsiella pneumoniae* controlled by cohorting patients and reinforcing infection control measures. *Infect Control Hosp Epidemiol* 29 (6): 517–524.

Lederman, E.R. and Crum, N.F. (2005). Pyogenic liver abscess with a focus on *Klebsiella pneumoniae* as a primary pathogen: an emerging disease with unique clinical characteristics. *Am J Gastroenterol* 100 (2): 322–331.

Li, P., Wu, D., Liu, K. et al. (2014). Investigation of antimicrobial resistance in *Escherichia coli* and enterococci isolated from Tibetan pigs. *PLoS One* 9 (4): e95623.

Lupo, A., Coyne, S., and Berendonk, T.U. (2012). Origin and evolution of antibiotic resistance: the common mechanisms of emergence and spread in water bodies. *Front Microbiol* 3: 18.

Lynch, J.P., Clark, N.M., and Zhanel, G.G. (2013). Evolution of antimicrobial resistance among *Enterobacteriaceae* (focus on extended spectrum β-lactamases and carbapenemases). *Expert Opin Pharmacother* 14 (2): 199–210.

Manzetti, S. and Ghisi, R. (2014). The environmental release and fate of antibiotics. *Mar Pollut Bull* 79 (1-2): 7–15.

Marinho, C.M., Santos, T., Gonçalves, A. et al. (2016). A decade-long commitment to antimicrobial resistance surveillance in Portugal. *Front Microbiol* 7: 1650.

Michael, C.A., Dominey-Howes, D., and Labbate, M. (2014). The antimicrobial resistance crisis: causes, consequences, and management. *Front Public Health* 2: 145.

Morrill, H.J., Pogue, J.M., Kaye, K.S., and LaPlante, K.L. (2015). Treatment options for carbapenem-resistant *Enterobacteriaceae* infections. *Open Forum Infect Dis* 2 (2): ofv050.

Moucheraud, C., Worku, A., Molla, M. et al. (2015). Consequences of maternal mortality on infant and child survival: a 25-year longitudinal analysis in Butajira Ethiopia (1987-2011). *Reprod Health* 12 (Suppl 1): S4–S4.

Moura, I., Torres, C., Silva, N. et al. (2013). Genomic description of antibiotic resistance in *Escherichia coli* and enterococci isolates from healthy Lusitano horses. *J Equine Vet* 33 (12): 1057–1063.

Okeke, I.N., Laxminarayan, R., Bhutta, Z.A. et al. (2005). Antimicrobial resistance in developing countries. Part I: recent trends and current status. *Lancet Infect Dis* 5 (8): 481–493.

Onori, R., Gaiarsa, S., Comandatore, F. et al. (2015). Tracking nosocomial *Klebsiella pneumoniae* infections and outbreaks by whole-genome analysis: small-scale Italian scenario within a single hospital. *J Clin Microbiol* 53 (9): 2861–2868.

Overdevest, I., Willemsen, I., Rijnsburger, M. et al. (2011). Extended-spectrum β-lactamase genes of *Escherichia coli* in chicken meat and humans, the Netherlands. *Emerg Infect Dis* 17 (7): 1216–1222.

Planta, M.B. (2007). The role of poverty in antimicrobial resistance. *J Am Board Fam Med* 20 (6): 533–539.

Poeta, P. (2006). Resistência a Antibióticos, Factores de Virulência e Bacteriocinas em Estirpes Comensais de Enterococcus spp. de Animais e Humanos. Dissertação de Doutoramento em Medicina Veterinária Vila Real: UTAD.

Poeta, P., Costa, D., Rodrigues, J., and Torres, C. (2006). Detection of genes encoding virulence factors and bacteriocins in fecal enterococci of poultry in Portugal. *Avian Dis* 50 (1): 64–68.

Poeta, P., Radhouani, H., Goncalves, A. et al. (2010). Genetic characterization of antibiotic resistance in enteropathogenic *Escherichia coli* carrying extended-spectrum beta-lactamases recovered from diarrhoeic rabbits. *Zoonoses Public Health* 57 (3): 162–170.

Potter, R.F., D'Souza, A.W., and Dantas, G. (2016). The rapid spread of carbapenem-resistant *Enterobacteriaceae*. *Drug Resist Updat* 29: 30–46.

Poulou, A., Voulgari, E., Vrioni, G. et al. (2012). Imported *Klebsiella pneumoniae* carbapenemase-producing *K. pneumoniae* clones in a Greek hospital: impact of infection control measures for restraining their dissemination. *J Clin Microbiol* 50 (8): 2618–2623.

Poulsen, M.N., Pollak, J., Bailey-Davis, L. et al. (2017). Associations of prenatal and childhood antibiotic use with child body mass index at age 3 years. *Obesity (Silver Spring)* 25 (2): 438–444.

Radhouani, H., Poeta, P., Goncalves, A. et al. (2012). Wild birds as biological indicators of environmental pollution: antimicrobial resistance patterns of *Escherichia coli* and enterococci isolated from common buzzards (*Buteo buteo*). *J Med Microbiol* 61 (Pt 6): 837–843.

Reardon, S. (2014). Antibiotic resistance sweeping developing world. *Nature* 509: 141–142.

Ridolfo, A.L., Rimoldi, S.G., Pagani, C. et al. (2016). Diffusion and transmission of carbapenem-resistant *Klebsiella pneumoniae* in the medical and surgical wards of a university hospital in Milan, Italy. *J Infect Public Health* 9 (1): 24–33.

Righi, E., Peri, A.M., Harris, P.N. et al. (2016). Global prevalence of carbapenem resistance in neutropenic patients and association with mortality and carbapenem use: systematic review and meta-analysis. *J Antimicrob Chemother* 72 (3): 668–677.

Rummukainen, M.-L., Mäkelä, M., Noro, A. et al. (2013). Assessing prevalence of antimicrobial use and infections using the minimal data set in Finnish long-term care facilities. *Am J Infect Control* 41: e35-7.

Ruppé, É., Woerther, P.-L., and Barbier, F. (2015). Mechanisms of antimicrobial resistance in Gram-negative bacilli. *Ann Intensive Care* 5: 21.

Saenz, Y., Zarazaga, M., Brinas, L. et al. (2001). Antibiotic resistance in *Escherichia coli* isolates obtained from animals, foods and humans in Spain. *Int J Antimicrob Agents* 18 (4): 353–358.

Schmiedel, J., Falgenhauer, L., Domann, E. et al. (2014). Multiresistant extended-spectrum β-lactamase-producing *Enterobacteriaceae* from humans, companion animals and horses in central Hesse, Germany. *BMC Microbiol* 14: 187.

Seni, J., Falgenhauer, L., Simeo, N. et al. (2016). Multiple ESBL-producing *Escherichia coli* sequence types carrying quinolone and aminoglycoside resistance genes circulating in companion and domestic farm animals in Mwanza, Tanzania, harbor commonly occurring plasmids. *Front Microbiol* 7: 142.

Shaikh, S., Fatima, J., Shakil, S. et al. (2015). Antibiotic resistance and extended spectrum beta-lactamases: types, epidemiology and treatment. *Saudi J. Biol. Sci.* 22 (1): 90–101.

Sharma, P. and Towse, A. (2011). *New Drugs to Tackle Antimicrobial Resistance: Analysis of EU Policy Options*. London, UK: Office of Health Economics.

Silva, N., Igrejas, G., Gonçalves, A., and Poeta, P. (2012). Commensal gut bacteria: distribution of *Enterococcus* species and prevalence of *Escherichia coli* phylogenetic groups in animals and humans in Portugal. *Ann Microbiol* 62 (2): 449–459.

Simoneit, C., Burow, E., Tenhagen, B.A., and Käsbohrer, A. (2015). Oral administration of antimicrobials increase antimicrobial resistance in *E. coli* from chicken: a systematic review. *Prev Vet Med* 118 (1): 1–7.

de Sosa, A.J., Byarugaba, D.K., Amabile, C. et al. (2010). *Antimicrobial Resistance in Developing Countries*. New York: Springer.

Spellberg, B. (2010). The antibacterial pipeline: why is it drying up, and what must be done about it? In: *Antibiotic Resistance: Implications for Global Health and Novel Intervention Strategies: Workshop Summary* For the Institute of Medicine, Forum on Antimicrobial Threats (ed. E.R. Choffnes, D.A. Relman and A. Mack). Washington, DC: The National Academies Press.

Spellberg, B., Blaser, M., Guidos, R. et al. (2011). Combating antimicrobial resistance: policy recommendations to save lives. *Clin Infect Dis* 52 Suppl 5: S397–428.

Su, L.-H., Chu, C., Cloeckaert, A., and Chiu, C.-H. (2008). An epidemic of plasmids? Dissemination of extended-spectrum cephalosporinases among *Salmonella* and other *Enterobacteriaceae*. *FEMS Immunol. Med. Microbiol.* 52 (2): 155–168.

Tan, S.Y. and Tatsumura, Y. (2015). Alexander Fleming (1881–1955): discoverer of penicillin. *Singapore Med J* 56 (7): 366–367.

Tang, Q., Song, P., Li, J. et al. (2016). Control of antibiotic resistance in China must not be delayed: the current state of resistance and policy suggestions for the government, medical facilities, and patients. *Biosci Trends* 10 (1): 1–6.

Tenover, F.C. (2006). Mechanisms of antimicrobial resistance in bacteria. *Am J Med* 119 (6): S62–S70.

Tzouvelekis, L.S., Markogiannakis, A., Psichogiou, M. et al. (2012). Carbapenemases in *Klebsiella pneumoniae* and other *Enterobacteriaceae*: an evolving crisis of global dimensions. *Clin Microbiol Rev* 25 (4): 682–707.

Ulstad, C.R., Solheim, M., Berg, S. et al. (2016). Carriage of ESBL/AmpC-producing or ciprofloxacin non-susceptible *Escherichia coli* and *Klebsiella* spp. in healthy people in Norway. *Antimicrob Resist Infect Control* 5: 57.

Valentin, L., Sharp, H., Hille, K. et al. (2014). Subgrouping of ESBL-producing *Escherichia coli* from animal and human sources: an approach to quantify the distribution of ESBL types between different reservoirs. *Int J Med Microbiol* 304 (7): 805–816.

Van Boeckel, T.P., Brower, C., Gilbert, M. et al. (2015). Global trends in antimicrobial use in food animals. *Proc Natl Acad Sci U S A* 112 (18): 5649–5654.

Van Duin, D., Kaye, K.S., Neuner, E.A., and Bonomo, R.A. (2013). Carbapenem-resistant *Enterobacteriaceae*: a review of treatment and outcomes. *Diagn Microbiol Infect Dis* 75 (2): 115–120.

Venhuis, B.J., Keizers, P.H., Klausmann, R., and Hegger, I. (2016). Operation resistance: a snapshot of falsified antibiotics and biopharmaceutical injectables in Europe. *Drug Test Anal* 8 (3–4): 398–401.

Ventola, C.L. (2015). The antibiotic resistance crisis. Part 1: causes and threats. *P&T* 40 (4): 277–283.

Viegas, S., Brandão, J., Taylor, H., and Viegas, C. (2015). Environmental microbiology for public health: capturing international developments in the field. *Res Microbiol* 166 (7): 555–556.

WHO (2016). Antimicrobial Resistance – a global epidemic. https://www.wto.org/english/news_e/news16_e/heal_29aug16_e.pdf (Ed F. sheet).

Woodford, N., Fagan, E.J., and Ellington, M.J. (2006). Multiplex PCR for rapid detection of genes encoding CTX-M extended-spectrum β-lactamases. *J Antimicrob Chemother* 57 (1): 154–5.

Wu, G., Day, M.J., Mafura, M.T. et al. (2013). Comparative analysis of ESBL-positive *Escherichia coli* isolates from animals and humans from the UK, The Netherlands and Germany. *PLoS One* 8 (9): e75392.

12

Extended-Spectrum-β-Lactamase and Carbapenemase-Producing Enterobacteriaceae in Food-Producing Animals in Europe

An Impact on Public Health?

Nuno Silva[1], *Isabel Carvalho*[2-6], *Carol Currie*[1], *Margarida Sousa*[2,3,4], *Gilberto Igrejas*[3,4], *and Patrícia Poeta*[2,5]

[1] *Moredun Research Institute, Pentlands Science Park, Penicuik, Scotland*
[2] *Department of Veterinary Sciences, University of Trás-os-Montes and Alto Douro, Vila Real, Portugal*
[3] *Department of Genetic and Biotechnology, University of Trás-os-Montes and Alto Douro, Vila Real, Portugal*
[4] *Functional Genomics and Proteomics Unit, University of Trás-os-Montes and Alto Douro, Vila Real, Portugal*
[5] *LAQV-REQUIMTE, Faculty of Science and Technology, New University of Lisbon, Monte da Caparica, Portugal*
[6] *MicroART- Antibiotic Resistance Team, University of Trás-os-Montes and Alto Douro, Vila real, Portugal*

12.1 Extended-Spectrum β-Lactamase

The potential contribution of food-producing animals to public health risks by extended-spectrum β-lactamase (ESBL)-producing bacteria is related to specific plasmid-mediated ESBL. The ESBL plasmid-encoded enzymes found in *Enterobacteriaceae*, frequently in *Escherichia coli*, *Klebsiella pneumoniae*, and *Salmonella*, confer resistance to a variety of ß-lactam antibiotics – including penicillins; second-, third-, and fourth-generation cephalosporins; and monobactams – by hydrolysis of these antibiotics. But usually these do not affect cephamycins (second-generation cephalosporins) and the carbapenems and are inhibited by β-lactamase inhibitors such as clavulanic acid, sulbactam, and tazobactam (Rawat and Nair 2010). Moreover, resistance to broad-spectrum cephalosporins can be due to overexpression of chromosomal or plasmid-mediated AmpC enzymes, encoded by genes such as bla_{CMY} and bla_{DHA} genes. Such enzymes also confer resistance to cephamycins and cannot be inhibited by the β-lactamase inhibitors (Smet et al. 2010).

Although there are a large number of genes that encode ESBL enzymes, not all are equally prevalent among human and animal bacteria. Until the 1990s, the majority of ESBLs identified in human clinical isolates were SHV or TEM types and, more than one decade later, the CTX-M enzymes have now become the most common genetic variant type of ESBL (Shaikh et al. 2015).

Antibiotic Drug Resistance, First Edition. Edited by José-Luis Capelo-Martínez and Gilberto Igrejas.

The most frequently reported ESBL subtypes in *Enterobacteriaceae* in food-producing animals within the European Union (EU) are CTX-M-1, CTX-M-14, CTX-M-15, CTX-M-2, TEM-52, and SHV-12 (Figure 12.1). A wide range of additional CTX-M variant types (CTX-M-1, -2, -3, -8, -9, -14, -15, -17, -18, -20, -32, -53) (EFSA 2013), TEM (TEM-20, -52, -106, -126), and SHV variants (SHV-2, -5, -12) (Chiaretto et al. 2008; Endimiani et al. 2012) have been detected in both food-producing animals and food in several European countries; CTX-M-1, CTX-M-2, and CTX-M-14 enzymes have been found associated with *E. coli* mainly from poultry (Briñas et al. 2003; Girlich et al. 2007; Smet et al. 2008; Costa et al. 2009; Jones-Dias et al. 2016). CTX-M-15, the most widely diffused enzyme among *Enterobacteriaceae* in humans, was only recently detected in *E. coli* from poultry, cattle, and pigs (Maciuca et al. 2015; Zurfluh et al. 2015; Falgenhauer et al. 2016; Muller et al. 2016).

The massive and indiscriminate use of antibiotics in veterinary medicine has contributed to the selection and spread of multidrug-resistant Gram-negative bacteria. Food-producing animals are considered important reservoirs of antibiotic-resistant *Enterobacteriaceae*, and their role on human health has drawn considerable global attention (Seiffert et al. 2013).

In animals the spread of these β-lactamases started to increase only in the last decade, and it is possible that these enzymes might be of human origin (Hernandez et al. 2005). Similarly, β-lactamases in humans can also be directly derived from animal origin. In fact, ESBL enzymes from food-producing animals, are also predominantly in human bacteria. In Europe, the CTX-M group is the most frequently detected ESBL in food animals (Ewers et al. 2012). The most prevalent enzymes in commensal and pathogenic *E. coli* from both humans and animals are CTX-M-14 and CTX-M-32 in Portugal (Costa et al. 2009); CTX-M-9, SHV-12, and CTX-M-14 in Spain (Briñas et al. 2005; Riano et al. 2009); and CTX-M-1 in France (Girlich et al. 2007). In addition, two studies carried out in 2011 from the Netherlands identified CTX-M-1 as the most prevalent ESBL type shared by human patients, healthy carriers, poultry, and retail chicken meat, suggesting recent cross-transmission between human and avian hosts (Leverstein-van Hall et al. 2011; Overdevest et al. 2011). This may indicate that there is somehow a similar epidemiology among food-producing animal and human *Enterobacteriaceae*. However, only a few studies described clear evidence of direct transmission of ESBL-producing *E. coli* isolates between food-producing animals and humans via the food chain (Seiffert et al. 2013) or by direct contact between humans and animals (de Been et al. 2014).

12.1.1 ESBL-Producing *Enterobacteriaceae* in Food Animals

12.1.1.1 Poultry

One of the first description of ESBL producing *Enterobacteriaceae* isolated from food-producing animals was descripted in poultry, Spain. *Escherichia coli*

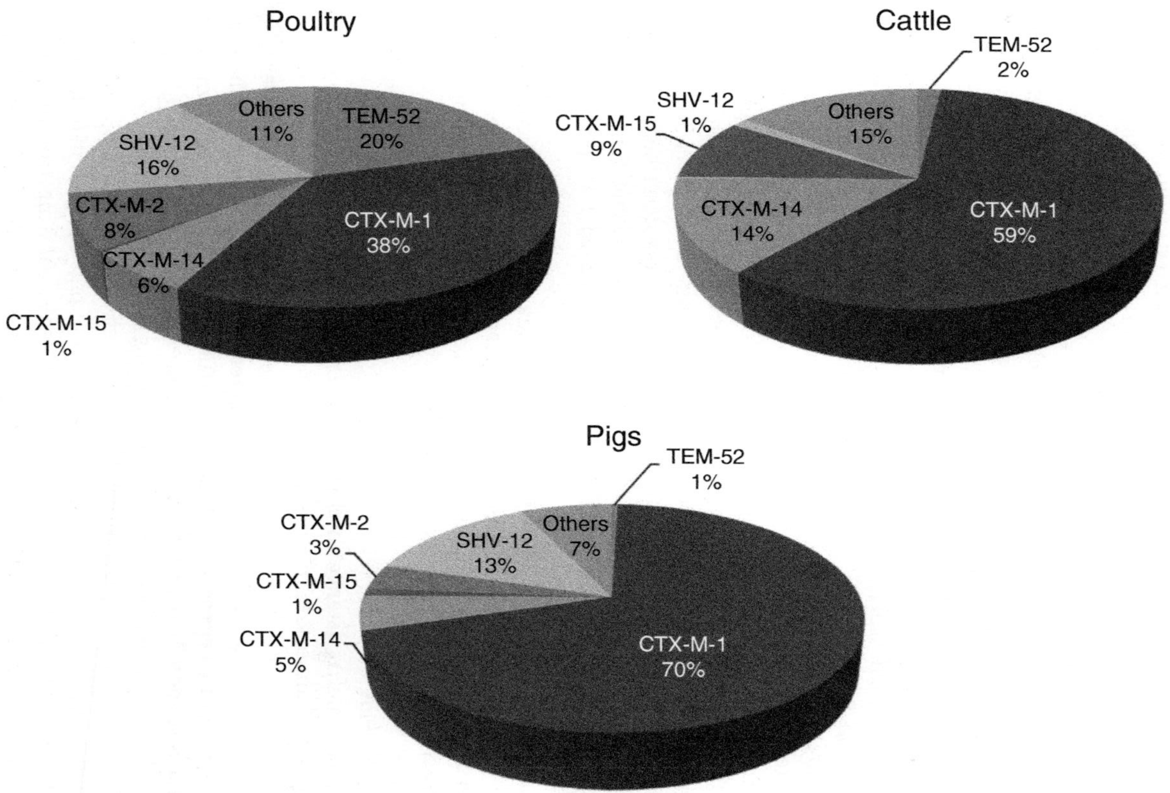

Figure 12.1 Schematic representation of studies reporting the presence of the most predominant extended-spectrum β-lactamase types among *Enterobacteriaceae* isolated from food-producing animals in Europe.

strains isolated from fecal samples of healthy chickens, between 2000 and 2001, were found to harbor CTX-M-14 and SHV-12 β-lactamases (Briñas et al. 2003). Subsequently, the occurrence of various ESBL types was described shortly after in pigs and cattle (Briñas et al. 2005; Meunier et al. 2006). During the last decade, the presence of ESBL-producing *E. coli* and *Salmonella* in animals and food has been increasingly reported in different European countries. Although these enzymes have been described in bacteria from all major food-producing animals, poultry and poultry products are most frequently reported to carry ESBL-producing *Enterobacteriaceae.*

Most of these studies are from European countries, with prevalence of ESBLs ranging between 0.5 and 94% (Bergenholtz et al. 2009; Cohen Stuart et al. 2012). High levels of bla_{ESBL} genes are observed in Portugal, Italy, Spain, the Netherlands, Belgium, and Germany. In Portugal, $bla_{\mathrm{TEM\text{-}52}}$, $bla_{\mathrm{CTX\text{-}M\text{-}14}}$, and $bla_{\mathrm{CTX\text{-}M\text{-}32}}$ were detected in cefotaxime-resistant *E. coli* obtained from fecal samples in broilers at slaughterhouses (Costa et al. 2009). Another study described *E. coli* isolates with high levels of bla_{ESBL}, obtained from healthy broilers in Italian farms, confirming the presence of $bla_{\mathrm{CTX\text{-}M\text{-}1}}$, $bla_{\mathrm{CTX\text{-}M\text{-}32}}$, and $bla_{\mathrm{SHV\text{-}12}}$ (Bortolaia et al. 2010). The presence of CTX-M-type genes was detected in a significant number of *E. coli* isolated from healthy chickens at poultry farms in Tenerife, Spain (Abreu et al. 2014). A recent study performed in the Netherlands showed the occurrence of $bla_{\mathrm{CTX\text{-}M\text{-}1}}$, $bla_{\mathrm{SHV\text{-}12}}$, and $bla_{\mathrm{TEM\text{-}52}}$ among *E. coli* isolated from poultry feces (Blaak et al. 2015). High rates and diversity of ESBL among cloacal *E. coli* was described in Belgian broiler farms (Smet et al. 2008) and in broiler chickens at slaughter in Germany (Reich et al. 2013), with TEM-52, SHV-12, and CTX-M-1 as the most often reported types.

According to these results, CTX-M-1 appears to be well disseminated in food-producing animals in most of the European countries mentioned. On the other hand, CTX-M-32 appears to be more associated with food-producing animals in Southern European countries.

12.1.1.2 Pigs

Antimicrobial resistance in commensal *Enterobacteriaceae* from pigs may also play an important role in the spread of resistance, constituting an important reservoir for these transmissible resistance genes. The occurrence of CTX-M-1-producing *E. coli* in pigs treated with ceftiofur in Danish farms was reported (Jørgensen et al. 2007). A study performed in different swine farms distributed in Spain showed the presence of SHV-12-, CTX-M-1-, CTX-M-9-, and CTX-M-14-encoding genes in *E. coli* isolates from fecal samples of healthy pigs, resistant to extended-spectrum cephalosporins (Escudero et al. 2010). In Portugal, about half of pigs sampled in a slaughterhouse revealed ESBL-producing *E. coli* isolates, showing the presence of CTX-M-1, CTX-M-9, CTX-M-14, CTX-M-32, and SHV-12 enzymes (Ramos et al. 2013). However, a recent study performed in Finland in different food-producing animals showed that no bla_{ESBL} gene-carrying isolates were found in pigs (Päivärinta et al. 2016).

12.1.1.3 Cattle

In many EU countries, ESBL-producing *E. coli* were detected in cattle. In the other food-producing animals, the $bla_{CTX\text{-}M\text{-}1}$ is the major ESBL gene found in cattle, presenting about 60% of all β-lactamase types. A study designed to assess the prevalence of ESBL-producing in dairy cows and beef cattle in Bavaria, Germany, showed that 32.8% of the animals contained ESBL-producing *E. coli*. Of these, more than 90% harbored CTX-M genes being the CTX-M group 1 the most frequently found group (Schmid et al. 2013).

Other European studies reported common occurrence of ESBL-producing *E. coli* in cattle in France (Madec et al. 2008), Denmark (DANMAP 2015), Switzerland (Geser et al. 2011), and the United Kingdom (Horton et al. 2011). In Europe the CTX-M-1 and -15 have been the predominant *bla* genes in western and northern European countries (Geser et al. 2011; Schmid et al. 2013; Carmo et al. 2014; Dahms et al. 2015), while the variant CTX-M-32 is predominant in food-producing animals in Southern Europe. CTX-M-9 group has been isolated from animals in Spain (Briñas et al. 2005), France (Valat et al. 2012), Switzerland (Geser et al. 2011), and the United Kingdom (Snow et al. 2011).

12.2 Carbapenemases

In Gram-negative bacteria, the main acquired mechanism leading to resistance or decreased susceptibility to carbapenem is due to the acquisition of exogenous genetic material containing gene(s) coding for carbapenemase production. The carbapenemases are β-lactamases with versatile hydrolytic capacities. They have the ability to hydrolyze penicillins, cephalosporins, monobactams, and carbapenems (Queenan and Bush 2007). Carbapenems are broad-spectrum β-lactam antimicrobials mostly used for the treatment of severe infections in humans and are considered a last resort of treatment for infections caused by highly drug-resistant Gram-negative bacteria (Papp-Wallace et al. 2011). Carbapenem resistance is much feared in clinical medicine since it results in loss of activity of nearly all β-lactam antibiotics. Additionally, given the high prevalence among the most clinically relevant carbapenem-resistant *Enterobacteriaceae* of co-resistance to a range of critically important classes of antimicrobial drugs, this leaves few therapeutic options (Magiorakos et al. 2012).

Based on their hydrolytic mechanism, carbapenemases are classified into two distinct groups, the serine carbapenemases, with an active site serine and, the metallo-β-lactamases (MBL) with zinc ions in the active site. The most frequently detected carbapenemase genes code for the production of class B metallo β-lactamases are *bla*$_{VIM}$, *bla*$_{IMP}$, and *bla*$_{NDM}$, class A β-lactamases *bla*$_{KPC}$, and class D β-lactamases *bla*$_{OXA-23, 40}$, $_{-48}$, and $_{-58}$ types (Nordmann et al. 2011). One of the milestones in the emergence of carbapenemases in *Enterobacteriaceae* was the detection of a novel carbapenemase, designated as *K. pneumoniae* carbapenemase (KPC) (Yigit et al. 2001).

Carbapenemases in *Enterobacteriaceae* are mainly found in *K. pneumoniae*, and in a lesser extent in *E. coli* and *Salmonella* and other enterobacterial species. The emergence of carbapenemases appeared mainly in Asia (Japan and India) and North America (United States). Nevertheless, the ESBL outbreaks occurred in many parts of the world, including Europe (Cantón et al. 2012). Although an increasing prevalence of carbapenemase-producing bacteria strains has been identified mainly in bacterial isolates from cases of human infection worldwide, carbapenemase-carrying bacteria have also been reported in food-producing animals (Guerra et al. 2014; Fischer et al. 2017) and their environment (Fischer et al. 2017). The first report of carbapenemase-producing isolates from livestock was made in 2012 in dairy cattle in France. *Acinetobacter* spp. from rectal swabs was described to harbor the *bla*$_{OXA-23}$ carbapenemase gene (Poirel et al. 2012).

Therefore, carbapenemases are now seen as a new and potentially emerging problem in food-producing animals. The prevalence of such resistance in bacteria from animals is largely unknown. To date, just a few publications have reported the presence of carbapenem resistance in bacteria from animals, including food-producing animals (pigs, poultry, bovines, and horses) and their environment in three different EU member states (Germany, France, and Belgium) (Fischer et al. 2012; Poirel et al. 2012; Smet et al. 2012). So far, no reports of such bacteria from animal-derived food products are available in Europe. However, carbapenemase-producing *Enterobacteriaceae* was only described in Germany, isolated from pig-fattening farms and broiler farms (Fischer et al. 2012, 2013, 2017; Falgenhauer et al. 2017). The isolates from the pig farms were from swine single feces or from the farm environment, and the poultry-related isolate was collected from a dust sample. In these studies, *Salmonella enterica subsp. enterica* and *E. coli* isolates were found to harbor the *bla*$_{VIM-1}$ gene, located on a class 1 integron (variable region with *bla*$_{VIM-1}$-*aac*A4-*aad*A1 gene cassettes) harbored by an incHI2 plasmid (Fischer et al. 2012, 2013). In addition to the VIM-1 carbapenemase, the isolates carried also the AmpC-encoding gene *bla*$_{ACC-1}$, which confers cephalosporin and cephamycin resistance and *str*A/B genes (streptomycin resistance). In a retrospective study, using isolates from the three carbapenemase-positive swine farms, an additional *Salmonella* and 35 new *E. coli* isolates producing carbapenemases was detected in one of the German swine farms cited above from different animals, farm environment, manure, and flies (Fischer et al. 2017). These isolates harboring the *bla*$_{VIM-1}$IncHI2 plasmids showed slightly decreased susceptibility to the carbapenems, according to the EUCAST epidemiological cut-offs (ECOFF) values, together with resistance to streptomycin, sulfonamides, and tetracycline.

In a study conducted in 2015 by the Danish Integrated Antimicrobial Resistance Monitoring and Research Programme (DANMAP), including randomly collected ESBL/AmpC *E. coli* isolates from healthy pigs and cattle at slaughterhouses and from domestically produced and imported broiler meat, pork and beef were tested for carbapenemase phenotype (meropenem, imipenem, and ertapenem susceptibility). None of the isolates showed reduced susceptibility to any of the three antimicrobials. Moreover, cephalosporin-resistant *E. coli* from pigs and pork, cattle and beef and, and poultry and broiler meat were tested for the presence of carbapenemase genes by whole genome sequencing and in silico analysis. None of the *E. coli* isolates from Danish animal and meat contained any known carbapenemase gene (DANMAP 2015). Furthermore, in two studies, fecal samples of pigs, cattle, and sheep and pooled fecal samples from poultry flocks were analyzed to determine the occurrence of carbapenemase-producing *Enterobacteriaceae* in Switzerland; no evidence was found (Stephan et al. 2013; Zurfluh et al. 2016).

On the other hand, in the United States, only recently, bacteria harboring plasmid-borne carbapenemase genes have been detected in livestock. Carbapenem-resistant *Enterobacteriaceae* carrying the metallo-β-lactamase gene $bla_{\text{IMP-27}}$ were recovered from the environment of a swine farrow-to-finish operation (Mollenkopf et al. 2017). However, no carbapenem resistance was detected from animal fecal samples.

12.3 Concluding Remarks

Food-producing animals are reservoirs of bacterial pathogens and may act as a vector for the transfer of resistant bacteria and antimicrobial resistance genes to humans. Although the magnitude of transmission of resistance from animal reservoirs to humans remains unknown, it is expected that the emergence and spread of β-lactamases (ESBLs and carbapenemases) from food-producing animals will have a significant impact on human health. Therefore, the detection of the existence and spread of these enzymes among members of the *Enterobacteriaceae* family in animal populations must be considered extremely important for the assessment of potential zoonotic risks. Thus, it is recommended that phenotypic and genotypic studies be carried out consistently in food-producing animals and their meat products, in order to perform a more accurate risk assessment concerning the spread of antimicrobial resistance, as well as on the mechanisms of linkage and transferability of β-lactam resistance determinants in natural environments. The evaluation of the possible impact of this resistance in food-producing animals for human health studies should not be limited to pathogenic bacteria, but must also include commensals because they may be a major reservoir of ESBLs and carbapenemases genes.

However, to interpret the entire data, it is necessary also to integrate information from the transmissible genetic material in zoonotic, commensal, and pathogenic bacteria from food-producing animals, food, and humans, as well as the data reported on antibiotic consumption in human and animals, in a harmonized way. Nevertheless, there are major gaps in surveillance and sharing of data about resistant bacteria that are transmitted through food-producing animals and the food chain. A multisectoral approach is needed to contain antimicrobial resistance in food-producing animals and the food chain.

International cooperation, in order to ensure the development and implementation of global strategies and measures designed to restrict the development and spread of ESBLs and carbapenemases-producing *Enterobacteriaceae*, should be developed. Disciplinary collaborations between World Health Organization (WHO), Food and Agriculture Organization (FAO), World Organisation for Animal Health (OIE), national and international surveillance networks, and researchers working on antimicrobial resistance in humans, animals, and the environment under the "One Health" approach are similarly needed at a global level to provide a coordinating platform for work in this area.

Moreover, because the main risk factors for selection and spread of resistant bacteria among animals are the excessive use of antimicrobial drugs, the prudent use of broad-spectrum cephalosporins (third and fourth generation) and the continued prohibition of the use of carbapenems in food-producing animals can be a simple and effective option in minimizing the further emergence and spread of multidrug-resistant bacteria, including carbapenem-resistant and ESBL-producing *Enterobacteriaceae*, via the food chain. As genes encoding ESBL and carbapenemase production are mostly plasmid-mediated and co-resistance may be an important issue in the spread of such resistance mechanisms, decreasing the frequency of use of antimicrobials in animal production in the EU in accordance with the use of guidelines is high priority. In addition, the improvement of animal husbandry environments, such as the biosecurity and hygienic conditions, and the implementation of alternative procedures to antimicrobials would decrease the need to use antimicrobials and the development of resistant bacteria in food-producing animals.

References

Abreu, R., Castro, B., Espigares, E. et al. (2014). Prevalence of CTX-M-type extended-spectrum β-lactamases in *Escherichia coli* strains isolated in poultry farms. *Foodborne Pathog. Dis.* 11 (11): 868–873.

Bergenholtz, R.D., Jorgensen, M.S., Hansen, L.H. et al. (2009). Characterization of genetic determinants of extended-spectrum cephalosporinases (ESCs) in

Escherichia coli isolates from Danish and imported poultry meat. *J. Antimicrob. Chemother.* 64 (1): 207–209.

Blaak, H., van Hoek, A.H., Hamidjaja, R.A. et al. (2015). Distribution, numbers, and diversity of ESBL-producing *E. coli* in the poultry farm environment. *PLoS One* 10 (8): e0135402.

Bortolaia, V., Guardabassi, L., Trevisani, M. et al. (2010). High diversity of extended-spectrum β-lactamases in *Escherichia coli* isolates from Italian broiler flocks. *Antimicrob. Agents Chemother.* 54 (4): 1623–1626.

Briñas, L., Moreno, M.A., Teshager, T. et al. (2005). Monitoring and characterization of extended-spectrum β-lactamases in *Escherichia coli* strains from healthy and sick animals in Spain in 2003. *Antimicrob. Agents Chemother.* 49 (3): 1262–1264.

Briñas, L., Moreno, M.A., Zarazaga, M. et al. (2003). Detection of CMY-2, CTX-M-14, and SHV-12 β-lactamases in *Escherichia coli* fecal-sample isolates from healthy chickens. *Antimicrob. Agents Chemother.* 47 (6): 2056–2058.

Cantón, R., Akova, M., Carmeli, Y. et al. (2012). Rapid evolution and spread of carbapenemases among *Enterobacteriaceae* in Europe. *Clin. Microbiol. Infect.* 18 (5): 413–431.

Carmo, L.P., Nielsen, L.R., da Costa, P.M., and Alban, L. (2014). Exposure assessment of extended-spectrum β-lactamases/AmpC β-lactamases-producing *Escherichia coli* in meat in Denmark. *Infect. Ecol. Epidemiol.* 4: 10.3402/iee.v4.22924.

Chiaretto, G., Zavagnin, P., Bettini, F. et al. (2008). Extended spectrum β-lactamase SHV-12-producing *Salmonella* from poultry. *Vet. Microbiol.* 128 (3–4): 406–413.

Cohen Stuart, J., van den Munckhof, T., Voets, G. et al. (2012). Comparison of ESBL contamination in organic and conventional retail chicken meat. *Int. J. Food Microbiol.* 154 (3): 212–214.

Costa, D., Vinue, L., Poeta, P. et al. (2009). Prevalence of extended-spectrum β-lactamase-producing *Escherichia coli* isolates in faecal samples of broilers. *Vet. Microbiol.* 138 (3–4): 339–344.

Dahms, C., Hubner, N.O., Kossow, A. et al. (2015). Occurrence of ESBL-producing *Escherichia coli* in livestock and farm workers in Mecklenburg-Western Pomerania, Germany. *PLoS One* 10 (11): e0143326.

DANMAP (2015). DANMAP 2015 – Use of antimicrobial agents and occurrence of antimicrobial resistance in bacteria from food animals, food and humans in Denmark. Last revised 6 October 2015.

de Been, M., Lanza, V.F., de Toro, M. et al. (2014). Dissemination of cephalosporin resistance genes between *Escherichia coli* strains from farm animals and humans by specific plasmid lineages. *PLoS Genet.* 10 (12): e1004776.

EFSA Panel on Biological Hazards (BIOHAZ) (2013). Scientific opinion on Carbapenem resistance in food animal ecosystems. *EFSA J.* 11 (12): 3501.

Endimiani, A., Rossano, A., Kunz, D. et al. (2012). First countrywide survey of third-generation cephalosporin-resistant *Escherichia coli* from broilers, swine, and cattle in Switzerland. *Diagn. Microbiol. Infect. Dis.* 73 (1): 31–38.

Escudero, E., Vinue, L., Teshager, T. et al. (2010). Resistance mechanisms and farm-level distribution of fecal *Escherichia coli* isolates resistant to extended-spectrum cephalosporins in pigs in Spain. *Res. Vet. Sci.* 88 (1): 83–87.

Ewers, C., Bethe, A., Semmler, T. et al. (2012). Extended-spectrum beta-lactamase-producing and AmpC-producing *Escherichia coli* from livestock and companion animals, and their putative impact on public health: a global perspective. *Clin. Microbiol. Infect.* 18 (7): 646–655.

Falgenhauer, L., Ghosh, H., Guerra, B. et al. (2017). Comparative genome analysis of IncHI2 VIM-1 carbapenemase-encoding plasmids of *Escherichia coli* and *Salmonella enterica* isolated from a livestock farm in Germany. *Vet. Microbiol.* 200: 114–117.

Falgenhauer, L., Imirzalioglu, C., Ghosh, H. et al. (2016). Circulation of clonal populations of fluoroquinolone-resistant CTX-M-15-producing *Escherichia coli* ST410 in humans and animals in Germany. *Int. J. Antimicrob. Agents* 47 (6): 457–465.

Fischer, J., Rodriguez, I., Schmoger, S. et al. (2012). *Escherichia coli* producing VIM-1 carbapenemase isolated on a pig farm. *J. Antimicrob. Chemother.* 67 (7): 1793–1795.

Fischer, J., Rodriguez, I., Schmoger, S. et al. (2013). *Salmonella enterica* subsp. *enterica* producing VIM-1 carbapenemase isolated from livestock farms. *J. Antimicrob. Chemother.* 68 (2): 478–480.

Fischer, J., San Jose, M., Roschanski, N. et al. (2017). Spread and persistence of VIM-1 carbapenemase-producing *Enterobacteriaceae* in three German swine farms in 2011 and 2012. *Vet. Microbiol.* 200: 118–123.

Geser, N., Stephan, R., Kuhnert, P. et al. (2011). Fecal carriage of extended-spectrum β-lactamase-producing *Enterobacteriaceae* in swine and cattle at slaughter in Switzerland. *J. Food Prot.* 74 (3): 446–449.

Girlich, D., Poirel, L., Carattoli, A. et al. (2007). Extended-spectrum β-lactamase CTX-M-1 in *Escherichia coli* isolates from healthy poultry in France. *Appl. Environ. Microbiol.* 73 (14): 4681–4685.

Guerra, B., Fischer, J., and Helmuth, R. (2014). An emerging public health problem: acquired carbapenemase-producing microorganisms are present in food-producing animals, their environment, companion animals and wild birds. *Vet. Microbiol.* 171 (3–4): 290–297.

Hernandez, J.R., Martinez-Martinez, L., Canton, R. et al. (2005). Nationwide study of *Escherichia coli* and *Klebsiella pneumoniae* producing extended-spectrum β-lactamases in Spain. *Antimicrob. Agents Chemother.* 49 (5): 2122–2125.

Horton, R.A., Randall, L.P., Snary, E.L. et al. (2011). Fecal carriage and shedding density of CTX-M extended-spectrum β-lactamase-producing *Escherichia coli* in cattle, chickens, and pigs: implications for environmental contamination and food production. *Appl. Environ. Microbiol.* 77 (11): 3715–3719.

Jones-Dias, D., Manageiro, V., Martins, A.P. et al. (2016). New class 2 Integron In2-4 among IncI1-positive *Escherichia coli* isolates carrying ESBL and PMAβ genes from food animals in Portugal. *Foodborne Pathog. Dis.* 13 (1): 36–39.

Jørgensen, C.J., Cavaco, L.M., Hasman, H. et al. (2007). Occurrence of CTX-M-1-producing *Escherichia coli* in pigs treated with ceftiofur. *J. Antimicrob. Chemother.* 59 (5): 1040–1042.

Leverstein-van Hall, M.A., Dierikx, C.M., Cohen Stuart, J. et al. (2011). Dutch patients, retail chicken meat and poultry share the same ESBL genes, plasmids and strains. *Clin. Microbiol. Infect.* 17 (6): 873–880.

Maciuca, I.E., Williams, N.J., Tuchilus, C. et al. (2015). High prevalence of *Escherichia coli*-producing CTX-M-15 extended-spectrum beta-lactamases in poultry and human clinical isolates in Romania. *Microb. Drug Resist.* 21 (6): 651–662.

Madec, J.Y., Lazizzera, C., Chatre, P. et al. (2008). Prevalence of fecal carriage of acquired expanded-spectrum cephalosporin resistance in *Enterobacteriaceae* strains from cattle in France. *J. Clin. Microbiol.* 46 (4): 1566–1567.

Magiorakos, A.P., Srinivasan, A., Carey, R.B. et al. (2012). Multidrug-resistant, extensively drug-resistant and pandrug-resistant bacteria: an international expert proposal for interim standard definitions for acquired resistance. *Clin. Microbiol. Infect.* 18 (3): 268–281.

Meunier, D., Jouy, E., Lazizzera, C. et al. (2006). CTX-M-1- and CTX-M-15-type β-lactamases in clinical *Escherichia coli* isolates recovered from food-producing animals in France. *Int. J. Antimicrob. Agents* 28 (5): 402–407.

Mollenkopf, D.F., Stull, J.W., Mathys, D.A. et al. (2017). Carbapenemase-producing *Enterobacteriaceae* recovered from the environment of a swine farrow-to-finish operation in the United States. *Antimicrob. Agents Chemother.* 61 (2): pii: e01298-16.

Muller, A., Stephan, R., and Nuesch-Inderbinen, M. (2016). Distribution of virulence factors in ESBL-producing *Escherichia coli* isolated from the environment, livestock, food and humans. *Sci. Total Environ.* 541: 667–672.

Nordmann, P., Naas, T., and Poirel, L. (2011). Global spread of carbapenemase-producing *Enterobacteriaceae*. *Emerg. Infect. Dis.* 17 (10): 1791–1798.

Overdevest, I., Willemsen, I., Rijnsburger, M. et al. (2011). Extended-spectrum β-lactamase genes of *Escherichia coli* in chicken meat and humans, the Netherlands. *Emerg. Infect. Dis.* 17 (7): 1216–1222.

Päivärinta, M., Pohjola, L., Fredriksson-Ahomaa, M., and Heikinheimo, A. (2016). Low occurrence of extended-spectrum beta-lactamase-producing *Escherichia coli* in Finnish food-producing animals. *Zoonoses Public Health* 63 (8): 624–631.

Papp-Wallace, K.M., Endimiani, A., Taracila, M.A., and Bonomo, R.A. (2011). Carbapenems: past, present, and future. *Antimicrob. Agents Chemother.* 55 (11): 4943–4960.

Poirel, L., Bercot, B., Millemann, Y. et al. (2012). Carbapenemase-producing *Acinetobacter* spp. in Cattle, France. *Emerg. Infect. Dis.* 18 (3): 523–525.

Queenan, A.M. and Bush, K. (2007). Carbapenemases: the versatile β-lactamases. *Clin. Microbiol. Rev.* 20 (3): 440–458.

Ramos, S., Silva, N., Dias, D. et al. (2013). Clonal diversity of ESBL-producing *Escherichia coli* in pigs at slaughter level in Portugal. *Foodborne Pathog. Dis.* 10 (1): 74–79.

Rawat, D. and Nair, D. (2010). Extended-spectrum β-lactamases in Gram negative bacteria. *J. Glob. Infect. Dis.* 2 (3): 263–274.

Reich, F., Atanassova, V., and Klein, G. (2013). Extended-spectrum β-lactamase- and AmpC-producing enterobacteria in healthy broiler chickens, Germany. *Emerg. Infect. Dis.* 19 (8): 1253–1259.

Riano, I., Garcia-Campello, M., Saenz, Y. et al. (2009). Occurrence of extended-spectrum β-lactamase-producing *Salmonella enterica* in northern Spain with evidence of CTX-M-9 clonal spread among animals and humans. *Clin. Microbiol. Infect.* 15 (3): 292–295.

Schmid, A., Hormansdorfer, S., Messelhausser, U. et al. (2013). Prevalence of extended-spectrum β-lactamase-producing *Escherichia coli* on Bavarian dairy and beef cattle farms. *Appl. Environ. Microbiol.* 79 (9): 3027–3032.

Seiffert, S.N., Hilty, M., Perreten, V., and Endimiani, A. (2013). Extended-spectrum cephalosporin-resistant Gram-negative organisms in livestock: an emerging problem for human health? *Drug Resist. Updat.* 16 (1–2): 22–45.

Shaikh, S., Fatima, J., Shakil, S. et al. (2015). Antibiotic resistance and extended spectrum beta-lactamases: types, epidemiology and treatment. *Saudi J. Biol. Sci.* 22 (1): 90–101.

Smet, A., Boyen, F., Pasmans, F. et al. (2012). OXA-23-producing *Acinetobacter* species from horses: a public health hazard? *J. Antimicrob. Chemother.* 67 (12): 3009–3010.

Smet, A., Martel, A., Persoons, D. et al. (2008). Diversity of extended-spectrum beta-lactamases and class C β-lactamases among cloacal *Escherichia coli* isolates in Belgian broiler farms. *Antimicrob. Agents Chemother.* 52 (4): 1238–1243.

Smet, A., Martel, A., Persoons, D. et al. (2010). Broad-spectrum β-lactamases among *Enterobacteriaceae* of animal origin: molecular aspects, mobility and impact on public health. *FEMS Microbiol. Rev.* 34 (3): 295–316.

Snow, L.C., Wearing, H., Stephenson, B. et al. (2011). Investigation of the presence of ESBL-producing *Escherichia coli* in the North Wales and West Midlands areas of the UK in 2007 to 2008 using scanning surveillance. *Vet. Rec.* 169 (25): 656.

Stephan, R., Sarno, E., Hofer, E. et al. (2013). Lack of evidence so far for carbapenemase-producing *Enterobacteriaceae* in food-producing animals in Switzerland. *Schweiz. Arch. Tierheilkd.* 155 (7): 417–419.

Valat, C., Auvray, F., Forest, K. et al. (2012). Phylogenetic grouping and virulence potential of extended-spectrum-β-lactamase-producing *Escherichia coli* strains in cattle. *Appl. Environ. Microbiol.* 78 (13): 4677–4682.

Yigit, H., Queenan, A.M., Anderson, G.J. et al. (2001). Novel carbapenem-hydrolyzing beta-lactamase, KPC-1, from a carbapenem-resistant strain of *Klebsiella pneumoniae*. *Antimicrob. Agents Chemother.* 45 (4): 1151–1161.

Zurfluh, K., Glier, M., Hachler, H., and Stephan, R. (2015). Replicon typing of plasmids carrying $bla_{\text{CTX-M-15}}$ among *Enterobacteriaceae* isolated at the environment, livestock and human interface. *Sci. Total Environ.* 521–522: 75–78.

Zurfluh, K., Hindermann, D., Nüesch-Inderbinen, M. et al. (2016). Occurrence and features of chromosomally encoded carbapenemases in Gram-negative bacteria in farm animals sampled at slaughterhouse level. *Schweiz. Arch. Tierheilkd.* 158 (6): 457–460.

Part IV

Therapeutic Strategy for Overcoming AR

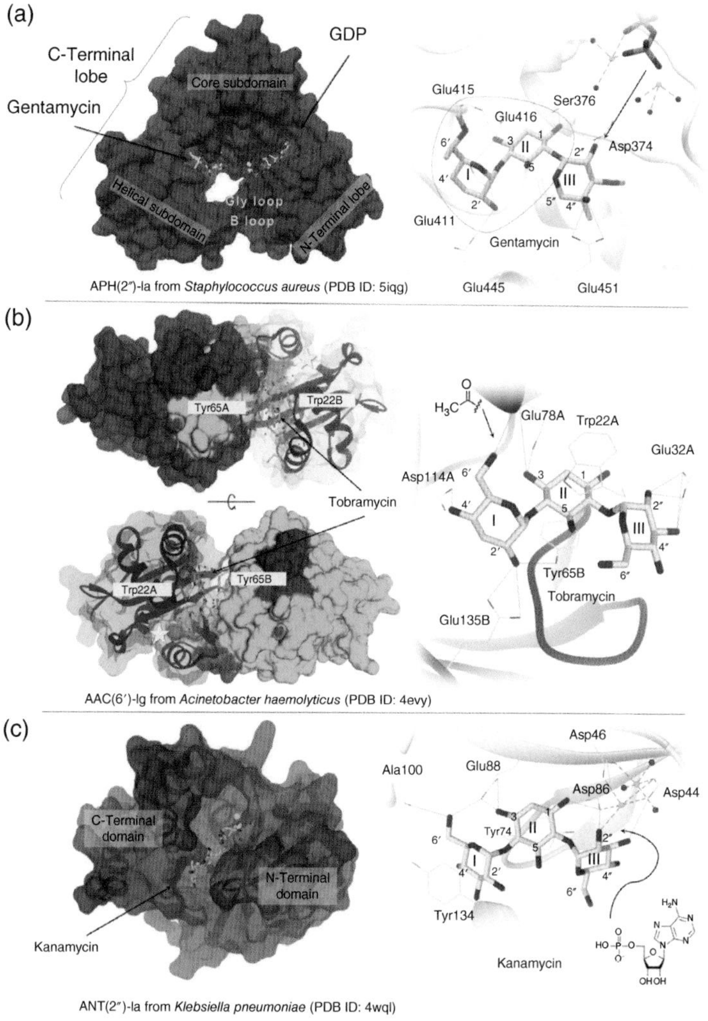

Figure 1.3 Representative crystal structures of the three types of AMEs and respective sites of aminoglycoside modification. (a) Structure of the aminoglycoside 2″-phosphotransferase APH(2″)-Ia from Staphylococcus aureus, in complex with gentamycin and GDP (Caldwell et al. 2016). The neamine-like moiety is circumscribed by a dashed blue line. (b) Dimeric structure of aminoglycoside 6′-N-acetyltransferase AAC(6′)-Ig from Acinetobacter haemolyticus in complex with tobramycin. Chains A and B are depicted in blue and red ribbons, respectively, while the yellow star marks the putative binding site of acetyl-CoA in each monomer (Stogios et al. 2017). (c) Structure of aminoglycoside nucleotidylyltransferase ANT(2″)-Ia from Klebsiella pneumoniae in complex with kanamycin (Cox et al. 2015). All protein structures are shown in surface representation. AGAs and GDP are represented as stick model; magnesium atoms and water molecules are represented as green and red spheres, respectively. Hydrogen bonds are represented by thin blue lines. Hydrogen atoms are omitted for simplicity. *Source:* All images were produced with program Chimera (Pettersen et al. 2004).

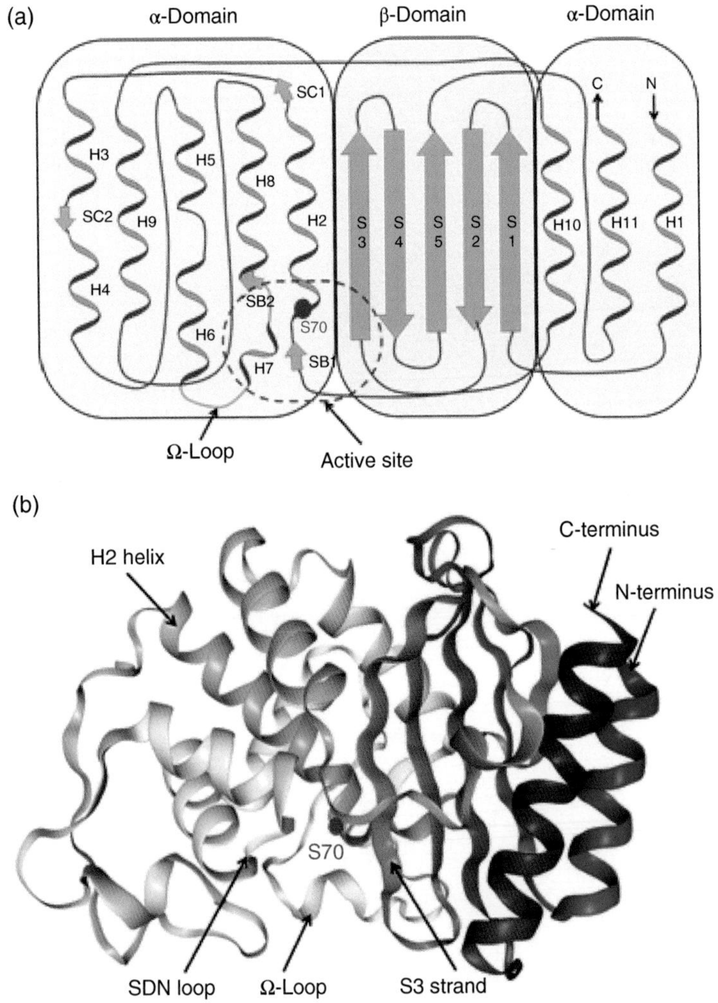

Figure 6.2 Schematic representation of the structural fold of β-lactamase TEM-1 consisting of three domains (a). The dotted line indicates the active site with catalytic S70. The presented scheme corresponds to the location of domains in the tertiary structure of β-lactamase TEM-1 (PDB ID 1ERO (Ness et al. 2000)) (b). The locations of the catalytic S70, H2 α-helix, S3 β-strand, and the Ω-loop are highlighted.

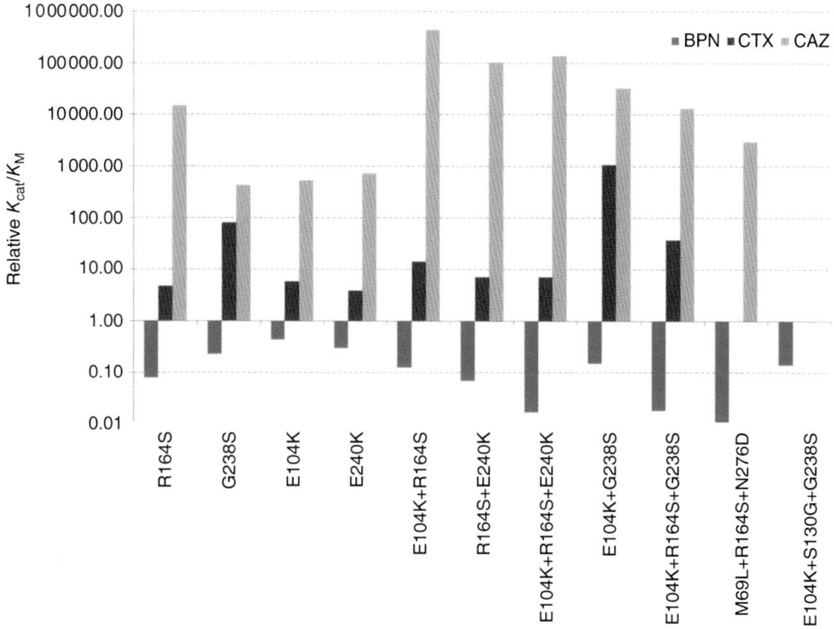

Figure 6.6 Relative catalytic efficiency (k_{cat}/K_M) of TEM ESBLs and CMT β-lactamases compared with that of TEM-1 against benzylpenicillin (BNP), cefotaxime (CTX), and ceftazidime (CAZ) (Sowek et al. 1991; Sirot et al. 1997; Neuwirth et al. 2001; Perilli et al. 2007; Robin et al. 2007; Palzkill 2018). A value larger than one indicates that the mutant is more active than TEM-1.

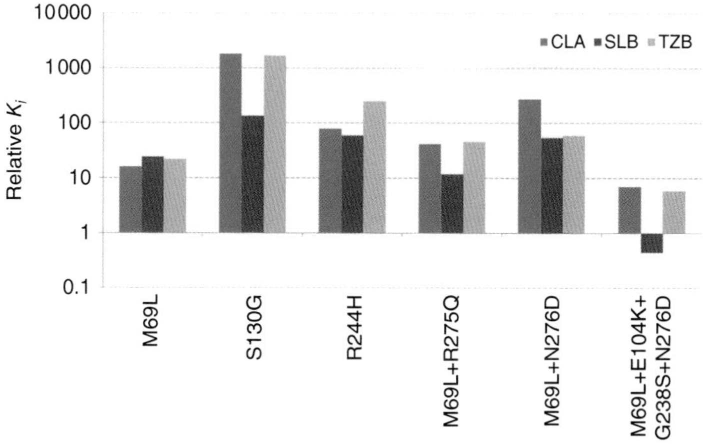

Figure 6.7 Relative Ki of IRT and CMT TEM-type β-lactamases compared with K_i of TEM-1 for inhibition by clavulanic acid (CLA), sulbactam (SLB), and tazobactam (TZB) (Sirot et al. 1997; Bermudes et al. 1999; Chaïbi et al. 1999). A value larger than one indicates that a complex of mutant enzyme with the inhibitor is less stable than TEM-1 with this inhibitor.

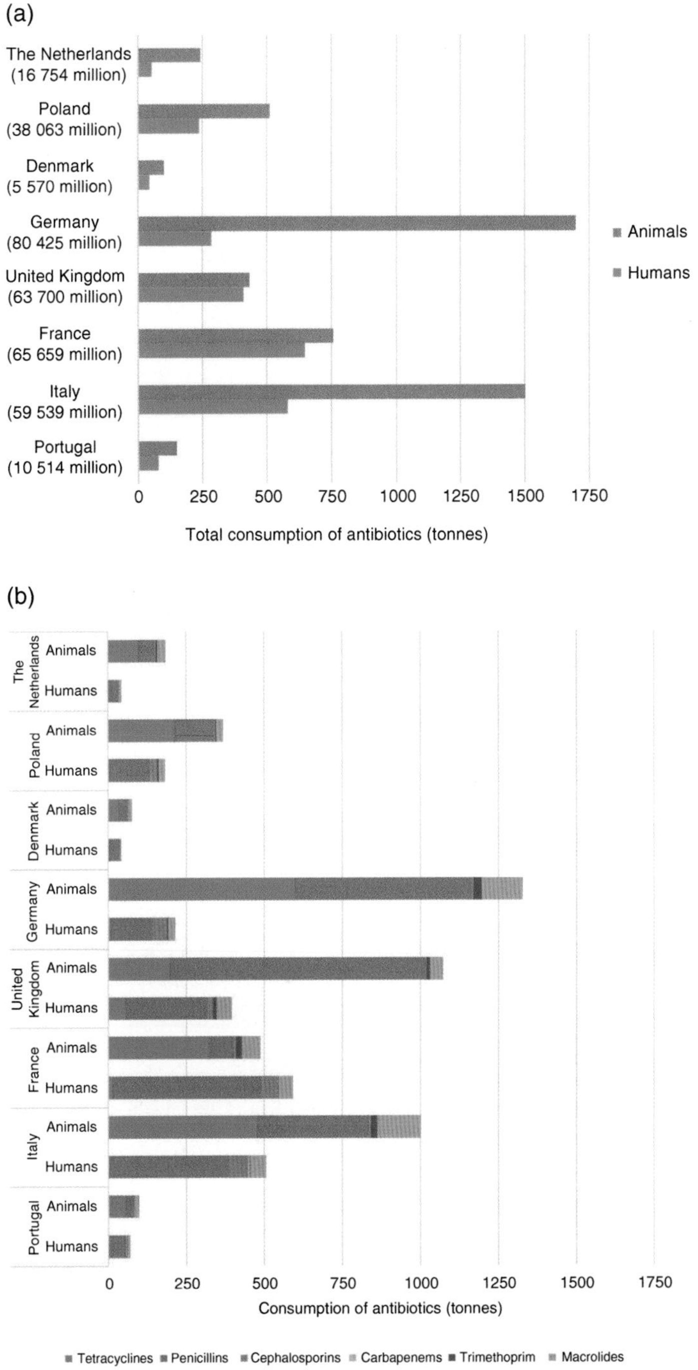

Figure 10.1 Consumption of antibiotics in humans and food-producing animals (a) by selected European countries in 2012 and (b) by class of antibiotics for each European country. Data of consumption in tonnes was recovered from ECDC, EFSA, and EMA report (ECDC/EFSA/EMA 2015). Data from human consumption includes the antibiotics used in the community and hospitals. Data from food-producing animal's consumption includes the use in pigs, cattle, and broilers. The countries' population size data are estimates from the World Bank for the year 2012 (http://data.worldbank.org/indicator/SP.POP.TOTL).

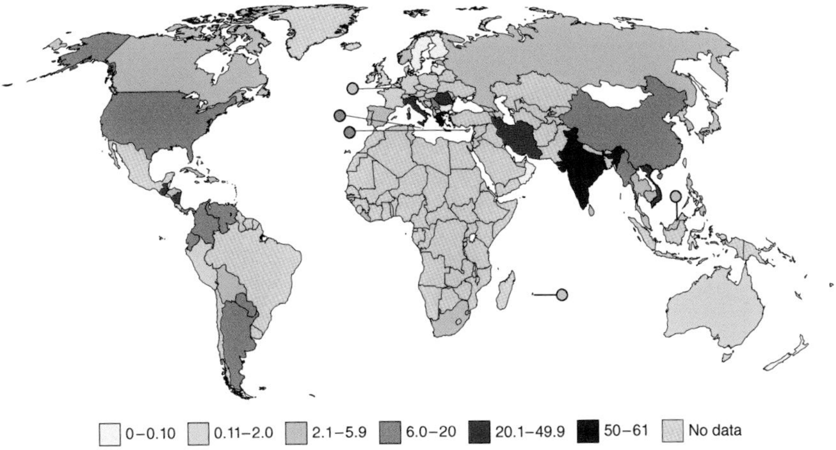

Figure 11.2 Wide distribution of carbapenem-resistant *Klebsiella pneumoniae*, by country. The frequency of prescription antibiotics varies greatly from state to state, suggesting areas where improvements in antibiotic prescription (fewer unnecessary prescriptions) would be most helpful. *Source*: Figure courtesy of the Center for Disease Dynamics, Economics & Policy (2015). Used with permission via Creative Commons license.

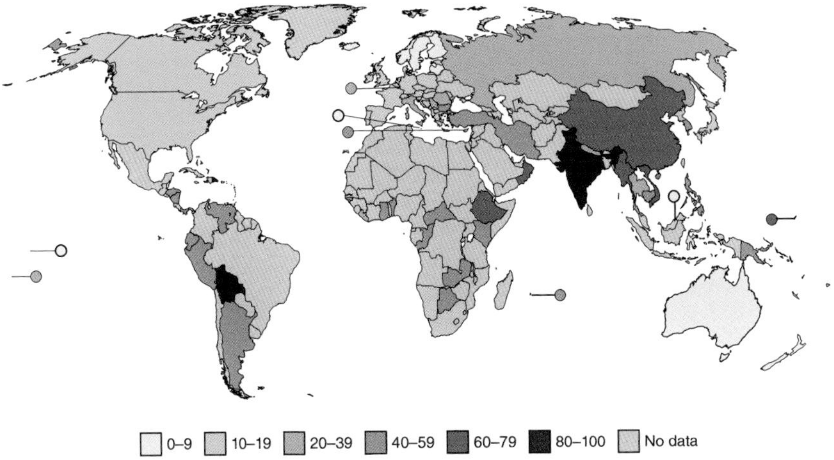

Figure 11.3 Wide distribution of extended-spectrum β-lactamase-producing *Escherichia coli*, by country (resistance to one or more of the following drugs were used: ceftazidime, ceftriaxone and cefotaxime). *Source*: Adapted from https://cddep.org/wp-content/uploads/2017/06/swa_edits_9.16.pdf.

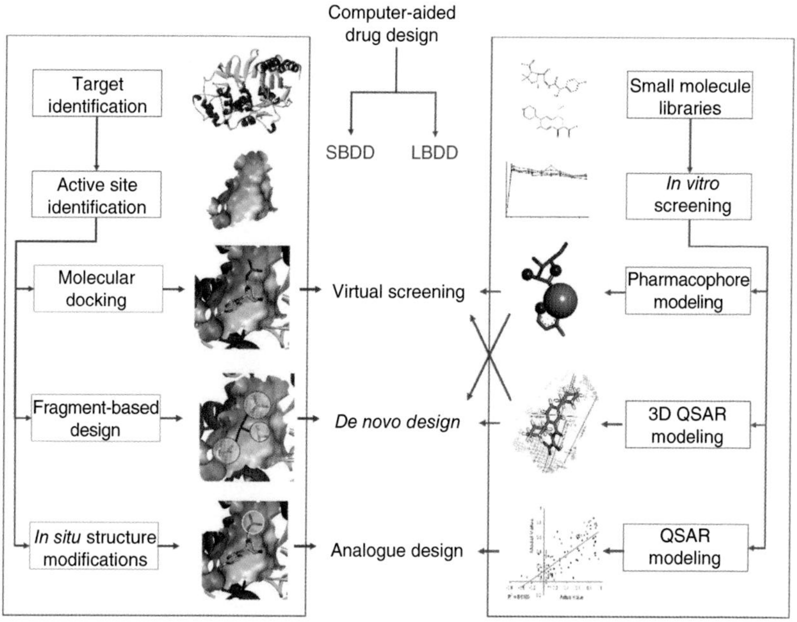

Figure 13.2 Overview of the methods used in the computer-aided design of antibiotics and inhibitors of antibiotic resistance mechanisms (SBDD, structure-based drug design; LBDD, ligand-based drug design; QSAR, quantitative structure–activity relationship; 3D, three-dimensional).

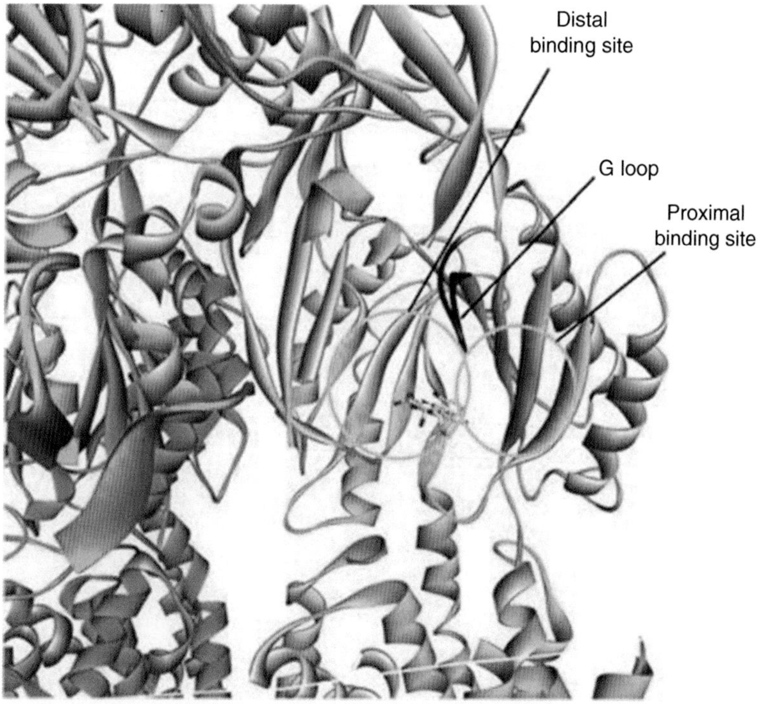

Figure 13.5 Tetracycline in the multi-binding sites within the binding protomer of the homotrimer model of the AcrB efflux pump. *Source:* Reprinted from Jamshidi et al. (2018) under Creative Commons license http://creativecommons.org/licenses/by/4.0.

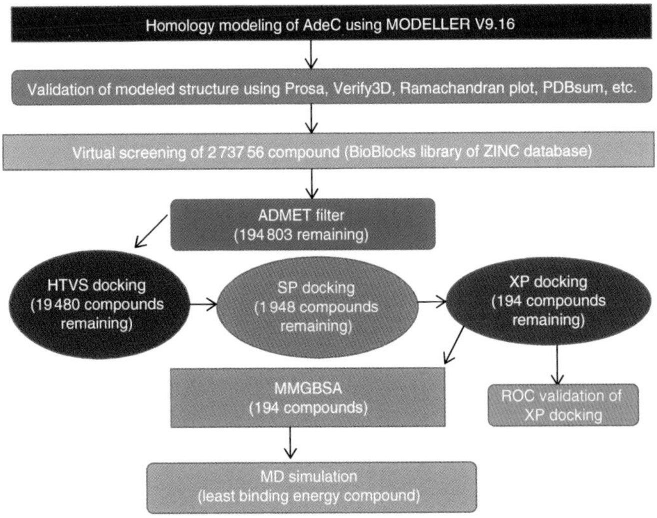

Figure 13.6 Schematic outline of the methods used for *in silico* inhibitor designed for AdeC. *Source:* Reprinted by permission from: Cell Biochemistry and Biophysics, Targeting Outer Membrane Protein Component AdeC for the Discovery of Efflux Pump Inhibitor against AdeABC Efflux Pump of Multidrug Resistant *Acinetobacter baumannii* (Verma et al. 2018). Copyright (2018) Springer.

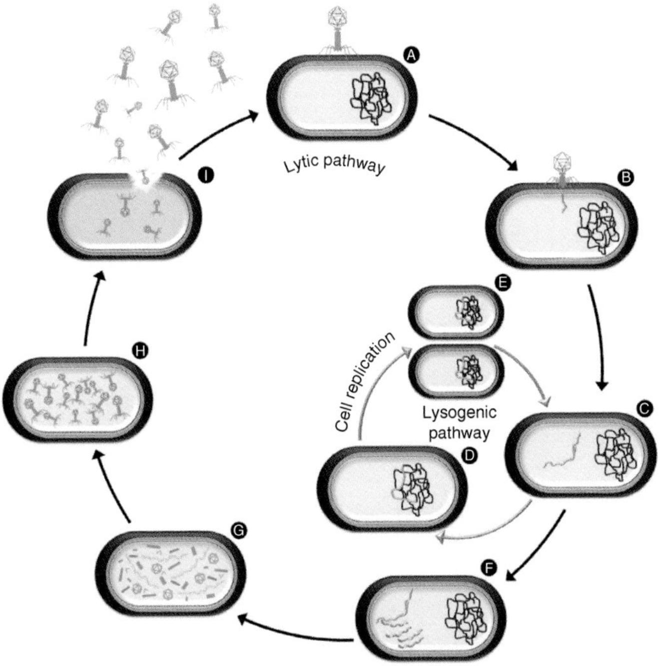

Figure 15.1 The phage replication cycle. Phage adsorbs to the bacterial cell (a) and injects its genome (b) to follow the lytic or the lysogenic pathway (c). In the lysogenic pathway, the genome integrates in the bacterial chromosome (d) and replicates with the cell (e). Continuing to the lytic pathway, the phage genes are translated (f) and proteins (including those composing the virus particle) are produced (g), assembled with the replicated phage genome (h) and the cell bursts to release the progeny (i) that will start a new cycle (a).

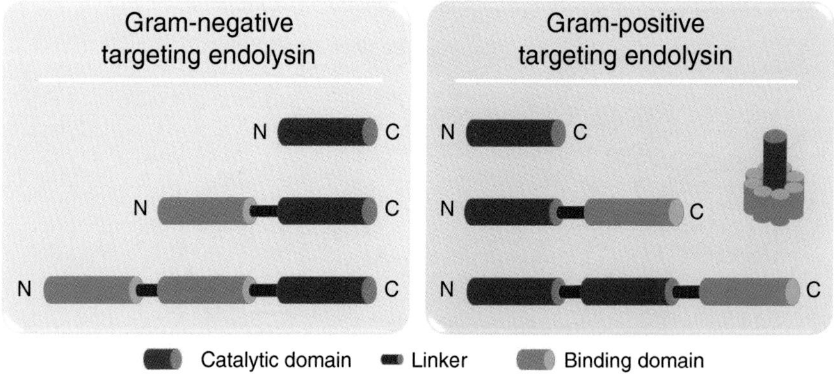

Figure 15.5 Different molecular arrangements possibilities of phage endolysins. On the left, architecture types of Gram-negative targeting endolysins with an inverted arrangement of that of Gram-positive targeting endolysins, with one catalytic domain at the C-terminus and one or more binding domains at the N-terminus. On the right, architecture types of Gram-positive phage endolysins with one or more catalytic domains at the N-terminus and one binding domain at the C-terminus. A multimeric structure with a single heavy chain (catalytic domain) and eight light chains (binding domain) has also been proved.

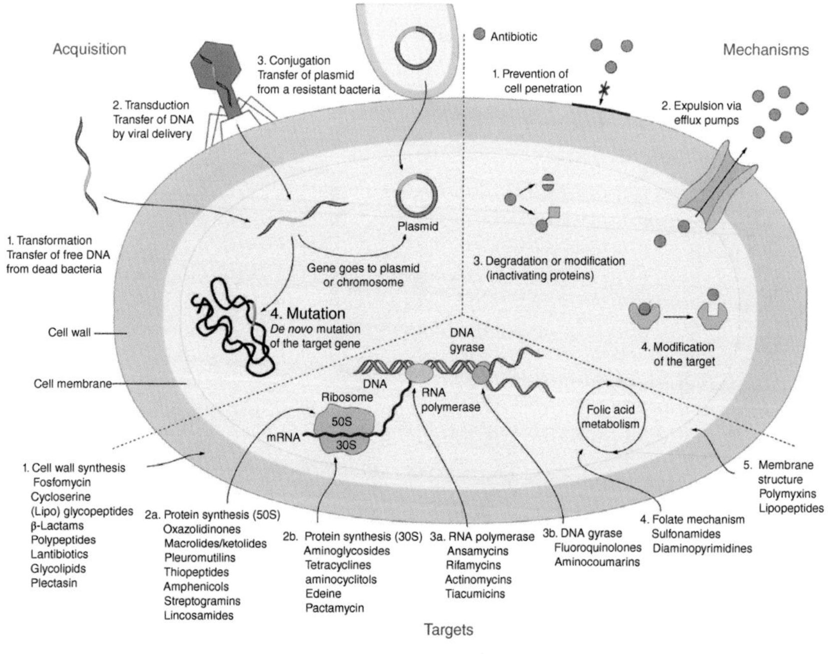

Scheme 16.6 Schematic illumination of the resistance acquisition pathways and the main mechanisms of resistance and the five main targets for antibiotics. *Source:* Reproduced from Chellat et al. (2016) with permission. Copyright 2016 Wiley-VCH Verlag GmbH & Co. KGaA.

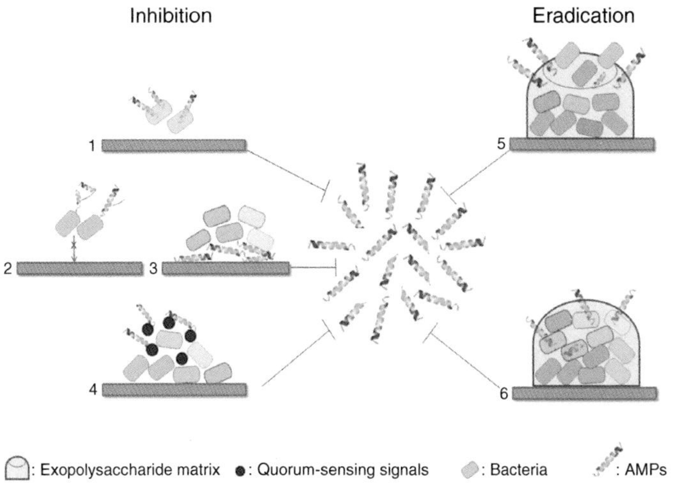

Figure 20.1 Mechanisms of anti-biofilm action of AMPs. (1) Inhibition of biofilm formation by targeting cells in the planktonic state. (2) Inhibition of adhesion on biotic surface reaching the regulatory pathways responsible for the production of adhesion defects, such as pili and flagellum. (3) Inhibition of adhesion by abiotic surface coating. (4) Inhibition of quorum sensing. (5) Eradication of mature biofilm by disruption of the extracellular matrix. (6) Eradication of the mature biofilm through the death of the adhered cells and incorporated into the extracellular matrix.

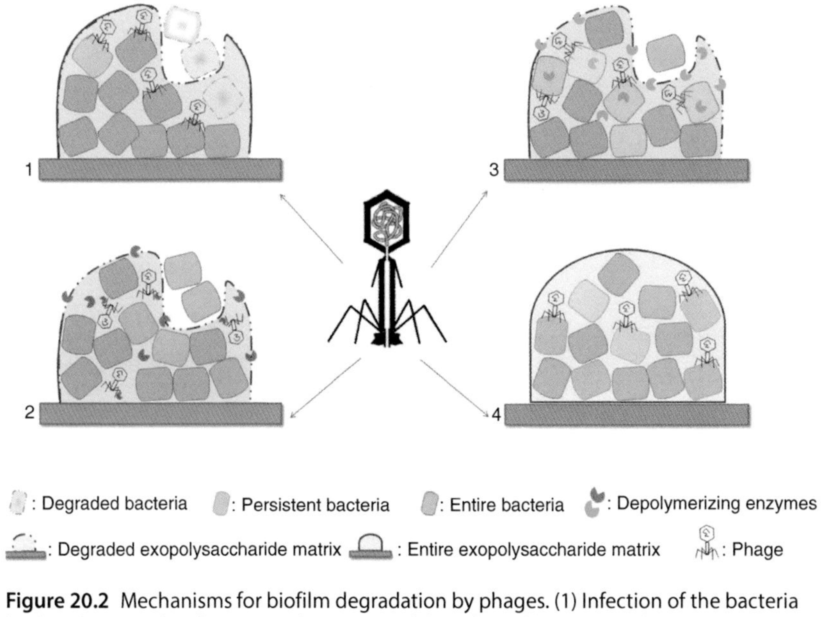

Figure 20.2 Mechanisms for biofilm degradation by phages. (1) Infection of the bacteria by the phage, with subsequent elimination of the infected bacteria and consequent degradation of EPS. (2) Degradation of EPS by depolymerization enzymes carried by the phage. (3) Degradation of EPS by depolymerization enzymes expressed by bacteria infected by the phage. (4) Infection of persistent bacteria by phage.

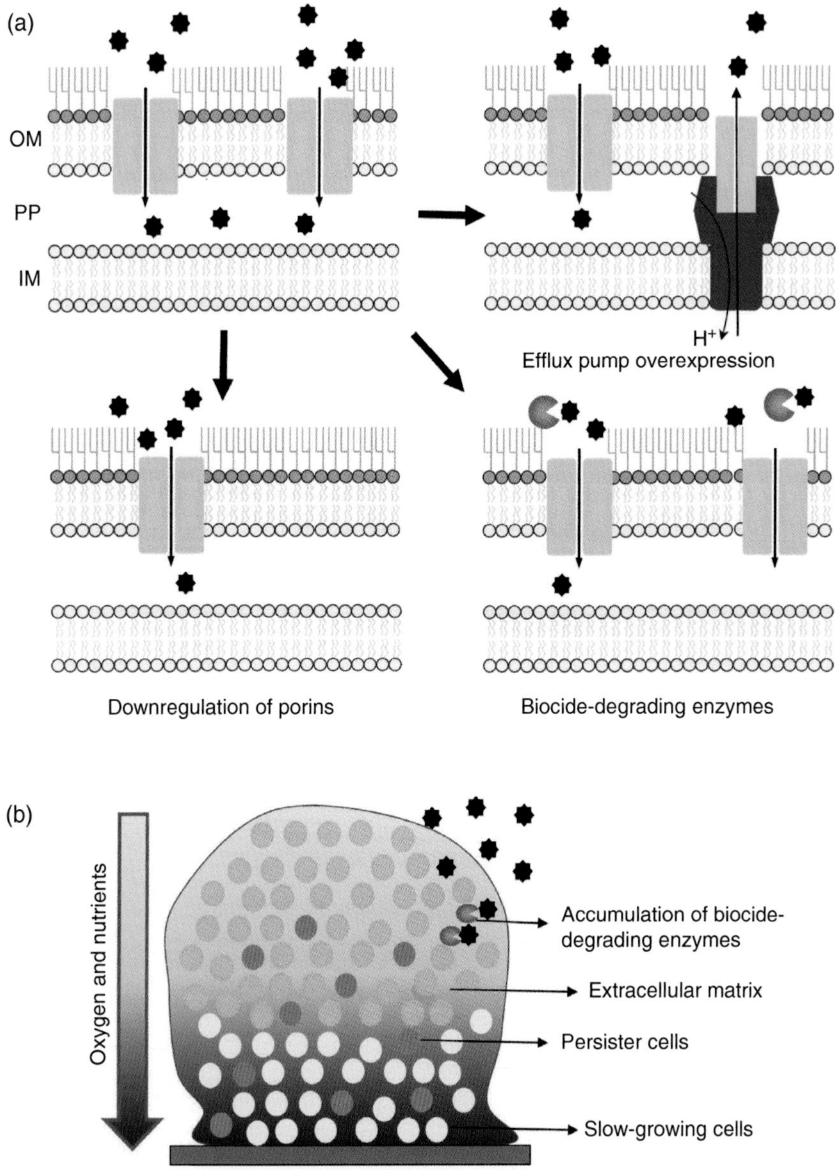

Figure 22.2 Mechanisms of adaptive resistance to disinfectants. (a) Examples of adaptations that confer decreased susceptibility to biocidal compounds in a Gram-negative bacterium: increased expression of degrading enzymes or efflux pumps and decreased expression of porin-encoding genes. (b) Examples of factors that increase resistance to disinfectants of bacterial biofilms. OM, outer membrane; PP, periplasm; IM, inner membrane.

13

AR Mechanism-Based Drug Design

Mire Zloh

UCL School of Pharmacy, University College London, London, UK

13.1 Introduction

Antibiotic agents currently used in the clinic disrupt biosynthesis of essential cell components (metabolites, proteins, peptidoglycans, and DNA) in actively growing bacteria and can also affect DNA replication and cell wall formation (Chopra et al. 2002; Kohanski et al. 2010). Although most of these drugs generally stimulate hydroxyl radical formation and induce a common oxidative damage cellular death pathway (Kohanski et al. 2007, 2008), they form interactions with different cellular targets. The discovery of antibiotic agents initially relied on serendipity and empirical screening of natural products for their ability to inhibit bacterial growth during "golden era" of antibiotic discovery. This period was followed by a period of rational drug design where new therapeutic agents were developed by modification of existing antibiotics mainly without knowledge of their site of action (Aminov 2010). Computer-aided molecular design approaches in combination with genomics-based identification of targets were further employed from the early 1990s, however with the limited success in the discovery of novel classes of antibiotics (Simmons et al. 2010; Brown and Wright 2016). The agents that entered the full development since 1995 were mostly from the already known classes of antibiotics (Bush 2012).

Bacteria's ability to respond to stresses and challenges by toxic compounds, which include antibiotics, results in resistance to therapies. Defense mechanisms can be developed from the presence of intrinsic genes in their genome that could generate a resistance phenotype and from the mutations in chromosomal genes (Davies and Davies 2010), as well as due to horizontal gene transfer (HGT) responsible for increased propagation of

Antibiotic Drug Resistance, First Edition. Edited by José-Luis Capelo-Martínez and Gilberto Igrejas.

resistance through bacterial populations (Nakamura et al. 2004). More often than not, resistance induced by an antibiotic leads to decrease of efficacy of other structurally unrelated antibiotics, resulting in the development of multidrug-resistant (MDR) bacterial strains. The specter of untreatable infections is becoming a real threat as the resistance was recently observed against the drugs of last resort, such as vancomycin and colistin (Boucher et al. 2009; Liu et al. 2016), and the demand for new antibiotic therapies is critical.

In the era of the antibiotic resistance, the development of the new antibiotics focuses on the new or unexplored targets present in resistant strains, while the alternative strategies include developing agents that target the mechanisms responsible for antibiotic resistance and failure of infection treatments

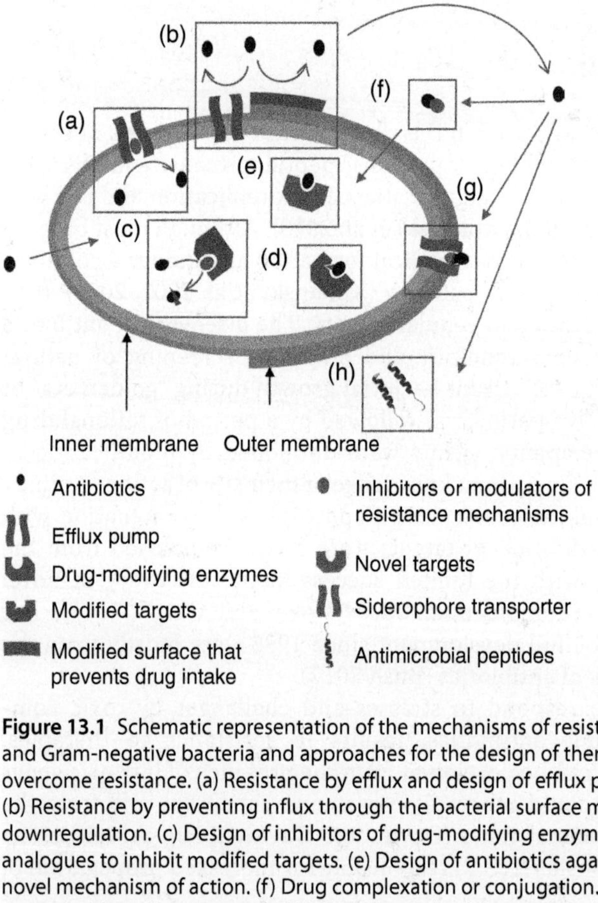

Figure 13.1 Schematic representation of the mechanisms of resistance in Gram-positive and Gram-negative bacteria and approaches for the design of therapeutic agents to overcome resistance. (a) Resistance by efflux and design of efflux pump inhibitors (EPI). (b) Resistance by preventing influx through the bacterial surface modification or porin downregulation. (c) Design of inhibitors of drug-modifying enzymes. (d) Design of drug analogues to inhibit modified targets. (e) Design of antibiotics against novel targets or with novel mechanism of action. (f) Drug complexation or conjugation. (g) Drug influx increase through siderophore receptors. (h) Use of antimicrobial peptides (AMP).

(Schillaci et al. 2017). The main reason behind antibiotic resistance are (Walsh 2000; Soon et al. 2011; Blair et al. 2015; González-Bello 2017):

- Removal of the antibiotic from the bacteria by membrane transporters named efflux pumps (Figure 13.1a).
- Decrease of drug uptake by modification of porin transporters and/or by modification of the surface charge of the Gram-negative bacteria and consequently reducing colistin binding (Figure 13.1b).
- Expression of enzymes that change or degrade antibiotics rendering them inactive (Figure 13.1c).
- Change of the structure binding site of the target biomolecule, so that antibiotics cannot exert its activity (Figure 13.1d).

While the pharmaceutical approaches and other alternative technologies to target antibiotic resistance were discussed elsewhere (Schillaci et al. 2017; Baker et al. 2018), the focus of this chapter is on the computer-aided drug design (CADD) of small molecules to overcome bacterial resistance. These approaches and methods are used with intention to develop new antibiotics by utilizing new targets, to discover new inhibitors of efflux pumps and drug-modifying enzymes, and to design novel antimicrobial peptides (AMPs).

13.2 Drug Design Principles

The drug design is a process of finding a molecule, known as a ligand or hit molecule, which has a potential to inhibit or enhance the activity of a targeted biomolecule through specific interaction and often tight binding (Figure 13.1c–e). This requires complementarity in shape and charges at the ligand–biomolecule interface. A hit molecule is subjected to limited optimization to develop a lead molecule with improved physicochemical properties and bioavailability. A lead molecule, promising drug candidate, can be considered for further process of drug development through optimization of its activity and metabolic half-life while minimizing side effects through iterative medicinal chemistry efforts during preclinical trials.

The plethora of the information available on molecular properties of potential drug candidates, their activities against different bacterial strains, mechanisms of action, target structures, and biochemical pathways can be optimally utilized to specifically interfere with bacterial cellular processes (Zloh and Kirton 2018). Molecules can be designed to interfere with the function of a biomolecular target that is associated with the survival of a microbial pathogen while designing out undesirable off-target interactions with the host. Approaches that can be utilized depend on the available information related to the selected target and pathogen of interest (Figure 13.2). Some general information and overview of methods used in CADD can be found elsewhere (Zhang 2016; Yu and MacKerell 2017).

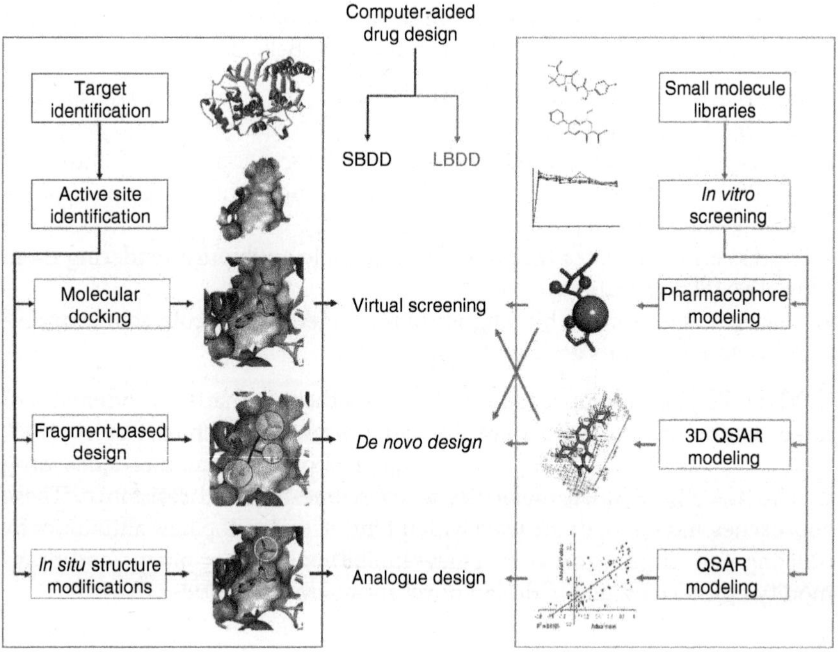

Figure 13.2 Overview of the methods used in the computer-aided design of antibiotics and inhibitors of antibiotic resistance mechanisms (SBDD, structure-based drug design; LBDD, ligand-based drug design; QSAR, quantitative structure–activity relationship; 3D, three-dimensional). *(See insert for color representation of the figure.)*

A knowledge of the three-dimensional (3D) structure of a biomolecule enables a structure-based drug design (SBDD) of candidates for the development of new therapeutic agents. These macromolecules should be validated as drug targets by proving that modulation of the target results in changing the disease states. A first stop source of 3D structures of biomolecules complexed with either drugs or active ligands is Protein Data Bank (PDB) website (Berman et al. 2000) that currently contains over 83 000 structures of bacterial macromolecules obtained by X-ray crystallography or NMR spectroscopy. Most of these entries provide information about the static structure of biomolecules with druggable sites or mutated active sites. For some proteins, there are multiple entries of the same biomolecule complexed with different ligands that can provide additional information about conformational flexibility of the site and the ability of target biomolecule to accommodate ligands of different sizes and shape. In the absence of the multiple entries, conformational flexibility of proteins and binding of different ligands can be explored using molecular dynamics simulation. Furthermore, homology modeling can be employed to predict the 3D structures of proteins without experimental structures. Once confirmed, their 3D structures can be used to identify small molecules that favorably bind into an active or an allosteric site by conducting virtual screening of small molecule libraries, modification of existing drug molecules, or *de novo* design of novel molecules (Simmons et al. 2010):

- Structure-based virtual screening (SBVS) for possible active molecules that forms favorable interactions with selected residues of a target. The SBVS relies on molecular docking of ligands from small molecule libraries into of a known or putative active site on the macromolecule.
- Structure-based *de novo* design provides opportunities to create new chemical entities with adequate molecular properties that could favorably bind into the active site and possibly have novel scaffolds (Hartenfeller and Schneider 2011; Cain et al. 2014).
- The availability of the 3D structure of a drug complexed with the target biomolecule provides a basis for *in silico* structural modifications of the drug molecule to improve the shape and charge complementarity on the binding interface.

On the contrary, ligand-based drug design (LBDD) is utilized when the 3D structure of the drug target is not known, but the effects of a series of molecules on a particular biological process or organism are known. Curation just one of the repositories of the biological activities of small molecules, the ChEMBL database (Gaulton et al. 2012), indicates that over 300 000 compounds were tested for antibacterial activity in cell-based assays, while 110 000 compounds were tested for their inhibitory activity of 500 proteins from different bacterial pathogens. In cases when selected compounds are known to bind into the same active site, the knowledge of their 3D structures enables the development of 3D quantitative structure–activity relationship (3D QSAR) and pharmacophore models, pharmacophore being a structural feature necessary for interaction with the target molecule. Both computer-aided approaches are suitable for lead identification and their further optimization into drug candidates (Acharya et al. 2011). Once the available information is used to generate and validate models, high-throughput virtual screening can be conducted to identify molecules with key molecular properties and/or pharmacophoric features that are essential for the desired biological activity. The identified hit molecules can be reintroduced into ligand-based models for the fine optimization of the structure to improve their efficacy and potency.

The key advantages of the *in silico* approaches to drug design lie in the possibility to generate and evaluate hypotheses before undertaking the extensive and resource demanding experimental projects. As the number of bacterial biomolecules with known 3D structures and availability of the information on the antibacterial activity of small molecules are on the rise, the increase of computational power provides opportunities for the development of new algorithms for the design of novel antibiotics and inhibitors of resistance. The analysis of such large and complex data sets increasingly requires the use of artificial intelligence methods to ensure the best outcome of the rational design of therapeutic agents.

13.3 Identification of Novel Targets and Novel Mechanisms of Action

A recent analysis of FDA-approved drugs indicates that only 198 biomolecules from various pathogens are targeted by antibiotics for the use in humans (Santos et al. 2016). This limited number of bacterial targets in combination with the significant decrease in number of approved antibiotics (Kinch et al. 2014; Kinch and Griesenauer 2018) that either belong to the previously known class or do act on known target class (Coussens et al. 2018) signifies the need to design molecules that would bypass resistance mechanisms by acting on novel bacterial targets. It is believed that such novel antibiotics would, therefore, reduce the likelihood of cross-resistance through the existing resistance mechanisms as they would inherently have a possibly different mode of action (Chopra et al. 2002). Furthermore, identifying novel targets in the bacterial resistant strains would be essential to address the issues related to failed treatments of infections by these pathogens.

The application of genome sequencing techniques can be used not only for the detection of pathogens and monitoring the spread of infection (Punina et al. 2015), but also it can reveal the capacity of pathogens to encode proteins, which potentially serve as targets for antibiotics. In combination with information that can be obtained from metabolomic, proteomic, and transcriptomic techniques, whole genome sequencing can identify those proteins that are essential for the pathogen survival (Fields et al. 2017) and possibly orthologue to those found in human. Identification of such proteins, especially those that do not belong to known classes of drug targets, would ensure that any novel antibiotic developed would have minimal side effects.

Genome sequence of multiple strains of *Staphylococcus aureus* revealed a core genome comprising 1441 genes with 239 genes conserved in all *S. aureus* strains. Interestingly, 94 of these genes are experimentally found to be essential for *Escherichia coli*, providing opportunities for developing inhibitors of proteins encoded by these genes as potential antibiotics (Bosi et al. 2016). A genome-wide protein interaction network, interactome, can further improve the understanding of essential biochemical pathways in bacteria and support identification of pathogen-specific proteins as putative drug targets. A protein interaction network built between proteins of a selected *S. aureus* strain, MRSA252, enabled identification of pyruvate kinase as an essential hub protein in MRSA interactome. The significant divergence of MRSA and human pyruvate kinase sequence was found in the domains responsible for the enzyme catalytic activity. These distinct sequence regions in the MRSA were selected as a target for the structure-based inhibitor design using X-ray structures of this bacterial enzyme (PDB entries: 3T07 and 3T0T). This unique pocket was utilized for *in silico* screening of small molecules from the ZINC database (Irwin and Shoichet 2005). Hit molecules with best inhibitory and antibacterial activities served as a starting point for a combination of fragment-based design and medicinal chemistry efforts to identify IS-130 scaffold analogues (Scheme 13.1) with low nanomolar minimum inhibitory concentrations (MIC). These novel compounds that inhibited pyruvate kinase were found to possess activity against both MRSA and multidrug-resistant *S. aureus* (MDRSA) strains and thus potentially could be developed into novel antibiotics (Zoraghi et al. 2011; Axerio-Cilies et al. 2012).

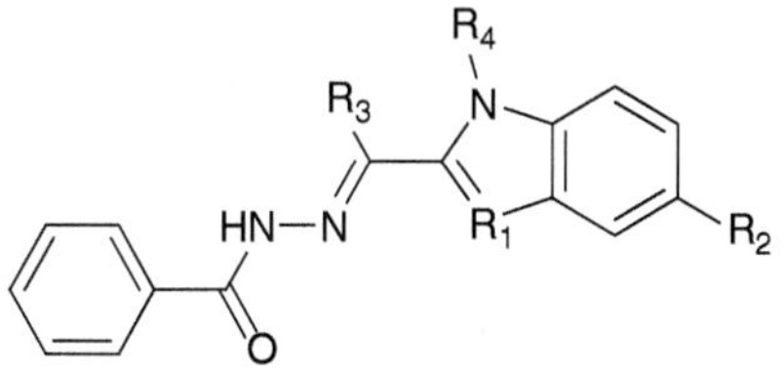

Scheme 13.1 IS-130 scaffold.

Computational tools for target identification are being developed by utilizing various databases and a wide range of approaches (chemoinformatics, omics technologies, systems networks, etc.) (Katsila et al. 2016). Although most of the approaches reviewed are focused on human targets, there are opportunities to extend some of these approaches to antibiotic discovery. Comparative genomics can not only identify bacterial proteins that do not exist in host organisms, but it also can unravel the differences in the physiology of sensitive bacterial cells and their resistant counterparts. This would allow identification of targets that would be specific to resistant bacterial strains. A stand-alone software, TiD, is developed to prioritize bacterial targets based on the subtractive channel analysis of genomes of pathogens, human, and gut microbes (Gupta et al. 2017). Information obtained on putative drug targets can be integrated with structural biology and molecular modeling techniques to facilitate structure-based drug discovery. One of the protocols that exemplified an integrated approach is used to identify a drug target in *Acinetobacter baumannii* strains (Figure 13.3). A comparative analysis of the genome of 35 from the NCBI genome database and subsequent proteome analysis enabled an identification of orthologous proteins against human proteome and, yet, part of the bacterial essential genome. Drug target prioritization was achieved by assessing proteins for suitability based on their localization and molecular properties, which resulted in the selection of KdsA protein for further extensive validation of this approach. The homology model of this protein was used for virtual screening and selection of ligands with best interactions with the target (Ahmad et al. 2018). Although the results of this study are not fully validated by *in vitro* experiments, this approach can serve as a guide in the design of antibiotic discovery experiments.

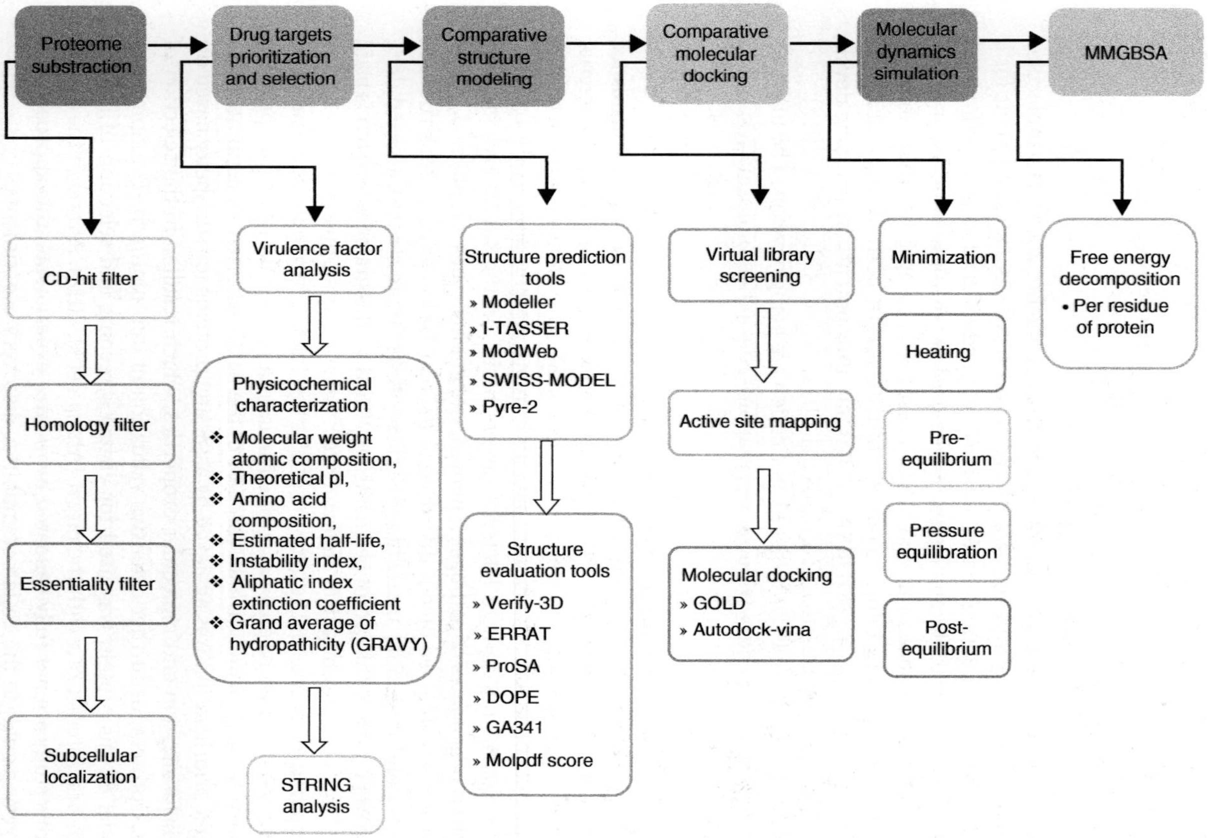

Figure 13.3 A complete step-by-step flow of different methodologies to identify inhibitors of a novel protein target from *Acinetobacter baumannii*. Source: Reprinted from Ahmad et al. (2018). Copyright (2018), with permission from Elsevier.

Furthermore, information on critical evolutionary trajectories that could be responsible for resistance can be obtained using whole genome sequencing. In turn, proteins or biochemical pathways within those trajectories could also provide candidates for the discovery of agents that target mechanisms of resistance and prevent the development of resistance. For example, *Pseudomonas aeruginosa*, evolved to colistin resistance in a laboratory environment using quantitative evolution pipeline, exhibits accumulation of mutations specific to resistance in the pmrAB two-component system. This confirmed the previously established link to the colistin resistance (Mehta et al. 2017). Such invaluable information can be utilized in not only the development of new inhibitors in *P. aeruginosa*, but this approach can be utilized in the discovery of mutations responsible for resistance in other bacterial strains.

It is not always that new target biomolecules should be identified to overcome the bacterial resistance. DNA gyrase, an essential bacterial enzyme, is a well-established and primary target for fluoroquinolones. The mutations in bacterial DNA gyrase result in quinoline resistance (Gruger et al. 2004), thus requiring a novel mechanism to utilize this bacteria-specific target. The 3D structure of *S. aureus* DNA gyrase can be utilized for discovery of compounds that target both Gram-positive and Gram-negative bacteria. A thiophene-based inhibitor ([*N*-(2-amino-1-phenylethyl)-4-(1H,2H,3H-pyrrolo[2,3-β]pyridin-3-yl)thiophene-2-carboxamide]) was found to be active in both Gram-positive and Gram-negative resistant strains by binding to the protein at a site remote from the DNA cleavage site. This previously unexploited pocket in gyrase can be used for the design of drug candidates against fluoroquinolone-resistant bacteria (Chan et al. 2017).

One of the possible future target and underexplored targets is RecA, a bacterial multifunctional protein essential to genetic recombination, DNA repair, and regulation of SOS response. RecA is a recombinase protein implicated in the bacterial drug resistance, survival, and pathogenicity (Pavlopoulou 2018), where the activation of bacterial SOS response is responsible for the development of antibiotic intrinsic and/or acquired resistance (Bellio et al. 2017), and therefore could be a potential drug target (Nautiyal et al. 2014).

The above demonstrates that there are considerable efforts in the discovery of novel targets, druggability potential of known proteins, or identification of new sites on known target proteins. There are challenges in utilizing such information and obtaining drug candidates interacting with these new target sites, which may be overcome a greater integration of available of the experimentally obtained information and computational resources.

13.4 Efflux Pump Inhibitors

Proteins present in membranes that remove antibiotics from bacteria, called efflux pumps, can be responsible for high levels of resistance especially if overexpressed. These efflux pumps reduce or restrict the intracellular concentrations of antibiotics and are often responsible for the failure of infection therapies. MDR efflux pumps transport a wide range of substrates that are structurally dissimilar, while other pumps can transport substrates with a degree of specificity toward a class of molecules or according to their physicochemical properties (Blair et al. 2015). The ability of pumps to extrude a wide range of drugs from its interior prevents the use of drug modification approaches to overcoming the resistance by efflux, and alternative approaches have to be employed. One approach would be to identify compounds that evade efflux pump or to develop inhibitors of efflux that would restore the efficacy of antibiotics (Figure 13.1a). Therefore, efflux pumps were considered as attractive targets where the inhibition by an efflux pump inhibitor (EPI), delivered together with an antibiotic drug, should result in the increased potency as well as the enhanced spectrum of activity of the antibiotic.

Early efforts in the discovery of such agents were focused on *in vitro* screening of antibiotics in combinations with natural products, other drugs, or synthetic analogues that resulted in some hit molecules that acted as broad-spectrum EPIs (Lomovskaya and Bostian 2006). EPIs such as MC-207,100 (Phe-Arg-ß-naphthylamide [PAßN]), reserpine, GG918, INF55, INF277, verapamil (Scheme 13.2), and their analogues never reached clinic due to their poor pharmacokinetic or toxicological profiles or unfavorable molecular properties.

Substrates that are more susceptible for efflux by the same efflux pump are assumed to have a common map of pharmacophore features despite their chemical structures dissimilarity, which may potentially be applied to EPI acting on the same pump. Initial analysis of the molecular similarity of structurally unrelated EPIs using MIPSIM software revealed that overall 3D distributions of their molecular interaction potentials overlap considerably with the similar spatial orientation of aromatic rings (Zloh and Gibbons 2004).

That indicates that the pharmacophore approach may be useful in EPI discovery. The *S. aureus* NorA efflux pump, responsible for resistance to the quinolone drugs, is considered as a good target for the development of EPIs, especially when the efflux is the only mechanism of resistance that affects the therapy. A set of compounds with the known inhibitory activities of this efflux pump served as a training set for developing a pharmacophore model (Figure 13.4). The training set included both active and inactive compounds while comprising nine different scaffolds to ensure that chemical diversity is considered during *in silico* screening for new EPI candidates.

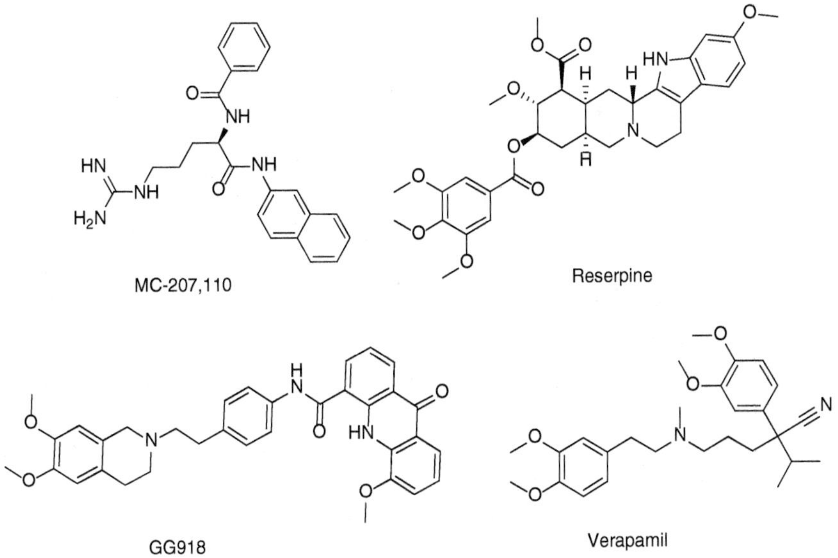

Scheme 13.2 Structures of some well-known EPIs.

FLAP, software based on the molecular interaction fields, was utilized to develop a model based on the training set and conduct a virtual screening of 300 000 compounds from the external set of commercially available compounds. Four molecules were selected by *in silico* screening and experimentally proven to be inhibitors of the NorA efflux pump. These four compounds belong to different chemical classes indicating the potential of this approach to discover novel hits where the overall molecular shape and properties may be crucial for their activity (Brincat et al. 2011).

Furthermore, the same software was used in a ligand-based approach to utilize propensity for further optimization of indole scaffold observed in INF55 (5-nitro-2-phenyl-(1H)-indole). A novel class of highly active NorA inhibitors was identified alongside to discovery that parts of the active molecules can be replaced with a molecular fragment similar in shape, size, and chemical properties without detrimental effects on the activity (Lepri et al. 2016; Buonerba et al. 2017).

Pharmacophore models of the NorA inhibition, developed using a different approach and Phase software, were also used to evaluate synergistic activities of antibiotics and non-antibiotic drugs. Such *in silico* drug repositioning would enable identification of NorA EPIs that are already approved for the use in the clinic and would have an acceptable toxicological profile by default. A training data set of 61 known NorA inhibitors was used to develop pharmacophore models validated by correct prediction of NorA inhibition for selected active

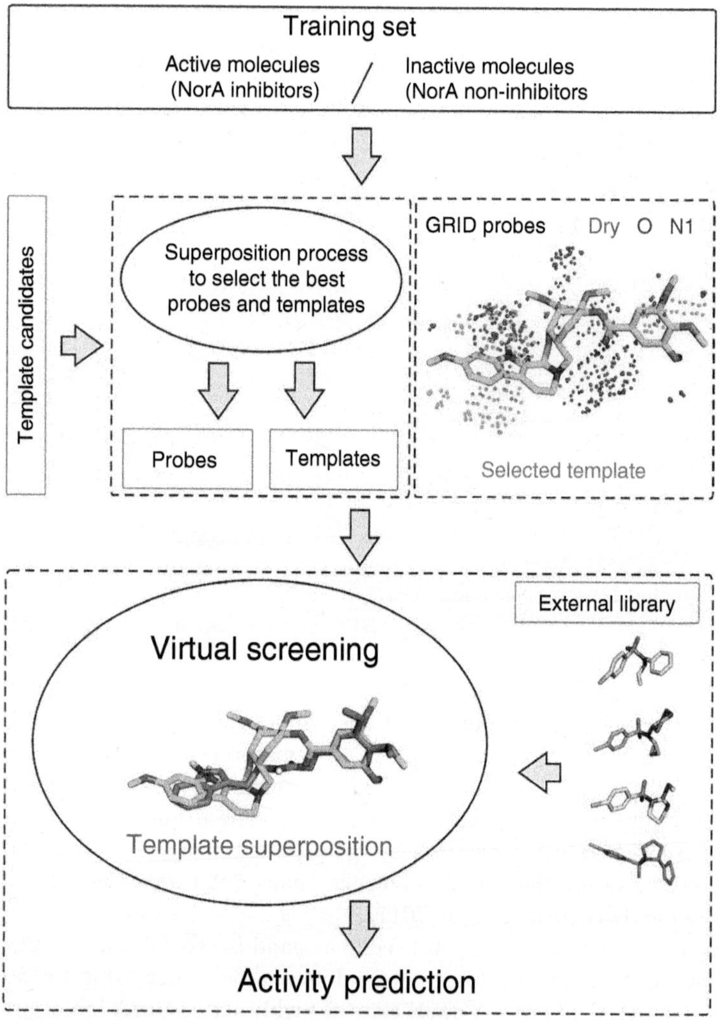

Figure 13.4 Scheme for the development of the pharmacophore model of the *Staphylococcus aureus* NorA pump inhibitors and subsequent *in silico* screening using FLAP software. Source: Reprinted with permission from Brincat et al. (2011). Copyright (2011) American Chemical Society.

and inactive molecules. Two best performing models were used to predict activities of 1620 approved drug molecules that allowed selection of seven candidates for *in vitro* testing. Dasatinib, gefitinib, and nicardipine exhibited a strong synergistic activity with ciprofloxacin against strains overexpressing NorA. These drugs may potentially be administered with ciprofloxacin for treatment of infections as long as the indications of the drug combinations are compatible, which would require in-depth investigation of possible drug–drug interactions (Astolfi et al. 2017).

The design of inhibitors of resistance by efflux in Gram-negative bacteria is further complicated by the presence of the outer membrane and consequent innate low permeability of small molecules. *Pseudomonas aeruginosa* exhibits a high level of antibiotic resistance to aminoglycosides and quinolones associated with the expression of the MexAB-OprM efflux pump. The potent inhibitors of this efflux pump were discovered in the late 1990s, such as MC-207,110 (Scheme 13.2), but due to its toxicity were used mainly as a research agent (Alibert-Franco et al. 2009).

However, limited success was achieved in generating pharmacophore or 3D QSAR models that would facilitate discovery of clinically relevant EPIs, despite extensive efforts in identifying a large number of hits by *in vitro* screening and generating their analogues. One of possible reasons for the difficulty to develop reliable ligand-based models, especially in the case of Gram-negative bacteria, may lie in the size of the binding site of the transporter and possible multiple binding sites for substrates and/or inhibitors as shown in the X-ray structure of AcrB pump from *E. coli* (Edward et al. 2003). Furthermore, there are possibilities for simultaneous binding of a substrate (tetracycline) and EPI (MC-207,110) at the proximal and distal binding sites (Figure 13.5) (Jamshidi et al. 2018). This allows a significant diversity of binding modes of EPI candidates that prevents identifying common structural features (pharmacophores) that are responsible for efflux pump inhibition.

The 3D structures of efflux pumps open opportunities to study possible binding modes of substrates (antibiotics) and inhibitors (EPIs). There are currently over one hundred entries in the Protein Data Bank with structural information about bacterial efflux pumps. The number of released structures is on the rise (45 new entries released since 2015). It is apparent that efflux pumps present in Gram-negative bacteria are of significant interest as there are over 50 entries that are related to *E. coli* and 14 entries that are related to proteins from *P. aeruginosa*. These structures together with 3D structures of other transporters provide the basis for homology modeling of other efflux pumps without experimentally known structures (Ramaswamy et al. 2017) and other computational analysis of their sequences, structures, and dynamic behavior (Travers et al. 2018).

The early efforts in structural studies of NorA were focused on explaining the mode of binding between a homology model of this efflux pump and molecules that successfully inhibited drug efflux. A set of 33 flavonoids, chosen by QSAR model on the basis of *in silico* predicted activity, were analyzed for binding ability by molecular docking into two possible active sites of the NorA homology model (Thai et al. 2015). Similarly, a homology model of NorA built using a 3D structure of *E. coli* lactose permease (PDBID entry: 1PV7) as template was used for docking of non-antibiotic agent ginsenoside 20(S)-Rh that

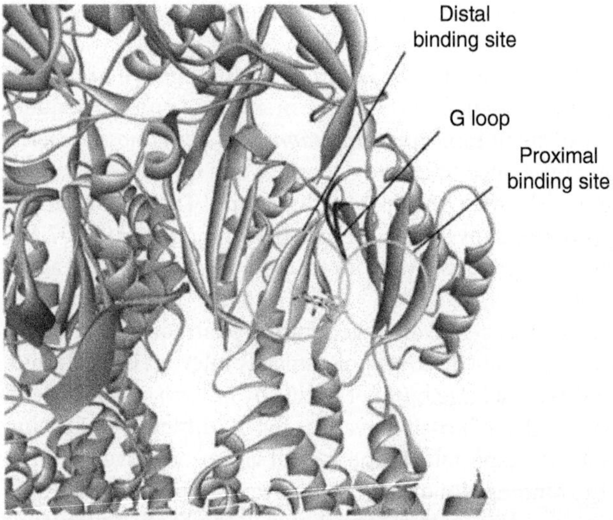

Figure 13.5 Tetracycline in the multi-binding sites within the binding protomer of the homotrimer model of the AcrB efflux pump. Source: Reprinted from Jamshidi et al. (2018) under Creative Commons license http://creativecommons.org/licenses/by/4.0. *(See insert for color representation of the figure.)*

acted as a potential EPI. A binding pocket was selected based on the experimentally known residues essential for NorA activity, and the binding modes of reserpine and potential EPI revealed key residues of the hydrophobic pocket and interactions with the protein, which included hydrogen bond formed with GLN51. The same approach was used to study the molecular mechanism of action of riparin synthetic analogues with proven ability to restore the activity of norfloxacin and ciprofloxacin against resistant *S. aureus*. It was revealed that most active riparin analogue fits into the hydrophobic cleft, which is the same binding site previously identified for other NorA inhibitors (Costa et al. 2016).

Further improvements of models obtained by the homology modeling included simulation of the solvated system comprising protein embedded in the lipid bilayer. Such molecular dynamics simulations include effects of the environment on the overall 3D structure of the protein and its active site. A homology model of NorA built using glycerol-3-phosphate transporter from *E. coli* (PDB entry: 1PW4) achieved the equilibrated conformation after 1.5 ns of simulation to provide a suitable target for virtual screening by molecular docking. A set of known inhibitors (including reserpine, verapamil, and omeprazole) and various substrates (including levofloxacin and ciprofloxacin) were docked into the active site, and the results clearly demonstrate that the reserpine has highest favorable binding energy though forming pi–pi interactions with PHE317 and hydrogen bonds with ARG324 and PHE129. Such validation of the NorA model enabled the virtual screening of selected ligands from the ZINC database, and 14 hit molecules with adequate molecular properties and acceptable predicted toxicity were identified as potential candidates for EPI development (Bhaskar et al. 2016).

The limited availability of the 3D structures of efflux pumps in Gram-positive bacteria may be a reason for limited use of the structure-based approach to design EPI for NorA efflux pump. On the contrary, the availability of the resistance-nodulation-division (RND) efflux pump structures provided a good basis for studying the molecular mechanism of efflux pump inhibition in Gram-negative bacteria and identification of novel EPIs.

A reduced model of acriflavine resistance protein B (AcrB) without transmembrane domains (PDB entry: 4DX5) and in the absence of other accessory proteins, such as AcrA and AcrZ, was used to evaluate the interactions with the compounds MBX2319 (a novel pyranopyridine EPI), D13-9001, 1-[1-naphthylmethyl]-piperazine, and MC-207,110. Molecular docking of these molecules into AcrB binding site was used to generate starting points for molecular dynamics simulations of complexes to consistently compare binding modes of these molecules and ability of the efflux pump to accommodate different substrates. It was confirmed that all inhibitors act most likely via competitive binding into active sites for substrates and possibly preventing conformational changes necessary for efflux to happen (Vargiu et al. 2014). Similarly, a set of 2-substituted benzothiazoles achieve a reversal in the antibacterial activity of ciprofloxacin with up to 10-fold lower MIC values. These compounds were docked into AcrB binding site (PDB entry: 2DRD) indicating that the tested compounds establish possible binding interactions with the phenylalanine-rich region in the distal binding site (Yilmaz et al. 2015).

However, to avoid a possibility of inhibitors to become substrates of the efflux pump, efflux substrate-like features could also be considered as an exclusion criterion to enrich the inhibitor identification process. A library of phytochemicals that did not contain pharmacophoric features of efflux pump substrates was screened using high-throughput docking against AcrB (PDB entry: 1T9Y) and MexB (PDB entry: 2V50) proteins. The two of the best hits from the screening, lanatoside C and daidzein, were validated as potential EPI by checkerboard synergy assay and ethidium bromide accumulation assay, achieving a good correlation between *in silico* screening and positive efflux inhibitory activity *in vitro* (Aparna et al. 2014).

Other components of the RND efflux pump, such as AdeB from the MDR strains of *A. baumannii*, were also considered as possible targets for EPI identification. A homology model was build using the AcrB multidrug efflux pump template (PDB entry: 1oy6) and used as a target for high-throughput virtual screening of 159 868 compounds. The compounds were filtered initially based on their predicted molecular properties and toxicological profile followed by two-stage docking process (standard and extra precision). This resulted in 123

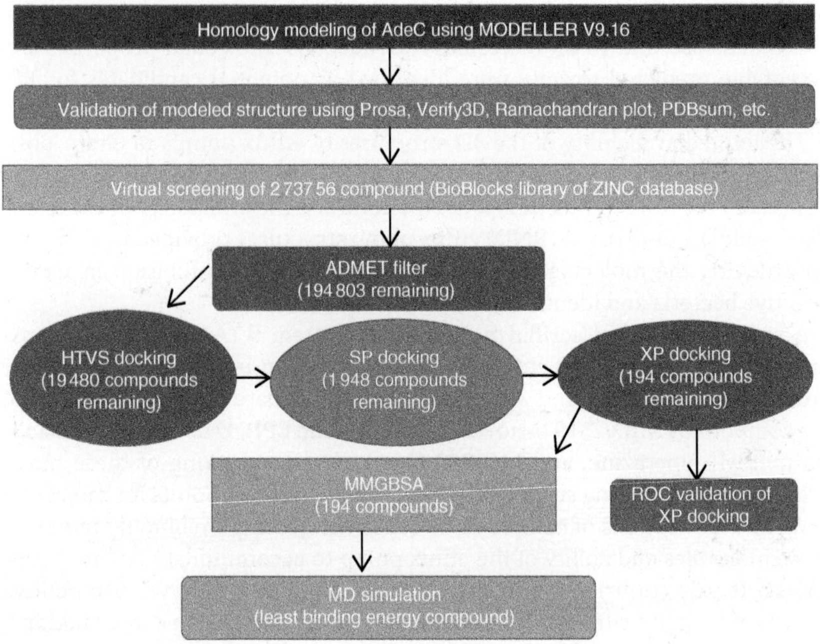

Figure 13.6 Schematic outline of the methods used for *in silico* inhibitor designing for AdeC. Source: Reprinted by permission from: Cell Biochemistry and Biophysics, Targeting Outer Membrane Protein Component AdeC for the Discovery of Efflux Pump Inhibitor against AdeABC Efflux Pump of Multidrug Resistant *Acinetobacter baumannii* (Verma et al. 2018). Copyright (2018) Springer. *(See insert for color representation of the figure.)*

compounds that could have an inhibitory potential toward AdeB. A most promising candidate (ZINC01155930) and its interaction with the pump were validated by molecular dynamics simulation (Verma et al. 2018). Using a similar approach (Figure 13.6), a molecule, ZINC77257599, was identified as a potential inhibitor of outer membrane protein component (AdeC) of efflux pump AdeABC of *A. baumannii* (Verma and Tiwari 2018).

The use of structural information on bacterial pumps is currently focused on elucidating the mechanisms of action of inhibitors of efflux that were discovered via *in vitro* screening assays. It is a combination of cellular, X-ray crystallographic, homology modeling, molecular docking, and molecular dynamics simulations that provides information on binding modes of hit molecules (Sjuts et al. 2016, 2017; Dwivedi et al. 2018; Atzori et al. 2019). The structure-based approaches to design or discover novel EPIs resulted in a low hit rate, especially when compared with successes in the design of enzyme inhibitors to treat other diseases, such as cancer. Further information on structures of efflux pumps is needed, but it would be essential to understand the atomistic description of the efflux process. As the molecular determinants that drive the functional rotation mechanism and corresponding conformational changes induced in AcrB are not fully understood, computational studies may be able to address this challenge (Neuberger et al. 2018; Vargiu et al. 2018). The current 3D models of tripartite assemblies (Wang et al. 2017a) and knowledge of putative channels for substrate entry in AcrB may provide the basis for rational design of EPIs as well as identification of new sites that can be targeted (Neuberger et al. 2018).

However, one of the key challenges in the use of EPIs is that these active molecules have to be combined with an antibiotic. To achieve an effective antibacterial therapy requires that the synergistic activity is maximized with minimized toxicity by ensuring compatibility of pharmacokinetics and mode of action of both compounds.

Adjuvant therapy has greater success in overcoming resistance by preventing drug degradation (β-lactam–β-lactamase inhibitor combination) than in developing a combination therapy to overcome drug efflux (antibiotic–EPI combination) (Domalaon et al. 2018). Computer-aided approaches in the rational design of adjuvant therapies for antibiotics are not explored to a greater extent, except using molecular dynamics to study various drug delivery systems (Albano et al. 2018). New strategies and approaches are thus required.

A concept of "escort" molecules was initially proposed based on the *in silico* prediction of intermolecular interactions between antibiotics and EPIs. These two molecules could complex in the cell-based assay medium, resulting in a noncovalently bound entity (Figure 13.7) that could bypass efflux pump (Zloh et al. 2004). Such complexes are not only larger than the antibiotic used, but they may change the molecular properties of an antibiotic and increase its permeability through the cell membrane (Zloh and Gibbons 2007). In an effort to repurpose FDA-approved drugs, an *in silico* protocol was developed for the identification of molecules that could restore the activity of antibiotics using intermolecular interaction energy and change of virtual log P as selection criteria. Initial studies revealed that two drugs (chlorpromazine and apomorphine) out of 10 selected based on the *in silico* screening had a weak to moderate potentiation activity as they reduced the MIC of norfloxacin two- and fourfold, respectively (Rahman et al. 2011).

Therefore to increase number of hit molecules that have chances to enter into clinic, the process of discovery of EPIs would benefit from the use of integration of SBDD, LBDD, and other approach. Such studies would not predict only binding of an inhibitor into active site by molecular docking, but would also predict the effects of binding on conformational changes and function of efflux pump by molecular dynamics; minimize off-target interactions within host by target prediction using PharmMapper (Wang 2017b), Pidgin (Mervin et al. 2015), and similar methods reviewed elsewhere (Huang et al. 2018; Zloh and Kirton 2018); and evaluate effects on biochemical pathways by using systems biology approaches.

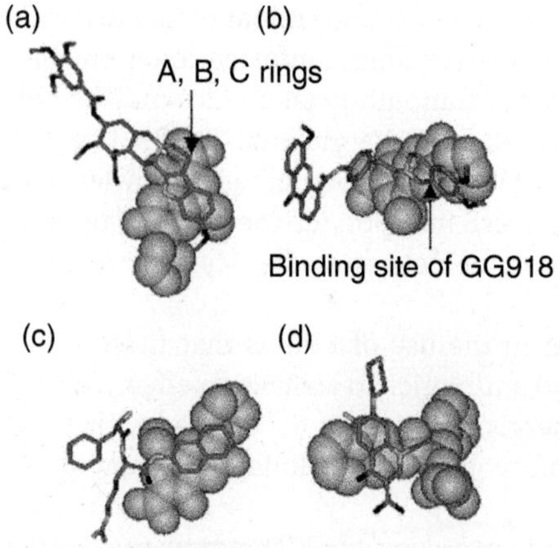

Figure 13.7 Highest interaction energy complexes between MDR inhibitors and substrates for bacterial MDR (substrate molecules are represented as spheres). (a) reserpine strongly binds norfloxacin at the same binding site for anticancer drugs, namely, rings A, B, and C. (b) Binding site of GG918 for norfloxacin is the same as for the topotecan complex. (c) The aromatic moiety of the MC-207,110 strongly binds the antibiotic levofloxacin. (d) The interaction between MC-002,595 and NPN involves π–π stacking of the aromatic rings. Source: Reprinted with permission from Zloh et al. (2004). Copyright (2004) Elsevier.

13.5 Design of Inhibitors of Drug-Modifying Enzymes

Bacterial resistance mechanism that is most clinically relevant is caused by the enzymatic modifications of antibiotics. To protect itself, bacteria can acquire and produce an enzyme that changes the chemical structure of an antibiotic and prevents its interaction with the target. Such an enzyme can degrade an antibiotic by hydrolysis (for example, β-lactamases) or add a chemical group(s) (for example, aminoglycoside-modifying enzymes) (Blair et al. 2015).

Such enzymes have distinct active sites that bind antibiotics, which may potentially be targeted by new molecular entities acting as inhibitors of such enzymes. These molecules would be suitable for developing combination therapies, where co-administration of an inhibitor with an antibiotic would restore its activity. These inhibitors generally mimic the shape and charges of antibiotic, which preferably bind into the enzyme binding site over antibiotic. A knowledge of the enzyme structure and its interactions with the antibiotics would, therefore, facilitate the design of effective adjuvants for combination therapies.

One of the mechanisms of resistance against β-lactam antibiotics is their degradation by one or more β-lactamases for which combination therapies are already developed and commercially exploited (Buynak 2006). Despite the progress in the development of β-lactamase inhibitors resulting in five used in the clinic, including clavulanic acid, tazobactam, and sulbactam, there are indications that pathogens are developing resistance to these inhibitors (van den Akker and Bonomo 2018). The emergence of resistance and the diversity of β-lactamases indicate the need for continued development of new inhibitors.

Currently, there are over 1200 entries in the Protein Data Bank related to β-lactamases, which provides a basis for structure-based design of their inhibitors. The availability of 3D structures provides opportunities for screening small molecules libraries or designing molecules with scaffolds that were not previously used in inhibitors and thus avoiding problems related to emerging resistance.

CTX-M class A extended-spectrum β-lactamases is one of such protein targets, especially as it poses a health threat due to causing failure of infection therapies. CTX-M-9 β-lactamase (PDB entry: 1YLY) was used as a target for virtual screening of molecules from ZINC lead-like database. This approach yielded a tetrazole-containing hit molecule amenable to further structural modifications to improve the potency by targeting hot spots important for binding, defined by two key residues ASP240 and PRO167. The synthesis and testing of the most promising analogue, found to have best *in silico* binding energies to the protein, has demonstrated ability to potentiate the activity of cefotaxime against CTX-M-9 *E. coli* strains. It has been postulated that tetrazole could target hot spots shared by other therapeutically important groups of β-lactamases and lead to a novel class of compounds that are effective against clinically relevant lactam-modifying enzyme (Nichols et al. 2012).

De novo design of β-lactamase inhibitors can result in their improved specificity toward certain classes. *In silico* fragment design was applied against zinc ion-dependent β-lactamases using New Delhi metallo-β-lactamase (NDM-1) structure (PDB entry: 2Q6X). In the first instance, hot spot regions of active site were identified using SPROUT software (Gillet et al. 1994) and key residues, TRP87 and LYS224/ARG228 (Figure 13.8). Rational design based on the structures of known inhibitors produced a lead compound **1**, with a scaffold amenable for structure-guided modifications. The designed analogues restored the activity of meropenem against *E. coli* species where NDM-1 is causing clinical problems. These inhibitors are potent (metal-β-lactamases) inhibitors but could also affect the function of some serine-β-lactamases (SBL) (Cain et al. 2018).

The example of enzymes that confer resistance by addition of chemical groups to the antibiotics is aminoglycoside-modifying enzymes. Such chemical modification prevents modified antibiotics to bind into the target site by steric hindrance. Depending on the type of modification, enzymes can be classified as *N*-acetyltransferases, *O*-nucleotidyltransferases, and *O*-phosphotransferases, which have a different propensity for clinically relevant aminoglycoside antibiotics (Zárate et al. 2018). Some of these enzymes can

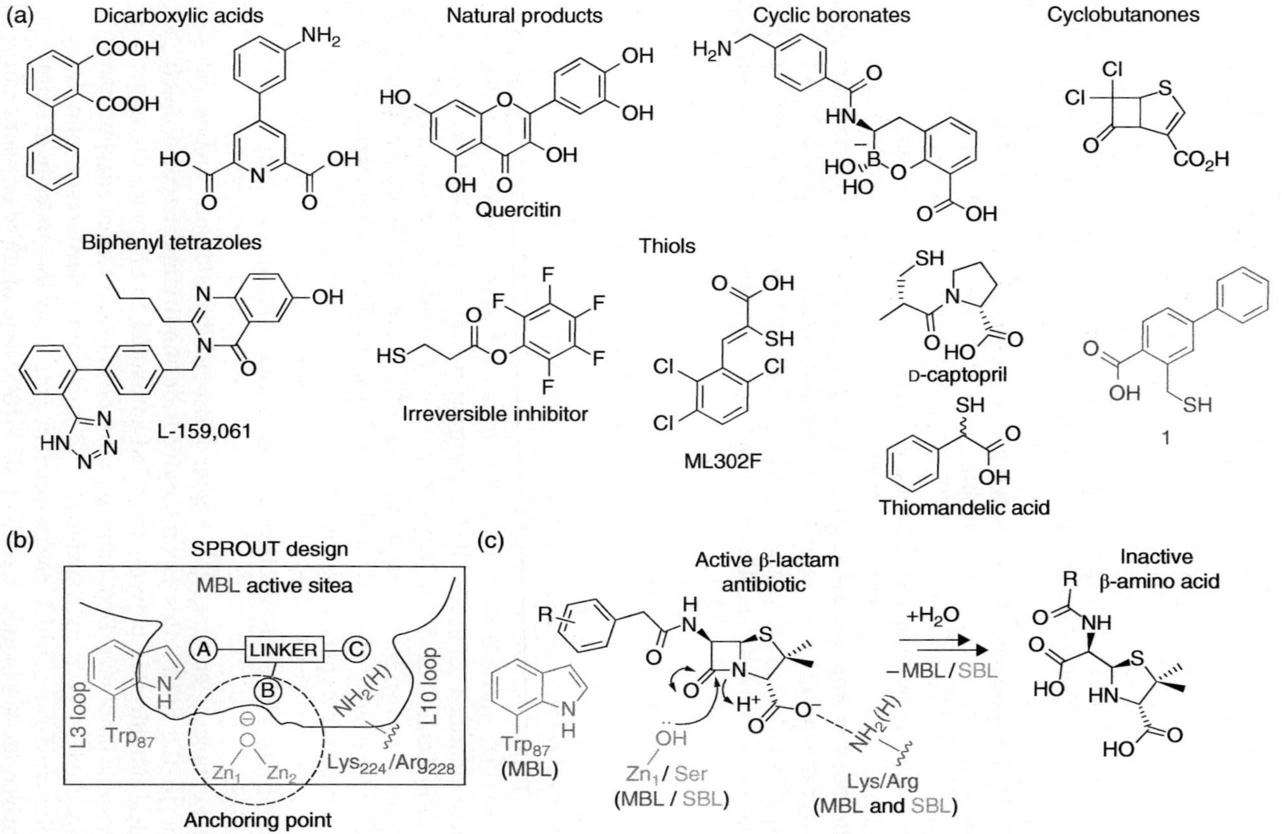

Figure 13.8 Structure-based design of broad-spectrum MBL inhibitors. (a) Known MBL inhibitors. (b) Key elements used in the SPROUT computational design process leading to the identification of putative broad-spectrum MBL inhibitors. (c) Outline mechanism for SBLs and MBLs as shown for a penicillin substrate. Source: Reprinted with permission from Cain et al. (2018). Copyright (2018) American Chemical Society.

inactive up to 15 aminoglycosides from different classes. Although there are over 110 structures in Protein Data Bank related to aminoglycoside-resistance enzymes, there are fewer studies on the structure-based design of their inhibitors (Chiem et al. 2016; Garzan et al. 2016; Parulekar and Sonawane 2018). This suggests that there are opportunities to development of effective and potent inhibitors of aminoglycoside-modifying enzymes using integrated approaches discussed earlier.

13.6 Antimicrobial Peptides

AMPs are generally short peptides with sequences that are less than 50 amino acids in length. Those molecules are often found in living organisms and part of a defense mechanism against bacterial infections (Ageitos et al. 2017). AMPs are being considered as an alternative to antibiotics or as a component of combination therapy with classic antibiotic (Figure 13.1h). These molecules are generally cationic and can adopt linear or cyclic structures. A judicious analysis of short AMPs (comprising <15 amino acids) with proven activity against Gram-negative bacteria has shown that these molecules form amphipathic surfaces via formation of either alpha helical or beta-turn structure (Passarini et al. 2018). However, there is a limited application of pure AMPs in the clinic due to their poor pharmacokinetic profiles or toxicity despite their observed promising activities against a range of different bacteria, but they can provide the basis for the design of novel modified peptides or peptidomimetic compounds. However, as the mechanism of action of AMP is commonly directed toward disruption of the membrane via various mechanisms or unknown (Ageitos et al. 2017; Kumar et al. 2018), the conventional structure-based design of peptides is not adequate, but other approaches are being employed.

Conformational studies of AMPs in the absence and presence of micelles are often conducted by NMR spectroscopy that provides not only structural information in a similar way as X-ray crystallography but can also provide information about their flexibility in different environments, particularly in the case of linear peptides (Vermeer et al. 2012). These conformational changes can be explored by using molecular dynamics simulations, which may provide information on their mechanism of action (Marquette and Bechinger 2018; Montales et al. 2018; Strandberg et al. 2018) and interactions with the membrane (Berglund et al. 2015; Balatti et al. 2017; Ulmschneider and Ulmschneider 2018). However, there is a need for continuous improvement and validation of methods and force fields used for conducting these types of simulations to ensure the reliable representation of the experimental observations (Wang et al. 2014). Although there are some successes in their modifications guided by molecular dynamics AMPs to change their activity (Waghu et al. 2018) or to reduce side effects (Dubovskii et al. 2018), the duration of molecular dynamics simulation needed to cover the timescale of conformational changes of peptides in the presence of membrane (order of several microseconds) is often bottleneck for efficient design of novel peptides using this approach. Therefore, in this section, some alternative methods to AMP design are considered.

As the structural diversity of AMPs with their conformational flexibility prevents the straightforward establishment of structure–activity relationships, testing a set of peptides with random sequences can provide a basis for developing a QSAR model by the amino acid composition of the most active peptides. Such a model created by using artificial neural networks enabled a development of quantitative and predictive *in silico* models of antibiotic activity that successfully predicted the activity of 100 000 virtual peptides (Cherkasov et al. 2009). The inclusion of the genetic algorithm heuristic search method improved further the identification of additional active peptides with greater efficiency, which was confirmed by *in vitro* testing of most promising candidates against several resistant bacterial strains (Fjell et al. 2011).

The use of machine learning approaches can be utilized to develop predictive models based on the results of short molecular dynamics simulations of AMPs (100 ns). The predicted 3D structures of AMPs extracted from the molecular dynamics simulation trajectories were used to calculate their 3D descriptors used to build random forest (RF) and support vector machine (SVM) models. These models were used to design novel AMPs, with proven antibacterial activities in *in vitro* experiments against several strains of bacteria (Liu et al. 2018).

The analysis of the sequence of AMPs is less demanding on computational resources. The arrangement and frequencies of amino acid residues in known AMP sequences carry information on the propensity of residue combinations to form particular structures. While the sequences can be used as strings in the development of predictive SVM models (Meher et al. 2017; Bhadra et al. 2018), it would be beneficial to understand the underlying principles and relationships behind particular arrangements of amino acids and their activities. A long short-term memory (LSTM) language model was developed that correctly deciphered the underlying grammar of AMP sequences. The application of this model has resulted in a lead molecule with potential to inhibit methicillin-resistant and carbapenem-resistant clinical isolates *in vitro* and *in vivo* activity (Nagarajan et al. 2018).

It is apparent that the further work on the design of novel peptides should encompass the prediction of their primary and tertiary structure. Therefore, as much the structural information available in the Protein Data Bank is needed, the information available in the AMP database (APD) (Wang et al. 2016), the collection of sequences and structures of antimicrobial peptides (CAMP) (Waghu et al. 2016), and the database of biofilm-active antimicrobial peptides (BaAMPs) (Di Luca et al. 2015) should be equally utilized.

13.7 Other Approaches to Overcome Bacterial Resistance

The outer membrane is one of the main hurdles for antibiotics overcome to reach targets in Gram-negative bacteria. The hijacking the nutrient-import transporters is one of a possible route for overcoming the impenetrability of this cell wall. The bacterial iron transport system is often targeted by conjugating antibiotics to siderophores, small molecules that bacteria use to transport iron (Schalk 2018). Siderophore conjugated to an antibiotic should still be recognized by the mechanism of siderophore recognition, with a suitable linker being stable in the extracellular environment and allowing the intracellular release of the antibiotic by enzyme action (Figure 13.1g). The 3D structures of siderophore receptors that are available in the Protein Data Bank should be utilized to design conjugates that have siderophore recognition motif accessible, but the features of recognized by efflux pumps should be masked or designed out.

Furthermore, quorum sensing, a regulation mechanism in bacteria that allows coordination of population density, relies on release and accumulation of signaling molecules in the environment. These signaling molecules interact with their target protein, which depends on the type and strain of the bacteria. The interactions that occur within quorum-sensing communication are similar to ligand–receptor interactions, suggesting that anti-quorum-sensing, anti-biofilm antibiotics could be designed using current standard pharmacologic principles (Raffa et al. 2005).

Such quorum-sensing inhibitors (QSIs) can effectively block QS and not only attenuate the virulence of *P. aeruginosa*, but it can also increase its susceptibility to both antibiotics and the immune system. Quorum-sensing receptor LasR (PDB entry: 2UV0) was considered as a target for an SVBS approach. A set of 3040 compounds were docked, resulting in a selection of 22 QSI candidates based on their docking scores and molecular masses. Five compounds were found to be able to inhibit QS-regulated gene expression in *P. aeruginosa* in a dose-dependent manner (Tan et al. 2013). Conversely to the structure-based approach, the search for an inhibitor of pqs quorum-sensing communication system characteristic for *P. aeruginosa* that controls virulence factor production and is involved in biofilm formation, an LBDD approach. A series of compounds targeting pqsR, the receptor of the pqs system, was designed by modifications of 2-heptyl-4-hydroxyquinoline. The testing of these compounds showed the reduction of virulence factor pyocyanin, but without reducing the viability of *P. aeruginosa*; therefore, they should not induce natural selection pressure and result in antibiotic resistance (Lu et al. 2012). These approaches can result in compounds that overcome the shortcomings of traditional antibiotics and may open new avenues for addressing the issue of antibiotic resistance (Kalia et al. 2019).

The impact of the infection can also be modulated by affecting the host signaling enzymes that are exploited by bacteria for their invasion, replication, or dissemination. Small molecule directed to act on host targets could potentiate innate immunity. This can be achieved by employing systems biology approaches, with key signaling pathways in the host–pathogen interplay identified as host proteins for therapeutic targeting. The 3D structures of such proteins could be modeled and used for the SBDD of host-directed small molecules that will regulate bacterial infection (Chiang et al. 2018).

13.8 Conclusion

The computer-aided design of therapeutic agents to fight antibiotic resistance did not fully achieve its full potential despite promising success stories. The time required for bacteria to equip itself with a defense mechanism against antibiotics is far shorter than for a new drug to treat infection to be discovered, to be developed, and to enter the clinic. Thus, in addition to developing new antibiotics, new strategies such as developing combination therapies, repurposing of already approved drugs, and design of drug delivery systems capable to specifically target pathogens are needed. A "clever" use of all available information on structures, activities, molecular properties, toxicities, plasma binding, and off-target interactions should enable utilization of modular workflows in machine learning methods and artificial intelligence approaches. Furthermore, "open-source" or a "community-based research" efforts, such as Community for Open Antimicrobial Drug Discovery (CO-ADD) initiative, should be supported and stimulated to enable our ability to address the global threat of antimicrobial resistance. The current progress in this field and increased availability of information and resources indicate that "golden era" for computer-aided antibiotic therapy design is coming.

References

Acharya, C., Coop, A., Polli, J.E., and Mackerell, A.D. Jr. (2011). Recent advances in ligand-based drug design: relevance and utility of the conformationally sampled pharmacophore approach. *Current Computer-Aided Drug Design* 7 (1): 10–22.

Ageitos, J.M., Sánchez-Pérez, A., Calo-Mata, P., and Villa, T.G. (2017). Antimicrobial peptides (AMPs): ancient compounds that represent novel weapons in the fight against bacteria. *Biochemical Pharmacology* 133: 117–138. https://doi.org/10.1016/j.bcp.2016.09.018.

Ahmad, S., Raza, S., Uddin, R., and Azam, S.S. (2018). Comparative subtractive proteomics based ranking for antibiotic targets against the dirtiest superbug: *Acinetobacter baumannii*. *Journal of Molecular Graphics and Modelling* 82: 74–92. https://doi.org/10.1016/j.jmgm.2018.04.005.

Albano, J.M.R., de Paula, E., and Pickholz, M. (2018). Molecular dynamics simulations to study drug delivery systems. In: *Molecular Dynamics* (ed. A. Vakhrushev). IntechOpen.

Alibert-Franco, S., Pradines, B., Mahamoud, A. et al. (2009). Efflux mechanism, an attractive target to combat multidrug resistant plasmodium falciparum and *Pseudomonas aeruginosa*. *Current Medicinal Chemistry* 16 (3): 301–317. https://doi.org/10.2174/092986709787002619.

Aminov, R.I. (2010). A brief history of the antibiotic era: lessons learned and challenges for the future. *Frontiers in Microbiology* 1: 134–134. https://doi.org/10.3389/fmicb.2010.00134.

Aparna, V., Dineshkumar, K., Mohanalakshmi, N. et al. (2014). Identification of natural compound inhibitors for multidrug efflux pumps of *Escherichia coli* and *Pseudomonas aeruginosa* using in silico high-throughput virtual screening and in vitro validation. *PLoS One* 9 (7): e101840. https://doi.org/10.1371/journal.pone.0101840.

Astolfi, A., Felicetti, T., Iraci, N. et al. (2017). Pharmacophore-based repositioning of approved drugs as novel *Staphylococcus aureus* NorA efflux pump inhibitors. *Journal of Medicinal Chemistry* 60 (4): 1598–1604. https://doi.org/10.1021/acs.jmedchem.6b01439.

Atzori, A., Malviya, V.N., Malloci, G. et al. (2019). Identification and characterization of carbapenem binding sites within the RND-transporter AcrB. *Biochimica et Biophysica Acta (BBA)-Biomembranes* 1861 (1): 62–74.

Axerio-Cilies, P., See, R.H., Zoraghi, R. et al. (2012). Cheminformatics-driven discovery of selective, nanomolar inhibitors for staphylococcal pyruvate kinase. *ACS Chemical Biology* 7 (2): 350–359. https://doi.org/10.1021/cb2003576.

Baker, S.J., Payne, D.J., Rappuoli, R., and De Gregorio, E. (2018). Technologies to address antimicrobial resistance. *Proceedings of the National Academy of Sciences* 115 (51): 12887–12895. https://doi.org/10.1073/pnas.1717160115.

Balatti, G.E., Ambroggio, E.E., Fidelio, G.D. et al. (2017). Differential interaction of antimicrobial peptides with lipid structures studied by coarse-grained molecular dynamics simulations. *Molecules* 22 (10): 1775.

Bellio, P., Di Pietro, L., Mancini, A. et al. (2017). SOS response in bacteria: inhibitory activity of lichen secondary metabolites against *Escherichia coli* RecA protein. *Phytomedicine* 29: 11–18. https://doi.org/10.1016/j.phymed.2017.04.001.

Berglund, N.A., Piggot, T.J., Jefferies, D. et al. (2015). Interaction of the antimicrobial peptide polymyxin B1 with both membranes of *E. coli*: a molecular dynamics study. *PLoS Computational Biology* 11 (4): e1004180.

Berman, H.M., Westbrook, J., Feng, Z. et al. (2000). The protein data bank. *Nucleic Acids Research* 28 (1): 235–242.

Bhadra, P., Yan, J., Li, J. et al. (2018). AmPEP: sequence-based prediction of antimicrobial peptides using distribution patterns of amino acid properties and random forest. *Scientific Reports* 8 (1): 1697. https://doi.org/10.1038/s41598-018-19752-w.

Bhaskar, B.V., Babu, T.M.C., Reddy, N.V., and Rajendra, W. (2016). Homology modeling, molecular dynamics, and virtual screening of NorA efflux pump inhibitors of *Staphylococcus aureus*. *Drug Design, Development and Therapy* 10: 3237–3252. https://doi.org/10.2147/DDDT.S113556.

Blair, J.M.A., Webber, M.A., Baylay, A.J. et al. (2015). Molecular mechanisms of antibiotic resistance. *Nature Reviews Microbiology* 13 (1): 42.

Bosi, E., Monk, J.M., Aziz, R.K. et al. (2016). Comparative genome-scale modelling of *Staphylococcus aureus* strains identifies strain-specific metabolic capabilities linked to pathogenicity. *Proceedings of the National Academy of Sciences* 113 (26): E3801–E3809. https://doi.org/10.1073/pnas.1523199113.

Boucher, H.W., Talbot, G.H., Bradley, J.S. et al. (2009). Bad bugs, no drugs: no ESKAPE! An update from the Infectious Diseases Society of America. *Clinical Infectious Diseases* 48 (1): 1–12.

Brincat, J.P., Carosati, E., Sabatini, S. et al. (2011). Discovery of novel inhibitors of the NorA multidrug transporter of *Staphylococcus aureus*. *Journal of Medicinal Chemistry* 54 (1): 354–365. https://doi.org/10.1021/jm1011963.

Brown, E.D. and Wright, G.D. (2016). Antibacterial drug discovery in the resistance era. *Nature* 529: 336. https://doi.org/10.1038/nature17042.

Buonerba, F., Lepri, S., Goracci, L. et al. (2017). Improved potency of indole-based NorA efflux pump inhibitors: from serendipity toward rational design and development. *Journal of Medicinal Chemistry* 60 (1): 517–523. https://doi.org/10.1021/acs.jmedchem.6b01281.

Bush, K. (2012). Improving known classes of antibiotics: an optimistic approach for the future. *Current Opinion in Pharmacology* 12 (5): 527–534. https://doi.org/10.1016/j.coph.2012.06.003.

Buynak, J.D. (2006). Understanding the longevity of the β-lactam antibiotics and of antibiotic/β-lactamase inhibitor combinations. *Biochemical Pharmacology* 71 (7): 930–940. https://doi.org/10.1016/j.bcp.2005.11.012.

Cain, R., Brem, J., Zollman, D. et al. (2018). In silico fragment-based design identifies subfamily B1 metallo-β-lactamase inhibitors. *Journal of Medicinal Chemistry* 61 (3): 1255–1260. https://doi.org/10.1021/acs.jmedchem.7b01728.

Cain, R., Narramore, S., McPhillie, M. et al. (2014). Applications of structure-based design to antibacterial drug discovery. *Bioorganic Chemistry* 55: 69–76. https://doi.org/10.1016/j.bioorg.2014.05.008.

Chan, P.F., Germe, T., Bax, B.D. et al. (2017). Thiophene antibacterials that allosterically stabilize DNA-cleavage complexes with DNA gyrase. *Proceedings of the National Academy of Sciences* 114 (22): E4492–E4500. https://doi.org/10.1073/pnas.1700721114.

Cherkasov, A., Hilpert, K., Jenssen, H. et al. (2009). Use of artificial intelligence in the design of small peptide antibiotics effective against a broad spectrum of highly antibiotic-resistant superbugs. *ACS Chemical Biology* 4 (1): 65–74. https://doi.org/10.1021/cb800240j.

Chiang, C.-Y., Uzoma, I., Moore, R.T. et al. (2018). Mitigating the impact of antibacterial drug resistance through host-directed therapies: current progress, outlook, and challenges. *mBio* 9 (1): https://doi.org/10.1128/mBio.01932-17.

Chiem, K., Jani, S., Fuentes, B. et al. (2016). Identification of an inhibitor of the aminoglycoside 6'-N-acetyltransferase type Ib [AAC(6')-Ib] by glide molecular docking. *Medchemcomm* 7 (1): 184–189. https://doi.org/10.1039/c5md00316d.

Chopra, I., Hesse, L., and O'Neill, A.J. (2002). Exploiting current understanding of antibiotic action for discovery of new drugs. *Journal of Applied Microbiology* 92 (s1): 4S–15S. https://doi.org/10.1046/j.1365-2672.92.5s1.13.x.

Costa, L.M., de Macedo, E.V., Oliveira, F.A.A. et al. (2016). Inhibition of the NorA efflux pump of *Staphylococcus aureus* by synthetic riparins. *Journal of Applied Microbiology* 121 (5): 1312–1322. https://doi.org/10.1111/jam.13258.

Coussens, N.P., Molinaro, A.L., Culbertson, K.J. et al. (2018). Better living through chemistry: addressing emerging antibiotic resistance. *Experimental Biology and Medicine (Maywood)* 243 (6): 538–553. https://doi.org/10.1177/1535370218755659.

Davies, J. and Davies, D. (2010). Origins and evolution of antibiotic resistance. *Microbiology and Molecular Biology Reviews* 74 (3): 417–433. https://doi.org/10.1128/MMBR.00016-10.

Di Luca, M., Maccari, G., Maisetta, G., and Batoni, G. (2015). BaAMPs: the database of biofilm-active antimicrobial peptides. *Biofouling* 31 (2): 193–199. https://doi.org/10.1080/08927014.2015.1021340.

Domalaon, R., Idowu, T., Zhanel, G.G., and Schweizer, F. (2018). Antibiotic hybrids: the next generation of agents and adjuvants against Gram-negative pathogens? *Clinical Microbiology Reviews* 31 (2): e00077-17. https://doi.org/10.1128/cmr.00077-17.

Dubovskii, P.V., Ignatova, A.A., Volynsky, P.E. et al. (2018). Improving therapeutic potential of antibacterial spider venom peptides: coarse-grain molecular dynamics guided approach. *Future Medicinal Chemistry* 10 (19): 2309–2322.

Dwivedi, G.R., Maurya, A., Yadav, D.K. et al. (2018). Synergy of clavine alkaloid 'chanoclavine' with tetracycline against multi-drug-resistant *E. coli*. *Journal of Biomolecular Structure and Dynamics* 37 (5): 1307–1325.

Edward, W.Y., McDermott, G., Zgurskaya, H.I. et al. (2003). Structural basis of multiple drug-binding capacity of the AcrB multidrug efflux pump. *Science* 300 (5621): 976–980.

Fields, F.R., Lee, S.W., and McConnell, M.J. (2017). Using bacterial genomes and essential genes for the development of new antibiotics. *Biochemical Pharmacology* 134: 74–86. https://doi.org/10.1016/j.bcp.2016.12.002.

Fjell, C.D., Jenssen, H., Cheung, W.A. et al. (2011). Optimization of antibacterial peptides by genetic algorithms and cheminformatics. *Chemical Biology and Drug Design* 77 (1): 48–56. https://doi.org/10.1111/j.1747-0285.2010.01044.x.

Garzan, A., Willby, M.J., Green, K.D. et al. (2016). Sulfonamide-based inhibitors of aminoglycoside acetyltransferase Eis abolish resistance to kanamycin in *Mycobacterium tuberculosis*. *Journal of Medicinal Chemistry* 59 (23): 10619–10628.

Gaulton, A., Bellis, L.J., Patricia Bento, A. et al. (2012). ChEMBL: a large-scale bioactivity database for drug discovery. *Nucleic Acids Research* 40 (Database issue): D1100–D1107. https://doi.org/10.1093/nar/gkr777.

Gillet, V.J., Newell, W., Mata, P. et al. (1994). SPROUT: recent developments in the de novo design of molecules. *Journal of Chemical Information and Computer Sciences* 34 (1): 207–217. https://doi.org/10.1021/ci00017a027.

González-Bello, C. (2017). Antibiotic adjuvants – a strategy to unlock bacterial resistance to antibiotics. *Bioorganic and Medicinal Chemistry Letters* 27 (18): 4221–4228. https://doi.org/10.1016/j.bmcl.2017.08.027.

Gruger, T., Nitiss, J.L., Anthony, M. et al. (2004). A mutation in *Escherichia coli* DNA gyrase conferring quinolone resistance results in sensitivity to drugs targeting eukaryotic topoisomerase II. *Antimicrobial Agents and Chemotherapy* 48 (12): 4495–4504. https://doi.org/10.1128/AAC.48.12.4495-4504.2004.

Gupta, R., Pradhan, D., Jain, A.K., and Rai, C.S. (2017). TiD: standalone software for mining putative drug targets from bacterial proteome. *Genomics* 109 (1): 51–57. https://doi.org/10.1016/j.ygeno.2016.11.005.

Hartenfeller, M. and Schneider, G. (2011). De novo drug design. In: *Chemoinformatics and Computational Chemical Biology* (ed. J. Bajorath), 299–323. Totowa, NJ: Humana Press.

Huang, G., Yan, F., and Tan, D. (2018). A review of computational methods for predicting drug targets. *Current Protein and Peptide Science* 19 (6): 562–572.

Irwin, J.J. and Shoichet, B.K. (2005). ZINC – a free database of commercially available compounds for virtual screening. *Journal of Chemical Information and Modeling* 45 (1): 177–182. https://doi.org/10.1021/ci049714+.

Jamshidi, S., Sutton, J.M., and Rahman, K.M. (2018). Mapping the dynamic functions and structural features of AcrB efflux pump transporter using accelerated molecular dynamics simulations. *Scientific Reports* 8 (1): 10470. https://doi.org/10.1038/s41598-018-28531-6.

Kalia, V.C., Patel, S.K.S., Kang, Y.C., and Lee, J.-K. (2019). Quorum sensing inhibitors as antipathogens: biotechnological applications. *Biotechnology Advances* 37 (1): 68–90. https://doi.org/10.1016/j.biotechadv.2018.11.006.

Katsila, T., Spyroulias, G.A., Patrinos, G.P., and Matsoukas, M.-T. (2016). Computational approaches in target identification and drug discovery. *Computational and Structural Biotechnology Journal* 14: 177–184. https://doi.org/10.1016/j.csbj.2016.04.004.

Kinch, M.S. and Griesenauer, R.H. (2018). 2017 in review: FDA approvals of new molecular entities. *Drug Discovery Today* 23 (8): 1469–1473. https://doi.org/10.1016/j.drudis.2018.05.011.

Kinch, M.S., Patridge, E., Plummer, M., and Hoyer, D. (2014). An analysis of FDA-approved drugs for infectious disease: antibacterial agents. *Drug Discovery Today* 19 (9): 1283–1287. https://doi.org/10.1016/j.drudis.2014.07.005.

Kohanski, M.A., Dwyer, D.J., Hayete, B. et al. (2007). A common mechanism of cellular death induced by bactericidal antibiotics. *Cell* 130 (5): 797–810. https://doi.org/10.1016/j.cell.2007.06.049.

Kohanski, M.A., Dwyer, D.J., and Collins, J.J. (2010). How antibiotics kill bacteria: from targets to networks. *Nature Reviews Microbiology* 8 (6): 423–435. https://doi.org/10.1038/nrmicro2333.

Kohanski, M.A., Dwyer, D.J., Wierzbowski, J. et al. (2008). Mistranslation of membrane proteins and two-component system activation trigger aminoglycoside-mediated oxidative stress and cell death. *Cell* 135 (4): 679–690. https://doi.org/10.1016/j.cell.2008.09.038.

Kumar, P., Kizhakkedathu, J.N., and Straus, S.K. (2018). Antimicrobial peptides: diversity, mechanism of action and strategies to improve the activity and biocompatibility in vivo. *Biomolecules* 8 (1): 4. https://doi.org/10.3390/biom8010004.

Lepri, S., Buonerba, F., Goracci, L. et al. (2016). Indole based weapons to fight antibiotic resistance: a structure–activity relationship study. *Journal of Medicinal Chemistry* 59 (3): 867–891. https://doi.org/10.1021/acs.jmedchem.5b01219.

Liu, S., Bao, J., Lao, X., and Zheng, H. (2018). Novel 3D structure based model for activity prediction and design of antimicrobial peptides. *Scientific Reports* 8 (1): 11189. https://doi.org/10.1038/s41598-018-29566-5.

Liu, Y.-Y., Yang, W., Walsh, T.R. et al. (2016). Emergence of plasmid-mediated colistin resistance mechanism MCR-1 in animals and human beings in China: a microbiological and molecular biological study. *The Lancet Infectious Diseases* 16 (2): 161–168.

Lomovskaya, O. and Bostian, K.A. (2006). Practical applications and feasibility of efflux pump inhibitors in the clinic – a vision for applied use. *Biochemical Pharmacology* 71 (7): 910–918. https://doi.org/10.1016/j.bcp.2005.12.008.

Lu, C., Kirsch, B., Zimmer, C. et al. (2012). Discovery of antagonists of PqsR, a key player in 2-Alkyl-4-quinolone-dependent quorum sensing in *Pseudomonas aeruginosa*. *Chemistry and Biology* 19 (3): 381–390. https://doi.org/10.1016/j.chembiol.2012.01.015.

Marquette, A. and Bechinger, B. (2018). Biophysical investigations elucidating the mechanisms of action of antimicrobial peptides and their synergism. *Biomolecules* 8 (2): 18.

Meher, P.K., Sahu, T.K., Saini, V., and Rao, A.R. (2017). Predicting antimicrobial peptides with improved accuracy by incorporating the compositional, physico-chemical and structural features into Chou's general PseAAC. *Scientific Reports* 7: 42362–42362. https://doi.org/10.1038/srep42362.

Mehta, H.H., Prater, A.G., and Shamoo, Y. (2017). Using experimental evolution to identify druggable targets that could inhibit the evolution of antimicrobial resistance. *The Journal of Antibiotics* 71: 279. https://doi.org/10.1038/ja.2017.108.

Mervin, L.H., Afzal, A.M., Drakakis, G. et al. (2015). Target prediction utilising negative bioactivity data covering large chemical space. *Journal of Cheminformatics* 7 (1): 51.

Montales, K.P., Wade, H.M., Figueroa, D.M. et al. (2018). Characterizing changes in antimicrobial peptide mechanism against different bacterial strains. *Biophysical Journal* 114 (3): 456a.

Nagarajan, D., Nagarajan, T., Roy, N. et al. (2018). Computational antimicrobial peptide design and evaluation against multidrug-resistant clinical isolates of bacteria. *Journal of Biological Chemistry* 293 (10): 3492–3509.

Nakamura, Y., Itoh, T., Matsuda, H., and Gojobori, T. (2004). Biased biological functions of horizontally transferred genes in prokaryotic genomes. *Nature Genetics* 36 (7): 760.

Nautiyal, A., Patil, K.N., and Muniyappa, K. (2014). Suramin is a potent and selective inhibitor of *Mycobacterium tuberculosis* RecA protein and the SOS response: RecA as a potential target for antibacterial drug discovery. *Journal of Antimicrobial Chemotherapy* 69 (7): 1834–1843. https://doi.org/10.1093/jac/dku080.

Neuberger, A., Dijun, D., and Luisi, B.F. (2018). Structure and mechanism of bacterial tripartite efflux pumps. *Research in Microbiology* 169 (7): 401–413. https://doi.org/10.1016/j.resmic.2018.05.003.

Nichols, D.A., Jaishankar, P., Larson, W. et al. (2012). Structure-based design of potent and ligand-efficient inhibitors of CTX-M class A β-lactamase. *Journal of Medicinal Chemistry* 55 (5): 2163–2172. https://doi.org/10.1021/jm2014138.

Parulekar, R.S. and Sonawane, K.D. (2018). Molecular modeling studies to explore the binding affinity of virtually screened inhibitor toward different aminoglycoside kinases from diverse MDR strains. *Journal of Cellular Biochemistry* 119 (3): 2679–2695. https://doi.org/10.1002/jcb.26435.

Passarini, I., Rossiter, S., Malkinson, J., and Zloh, M. (2018). In silico structural evaluation of short cationic antimicrobial peptides. *Pharmaceutics* 10 (3): 72.

Pavlopoulou, A. (2018). RecA: a universal drug target in pathogenic bacteria. *Frontiers in Bioscience (Landmark Edition)* 23: 36–42.

Punina, N.V., Makridakis, N.M., Remnev, M.A., and Topunov, A.F. (2015). Whole-genome sequencing targets drug-resistant bacterial infections. *Human Genomics* 9 (1): 19. https://doi.org/10.1186/s40246-015-0037-z.

Raffa, R.B., Iannuzzo, J.R., Levine, D.R. et al. (2005). Bacterial communication ("quorum sensing") via ligands and receptors: a novel pharmacologic target for the design of antibiotic drugs. *Journal of Pharmacology and Experimental Therapeutics* 312 (2): 417–423. https://doi.org/10.1124/jpet.104.075150.

Rahman, S.S., Simovic, I., Gibbons, S., and Zloh, M. (2011). In silico screening for antibiotic escort molecules to overcome efflux. *Journal of Molecular Modeling* 17 (11): 2863–2872.

Ramaswamy, V.K., Cacciotto, P., Malloci, G. et al. (2017). Computational modelling of efflux pumps and their inhibitors. *Essays in Biochemistry* 61 (1): 141–156.

Santos, R., Ursu, O., Anna, G. et al. (2016). A comprehensive map of molecular drug targets. *Nature Reviews Drug Discovery* 16: 19–34. https://doi.org/10.1038/nrd.2016.230.

Schalk, I.J. (2018). Siderophore-antibiotic conjugates: exploiting iron uptake to deliver drugs into bacteria. *Clinical Microbiology and Infection* 24 (8): 801–802. https://doi.org/10.1016/j.cmi.2018.03.037.

Schillaci, D., Spanò, V., Parrino, B. et al. (2017). Pharmaceutical approaches to target antibiotic resistance mechanisms. *Journal of Medicinal Chemistry* 60 (20): 8268–8297. https://doi.org/10.1021/acs.jmedchem.7b00215.

Simmons, K.J., Chopra, I., and Fishwick, C.W.G. (2010). Structure-based discovery of antibacterial drugs. *Nature Reviews Microbiology* 8: 501. https://doi.org/10.1038/nrmicro2349, https://www.nature.com/articles/nrmicro2349#supplementary-information.

Sjuts, H., Vargiu, A.V., Kwasny, S.M. et al. (2016). Molecular basis for inhibition of AcrB multidrug efflux pump by novel and powerful pyranopyridine derivatives. *Proceedings of the National Academy of Sciences* 113 (13): 3509–3514.

Sjuts, H., Vargiu, A.V., Kwasny, S.M. et al. (2017). Molecular insights into compound-transporter interactions: the case of inhibitors of Gram-negative bacteria efflux pumps. *Biophysical Journal* 112 (3): 274a.

Soon, R.L., Nation, R.L., Cockram, S. et al. (2011). Different surface charge of colistin-susceptible and -resistant *Acinetobacter baumannii* cells measured with zeta potential as a function of growth phase and colistin treatment. *Journal of Antimicrobial Chemotherapy* 66 (1): 126–133. https://doi.org/10.1093/jac/dkq422.

Strandberg, E., Zerweck, J., Wadhwani, P. et al. (2018). Molecular mechanism of synergy between the antimicrobial peptides PGLa and magainin 2 in membranes. *Biophysical Journal* 114 (3): 452a–453a.

Tan, S.Y.-Y., Chua, S.-L., Chen, Y. et al. (2013). Identification of five structurally unrelated quorum-sensing inhibitors of *Pseudomonas aeruginosa* from a natural-derivative database. *Antimicrobial Agents and Chemotherapy* 57 (11): 5629–5641. https://doi.org/10.1128/aac.00955-13.

Thai, K.-M., Ngo, T.-D., Phan, T.-V. et al. (2015). Virtual screening for novel *Staphylococcus aureus* NorA efflux pump inhibitors from natural products. *Medicinal Chemistry* 11 (2): 135–155.

Travers, T., Wang, K.J., Lopez, C.A., and Gnanakaran, S. (2018). Sequence-and structure-based computational analyses of Gram-negative tripartite efflux pumps in the context of bacterial membranes. *Research in Microbiology* 169 (7–8): 414–424.

Ulmschneider, J.P. and Ulmschneider, M.B. (2018). Molecular dynamics simulations are redefining our view of peptides interacting with biological membranes. *Accounts of Chemical Research* 51 (5): 1106–1116.

van den Akker, F. and Bonomo, R.A. (2018). Exploring additional dimensions of complexity in inhibitor design for serine β-lactamases: mechanistic and intra- and inter-molecular chemistry approaches. *Frontiers in Microbiology* 9: 622.

Vargiu, A.V., Ruggerone, P., Opperman, T.J. et al. (2014). Molecular mechanism of MBX2319 inhibition of Escherichia coli AcrB multidrug efflux pump and comparison with other inhibitors. *Antimicrobial Agents and Chemotherapy* 58 (10): 6224–6234. https://doi.org/10.1128/aac.03283-14.

Vargiu, A.V., Ramaswamy, V.K., Malloci, G. et al. (2018). Computer simulations of the activity of RND efflux pumps. *Research in Microbiology* 169 (7): 384–392. https://doi.org/10.1016/j.resmic.2017.12.001.

Verma, P., Tiwari, M., and Tiwari, V. (2018). In silico high-throughput virtual screening and molecular dynamics simulation study to identify inhibitor for AdeABC efflux pump of *Acinetobacter baumannii. Journal of Biomolecular Structure and Dynamics* 36 (5): 1182–1194. https://doi.org/10.1080/07391102.2017.1317025.

Verma, P. and Tiwari, V. (2018). Targeting outer membrane protein component AdeC for the discovery of efflux pump inhibitor against AdeABC efflux pump of multidrug resistant *Acinetobacter baumannii. Cell Biochemistry and Biophysics* 76 (3): 391–400.

Vermeer, L.S., Lan, Y., Abbate, V. et al. (2012). Conformational flexibility determines selectivity and anti-bacterial,-Plasmodium and-cancer potency of cationic α-helical peptides. *Journal of Biological Chemistry* 287 (41): 34120–34133.

Waghu, F.H., Joseph, S., Ghawali, S. et al. (2018). Designing antibacterial peptides with enhanced killing kinetics. *Frontiers in Microbiology* 9: 325.

Waghu, F.H., Barai, R.S., Gurung, P., and Idicula-Thomas, S. (2016). CAMPR3: a database on sequences, structures and signatures of antimicrobial peptides. *Nucleic Acids Research* 44 (D1): D1094–D1097. https://doi.org/10.1093/nar/gkv1051.

Walsh, C. (2000). Molecular mechanisms that confer antibacterial drug resistance. *Nature* 406 (6797): 775.

Wang, G., Li, X., and Wang, Z. (2016). APD3: the antimicrobial peptide database as a tool for research and education. *Nucleic Acids Research* 44 (D1): D1087–D1093. https://doi.org/10.1093/nar/gkv1278.

Wang, X., Shen, Y., Wang, S. et al. (2017b). PharmMapper 2017 update: a web server for potential drug target identification with a comprehensive target pharmacophore database. *Nucleic Acids Research* 45 (W1): W356–W360.

Wang, Y., Zhao, T., Wei, D. et al. (2014). How reliable are molecular dynamics simulations of membrane active antimicrobial peptides? *Biochimica et Biophysica Acta (BBA) – Biomembranes* 1838 (9): 2280–2288. https://doi.org/10.1016/j.bbamem.2014.04.009.

Wang, Z., Fan, G., Hryc, C.F. et al. (2017a). An allosteric transport mechanism for the AcrAB-TolC multidrug efflux pump. *eLife* 6: e24905.

Yilmaz, S., Altinkanat-Gelmez, G., Bolelli, K. et al. (2015). Binding site feature description of 2-substituted benzothiazoles as potential AcrAB-TolC efflux pump inhibitors in *E. coli*. *SAR and QSAR in Environmental Research* 26 (10): 853–871. https://doi.org/10.1080/1062936X.2015.1106581.

Yu, W. and MacKerell, A.D. Jr. (2017). Computer-aided drug design methods. *Methods in Molecular Biology (Clifton, N.J.)* 1520: 85–106. https://doi.org/10.1007/978-1-4939-6634-9_5.

Zárate, S., De la Cruz Claure, M., Benito-Arenas, R. et al. (2018). Overcoming aminoglycoside enzymatic resistance: design of novel antibiotics and inhibitors. *Molecules* 23 (2): 284.

Zhang, W. (2016). *Computer-Aided Drug Discovery*. Springer.

Zloh, M. and Gibbons, S. (2004). Molecular similarity of MDR inhibitors. *International Journal of Molecular Sciences* 5 (2): 37.

Zloh, M. and Gibbons, S. (2007). The role of small molecule: small molecule interactions in overcoming biological barriers for antibacterial drug action. *Theoretical Chemistry Accounts* 117 (2): 231–238.

Zloh, M., Kaatz, G.W., and Gibbons, S. (2004). Inhibitors of multidrug resistance (MDR) have affinity for MDR substrates. *Bioorganic and Medicinal Chemistry Letters* 14 (4): 881–885.

Zloh, M. and Kirton, S.B. (2018). The benefits of in silico modeling to identify possible small-molecule drugs and their off-target interactions. *Future Medicinal Chemistry* 10 (4): 423–432. https://doi.org/10.4155/fmc-2017-0151.

Zoraghi, R., See, R.H., Axerio-Cilies, P. et al. (2011). Identification of pyruvate kinase in methicillin-resistant *Staphylococcus aureus* as a novel antimicrobial drug target. *Antimicrobial Agents and Chemotherapy* 55 (5): 2042–2053. https://doi.org/10.1128/aac.01250-10.

14

Antibiotics from Natural Sources

David J. Newman

Newman Consulting LLC, Wayne, PA, USA

14.1 Introduction

14.1.1 The Origin of Microbial Resistance Gene Products

It is often thought, even today in some areas of the scientific press, that antibiotic resistance began once penicillin and the sulfonamides were in general use in the middle to late 1930s in the latter case and then in the early part of WWII with penicillins G (**1**) and V (**2**). That this was not the case was shown by the work of D'Costa et al. published in a *Nature* paper in 2011 (D'Costa et al. 2011). In this report, the authors demonstrated that antibiotic resistance is ancient after a metagenomic analyses of authenticated ancient DNA collected from sediments from 30 000-year-old Beringian permafrost in the Yukon, as well as the subsequent identification of a diverse collection of genes encoding resistance to β-lactam, tetracycline, and glycopeptide antibiotics (GPAs). These resistant determinants included, but were not limited to, the ribosomal protection protein *TetM*, which confers resistance to tetracycline antibiotics; the D-Ala-D-Ala dipeptide hydrolase *VanX*, a component of the vancomycin resistance operon; the aminoglycoside-modifying acetyltransferase *AAC*(*3*); a penicillin-inactivating β-lactamase *Bla* (a member of the large TEM β-lactamase grouping); and the ribosome methyltransferase *Erm*, which blocks macrolide, lincosamide, and type B streptogramin antibiotics from binding to the ribosome. Thus "biochemical warfare" between microbes has been going on for eons before the discovery of naturally occurring antibiotics. Further commentary was published the next year by Wright and Poinar (2012).

Thus the phenomenon of resistance is a very old one and it is reasonable to assume that protective measures led to evolution and subsequent production of naturally occurring microbial products that would overcome some of these

Antibiotic Drug Resistance, First Edition. Edited by José-Luis Capelo-Martínez and Gilberto Igrejas.

protective measures, and thus it has been the "task" of microbiologists and chemists to find these "new but old" agents and/or to modify these agents that are known to overcome these protective measures.

14.2 Organization of the Following Sections

I will briefly discuss active agents based on cyclic peptides, including the glycopeptides (vancomycin class) and lipopeptides, β-lactam chemistry and microbiology, and the work on aminoglycosides, tetracyclines, and macrolides using erythromycin and others as examples. The following sections will also cover, in specific cases, the chemical modifications that led to molecules or current combinations that are designed to overcome resistance to specific classes, and will include certain examples of molecules that have been designed to overcome resistance but are not yet approved drugs though may be at any level from biological testing to late clinical trials. This is primarily due to the size and cost of clinical trials of antibiotics that have to first demonstrate safety and then their superiority to those in current use. Thus the number of patients is in the thousands in phase III trials. Finally, the use of "Omics" methods to locate potential antibiotics, either single agents or biogenetic classes, will be discussed, as this area is the "future source of NP-based antibiotics."

14.3 Peptidic Antibiotics (Both Cyclic and Acyclic)

14.3.1 Tyrocidines, Gramacidins, and Derivatives

The first antibacterial natural product to go into clinical use was not penicillin as most people think, but the cyclic peptide tyrocidine (**3**) and a derivative of "gramicidin" that was identified in a 1940 report (Hotchkiss and Dubois 1940), with the structure of tyrocidine A reported in 1952 (Battersby and Craig 1952). A chemical relative of "gramicidin" reported by Hotchkiss and Dubois was possibly the second antibiotic series to begin clinical use, without formal approval as a drug entity in the modern sense. It was not as what most people in the West seem to think, the aminoglycosides as exemplified by streptomycin (**4**), but was gramicidin S (**5**), first used by the USSR for battlefield injuries in 1943 (Gause and Brazhnikova 1944). Further reports in the same basic time frame from the then USSR covered more clinical reports (Belozersky and Passhina 1944; Sergiev 1944). Then in 1946, Gause discussed the large-scale production and chemistry of this agent (Gause 1946). However, though these molecules including tyrocidine were not used significantly in the West due to various toxicity problems, they did demonstrate that cyclic peptides worked as antibiotics. By using modern chemical methods, derivatives of this basic structure have potential utility and are now being actively explored. Thus Wan et al.

(2018) reported that by using β,γ-diaminoacids as a β-turn mimic, they could synthesize analogues (**6** and **7**) that minimized the hemolytic activity effectively to less than 1% compared with 74–83% at comparable concentrations to GS, though the antibiotic activity was now between 2 and 8 times less effective against comparable microbes. However, the essential point is that these analogues are effectively non-hemolytic, thus expanding their potential for use other than as topical medications.

14.3.2 Streptogramins and Derivatives: Cyclic Peptides

Relatively recently, modifications of other peptide-based compounds from natural sources resulted in potential antibiotics, in particular the reassignment for human use of a class of agents (streptogramins) that were initially used solely for agricultural purposes. These compounds were divided into two distinct subgroups, group A (23-membered macrocyclic polyketide/non-ribosomal peptide hybrids) and group B (19-membered cyclic depsipeptides), which had a synergistic relationship. However, it was not until 1999 that the slightly modified compounds known as dalfopristin (**8**) (from virginiamycin M1; **9**), and quinupristin (**10**) (from virginiamycin S1; **11**) in a 70 : 30 mixture began clinical use as an IV formulation against multidrug resistance caused by vancomycin-resistant *Enterococcus faecium*. These compounds were produced via semisynthesis of the natural products, but in 2017, Li and Seiple (2017) published a modular and scalable synthesis of, in particular, virginiamycin M1 and maduramycin (structure not shown), as virginiamycin M1 has a dehydroproline moiety that may serve as a "handle" for the installation of side chains that could improve water solubility, as in the case of the semisynthesis of dalfopristin. The 2018 review by Luther et al. (2018) should also be consulted as it gives excellent coverage of the streptogramins and related compounds still under preclinical and early clinical studies.

14.3.3 Arylomycins (Lipopeptide and Modification, Preclinical)

Among the agents covered by Luther et al. are lipopeptides that target signal peptidase I, thus adding yet another bacterial target to the well-known earlier ones. None of these agents are yet in clinical trials, including the modified arylomycins, originally isolated from the soil bacterium *Streptomyces* Tü 6075 (Holtzel et al. 2002; Schimana et al. 2002) and shown to be active against the signal peptidase by workers at Lilly (Kulanthaivel et al. 2004). Genentech used the arylomycin structure (**12**) and modified it to produce G0775 (**13**) that covalently binds to a lysine in the *LepB* target as described in the recent *Nature* paper by Smith et al. (2018). Although these agents are still at the preclinical stage, their target is one that few agents have been tested against; therefore the possibilities are significant, and Genentech is the only major US-based company still "actively looking" for natural product-based antibiotics, with an active program with Lodo Therapeutics (see genetic methods below).

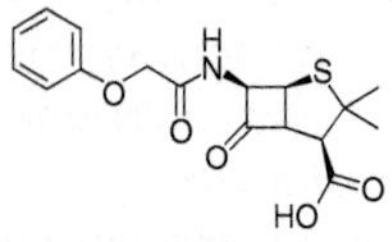
1; Penicillin V

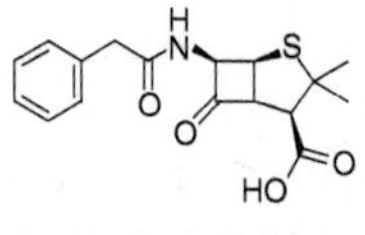
2; Penicillin G

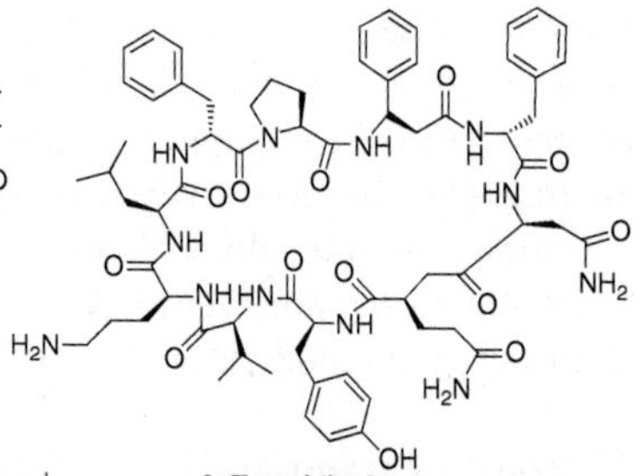
3; Tyrocidin A

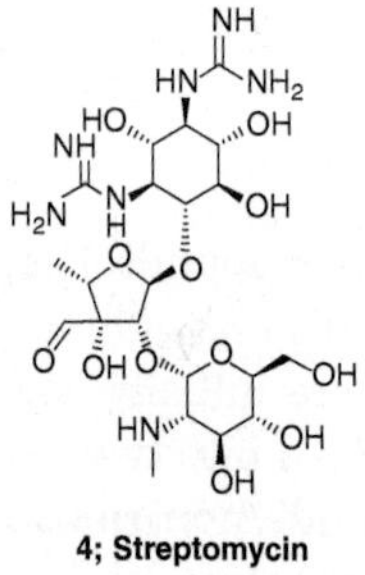
4; Streptomycin

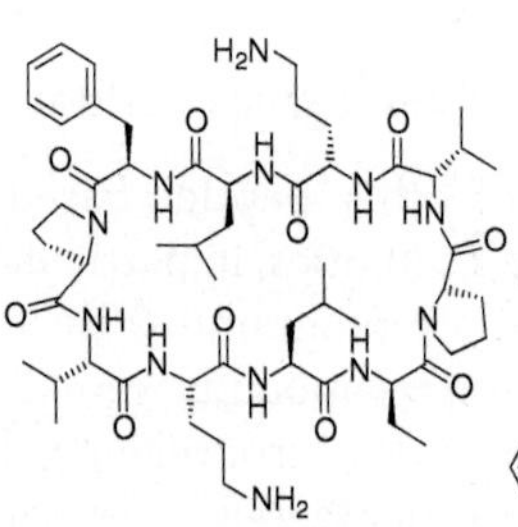
5; Gramicidin S

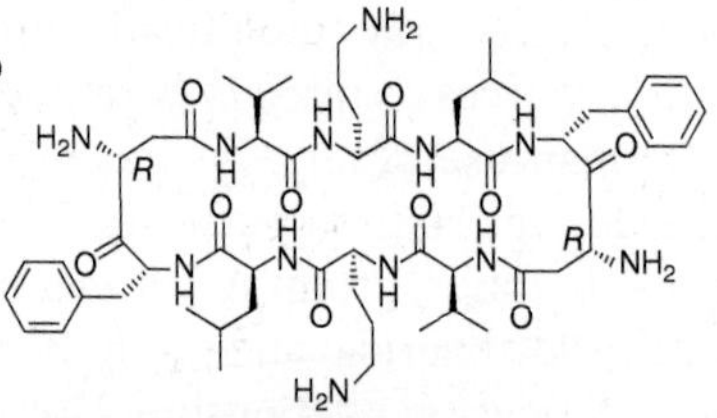
6; Analogue A (both *R*)

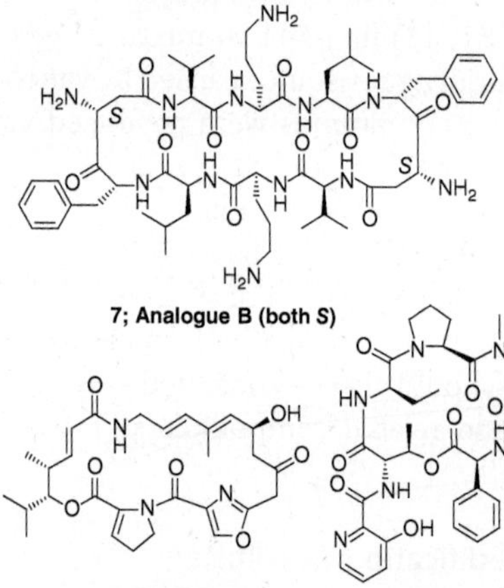
7; Analogue B (both *S*)

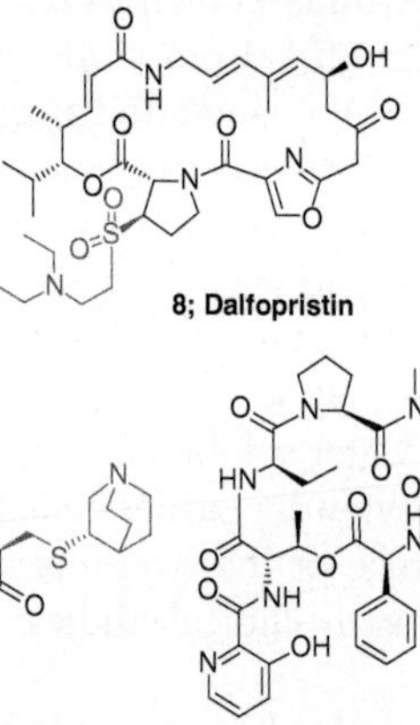
8; Dalfopristin

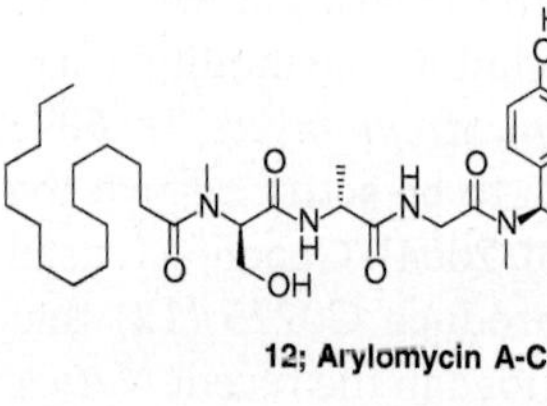
9; Virginiamycin M1

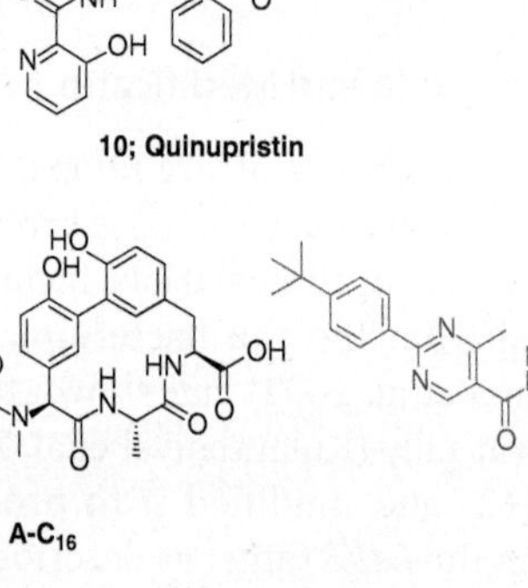
10; Quinupristin

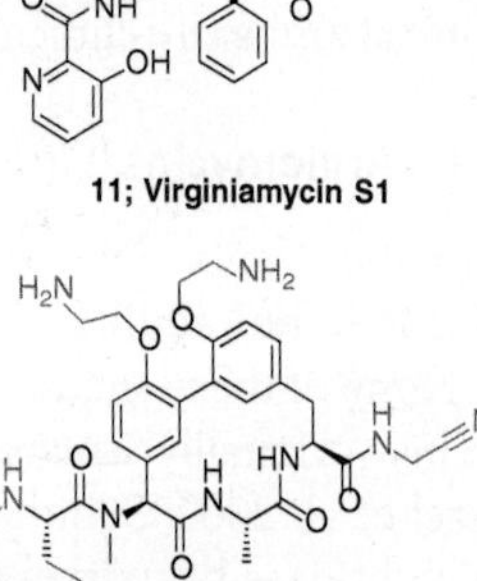
11; Virginiamycin S1

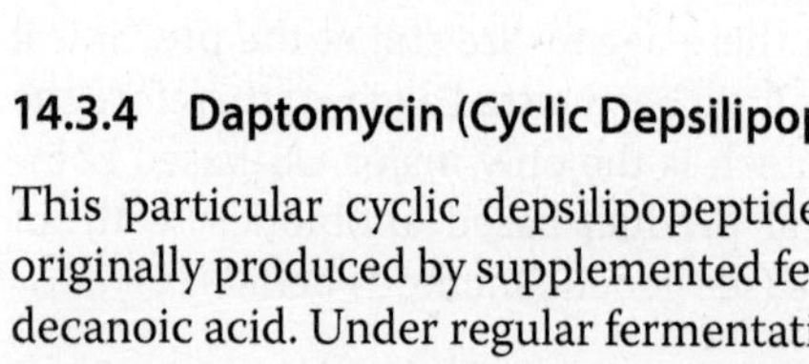
12; Arylomycin A-C_{16}

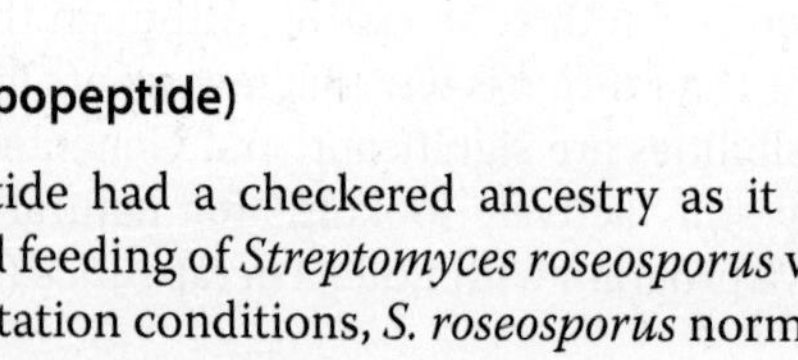
13; G0775

14.3.4 Daptomycin (Cyclic Depsilipopeptide)

This particular cyclic depsilipopeptide had a checkered ancestry as it was originally produced by supplemented feeding of *Streptomyces roseosporus* with decanoic acid. Under regular fermentation conditions, *S. roseosporus* normally produced a complex of lipopeptides (A21978C) with different long-chain fatty acid tails, but by feeding the straight C10 fatty acid, daptomycin was produced in quantity. Inspection of the structure of daptomycin (**14**) shows that the peptide portion of daptomycin contains 13 amino acids (10 with *L* and 3 with *D* stereochemistry), and the ring is closed by an ester bond between the terminal kynurenine (Kyn) and the hydroxyl group of Thr. This lipodepsipeptide and some of its close chemical relatives, have an obligate requirement for Ca^{2+} to be active and require a level of 1.25 mM for optimal activity, which is equivalent to the normal levels in human serum. This agent is active against MRSA and to some of other resistant Gram-positive microbes but has no activity against Gram-negative bacteria. Full details up through 2005 were presented in an excellent review by Baltz et al., which is worth studying for the early details on this class of compounds as well (Baltz et al. 2005).

14.3.4.1 Analogues of Daptomycin

What are very interesting are the recent articles by Ghosh et al. (2017) covers the medicinal chemistry involved in linking an analogue (**15**) of the siderophore fimsbactin (**16**) from *Acinetobacter baumannii* to daptomycin generating a conjugate that had significant *in vitro* and *in vivo* activities against multidrug-resistant strains of *A. baumannii*. Then in 2018, the same group generated bis- and tris-siderophore derivatives linked to daptomycin (**17**; **18**) (Ghosh et al. 2018) that were active against β-lactam-resistant *A. baumannii* strains. Thus, a relatively simple modification of a natural siderophore coupled to a complex nominally Gram-positive only active natural product, extended activity into the Gram-negative arena. Hopefully other similar modifications will extend the activity of daptomycin in the future, as it is one of the more complex but soluble antibiotics from nature. Finally, the recent review by Heidary et al. (2018) gives an excellent overview of the current usage of this molecule in clinical settings.

14.3.5 Colistins (Cyclic Peptides with a Lipid Tail)

Another group of agents that has had a rebirth due to major increases in resistance in Gram-negative microbes are the "ancient compounds" known as the colistins. These were used clinically in the 1960s but were withdrawn due to reports of nephrotoxicity and neurological problems in a large number of patients. With the recent emergence of multidrug-resistant bacteria and new and more stringent dosing regimens, colistin (**19**), a mixture of colistins A and B and also known as polymyxin E, has experienced a revival in clinical use as a last resort antibiotic against *Pseudomonas aeruginosa, A. baumannii,* and *Klebsiella pneumoniae* (part of the so-called ESKAPE pathogen series). One problem is the discovery of a transferable resistant plasmid *mcr-1* that has moved from the *Enterobacter* to other members of the ESKAPE pathogen series. This means that even this antibiotic of last resort now has resistance problems (Zhao et al. 2017; Machado et al. 2018).

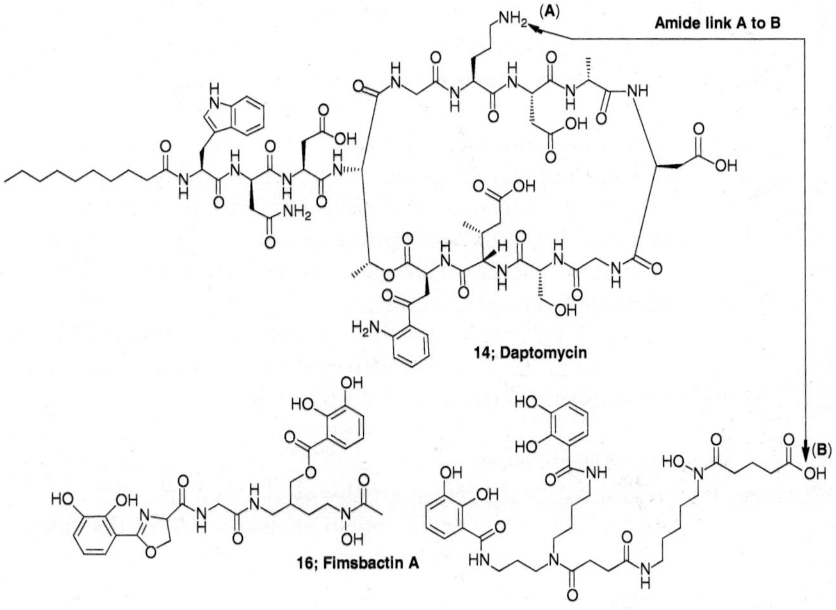

14; Daptomycin

16; Fimsbactin A

15; Synthetic Fimsbactin A analogue

17; *bis*-Catechol

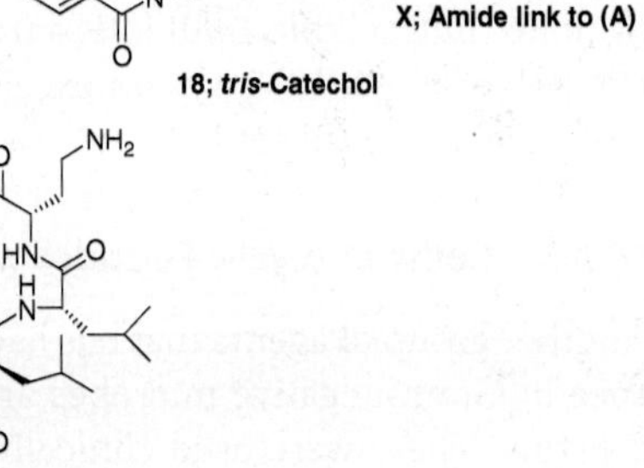

18; *tris*-Catechol

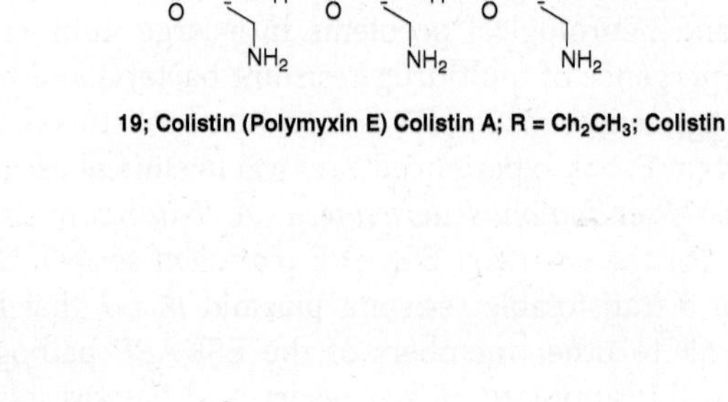

19; Colistin (Polymyxin E) Colistin A; R = CH_2CH_3; Colistin B; R = CH_3

14.3.6 Glycopeptides

14.3.6.1 Vancomycin and Chemical Relatives

The antibiotics based on peptides that have had the longest "continuous usage," are those in the vancomycin class, normally known as GPAs. Vancomycin (**20**) was first introduced into medicine in the middle to late 1950s by Lilly (Griffith 1981). Although in clinical use for over 20 years, it was not until 1982 that the full structure was identified when Harris and Harris correctly identified an asparagine residue in the antibiotic (Harris and Harris 1982). Subsequently, this molecule was used mainly as a "treatment of last resort" when resistance to methicillin in *Staphylococcus aureus* arose, giving rise in due course to the soubriquet MRSA for this resistant organism. Relatively recently, in order to help overcome the side effects of the normal IV administration of vancomycin, Shire Pharmaceuticals marketed an oral formulation "Vancocin Pulvules®." This can be considered a product of "pharmaceutics manipulation of its physical properties" in order to produce an orally active compound, but it does not alter the basic structure of the molecule itself.

As with all antibiotics, resistance to vancomycin developed, and for a considerable amount of time, the reason for this resistance was not known. Vancomycin and similar molecules predominately function as antibiotics by binding to the L-Lys-D-Ala-D-Ala-COOH terminals of the cross-links in the Gram-positive cell wall. In the late 1970s, the author (and colleagues) then at Smith, Kline & French (SK&F) devised a very simple screen for discovering vancomycin-like glycopeptides, by initially competing their activity on a simple disc assay on a test plate of the well-known *S. aureus* strain 209P. They used a sacculus preparation from the Gram-positive bacterium *Bacillus subtilis*, as the cost at that time of acetyl-L-Lys-D-Ala-D-Ala-COOH was prohibitive. This was published in addition to using the tripeptide method, many years later as SK&F ceased antibacterial discovery in 1985 (Rake et al. 1986). This resulted in the discovery and then subsequent development of the aridicins (Shearer et al. 1985). Subsequent work demonstrated that the *VanR* phenotype was simply due to a change in the terminal D-Ala residue to D-Lactate in the *vanA*, *vanB*, and *vanD* or to D-Ser in the *vanC*, *vanE*, and *vanG* phenotypes.

The rise of the vancomycin-resistance phenotype in clinical practice, was such that new antibiotics were required to deal with it, as one now had an increase in infections where MRSA was also exhibiting resistance to vancomycin. The D-Lac modification increased resistance by roughly 1000-fold, with the D-Ser modification causing roughly 140-fold increase in resistance compared to the base strain.

In the last few years, three semisynthetic GPAs entered clinical use in the United States and other countries: telavancin (**21**), which was approved in 2009, derived from vancomycin, as can be seen by comparing structure (**20**) with structure (**21**). In 2014, two more were approved by the FDA, dalbavancin (**22**) derived from a teicoplanin-like compound that was part of the A40926 natural product complex and oritavancin (**23**) derived from the natural product, chloroeremomycin. One other GPA, teicoplanin (structure not shown), is composed of a mixture of five compounds, TA_2–1 to TA_2–5, which differ in their fatty acid side chains, and the presence of TA_3–1, which is deacylated, was launched in 1988 in Italy and is still used extensively in Europe, with an oral formulation reported to have efficacy in geriatric patients with *Clostridium difficile* infections in Serbia in 2015 (Popovic et al. 2015) and in France in 2017 (Davido et al. 2017). Teicoplanin does maintain resistance against strains with the *vanB* operon, and there are strains with resistance against teicoplanin that retain sensitivity to vancomycin, but more usually strains are resistant to both.

14.3.6.2 Synthetic Modifications of Vancomycin

The major group involved from a total synthesis perspective is that of Boger's at the Scripps Research Institute in San Diego. This group has optimized the basic structure of vancomycin as shown by their reports over the last 10 plus years on the total syntheses of very interesting and active variations (Xie et al. 2011; Okano et al. 2014, 2015) coupled to earlier work on the total synthesis of the ristocetin aglycone (Crowley et al. 2004). All of these modifications were made by total synthesis, not by modifying the natural product. Recently, the Boger group reported the synthesis of a derivative of vancomycin that following subtle modifications, including the conversion of an amide carbonyl in the cyclic peptide component to a methylene, yielded an exceptionally potent antibiotic (**24**) with activity against both D-Ala-D-Ala and D-Ala-D-Lac end groups, increased membrane permeability, and more than a 1000-fold increase in activity over vancomycin, and no significant resistance after 25 passages (Okano et al. 2017a). Inspection of the five structures (**20–24**) demonstrates that relatively small structural changes significantly influence the bioactivity/resistance profile and membrane permeability. More details are given in three excellent recent reports (Okano et al. 2017a, b; Yang et al. 2017), including the significant synthetic challenges involved. The further development of this molecule should be extremely interesting.

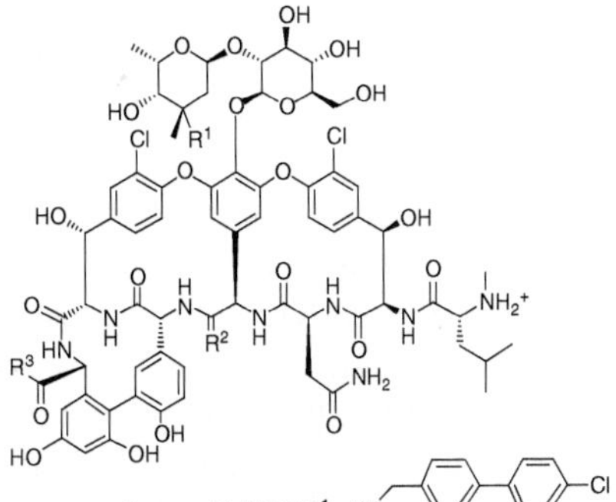

20; Vancomycin, $R^1 = NH_2$ 24; Boger R^1 = HN
20. Vancomycin, $R^2 = O$ 24;.Boger $R^2 = H_2$
20. Vancomycin, $R^3 = OH$ 24; Boger $R^3 = -NH(CH_2)_3N(CH_3)_3^+$

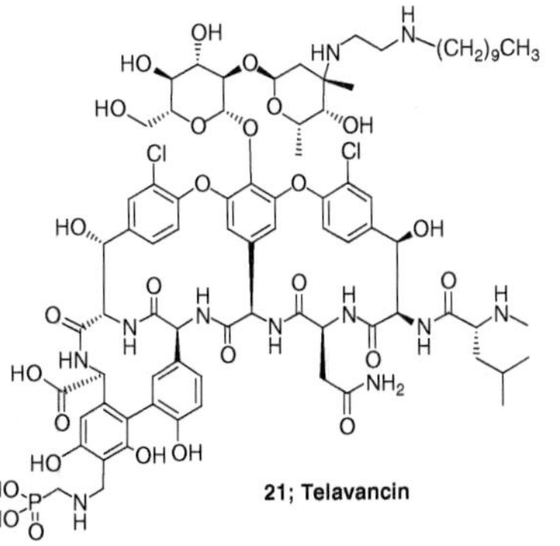

21; Telavancin

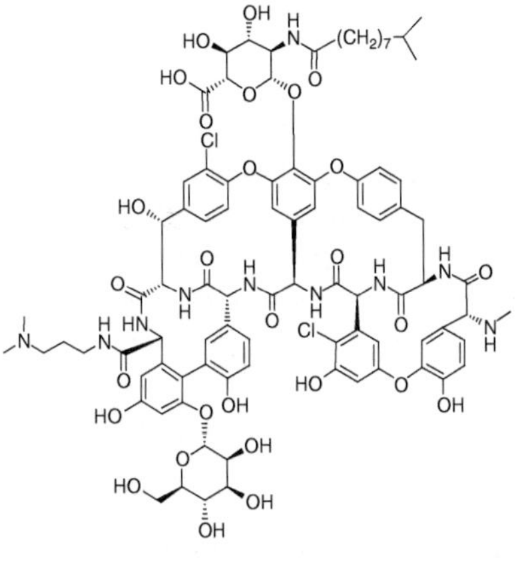

22; Dalbavancin

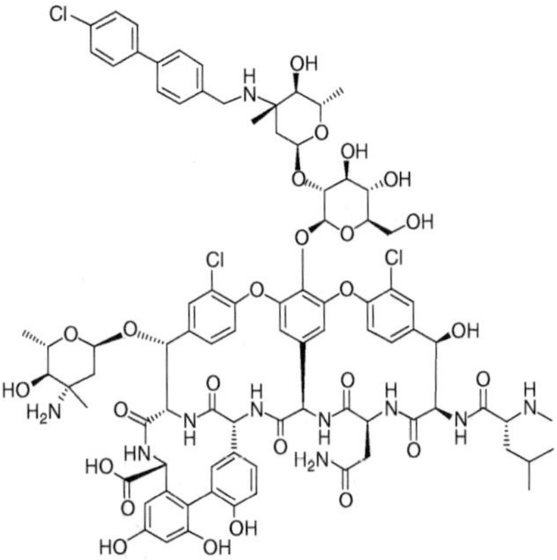

23; Oritavancin

14.3.7 Host Defense Peptides

14.3.7.1 Magainins and Derivatives

The peptidic molecules in the previous sections do not include peptides such as the magainins. Linear peptides using L amino acids with the base molecule shown in structure (**25**) and their descendants. These are part of a vast number of "host defense peptides" that all vertebrate animals secrete as cationic peptides, usually around 20–35 residues of regular amino acids that attack "microbial invaders." The history of the frog-derived peptides known as magainins were given in a series of papers by Zasloff (the discoverer) (Zasloff 1987) followed by significant discussion by the company formed to develop this agent (Berkowitz et al. 1990) and then a more recent book chapter by Zasloff (2016).

Although magainins were not approved as drugs, there are molecules based upon them that are in current clinical trials, and in one case, expressing the required peptide plectasin (structure not given) using *Escherichia coli* to produce the mushroom-based defensin. Synergistic activity was reported in 2015 for this agent against MRSA by Hu et al., but currently no clinical trials are reported (Hu et al. 2015). Conversely, pexiganan (**26**) a slight modification of the natural product magainin 2, is in phase III clinical trials for infected diabetic foot ulcers, with *in vitro* data published in 2015 by Flamm et al. (2015).

14.3.7.2 Synthetic Variations Using the "Defensin Concept"

Rather than working with the magainin-style of molecule, in order to overcome some of the pharmacological problems with the "defensins," the DeGrado group reported in 2002 the synthesis and characterization of a novel family of arylamide amphipathic anionic polymers that exhibited many of the physical and biological properties of antimicrobial peptides, such as the magainins (Tew et al. 2002). These mimetics held an advantage as they could easily be synthesized from inexpensive monomers via classical polymerization techniques, yielding agents that were resistant to proteolytic hydrolysis. By following refinements using structure–activity relationships, the potency against Gram-positive bacteria was optimized while maintaining minimal lytic activity against human erythrocytes. Choi et al. reported the success of these techniques and the synthesis of brilacidin (**27**) (Choi et al. 2009). In solution, the molecule has a planar secondary structure, with the guanidinyl and pyridinyl groups positioned on one edge of the scaffold and trifluoromethane groups on another. As reported in 2014 by the DeGrado group, brilacidin selectively damages the microbial membrane in a manner comparable to linear antimicrobial peptides and the lipopeptide daptomycin (Mensa et al. 2014). Currently the molecule is in phase II clinical trials for a variety of Gram-positive infections.

Though not derived from natural products, other than using the underlying principles involved in the mechanism of the antimicrobial peptides, in the middle of 2018, a Spanish academic group published an interesting report on synthetic cationic and non-peptidic small molecules. These molecules (structures not shown) had a total of six positive charges on one side, due to two aminophenol residues, and a long aliphatic tail on the other. The molecules demonstrated significant antimicrobial activity against Gram-positive bacteria including multidrug-resistant strains but were inactive against pathogenic Gram-negative strains. It will be interesting to see how these molecules progress in due course (Jiménez et al. 2018).

H_2N-Gly-Ile-Gly-Lys-Phe-Leu-His-Ser-Ala-Gly-Lys-Phe-Gly-Lys-Ala-Phe-Val-Gly-Glu-Ile-Met-Lys-Ser-CO_2H

25; Magainin 1

H_2N-Gly-Ile-Gly-Lys-Phe-Leu-Lys-Lys-Ala-Lys-Lys-Phe-Gly-Lys-Ala-Phe-Val-Lys-Ile-Leu-Lys-Lys-NH_2

26; Pexiganan

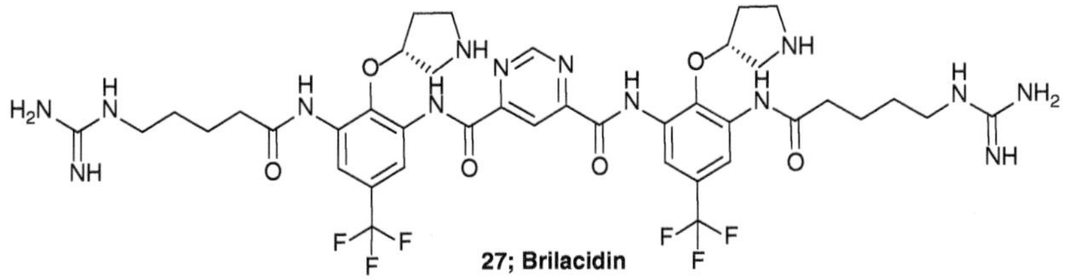

27; Brilacidin

14.4 β-Lactams: Development, Activities, and Chemistry

From the basic penicillin backbone (as in **1** and **2**), one can argue that we are now at the fifth generation of these agents, particularly in the case of the cephalosporins as exemplified by the cephalosporin prodrug ceftaroline fosamil (**28**). This compound is effective against Gram-positive and Gram-negative bacteria, including *P. aeruginosa* (though not all strains), and has increased affinity for the PBP2a, the gene product that mediates methicillin resistance in *Staphylococci*.

The original cephalosporin, cephalosporin C (**29**) was initially reported by Brotzu (1948) from a discovery made in 1945 of a bioactive fungus from Sardinian waters. This fungus was identified as *Cephalosporium acremonium* in the 1948 report, and that year, a culture was sent to Oxford University for further investigation. Over the next 20 years, active components, such as cephalosporins C (**29**) and N (now penicillin N; **30**), were identified. Cephalosporin C (**29**) had a wide range of bioactivities against a number of penicillin-resistant and -sensitive strains of *S. aureus* (Newton and Abraham 1955), and although its bioactivities were weak, cephalosporin C (**29**) was not susceptible to the then known β-lactamase-producing strains (Abraham and Newton 1956; Newton and Abraham 1956) and therefore had potential bioactivity against penicillin-resistant microbes, which had been reported soon after the initial usage of penicillin in the middle 1940s. These results encouraged Lilly to expand upon the 7-aminocephalosporanic acid scaffold to produce many cephalosporins to treat penicillin-resistant infections, and today, as mentioned above, there are now five generations of these agents. The 2013 review by Fernandes et al. (2013), though dated, gives an excellent overview of the various "chemical classes" of these agents and an introduction to their chemistry. The total number of such agents that have been synthesized and then tested *in vitro* against microbes is not currently calculable, but in the heyday of β-lactam synthesis and testing in the early to middle 1980s, figures in excess of 25000 compounds were provided from literature analyses (J.R.E. Hoover SK&F Labs, personal communication).

14.4.1 Combinations with β-Lactamase Inhibitors

One method of overcoming β-lactamases was introduced in the late 1970s to early 1980s: the use of a combination of penicillins (and later cephalosporins) and specific inhibitors. The earliest one was sulbactam (structure not shown), but then the natural product clavulanic acid (**31**) was introduced, initially with ampicillin but then as a fixed dose combination with amoxicillin, which is still in use. Three combinations of inhibitor and β-lactam-based antibiotics have been approved: two in 2015, ceftolozane/tazobactam (**32, 33**), an unusual cephalosporin and penicillin combination, and the cephalosporin ceftazidime and the synthetic inhibitor avibactam (**34, 35**). The penem meropenem (**36**), which was based upon the natural product thienamycin (structure not shown), the first of the "penem" class of antibiotics, was combined with the synthetic boronic inhibitor vaborbactam (**37**) and launched in 2017.

Thus, more than 70 years after penicillin first went into clinical but not yet commercial use, the chemical skeleton is still in use and there are at least three variations still under clinical studies as of the end of 2018.

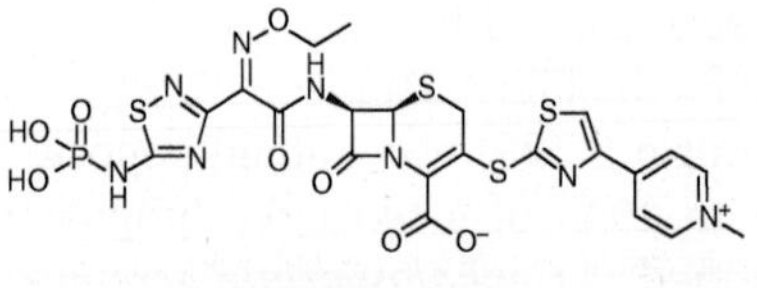

28; Ceftaroline fosamil

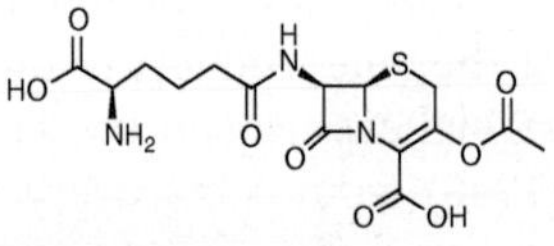

29; Cephalosporin C

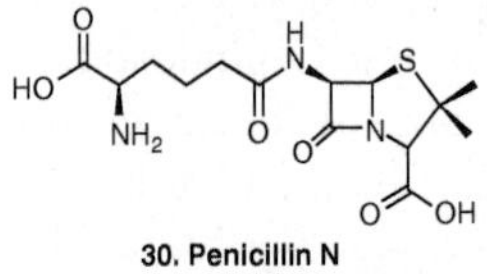

30. Penicillin N

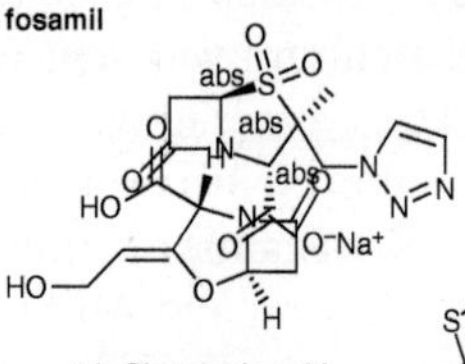

31; Clavulanic acid

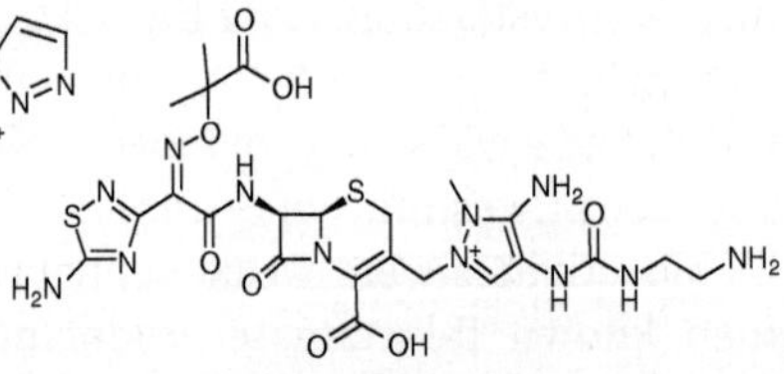

32; Cefloozane

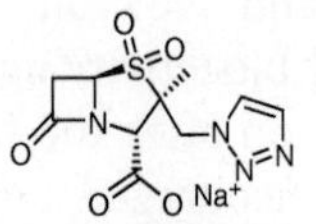

33; Tazobactam

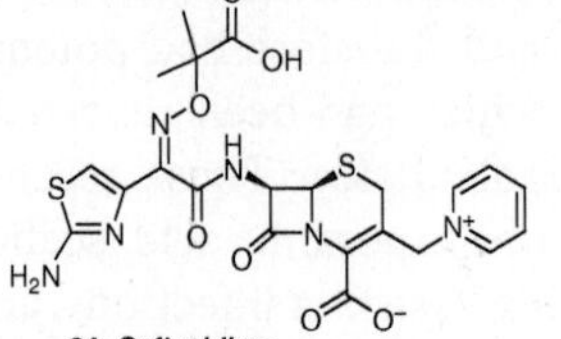

34; Ceftazidime

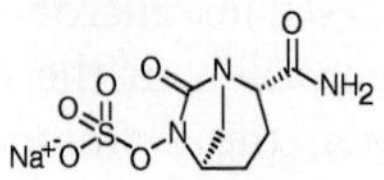

35; Avibactam sodium

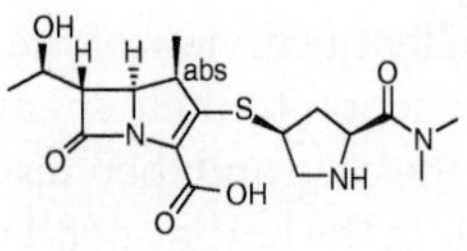

36; Meropenem

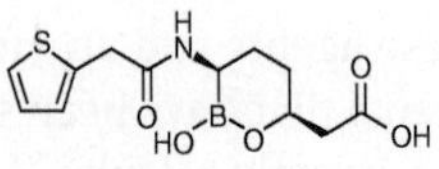

37; Vaborbactam

14.5 Aminoglycosides

14.5.1 Streptomycin

Streptomycin (**4**), the first of the aminoglycosides, was often considered in the West to be the second antibiotic to begin clinical use (during WWII) after penicillin G/V; however, as shown earlier, the use of gramicidin S predated it. It did begin use in wounded soldiers in the late 1944 to early 1945 time frame. Its discovery was formally reported by the Waksman group in 1944 (Schatz et al. 1944) and it was not until 30 years later that its formal synthesis via dihydrostreptomycin was published by Umezawa et al. (1974a, b). The aminoglycosides were used in large quantities and over the years, different variations of the basic structure either were found from nature or were the product of slight chemical modifications to overcome the aminoglycoside-modifying enzymes (AMEs), plasmid-borne enzymes that *N*-acetylate, or *O*-adenylate or *O*-phosphorylate on specific hydroxyl groups in the molecules. Most of the naturally occurring aminoglycosides were discovered between the 1940s and 1970s and classified into three basic groups: streptidine containing (**4**); 2-deoxystreptamine containing (neomycins, paromomycins, kanamycins, gentamicins, tobramycin (**38**), sisomicin, ribostamycins, lividomycins, butirosins, and verdamicin). In addition, there were novel ones other than the two earlier classes, hygromycins, spectinomycin, apramycin, and fortimicins (structures not shown). A very recent paper demonstrated that even 44 years after the introduction of tobramycin in 1974, the compound is still very useful in combatting biofilm-bound *P. aeruginosa* in clinical settings (Müsken et al. 2018).

14.5.2 Plazomicin

The slight modification of sisomicin (**39**), which was originally reported in 1970 from *Micromonospora inyoensis* (Weinstein et al. 1970) that produced the semisynthetic plazomicin (**40**), was described in detail by Galani (2014) with the agent being approved by the US FDA in 2018 for treatment of complex urinary tract infections such as those caused by carbapenem-resistant *Enterobacteriaceae* (CRE) and extended-spectrum beta-lactamase (ESBL)-producing *Enterobacteriaceae*.

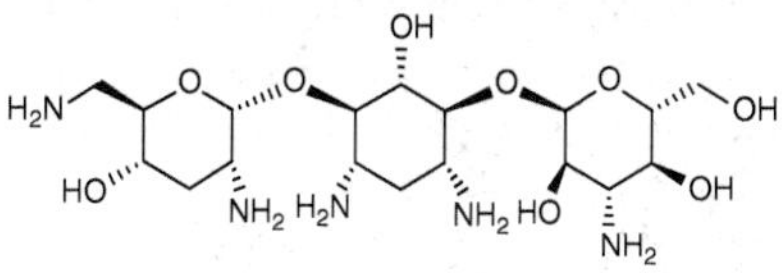

38; Tobramicin

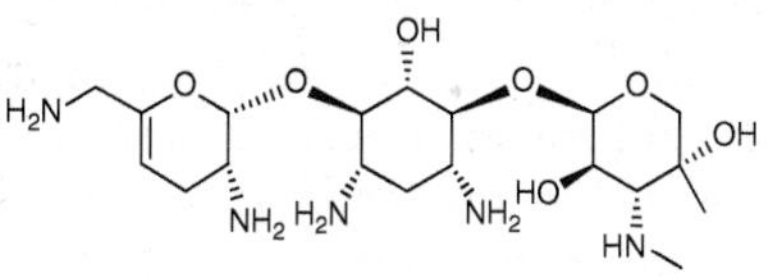

39; Sisomicin

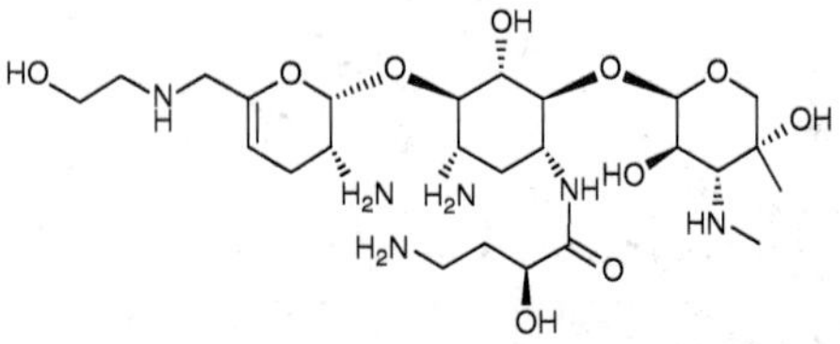

40; Plazomicin

14.6 Early Tetracyclines: Aureomycin and Terramycin

In 1948, Duggar reported the discovery of aureomycin, subsequently identified as chlortetracycline (**41**) by Woodward's group, who also identified terramycin as oxytetracycline (**42**). The early history of the tetracyclines was covered in an excellent article in 2001 by Nelson (2001) in the book *Tetracyclines in Biology, Chemistry and Medicine,* also edited by Nelson. Another chapter in this book by Schneider (2001) discussed the metal-binding characteristics of tetracyclines, a discussion that has become relevant in the last few years.

What is also of significant interest is that the tetracycline nucleus was probably the first drug "structure" that led to the identification of "active transport" as a mechanism of removal of a drug from its target, in addition to "ribosome protection." A very recent discussion of the various mechanisms has been published by Markley and Wencewicz (2018), which should be consulted to demonstrate the many interlocking systems that bacteria use to defend themselves against this class of molecules. Another interesting review, in a journal that few microbiologists would probably read, is the discussion by Guerra et al. (2016) in *Coordination Chemistry Reviews,* where they demonstrate that using a platinum or palladium complex with tetracycline can overcome the *tetX* and other transporters, leading to abrogation of the resistance in microbes resistant to tetracyclines (Guerra et al. 2016).

14.6.1 Semisynthetic Tetracyclines from 2005

Based on the previous sections and, in particular, the resistance profiles that have arisen since the late 1940s, one might think that this particular base structure is not one to continue to study for new antibiotics. However, in 2005 Lederle (Wyeth and then Pfizer) had tigecycline (**43**) approved by the FDA. This is effectively a derivative of the slightly modified tetracycline, minocycline (**44**), and is a "glycylcycline." The original and subsequent papers covering this compound are worth reading as what was effectively a relatively simple modification, led to a new and active tetracycline (Sum et al. 1993; Petersen et al. 1999; Hunter and Castaner 2001).

Then in the middle to late 2018, three new semisynthetic tetracyclines were approved. On 27 August, the US FDA approved eravacycline (**45**) (Thakare et al. 2018) for the treatment of complicated abdominal infections and complicated urinary tract infections, with approval in the EU a month later. A recent paper discusses in detail the synthetic processes to produce this compound in bulk, which makes interesting reading when compared to the usual reports of isolating a few milligrams (Ronn et al. 2013). Then on 1 October, the US FDA approved sarecycline (**46**) for the oral treatment of acne vulgaris in patients nine years old or older. One has yet to observe whether this agent will also induce staining of adolescent's teeth due to chronic usage, as with the early tetracyclines. The next day, 2 October, omadacycline (**47**) was approved by the US FDA for treatment of acute bacterial skin and skin structure infections (ABSSSI); the EU is yet to provide approval (Cho et al. 2018).

Thus 70 years after the first report of aureomycin, the basic nucleus of that drug is still a viable entity for the semisynthesis of new molecules. In addition, a simple search of the NIH clinical trials database on 1 January 2019, for doxycycline, which was approved in 1967, shows over 200 listed trials of which 50 plus are at phase IV. Thus, old compounds have potential new uses.

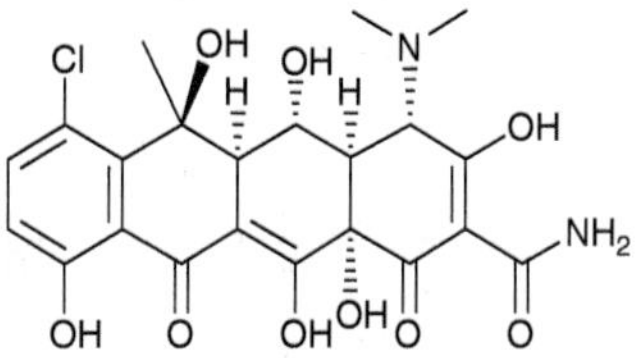

41; Chlortetracycline

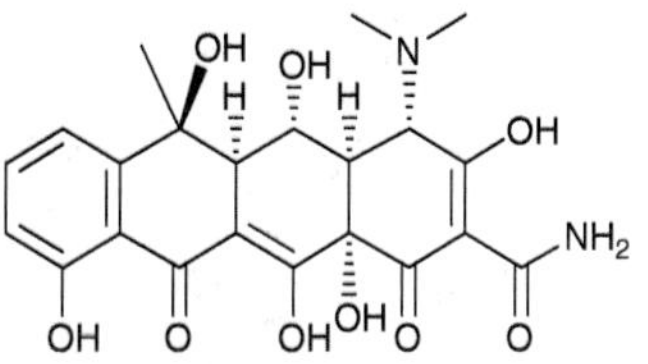

42; Oxytetracylcine

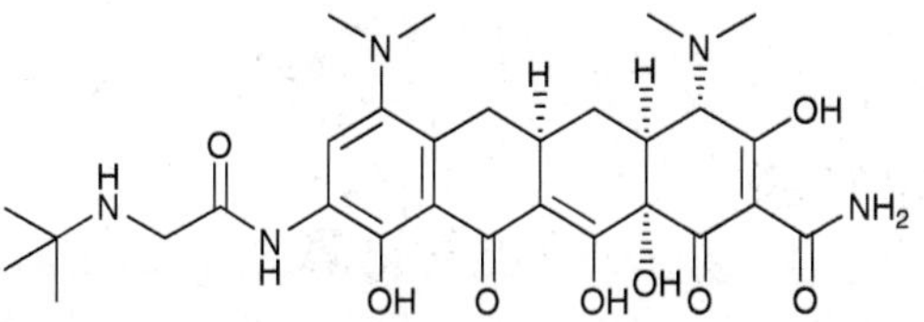

43; Tigecycline

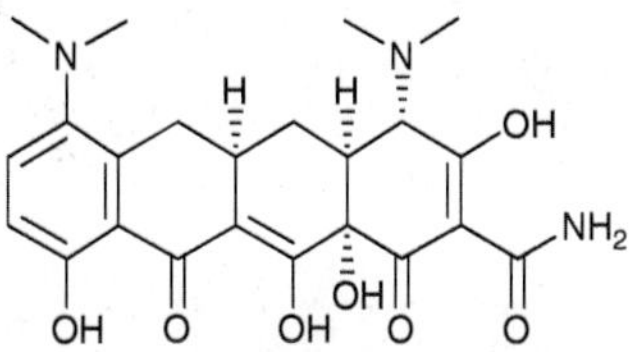

44; Minocycline

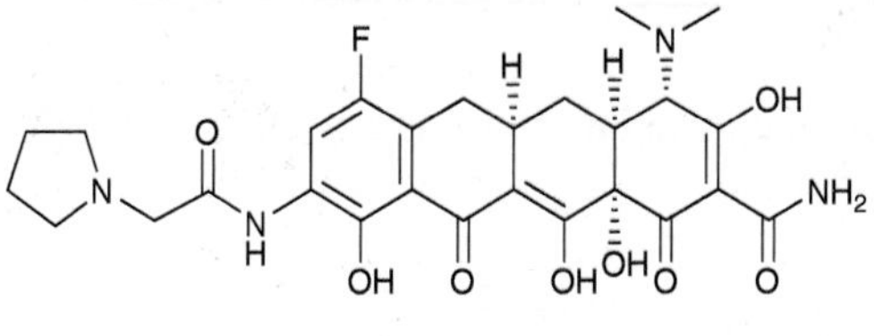

45; Eravacycline

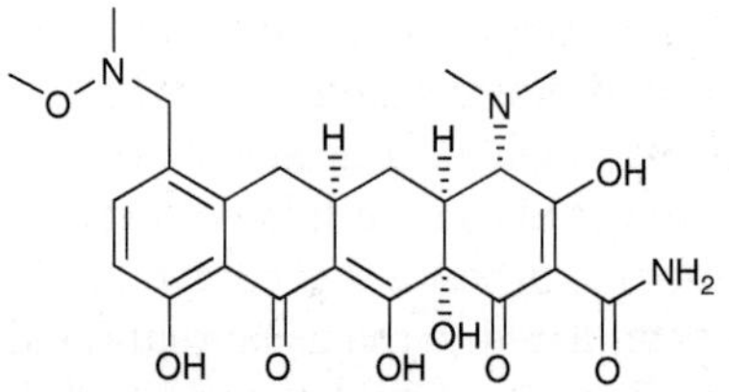

46; Sarecycline

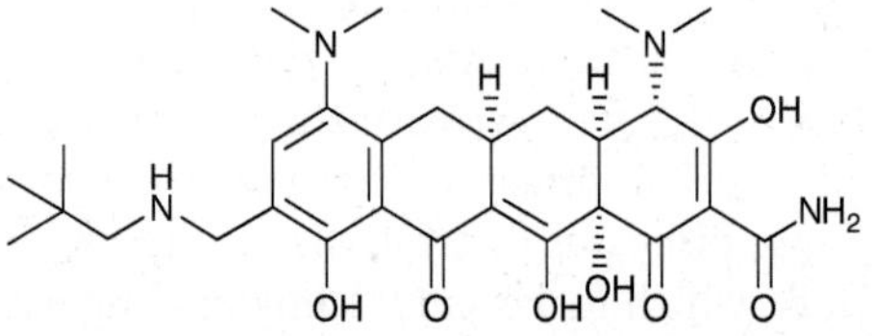

47; Omadacycline

14.7 Erythromycin Macrolides

The basic compound erythromycin (**48**) was first reported by McGuire et al. (1952) following its discovery three years earlier. Its biosynthesis and *in vitro* establishment of the biosynthetic process were well described by workers at Abbott and led to significant attempts in the early days of what might be considered the "genomic revolution," to produce modified macrolide ring structures and use these to obtain novel agents. Though the process worked well on a small scale, attempts at scale up were not successful. *A case of being too early even though the ideas were feasible.*

The basic molecule and early variants rapidly met resistance, usually from one or more of the following: macrolide efflux, rRNA methylation, or ribosomal mutation. Macrolides and ketolides bind to the 23S rRNA of the 50S ribosomal subunit, which blocks the peptide exit channel and stimulates the dissociation of peptidyl-tRNA from the ribosome during the translocation process. In addition, these molecules also had the capability to interact with the precursors of a partially assembled 50S subunit, plus being capable of inhibiting the complete formation of bacterial ribosomes with the unassembled precursor particles undergoing nucleolytic degradation.

14.7.1 Recent Semisynthetic Macrolides

Although no new macrolides that were produced directly from nature have been utilized in the past 30 years, there have been some significant "modified macrolides" that have been successful. Only three will be mentioned, mainly because, even years after their original approval, they are still under active development in major countries. These will be discussed individually.

The first is clarithromycin (**49**), which was launched in 1990 in Europe and in 1991 in the United States, and it can be generically named as 6-*O*-methyl erythromycin A, with its initial synthesis reported in 1984 (Morimoto et al. 1984). Currently over 100 trials at phase IV are listed in the NIH website with one, NCT03516669, actively recruiting for study on a treatment for *Helicobacter pylori* infection in Taiwan, while others at phase III are recruiting in Egypt and Korea – just to give examples of the areas of the world that are currently involved.

The second is azithromycin (**50**). A very interesting modification was made by Pliva, in what was then Yugoslavia, wherein the 14-ring base macrolide was converted to a 15-ring azalide, chemically named as 9-deoxo-9*a*-aza-9*a*-methyl-9*a*-homoerythromycin A. It was put into clinical use in 1988 in Yugoslavia and was then licensed to Pfizer and approved in the United States in 1991. As with clarithromycin, the NIH clinical trial database currently has 100 studies listed at phase IV, with 81 at phase II, including trials directed at pulmonary tuberculosis and potential use for treatment post tick bites.

Solithromycin (**51**), which is currently at phase III in Japan for the treatment of community acquired bacterial pneumonia (CABP), is possibly the most "bioactive" erythromycin-derived molecule, but it was pulled from studies in the EU and the United States due to problems associated with the formulation and delivery systems in 2016. The compound was then licensed to Fujifilm Toyama Chemical, and as mentioned above, it is currently at Phase III in Japan, with code numbers JapicCTI-163438, JapicCTI-163439, JapicCTI-163467, and JapicCTI-173733. So in due course, this potent antibiotic might be approved for use.

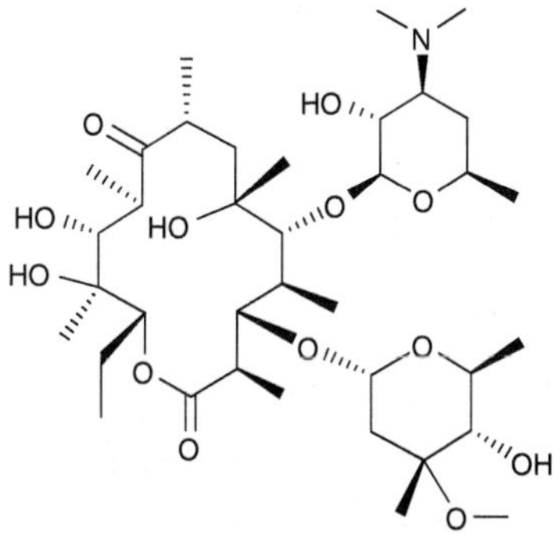

48; Erythromycin

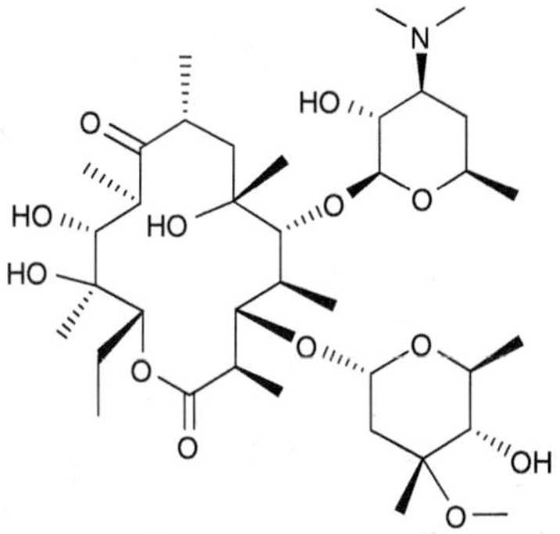

49; Clarithromycin

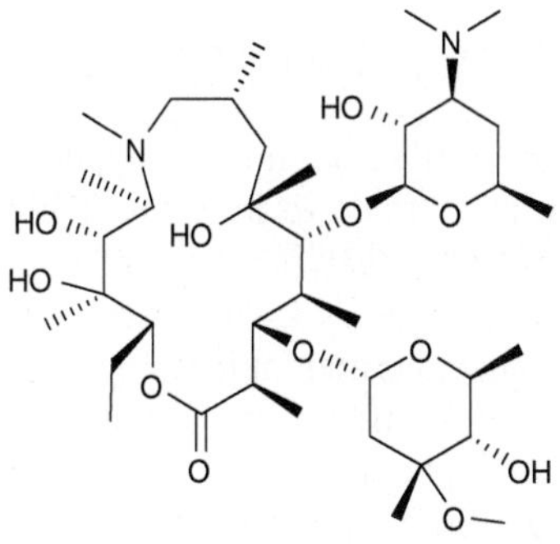

50; Azithromycin

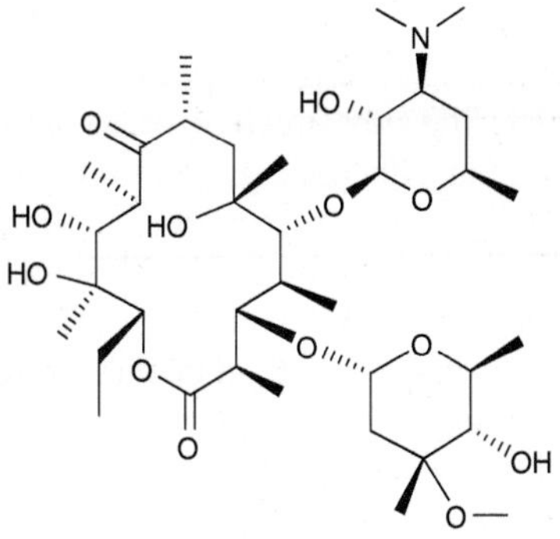

51; Solithromycin (Phase III Japan)

14.8 Current Methods of "Discovering Novel Antibiotics"

14.8.1 Introduction

As can be seen from the earlier discussions, companies and, to some extent, governments and academic organizations have spent very significant time and money in "incremental adjustments to known chemical skeletons" and have been quite successful in doing so, but in most cases, the molecules that have been approved, or are in late-stage clinical trials, are now relatively narrow-spectrum antibiotics. It is interesting that the molecules that are now being "modified" are "chemical cousins" of the base molecules that in the 1ate 1960s to middle 1980s were usually discarded by pharmaceutical management due to being narrow spectrum (either Gram-positive or Gram-negative only, or, in certain cases, only one particular genus and species).

The current methods that are being applied can be broadly subdivided into the following categories:

1) Genomics (whole cell).
2) Isolated genomics (cell free).
3) New sources (including older ones that are being reinvestigated).
4) Baiting for microbes.
5) Use of elicitors, either chemical or biological.

14.8.2 Initial Rate-Limiting Step (Irrespective of Methods)

The initial overriding principle, in all searching using old or new techniques for novel or even old known agents, is the rapid identification of any candidate, followed by the use of a decision tree by natural product chemists regarding the "value" from a chemical viewpoint as to whether a compound should be worked on or has been seen before. This process is called "dereplication," formerly known as "classification," and the faster that a chemist can decide as to whether the compound is known, or from a class that they have no interest in, the faster the subsequent process of discovery.

What has effectively revolutionized this aspect of the search is the application of mass spectroscopy to separation methods, coupled to freely available databases. This melding of techniques has permitted researchers in academia or government (who publish relatively rapidly) and users in industry who usually do not, other than in patents, to make a conscious and rapid decision as to whether or not to pursue a particular isolate/fraction from a chemical aspect. Later work with microbiologists would then give sufficient information to make decisions as to "go or no go" depending upon spectrum of activity, perhaps rapid *in vivo* assays.

14.8.3 Genomic Analyses of Whole Microbes

Possibly the two most influential papers in the last few years showing the possibilities for using genomic information described the total sequences of *Streptomyces coelicolor* (Challis 2014) and *Streptomyces avermitilis* (Ikeda et al. 2014). Though neither of these papers were the originals from the Hopwood group and the Omura group, respectively, they show the potential that could be realized from studies of the complete genome of an antibiotic-producing microbe. What was surprising in the initial studies was the number of potential biosynthetic gene clusters (BGCs) that were present but not "activated" under normal conditions. Inspection of these two papers demonstrated how this information led to the spread of information once the genetic sequences of microbes were regularly analyzed.

14.8.3.1 Current Processes

When the cost of complete and accurate sequencing of microbes dropped to somewhere in the order of US$5000 (it is now close to US$600–1000), it became reasonable for well-funded academic groups, particularly in the United States, to couple their own genomic studies on microbes isolated from varied sources to techniques known from the very early days of using varied fermentation conditions or, in some cases, to express the relevant BGC in a surrogate host and analyze the results using current mass spectral techniques.

Significant recent papers that cover these types of processes are as follows, though due to page restraints, they will not be analyzed in any detail. For studies related to activation of silent gene clusters, the following are worth consulting: Rutledge and Challis (2015), Ziemert et al. (2016), Adamek et al. (2017), Genilloud (2018), and Hug et al. (2018). In particular, as it was an early published example of sequencing and analyzing data from a large collection of actinomycetes at the USDA's old NRRL group, the paper by Doroghazi et al., should definitely be consulted, as it includes data from over 300 previously unpublished streptomycete genomes and shows the potential power of mass spectral information (Doroghazi et al. 2014).

Subsequent operations, particularly by the mass spectroscopy groups at the University of California, San Diego, working with researchers at the Scripps Institute of Oceanography, have led to the establishment of web-based information systems that cover not only microbes but also members of the Eukaryota. As in the previous section, these are not annotated but they are definitely worth reading and using (Wang et al. 2016; Aksenov et al. 2017; Gonzalez et al. 2018; Mohimani et al. 2018; Nothias et al. 2018).

14.8.4 Isolated Genomics

Are there academic groups and their spin-offs, aside from those utilizing very large-scale genomic analyses such as Warp Drive, Bio, the Cambridge biotech launched six years ago by Third Rock Ventures in partnership with the French drug giant Sanofi, that are aiming to identify nominal biosynthetic gene clusters related to potential antibiotic structures, working in this area?

The answer is definitely yes. One group that decided not to follow the whole genome route is led by Sean Brady at the Rockefeller University. In 2014, the Brady group published an article on metagenomic discovery methods in *Current Opinion in Microbiology* (Charlop-Powers et al. 2014). In this paper they described the then current methods for using metagenomic studies to obtain information without isolation. Then in 2016, they demonstrated that parks in New York City were a prolific source of natural product biosynthetic diversity (Charlop-Powers et al. 2016) and that they could find gene encoding in the biosynthesis of nonribosomal peptides and polyketides, which included a potentially novel calcium-dependent antibiotic (CDA).

This initial discovery of a CDA was followed up by the Brady group in a report in 2018, where they demonstrated the novel natural products, malacidin A (**52**) and B (**53**), following heterologous expression in *Streptomyces albus* (Hover et al. 2018). These are dissimilar in structure to the known CDA lipopeptide daptomycins and did not contain the Ca-binding motif, "Asp-X-Asp-Gly," but still required Ca^{2+} for activity against Gram-positive bacteria.

Using a different approach, this time using data from specific biosynthetic clusters and then using small peptide synthetic techniques, the same group reported in 2018 (Chu et al. 2018) further work from a discovery that they first reported in 2016 (Chu et al. 2016). In that initial report, they published a process (working from metagenomic data) that they named "syn-BNP". This acronym stands for "synthetic–bioinformatic natural products." Syn-BNPs are not intended to be exact copies of the isolated NPs but are analogues that structurally resemble the original agent; thus, using the lexicon of Newman and Cragg, these would be "NDs" (Newman and Cragg 2016). The base humimycin A structure (not shown) came from a study of the human microbiome and though not an antibiotic in its own right, it is in fact a "flippase" inhibitor that, when co-administered with a β-lactam such as carbenicillin (**54**), can overcome, to some extent, MRSA and *Enterococcus faecalis* (VRE-resistant) microbes. When the Brady group modified the *N*-acylated 7-mer linear peptide (humimycin A, 1S, **55**) by systematically changing the amino acid sequence and "adjusting the *N*-acyl group)," they developed humimycin 17S (**56**) that when co-administered with carbenicillin, inhibited the growth of highly resistant MRSA strains, including some that were also vancomycin resistant (Chu et al. 2018).

A second series of compounds that this time "descended" from studies from a bioinformatic analysis of nonribosomal peptide synthetase gene clusters led to the "syn-BNP-derived" paenimucillin A (**57**) and B (**58**) (Vila-Farres et al. 2018). Initially these were effectively only Gram-positive active, but by modification of the pair, paenimucillin C (**59**) was synthesized and using a rat open cutaneous wound model, infected with the ESKAPE pathogen *A. baumannii* showed no reinfection after twice daily treatment. It was recently announced that Genentech has partnered with Lodo Therapeutics, a company formed by Sean Brady with the aim of further expanding the possibilities of this type of "isolated genomics studies."

14.8.5 New Sources (and Old Ones?) for Investigation

What is of significance is that almost all of the work reported above on BGC analyses and expression deal with products from Gram-positive organisms (mainly from the actinomycetales). However, as mentioned by the Challis group in an excellent recent review in *Natural Product Reports*, Gram-negative organisms also produce bioactive secondary metabolites, so this major group of organisms should not be forgotten (Masschelein et al. 2017). To add to this, it is usually forgotten that fungi are also very prolific producers of bioactive metabolites because cephalosporins were isolated from fungi. In 2017, the Keller group at Wisconsin published an excellent review demonstrating a "scalable platform" to identify fungal secondary metabolites and their gene clusters. This is a paper that should definitely be read (Clevenger et al. 2017).

In addition, it would be remiss to ignore the very interesting work that has been performed over the years by the Cichewicz group in Oklahoma, investigating molecules produced by organisms from "strange places" (Motley et al. 2017). Currently the group has over 5000 fungi collected and fermented together with a significant number of novel compounds that are actively investigated. A substantial number are direct isolation, but some are from use of epigenetic modifiers.

14.8.6 "Baiting" for Microbes

So what is meant by this phrase? A technique used in working with fungi and some other microbes is to effectively establish a procedure that permits microbes that are present, usually in soils and muds, to create colonies that can then be seen. These are quite different from the "as yet uncultivatable" microbes that require isolation of a single cell, followed by expression of the total genome from that one microbe (including some that will never be isolated as free microbes). Further information on the latter sources was presented by Newman in a recent review (Newman 2018).

However, the major recent discovery was that of the peptide texiobactin (**60**) reported by the Lewis group in 2015 (Ling et al. 2015). They devised a technique that was a variation on earlier "baiting techniques," such as the single cell fermentation systems used by "One Cell" and then by Diversa in the mid-1990s. This was brought up to date using miniaturization systems, and their system is now known as the "iChip." Using their system, they obtained from a previously unknown soil microbe, the peptide known as texiobactin (**60**). As of early January 2019, their original paper has over 750 citations in the Scopus database and has effectively generated "an industry" in synthesizing variations on the initial structure including agents active both *in vitro* and *in vivo* against multiply resistant microbes (Guo et al. 2018; Parmar et al. 2018; Zong et al. 2018).

Obviously, the parent molecule or a closely related one made by synthesis will undergo clinical trials in due course, as in the original paper, the Lewis group reported no resistance evolving even after 25 passages. This is an example of where it may be more difficult to express the BGC in a surrogate host than to use modern synthetic techniques. Such techniques as shown earlier in the discussion of the Brady group can permit production on a significant scale under the required cGMP conditions for clinical trials. The important lesson however is that without the initial active lead, no chemistry would exist.

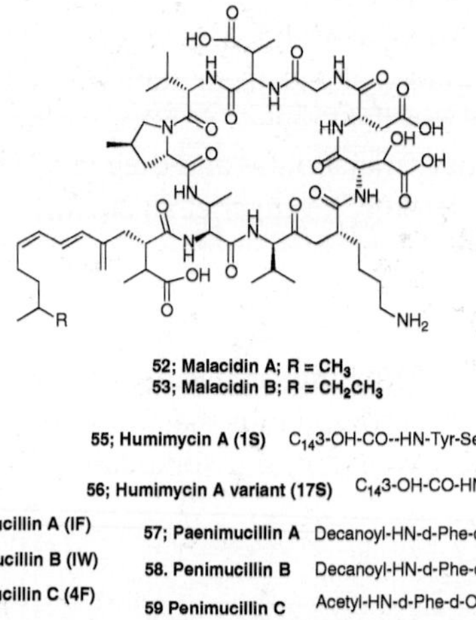

52; Malacidin A; R = CH_3
53; Malacidin B; R = CH_2CH_3

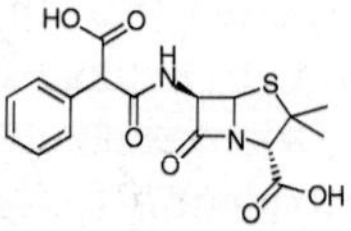

54; Carbenicillin

55; Humimycin A (1S) C_{14}3-OH-CO--HN-Tyr-Ser-d-Tyr-Phe-D-Thr-Val-Val-CO_2H

56; Humimycin A variant (17S) C_{14}3-OH-CO-HN-Trp-Ser-d-Trp-Trp-d-Thr-Val-Val-CO_2H

4a; Paenimucillin A (IF) **57; Paenimucillin A** Decanoyl-HN-d-Phe-d-Orn-d-Ile-d-Ser-d-Phe-Ser-Ser-d-Orn-Phe-Ser-Val-d-Ile-Orn-CO_2H

4b; Paenimucillin B (IW) **58. Penimucillin B** Decanoyl-HN-d-Phe-d-Orn-d-Ile-d-Ser-d-Trp-Ser-Ser-d-Orn-Trp-Ser-Val-d-Ile-Orn-CO_2H

4c; Paenimucillin C (4F) **59 Penimucillin C** Acetyl-HN-d-Phe-d-Orn-d-Ile-d-Ser-d-Phe-Ser-Ser-d-Orn-Phe-Ser-Val-d-Ile-Orn-CO_2H

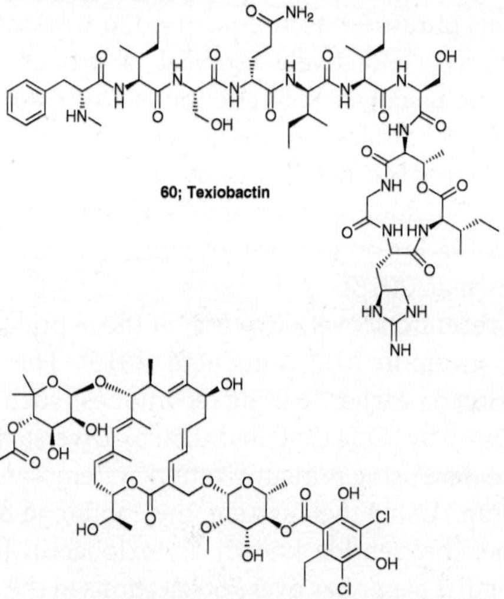

60; Texiobactin

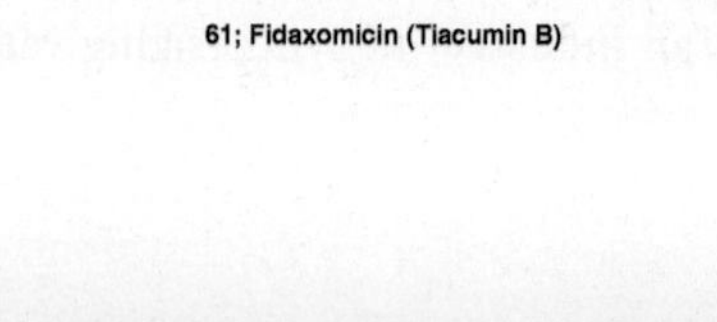

61; Fidaxomicin (Tiacumin B)

14.8.7 Use of Elicitors

The concept that microbes “talk to each other chemically” is well established, but what is not so well known is that these signals occur between all “types of organisms,” including talk between *Homo sapiens* and the microbiome. Two excellent papers demonstrating this type of interspecies “talk” are the publication by Kenny and Balskus, in *Chemical Society Reviews* (Kenny and Balskus 2018), and the review by Pupo’s group in Brazil in the same journal showing plant–microbe interactions (Chagas et al. 2018).

There have been a number of excellent reviews showing how the addition of epigenetic modifiers will significantly alter the metabolome of microbes, but perhaps the “piece de resistance” in this topic was the 2016 report by the Seyedsayamdost group at Princeton, on the very large number of previously unknown molecules discovered when the Gram-negative microbe *Burkholderia thailandensis* E264 was treated with trimethoprim. This treatment yielded over 100 previously unknown metabolites (Okada et al. 2016). In the following year, the same group published an excellent review on expression of silent gene clusters by small molecules (Okada and Seyedsayamdost 2017) with another review on the same topic in 2018 (Mao et al. 2018).

Though not using defined small molecules, the 2013 paper by Rateb et al. demonstrated the potential for co-culture of fungi with bacteria in order to produce previously unknown metabolites (Rateb et al. 2013), and an earlier example showing how bacterial–fungal interactions produce different molecules from the individually fermented microbes was given in the 2011 paper by Zuck et al. (2011).

14.9 Conclusions

Natural product-derived metabolites, either as the natural product or as modifications usually to improve pharmacodynamics and/or to increase bioactivity and/or reduce toxicity, are still alive and well as leads to and even in recent cases, the antibiotic itself. Though not mentioned in the text until now, the antibiotic fidaxomicin (**61**), though having a very long and convoluted progression to clinical use, was approved in 2011 by the FDA as a treatment for *C. difficile* infections. Little to no resistance to this molecule has been observed.

However, in the post-genomics age, which is where we are now, the initial euphoria about the potential to express BGCs once identified as a result of modern rapid and cheap sequencing; “new generation sequencing,” as it is termed, has not yet proven to be a viable source of leads to develop directly. However, the use of genomic and metabolomic techniques to identify “isolated genomic traces” (for want of a better term) as practiced by the Brady group, the use of “elicitors” be they natural or chemically derived and the interplay of mixed cultures, bodes well for the discovery and development of *as yet unrecognized agents* that may well be more resistant in the constant battle between microbes and antibiotics. The "reach back" to agents such as the colistins and the other peptide-based molecules may also prove to be of significant value.

14.9.1 Funding?

I have left the major hurdle to last, and that is how to obtain sufficient funding to find and then develop these compounds. Currently, the approval agencies in the West require large numbers of patients for clinical trials at the phase III level. These are designed to demonstrate that the agent under consideration is better than the alternatives that are already available. With the advent of agents that are only active against specific infections rather than being broad spectrum, it behooves the approval agencies to rethink their requirement and to realize that antibiotics, by their very nature, have a low return on investment, which is not what the major pharmaceutical companies wish to hear and may well account tor why the major companies with the exception of Genentech/Roche have effectively abandoned the search.

What also adds fuel to that fire is that no de novo combinatorially derived chemical compound has even made it to phase I trials as an antibiotic. This does not include agents based upon a natural product skeleton. Thus, the massive sums of money expended on the production of "regular" combi-chem libraries have not worked as generators of antibiotic leads. One should read the 2007 review from GSK authors demonstrating how 70 isolated targets and their combi-chem collection were used; none reached preclinical stages (Payne et al. 2007).

14.9.2 The "Take-Home Lesson"

Natural product or compounds based thereon do work, unfortunately currently there are few companies (medium or small) left to search for these needed items.

References

Abraham, E.P. and Newton, G.G.F. (1956). Experiments on the degradation of cephalosporin C. *Biochem. J.* 62: 658–665.

Adamek, M., Spohn, M., Stegmann, E., and Ziemert, N. (2017). Mining bacterial genomes for secondary metabolite gene clusters. In: *Antibiotics: Methods and Protocols*, Methods in Molecular Biology, vol. 1520 (ed. P. Sass), 23–47. New York: Springer.

Aksenov, A.A., da Silva, R., Knight, R. et al. (2017). Global chemical analysis of biology by mass spectrometry. *Nat. Rev. Chem.* 1: 0054. https://doi.org/10.1038/s41570-017-0054.

Baltz, R.H., Miao, V., and Wrigley, S.K. (2005). Natural products to drugs: daptomycin and related lipopeptide antibiotics. *Nat. Prod. Rep.* 22: 717–741.

Battersby, A.R. and Craig, L.C. (1952). The chemistry of tyrocidine. I. Isolation and characterization of a single peptide. *J. Chem. Soc.* 74: 4019–4023.

Belozersky, A.N. and Passhina, T.S. (1944). Chemistry of gramacidin S. *Lancet* 716–717.

Berkowitz, B.A., Bevins, C.L., and Zasloff, M.A. (1990). Magainins: a new family of membrane-active host defense peptides. *Biochem. Pharmacol.* 39: 625–629.

Brotzu, G. (1948). Ricerche su di un nuovo antibiotico. Lav. Inst. Igiene Cagliari1-11.

Chagas, F.O., de Cassia Pessotti, R., Caraballo-Rodriguez, A.M., and Pupo, M.T. (2018). Chemical signaling involved in plant-microbe interactions. *Chem. Soc. Rev.* 47: 1652–1704.

Challis, G.L. (2014). Exploitation of the *Streptomyces coelicolor* A3(2) genome sequence for discovery of new natural products and biosynthetic pathways. *J. Ind. Microbiol. Biotechnol.* 41: 219–232. https://doi.org/10.1007/s10295-013-1383-2.

Charlop-Powers, Z., Milshteyn, A., and Brady, S.F. (2014). Metagenomic small molecule discovery methods. *Curr. Opin. Microbiol.* 19: 70–75.

Charlop-Powers, Z., Pregitzer, C.C., Lemetre, C. et al. (2016). Urban park soil microbiomes are a rich reservoir of natural product biosynthetic diversity. *Proc. Natl. Acad. Sci. U. S. A.* 113: 14811–14816.

Cho, J.C., Childs-Kean, L.M., Zmarlicka, M.T., and Crotty, M.P. (2018). Return of the tetracyclines: Omadacycline, a novel aminomethylcycline antimicrobial. *Drugs Today* 54: 209–217.

Choi, S., Isaacs, A., Clements, D. et al. (2009). De novo design and in vivo activity of conformationally restrained antimicrobial arylamide foldamers. *Proc. Natl. Acad. Sci. U. S. A.* 106: 6968–6973.

Chu, J., Vila-Farres, X., Inoyama, D. et al. (2016). Discovery of MRSA active antibiotics using primary sequence from the human microbiome. *Nat. Chem. Biol.* 12: 1004–1006.

Chu, J., Vila-Farres, X., Inoyama, D. et al. (2018). Human microbiome inspired antibiotics with improved β-lactam synergy against MDR *Staphylococcus aureus*. *ACS Infect. Dis.* 4: 33–38.

Clevenger, K.D., Bok, J.W., Ye, R. et al. (2017). A scalable platform to identify fungal secondary metabolites and their gene clusters. *Nat. Chem. Biol.* 13: 895–901. https://doi.org/10.1038/nchembio.2408.

Crowley, B.M., Mori, Y., McComas, C.C. et al. (2004). Total synthesis of the ristocetin aglycon. *J. Am. Chem. Soc.* 126: 4310–4317.

Davido, B., Leplay, C., Bouchand, F. et al. (2017). Oral teicoplanin as an alternative first-line regimen in *Clostridium difficile* infection in elderly patients: a case series. *Clin. Drug Investig.* 37: 699–703.

D'Costa, V.M., King, C.E., Kalan, L. et al. (2011). Antibiotic resistance is ancient. *Nature* 477: 457–461.

Doroghazi, J.R., Albright, J.C., Goering, A.W. et al. (2014). A roadmap for natural product discovery based on large-scale genomics and metabolomics. *Nat. Chem. Biol.* 10: 963–968. https://doi.org/10.1038/nCHeMBIO.1659.

Fernandes, R., Amador, P., and Prudêncio, C. (2013). β-Lactams: chemical structure, mode of action and mechanisms of resistance. *Rev. Med. Microbiol.* 24: 7–17.

Flamm, R.K., Rhomberg, P.R., Simpson, K.M. et al. (2015). *In vitro* spectrum of pexiganan activity when tested against pathogens from diabetic foot infections and with selected resistance mechanisms. *Antimicrob. Agents Chemother.* 59: 1751–1754. https://doi.org/10.1128/AAC.04773-14.

Galani, I. (2014). Plazomicin. *Drugs Future* 39: 25–35. https://doi.org/10.1358/dof.2014.39.1.2095267.

Gause, G.F. (1946). Gramicidin S. *Lancet* 46–47.

Gause, G.F. and Brazhnikova, M.G. (1944). Gramicidin S and its use in the treatment of infected wounds. *Nature* 154: 703.

Genilloud, O. (2018). Mining actinomycetes for novel antibiotics in the omics era: are we ready to exploit this new paradigm? *Antibiotics* 7: 85. https://doi.org/10.3390/antibiotics7040085.

Ghosh, M., Lin, Y.-M., Miller, P.A. et al. (2018). Siderophore conjugates of daptomycin are potent Iinhibitors of carbapenem resistant strains of *Acinetobacter baumannii*. *ACS Infect. Dis.* 4: 1529–1535.

Ghosh, M., Miller, P.A., Mollmann, U. et al. (2017). Targeted antibiotic delivery: selective siderophore conjugation with daptomycin confers potent activity against multidrug resistant *Acinetobacter baumannii* both *in vitro* and *in vivo*. *J. Med. Chem.* 60: 4577–4583.

Gonzalez, A., Navas-Molina, J.A., Kosciolek, T. et al. (2018). Qiita: rapid, web-enabled microbiome meta-analysis. *Nat. Methods* 15: 796–798. https://doi.org/10.1038/s41592-018-0141-9.

Griffith, R.S. (1981). Introduction to vancomycin. *Rev. Infect. Dis.* 3: S200–S204.

Guerra, W., Silva-Caldeira, P.P., Terenzi, H., and Pereira-Maia, E.C. (2016). Impact of metal coordination on the antibiotic and non-antibiotic activities of tetracycline-based drugs. *Coord. Chem. Rev.* 327–328: 188–199.

Guo, C., Mandalapu, D., Ji, X. et al. (2018). Chemistry and biology of teixobactin. *Chem. A Eur. J.* 24: 5406–5422.

Harris, C.M. and Harris, T.M. (1982). Structure of the glycopeptide antibiotic vancomycin. Evidence for an asparagine residue in the peptide. *J. Am. Chem. Soc.* 104: 4293–4295.

Heidary, M., Khosravi, A.D., Khoshnood, S. et al. (2018). Daptomycin. *J. Antimicrob. Chemother.* 73: 1–11.

Holtzel, A., Schmid, D.G., Nicholson, G.J. et al. (2002). Arylomycins A and B, new biaryl-bridged lipopeptide antibiotics produced by *Streptomyces* sp. Tu 6075. II. Structure elucidation. *J. Antibiot.* 55: 571–577.

Hotchkiss, R.D. and Dubois, R.J. (1940). Fractionation of the bacterial agent from cultures of a soil bacillus. *J. Biol. Chem.* 132: 791–792.

Hover, B.M., Kim, S.-H., Katz, M. et al. (2018). Culture-independent discovery of the malacidins as calcium-dependent antibiotics with activity against multidrug-resistant Gram-positive pathogens. *Nat. Microbiol.* 3: 415–422.

Hu, Y., Liu, A., Vaudrey, J. et al. (2015). Combinations of β-lactam or aminoglycoside antibiotics with plectasin are synergistic against methicillin-sensitive and methicillin-resistant *Staphylococcus aureus*. *PLoS One* 10: e0117664. https://doi.org/10.1371/journal.pone.0117664.

Hug, J.J., Bader, C.D., Remškar, M. et al. (2018). Concepts and methods to access novel antibiotics from actinomycetes. *Antibiotics* 7: 44. https://doi.org/10.3390/antibiotics7020044.

Hunter, P.A. and Castaner, J. (2001). GAR-936. Tetracycline antibiotic. *Drugs Future* 26: 851–858. https://doi.org/10.1358/dof.2001.026.09.634733.

Ikeda, H., Shin-ya, K., and Omura, S. (2014). Genome mining of the *Streptomyces avermitilis* genome and development of genome-minimized hosts for heterologous expression of biosynthetic gene clusters. *J. Ind. Microbiol. Biotechnol.* 41: 233–250. https://doi.org/10.1007/s10295-013-1327-x.

Jiménez, A., García, P., de la Puente, S. et al. (2018). A novel class of cationic and non-peptidic small molecules as hits for the development of antimicrobial agents. *Molecules* 23: E1513. https://doi.org/10.3390/molecules23071513.

Kenny, D.J. and Balskus, E.P. (2018). Engineering chemical interactions in microbial communities. *Chem. Soc. Rev.* 47: 1705–1729.

Kulanthaivel, P., Kreuzman, A.J., Strege, M.A. et al. (2004). Novel lipoglycopeptides as inhibitors of bacterial signal peptidase I. *J. Biol. Chem.* 279: 36250–36258.

Li, Q. and Seiple, I.B. (2017). Modular, scalable synthesis of group A streptogramin antibiotics. *J. Am. Chem. Soc.* 139: 13304–13307. https://doi.org/10.1021/jacs.7b08577.

Ling, L.L., Schneider, T., Peoples, A.J. et al. (2015). A new antibiotic kills pathogens without detectable resistance. *Nature* 517: 455–459.

Luther, A., Bisang, C., and Obrecht, D. (2018). Advances in macrocyclic peptide-based antibiotics. *Bioorg. Med. Chem.* 26: 2850–2858.

Machado, D., Antunes, J., Simões, A. et al. (2018). Contribution of efflux to colistin heteroresistance in a multidrug resistant *Acinetobacter baumannii* clinical isolate. *J. Med. Microbiol.* 67: 740–749. https://doi.org/10.1099/jmm.0.000741.

Mao, D., Okada, B.K., Wu, Y. et al. (2018). Recent advances in activating silent biosynthetic gene clusters in bacteria. *Curr. Opin. Microbiol.* 46: 156–163. https://doi.org/10.1016/j.mib.2018.05.001.

Markley, J.L. and Wencewicz, T.A. (2018). Tetracycline-inactivating enzymes. *Front. Microbiol.* 9: 1058. https://doi.org/10.3389/fmicb.2018.01058.

Masschelein, J., Jenner, M., and Challis, G.L. (2017). Antibiotics from Gram-negative bacteria: a comprehensive overview and selected biosynthetic highlights. *Nat. Prod. Rep.* 34: 712–783. https://doi.org/10.1039/c7np00010c.

McGuire, J.M., Bunch, R.L., Anderson, R.C. et al. (1952). Ilotycin, a new antibiotic. *Antibiot. Chemother.* 2: 281–283.

Mensa, B., Howell, G.L., Scott, R., and DeGrado, W.F. (2014). Comparative mechanistic studies of brilacidin, daptomycin, and the antimicrobial peptide LL16. *Antimicrob. Agents Chemother.* 58: 5136–5145. https://doi.org/10.1128/AAC.02955-14.

Mohimani, H., Cao, L., Gurevich, A. et al. (2018). Dereplication of microbial metabolites through database search of mass spectra. *Nat. Commun.* 9: 4035. https://doi.org/10.1038/s41467-018-06082-8.

Morimoto, S., Takahashi, Y., Watanabe, Y., and Omura, S. (1984). Chemical modification of erythromycins. I. Synthesis and antibacterial activity of 6-*O*-methylerythromycins A. *J. Antibiot.* 37: 187–189.

Motley, J.L., Stamps, B.W., Mitchell, C.A. et al. (2017). Opportunistic sampling of roadkill as an entry point to accessing natural products assembled by bacteria associated with nonanthropoidal mammalian microbiomes. *J. Nat. Prod.* 80: 598–608.

Müsken, M., Pawar, V., Schwebs, T. et al. (2018). Breaking the vicious cycle of antibiotic killing and regrowth of biofilm-residing *Pseudomonas aeruginosa*. *Antimicrob. Agents Chemother.* 62: e01635-18. https://doi.org/10.1128/AAC.01635-18.

Nelson, M.L. (2001). The chemistry and cellular biology of the tetracyclines. In: *Tetracyclines in Biology, Chemistry and Medicine* (ed. M. Nelson, W. Hillen and A.A. Greenwald), 3–63. Switzerland: Birkauser Verlag.

Newman, D.J. (2018). From large-scale collections to the potential use of genomic techniques for supply of drug candidates. *Front. Mar. Sci.* 5: 401. https://doi.org/10.3389/fmars.2018.00401.

Newman, D.J. and Cragg, G.M. (2016). Natural products as sources of new drugs from 1981 to 2014. *J. Nat. Prod.* 79: 629–661. https://doi.org/10.1021/acs.jnatprod.5b01055.

Newton, G.G.F. and Abraham, E.P. (1955). Cephalosporin C, a new antibiotic containing sulphur and D-α-aminoadipic acid. *Nature* 175: 548. https://doi.org/10.1038/175548a0.

Newton, G.G.F. and Abraham, E.P. (1956). Isolation of cephalosporin C, a penicillin-like antibiotic containing D-α-aminoadipic acid. *Biochem. J.* 62: 651–658.

Nothias, L.-F., Nothias-Esposito, M., da Silva, R. et al. (2018). Bioactivity-based molecular networking for the discovery of drug leads in natural product bioassay-guided fractionation. *J. Nat. Prod.* 81: 758–767. https://doi.org/10.1021/acs.jnatprod.7b00737.

Okada, B.K. and Seyedsayamdost, M.R. (2017). Antibiotic dialogues: induction of silent biosynthetic gene clusters by exogenous small molecules. *FEMS Microbiol. Rev.* 41: 19–33. https://doi.org/10.1093/femsre/fuw035.

Okada, B.K., Wu, Y., Mao, D. et al. (2016). Mapping the trimethoprim-induced secondary metabolome of *Burkholderia thailandensis*. *ACS Chem. Biol.* 11: 2124–2130. https://doi.org/10.1021/acschembio.6b00447.

Okano, A., Isley, N.A., and Boger, D.L. (2017a). Peripheral modifications of [Ψ[CH2NH]Tpg4]vancomycin with added synergistic mechanisms of action provide durable and potent antibiotics. *Proc. Natl. Acad. Sci. U. S. A.* E5052–E5061. https://doi.org/10.1073/pnas.1704125114.

Okano, A., Isley, N.A., and Boger, D.L. (2017b). Total syntheses of vancomycin-related glycopeptide antibiotics and key analogues. *Chem. Rev.* 117: 11952–11993.

Okano, A., Nakayama, A., Schammel, A.W., and Boger, D.L. (2014). Total synthesis of [Ψ[C(=NH)NH]Tpg^4]vancomycin and its (4-chlorobiphenyl)methyl derivative: impact of peripheral modifications on vancomycin analogues redesigned for dual D-Ala-D-Ala and D-Ala-D-Lac binding. *J. Am. Chem. Soc.* 136: 13522–13525.

Okano, A., Nakayama, A., Wu, K. et al. (2015). Total syntheses and initial evaluation of [Ψ[C(=S)NH]Tpg4]vancomycin, [Ψ[C(=NH)NH]Tpg4]vancomycin, [Ψ[CH2NH]Tpg4]vancomycin, and their (4-chlorobiphenyl)methyl derivatives: synergistic binding pocket and peripheral modifications for the glycopeptide antibiotics. *J. Am. Chem. Soc.* 137: 3693–3704.

Parmar, A., Lakshminarayanan, R., Iyer, A. et al. (2018). Design and syntheses of highly potent teixobactin analogues against *Staphylococcus aureus*, methicillin-resistant *Staphylococcus aureus* (MRSA), and vancomycin-resistant *Enterococci* (VRE) *in vitro* and *in vivo*. *J. Med. Chem.* 61: 2009–2017.

Payne, D.J., Gwynn, M.N., Holmes, D.J., and Pompliano, D.L. (2007). Drugs for bad bugs: confronting the challenges of antibacterial discovery. *Nat. Rev. Drug Discov.* 6: 28–39.

Petersen, P.J., Jacobus, N.V., Weiss, W.J. et al. (1999). In vitro and in vivo antibacterial activities of a novel gycylcycline, the 9-t-butylglycylamido derivative of minocycline (GAR-936). *Antimicrob. Agents Chemother.* 43: 738–744.

Popovic, N., Korac, M., Nesic, Z. et al. (2015). Oral teicoplanin for successful treatment of severe refractory *Clostridium difficile* infection. *J. Infect. Dev. Ctries.* 9: 1062–1067.

Rake, J.B., Gerber, R., Mehta, R.J. et al. (1986). Glycopeptide antibiotics: a mechanism-based screen employing a bacterial cell wall receptor mimetic. *J. Antibiot.* 39: 58–67.

Rateb, M.E., Hallyburton, I., Houssen, W.E. et al. (2013). Induction of diverse secondary metabolites in *Aspergillus fumigatus* by microbial co-culture. *RSC Adv.* 3: 14444–14450. https://doi.org/10.1039/c3ra42378f.

Ronn, M., Zhu, Z., Hogan, P.C. et al. (2013). Process R&D of eravacycline: the first fully synthetic fluorocycline in clinical development. *Org. Process Res. Dev.* 17: 838–845.

Rutledge, P.J. and Challis, G.L. (2015). Discovery of microbial natural products by activation of silent biosynthetic gene clusters. *Nat. Rev. Microbiol.* 13: 509–523. https://doi.org/10.1038/nrmicro3496.

Schatz, A., Bugie, E., and Waksman, S.A. (1944). Streptomycin, a new substance exhibiting antibiotic activity against Gram-positive and Gram-negative bacteria. *Proc. Soc. Exp. Biol. Med.* 55: 66–69.

Schimana, J., Gebhardt, K., Holtzel, A. et al. (2002). Arylomycins A and B, new biaryl-bridged lipopeptide antibiotics produced by *Streptomyces* sp. Tu 6075. I. Taxonomy, fermentation, isolation and biological activities. *J. Antibiot.* 55: 565–570.

Schneider, S. (2001). Proton and metal ion binding of tetracyclines. In: *Tetracyclines in Biology, Chemistry and Medicine* (ed. M. Nelson, W. Hillen and A.A. Greenwald), 65–104. Switzerland: Birkhauser Verlag.

Sergiev, P.G. (1944). Clinical use of Gramicidin S. *Lancet* 717–718.

Shearer, M.C., Actor, P., Bowie, B.A. et al. (1985). Aridicins, novel glycopeptide antibiotics. I. Taxonomy, produciton and biological activity. *J. Antibiot.* 38: 555–560.

Smith, P.A., Koehler, M.F.T., Girgis, H.S. et al. (2018). Optimized arylomycins are a new class of Gram-negative antibiotics. *Nature* 561: 189–194. https://doi.org/10.1038/s41586-018-0483-6.

Sum, P.E., Lee, V.J., Testa, R.T. et al. (1993). Glycylcyclines. I. A new generation of potent antibacterial agents through modification of 9-aminotetracyclines. *J. Med. Chem.* 37: 184–188.

Tew, G.N., Liu, D., Chen, B. et al. (2002). *De novo* design of biomimetic antimicrobial polymers. *Proc. Natl. Acad. Sci. U. S. A.* 99: 5110–5114.

Thakare, R., Dasgupta, A., and Chopra, S. (2018). Eravacycline for the treatment of patients with bacterial infections. *Drugs Today* 54: 245–254.

Umezawa, S., Takahashi, Y., Usi, T., and Tsuchiya, T. (1974a). Total synthesis of streptomycin. *J. Antibiot.* 28: 997–999.

Umezawa, S., Tsuchiya, T., Yamasaki, T. et al. (1974b). Total synthesis of of dihydrostreptomycin. *J. Am. Chem. Soc.* 96: 920–921.

Vila-Farres, X., Chu, J., Ternei, M.A. et al. (2018). An optimized synthetic-bioinformatic natural product antibiotic sterilizes multidrug-resistant *Acinetobacter baumannii*-infected wounds. *mSphere* 3: e00528-17.

Wan, Y., Stanovych, A., Gori, D. et al. (2018). β,γ-diamino acids as building blocks for new analogues of Gramicidin S: synthesis and biological activity. *Eur. J. Med. Chem.* 149: 122–128.

Wang, M., Carver, J.J., Phelan, V.V. et al. (2016). Sharing and community curation of mass spectrometry data with Global Natural Products Social Molecular Networking. *Nat. Biotechnol.* 34: 828–837. https://doi.org/10.1038/nbt.3597.

Weinstein, M.J., Marquez, J.A., Testa, R.T. et al. (1970). Antibiotic 6640, a new *Micromonospora*-produced aminoglycoside antibiotic. *J. Antibiot.* 23: 551–554.

Wright, G.D. and Poinar, H. (2012). Antibiotic resistance is ancient: implications for drug discovery. *Trends Microbiol.* 20: 157–159.

Xie, J., Pierce, J.G., James, R.C. et al. (2011). A redesigned vancomycin engineered for dual D-Ala-D-Ala and D-Ala-D-Lac binding exhibits potent antimicrobial activity against vancomycin-resistant bacteria. *J. Am. Chem. Soc.* 133: 13946–13949.

Yang, S., Sankar, K., Skepper, C.K. et al. (2017). Total synthesis of a key series of vinblastines modified at C4 that define the importance and surprising trends in activity. *Chem. Sci.* 8: 1560–1569.

Zasloff, M. (1987). Magainins, a class of antimicrobial peptides from *Xenopus* skin: isolation, characterization of two active forms, and partial cDNA sequence of a precursor. *Proc. Natl. Acad. Sci. U. S. A.* 84: 5449–5453.

Zasloff, M. (2016). Antimicrobial peptides: do they have a future as therapeutics? In: *Antimicrobial Peptides; Role in Human Health and Disease* (ed. J. Harder and J.M. Schroder), 147–154. Cham: Springer.

Zhao, F., Feng, Y., Lü, X. et al. (2017). IncP plasmid carrying colistin resistance gene *mcr-1* in *Klebsiella pneumoniae* from hospital sewage. *Antimicrob. Agents Chemother.* 61: e02229-16. https://doi.org/10.1128/AAC.02229-16.

Ziemert, N., Alanjary, M., and Weber, T. (2016). The evolution of gemone mining in microbes: a review. *Nat. Prod. Rep.* 33: 988–1005. https://doi.org/10.1039/C6NP00025H.

Zong, Y., Sun, X., Gao, H. et al. (2018). Developing equipotent teixobactin analogues against drug resistant bacteria and discovering a hydrophobic interaction between lipid II and teixobactin. *J. Med. Chem.* 61: 3409–3421.

Zuck, K.M., Shipley, S., and Newman, D.J. (2011). Induced produciton of *N*-formyl alkaloids from *Aspergillus fumigatus* by co-culture with *Streptomyces peucetius*. *J. Nat. Prod.* 74: 1653–1657. https://doi.org/10.1021/np200255f.

15

Bacteriophage Proteins as Antimicrobials to Combat Antibiotic Resistance

Hugo Oliveira, Luís D. R. Melo, and Sílvio B. Santos

Centre of Biological Engineering (CEB), Laboratório de Investigação em Biofilmes Rosário Oliveira (LIBRO), University of Minho, Braga, Portugal

15.1 Introduction

Sir Alexander Fleming, during his Nobel Prize acceptance lecture in 1945, already predicted the possibility of bacterial antibiotic resistance development, something that could be the result of negligent use of antibiotics by man (Fleming 1945).

Currently, antibiotic resistance is a massive public health challenge. According to the Centers for Disease Control and Prevention (CDC) in the United States, at least two million people get an antibiotic-resistant infection that causes at least 23 000 deaths annually. When bacteria develop mechanisms to survive and grow in the presence of the drugs that were designed to kill them, they are considered resistant. The problem worsens when these bacteria cause infections that become difficult or even impossible to treat due to their resistance. This leads to a high morbidity and mortality, as well as a prolongation of hospitalization, which ultimately increases the financial costs dispended for each patient. Also, the infected patient will have a higher workplace absenteeism, which leads to innumerous socioeconomic consequences. And yet to be determined are the costs related to psychological and emotional changes caused by pain and suffering (among other factors) (Scott 2009).

In 2017, the World Health Organization (WHO) issued a list of priority pathogens resistant to antibiotics, encouraging both the scientific community and pharmaceutical industries to develop new antimicrobials to combat these antimicrobial-resistant pathogens (WHO 2017). Unfortunately, the pipeline for the development of new antibacterial drugs is now virtually empty, particularly for the treatment of Gram-negative enteric bacteria (WHO 2014). This, coupled with the common appearance of resistance shortly after the

Antibiotic Drug Resistance, First Edition. Edited by José-Luis Capelo-Martínez and Gilberto Igrejas.

introduction of a new antibiotic and with the regulatory constraints to their use, requires the development of effective alternatives, a research that is still in the early stages (WHO 2014; Ruer et al. 2015).

During the last decades, bacteriophages (phages) gained attention in the West as alternatives to antibiotics. Phages are viruses that infect bacteria. Compared to antibiotic therapy, the use of phages has become very appealing: (i) they show great specificity and efficiency in lysing the host bacteria, without affecting the natural flora of the patient; (ii) phages are effective against antibiotic-resistant bacteria (Loc-Carrillo and Abedon 2011); (iii) as viruses of bacteria they are harmless for humans and animals; (iv) they replicate only when the host bacterium is present, allowing to increase their concentration where they are needed; (v) the costs of production and development are far lower than that of antibiotics (Loc-Carrillo and Abedon 2011); (vi) they are the most abundant entities on earth rendering easy to find a phage infecting a particular host. However, many regulatory hurdles still exist as regards the use of phages due to their viral nature, and still, bacteria can also acquire resistance to phages (Labrie et al. 2010).

Current advances in sequencing technologies and DNA manipulation introduce more insights on phage-derived proteins used for bacterial control in food, agriculture, and health (Santos et al. 2018). Though some applications involving phage-encoded proteins can be performed by the phage itself, the use of their proteins has advantages considering the regulatory issues and public acceptance. Phages have evolved for billions of years to recognize, multiply, and kill their hosts in a very competent and efficient manner. To understand the role of phage proteins and their potential as antimicrobials, it is important to depict the phage replication cycle (Figure 15.1).

The phage life cycle starts with the encounter of the phage and its host. Phages interact with bacterial cell surface through adsorption. In this process, phage proteins specifically recognize receptor structures on the bacterial cell. At this time point, phages may need the action of phage enzymes, such as depolymerases that degrade host capsular structures, to gain access to their receptors (Figure 15.1a) (Pires et al. 2016).

After meeting and recognizing the host, phages eject their DNA into the host cell in a process facilitated by phage enzymes, namely, VALs, which are part of the phage structure. These enzymes produce pores in the bacterial peptidoglycan, allowing the passage of the phage genetic material (Figure 15.1b) (Kutter and Sulakvelidze 2005). When the nucleic acid is inside the host cell, a decision has to be made depending on the phage nature: lysogenic or lytic pathway (Figure 15.1c). In the case of a temperate phage, the lysogenic pathway may be chosen leading to the integration of the phage nucleic acid into the bacterial chromosome as a prophage (Figure 15.1d). The phage will replicate along with the bacterial genetic material passing to the bacterial daughters (Figure 15.1e) until a specific stimulus triggers the lytic pathway. When this happens, or after

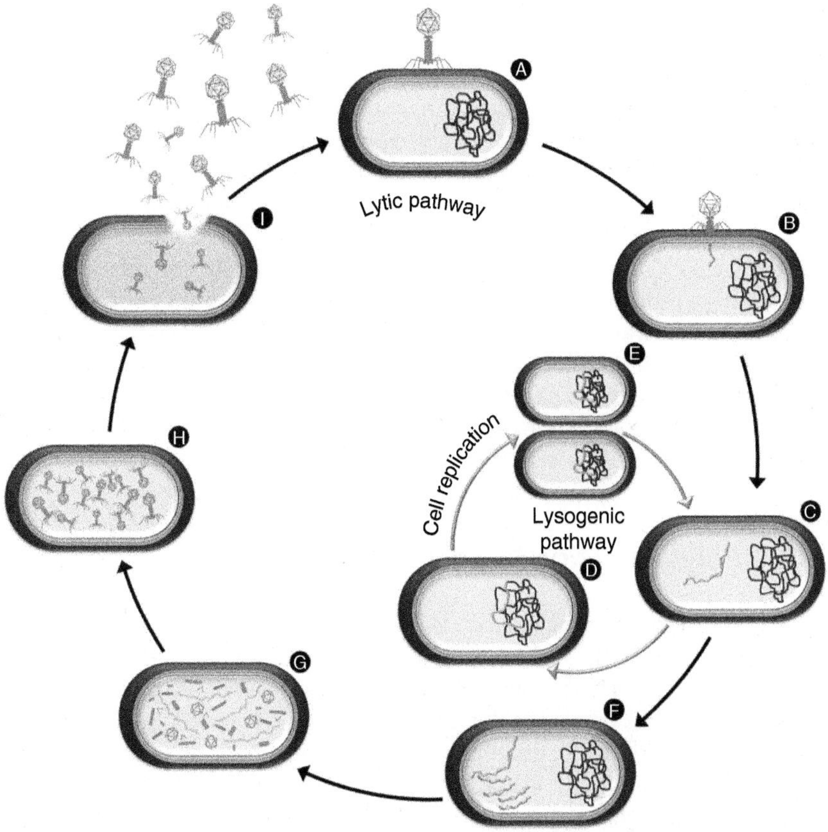

Figure 15.1 The phage replication cycle. Phage adsorbs to the bacterial cell (a) and injects its genome (b) to follow the lytic or the lysogenic pathway (c). In the lysogenic pathway, the genome integrates in the bacterial chromosome (d) and replicates with the cell (e). Continuing to the lytic pathway, the phage genes are translated (f) and proteins (including those composing the virus particle) are produced (g), assembled with the replicated phage genome (h) and the cell bursts to release the progeny (i) that will start a new cycle (a). *(See insert for color representation of the figure.)*

DNA ejection of a virulent phage, early genes are immediately expressed (Figure 15.1f). These genes are responsible for host takeover by hijacking its cellular machinery, redirecting it to replicate the phage genome. Virion accessory and structural proteins are thus expressed and the phage genetic material is packaged into the empty capsids (Figure 15.1g). All the phage pieces that were expressed separately are now assembled and the new progeny is ready to start a new cycle (Figure 15.1h). However, phage particles are still trapped within the host cell. While the phage is replicated inside the cell, two other phage-encoded proteins are also expressed – the holin and the endolysin. The holin opens the pores in the cytoplasmic membrane, allowing endolysin to reach and cleave specific peptidoglycan bonds, causing the disruption of the cell wall structure. As a consequence, the internal osmotic pressure is affected and the cell is disrupted by through hypotonic lysis through a process called “lysis from within” (Figure 15.1i). In tailed phages, the endolysin-mediated lysis is generally preceded by the holin action. This means that while endolysins accumulate in the cytosol, holins oligomerize to permeabilize the cytoplasm, exposing the peptidoglycan to the endolysins (Wang et al. 2000).

After the action of these two enzymes (spanins may also be required), the mature phage progeny particles are able to find new hosts to start new infection cycles (Figure 15.1a). All these proteins involved in the phage lytic cycle have their specific and indispensable role in the process of bacteria predation improved through millions of years of evolution, and their function can thus be used to combat the antibiotic resistance crisis. We will describe in the next sections the proteins that present the higher antimicrobial potential (Figure 15.2).

15.2 Polysaccharide Depolymerases

To initiate infection, phages need to recognize specific ligands on the host bacterial cell surface through receptor binding proteins (Figure 15.1a); in some cases, phage recognition involves enzymatic degradation of extracellular bacterial polysaccharides present in the K (capsule) and O (outermost portion of the lipopolysaccharides [LPS], usually present in Gram-negative bacteria) antigens (Bertozzi Silva et al. 2016). While some bacteria have evolved to produce a surrounding capsule that mask and hide these receptors, phages co-evolved to encode enzymes that specifically recognize and degrade those capsules (using it as their primary receptor) exposing the second receptor and allowing an irreversible binding and consequent infection (Bertozzi Silva et al. 2016). These enzymatic functions are accomplished by the phage-encoded depolymerases (Figure 15.2), which recognize and degrade the bacterial surface polysaccharides (K- and O-antigens) as well as exopolysaccharides (EPS).

In our current knowledge, only a few phages encode polysaccharide depolymerases. The presence of these enzymes was first described in 1956 after the observation of growing haloes surrounding the phage plaques during the spot-on-lawn method (Adams and Park 1956).

Depolymerases are integrated at the virion particle, usually associated with the phage tail spikes (Figure 15.2). Therefore, these molecules have high genetic plasticity allowing degradation of a vast variability of polymorphic O- and K-antigens of their hosts. For instance, *Escherichia coli* has 186 O and 80 K forms defined by serology (Whitfield and Roberts 1999), which means that in nature, phages infecting *E. coli* should encode depolymerases to recognize and degrade all these antigens.

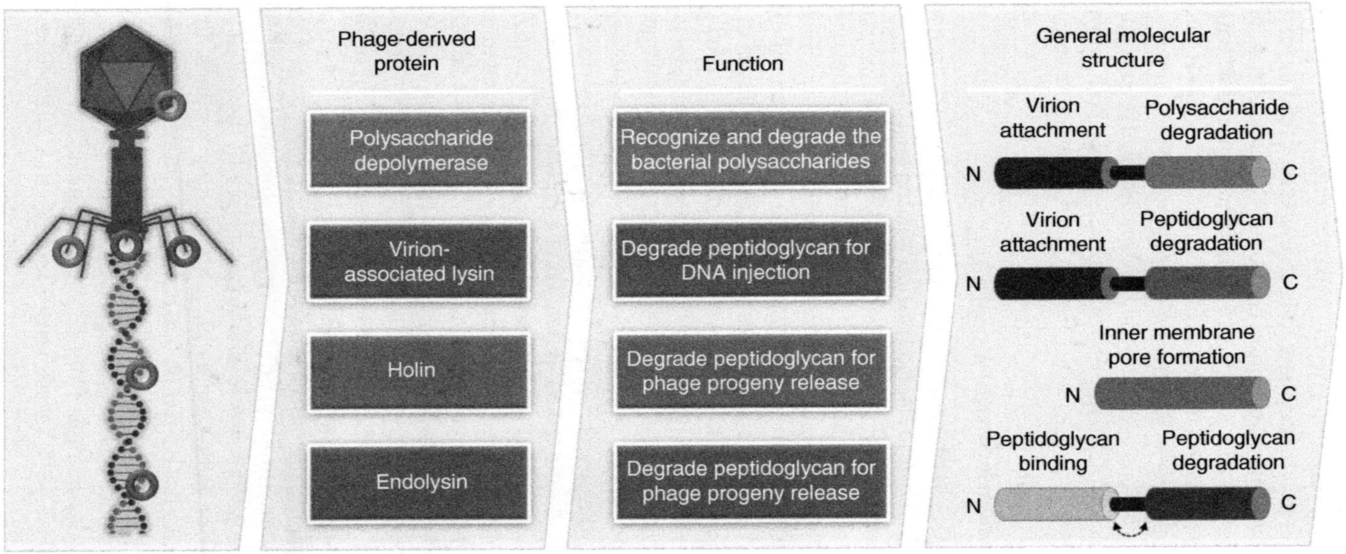

Figure 15.2 Phage-derived proteins with high potential as antimicrobials, their location in the phage particle, function, and general structure.

Bacterial polysaccharides are a diverse and important virulence factor, facilitating bacterial colonization, evasion, and protection from microbial defenses. Phage-encoded depolymerases are thus an important field of research as antivirulent agents or as bio-recognition elements (Yamaguchi et al. 2000; Spinosa et al. 2007; Geisinger and Isberg 2015). Therefore, special attention has been given to the use of depolymerases to recognize and remove these polymers from the bacterial surface, identifying the pathogens or attenuating their virulence, respectively. Herein, we provide a comprehensive review on the diversity of depolymerase structures, enzymatic activities, and biotechnological applications.

15.2.1 Depolymerase Structure

To date, all characterized recombinant depolymerases have been associated to the virion particle. Their preferred location is at the tail spikes or tail fibers (e.g. *Pseudomonas* phage LKA1, *Acinetobacter* phage vB_ApiP_P1) (Oliveira et al. 2017; Olszak et al. 2017), but depolymerases can also be attached to the preneck appendage proteins, as demonstrated with the *Staphylococcus* phage vB_SepiS-φIPLA7 (Gutiérrez et al. 2015). This information is also corroborated by a bioinformatics study (Pires et al. 2016). The analysis of 143 phages, with complete genomes infecting 24 bacterial genera, identified 160 putative depolymerases, where the majority were encoded in genes associated with the abovementioned structural functions. Interestingly, it was predicted that phages can encode more than one depolymerase gene to increase their host range. This hypothesis was recently validated with the *Klebsiella pneumoniae* phages K5-2, K5-4, and ϕK64-1 harboring multiple depolymerases (Hsieh et al. 2017; Pan et al. 2017). The former is the best depolymerase-equipped phage example. This giant (346 602 bp genome) virus has eight experimental validated genes encoding tail fibers, tails spikes, or lyases with depolymerase activity specific toward distant capsules types (K1, K11, K21, K25, K30/K69, K35, K64, KN4, and KN5).

Structural analysis has shown that most tail-associated depolymerases form elongated homodimers (e.g. *Pseudomonas* phage LKA1) (Olszak et al. 2017), but tetramer shapes have also been found (e.g. *E. coli* phage 63D) (Machida et al. 2000) ideal for virion cell puncturing. The elongated shape made of beta-sheets allows the proteins to scan a larger surface of the polysaccharide and expose its intra- or inter-subunit substrate binding residues. In agreement, phage P22 tail spike has been shown to bind to the entire *Salmonella* O-antigen polysaccharide layer (Baxa et al. 1996). Furthermore, the highly interwoven β-sheets of depolymerases favor a high stability toward temperature. So far, all characterized depolymerases from *K. pneumoniae* and *E. coli* phages melt at temperatures above 65 °C (Majkowska-Skrobek et al. 2016; Guo et al. 2017). This feature is likely a consequence of the evolution of depolymerases to withstand difficult extracellular environments to preserve phage infectivity and therefore ensure phage survival.

15.2.2 Depolymerase Classification

Phage-derived depolymerases are enzymatically diverse proteins as response to a diversified list of bacterial polysaccharides in this constant phage–bacteria arms race. According to their general mode of action, depolymerases can be divided into two main classes: hydrolases or lyases (Figure 15.3).

Both result in the specific digestion of the polymers into soluble oligosaccharides via distinct enzymatic paths. The hydrolases catalyze the hydrolysis of O- or K- antigen polymers through cleavage of glycosidic bonds with the consumption of a water molecule. These are subdivided into sialidases, rhamnosidases, and peptidases (Davies and Henrissat 1995; Pires et al. 2016). Sialidases are enzymes that hydrolyze the α-linkage of the terminal sialic acids and are a well-characterized group of enzymes in phages. They are mostly found in phages targeting *E. coli* K1- and K92-capsule types (e.g. K1A, K1E, K1F, 63D, and φ92) (Machida et al. 2000; Stummeyer et al. 2006; Schwarzer et al. 2015). Rhamnosidases have been found to cleave the α-1,3 *O*-glycosidic bonds of the *Salmonella* O-antigens. Consistently, rhamnosidases enzymatic domains are well characterized in phage genomes infecting *Salmonella* (e.g. Epsilon15, KB1, P27, 9NA, P22) (Wollin et al. 1981; Baxa et al. 1996; Guichard et al. 2013) and closely related *Shigella* (e.g. Sf6) species (Chua et al. 1999). Peptidases are

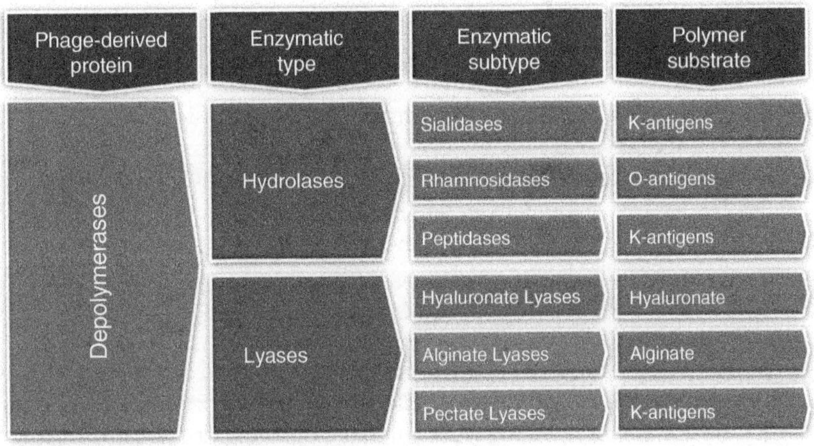

Figure 15.3 Distribution of the different polysaccharide depolymerase classes. Only the depolymerase classification types for which experimental validation has been reported are presented. Although peptidases do not degrade polysaccharides they were included here due to their degrading activity on bacterial capsules (those formed by polypeptides).

an exception to the other depolymerases since they not degrade sugars, but instead catalyze the hydrolysis of peptide bonds (polypeptide depolymerases). One example is found in the *Bacillus* phage φNIT1 enzyme, which proved to digest the poly-γ-glutamate-rich capsular polypeptides (Kimura and Itoh 2003). The xylosidase, dextranase, and levanase group of enzymes hydrolyze xylan, dextran, and levan sugars present in bacterial species, respectively. These last three subclasses of enzymes have only been predicted by bioinformatics studies, lacking experimental validation (Pires et al. 2016).

The other main class of depolymerases are the polysaccharide lyases that cleave 1,4 glycosidic bonds via β-elimination mechanism. Three different subgroups of phage depolymerases are found: hyaluronate, alginate, and pectate lyases. Hyaluronate and alginate lyases are a group of enzymes capable of digesting the hyaluronate polymer found in several organisms and the polysaccharide alginate synthetized by bacteria, respectively. For instance, alginate is produced by mucoid *Pseudomonas aeruginosa* strains that are important pathogens in patients with cystic fibrosis.

Lastly, pectate lyases are pectolytic enzymes predicted to degrade the galacturonic acid, one of the major constituents of bacterial polysaccharides. Pectate lyases are one of the most well-known enzymatic group found in phage depolymerases infecting mostly *K. pneumoniae* and *Acinetobacter baumannii* and found to sometimes cleave two capsule types (Majkowska-Skrobek et al. 2016; Hsieh et al. 2017; Oliveira et al. 2017, 2018; Pan et al. 2017; Popova et al. 2017).

15.2.3 Depolymerase Activity Assessment

The *in vitro* assessment of depolymerase enzymatic activity can be accomplished both qualitatively and quantitatively. The latter is more challenging due to the difficulties in separating the K- and O-antigen polymers from the remaining bacterial EPS. Generally, the methods can be divided into biological, biochemical, and physical.

The most common biological method used is the spot-on-lawn (Oliveira et al. 2017). Due to its simplicity and inexpensive implementation, it is usually the first assay used to assess the depolymerase activity. By dropping an enzyme solution into a bacterial host spread on agar plates using the overlay technique, the presence of a halo after a few hours is indicative of depolymerase activity.

The biochemical methods allow an estimation of the enzymatic activity. The turbidimetry assay is the most popular biochemical method and several experimental variants exist. The 3,5-Dinitrosalicylic acid (DNS) assay is used to quantify the number of reducing ends generated upon digestion of the polysaccharide polymer. If combined with glucose as a standard, it is possible to correlate the activity as glucose-reducing end equivalents (Lee et al. 2017). The bicinchoninic acid assay is an analogous method that can also be used to determine the reducing ends in a digested EPS solution (Hernandez-Morales et al.

2018). The cetylpyridinium chloride assay enables to quantify the enzymatic activity by the precipitation of digested polysaccharides and monitorization of turbidity decrease (Majkowska-Skrobek et al. 2016). Other methods include the examination of periodate oxidation (Kwiatkowski et al. 1983), measurement of uronic acid release (Glonti et al. 2010), or viscosity of polysaccharide solutions (Hernandez-Morales et al. 2018).

Finally, physical methods such as polyacrylamide gel electrophoresis can separate and compare the digested and nondigested polymer profile after staining with alcian blue/silver or methylene blue (Clarke et al. 2000; Barbirz et al. 2008; Majkowska-Skrobek et al. 2016). Nevertheless, they only allow visualization and not quantification of the enzymatic activity. Fluorophore-assisted carbohydrate electrophoresis and capillary electrophoresis-based methods coupled with biochemistry techniques (e.g. nuclear magnetic resonance spectroscopy, matrix-assisted laser desorption/ionization, and electromagnetic inspection scanner) are powerful methods to determine the substrate cleavage site (Thurow et al. 1974; Linnerborg et al. 2001; Lee et al. 2017). Other still not well disseminated techniques in the depolymerase field are optical (e.g. negative staining with congo red and nigrosine) and electronic microscopy (e.g. atomic force microscopy, confocal laser scanning microscopy, transmission electron microscopy, and scanning electron microscopy), which offers the possibility to visualize the degradation of the polysaccharide content of the bacterial surface and the biofilm slime.

15.2.4 Depolymerases as Antimicrobials

The ability of depolymerases to specifically recognize, modify, or remove bacterial extracellular polysaccharides that promote virulence renders them attractive enzymes for the control and detection of bacterial pathogens (Table 15.1).

The use of depolymerases as antimicrobials has been mostly explored as anti-virulence agents, favoring the host immune system. Although they are not able to lyse the cells, depolymerases can strip cells from their polymeric coats exposing them to microbial defenses. This has been proven successful in models of *Galleria mellonella* caterpillar, murine sepsis, and human serum. To assess the anti-virulence properties of depolymerases, *G. mellonella* has been used as an intermedium and inexpensive animal model before using other *in vivo* models.

In one study, O5-speficic depolymerase treatments could significantly prolong larvae lifespan from deadly *P. aeruginosa* infections (Olszak et al. 2017). Enzyme pretreatments extended the survival of about 20% caterpillars up to 72 hours, and injections 1 hour after bacterial challenge also resulted in an even greater survival rate of 35% of larvae, compared with the control groups. Likewise, it was observed that there was an increase of the survival rates in

Table 15.1 Depolymerase as antimicrobials.

Source phage	Target	Model	Main achievements	References
Pseudomonas phage LKA1	*Pseudomonas aeruginosa*	*In vivo* – *Galleria mellonella*	O5-depolymerase treatment rescued 35% of larvae	(Olszak et al. (2017)
Klebsiella phage KP36	*Klebsiella pneumoniae*	*In vivo* – *G. mellonella*	K36-depolymerase treatment rescued 43% of larvae	Majkowska-Skrobek et al. (2016)
Escherichia phages K1E, K1F, and K1H	*Escherichia coli*	*In vivo* – mice	K1F, K1H, K5, and K30- depolymerases rescued mice, but K1E-depolymerase did not	Lin et al. (2017)
Salmonella phage P22	*Salmonella enterica* serovar *typhimurium*	*In vivo* – chickens	O-depolymerase that recognize serotypes A, B, and D1 significantly reduced (at least tenfold) *Salmonella* in the cecum and its further penetration into internal organs	Waseh et al. (2010)
Klebsiella phage NTUH-K2044-K1-1	*K. pneumoniae*	*In vivo* – mice	K1-depolymerase fully protected mice from a deadly infection	Lin et al. (2014)
Klebsiella phage K64-1	*K. pneumoniae*	*In vivo* – mice	K64-depolymerase protected 20–100% of infected mice, depending on postinfection times	Pan et al. (2015)
Escherichia phage E	*E. coli*	*In vivo* – rat	K1-depolymerase protected more than 30% of rats comparatively to the control group	Mushtaq et al. (2004)
Escherichia phage E	*E. coli*	*In vivo* – rat	K1-depolymerase prevented death in at least 80% of infected animals	Mushtaq et al. (2005)

Escherichia phage E	*E. coli*	*In vivo* – rat	K1-depolymerase prevented death in 95% of infected animals	Zelmer et al. (2010)
Pasteurella phage PHB20	*Pasteurella multocida*	*In vivo* – mice	KA-depolymerase multiple-dose treatment rescued more than 60% of challenged mice	Chen et al. (2018)
Staphylococcus phage phiIPLA7	*Staphylococcus aureus*	*In vitro*	EPS-depolymerase achieved a maximum removal (>90%) of biofilm-attached cells	Gutierrez et al. (2015)
Erwinia phage phiEa1h.	*Erwinia amylovora*	*In vivo*	EPS-depolymerase expressed in transgenic pear reduced *E. amylovora* in a greenhouse	Malnoy et al. (2005)
Chimera	*E. coli* O157:H7	*In vitro*	O157-depolymerse with a R-type pyocin gain a broad range of antimicrobial activity against *E. coli* O157: H7 isolates	Scholl et al. (2009)

wax moths infected with *K. pneumoniae* pretreated with K46-specific depolymerase applied five minutes after bacterial inoculation (Majkowska-Skrobek et al. 2016). More convincing data comes from murine and poultry models. The few studies available have targeted only Gram-negative pathogens, including *E. coli* (Mushtaq et al. 2005; Lin et al. 2017), *Salmonella enterica* (Waseh et al. 2010), K. pneumoniae (Lin et al. 2014; Pan et al. 2015; Majkowska-Skrobek et al. 2016), and more recently *Pasteurella multocida* (Chen et al. 2018).

Escherichia coli is a well-known human commensal organism, but specific serotypes such as K1 and K5 capsules are frequently associated with extra-intestinal infections (Orskov et al. 1977). Animal experiments were conducted using different variants of K1-, K5-, or K30-specific depolymerases in *E. coli*-infected mice (Lin et al. 2017). Treatment doses (2–20 μg per mice) intramuscularly injected in the right thigh 30 minutes after bacterial challenge caused a different outcome. While most depolymerases did protect well mice from deadly infections, one K1E enzyme did not. As this K1E did sensitize the pathogen on serum in a way similar to K1F and K1H enzymes, the *in vivo* failure of this particular enzyme was explained by its inability to form trimers. Furthermore, to assess the potential acute toxicity, higher doses of enzyme (100 μg per mice) were tested with no significant alterations on survival rate, behavior, and body weight gains for five days, compared to the mock treated group. Three analogous studies proved the success of K1-specific depolymerases administrated intraperitoneally to neonatal rats by reducing the mortality rates (Mushtaq et al. 2004; Mushtaq et al. 2005; Zelmer et al. 2010).

Salmonella enterica serovar Typhimurium is a major cause of morbidity and mortality in poultry animals, posing serious economical burdens. In one study, chick gavage with three doses of O-degrading depolymerase (30 $\mu g\,\mu l^{-1}$) at 18, 42, and 66 hours postinfection significantly reduced *Salmonella* colonization in the gut and further penetration into internal organs (Waseh et al. 2010). The O-antigen was proposed to be essential for bacterial colonization in the animal gut.

The multidrug-resistant *K. pneumoniae* is associated with several hospital-acquired infections (e.g. ventilator-associated pneumonia, bloodstream, urinary tract, and soft tissue infections) and an important vehicle of extended-spectrum β-lactamases resistance genes dissemination (Vading et al. 2018). Therefore, several independent studies using bacteremic mice were developed to test the anti-virulent effect of depolymerases *in vivo*. In one study, researchers isolated a phage depolymerase that targets K1, a common capsular type in bacteria causing liver abscesses in Taiwan (Fung et al. 2002). Preclinical evaluation of the K1-specific depolymerase injected intraperitoneally (25 μg per mice) 30 minutes after a deadly bacterial challenge could rescue all mice (Lin et al. 2014). In another study, the same authors found the capsular type K64 to be the most predominant type among carbapenem-resistant *K. pneumoniae* strains isolated from Taiwanese hospitals (Pan et al. 2015). Applying a

K64-specific depolymerase successfully enhanced the serum and neutrophil killing *in vitro* and also fully protected K64 *K. pneumoniae*-inoculated mice through depolymerase intraperitoneal injections (18.75 μg per mice) one hour after challenge. Nevertheless, the survival rate significantly decreased to almost 20% when added 8 hours later.

The zoonotic bacteria *P. multocida* is the causative agent of various diseases in intensive animal and can be transmitted to humans usually followed by bites or scratches from domestic animals (Giordano et al. 2015). The capsule A-specific recombinant depolymerase from phage PHB20 that specifically infects *P. multicida* capsular A strains successfully reduced 3.5–4.5 log units of cells in serum (Chen et al. 2018). Moreover, the enzyme successfully rescued more than 60% of challenged mice with a multiple-dose treatment intravenous injected 13 hours postinfection, from an otherwise deadly infection. Also, no pathological changes were observed when compared with the control group. Taken together, depolymerases demonstrated to be efficient in reducing bacterial virulence *in vivo* in animal models by enhancing killing by complement, neutrophils and macrophages.

Other possibilities on the use of depolymerases as antimicrobials have arisen. They have been used to enzymatically degrade and disperse biofilms of *S. aureus* (Gutiérrez et al. 2015) or to be expressed in transgenic pear to reduce *Erwinia amylovora,* the causative agent of fire blight (Malnoy et al. 2005; Flachowsky et al. 2008). Using genetic engineering, depolymerases were incorporated into phage genomes to express them during the phage lytic cycle. This was first demonstrated by engineering phages to express dispersin B, a depolymerase from bacterial origin (Lu and Collins 2007). The synthetic phage was able to reduce biofilms by 4.5 log units, twofold better than the nonenzymatic wild-type phage. This concept was recently translated to an *E. amylovora* phage (Born et al. 2017). While the wild-type phage exhibited poor lytic activity, the engineered phage Y2::*dpoL1-C* had an enhanced performance by combining the capsule–depolymerase activity and the phage lytic activity. This synthetic phage could also significantly reduce the ability of *E. amylovora* to colonize the surface of detached flowers. Finally, the possibility to construct chimeric proteins with enhanced performance is an emerging field of research on phage proteins. One innovative experiment introduced a lytic activity to a depolymerase via fusion with the R-type pyocin (Scholl et al. 2009). This engineered depolymerase became specific and was able to efficiently kill *E. coli* O157:H7 isolates.

15.2.5 Remarks on Depolymerases

The biotechnological application of phage depolymerases is currently unexplored when compared to other phage proteins like endolysins. The first proteins have been characterized from *E. coli*-infecting phages. Since then, many other protein homologs from diversified hosts have been shown to control infectious diseases with the cooperation of the host complement lytic action, with many evidences proved in murine models. Fewer studies have also shown that they could also be used to degrade EPS and compose new anti-biofilm agents. Overall, depolymerases proved their potential as alternatives to antibiotics to control multidrug-resistant bacterial pathogens. Moreover, due to their high specificity and ability to discriminate the bacterial polysaccharides, they are also expected not to interfere with the normal flora.

15.3 Peptidoglycan-Degrading Enzymes

During their replication cycle, the genome and the progeny particles of tailed phages need to cross the bacterial cell. The first needs to get in (Figure 15.1b) and the second has to get out (Figure 15.1i). In both cases, phages need to degrade the peptidoglycan (a polymer also known as murein), the major component of the bacteria cell wall and the responsible for the structural integrity and shape of the cell. To accomplish that, they use different enzymes that are specialized for each case (Figure 15.2) (Oliveira et al. 2018). In the first case, to facilitate injection of their genome, phages use VALs, enzymes anchored to the phage particle that promote a local degradation of the peptidoglycan. This peptidoglycan digestion must not be harsh to prevent premature death of the host cells. In the second case, phages use endolysins, late proteins synthesized as soluble molecules that degrade the peptidoglycan to a large extent from within the cells, causing lysis of the cells, allowing the release of the phage progeny (Schmelcher et al. 2012).

Besides degrading the same polymer, VALs and endolysins face a diverse and changeable composition of the peptidoglycan. Peptidoglycan is composed of several chains of alternating residues of *N*-acetylmuramic acid (MurNAc) and *N*-acetylglucosamine (GlcNAc), connected by β-1,4 glycosidic bonds, linked to a short stem tetrapeptide. Whereas the carbohydrate backbone is conserved in all bacteria, the peptide moiety made of L- and D-amino acids is only conserved in Gram-negative organisms (Silhavy et al. 2010). In Gram-positive bacteria, it is considerably more diverse in terms of length and composition. As a response to this variability, VALs and endolysins have evolved to recognize and degrade all peptidoglycan types found in phage hosts. Their diversity, enzymatic activity, and biotechnological applications are discussed in the next sections.

Different enzymatic catalytic domains (ECDs) can cleave different peptidoglycan bonds and, depending on the bond that they cleave, they have different classifications (Figure 15.4) (Oliveira et al. 2013). Glycosidases cleave the glycan component at the reducing end of GlcNAc as the *N*-acetyl-β-D-glucosaminidases, or at the reducing end of MurNAc as the *N*-acetyl-β-D-muramidases (often called lysozymes or muramidases) and the lytic transglycosidases. The

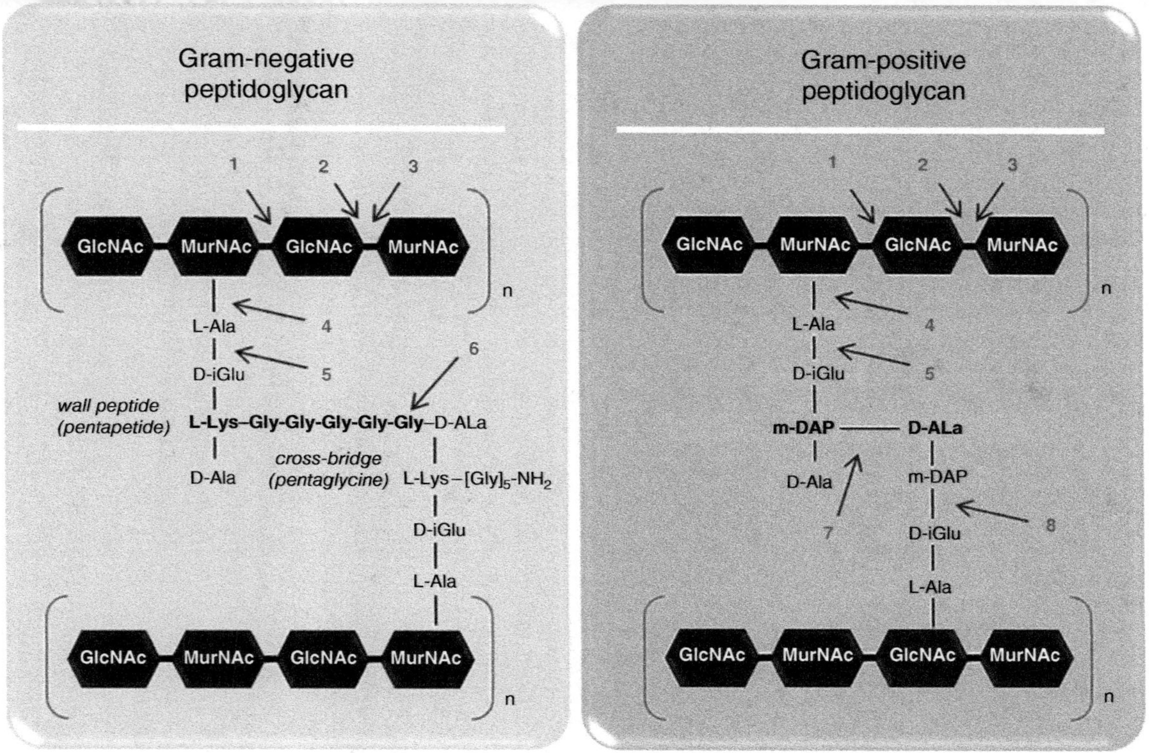

Figure 15.4 Illustration of the peptidoglycan bonds cleaved by enzymatic catalytic domains (ECDs) and their classification. Differences between Gram-negative and Gram-positive bacteria peptidoglycan are shown. In bold, the typical differences observed at the cross-bridge are indicated. Possible target sites of phage-derived peptidoglycans are shown: (1) *N*-acetyl-β-D-glucosaminidase; (2) *N*-acetyl-β-D-muramidase; (3) lytic transglycosylase; (4) *N*-acetylmuramoyl-L-alanine amidase; (5) L-alanoyl-D-glutamate endopeptidase; (6) D-alanyl-glycyl endopeptidase; (7) D-alanine-D-meso-DAP endopeptidase; and (8) D-glutamyl-m-DAP endopeptidase. *Source*: Adapted from Oliveira et al. (2012).

amidases hydrolyze the bonds between MurNAc and the L-alanine, which separate the glycan strand from the stem peptide. Finally, peptidases degrade the links of the peptide stem, which sometimes serve as crosslinks between the cell wall (Oliveira et al. 2012). Some proteins, which have the cysteine–histidine-dependent amidohydrolase/peptidase (CHAP) as a catalytic domain, have been shown to act both as peptidases and amidases (Navarre et al. 1999; Baker et al. 2006; Nelson et al. 2006).

15.3.1 Virion-Associated Lysins (VALs)

In literature, VALs are also known as virion-associated peptidoglycan hydrolases (VAPGHs) (Rodriguez-Rubio et al. 2013b), tail-associated muralytic enzymes (TAMEs) (Paul et al. 2011), and structural lysins or exolysins (Oliveira et al. 2013), which have been discouraged for being misleading terms.

VALs are frequently associated with the phage structural proteins. They are found in many phages mostly associated not only with tail-associated proteins (e.g. *Lactococcus lactis* phage TP901-1) but also with tape measure proteins (e.g. *Staphylococcus aureus* phage φIPLA35), baseplates (e.g. *E. coli* phage T4), and internal capsid proteins (e.g. *E. coli* phage T7) (Lavigne et al. 2006; Nishima et al. 2011). Besides being a small part of a larger structural protein, VALs have been successfully cloned and recombinantly produced in *E. coli* expression systems. Nevertheless, compared to endolysins, VALs have been much less explored in terms of structure, muralytic, and antibacterial activity. First studies started with the characterization of the muratlytic activity of VALs from *L. lactis* phages Tuc2009 and TP901-1 and *Bacillus subtilis* phage φ29, showing their ability to digest the peptidoglycan during the initial phage infection cycle (Kenny et al. 2004; Sudiarta et al. 2010; Mahony et al. 2016). Later studies have validated the muralytic activity of VALs against *S. aureus* and *Pseudomonas* hosts (Caldentey and Bamford 1992; Briers et al. 2008; Lavigne et al. 2006).

15.3.1.1 VAL Structure

While VALs from phages infecting Gram-negative bacteria harbor a single catalytic domain, those from Gram-positive background often encode two enzymatic active domains and with distinct organizations. For instance, *S. aureus* φMR11 VAL was proven to have a bifunctional lytic function by cloning and testing their CHAP and lysozyme individual catalytic domains (Rashel et al. 2008; Briers et al. 2014). A likely explanation for the multi-domain nature of these VALs may be to offer enhanced enzymatic degradation of the thicker peptidoglycan found in Gram-positive host cells.

There are several features that distinguish VALs from endolysins. Structurally, studies have shown that VALs (37–252 kDa) are larger than endolysins (15–40 kDa), with the smallest and biggest being the *Pseudoalteromonas* phage PM2 and *Bacillus* SP-beta prophage VALs, respectively (Kivela et al. 2004;

Sudiarta et al. 2010). Also, they form monomers (e.g. *Pseudomonas* phage φ6) (Caldentey and Bamford 1992), dodecamer (e.g. *Salmonella* phage P22) (Tang et al. 2011), and oligomers. The latter are the most reported state in the virus particle. Biologically, VALs are proteins that digest the peptidoglycan of actively growing cells from the outside, and endolysins digest the peptidoglycan from within of an already dead cell by the pre-action of holins. Enzymatically, VALs usually possess domains conferring glycosidase or endopeptidase activity that are shared with endolysins, being the latter often observed in Gram-positive phage VALs. Interestingly, the endopeptidase (M23 family) is almost exclusively found on VALs. This domain is also found by analogous bacterial enzymes (e.g. lysostaphin and enterolysin A) that also degrade the cell wall from outside, suggesting being more prepared to degrade the peptidoglycan from actively growing cells (Khan et al. 2013). The absence of an identifiable binding domain is another distinctive feature of VALs. The approach of these proteins to the cell wall is ensured by other embedded subdomains found in the same open reading frame where VALs are found (e.g. tail fiber, tail spikes). VALs have also been found to be highly thermostable, which contrasts with the mesophilic properties of most endolysins. For instance, the *P. aeruginosa* φKMV VAL has the ability to resist temperatures up to 100 °C and even to autoclaving (Lavigne et al. 2004). This is probably explained by the fact that being anchored to the phage particle, these proteins have evolved to endure harsh external conditions where phage inhabits, ensuring their infectivity and survival. Therefore, VALs may find several biotechnological applications when thermic processing is applied.

Moreover, the role of VALs goes beyond what was previously anticipated. Some VALs (e.g. *S. aureus* phage φ11 and *E. coli* phage T7) are reported to be dispensable for phage infection and to degrade highly cross-linked peptidoglycan, which is found in stationary phase cells (Moak and Molineux 2000; Rodriguez-Rubio et al. 2013c; Stockdale et al. 2013). These observations suggest that VALs offer competitive advantages to degrade the peptidoglycan in the initial step of phage infection under specific physiological conditions.

15.3.1.2 VALs as Antimicrobials

The natural lytic activity of VALs causes "lysis from without," a phenomenon that was first documented in 1940. This phenomenon occurs when several phages are drilling holes in the peptidoglycan of the same cell through enzymatic degradation of the VALs, causing premature death of the host. After this observation, VALs have emerged as novel antibacterial agents being mostly applied to kill *S. aureus* (Table 15.2). For instance, *S. aureus* phage φIPLA88 and P68 VALs demonstrated to kill 99% of methicillin-resistant *S. aureus* (MRSA) strains 20 minutes after exposure (Takac and Blasi 2005; Rodríguez et al. 2011). The thermostability of VALs confers them high potential to control undesirable bacteria for example in pasteurized milk or other dairy products

Table 15.2 Virion-associated lysins (VALs) as antimicrobials.

Source phage	Target	Model	Main achievements	References
Staphylococcus phage phiIPLA88	*Staphylococcus aureus*	*In vitro*	HydH5 and its derivative fusion proteins killed MRSA and inhibited their growth in whole, skimmed, and pasteurized milk	Rodriguez-Rubio et al. (2013a)
Staphylococcus phage P68	*S. aureus*	*In vitro*	Protein 17 displays activity against MRSA strains that were resistant to the phage	Takac and Blasi (2005)
S. aureus phage phi MR11	*S. aureus*	*In vitro*	Both VAL domains (cysteine-histidine-dependent amidohydrolase/peptidase [CHAP] and lysozyme) individually isolated could efficiently kill *S. aureus* cells	Rashel et al. (2008)
Staphylococcus phage phiIPLA88	*S. aureus*	*In vitro*	HydH5 exhibits an enhanced staphylolytic activity and synergizes with an endolysin to kill *S. aureus* cells	Rodriguez-Rubio et al. (2012)
Chimeric protein	*S. aureus*	*In vitro*	P128 synergizes with an antibiotic to reduce *S. aureus* biofilms	Poonacha et al. (2017)
Chimeric protein	*Enterococcus faecalis*	*In vitro*	EC300 had a superior antibacterial effect than the parental enzymes, having a robust antibacterial activity against actively dividing cells	Proenca et al. (2015)
Pseudomonas phage phiKMV	*Pseudomonas aeruginosa*	*In vitro*	VAL (gp36) was able to have muralytic activity and to be highly thermoresistant	Briers et al. (2006)
Chimeric protein	*S. aureus*	*In vitro*	Ply187AN-KSH3b displayed a tenfold increase in antimicrobial activity than the parental enzyme	Mao et al. (2013)
Chimeric protein	*S. aureus*	*In vitro*	Ply187N-V12C harboring a non-SH3 binding domain extended the lytic activity against staphylococci and streptococci	Dong et al. (2015)

that need to be heat-treated. HydDH5 VAL was shown to prevent the *S. aureus* growth in raw (whole and skimmed) and pasteurized milk (Rodriguez-Rubio et al. 2013a).

Other studies have also demonstrated the individual activity of the catalytic domains of the *S. aureus* phage φMR11 VAL (Rashel et al. 2008). Authors showed that both the lysozyme and the amidase domains could cause efficient lysis of *S. aureus* cells. Moreover, both proteins demonstrated to have a broader lytic activity when compared to the parental phage. Additionally, there are evidences that VALs can synergize with endolysins and antibiotics, as shown against *S. aureus* (Rodriguez-Rubio et al. 2012; Poonacha et al. 2017).

In contrast to Gram-positive bacteria, where VAL have been used against intact cells, in Gram-negative bacteria, they have been only applied against outer membrane (OM)-permeabilized cells treated with chloroform, to assess their muralytic activity. The majority of the studies have validated the muralytic activity of VALs from *P. aeruginosa* infecting phages φKMV, φ6, and φKZ (Caldentey and Bamford 1992; Briers et al. 2006, 2008), demonstrating that they present a broad lytic spectrum. This can be attributed to the conserved structure of the Gram-negative bacterial peptidoglycan (chemotype A1γ), which contrasts with the diversity of chemotypes found in Gram-positive bacteria (Schleifer and Kandler 1972).

15.3.1.2.1 Protein Engineering of VALs

While some interesting data was validated using native VALs, the most exciting results come from chimeric proteins that often harbor domains from both endolysins and VALs to generate proteins with higher lytic activity (Sao-Jose 2018). Taking advantage of their thermoresistant properties, VALs have been engineered to contain heterologous binding domains to further enhance its antibacterial activity. HydH5 chemolysin harboring one of the *S. aureus* phage φIPLA88 VAL catalytic domains (CHAP) was fused to a bacterial lysostaphin binding domain (SH3b type). The chimera exhibited an enhanced staphylolytic activity and tolerance to temperatures, retaining some lytic activity after pasteurization treatment and after storage at 4 °C for three days (Rodriguez-Rubio et al. 2012). P128 chemolysin, which is a truncated version of the *S. aureus* phage K VAL (endopeptidase) fused with the staphylococcal cell wall binding SH3b domain of lysostaphin, is another example of success (Paul et al. 2011; Vipra et al. 2012). This engineered protein presented a 100-fold increase in the antibacterial activity and a more effective anti-biofilm activity compared to the isolated VAL catalytic domain and better thermostability than lysostaphin. A *S. aureus* phage 187 lytic protein has also been used to construct several chemolysins. Although being presented as an endolysin, its genetic organization and high similarity to other VALs make Ply187 likely a VAL too (Loessner et al. 1999). Ply187AN-KSH3b is a protein that combines the *S. aureus* phage ply187 VAL catalytic domain (endopeptidase) to a *S. aureus* phage K binding domain

Chimeric protein	*S. aureus*	*In vivo* – mice	ClyF rescued 100% of mice in a bacteremia model. ClyF also reduced skin colonization by approximately 1.5 and 3.3 log units in a burn wound model	Yang et al. (2017)
Chimeric protein	*S. aureus*	*In vivo* – mice	ClyH rescued more than 66.7% of mice from a deadly bacteremia infection	Yang et al. (2014)
Chimeric protein	*S. aureus*	*In vivo* – mice	Ply187AN-KSH3b significantly reduced bacterial burden and the endophthalmitis outcome	Singh et al. (2014)
Chimeric protein	*S. aureus*	*In vivo* – rat	P128 formulated as a hydrogel effectively decolonized rat nares of *S. aureus* USA300 in an experimental mode	Paul et al. (2011)

Research examples and main results.

(SH3b-type) which resulted in a tenfold increase in antimicrobial activity compared to the wild-type ply187 VAL, being also more effective than the truncated versions, and to the phage K endolysin itself (Mao et al. 2013). Resorting to a random screening of catalytic and binding domains, ClyF (derived from the Ply187 VAL and PlySs2 endolysin) emerged as a protein with better thermostability and activity properties against planktonic and biofilm MRSA *in vitro* (Yang et al. 2017). In some cases, genetic engineering of proteins has altered the lytic spectrum. This has been done by adding non-SH3b binding domains to Ply187, originating ClyF, Ply187AN-V12C, and ClyH chemolysins. ClyF resulted in a protein with a 3.7–13.6-fold increase in lytic activity against *S. aureus* and extended activity toward *Streptococcus sobrinus*, which ply187 VAL could not lyse before (Yang et al. 2014). Similar results were obtained with Ply187AN-V12C, which in this case the activity was extended to the cow mastitis main pathogen, *Streptococcus dysgalactiae* (Dong et al. 2015). Following the idea that VALs are better prepared to degrade the peptidoglycan of an actively growing cell, a chimeric protein was constructed to kill pathogens in conditions other than buffers. EC300 resulted from a fusion of a VAL catalytic domain and the binding domain of the cognate Lys170 endolysin both produced from the *Enterococcus faecalis* phage F170/08 (Proença et al. 2015). While both VAL and endolysin produced by phage F170/08 could lyse *E. faecalis* suspended in a nutrient-depleted buffered solutions, the chimera EC300 could kill cells in rich media, which promotes robust bacterial growth. The EC300 superior antibacterial effect was also validated in a vast panel of multidrug-resistant *S. aureus* strains.

15.3.1.2.2 In Vivo *Trials of VALs*

Opposite to phage endolysins, only a few studies have applied VALs *in vivo*. To our knowledge, only P128 and abovementioned ply187-derived proteins (ClyH, Ply187AN-V12C, and ClyF, if considered as VALs) were reported to be effective antibacterial agents in a murine model (Paul et al. 2011). As these proteins derived from *S. aureus*-infecting phages, only their staphylolytic activity was evaluated. The ability of *S. aureus* to elicit topical or systemic inflammatory response syndrome is mediated by the various infections associated with this pathogen, from the skin and soft skin tissue to necrotizing pneumonia and sepsis. Treatment of these infections has become more challenging since the appearance of multidrug-resistant strains, such as MRSA. Therefore, VALs have been tested in models reproducing these diseases, namely, respiratory (Paul et al. 2011) and skin infections (Yang et al. 2017), endophthalmitis (Singh et al. 2014), and bacteremia (Yang et al. 2014). The chimeric P128 protein showed to have a stronger anti-staphylococcal activity and an effective decolonizing action in rat nares (Paul et al. 2011). Mice challenged with MRSA strain were intranasally treated twice daily over three days, with P128 formulated as a hydrogel, a placebo gel, or a mupirocin nasal (30 mg per dose, 2% mupirocin ointment, GlaxoSmithKline). The median bacterial counts recovered in the P128 hydrogel-treated group was found to present a reduction of more than 2 log units than that of the control groups. Safety use and efficacy of P128 is now studied by GangaGen as a topical treatment in volunteers and patients who are nasal carriers of *S. aureus*, on a combined Phase I and Phase II clinical trials (ClinicalTrials.gov Identifier NCT01746654), respectively. Ply187AN-KSH3b was used to study for the first time a mouse model of *S. aureus* endophthalmitis, an infection of the intraocular cavities, which occur after ocular surgery or trauma (Singh et al. 2014). Eyes treated with Ply187AN-KSH3b with an intravitreal injection (1 μg per eye of mouse) 6 or 12 hours postinfection significantly reduced bacterial burden and the endophthalmitis outcome in mice. Furthermore, protein treatment maintained the normal retinal function and reduced inflammatory cytokines and neutrophil infiltration in the eyes. ClyH was tested in a *S. aureus*-induced bacteremia (Yang et al. 2014). Treatments with ClyH injected intraperitoneally three hours postinfection rescued more than 66.7% of mice from a deadly infection. Furthermore, daily injections of ClyH did not cause harmful effects. Finally, ClyF chemolysin was tested in a bacteremia and burned wound mice models (Yang et al. 2017). In the bacteremia model, ClyF intraperitoneal injected rescued up to 100% of mice from an otherwise deadly infection. No mice died when exposed to higher or repeated doses for five continuous days, which attest for its nontoxicity. In the mice burn wound infection model, topically applied ClyF in a single or double dose or double could reduce the skin bioburden by approximately 1.5 and 3.3 log units of *S. aureus* colonization. The histological analysis showed that ClyF-treated mice exhibited less *S. aureus* colonization below the epidermis of the skin compared to the mocked treated groups.

15.3.1.3 Remarks on VALs

The demonstration of peptidoglycan-degrading activity was first spotted with VALs by the "lysis from without" phenomenon observed in 1940, i.e. when an infection is aborted by a high number of phages that are puncturing the sane cell wall through enzymatic degradation, leading to premature cell death. These peptidoglycan-degrading enzymes have sparked the interested of many researchers to find novel antibacterial agents. Since their discovery, VALs have demonstrated to have better thermostatibility properties when compared to endolysins and differentiate from antibiotics by having (i) distinct mechanism of action, (ii) high specificity and activity against multidrug-resistant pathogens, and (iii) low probability of resistance development. Especially, the engineered VALs represent a new class of enzybiotics that could be custom designed to solve or improve antibacterial traits against emerging antimicrobial resistance.

15.3.2 Gram-Positive Targeting Endolysins

The exposed peptidoglycan in Gram-positive bacteria allow for the endolysins that target these bacteria to present activity when applied exogenously, causing lysis of the cell through a process also known as "lysis from without." This process is described as quick and effective characteristics that make endolysins an interesting therapeutic approach (Loessner 2005; Fischetti 2010).

15.3.2.1 Gram-Positive Targeting Endolysin Structure

While the molecular weight of antibiotics is typically low, ranging between 0.3 and 1.6 kDa, endolysins are usually larger than 25 kDa. Gram-positive and *Mycobacterium* endolysins display, mostly, a modular organization with at least two different domains with different functions: an ECD responsible for peptidoglycan hydrolysis and a cell binding domain (CBD) for substrate recognition (Borysowski et al. 2006; Nelson et al. 2012; Payne and Hatfull 2012). Moreover, both domains are usually connected by a linker (Figure 15.5).

Generally, each endolysin is composed by only one ECD; however some possess two ECDs with different cleavage sites (Oliveira et al. 2013). The specificity of endolysins is provided by the CBDs that bind non-covalently to specific conserved ligands or polymers mostly located at the cell wall (Loessner et al. 2002; Chang and Ryu 2017). The affinity of CBDs to their ligands has been compared to mature antibody–epitope binding, which explains the quick

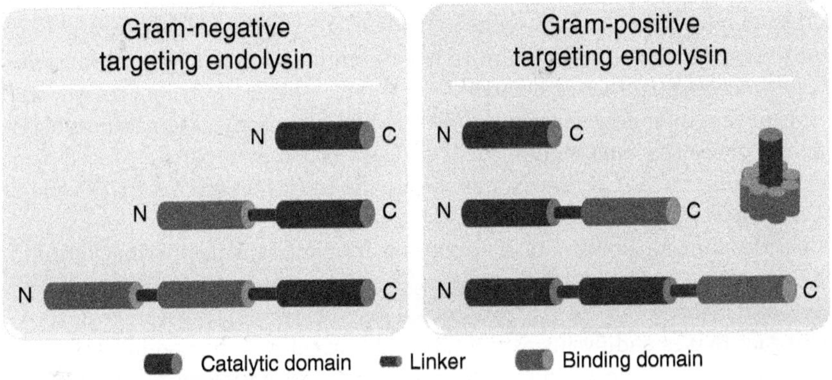

Figure 15.5 Different molecular arrangements possibilities of phage endolysins. On the left, architecture types of Gram-negative targeting endolysins with an inverted arrangement of that of Gram-positive targeting endolysins, with one catalytic domain at the C-terminus and one or more binding domains at the N-terminus. On the right, architecture types of Gram-positive phage endolysins with one or more catalytic domains at the N-terminus and one binding domain at the C-terminus. A multimeric structure with a single heavy chain (catalytic domain) and eight light chains (binding domain) has also been proved. *(See insert for color representation of the figure.)*

manner that endolysins identify their targets (Kretzer et al. 2007). Moreover, it was demonstrated with the pneumococcal endolysin Cpl-1 that the presence of the CBD also helps to position the ECD (Hermoso et al. 2003). Theoretically, the presence of the CBD will limit endolysin diffusion after the lytic cycle due to its high affinity with the peptidoglycan, maintaining the endolysin bound to the peptidoglycan even after its digestion. This prevents the diffusion of the endolysin and thus the lysis of uninfected cells, which can thus be performed by the newly formed phage particles, ensuring phage survival.

15.3.2.2 Gram-Positive Targeting Endolysins as Antimicrobials

In vitro results have demonstrated the tremendous potential of endolysins against different pathogens including *Listeria monocytogenes* (Loessner et al. 2002), *S. aureus* and *Staphylococcus epidermidis* (Gutiérrez et al. 2014), *Bacillus anthracis* and *Bacillus cereus* (Schuch et al. 2002), *E. faecalis* and *Enterococcus faecium* (Proença et al. 2012), and *Streptococcus pneumoniae, Streptococcus agalactiae,* and *Streptococcus pyogenes* (Loeffler et al. 2001; Nelson et al. 2001; Cheng et al. 2005) (Table 15.3). In contrast to antibiotics, endolysin's specificity is limited to the phage host from whom they were isolated, allowing them to kill only the target bacterium with very low or no effect on commensal flora. This means that endolysins derived from *Enterococcus* phages will lyse only enterococcal strains (Proença et al. 2012). To date, the *B. anthracis* endolysin PlyG has been reported as the most specific endolysin, as it binds to a specific polysaccharide that is usually found only in *B. anthracis* strains (Schuch et al. 2013). On the other hand, the endolysin PlySs2 from a *Streptococcus suis* phage has the broadest lytic spectra able to lyse different species of *Streptococcus* and *Staphylococcus* (Gilmer et al. 2013).

Interestingly, endolysins were proven to be highly active against both exponential phase cells and stationary phase cells (Melo et al. 2018). Moreover, it was demonstrated that they can penetrate the thick polysaccharide capsule of *B. anthracis* (Yang et al. 2012).

Besides the high specificity of endolysins from a Gram-positive background, they were also found to present a very fast antimicrobial activity. The endolysin PlyC eradicated a culture of *S. pyogenes* in just five seconds (Nelson et al. 2001). This fact makes endolysins as the faster biological lytic compound acting on bacterial cells, an effect just comparable to chemical antiseptics.

Due to their antimicrobial properties, endolysin applications have been gaining more relevance in the last two decades. There are several reports of the *in vitro* potential of different endolysins against relevant pathogens, such as MRSA (O'Flaherty et al. 2005), *S. pneumoniae* (Blázquez et al. 2016), and vancomycin-resistant *E. faecalis* and *E. faecium* (Yoong et al. 2004). Despite the success against planktonic cells, their effect against sessile cells is considered of utmost importance in healthcare facilities. The bacterial biofilm formation in medical devices and other clinical environments is of extreme relevance, since

Table 15.3 Gram-positive targeting endolysins as antimicrobials.

Source phage	Target	Model	Main achievements	References
Staphylococcus aureus phage vB_SauS-phiIPLA88	*S. aureus* and *Staphylococcus epidermidis*	*In vitro*	LysH5 reduced biofilm cell counts by 1–3 log units	Gutiérrez et al. (2014)
Bacillus anthracis phage γ	*Bacillus* sp.	*In vivo* – mice	PlyG administered i.p. rescue 76.9 of mice	Schuch et al. (2002)
Enterococcus faecalis phages F168/08 and F170/08	*E. faecalis*	*In vitro*	Lys168 or Lys170 lysed ≥98% of the initial CFU per ml in 90 minutes	Proença et al. (2012)
Streptococcus phage C1	Group A *streptococcus*	*In vivo* – mice	Lysin could prevent and control bacterial colonization in mice	Nelson et al. (2001)
Streptococcus phage NCTC 11261	*Streptococcus agalactiae*	*In vivo* – mice	PlyGBS reduced bacterial charge in the vagina by 3 log units	Cheng et al. (2005)
Streptococcus phage Dp-1	*Streptococcus pneumoniae*	*In vivo* – mice	Pal administration reduced bacteria do undetectable levels	Loeffler et al. (2001)
B. anthracis phage γ	*B. anthracis*	*In vivo* – mice	PlyG lead to 66–100% of mice survival depending on the time of administration	Schuch et al. (2013)
Streptococcus suis strain 89/1591 (prophage)	*Streptococcus pyogenes and S. aureus*	*In vivo* – mice	PlySs2 administration resulted in 92% mice survival from mixed infection	Gilmer et al. (2013)
S. aureus phage vB_SauM-LM12	*S. aureus*	*In vitro*	E-LM12 reduced exponential phase and stationary phase cell by 3–4.5 log units in two hours	Melo et al. (2018)
Staphylococcus phage K	*S. aureus*	*In vitro*	99% reduction in staphylococcal cell counts was observed one hour after the addition of LysK	O'Flaherty et al. (2005)

(*Continued*)

Table 15.3 (Continued)

Source phage	Target	Model	Main achievements	References
Chimeric protein	*S. pneumoniae*	*In vivo* – zebrafish embryos	PL3 fully protect embryos from infection	Blázquez et al. (2016)
Enterococcus phage Φ1	*E. faecalis*	*In vitro*	PlyV12 reduced more than 4 log units cell counts in 15 minutes	Yoong et al. (2004)
Staphylococcus phage GRCS	*Staphylococcus* sp.	*In vitro*	PlyGRCS decrease biofilm biomass (c. 50%) in one hour	Melo et al. (2018) and Linden et al. (2015)
Eight different endolysins	*S. aureus*	*In vivo* – mice	Endolysins 80α, phi11, LysK, 2638A, and WMY protected 100% of the animals from death	Schmelcher et al. (2015)
Chimeric protein	*S. aureus*	*In vitro*	ClyH had an MBEC inferior to several antibiotics	Yang et al. (2014)
Staphylococcus phage MR-10	*S. aureus*	*In vitro*	Approximately 1 log unit reduction on biofilms	Chopra et al. (2015)
Staphylococcus phage K	*S. aureus*	*In vitro*	LysK act synergistically with a depolymerase in dynamic biofilms reducing the number of cells in 2.5 log units	Olsen et al. (2018)
Chimeric protein	*S. aureus*	*In vivo* – mice	ClyS reduced by 3 log units the bacterial load	Pastagia et al. (2011)
Streptococcus phage Cp-7	*S. pyogenes* and *S. pneumoniae*	*In vivo* – zebrafish embryos	Engineered lysin (Cpl-7S) rescue 95–99% of embryos	Díez-Martínez et al. (2013)
Streptococcus phage Cp-1	*S. pneumoniae*	*In vivo* – mice	Cpl-1 reduced nasopharynx colonization prolonging mice survival	Loeffler et al. (2003)

Chimeric protein	*S. pneumoniae*	*In vivo* – mice	Cpl-711 increase mice survival 45–100% in a concentration-dependent manner	Diez-Martinez et al. (2015)
Streptococcus phage Cp-1	*S. pneumoniae*	*In vivo* – CD1 mice	Cpl-1 combination with daptomyxin resulted in ≥80% survival	Vouillamoz et al. (2013)
S. pyogenes strain MGAS315 (prophage)	*S. pyogenes*	*In vivo* – mice	PlyPy increased mice survival at least 90%	Lood et al. (2014)
Streptococcus dysgalactiae strain SK1249 (prophage)	*S. agalactiae*	*In vivo* – mice	PlySK1249 consecutive doses rescued 60% of animals	Oechslin et al. (2013)
Staphylococcus phage ϕMR11	*S. aureus*	*In vivo* – mice	MV-L fully protected mice from death	Rashel et al. (2007)
Staphylococcus phage GH15	*S. aureus*	*In vivo* – mice	LysGH15 rescued 40–100% of mice in a concentration dependent manner	Gu et al. (2011)
Staphylococcus phage SAP-1	*S. aureus*	*In vitro*	SAL-1 had a twofold lower MIC than the well-known LysK	Jun et al. (2011)
Staphylococcus phage SAP-1	*S. aureus*	*In vivo* – mice	SAL-1 formulation (SAL200) rescued at least 93.3% of mice	Jun et al. (2013)
Staphylococcus phage SAP-1	*S. aureus*	*In vivo* – rats	SAL200 shown no toxicity signals in a safety assessment	Jun et al. (2014)
Chimeric proteins	*S. aureus*	*In vitro*	Lys168-87 and Lys170-87 have increased solubility than parental enzymes and reduced bacterial counts by about 99%	Fernandes et al. (2012)
Chimeric protein	*S. aureus*	*In vitro*	Ply187AN-KSH3b MIC is lower than parental enzymes	Mao et al. (2013)
Chimeric protein	*Listeria monocytogenes*	*In vitro*	EAD118-CBDPSA protein exhibited around threefold higher lytic activity than parental endolysins	Schmelcher et al. (2011)
Chimeric protein	*S. aureus*	*In vivo* – mice	ClyS and oxacillin acted synergistically and prevented 80% of mice septic death	Daniel et al. (2010)

Research examples and main results.

biofilm cells are described as more tolerant to antibiotics than their planktonic counterparts. This is mainly due to antibiotics' difficulty in penetrating the complex biofilm matrix and the targeting of slow-growing cells as several antibiotics can only target active cells.

The endolysins PlyGRCS and E-LM12 were shown to be active against *S. aureus* stationary phase cells and biofilms (Linden et al. 2015; Melo et al. 2018). Schmelcher et al. compared the effect of eight endolysins and lysostaphin against MRSA static biofilms, and LysK was shown to be the more active (Schmelcher et al. 2015). The anti-biofilm effect of the chimeric endolysin ClyH was shown to be strain dependent and also varied through biofilm maturation stage. In this sense, the removal of more mature biofilms required a longer treatment than younger biofilms (Yang et al. 2014). Despite the promising results, it is not easy to achieve biofilm eradication using just endolysins, but the solution might be to combine endolysins with other antimicrobials. The combination of endolysin MR-10 and minocycline could significantly reduce biofilms with different maturation times (Chopra et al. 2015). More recently, Olsen et al. combined LysK with a bacterial depolymerase and demonstrated a synergistic effect of their combination both in static and dynamic biofilm models (Olsen et al. 2018).

15.3.2.2.1 **In Vivo** ***Trials of Gram-Positive Targeting Endolysins***

The effect of several endolysins has also been assessed in different animal models, namely, against *Streptococcus*, *Staphylococcus*, *B. anthracis*, and *E. faecalis* (reviewed in [Oliveira et al. 2018]). Although the majority of the studies are performed in models of systemic infection, there are few studies in different models. The highly active PlyC endolysin that showed for the first time the enormous *in vivo* potential of phage endolysins by specifically eradicating a group of A streptococci within seconds was tested on a mice model for pharyngitis infection. The use of PlyC in nanograms via oral and nasal administration could successfully prevent (70% survival) and treat (100% survival) mice heavily infected with streptococci (Nelson et al. 2001). In another study, the chimeric endolysin ClyS was applied topically on a mouse skin infected with *S. aureus* and presented better results than mupirocin, the standard topical antibiotic used (Pastagia et al. 2011).

In bacteremia models, animals are challenged with deadly doses of bacteria, and few hours later, endolysins are administered. A chimeric lysin targeting streptococci significantly increased the survival rate on a zebrafish model (Díez-Martínez et al. 2013). On a mice model, an intravenous administration of endolysin Cpl-1 reduced 3 log units of the number of culturable *S. pneumoniae* in the bloodstream, when administered 10 hours postinfection (Loeffler et al. 2003). Using another route of administration (intraperitoneal injection), a single injection of a chimeric enzyme administered one hour after challenge improved mice survival in 50% (Diez-Martinez et al. 2015). Recently, Vouillamoz et al. reported that a combination of Cpl-1 and daptomycin showed synergism in a pneumococcal bacteremia mouse model, increasing their survival rate to 95% (Vouillamoz et al. 2013). Lysins were also proven to be efficient against bacteremic mice models of other streptococci. Intraperitoneal injections of PlyC, PlySs2, and PlyPy have proven to rescue the majority of mice from lethal *S. pyogenes* infections (Nelson et al. 2001; Gilmer et al. 2013; Lood et al. 2014). Likewise, a single intraperitoneal administration of prophage-derived endolysin PlySK1249, which is active against *S. agalactiae*, prolonged mice survival (Oechslin et al. 2013). This treatment became curative when it was increased to three injections on the first day of infection.

The *B. anthracis* phage endolysin PlyG, which has been shown *in vitro* to be active against both vegetative cells and germinating spores in several *Bacillus* sp., was administered in mice infected with lethal *B. anthracis* bacteremia, rescued 65% of mice, and prolonged the survival of the mice that did not recover (Schuch et al. 2002).

Due to its clinical significance, *S. aureus* has been the most targeted species for endolysin testing in animal models of infection. Endolysins have proven to reduce *S. aureus* cell counts in the bloodstream and increased rodent's survival from septic septicemia. A single intraperitoneal injection of MV-L endolysin 30 minutes after bacterial challenge rescued 100% of mice (Rashel et al. 2007). LysGH15 extended this effect until one hour after MRSA challenging (Gu et al. 2011). In the recent study where the efficacy of eight endolysins was compared with that of lysostaphin (and two antibiotics [Schmelcher et al. 2015]), a mice model infected with MRSA was also tested. In this model, five of the endolysins, lysostaphin, and vancomycin rescued 100% of the mice. A stabilized formula containing SAL-1 (the phage SAP-1 endolysin that has shown *in vitro* activity against MRSA biofilms [Jun et al. 2011]) also showed to significantly reduce the bacterial counts in the bloodstream and prolonged the mice viability on a MRSA infection model (Jun et al. 2013). Furthermore, this formula was administered for two weeks on dogs and the results showed no alterations on their physical and vital signs (Jun et al. 2014). Moreover, the repeated exposure to this endolysin elicited an immune response, which might be advantageous in the killing of infective cells.

It is expected that the first endolysin-based antimicrobials to get into the market will target the multidrug-resistant Gram-positive *S. aureus* pathogen. Recently, the pharmacokinetics, pharmacodynamics, and tolerance of SAL200 were assessed. SAL200 is a formulation containing the recombinant phage endolysin SAL-1 as its active pharmaceutical ingredient. A single intravenous injection of increasing doses of SAL200 (0.1, 0.3, 1, 3, and 10 $mg\,kg^{-1}$) was administered in healthy male volunteers and generally, the endolysin was well tolerated with no serious adverse effect noticed. Moreover, no clinically significant values were observed in different patient analysis (Jun et al. 2017).

Currently under development is a double-blinded and randomized superiority trial with a parallel group, designed to test the endolysin efficacy on patients with atopic dermatitis. In that trial with 100 patients, a placebo or an engineered endolysin named Staphefekt was administered in a cetomacrogol-based cream for 12 weeks. The patient recruitment was expected to be finished in August 2017, and the results are not published yet (Totté et al. 2017).

15.3.2.2.2 Protein Engineering of Gram-Positive Targeting Endolysins

After millions of years of coevolution, it can be said that endolysins were optimized to lyse the cells from within. However, there is still room to improve endolysins' activity from without. The huge diversity of enzymatic activities and different binding affinities opens the possibility to genetically engineer novel lysins with different binding and catalytic domains (Schmelcher et al. 2012). Domains can thus be switched to generate chimeric endolysins with different catalytic activities and binding specificities. This *a la carte* designing of endolysins can lead to the creation of proteins with enhanced functional properties, namely, efficacy, specificity, solubility, and thermostability (Fenton et al. 2010). Several chimerical endolysins were constructed to solve the solubility problems associated with Gram-positive pathogens (Daniel et al. 2010; Schmelcher et al. 2011; Fernandes et al. 2012; Mao et al. 2013). For instance, the chimeric endolysin ClyS was generated by the fusion of the catalytic domain of *Staphylococcus hyicus* phage Twort to the CBD of the *S. aureus* phage φNM3 endolysin. ClyS was shown to be highly soluble and very active in protecting mice against *S. aureus*-induced sepsis (Daniel et al. 2010). In 2006, Donovan et al. fused both the truncated and the entire versions of *S. agalactiae* phage B30 endolysin to the mature lysostaphin, and the resulting chimeras were active both on staphylococcal and streptococcal strains (Donovan et al. 2006).

Several fusions of ECD derived from the streptococcal phage lSA2 endolysin, with the *Staphylococcus* SH3 CBD, led to an increased activity against staphylococci while maintaining the streptolytic activity (Becker et al. 2009). Moreover, the duplication of a CBD in a single construct can increase the binding affinity, even at higher salt concentrations (Schmelcher et al. 2011). On the other hand, the elimination of the CBD from endolysins can increase their lytic spectra (Low et al. 2011; Mayer et al. 2011).

Taken together, these findings open new opportunities to customize and optimize different endolysins properties to create more potent proteins against a pathogen of interest (Gerstmans et al. 2018).

15.3.2.2.3 Resistance Development of Gram-Positive Targeting Endolysins

So far, no resistance mechanisms or resistant phenotypes emerged after exposure to low endolysin concentrations (Loeffler et al. 2001; Schuch et al. 2002; Pastagia et al. 2011). Even after 40 cycles of selection of streptococcal colonies that were picked in the border of lysis zone on a lawn, no resistant bacteria were detected (Loeffler et al. 2001). When *S. aureus* cells were exposed to increasing concentrations of a specific endolysin (ClyS) or to mupirocin, after eight days, low levels of resistance were only detected with the antibiotic treatment (Pastagia et al. 2011). In another study, an endolysin-susceptible *B. cereus* strain was randomly mutated with methanesulfonic acid ethyl ester, and no endolysin-resistant mutants emerged, while an increase on streptomycin and novobiocin resistance was observed (Schuch et al. 2002).

This lack of resistance may be a consequence of endolysin evolution to target bonds and/or molecules that are essential for bacterial viability (Fischetti 2005). While they are essential for cell viability, losing or modifying those ligands will likely result in cell death, which can cause the inability to isolate resistant mutants (Schuch et al. 2002).

Since resistance to other peptidoglycan hydrolases, namely, human lysozyme (Guariglia-Oropeza and Helmann 2011; Visweswaran et al. 2011) and lysostaphin (DeHart et al. 1995; Grundling et al. 2006) was described, it was theorized that bacteria could also acquire resistance mechanisms against endolysins, namely, by modifications within the pentaglycine bridge (Schmelcher et al. 2012). Hypothetically, even if some resistant strains start to emerge, the use of endolysins with different ECDs or endolysin cocktails may counteract the probability of resistance (Fenton et al. 2010). To avoid the possible resistance issue, a protein was created by combining three different ECDs with different cleavage sites (amidase, CHAP and M23 endopeptidase) that showed to be extremely active against staphylococcal strains (Donovan et al. 2009).

15.3.2.2.4 Immune Response to Gram-Positive Targeting Endolysins

For millions of years, humans have been exposed to phages and, therefore, to endolysins, and no apparent consequences are visible (Reyes et al. 2010). The highly conserved peptidoglycan cleavage sites and endolysin's specificity suggest that no eukaryotic cell will be affected. Moreover, the therapeutic administration of endolysins (topical or systemic) did not produce any relevant side effects so far (Gutiérrez et al. 2018).

An important issue in therapy is the potential immune response that the antimicrobial may elicit. Considering the protein nature of endolysins, it is expected that antibodies against endolysins can be produced. Mice injected with the pneumococcal endolysin Cpl-1 elicited IgG production; however, this did not affect the endolysin activity, still enabling mice recovery with no adverse effects (Loeffler et al. 2003). Although, in this same study, the hyperimmune serum of a rabbit model treated with endolysin reduced the enzyme efficacy *in vitro* (Loeffler et al. 2003). The inflammatory effect of Cpl-1 was compared with amoxicillin and the results showed to be similar (Witzenrath et al. 2009). On the other hand, in an endocarditis rat model, Cpl-1 led to a larger production of several cytokines than vancomycin (Entenza et al. 2005). These inconsistencies between different models are yet to be explained, but it should be noted that there is a set of different variables in the studies, namely, the animal species, endolysin dosage, and route of administration.

Several other studies using different endolysins and different bacteria also reported the development of anti-endolysin antibodies. Since endolysins from a Gram-positive background remain bound to the peptidoglycan after bacterial lysis, the inflammatory response raised by the cellular debris can explain this antibody production (Pastagia et al. 2013). Despite that, the majority of the studies reported no adverse side effects nor enzyme inactivation (reviewed in [Pastagia et al. 2013; Gerstmans et al. 2018]), something that might be explained by the high binding affinity and the fast kinetics of endolysins (Schmelcher et al. 2012).

15.3.2.3 Remarks on Gram-Positive Targeting Endolysins

The intrinsic muralytic activity of endolysins from a Gram-positive background together with high specificity and many other desired characteristics makes them the perfect antimicrobial. Their high potential to control pathogenic Gram-positive bacteria *in vitro* and *in vivo* has made them the most explored phage-encoded proteins so far. Their use in therapies remains very promising especially for topical applications where the human immune and cytokine responses decline. In fact, some endolysin-based products to treat skin infections cause by MRSA are very close to get into the market.

Despite the natural potential of endolysins, considering their modular structure, there is still a large room to improve their properties through protein engineering, allowing the creation of engineered endolysins that can probably target any Gram-positive pathogen with superior results. If optimized, as suggested by the recent results, systemic applications will probably be the next step. Furthermore, the use of endolysins to disinfect different medical devices, such as catheters, valves, and prostheses, is predicted and very likely.

Considering the endolysins specificity, the administration of lysins requires the identification of the etiological agent. Nevertheless, the improvements achieved on diagnostic tests and the desired rational for the use of any antimicrobial will probably facilitate the use of endolysins for therapy in the near future (Ince and McNally 2009; Pastagia et al. 2013).

An important fact to consider on the use of endolysins as antimicrobials is that after endolysin treatment, resensitization to obsolete antibiotics occurs (Daniel et al. 2010), something that empowers the therapeutic use of endolysins and their potential against antibiotic-resistant bacteria.

15.3.3 Gram-Negative Targeting Endolysins

The cell wall of Gram-negative bacteria presents some structural differences from that of Gram-positive bacteria. While Gram-positive bacterial cells have their peptidoglycan exposed from the outside, Gram-negative cells possess an OM involving the whole cell and isolating the peptidoglycan from the external environment. Such structure avoids the contact between endolysins released from the burst of phage-infected bacterial cells and the peptidoglycan of phage-uninfected and susceptible cells. Although the natural mode of action of endolysins from within is similar in Gram-positive and Gram-negative bacteria, the presence of the OM in Gram-negative bacteria has driven evolution of endolysins with different structures depending on the bacteria type that their source phage infects (Santos et al. 2018).

15.3.3.1 Gram-Negative Targeting Endolysin Structure

Endolysins encoded on phages infecting Gram-negative bacteria usually present a globular structure (in opposite with the typical modular structure of endolysins from phages infecting Gram-positive bacteria) composed of a single catalytic domain (Figure 15.5). Even so, rare cases exist of endolysins from a Gram-negative background that possess an experimentally confirmed modular structure composed of a binding and a catalytic domain. In these cases, the organization of the catalytic and binding domains is inverted when compared to the endolysins from a Gram-positive background, with the CBD at the N-terminal and the catalytic domain at the C-terminal (Briers et al. 2007; Fokine et al. 2008). The presence of such a binding domain confers to the modular endolysins a higher enzymatic activity, which can surpass more than 100 times that of T4 lysozyme, probably by keeping the enzyme in close proximity to its peptidoglycan substrate (Walmagh et al. 2012).

15.3.3.2 Gram-Negative Targeting Endolysins as Antimicrobials

On the WHO global priority pathogen list of antibiotic-resistant bacteria that pose the greatest threat to human health, only Gram-negative bacteria were classified as priority 1 (Critical), composed of *A. baumannii*, *P. aeruginosa*, and *Enterobacteriaceae* (WHO 2017).

However, research on the use of endolysins as antimicrobials through external addition has been almost exclusively focused on Gram-positive bacteria due to their exposed peptidoglycan layer. The existence of an OM in Gram-negative bacteria constitutes a physical protective barrier against the activity of endolysins from the outside hindering their contact with the peptidoglycan, which impairs their use as antimicrobials (*A. baumannii* seems to be an exception) (Fenton et al. 2010; Fischetti 2010; Lai et al. 2011; Schmelcher et al. 2012; Gerstmans et al. 2016; Oliveira et al. 2016). Consequently, the successful application of these enzymes in Gram-negative bacteria demands for a strategy that allows them to translocate the OM.

15.3.3.2.1 Combination of Gram-Negative Targeting Endolysins with OM Permeabilizers

One of the strategies is the use of substances that permeabilize the bacterial OM allowing the endolysin to get access to the peptidoglycan. Evaluation of different substances with OM permeabilization potential was assessed in their ability to enable the activity of the endolysin EL188 on antibiotic-resistant *P. aeruginosa* strains. The use of endolysin EL188 *per se* exhibited no antibacterial activity, but the combination of EL188 with ethylenediaminetetraacetic acid (EDTA) led to a reduction on the cell counts of *P. aeruginosa* of up to 4 log units in just 30 minutes, in a strain-dependent manner (Briers et al. 2011). Such study demonstrated that the application of endolysins therapeutically is not limited to Gram-positive bacteria and holds great potential against Gram-negative pathogens (Table 15.4).

Other examples include the *Salmonella* phage endolysin SPN1S. This endolysin did not affect the number of cell counts without EDTA; but after its addition, a dramatic reduction in bacterial viability was observed reaching a reduction of more than 4 log units. The lytic activity varied not only with the endolysin concentration but also with the EDTA concentration and the reaction time (Lim et al. 2012). Similarly, the *Salmonella* phage endolysin gp110 combined with EDTA reduced the number of cell counts of *P. aeruginosa* PAO1 in roughly 3 log units (but only 0.38 log units in the case of *S.* Typhimurium LT2 cells) (Rodríguez-Rubio et al. 2016).

The characterization of five new endolysins from a Gram-negative background with a globular structure targeting *Burkholderia cepacia, E. coli, S. enterica, and K. pneumoniae* showed that all five endolysins present a muralytic activity on the peptidoglycan of several Gram-negative bacterial species when applied exogenously in the presence of the OM permeabilizer (OMP) EDTA. Reductions of up to 3 log units on the cell counts were observed depending on the endolysin. Moreover, the results demonstrated that the muralytic activity of the endolysins depends on the type of the chemical bond that is cleaved within the peptidoglycan, with transglycosylases showing a substantially higher antibacterial activity compared to the amidases (Walmagh et al. 2013).

The ability of EDTA to allow the activity of endolysins seems to be dependent on the type and structure of the OM and its ability to remove the stabilizing divalent Mg^{2+} and Ca^{2+} cations present in the LPS layer of the Gram-negative bacteria (Walmagh et al. 2012). The differences on the LPS composition turn the *Pseudomonas* genus much more prone to the muralytic activity of the externally added endolysins than the *Enterobacteriaceae* (e.g. *E. coli* and *S.* Typhimurium), demonstrating that the positive effect of EDTA is strain/species dependent (Knirel et al. 2006).

Besides the good results obtained with the combined action of endolysins and EDTA, it is important to highlight that EDTA has a noticeable toxicity to biological membranes and is able to inhibit blood coagulation. Consequently, its use must be restricted to topical applications or to control Gram-negative bacteria in foodstuff (Love et al. 2018; Bolla et al. 2011). Despite its strain-dependent activity and its restricted topical/foodstuff application, EDTA still is the most well-known and used OMP and it has been used to assess the

Table 15.4 Gram-negative targeting endolysins as antimicrobials.

Source phage	Target	Model	Main achievements	References
Pseudomonas phage EL	*Pseudomonas aeruginosa*	*In vitro* – cell counts	Reductions >4 log units with ethylenediaminetetraacetic acid (EDTA) in 30 minutes in a strain-dependent manner	Briers et al. (2011)
Salmonella Typhimurium phage SPN1S	*Escherichia coli*	*In vitro* – cell counts	Reduction >4 log units with EDTA. Lytic activity varied with endolysin and EDTA concentration and reaction time. Active on *Salmonella* and *Pseudomonas*	Lim et al. (2012)
Salmonella phage 10	*P. aeruginosa* and *Salmonella* Typhimurium LT2	*In vitro* – cell counts	Roughly 3 log units reduction on *P. aeruginosa* cells. Only 0.38 log unit on *S. typhimurium* LT2 cells. Both results in the presence of EDTA	Rodríguez-Rubio et al. (2016)
Five endolysins	*P. aeruginosa*	*In vitro* – cell counts	Reductions of up to 3 log units on the cell counts were observed depending on the endolysin	Walmagh et al. (2013)
Salmonella phage φ68	Panel of Gram-negative bacteria	*In vitro* – cell counts	Citric and malic acids broadened activity to 9/11 genus tested (only on *Pseudomonas* with EDTA). Reductions of 5 log units on *Pseudomonas* and *Salmonella* and 3 log units on other genera. Reductions of 1 log unit on stationary phase and biofilm cells	Oliveira et al. (2014)
Salmonella phage S-394	*E. coli*	*In vitro* – cell counts	Reduction on the cell counts of up to 4 log units with EDTA and PGLa	Legotsky et al. (2014)

(*Continued*)

Table 15.4 (Continued)

Source phage	Target	Model	Main achievements	References
E. coli T5 phage	*E. coli*	*In vitro* – cell counts	Reduction of 4 log units on stationary cells with polymyxin B or chlorhexidine. With poly-L-lysine reduced 5 log units	Shavrina et al. (2016)
E. coli T4 and lambda phages	*P. aeruginosa, Shigella flexneri,* and *Yersinia enterocolitica*	*In vitro* – cell counts	Reductions of up to 3 log units when combined with high pressure	Nakimbugwe et al. (2006)
E. coli lambda phage	*E. coli* and *S.* Typhimurium	*In vitro* – cell counts	With high pressure, reductions of 6.5 log units in banana juice and skimmed milk (more than the HEWL)	Nakimbugwe et al. (2006)
P. aeruginosa phages φKZ and EL	*P. aeruginosa*	*In vitro* – cell counts	With high pressure, a reduction of 4 log units was observed compared to the endolysins alone. Endolysins can reduce the required pressure level	Briers et al. (2008)
Pseudomonas fluorescens phage OBP, *Salmonella* Enteritidis phage PVP-SE1 and *Pseudomonas chlororaphis* phage 201φ2-1	*Pseudomonas* and *E. coli*	*In vitro* – cell counts	Reduction on the bacterial counts of up to 4 log units with EDTA	Walmagh et al. (2012)
P. fluorescens phage OBP and *S.* Enteritidis phage PVP-SE1	*P. aeruginosa, Acinetobacter Baumannii, E. coli,* and *S. Typhimurium*	*In vitro* – cell counts	Reduction of up to 3.41 log units, with EDTA increased to more than 5 log units	Briers et al. (2014)

Chimeric protein	*P. aeruginosa* and *A. baumannii*	*In vitro* – cell counts and microfluidic flow cell	Eradication of large bacterial inocula (including persister and resistant cells) without hemolysis or cytotoxicity. No cross-resistance with antibiotics	Briers et al. (2014) and Defraine et al. (2016)
Chimeric protein	*E. coli, Shigella sonnei,* and *P. aeruginosa*	*In vitro* – cell counts	Reductions of up to 1 log unit	Zampara et al. (2018)
A. baumannii phage ϕAB2	*A. baumannii*	*In vitro* – optical density	Reduction on the optical density of a suspension of viable cells	Lai et al. (2011)
A. baumannii prophage	*A. baumannii*	*In vitro* – cell counts	Viability reduction of exponentially growing cells >3 log units and of stationary-phase cells >1 log unit. Reduction in biofilm cells in 1.6 log units (marked reduction in total biofilm biomass on catheters)	Lood et al. (2015)
Acinetobacter phage vb_AbaP_CEB1	*A. baumannii*	*In vitro* – cell counts	Reduction in cell counts of 2 log units. Antibacterial activity was improved and broadened with citric and malic acid, with reductions >5 log units	Oliveira et al. (2016)
P. aeruginosa phage	*Acinetobacter, E. coli, Klebsiella pneumoniae, and P. aeruginosa*	*In vitro* – cell counts	Up to 4 log units of an exponential growing suspension with log 8 cells of *P. aeruginosa* were killed. Antimicrobial activity also observed on *E. coli* and *K. pneumoniae*. Biofilms were disrupted	Guo et al. (2017)
S. Typhimurium phage SPN9CC	*Salmonella, Pseudomonas, E. coli*	*In vitro* – cell counts	Reduction of 2 log units in one hour in the number of viable *E. coli* cells in suspension. EDTA had no effect	Lim et al. (2014)

(*Continued*)

Table 15.4 (Continued)

Source phage	Target	Model	Main achievements	References
Citrobacter freundii phage CfP1	*C. freundii* and *Citrobacter koseri*	*In vitro* – cell counts	Reductions on the cell counts above 4 log units. Addition of EDTA resulted in the eradication of some strains	Oliveira et al. (2016)
A. baumannii prophage	*A. baumannii*	*In vivo* – catheters implanted in mice/ mice bacteremia	Reduction in the number of viable cells in 2 log units in catheters. Fifty percent of the mice with bacteremia were rescued (10% in the control group)	Lood et al. (2015)
Salmonella phage PVP-SE1	*P. aeruginosa*	*In vivo* – *Caenorhabditis elegans* nematode	A 63% of survival was observed	Briers et al. (2014)
Chimeric protein	*P. aeruginosa*	*In vivo* – dog	Two dog otitis refractory to antibiotics were healed and no relapses were reported	Briers and Lavigne (2015)

Research examples and main results.

antimicrobial activity of the majority of endolysins from a Gram-negative background. Consequently, this OMP can be used to compare the muralytic activity of the different Gram-negative phage endolysins.

The *Salmonella* phage endolysin Lys68 was tested against 11 different bacterial genera and its combination with EDTA allowed its activity only on *Pseudomonas* species. However, combination of Lys68 with the weak citric and malic acids broadened its antibacterial activity to 9 of the 11 genus, with reductions on the bacterial cell counts of up to 5 log units on *Pseudomonas* and 3 log units on the other genera within 30 minutes. By increasing the incubation time to two hours, the reductions on the cell counts of *Salmonella* can go up to 5 log units with citric acid. Moreover, such combination was also active on stationary-phase cells and bacterial biofilms, reducing the cell counts by approximately 1 log unit (Oliveira et al. 2014). The destabilization of the OM integrity by the acidity of weak acids combined with the muralytic activity of endolysins resistant at low pH thus seems to be a feasible approach to control Gram-negative bacteria, especially other than *Pseudomonas sp.*

The permeabilizing activity of poly-L-lysine and poly-L-arginine with different lengths was also studied and poly-L-arginine with molecular weight distribution from 5 to 15 kDa was the most active permeabilizing agent, similar to the cationic antimicrobial peptide PGLa. However, the use of poly-L-arginine or PGLa with the Lys394 *Salmonella* phage endolysin had no significant antimicrobial activity on *E. coli* planktonic cells. Identical results were obtained with EDTA and Lys394, but when Lys394 was combined with EDTA and PGLa, a reduction on the cell counts of up to 4 log units was obtained in 30 minutes, showing a synergistic effect between these peptides (Legotsky et al. 2014).

Different cationic OMPs were assessed in combination with the *E. coli* T5 phage endolysin on *E. coli* stationary cells and the results were promising. In this case, the poly-L-lysine combined with the endolysin produced a reduction in the bacterial counts of 4 log units and when the endolysin was combined with polymyxin B or chlorhexidine, the reduction was raised up to 5 log units (Shavrina et al. 2016).

15.3.3.2.2 Combination of Gram-Negative Targeting Endolysins with High Pressure

High-pressure treatment is a technology with application in the food industry and has been demonstrated to transiently permeabilize the OM of Gram-negative bacteria (Kalchayanand et al. 1994; Hauben et al. 1996; García-Graells et al. 1999; Masschalck et al. 2001; Nakimbugwe et al. 2006) broadening the antimicrobial range of different compounds (Briers et al. 2008) in a process termed pressure-assisted self-promoted uptake (Masschalck et al. 2000). The combination of this treatment with T4 *E. coli* phage endolysin and the lambda-phage endolysin allowed the reduction of cell counts of *P. aeruginosa*, *Shigella flexneri*, and *Yersinia enterocolitica* of up to 3 log units (Nakimbugwe et al.

2006) showing its potential application in foodstuff. The efficiency of the lambda-phage endolysin was also shown to be more effective than the hen egg white lysozyme (when combined with high pressure) in two food matrices (banana juice and skimmed milk) with reductions on the bacterial loads of *E. coli* and *S.* Typhimurium of up to 6.5 log units (Nakimbugwe et al. 2006).

The synergistic bactericidal effect between high hydrostatic pressure and the muralytic activity of the *Pseudomonas* bacteriophage-encoded endolysins KZ144 and EL188 was also observed with a reduction of up to 4 log units on the cell counts of *P. aeruginosa* when compared to the use of the endolysins alone. Moreover, the concomitant use of endolysins can reduce the pressure level required to achieve the desired reduction of undesired and pathogenic bacteria, making high pressure a more cost-effective treatment (Briers et al. 2008).

15.3.3.2.3 The Rare Modular Structure of Gram-Negative Targeting Endolysins

Three new endolysins, unique among endolysins from a Gram-negative background for their modular structure, presenting an N-terminal cell wall binding domain, showed a strong muralytic activity on the peptidoglycan of a broad range of Gram-negative bacteria. The combination of EDTA with each of the three endolysins – OBPgp279 from the *Pseudomonas fluorescens* phage OBP, PVP-SE1gp146 from the *S. enterica* serovar Enteritidis phage PVP-SE1, and 201φ2-1gp229 from the *Pseudomonas chlororaphis* phage 201φ2-1 – allowed for a reduction on the bacterial counts of up to 4 log units (Walmagh et al. 2012). The *Pseudomonas* phage endolysins KZ144 and EL188 also showed to present a modular structure and a high muralytic activity, 143 and 69 times higher compared with the broad-spectrum hen egg white lysozyme (Briers et al. 2007).

The endolysin structure seems to influence the antibacterial activity of Gram-negative endolysins (Walmagh et al. 2012, 2013). Such a strong muralytic activity and bacterial spectrum of Gram-negative phage endolysins seems to be improved by the presence of the cell wall binding domain (Briers et al. 2007; Walmagh et al. 2012). The strong affinity of the N-terminal peptidoglycan binding domains can provide increased enzyme–substrate proximity, which can explain the exceptionally high activity of Gram-negative modular endolysins (Briers et al. 2007; Briers et al. 2009). Enzymes presenting a modular structure (and consequently a higher reduction and a broader host range) should thus be favored to globular endolysins, since they have a higher potential to control these pathogens.

The important role of the endolysin binding domains on the muralytic activity of these enzymes was explored by combining the binding domain of endolysin KZ144 with the peptidoglycan hydrolase domain KMV36C of *P. aeruginosa* bacteriophage φKMV. Such combination increased the activity of the native catalytic domain by threefold (Briers et al. 2009) and demonstrated that the construction of new modular phage endolysins targeting Gram-negative bacteria is a rational approach to improve their specific activity and gives rise to promising antibacterials.

The improved activity of modular endolysins over the globular ones raises a question: Why only some endolysins from a Gram-negative origin evolved to a modular design, if it provides such a benefit (Briers et al. 2009)? The presence of an OM on Gram-negative bacteria that impairs lysis of new potential host cells by the released endolysins from previous burst phage-infected cells and the much less thick peptidoglycan of Gram-negative bacteria (which does not require such a high muralytic activity as the thicker peptidoglycan of Gram-positive bacteria) do not seem to imprint enough selective pressure to maintain a CBD (and thus a larger modular enzyme) on endolysins from a Gram-negative background.

15.3.3.2.4 Protein Engineering of Gram-Negative Targeting Endolysins

The use of endolysins against Gram-negative in a clinical and therapeutic context requires a product that presents no toxicity against eukaryotic cells, most likely protein based. Consequently, protein engineering is a promising strategy to build new endolysin-based chimeras that can be used as antibacterials on Gram-negative pathogens. With this in mind, a new class of antibacterials called Artilysin® was created. These chimeras are created through the fusion of an LPS-destabilizing peptide to either the N- or C-terminus of endolysins, without affecting the endolysin secondary and tertiary structures. The ability of these peptides to destabilize the LPS is based on their amphipathic or polycationic properties, allowing the endolysin to get access to the peptidoglycan and kill the bacterial cells (Gerstmans et al. 2016).

This concept was tested with the most active endolysins, those presenting a modular structure, to assess if Artilysins could further optimize their activity. Therefore, different peptides were fused to the endolysins OBPgp279 and PVP-SE1gp146 and challenged against multidrug-resistant strains of *P. aeruginosa*, *E. coli*, and *S.* Typhimurium, which resulted in an increased antibacterial activity of the parent endolysins, reaching a reduction of 2.61 log units in just 30 minutes on the cell counts of *P. aeruginosa*. However, no significant reduction was observed on *E. coli* and *S. enterica*, showing that the sensitivity to Artilysins may be dependent on the LPS structure. By optimizing the length of the peptide linker between the peptide and the endolysin, the reduction was increased to 3.41 log units (Briers et al. 2014).

The mechanism by which Artilysin destabilizes the OM is different from that of EDTA since the synergy with EDTA is not compromised. The addition of EDTA further improved the chimera activity up to more than 5 log unit reduction on the cell counts of *P. aeruginosa* (Briers et al. 2014).

A new Artilysin created through the fusion of SMAP-29 (the 29-amino-acid sheep myeloid antimicrobial peptide) to the N-terminus of the modular endolysin KZ144 was shown to have a high and fast bactericidal effect on multidrug-resistant strains of *P. aeruginosa* and *A. baumannii*. The bactericidal activity of this Artilysin, Art-175, is instantaneous resulting in a complete elimination of large inocula of *P. aeruginosa* and *A. baumannii*. Although SMAP-39 presents hemolytic activity, Art-175 did not produce hemolysis on erythrocytes or cytotoxicity on L-929 mouse connective tissue fibroblasts, showing a different bactericidal effect from SMAP-29 (Briers et al. 2014; Defraine et al. 2016).

Art-175 did not present cross-resistance with the resistance mechanisms of 21 therapeutically used antibiotics on *P. aeruginosa* or other first-line antibiotic classes (aminoglycosides, quinolones, and carbapenems) on *A. baumannii*. These multidrug-resistant strains were sensitive to Art-175 to the same extent as antibiotic-sensitive strains, which demonstrates the potential usefulness of Artilysins on the antibiotic resistance global problem (Briers et al. 2014; Defraine et al. 2016).

Also, no resistance development was observed for these bacteria after serial exposure to subinhibitory doses of Art-175. Remarkably, this chimera was found to be rapid and efficient in killing persister cells, dormant variants of regular cells that form stochastically in microbial populations and are highly tolerant to antibiotics and responsible for chronic infections (Lewis 2010; Briers et al. 2014; Defraine et al. 2016). This is possible because Art-175, similar to natural endolysins, does not require an active metabolism for its activity as it actively degrades the peptidoglycan layer.

Innolysins represent a new and different concept for engineering endolysins allowing their passage through the OM of Gram-negative bacteria. In this new concept, the enzymatic activity of endolysins is combined with the binding capacity of phage receptor binding proteins, and it was demonstrated through the fusion of the phage T5 endolysin and the receptor binding protein Pb5 that irreversibly binds to the phage receptor FhuA (which is involved in ferrichrome transport in *E. coli*). This chimera could reduce the number of cell counts of *E. coli*, *Shigella sonnei*, and *P. aeruginosa* up to 1 log unit (Zampara et al. 2018). A similar approach was already used through the fusion of the T4 endolysin and the FyuA binding domain of pesticin (FyuA belongs to the family of TonB-dependent transporters). Such hybrid protein was able to specifically kill *Yersinia* and pathogenic *E. coli* strains (Lukacik et al. 2012).

Protein engineering is thus a very promising approach to create novel endolysin-based antimicrobials that allow the use of endolysins to treat pathogenic Gram-negative bacteria.

15.3.3.2.5 Naturally OM Surpassing Gram-Negative Targeting Endolysins

From the growing number of phage endolysins targeting Gram-negative bacteria that have been identified and studied in the last years, some were found to present an innate antibacterial activity against Gram-negative strains without the need of using an OMP. This happens with a high frequency with endolysins from phages infecting the problematic *A. baumannii* and when this specie is the target. The endolysin LysAB2 was able to reduce the optical density of a suspension of viable cells of *A. baumannii* (Lai et al. 2011). A higher activity was observed with the globular endolysin PlyF307, reducing the viability of exponentially growing cells by more than 3 log units and the viability of stationary-phase bacteria by more than 1 log unit. In addition, PlyF307 was able to reduce in 1.6 log units the number of *A. baumannii* cells in biofilm with a marked reduction in total biofilm biomass on catheters (Lood et al. 2015).

The globular *A. baumannii* phage endolysin ABgp46 also showed innate and specific antimicrobial activity against multidrug-resistant *A. baumannii* strains, reducing the cell counts up to 2 log units. The antibacterial activity was improved and broadened in the presence of citric and malic acid achieving reductions over 5 log units to values below the detection limit (Oliveira et al. 2016).

The T4 phage endolysin was the first described to present antimicrobial activity on Gram-negative cells without the use of OMP. Nevertheless, the antimicrobial effect was devoid of enzymatic activity and attributed to the positively charged amphipathic nature of the endolysin C-terminal (Düring et al. 1999).

An intrinsic and native destabilization of the OM of *P. aeruginosa* was found to be a characteristic of the *P. fluorescens* phage endolysin OBPgp279, which reduced the number of cell counts of that bacterium in 1 log unit in the absence of any OMP (Walmagh et al. 2012). A higher antimicrobial activity was observed with the *P. aeruginosa* phage endolysin LysPA26, which could kill up to 4 log units of an exponential growing suspension with 10^8 cells of *P. aeruginosa*. The antimicrobial activity was also effective against *E. coli* and *K. pneumoniae*. Moreover, LysPA26 could disrupt *P. aeruginosa* biofilms in a concentration-dependent manner (Guo et al. 2017).

The addition of the *Salmonella* Typhimurium phage endolysin SPN9CC to a suspension of viable *E. coli* cells reduced its number in 2 log units in just one hour. Under these same conditions, adding EDTA could not increase the antibacterial activity of the endolysin (Lim et al. 2014).

An endolysin from the Maltocin P28 phage tail-like bacteriocin (probably a VAL) produced by *Stenotrophomonas maltophilia*, called P28, presents intrinsic antibacterial activity not only on Gram-negative but also on Gram-positive bacteria. While reductions on viable cell numbers of Gram-positive strains were roughly 2 log units, the viable cell numbers of the Gram-negative strains *Klebsiella mobilis*, *Xanthomonas campestris*, and *S. flexneri* were remarkably reduced from 3 to 5 log units (Dong et al. 2015).

Reductions on the cell counts above 4 log units on highly resistant *Citrobacter freundii* and *Citrobacter koseri* isolates were obtained with the *Citrobacter* phage endolysin CfP1, without any OMP. This strong antibacterial activity could be improved by the addition of EDTA, which led to the eradication of some of the strains (Oliveira et al. 2016).

As we see this spontaneous activity of endolysins against Gram-negative bacteria (that have a protective OM), one important observation should be made. All these studies have tested endolysins against pathogens that were suspended in low ionic strength buffers, which confer a lower stabilization of the bacterial OMs and subsequently enhance the antibacterial activity of endolysins that otherwise could be absent. Therefore, as the tested conditions of endolysins seem to be very restricted, one can expect that the efficacy of these enzymes may not be translated to other *in vitro* environmental conditions and also to *in vivo*.

15.3.3.2.6 Gram-Negative Targeting Endolysins with Enhanced Thermoresistance

The heat treatments commonly used in the food processing industry require that the antibacterials, when needed to be added in the process, are thermoresistant or their activity will be compromised. A few number of endolysins targeting Gram-negative bacteria, especially those targeting problematic bacteria in foodstuff, were found to present different degrees of resistance to high temperatures. The endolysin PVP-SE1gp146, encoded on the *Salmonella* phage PVP-SE1, was the first reported thermoresistant Gram-negative phage endolysin. This endolysin could maintain its activity at temperatures up to 90 °C (Walmagh et al. 2012). The endolysin retained its full activity even after being incubated for 1 hour at 80 °C and was only completely inactivated after 40 minutes at 100 °C. Similarly, another *Salmonella* phage endolysin, Lys68, could only be inactivated when incubated for 30 minutes at 100 °C, maintaining 54.7% of its residual activity at 80 °C (Oliveira et al. 2014). Circular dichroism analysis demonstrated the ability of Lys68 to refold into its original conformation upon thermal denaturation.

The *Pseudomonas* phage endolysin LysPA26 was able to retain its full activity at 50 °C and still has residual activity against OM-permeabilized cells under heat treatment at 100 °C (Guo et al. 2017). Endolysins KZ144 and EL188 present high activity up to 50 °C, but at 60 °C the activity of both endolysins is completely abolished (Briers et al. 2007). As such, the thermoresistance of these two endolysins is very limited.

To some of these endolysins, it was shown that they are able to rapidly refold after unfolding due to high temperatures (Oliveira et al. 2014). This refolding ability, coupled with a stable conformational structure, can be the reason for the high thermoresistance presented by these Gram-negative phage endolysins (Guo et al. 2017).

The high thermoresistance characteristic of these endolysins demonstrates their potential use as antibacterial components in hurdle technology for food preservation (Walmagh et al. 2012), increasing food safety and decreasing the use of chemical preservatives and the intensity of heat treatments, improving food quality and nutritional values. As a result, the use of endolysins in food has the potential to decrease the number of foodborne diseases and consequently to decrease the use of antibiotics to treat such diseases, which will contribute to combat the antimicrobial resistance crisis.

15.3.3.2.7 In Vivo *Trials of Gram-Negative Targeting Endolysins*

Understanding the true therapeutic potential of endolysins and its use on a clinical context requires the *in vivo* study. If the number of identified and studied phage endolysins that have antimicrobial activity on Gram-negative bacteria without hemolysis or cytotoxicity on eukaryotic cells is relatively scarce, the number of *in vivo* trials is diminutive.

Acinetobacter baumannii is especially problematic due to its ability to form biofilms on medical implants and catheters, something that contributes to its persistence and resistance to antimicrobials. With this in mind, catheters colonized with two-day-old *A. baumannii* biofilms were implanted subcutaneously in the backs of mice and challenged with the PlyF307 endolysin (two doses within four hours). The endolysin was able to reduce the number of viable cells in 2 log units. Considering these positive results, the endolysin was further assessed for its ability to work systemically and to rescue mice from lethal bacteremia caused by *A. baumannii*, and with only one dose of PlyF307, 50% of the mice were rescued, while only 10% of the control group survived to this highly lethal dose of *A. baumannii* (Lood et al. 2015).

A human neonatal foreskin keratinocyte model composed of a confluent keratinocyte monolayer infected with *P. aeruginosa* was used to evaluate PVP-SE1gp146 and LoGT-008 (Artilysin variant of PVP-SE1gp146). Both antimicrobials were able to fully protect the monolayer from the otherwise *P. aeruginosa* cytotoxic effects, an effect that was associated with a drastic decrease in the bacterial load (Briers et al. 2014). Consequently, LoGT-008 was further assessed for its efficacy *in vivo* on a *Caenorhabditis elegans* nematode gut infection model, which resulted in a survival rate of only 10%. When Artilysin was used, a protective effect superior to that of ciprofloxacin was observed (63% versus 45% survival) demonstrating its therapeutic potential application. The native PVP-SE1gp146 endolysin also rescued 40% of the infected worms when EDTA was added, a result similar to that of ciprofloxacin (Briers et al. 2014).

Two dogs with otitis that could not be healed with antibiotics for a prolonged time were then treated with Art-085 (Artilysin composed of the endolysin wild-type KZ144 fused to SMAP-29 at its N-terminus). Treatment with Art-085 clearly increased the dog's well-being and healed the dog's ears. Moreover, no relapses were reported (Briers and Lavigne 2015).

15.3.3.3 Remarks on Gram-Negative Targeting Endolysins

As we have seen, many are the advantages of endolysins as new and effective antimicrobials: their mode of action based on enzymatic peptidoglycan degradation that does not require an active metabolism with the consequent ability to kill persister and stationary cells; their effectiveness on the problematic biofilm arrangement of pathogenic bacteria; the absence of cross-resistance with antibiotic resistance mechanisms and thus their effectiveness on multidrug-resistant strains; and the immutable nature of peptidoglycan that turns highly unlikely the development of resistance, among many others. Even so, the existence of an OM in Gram-negative cells has impaired their use on these bacteria and their efficient use will almost always require the use of OMPs. We showed here that some strategies have been developed, turning the use of endolysins effective and efficient against problematic and resistant pathogenic Gram-negative bacteria. Considering the strategies presented, we believe that those based on protein engineering will rise and be established as the most promising strategy for therapeutic application since it will result in a completely protein-based product, most likely biodegradable, and thus with low probability to accumulate in the environment and with a lower probability of inducing resistance in bacteria, contrasting with many persisting antibiotics.

15.4 Holins

Both Gram-positive and Gram-negative bacteria present an inner membrane in their cell structure that separates and impairs the contact of the intracellular content with the peptidoglycan of the cell wall. Consequently, the endolysin that accumulates during phage replication to produce cell lysis and the release of the progeny phages usually cannot have access to its substrate per se. It is here where holins exert their main function. Holins are small hydrophobic membrane proteins that accumulate in the membrane. After reaching a genetically defined threshold concentration, they are triggered to form holes. These "time-programmed" holes are usually large enough to allow the passage of the fully folded endolysins and its contact with the peptidoglycan. The consequences are the digestion of the peptidoglycan and the bursting of the cell with the release of the phage progeny (Figure 15.1i) (Smith et al. 1998; Dewey et al. 2010; Fernandes and São-José 2018).

15.4.1 Holin Structure

Holins constitute one of the most diverse functional groups, and 52 families of established or putative holins have been identified. Experimental topological analyses suggest that these small proteins span the cytoplasmic membrane from one to four times as transmembrane α-helical segments. Within the same family, the protein size and topology are usually conserved and they have been identified in phages infecting both Gram-negative and Gram-positive bacteria (Reddy and Saier 2013; Saier and Reddy 2015). The high diversity in holin sequences makes it difficult to identify them although, they share some common characteristics: the presence of the said transmembrane domains; the highly charged and hydrophilic C-terminal domains; and the proximity of their encoding gene to that of the endolysin (Saier and Reddy 2015).

The function of these proteins is not yet well defined. Although it is clear that they confer access of endolysins to the bacterial peptidoglycan, it is not clear if that access is made through secretion, leakage, or membrane lysis, something that may depend on the holin type (Saier and Reddy 2015).

15.4.2 Holins as Antimicrobials

Besides their ancillary function in bacterial lysis, holins can cause cell death independent of endolysins and, unlike these, have an unspecific broad-spectrum antibacterial activity against both Gram-positive and Gram-negative bacteria. The holin HolGH15 from the *S. aureus* phage GH15 showed efficient antimicrobial activity on *S. aureus* and also on *L. monocytogenes*, *B. subtilis*, *P. aeruginosa*, *K. pneumoniae*, and *E. coli*. The antimicrobial activity was most notorious on *Staphylococcus* and *Listeria*, with reductions close to 5 log units. Importantly, HolGH15 did not induce erythrocyte rupture on defibrinated rabbit blood, showing that this protein that exerts its function on biological membranes somehow unexpectedly does not cause hemolysis (Jiang et al. 2016).

A metagenomic study of a goat skin surface identified a holin-like protein shown to complement the holin function in a lysis-defective bacteriophage lambda. The 34-amino-acid protein named Tmp1 presented antimicrobial activity on a panel of Gram-positive bacteria, but not on Gram-negative bacteria. By using error-prone PCR, four mutants of Tmp1 were obtained with an increased spectrum of antibacterial activity that also includes Gram-negative bacteria, with reductions in the number of viable bacteria of up to 4 log units. The increased antibacterial activity seems to be related with an increased hydrophobicity (independent of the protein net charge), but still did not show hemolytic activity. This study, besides demonstrating the antimicrobial potential of holins, also shows that they can be improved through protein engineering (Rajesh et al. 2011).

Strangely, the *S. suis* bacteriophage holin HolSMP causes weak lysis of *S. aureus* and *B. subtilis* cells but not of the phage host (Shi et al. 2012). Also surprising, these two strains were insensitive to its cognate endolysin LySMP. The combination of these two phage lytic proteins, the holin HolSMP and the endolysin LySMP, extended the spectrum of both proteins in a synergistic effect since the combination presented antibacterial activity not only on strains that were sensitive to any of the lytic proteins but also on many other drug-resistant strains that were insensitive to each of the proteins alone (Shi et al. 2012). Such results demonstrate that the combined use of a holin and an endolysin presents a higher potential as antibacterials against drug-resistant strains.

15.4.3 Remarks on Holins

The small number of studies on the antibacterial activity of holins suggest that they have potential against many pathogenic bacteria, including multidrug-resistant bacteria. The antibacterial properties seem to be related with their characteristic transmembrane domains and hydrophobicity, but further studies need to be carried out to understand their mode of action, especially to understand how these proteins that exert their natural action on biological membranes interact with the cell wall of Gram-positive bacteria that possess peptidoglycan as their external layer.

Activity improvement was shown to be a possible approach that can increase the range of sensitive bacteria, but holin unspecificity can be a drawback to their use as antimicrobials, limiting their application to surface disinfection or foodstuff sterilization. In cases where specificity is required, new strategies have to be designed and can include their fusion with other proteins that specifically bind to the target bacteria. This seems a promising approach since fusions have been shown not to affect the holin activity (White et al. 2011). Proteins with potential to confer specificity to holins include the CBDs of endolysins or phage recognition binding proteins, both from phage origin.

The potential of holins as antimicrobials has been impaired by their hydrophobic, insoluble, and toxic nature making difficult their cloning and expression in high yields that are required for their study. Consequently, exploring all their potential as antimicrobials will depend on the technological advances to solve these issues.

15.5 Final Considerations

The current and serious global crisis of antimicrobial resistance has been caused by incorrect use and abuse of antibiotics leading the humankind to a post-antibiotic era where small injuries or infections may result in death. This is demonstrated by the common appearance of bacterial strains showing resistance to most or even all existing classes of antibiotics, a fact already warned by the WHO.

Driven by this crisis and the difficulty in discovering new antibiotics, research on alternatives has increased considerably, particularly on bacteriophage proteins with antimicrobial activity (mainly endolysins) and their possible applications. Research on this field has undergone a significant acceleration in the last decade. The potential of phage enzymes is so obvious that a recent review on possible alternatives to antibiotics (Czaplewski et al. 2016) pointed phage endolysins as the therapeutics with the greatest potential considering their high clinical impact and high technical feasibility. But the use of phage lytic proteins is not limited to medicine; their antimicrobial potential has also been demonstrated in the veterinary and food sectors, agricultural and environmental control, and abiotic surfaces as disinfectants.

In addition to the natural antimicrobial potential of phage-derived proteins, protein engineering and synthetic biology of these proteins created new chimeras with improved and new properties providing even more efficient antimicrobials. The modular structure of endolysins has facilitated their improvement, but fusions of globular phage proteins and mutagenesis or direct evolution has also been proven efficient in optimizing and broadening their possible applications. The increasing knowledge on bioinformatics and proteomics will provide, in the near future, new opportunities to engineer and design proteins *a la carte*, meeting the specific requirements for a particular application. Some of the properties that were already improved include specificity, host range, fastness, antimicrobial activity, and enhanced thermostability. There is a huge number of possibilities, and this number increases every time that a new phage protein is discovered and is biochemically and structurally characterized.

We have provided here a number of examples that demonstrates the antimicrobial potential of phage proteins *in vitro* and *in vivo*. The potential is real, and a proof of that is the existence of phage–endolysin-based formulations in phase I and phase II human clinical trials against the problematic and multiresistant pathogenic *S. aureus*.

Although the success is more notorious for endolysins targeting Gram-positive pathogens, those targeting Gram-negative bacteria are also emerging. An example is Medolysin (from Lysando), a wound care spray based on Artilysin with bactericidal activity, which is on preclinical trials (Gerstmans et al. 2018). It is important to note that until recently the therapeutic use of endolysins targeting Gram-negative bacteria was unthinkable. The VALs have also found their place, mainly on the food industry due to their high thermoresistance, but many other applications can be envisaged. In the case of depolymerases and holins, research is still in the beginning and there is still a long way to go; but considering the evidences, the expectations are exciting.

Phage-derived proteins compose thus excellent alternatives to antibiotics due to the following: (i) their efficient antimicrobial activity on bacteria, producing rapid lysis of the cells and rapid reductions on cell counts; (ii) a unique mode of action that does not require an active metabolic cell as antibiotics do, allowing their activity on persister and biofilm cells; (iii) their specificity, which makes them active only on the target bacteria, avoiding interference with the normal flora; (iv) the ability to combine synergistically with other antimicrobials (including antibiotics); (v) the possibility to easily engineer through domain shuffling and fusion with other proteins; (vi) the absence of cross-resistance with existing antibiotic resistance mechanisms, thus making them efficient against multidrug-resistant strains; (vii) no resistance development, something that can be attributed to the immutable nature of peptidoglycan; (viii) their biodegradability due to their proteinaceous nature and consequently their non-accumulation in nature, reducing the risk to induce resistant bacteria in the environment (in contrast with antibiotics); (ix) phages are the most abundant and diverse biological entities on earth, thus representing an almost infinite source of new proteins with different properties that can be used to target almost any bacterial pathogen.

It is thus easy to conclude that phage-derived proteins will play, in a near future, a predominant role in the combat to the established antimicrobial crisis.

References

Adams, M.H. and Park, B.H. (1956). An enzyme produced by a phage-host cell system. II. The properties of the polysaccharide depolymerase.

Baker, J.R., Liu, C., Dong, S., and Pritchard, D.G. (2006). Endopeptidase and glycosidase activities of the bacteriophage B30 lysin. *Appl. Environ. Microbiol.* 72: 6825–6828.

Barbirz, S., Müller, J.J., Uetrecht, C. et al. (2008). Crystal structure of *Escherichia coli* phage HK620 tailspike: podoviral tailspike endoglycosidase modules are evolutionarily related. *Mol. Microbiol.* 69: 303–316.

Baxa, U., Steinbacher, S., Miller, S. et al. (1996). Interactions of phage P22 tails with their cellular receptor, Salmonella O-antigen polysaccharide. *Biophys. J.* 71: 2040–2048.

Becker, S.C., Foster-Frey, J., Stodola, A.J. et al. (2009). Differentially conserved staphylococcal SH3b_5 cell wall binding domains confer increased staphylolytic and streptolytic activity to a streptococcal prophage endolysin domain. *Gene* 443: 32–41.

Bertozzi Silva, J., Storms, Z., and Sauvageau, D. (2016). Host receptors for bacteriophage adsorption. *FEMS Microbiol. Lett.* 363: fnw002.

Blázquez, B., Fresco-Taboada, A., Iglesias-Bexiga, M. et al. (2016). PL3 amidase, a tailor-made lysin constructed by domain shuffling with potent killing activity against pneumococci and related species. *Front. Microbiol.* 7: 1156.

Bolla, J.-M., Alibert-Franco, S., Handzlik, J. et al. (2011). Strategies for bypassing the membrane barrier in multidrug resistant Gram-negative bacteria. *FEBS Lett.* 585: 1682–1690.

Born, Y., Fieseler, L., Thöny, V. et al. (2017). Engineering of bacteriophages Y2:: *dpoL1-C* and Y2:: *luxAB* for efficient control and rapid detection of the fire blight pathogen, *Erwinia amylovora*. *Appl. Environ. Microbiol.* 83: pii: e00341-17.

Borysowski, J., Weber-Dabrowska, B., and Górski, A. (2006). Bacteriophage endolysins as a novel class of antibacterial agents. *Exp. Biol. Med. (Maywood).* 231: 366–377.

Briers, Y., Cornelissen, A., Aertsen, A. et al. (2008). Analysis of outer membrane permeability of *Pseudomonas aeruginosa* and bactericidal activity of endolysins KZ144 and EL188 under high hydrostatic pressure. *FEMS Microbiol. Lett.* 280: 113–119.

Briers, Y. and Lavigne, R. (2015). Breaking barriers: expansion of the use of endolysins as novel antibacterials against Gram-negative bacteria. *Future Microbiol.* 10: 377–390.

Briers, Y., Lavigne, R., Plessers, P. et al. (2006). Stability analysis of the bacteriophage ϕKMV lysin gp36C and its putative role during infection. *Cell. Mol. Life Sci.* 63: 1899–1905.

Briers, Y., Miroshnikov, K., Chertkov, O. et al. (2008). The structural peptidoglycan hydrolase gp181 of bacteriophage ϕKZ. *Biochem. Biophys. Res. Commun.* 374: 747–751.

Briers, Y., Schmelcher, M., Loessner, M.J. et al. (2009). The high-affinity peptidoglycan binding domain of *Pseudomonas phage* endolysin KZ144. *Biochem. Biophys. Res. Commun.* 383: 187–191.

Briers, Y., Volckaert, G., Cornelissen, A. et al. (2007). Muralytic activity and modular structure of the endolysins of *Pseudomonas aeruginosa* bacteriophages? KZ and EL. *Mol. Microbiol.* 65: 1334–1344.

Briers, Y., Walmagh, M., Grymonprez, B. et al. (2014). Art-175 is a highly efficient antibacterial against multidrug-resistant strains and persisters of *Pseudomonas aeruginosa*. *Antimicrob. Agents Chemother.* 58: 3774–3784.

Briers, Y., Walmagh, M., and Lavigne, R. (2011). Use of bacteriophage endolysin EL188 and outer membrane permeabilizers against *Pseudomonas aeruginosa*. *J. Appl. Microbiol.* 110: 778–785.

Briers, Y., Walmagh, M., Van Puyenbroeck, V. et al. (2014). Engineered endolysin-based 'Artilysins' to combat multidrug-resistant gram-negative pathogens. *MBio* 5: e01379-14.

Caldentey, J. and Bamford, D.H. (1992). The lytic enzyme of the Pseudomonas phage Phi-6: purification and biochemical-characterization. *Biochim. Biophys. Acta* 1159: 44–50.

Chang, Y. and Ryu, S. (2017). Characterization of a novel cell wall binding domain-containing *Staphylococcus aureus* endolysin LysSA97. *Appl. Microbiol. Biotechnol.* 101: 147–158.

Chen, Y., Sun, E., Yang, L. et al. (2018). Therapeutic application of bacteriophage PHB02 and its putative depolymerase against *Pasteurella multocida* capsular Type A in mice. *Front. Microbiol.* 9: 1678.

Cheng, Q., Nelson, D., Zhu, S., and Fischetti, V.A. (2005). Removal of group B streptococci colonizing the vagina and oropharynx of mice with a bacteriophage lytic enzyme. *Antimicrob. Agents Chemother.* 49: 111–117.

Chopra, S., Harjai, K., and Chhibber, S. (2015). Potential of sequential treatment with minocycline and *S. aureus* specific phage lysin in eradication of MRSA biofilms: an in vitro study. *Appl. Microbiol. Biotechnol.* 99: 3201–3210.

Chua, J.E., Manning, P.A., and Morona, R. (1999). The Shigella flexneri bacteriophage Sf6 tailspike protein (TSP)/endorhamnosidase is related to the bacteriophage P22 TSP and has a motif common to exo- and endoglycanases, and C-5 epimerases. *Microbiology* 145 (Pt 7): 1649–1659.

Clarke, B.R., Esumeh, F., and Roberts, I.S. (2000). Cloning, expression, and purification of the K5 capsular polysaccharide lyase (KflA) from coliphage K5A: evidence for two distinct K5 lyase enzymes. *J. Bacteriol.* 182: 3761–3766.

Czaplewski, L., Bax, R., Clokie, M. et al. (2016). Alternatives to antibiotics: a pipeline portfolio review. *Lancet Infect. Dis.* 16: 239–251.

Daniel, A., Euler, C., Collin, M. et al. (2010). Synergism between a novel chimeric lysin and oxacillin protects against infection by methicillin-resistant *Staphylococcus aureus*. *Antimicrob. Agents Chemother.* 54: 1603–1612.

Davies, G. and Henrissat, B. (1995). Structures and mechanisms of glycosyl hydrolases. *Structure* 3: 853–859.

Defraine, V., Schuermans, J., Grymonprez, B. et al. (2016). Efficacy of artilysin Art-175 against resistant and persistent *Acinetobacter baumannii*. *Antimicrob. Agents Chemother.* 60: 3480–3488.

DeHart, H.P., Heath, H.E., Heath, L.S. et al. (1995). The lysostaphin endopeptidase resistance gene (epr) specifies modification of peptidoglycan cross bridges in *Staphylococcus simulans* and *Staphylococcus aureus*. *Appl. Environ. Microbiol.* 61: 1475–1479.

Dewey, J.S., Savva, C.G., White, R.L. et al. (2010). Micron-scale holes terminate the phage infection cycle. *Proc. Natl. Acad. Sci.* 107: 2219–2223.

Díez-Martínez, R., de Paz, H., Bustamante, N. et al. (2013). Improving the lethal effect of Cpl-7, a pneumococcal phage lysozyme with broad bactericidal activity, by inverting the net charge of its cell wall-binding module. *Antimicrob. Agents Chemother.* 57: 5355–5365.

Diez-Martinez, R., De Paz, H.D., Garcia-Fernandez, E. et al. (2015). A novel chimeric phage lysin with high in vitro and in vivo bactericidal activity against *Streptococcus pneumoniae*. *J. Antimicrob. Chemother.* 70: 1763–1773.

Dong, Q., Wang, J., Yang, H. et al. (2015). Construction of a chimeric lysin Ply187N-V12C with extended lytic activity against staphylococci and streptococci. *Microb Biotechnol* 8: 210–220.

Dong, H., Zhu, C., Chen, J. et al. (2015). Antibacterial activity of *Stenotrophomonas maltophilia* endolysin P28 against both Gram-positive and Gram-negative bacteria. *Front. Microbiol.* 6: 1299.

Donovan, D.M., Becker, S.C., Dong, S. et al. (2009). Peptidoglycan hydrolase enzyme fusions for treating multi-drug resistant pathogens. *Biotech Int.* 21: 6–10.

Donovan, D.M., Dong, S., Garrett, W. et al. (2006). Peptidoglycan hydrolase fusions maintain their parental specificities. *Appl. Environ. Microbiol.* 72: 2988–2996.

Düring, K., Porsch, P., Mahn, A. et al. (1999). The non-enzymatic microbicidal activity of lysozymes. *FEBS Lett.* 449: 93–100.

Entenza, J.M., Loeffler, J.M., Grandgirard, D. et al. (2005). Therapeutic effects of bacteriophage Cpl-1 lysin against *Streptococcus pneumoniae* endocarditis in rats. *Antimicrob. Agents Chemother.* 49: 4789–4792.

Fenton, M., Ross, P., McAuliffe, O. et al. (2010). Recombinant bacteriophage lysins as antibacterials. *Bioeng. Bugs* 1: 9–16.

Fernandes, S., Proença, D., Cantante, C. et al. (2012). Novel chimerical endolysins with broad antimicrobial activity against methicillin-resistant *Staphylococcus aureus*. *Microb. Drug Resist.* 18: 333–343.

Fernandes, S. and São-José, C. (2018). Enzymes and mechanisms employed by tailed bacteriophages to breach the bacterial cell barriers. *Viruses* 10: 396.

Fischetti, V.A. (2005). Bacteriophage lytic enzymes: novel anti-infectives. *Trends Microbiol.* 13: 491–496.

Fischetti, V.A. (2010). Bacteriophage endolysins: a novel anti-infective to control Gram-positive pathogens. *Int J Med Microbiol* 300: 357–362.

Flachowsky, H., Richter, K., Kim, W.-S. et al. (2008). Transgenic expression of a viral EPS-depolymerase is potentially useful to induce fire blight resistance in apple. *Ann. Appl. Biol.* 153: 345–355.

Fleming, A. (1945). *Penicillin: Nobel Lecture* 83–93. The Nobel Foundation. https://www.nobelprize.org/uploads/2018/06/fleming-lecture.pdf (accessed 7 January 2019).

Fokine, A., Miroshnikov, K.A., Shneider, M.M. et al. (2008). Structure of the bacteriophage ϕKZ lytic transglycosylase gp144. *J. Biol. Chem.* 283: 7242–7250.

Fung, C.-P., Chang, F.-Y., Lee, S.-C. et al. (2002). A global emerging disease of *Klebsiella pneumoniae* liver abscess: is serotype K1 an important factor for complicated endophthalmitis? *Gut* 50: 420–424.

García-Graells, C., Masschalck, B., and Michiels, C.W. (1999). Inactivation of *Escherichia coli* in milk by high-hydrostatic-pressure treatment in combination with antimicrobial peptides. *J. Food Prot.* 62: 1248–1254.

Geisinger, E. and Isberg, R.R. (2015). Antibiotic modulation of capsular exopolysaccharide and virulence in *Acinetobacter baumannii*. *PLOS Pathog.* 11: e1004691.

Gerstmans, H., Criel, B., and Briers, Y. (2018). Synthetic biology of modular endolysins. *Biotechnol. Adv.* 36: 624–640.

Gerstmans, H., Rodriguez-Rubio, L., Lavigne, R., and Briers, Y. (2016). From endolysins to Artilysin(R)s: novel enzyme-based approaches to kill drug-resistant bacteria. *Biochem. Soc. Trans.* 44: 123–128.

Gilmer, D.B., Schmitz, J.E., Euler, C.W., and Fischetti, V.A. (2013). Novel bacteriophage lysin with broad lytic activity protects against mixed infection by *Streptococcus pyogenes* and methicillin-resistant *Staphylococcus aureus*. *Antimicrob. Agents Chemother.* 57: 2743–2750.

Giordano, A., Dincman, T., Clyburn, B.E. et al. (2015). Clinical features and outcomes of *Pasteurella multocida* infection. *Medicine (Baltimore).* 94: e1285.

Glonti, T., Chanishvili, N., and Taylor, P.W. (2010). Bacteriophage-derived enzyme that depolymerizes the alginic acid capsule associated with cystic fibrosis isolates of *Pseudomonas aeruginosa*. *J. Appl. Microbiol.* 108: 695–702.

Grundling, A., Missiakas, D.M., and Schneewind, O. (2006). *Staphylococcus aureus* mutants with increased lysostaphin resistance. *J. Bacteriol.* 188: 6286–6297.

Gu, J., Xu, W., Lei, L. et al. (2011). LysGH15, a novel bacteriophage lysin, protects a murine bacteremia model efficiently against lethal methicillin-resistant *Staphylococcus aureus* infection. *J. Clin. Microbiol.* 49: 111–117.

Guariglia-Oropeza, V. and Helmann, J.D. (2011). Bacillus subtilis σ(V) confers lysozyme resistance by activation of two cell wall modification pathways, peptidoglycan O-acetylation and D-alanylation of teichoic acids. *J. Bacteriol.* 193: 6223–6232.

Guichard, J.A., Middleton, P.C., and McConnell, M.R. (2013). Genetic analysis of structural proteins in the adsorption apparatus of bacteriophage epsilon 15. *World J Virol* 2: 152–159.

Guo, M., Feng, C., Ren, J. et al. (2017). A novel antimicrobial endolysin, LysPA26, against *Pseudomonas aeruginosa*. *Front. Microbiol.* 8: 293.

Guo, Z., Huang, J., Yan, G. et al. (2017). Identification and characterization of Dpo42, a novel depolymerase derived from the *Escherichia coli* phage vB_EcoM_ECOO78. *Front. Microbiol.* 8: 1460.

Gutiérrez, D., Briers, Y., Rodríguez-Rubio, L. et al. (2015). Role of the pre-neck appendage protein (Dpo7) from phage vB_SepiS-phiIPLA7 as an anti-biofilm agent in staphylococcal species. *Front. Microbiol.* 6: 1315.

Gutiérrez, D., Fernández, L., Rodríguez, A., and García, P. (2018). Are phage lytic proteins the secret weapon to kill *Staphylococcus aureus*? *MBio* 9: pii: e01923-17.

Gutiérrez, D., Ruas-Madiedo, P., Martínez, B. et al. (2014). Effective removal of staphylococcal biofilms by the endolysin LysH5. *PLoS One* 9: e107307.

Hauben, K.J.A., Wuytack, E.Y., Soontjens, C.C.F., and Michiels, C.W. (1996). High-pressure transient sensitization of *Escherichia coli* to lysozyme and nisin by disruption of outer-membrane permeability. *J. Food Prot.* 59: 350–355.

Hermoso, J.A., Monterroso, B., Albert, A. et al. (2003). Structural basis for selective recognition of pneumococcal cell wall by modular endolysin from phage Cp-1. *Structure* 11: 1239–1249.

Hernandez-Morales, A.C., Lessor, L.L., Wood, T.L. et al. (2018). Genomic and biochemical characterization of *Acinetobacter* podophage Petty reveals a novel lysis mechanism and tail-associated depolymerase activity. *J. Virol.* 92: pii: e01064-17.

Hsieh, P.-F., Lin, H.-H., Lin, T.-L. et al. (2017). Two T7-like bacteriophages, K5-2 and K5-4, each encodes two capsule depolymerases: isolation and functional characterization. *Sci. Rep.* 7: 4624.

Ince, J. and McNally, A. (2009). Development of rapid, automated diagnostics for infectious disease: advances and challenges. *Expert Rev. Med. Devices* 6: 641–651.

Jiang, H., Li, X., Hu, L. et al. (2016). Identification and characterization of HolGH15: the holin of *Staphylococcus aureus* bacteriophage GH15. *J. Gen. Virol.* 97: 1272–1281.

Jun, S.Y., Jang, I.J., Yoon, S. et al. (2017). Pharmacokinetics and tolerance of the phage endolysin-based candidate drug SAL200 after a single intravenous administration among healthy volunteers. *Antimicrob. Agents Chemother.* 61: pii: e02629-16.

Jun, S.Y., Jung, G.M., Son, J.-S. et al. (2011). Comparison of the antibacterial properties of phage endolysins SAL-1 and LysK. *Antimicrob. Agents Chemother.* 55: 1764–1767.

Jun, S.Y., Jung, G.M., Yoon, S.J. et al. (2013). Antibacterial properties of a pre-formulated recombinant phage endolysin, SAL-1. *Int. J. Antimicrob. Agents* 41: 156–161.

Jun, S.Y., Jung, G.M., Yoon, S.J. et al. (2014). Preclinical safety evaluation of intravenously administered SAL200 containing the recombinant phage endolysin SAL-1 as a pharmaceutical ingredient. *Antimicrob. Agents Chemother.* 58: 2084–2088.

Kalchayanand, N., Sikes, T., Dunne, C.P., and Ray, B. (1994). Hydrostatic pressure and electroporation have increased bactericidal efficiency in combination with bacteriocins. *Appl. Environ. Microbiol.* 60: 4174–4177.

Kenny, J.G., McGrath, S., Fitzgerald, G.F., and van Sinderen, D. (2004). Bacteriophage Tuc2009 encodes a tail-associated cell wall-degrading activity. *J. Bacteriol.* 186: 3480–3491.

Khan, H., Flint, S.H., and Yu, P.L. (2013). Determination of the mode of action of enterolysin A, produced by *Enterococcus faecalis* B9510. *J. Appl. Microbiol.* 115: 484–494.

Kimura, K. and Itoh, Y. (2003). Characterization of poly-gamma-glutamate hydrolase encoded by a bacteriophage genome: possible role in phage infection of *Bacillus subtilis* encapsulated with poly-gamma-glutamate. *Appl Env. Microbiol* 69: 2491–2497.

Kivela, H.M., Daugelavicius, R., Hankkio, R.H. et al. (2004). Penetration of membrane-containing double-stranded-DNA bacteriophage PM2 into Pseudoalteromonas hosts. *J. Bacteriol.* 186: 5342–5354.

Knirel, Y.A., Bystrova, O.V., Kocharova, N.A. et al. (2006). Conserved and variable structural features in the lipopolysaccharide of *Pseudomonas aeruginosa*. *J. Endotoxin Res.* 12: 324–336.

Kretzer, J.W., Lehmann, R., Schmelcher, M. et al. (2007). Use of high-affinity cell wall-binding domains of bacteriophage endolysins for immobilization and separation of bacterial cells. *Appl. Environ. Microbiol.* 73: 1992–2000.

Kutter, E. and Sulakvelidze, A. (2005). *Bacteriophages: Biology and Applications*. CRC Press.

Kwiatkowski, B., Boschek, B., Thiele, H., and Stirm, S. (1983). Substrate specificity of two bacteriophage-associated endo-N-acetylneuraminidases. *J. Virol.* 45: 367–374.

Labrie, S.J., Samson, J.E., and Moineau, S. (2010). Bacteriophage resistance mechanisms. *Nat. Rev. Microbiol.* 8: 317–327.

Lai, M.-J., Lin, N.-T., Hu, A. et al. (2011). Antibacterial activity of *Acinetobacter baumannii* phage ϕAB2 endolysin (LysAB2) against both gram-positive and gram-negative bacteria. *Appl. Microbiol. Biotechnol.* 90: 529–539.

Lavigne, R., Briers, Y., Hertveldt, K. et al. (2004). Identification and characterization of a highly thermostable bacteriophage lysozyme. *Cell Mol Life Sci* 61: 2753–2759.

Lavigne, R., Noben, J.P., Hertveldt, K. et al. (2006). The structural proteome of *Pseudomonas aeruginosa* bacteriophage phiKMV. *Microbiology* 152: 529–534.

Lee, I.-M., Tu, I.-F., Yang, F.-L. et al. (2017). Structural basis for fragmenting the exopolysaccharide of *Acinetobacter baumannii* by bacteriophage ΦAB6 tailspike protein. *Sci. Rep.* 7: 42711.

Legotsky, S.A., Vlasova, K.Y., Priyma, A.D. et al. (2014). Peptidoglycan degrading activity of the broad-range Salmonella bacteriophage S-394 recombinant endolysin. *Biochimie* 107: 293–299.

Lewis, K. (2010). Persister cells. *Annu. Rev. Microbiol.* 64: 357–372.

Lim, J.-A., Shin, H., Heu, S., and Ryu, S. (2014). Exogenous lytic activity of SPN9CC endolysin against gram-negative bacteria. *J. Microbiol. Biotechnol.* 24: 803–811.

Lim, J.-A., Shin, H., Kang, D.-H., and Ryu, S. (2012). Characterization of endolysin from a Salmonella Typhimurium-infecting bacteriophage SPN1S. *Res. Microbiol.* 163: 233–241.

Lin, T.-L., Hsieh, P.-F., Huang, Y.-T. et al. (2014). Isolation of a bacteriophage and its depolymerase specific for K1 capsule of *Klebsiella pneumoniae*: implication in typing and treatment. *J. Infect. Dis.* 210: 1734–1744.

Lin, H., Paff, M.L., Molineux, I.J., and Bull, J.J. (2017). Therapeutic application of phage capsule depolymerases against K1, K5, and K30 capsulated *E. coli* in mice. *Front. Microbiol.* 8: 2257.

Linden, S.B., Zhang, H., Heselpoth, R.D. et al. (2015). Biochemical and biophysical characterization of PlyGRCS, a bacteriophage endolysin active against methicillin-resistant *Staphylococcus aureus*. *Appl. Microbiol. Biotechnol.* 99: 741–752.

Linnerborg, M., Weintraub, A., Albert, M.J., and Widmalm, G. (2001). Depolymerization of the capsular polysaccharide from *Vibrio cholerae* O139 by a lyase associated with the bacteriophage JA1. *Carbohydr. Res.* 333: 263–269.

Loc-Carrillo, C. and Abedon, S.T. (2011). Pros and cons of phage therapy. *Bacteriophage* 1: 111–114.

Loeffler, J.M., Djurkovic, S., and Fischetti, V.A. (2003). Phage lytic enzyme Cpl-1 as a novel antimicrobial for pneumococcal bacteremia. *Infect. Immun.* 71: 6199–6204.

Loeffler, J.M., Nelson, D., and Fischetti, V.A. (2001). Rapid killing of *Streptococcus pneumoniae* with a bacteriophage cell wall hydrolase. *Science* 294: 2170–2172.

Loessner, M.J. (2005). Bacteriophage endolysins: current state of research and applications. *Curr. Opin. Microbiol.* 8: 480–487.

Loessner, M.J., Gaeng, S., and Scherer, S. (1999). Evidence for a holin-like protein gene fully embedded out of frame in the endolysin gene of *Staphylococcus aureus* bacteriophage 187. *J. Bacteriol.* 181: 4452–4460.

Loessner, M.J., Kramer, K., Ebel, F., and Scherer, S. (2002). C-terminal domains of *Listeria monocytogenes* bacteriophage murein hydrolases determine specific recognition and high-affinity binding to bacterial cell wall carbohydrates. *Mol. Microbiol.* 44: 335–349.

Lood, R., Raz, A., Molina, H. et al. (2014). A highly active and negatively charged *Streptococcus pyogenes* lysin with a rare D-alanyl-L-alanine endopeptidase activity protects mice against streptococcal bacteremia. *Antimicrob. Agents Chemother.* 58: 3073–3084.

Lood, R., Winer, B.Y., Pelzek, A.J. et al. (2015). Novel phage lysin capable of killing the multidrug-resistant gram-negative bacterium *Acinetobacter baumannii* in a mouse bacteremia model. *Antimicrob. Agents Chemother.* 59: 1983–1991.

Love, M., Bhandari, D., Dobson, R., and Billington, C. (2018). Potential for bacteriophage endolysins to supplement or replace antibiotics in food production and clinical care. *Antibiotics* 7: 17.

Low, L.Y., Yang, C., Perego, M. et al. (2011). Role of net charge on catalytic domain and influence of cell wall binding domain on bactericidal activity, specificity, and host range of phage lysins. *J. Biol. Chem.* 286: 34391–34403.

Lu, T.K. and Collins, J.J. (2007). Dispersing biofilms with engineered enzymatic bacteriophage. *Proc. Natl. Acad. Sci.* 104: 11197–11202.

Lukacik, P., Barnard, T.J., Keller, P.W. et al. (2012). Structural engineering of a phage lysin that targets Gram-negative pathogens. *Proc. Natl. Acad. Sci.* 109: 9857–9862.

Machida, Y., Miyake, K., Hattori, K. et al. (2000). Structure and function of a novel coliphage-associated sialidase. *FEMS Microbiol. Lett.* 182: 333–337.

Mahony, J., Alqarni, M., Stockdale, S. et al. (2016). Functional and structural dissection of the tape measure protein of lactococcal phage TP901-1. *Sci. Rep.* 6: 36667.

Majkowska-Skrobek, G., Łątka, A., Berisio, R. et al. (2016). Capsule-targeting depolymerase, derived from Klebsiella KP36 phage, as a tool for the development of anti-virulent strategy. *Viruses* 8: 324.

Malnoy, M., Faize, M., Venisse, J.-S. et al. (2005). Expression of viral EPS-depolymerase reduces fire blight susceptibility in transgenic pear. *Plant Cell Rep.* 23: 632–638.

Mao, J., Schmelcher, M., Harty, W.J. et al. (2013). Chimeric Ply187 endolysin kills *Staphylococcus aureus* more effectively than the parental enzyme. *FEMS Microbiol. Lett.* 342: 30–36.

Masschalck, B., García-Graells, C., Van Haver, E., and Michiels, C.W. (2000). Inactivation of high pressure resistant *Escherichia coli* by lysozyme and nisin under high pressure. *Innov. Food Sci. Emerg. Technol.* 1: 39–47.

Masschalck, B., Van Houdt, R., Van Haver, E.G.R., and Michiels, C.W. (2001). Inactivation of Gram-negative bacteria by lysozyme, denatured lysozyme, and lysozyme-derived peptides under high hydrostatic pressure. *Appl. Environ. Microbiol.* 67: 339–344.

Mayer, M.J., Garefalaki, V., Spoerl, R. et al. (2011). Structure-based modification of a *Clostridium difficile*-targeting endolysin affects activity and host range. *J. Bacteriol.* 193: 5477–5486.

Melo, L.D.R., Brandão, A., Akturk, E. et al. (2018). Characterization of a new *Staphylococcus aureus* Kayvirus harboring a lysin active against biofilms. *Viruses* 10: 182.

Moak, M. and Molineux, I.J. (2000). Role of the Gp16 lytic transglycosylase motif in bacteriophage T7 virions at the initiation of infection. *Mol Microbiol* 37: 345–355.

Mushtaq, N., Redpath, M.B., Luzio, J.P., and Taylor, P.W. (2004). Prevention and cure of systemic *Escherichia coli* K1 infection by modification of the bacterial phenotype. *Antimicrob. Agents Chemother.* 48: 1503–1508.

Mushtaq, N., Redpath, M.B., Luzio, J.P., and Taylor, P.W. (2005). Treatment of experimental *Escherichia coli* infection with recombinant bacteriophage-derived capsule depolymerase. *J. Antimicrob. Chemother.* 56: 160–165.

Nakimbugwe, D., Masschalck, B., Anim, G., and Michiels, C.W. (2006). Inactivation of gram-negative bacteria in milk and banana juice by hen egg white and lambda lysozyme under high hydrostatic pressure. *Int. J. Food Microbiol.* 112: 19–25.

Nakimbugwe, D., Masschalck, B., Atanassova, M. et al. (2006). Comparison of bactericidal activity of six lysozymes at atmospheric pressure and under high hydrostatic pressure. *Int. J. Food Microbiol.* 108: 355–363.

Navarre, W.W., Ton-That, H., Faull, K.F., and Schneewind, O. (1999). Multiple enzymatic activities of the murein hydrolase from staphylococcal phage phi11. Identification of a D-alanyl-glycine endopeptidase activity. *J. Biol. Chem.* 274: 15847–15856.

Nelson, D., Loomis, L., and Fischetti, V.A. (2001). Prevention and elimination of upper respiratory colonization of mice by group A streptococci by using a bacteriophage lytic enzyme. *Proc. Natl. Acad. Sci.* 98: 4107–4112.

Nelson, D.C., Schmelcher, M., Rodriguez-Rubio, L. et al. (2012). Endolysins as antimicrobials. *Adv Virus Res* 83: 299–365.

Nelson, D., Schuch, R., Chahales, P. et al. (2006). PlyC: a multimeric bacteriophage lysin. *Proc. Natl. Acad. Sci.* 103: 10765–10770.

Nishima, W., Kanamaru, S., Arisaka, F., and Kitao, A. (2011). Screw motion regulates multiple functions of T4 phage protein gene product 5 during cell puncturing. *J Am Chem Soc* 133: 13571–13576.

Oechslin, F., Daraspe, J., Giddey, M. et al. (2013). *In vitro* characterization of PlySK1249, a novel phage lysin, and assessment of its antibacterial activity in a mouse model of *Streptococcus agalactiae* bacteremia. *Antimicrob. Agents Chemother.* 57: 6276–6283.

O'Flaherty, S., Coffey, A., Meaney, W. et al. (2005). The recombinant phage lysin LysK has a broad spectrum of lytic activity against clinically relevant staphylococci, including methicillin-resistant *Staphylococcus aureus*. *J. Bacteriol.* 187: 7161–7164.

Oliveira, H., Azeredo, J., Lavigne, R., and Kluskens, L.D. (2012). Bacteriophage endolysins as a response to emerging foodborne pathogens. *Trends Food Sci. Technol.* 28: 103–115.

Oliveira, H., Costa, A.R., Ferreira, A. et al. (2018). Functional analysis and anti-virulent properties of a new depolymerase from a myovirus that infects *Acinetobacter baumannii* capsule K45. *J. Virol.* 93 (4): pii: e01163-18.

Oliveira, H., Costa, A.R., Konstantinides, N. et al. (2017). Ability of phages to infect *Acinetobacter calcoaceticus-Acinetobacter baumannii* complex species through acquisition of different pectate lyase depolymerase domains. *Environ. Microbiol.* 19: 5060–5077.

Oliveira, H., Melo, L.D.R., Santos, S.B. et al. (2013). Molecular aspects and comparative genomics of bacteriophage endolysins. *J. Virol.* 87: 4558–4570.

Oliveira, H., Pinto, G., Oliveira, A. et al. (2016). Characterization and genome sequencing of a *Citrobacter freundii* phage CfP1 harboring a lysin active against multidrug-resistant isolates. *Appl. Microbiol. Biotechnol.* 100: 10543–10553.

Oliveira, H., São-José, C., and Azeredo, J. (2018). Phage-derived peptidoglycan degrading enzymes: challenges and future prospects for in vivo therapy. *Viruses* 10: 292.

Oliveira, H., Thiagarajan, V., Walmagh, M. et al. (2014). A thermostable Salmonella phage endolysin, Lys68, with broad bactericidal properties against Gram-negative pathogens in presence of weak acids. *PLoS One* 9: e108376.

Oliveira, H., Vilas Boas, D., Mesnage, S. et al. (2016). Structural and enzymatic characterization of ABgp46, a novel phage endolysin with broad anti-gram-negative bacterial activity. *Front. Microbiol.* 7: 208.

Olsen, N., Thiran, E., Hasler, T. et al. (2018). Synergistic removal of static and dynamic *Staphylococcus aureus* biofilms by combined treatment with a bacteriophage endolysin and a polysaccharide depolymerase. *Viruses* 10: 438.

Olszak, T., Shneider, M.M., Latka, A. et al. (2017). The O-specific polysaccharide lyase from the phage LKA1 tailspike reduces *Pseudomonas virulence. Sci. Rep.* 7: 16302.

Orskov, I., Orskov, F., Jann, B., and Jann, K. (1977). Serology, chemistry, and genetics of O and K antigens of *Escherichia coli*. *Bacteriol. Rev.* 41: 667–710.

Pan, Y.J., Lin, T.L., Chen, C.C. et al. (2017). Klebsiella phage ΦK64-1 encodes multiple depolymerases for multiple host capsular types. *J Virol* 91: pii: e02457-16.

Pan, Y.-J., Lin, T.-L., Lin, Y.-T. et al. (2015). Identification of capsular types in carbapenem-resistant *Klebsiella pneumoniae* strains by *wzc* sequencing and implications for capsule depolymerase treatment. *Antimicrob. Agents Chemother.* 59: 1038–1047.

Pastagia, M., Euler, C., Chahales, P. et al. (2011). A novel chimeric lysin shows superiority to mupirocin for skin decolonization of methicillin-resistant and -sensitive *Staphylococcus aureus* strains. *Antimicrob. Agents Chemother.* 55: 738–744.

Pastagia, M., Schuch, R., Fischetti, V.A., and Huang, D.B. (2013). Lysins: the arrival of pathogen-directed anti-infectives. *J. Med. Microbiol.* 62: 1506–1516.

Paul, V.D., Rajagopalan, S.S., Sundarrajan, S. et al. (2011). A novel bacteriophage Tail-Associated Muralytic Enzyme (TAME) from Phage K and its development into a potent antistaphylococcal protein. *BMC Microbiol.* 11: 226.

Payne, K.M. and Hatfull, G.F. (2012). Mycobacteriophage endolysins: diverse and modular enzymes with multiple catalytic activities. *PLoS One* 7: e34052.

Pires, D.P., Oliveira, H., Melo, L.D.R. et al. (2016). Bacteriophage-encoded depolymerases: their diversity and biotechnological applications. *Appl. Microbiol. Biotechnol.* 100: 2141–2151.

Poonacha, N., Nair, S., Desai, S. et al. (2017). Efficient killing of planktonic and biofilm-embedded coagulase-negative staphylococci by bactericidal protein P128. *Antimicrob. Agents Chemother.* 61.

Popova, A., Lavysh, D., Klimuk, E. et al. (2017). Novel Fri1-like viruses infecting *Acinetobacter baumannii*—vB_AbaP_AS11 and vB_AbaP_AS12—characterization, comparative genomic analysis, and host-recognition strategy. *Viruses* 9: 188.

Proença, D., Fernandes, S., Leandro, C. et al. (2012). Phage endolysins with broad antimicrobial activity against *Enterococcus faecalis* clinical strains. *Microb. Drug Resist.* 18: 322–332.

Proença, D., Leandro, C., Garcia, M. et al. (2015). EC300: a phage-based, bacteriolysin-like protein with enhanced antibacterial activity against *Enterococcus faecalis*. *Appl Microbiol Biotechnol* 99: 5137–5149.

Rajesh, T., Anthony, T., Saranya, S. et al. (2011). Functional characterization of a new holin-like antibacterial protein coding gene tmp1 from goat skin surface metagenome. *Appl. Microbiol. Biotechnol.* 89: 1061–1073.

Rashel, M., Uchiyama, J., Takemura, I. et al. (2008). Tail-associated structural protein gp61 of *Staphylococcus aureus* phage phi MR11 has bifunctional lytic activity. *FEMS Microbiol. Lett.* 284: 9–16.

Rashel, M., Uchiyama, J., Ujihara, T. et al. (2007). Efficient elimination of multidrug-resistant *Staphylococcus aureus* by cloned lysin derived from bacteriophage ϕMR11. *J. Infect. Dis.* 196: 1237–1247.

Reddy, B.L. and Saier, M.H. (2013). Topological and phylogenetic analyses of bacterial holin families and superfamilies. *Biochim. Biophys. Acta Biomembr.* 1828: 2654–2671.

Reyes, A., Haynes, M., Hanson, N. et al. (2010). Viruses in the faecal microbiota of monozygotic twins and their mothers. *Nature* 466: 334–338.

Rodríguez, L., Martínez, B., Zhou, Y. et al. (2011). Lytic activity of the virion-associated peptidoglycan hydrolase HydH5 of *Staphylococcus aureus* bacteriophage vB_SauS-phiIPLA88. *BMC Microbiol.* 11: 138.

Rodríguez-Rubio, L., Gerstmans, H., Thorpe, S. et al. (2016). DUF3380 domain from a Salmonella phage endolysin shows potent n-acetylmuramidase activity. *Appl. Environ. Microbiol.* 82: 4975–4981.

Rodriguez-Rubio, L., Martinez, B., Donovan, D.M. et al. (2013a). Potential of the virion-associated peptidoglycan hydrolase HydH5 and its derivative fusion proteins in milk biopreservation. *PLoS One* 8: e54828.

Rodriguez-Rubio, L., Martinez, B., Donovan, D.M. et al. (2013b). Bacteriophage virion-associated peptidoglycan hydrolases: potential new enzybiotics. *Crit Rev Microbiol* 39: 427–434.

Rodriguez-Rubio, L., Martinez, B., Rodriguez, A. et al. (2012). Enhanced staphylolytic activity of the *Staphylococcus aureus* bacteriophage vB_SauS-phiIPLA88 HydH5 virion-associated peptidoglycan hydrolase: fusions, deletions, and synergy with LysH5. *Appl Env. Microbiol* 78: 2241–2248.

Rodriguez-Rubio, L., Quiles-Puchalt, N., Martinez, B. et al. (2013c). The peptidoglycan hydrolase of *Staphylococcus aureus* bacteriophage 11 plays a structural role in the viral particle. *Appl Env. Microbiol* 79: 6187–6190.

Ruer, S., Pinotsis, N., Steadman, D. et al. (2015). Virulence-targeted antibacterials: concept, promise, and susceptibility to resistance mechanisms. *Chem Biol Drug Des.* 86 (4): 379–399. https://doi.org/10.1111/cbdd.12517.

Saier, M.H. and Reddy, B.L. (2015). Holins in bacteria, eukaryotes, and archaea: multifunctional xenologues with potential biotechnological and biomedical applications. *J. Bacteriol.* 197: 7–17.

Santos, S.B., Costa, A.R., Carvalho, C. et al. (2018). Exploiting bacteriophage proteomes: the hidden biotechnological potential. *Trends Biotechnol.* 36: 966–984.

Sao-Jose, C. (2018). Engineering of phage-derived lytic enzymes: improving their potential as antimicrobials. *Antibiot.* 7: pii: E29.

Schleifer, K.H. and Kandler, O. (1972). Peptidoglycan types of bacterial cell walls and their taxonomic implications. *Bacteriol. Rev.* 36: 407–477.

Schmelcher, M., Donovan, D.M., and Loessner, M.J. (2012). Bacteriophage endolysins as novel antimicrobials. *Future Microbiol.* 7: 1147–1171.

Schmelcher, M., Shen, Y., Nelson, D.C. et al. (2015). Evolutionarily distinct bacteriophage endolysins featuring conserved peptidoglycan cleavage sites protect mice from MRSA infection. *J. Antimicrob. Chemother.* 70: 1453–1465.

Schmelcher, M., Tchang, V.S., and Loessner, M.J. (2011). Domain shuffling and module engineering of *Listeria phage* endolysins for enhanced lytic activity and binding affinity. *Microb. Biotechnol.* 4: 651–662.

Scholl, D., Cooley, M., Williams, S.R. et al. (2009). An engineered R-type pyocin is a highly specific and sensitive bactericidal agent for the food-borne pathogen *Escherichia coli* O157:H7. *Antimicrob. Agents Chemother.* 53: 3074–3080.

Schuch, R., Nelson, D., and Fischetti, V.A. (2002). A bacteriolytic agent that detects and kills *Bacillus anthracis*. *Nature* 418: 884–889.

Schuch, R., Pelzek, A.J., Raz, A. et al. (2013). Use of a bacteriophage lysin to identify a novel target for antimicrobial development. *PLoS One* 8: e60754.

Schwarzer, D., Browning, C., Stummeyer, K. et al. (2015). Structure and biochemical characterization of bacteriophage phi92 endosialidase. *Virology* 477: 133–143.

Scott, R.D. (2009). *The Direct Medical Costs of Healthcare-Associated Infections in U.S. Hospitals and the Benefits of Prevention*. doi:CS200891-A. https://www.cdc.gov/hai/pdfs/hai/scott_costpaper.pdf (accessed 7 May 2019).

Shavrina, M.S., Zimin, A.A., Molochkov, N.V. et al. (2016). *In vitro* study of the antibacterial effect of the bacteriophage T5 thermostable endolysin on *Escherichia coli* cells. *J. Appl. Microbiol.* 121: 1282–1290.

Shi, Y., Li, N., Yan, Y. et al. (2012). Combined antibacterial activity of phage lytic proteins holin and lysin from *Streptococcus suis* bacteriophage SMP. *Curr. Microbiol.* 65: 28–34.

Silhavy, T.J., Kahne, D., and Walker, S. (2010). The bacterial cell envelope. *Cold Spring Harb. Perspect. Biol.* 2: a000414.

Singh, P.K., Donovan, D.M., and Kumar, A. (2014). Intravitreal injection of the chimeric phage endolysin ply187 protects mice from *Staphylococcus aureus* endophthalmitis. *Antimicrob. Agents Chemother.* 58: 4621–4629.

Smith, D.L., Chang, C.Y., and Young, R. (1998). The lambda holin accumulates beyond the lethal triggering concentration under hyperexpression conditions. *Gene Expr.* 7: 39–52.

Spinosa, M.R., Progida, C., Tala, A. et al. (2007). The Neisseria meningitidis capsule is important for intracellular survival in human cells. *Infect. Immun.* 75: 3594–3603.

Stockdale, S.R., Mahony, J., Courtin, P. et al. (2013). The lactococcal phages Tuc2009 and TP901-1 incorporate two alternate forms of their tail fiber into their virions for infection specialization. *J Biol Chem* 288: 5581–5590.

Stummeyer, K., Schwarzer, D., Claus, H. et al. (2006). Evolution of bacteriophages infecting encapsulated bacteria: lessons from *Escherichia coli* K1-specific phages. *Mol. Microbiol.* 60: 1123–1135.

Sudiarta, I.P., Fukushima, T., and Sekiguchi, J. (2010). Bacillus subtilis CwlP of the SP-beta prophage has two novel peptidoglycan hydrolase domains, muramidase and cross-linkage digesting DD-endopeptidase. *J. Biol. Chem.* 285: 41232–41243.

Takac, M. and Blasi, U. (2005). Phage P68 virion-associated protein 17 displays activity against clinical isolates of *Staphylococcus aureus. Antimicrob. Agents Chemother.* 49: 2934–2940.

Tang, J., Lander, G.C., Olia, A.S. et al. (2011). Peering down the barrel of a bacteriophage portal: the genome packaging and release valve in p22. *Structure* 19: 496–502.

Thurow, H., Niemann, H., Rudolph, C., and Stirm, S. (1974). Host capsule depolymerase activity of bacteriophage particles active on Klebsiella K20 and K24 strains. *Virology* 58: 306–309.

Totté, J., de Wit, J., Pardo, L. et al. (2017). Targeted anti-staphylococcal therapy with endolysins in atopic dermatitis and the effect on steroid use, disease severity and the microbiome: study protocol for a randomized controlled trial (MAAS trial). *Trials* 18: 404.

Vading, M., Nauclér, P., Kalin, M., and Giske, C.G. (2018). Invasive infection caused by *Klebsiella pneumoniae* is a disease affecting patients with high comorbidity and associated with high long-term mortality. *PLoS One* 13: e0195258.

Vipra, A.A., Desai, S.N., Roy, P. et al. (2012). Antistaphylococcal activity of bacteriophage derived chimeric protein P128. *BMC Microbiol.* 12: 41.

Visweswaran, G.R.R., Dijkstra, B.W., and Kok, J. (2011). Murein and pseudomurein cell wall binding domains of bacteria and archaea: a comparative view. *Appl. Microbiol. Biotechnol.* 92: 921–928.

Vouillamoz, J., Entenza, J.M., Giddey, M. et al. (2013). Bactericidal synergism between daptomycin and the phage lysin Cpl-1 in a mouse model of pneumococcal bacteraemia. *Int. J. Antimicrob. Agents* 42: 416–421.

Walmagh, M., Boczkowska, B., Grymonprez, B. et al. (2013). Characterization of five novel endolysins from Gram-negative infecting bacteriophages. *Appl. Microbiol. Biotechnol.* 97: 4369–4375.

Walmagh, M., Briers, Y., dos Santos, S.B. et al. (2012). Characterization of modular bacteriophage endolysins from *Myoviridae phages* OBP, 201φ2-1 and PVP-SE1. *PLoS One* 7: e36991.

Wang, I.-N., Smith, D.L., and Young, R. (2000). Holins: the protein clocks of bacteriophage infections. *Annu. Rev. Microbiol.* 54: 799–825.

Waseh, S., Hanifi-Moghaddam, P., Coleman, R. et al. (2010). Orally administered P22 phage tailspike protein reduces salmonella colonization in chickens: prospects of a novel therapy against bacterial infections. *PLoS One* 5: e13904.

White, R., Chiba, S., Pang, T. et al. (2011). Holin triggering in real time. *Proc. Natl. Acad. Sci.* 108: 798–803.

Whitfield, C. and Roberts, I.S. (1999). Structure, assembly and regulation of expression of capsules in *Escherichia coli. Mol. Microbiol.* 31: 1307–1319.

WHO (2014). WHO's first global report on antibiotic resistance reveals serious, worldwide threat to public health.

WHO (2017). Global priority list of antibiotic-resistant bacteria to guide research, discovery, and development of new antibiotics. *WHO.*

Witzenrath, M., Schmeck, B., Doehn, J.M. et al. (2009). Systemic use of the endolysin Cpl-1 rescues mice with fatal pneumococcal pneumonia. *Crit. Care Med.* 37: 642–649.

Wollin, R., Eriksson, U., and Lindberg, A.A. (1981). Salmonella bacteriophage glycanases: endorhamnosidase activity of bacteriophages P27, 9NA, and KB1. *J Virol* 38: 1025–1033.

Yamaguchi, K., Uchida, K., Tsujimoto, S. et al. (2000). Role of bacterial capsule in local and systemic inflammatory responses of mice during pulmonary infection with *Klebsiella pneumoniae. J. Med. Microbiol.* 49: 1003–1010.

Yang, H., Wang, D.-B., Dong, Q. et al. (2012). Existence of separate domains in lysin PlyG for recognizing *Bacillus anthracis* spores and vegetative cells. *Antimicrob. Agents Chemother.* 56: 5031–5039.

Yang, H., Zhang, Y., Huang, Y. et al. (2014). Degradation of methicillin-resistant *Staphylococcus aureus* biofilms using a chimeric lysin. *Biofouling* 30: 667–674.

Yang, H., Zhang, H., Wang, J. et al. (2017). A novel chimeric lysin with robust antibacterial activity against planktonic and biofilm methicillin-resistant *Staphylococcus aureus. Sci Rep* 7: 40182.

Yang, H., Zhang, Y., Yu, J. et al. (2014). Novel chimeric lysin with high-level antimicrobial activity against methicillin-resistant *Staphylococcus aureus* in vitro and in vivo. *Antimicrob Agents Chemother* 58: 536–542.

Yoong, P., Schuch, R., Nelson, D., and Fischetti, V.A. (2004). Identification of a broadly active phage lytic enzyme with lethal activity against antibiotic-resistant *Enterococcus faecalis* and *Enterococcus faecium. J. Bacteriol.* 186: 4808–4812.

Zampara, A., Sørensen, M.C.H., Grimon, D. et al. (2018). Innolysins: a novel approach to engineer endolysins to kill Gram-negative bacteria. *bioRxiv* 408948. https://doi.org/10.1101/408948.

Zelmer, A., Martin, M.J., Gundogdu, O. et al. (2010). Administration of capsule-selective endosialidase E minimizes upregulation of organ gene expression induced by experimental systemic infection with *Escherichia coli* K1. *Microbiology* 156: 2205–2215.

16

Antibiotic Modification Addressing Resistance

Haotian Bai and Shu Wang

Institute of Chemistry, Chinese Academy of Sciences, Beijing, P. R. China

Antibiotic-resistant pathogens are becoming prevalent in hospitals and communities (Fischbach and Walsh 2009). Thus, not only new antibiotics but also new classes of antibiotics are needed to fight the challenges caused by drug-resistant pathogens while the developing progress is much slower than expected until now. Most of the antibiotics currently in use are accidentally discovered, after which their mechanisms of action are elucidated gradually. To directly counter the antibiotic resistance and address the urgent need for new antibiotics, more rational approaches to modification and development of drug are clearly desired. It is a long history of utilizing chemical modification approach to improve drug activity, inhibit the growth of drug-resistant bacteria, and treat infectious diseases. Considering these unique advantages and distinguished effects, we will summarize several commonly used antibiotic modification methods for countering antibiotic resistance in this section. Additionally, we will also focus on some latest antibiotic modification concepts and ideas. We believe that the following rich content could inspire more scientists to devote themselves to antibiotic modification addressing resistance.

16.1 Chemical Synthesis of New Antibiotics

Members of each class antibiotic always owned an identical core structure. Therefore, from the traditional perspective of organic chemistry, it was a commonly used pathway to synthesize and screen new bactericidal compounds via modifying the major molecular skeleton of antibiotics. For example, erythromycin, as the first representative of macrolide antibiotics, was isolated in 1952 from the *Saccharopolyspora erythraea* and then was widely used for the

Antibiotic Drug Resistance, First Edition. Edited by José-Luis Capelo-Martínez and Gilberto Igrejas.

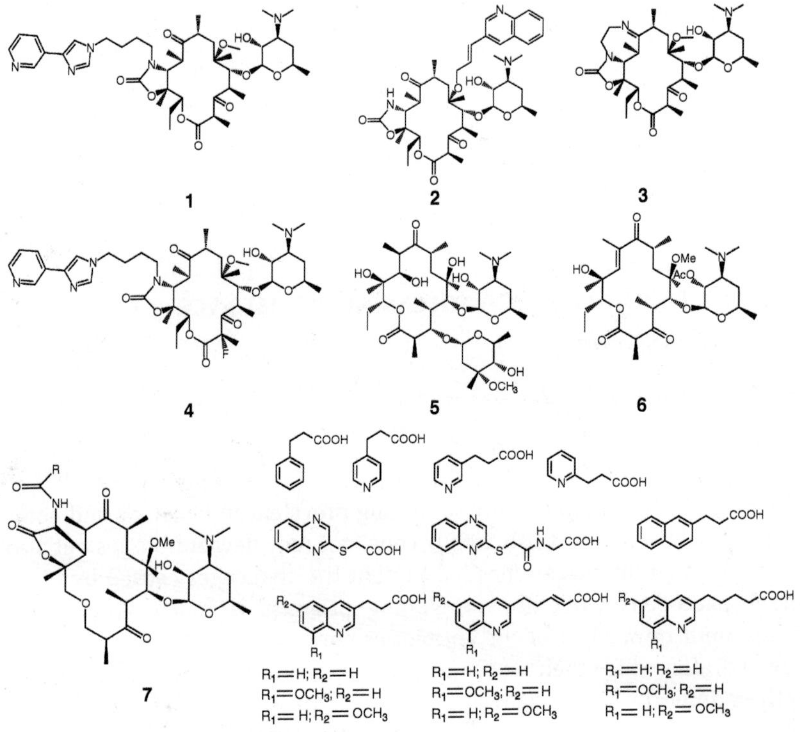

Scheme 16.1 Chemical structures of (**1**)–(**7**).

treatment of Gram-positive bacterial infections in the clinic. But with the emerging of macrolide-resistant bacteria, these initial macrolides are no longer able to meet the requirements of clinical therapy. Thus, as shown in Scheme 16.1, Structures (**1**)–(**4**), a series of ketolides belonging to the macrolide group, were successively synthesized since 1995, including telithromycin, cethromycin, and so on (Jiang et al. 2013).

Inspired by the outstanding previous works, Pavlovic's group has investigated the modification method of macrolide antibiotics for several years (Pavlovic et al. 2010; Pavlovic and Mutak 2011). In 2014, they first used the classic erythromycin as the raw material and constructed the compounds (**5**) and (**6**) according to the published reports (Baker et al. 1988). After oriented C-12 modification, they prepared a set of new ketolides (Structure **7**) (Pavlovic et al. 2014). By treating with different drug-resistant pathogens, all the obtained compounds exhibited great effectiveness against the selected Gram-positive bacteria, including the inducibly resistant *Staphylococcus aureus* and constitutively resistant and efflux-resistant *Staphylococcus pneumoniae* and *Staphylococcus pyogenes*.

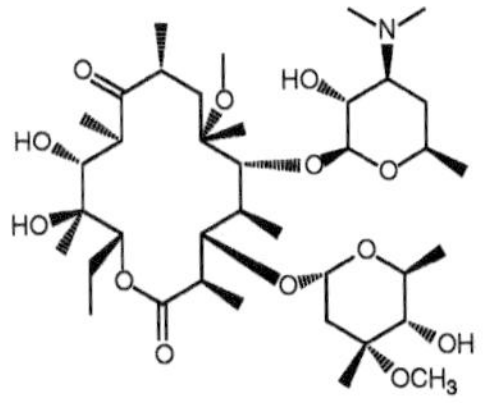

8

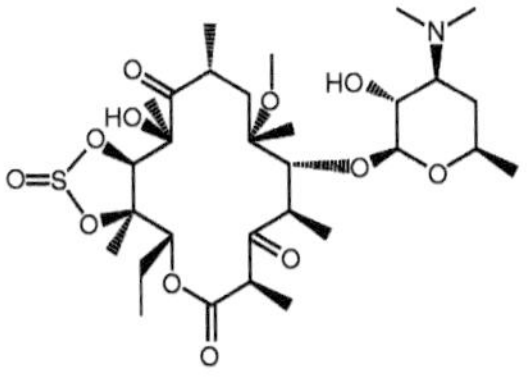

9

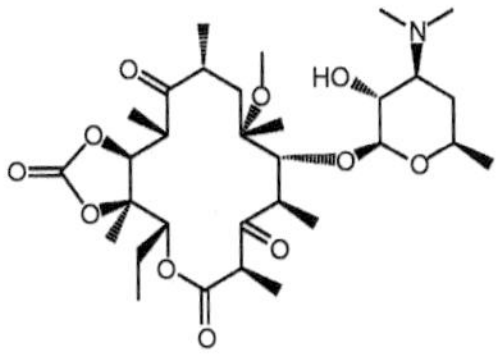

10

Scheme 16.2 Chemical structures of (**8**)–(**10**).

For the same goal of developing new generation macrolide antibiotics against macrolide-resistant pathogens, You's group also developed some ketolide derivatives and industrialized intermediates (Zhao et al. 2003; Wei and You 2006). In 2013, they used the clarithromycin (Structure **8**) as the starting material and obtained the antibacterial compounds (**9**) and (**10**) (Jiang et al. 2013). By evaluating the antibacterial activity, both of them showed outstanding killing ability comparing with the commercialized clarithromycin, and the minimum inhibitory concentration (MIC) against *Staphylococcus epidermidis* was markedly decreased from 64 to 8 $\mu g\,ml^{-1}$ and 0.0625 $\mu g\,ml^{-1}$. These two above reports not only illustrated effective potential antibiotics but also guided promising ketolide frameworks. More importantly, they demonstrate once again that the semisynthetic modification strategy of existing antibiotics is a rational method for improving drug effects (Scheme 16.2).

Glycopeptide antibiotics (GPAs) are another kind of key antibacterial drugs for fighting with drug-resistant pathogens. As a mainstream drug among them, vancomycin (**11**) is used for treating serious and life-threatening Gram-positive infections, when other common antibiotics are no long antibacterial responsive. Thus, vancomycin is also well known as the last resort for curing methicillin-resistant *S. aureus* (MRSA) (McComas et al. 2003). However, only 30 years after its clinical introduction, the vancomycin-resistant enterococci (VRE) was discovered and reported (Kaplan et al. 1988). Thus, many chemists have been attached in the modification of vancomycin for singling out potent vancomycin derivatives in dealing with drug resistance. Some of these products have been successfully industrialized and applied in clinical therapy, such as chloroeremomycin, teicoplanin, telavancin, oritavancin, and dalbavancin (Structures **12–16**) (Blaskovich et al. 2018; Jiang et al. 2018) (Scheme 16.3).

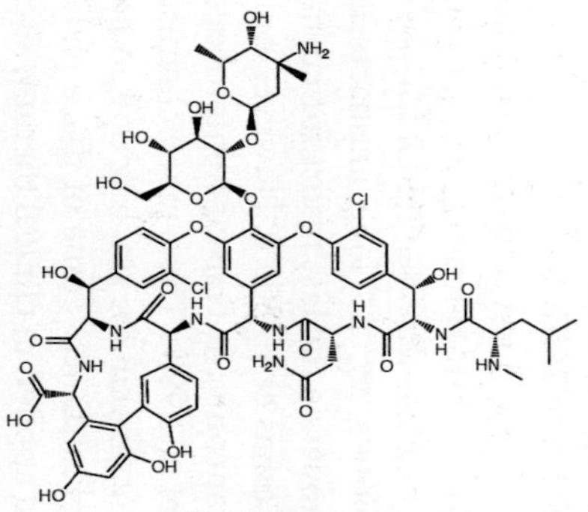

11 Vancomycin

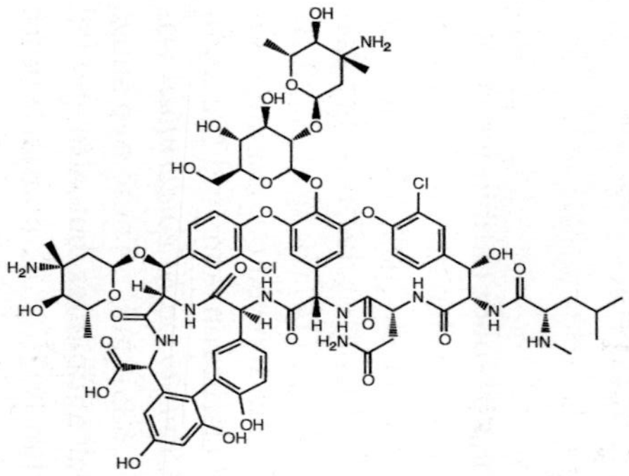

12 Chloroeremomycin

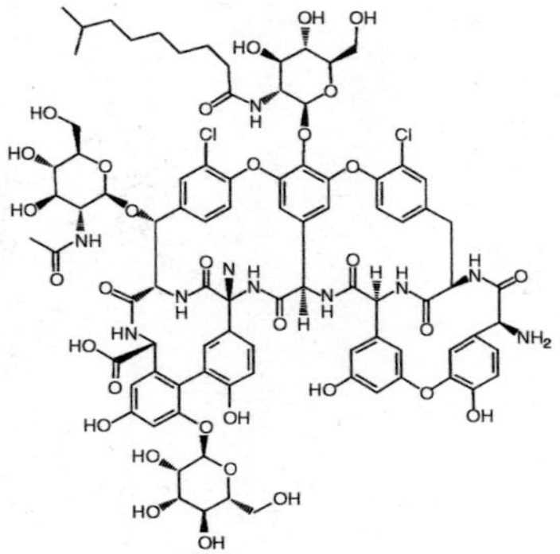

13 Teicoplanin

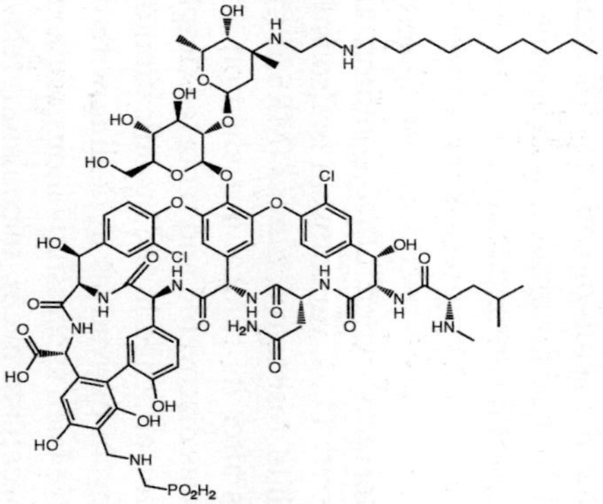

14 Telavancin

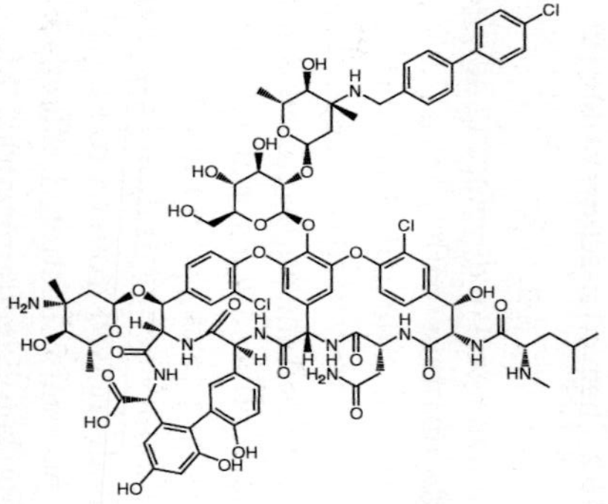

15 Oritavancin

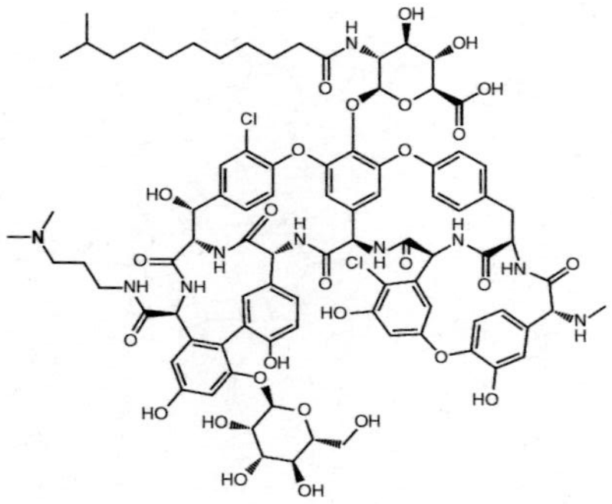

16 Dalbavancin

Scheme 16.3 Chemical structures of (**11**)–(**16**), including vancomycin, chloroeremomycin, teicoplanin, telavancin, oritavancin, and dalbavancin.

In addition, other new antibacterial compounds based on vancomycin modification were also exhibited prominent biological activity against resistant pathogens. In 2000, Daniel Kahne and co-workers investigated the role of hydrophobic substituents in sugar moiety of vancomycin for overcoming drug resistance (Kerns et al. 2000). The MIC assays demonstrated that all the resultant products 14 have better bactericidal activity toward both sensitive and resistant strains. Comparing with the original vancomycin, the biological activity of these derivatives with hydrophobic groups increased by 50–100 times against sensitive *Enterococcus faecium* and by 15–200 times against resistant *E. faecium*. It is proved that the hydrophobic modification strategy is one of the rational approaches for improving vancomycin activity. From the above point of primary synthetic chemistry, modifying of clinical antibiotics is a classical and appropriate method for obtaining counterpart derivatives. By screening active ones from them, this strategy has applied many highly effective new antibiotics. These listed proper examples demonstrated that the semisynthetic modification strategy is advantageous in treating drug-resistant pathogens. But we should also clearly understand that not all the modified antibiotic derivatives could obtain a better biological function, so the semisynthetic method is a difficult and unpredicted way (Scheme 16.4).

With the specific study on the sterilization mechanism, more directional modification strategies have been introduced for promoting vancomycin activity. For example, in order to prevent the bacterial cell wall maturation, it has been demonstrated that vancomycin must form specific vancomycin homodimers driven by hydrogen bonding and hydrophobic interaction (Mackay et al. 1994a, b) firstly and then selectively contact with the terminal of *N*-acyl-D-Ala-D-Ala (Sheldrick et al. 1978; Perkins 1982). It also has been proved that one of the resistance mechanisms of vancomycin is that the terminal D-Ala-D-Ala is replaced by D-Ala-D-Lac. This little mutation resulted in the decrement of the binding ability between the original vancomycin and peptidoglycan. Fortunately, the constructed vancomycin dimers could bind with the mutated D-Ala-D-Lac (Bugg et al. 1991). Thus, it was a predictable approach that constructing vancomycin dimers to improve the affinity between vancomycin and original/mutated peptidoglycan sit, regaining the bactericidal activity of vancomycin. In 2003, Mu's group synthesized a kind of vancomycin dimer (**18**) bearing disulfide bonds (Mu et al. 2004). They also verified that the vancomycin dimer had great antibacterial activity against some selected pathogens. In 2015, Haldar's group prepared the vancomycin aglycon dimers (**19**), bis(vancomycin aglycon)carboxamides (Yarlagadda et al. 2015). The polyamine linker has a variable hydrophobicity, positive charge, and primary amine group, which could couple to the carboxyl group of vancomycin aglycon. The final product also recovered the vancomycin activity against MRSA and VRE.

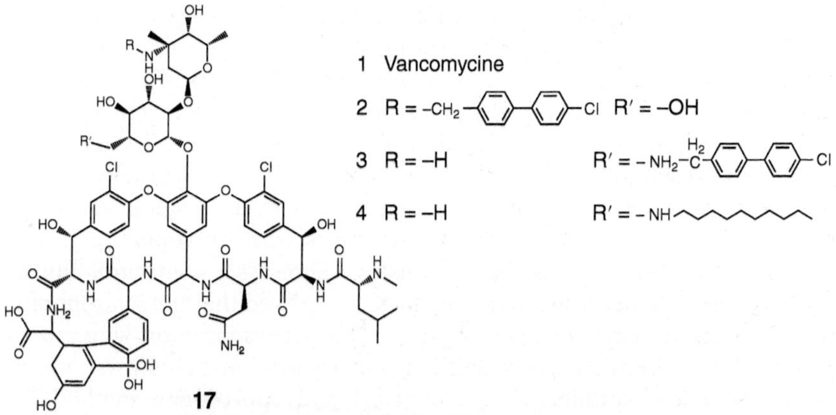

Intact glycopeptide	MIC (μg/ml^{-1})	
	Sensitive	Resistant
1	2	512
2	<0.03	2
3	<0.03	16
4	0.06	32

Scheme 16.4 The chemical structure of (**17**) of the synthesized glycopeptide antibiotics and their MIC values against sensitive and resistant pathogens.

Its MIC value against VanA-resistant *E. faecium* was only 1 mg ml^{-1}, and the bactericidal activity was 300-fold higher than initial vancomycin. In 2018, Sun's group linked two demethylvancomycin molecules by the N-terminal tails and constructed a range of dimeric derivatives (**20**) (Jiang et al. 2018). The biological assays were produced by standard broth microdilution susceptibility tests, and the results showed that all these three dimerization products also had excellent bactericidal activity against VRE. By further detecting the MIC values of these three products, they speculated that the activity recovery was possibly attributed to the bulky and rigid substituent on the linker, via weakening the activities between compound and VRE. Proved by these above researches, it is one of the probable strategies that synthetic dimers of vancomycin increase bactericidal activity to eliminate drug-resistant pathogens. Therefore, it will bring us more enlightenment of antibiotic modification in the future, with the further development of molecular biology and the study of microbial resistant mechanisms (Scheme 16.5).

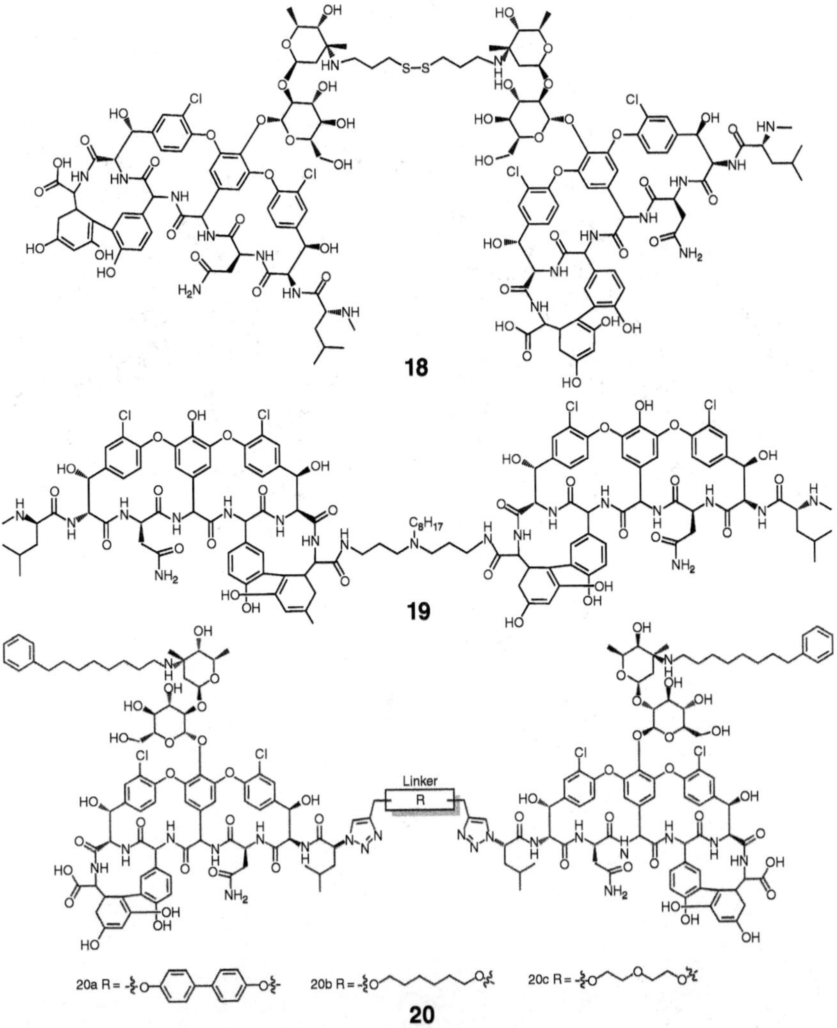

Scheme 16.5 Chemical structures of (**18**)–(**20**).

16.2 Antibiotic Modification with Targeted Groups

The in-depth exploring of the mechanisms of antibiotic resistance has recreated new possibilities for targeted antibiotic modification. Scientists proved that the antibiotic resistance can occur through three general pathways (Scheme 16.6): (i) prevention of the interaction between drug and related targets, (ii) efflux of the antibiotic outside the pathetic cell, and (iii) direct

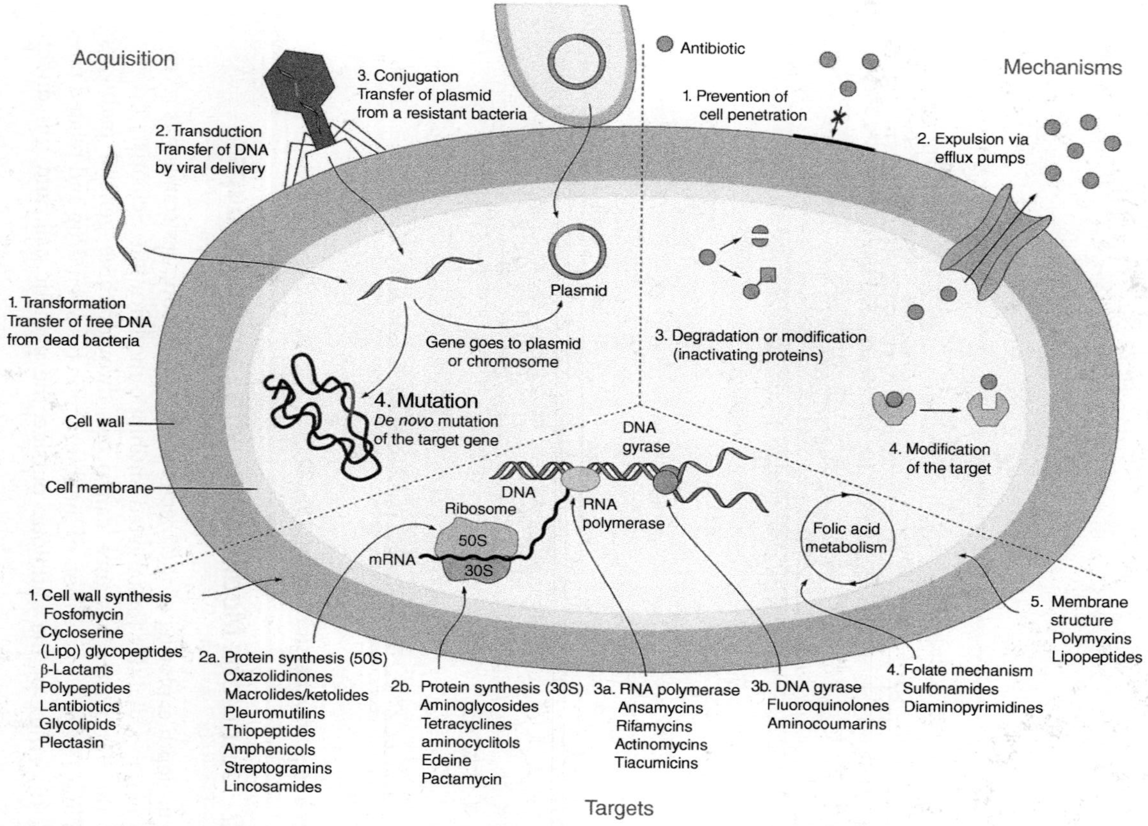

Scheme 16.6 Schematic illumination of the resistance acquisition pathways and the main mechanisms of resistance and the five main targets for antibiotics. Source: Reproduced from Chellat et al. (2016) with permission. Copyright 2016 Wiley-VCH Verlag GmbH & Co. KGaA. *(See insert for color representation of the figure.)*

destruction or modification of compounds (Wright 2005; Chellat et al. 2016). Inspired by these explored antibiotic resistance mechanisms, more and more researchers have been beginning to target modification on existing antibiotics and have obtained many exciting achievements.

For improving the binding ability between antibiotics and cell walls, the bactericidal compounds (**21**) and (**22**) were synthesized by Walsh (Lin and Walsh 2004). In this work, the carbohydrate was utilized for modifying the core structure of tyrocidine, which was a kind of nonribosomal peptide without specific target. The antibacterial and hemolytic assays demonstrated that the obtained Tyc4pg-14 and Tyc4pg-15 had improved the therapeutic index, compared with the natural tyrocidine (Scheme 16.7).

In order to realize the purpose of targeted modification, Sanjeev Mariathasan et al. used anti-*S. aureus* antibody to covalently link with highly efficacious antibiotics and improved the binding ability and targeting of the selected antibacterial drugs (Scheme 16.8) (Lehar et al. 2015). By further introducing the intracellular protease-sensitive linker, the novel antibody–antibiotic conjugate could be activated only by the phagolysosomal protease, and the released active group could efficiently eliminate intracellular *S. aureus*. A virulent subset of *S. aureus* was selected as the model pathogens, because it could establish intracellular infection even in the presence of vancomycin. After treated with these model bacteria, it was confirmed that the antibody–antibiotic conjugate

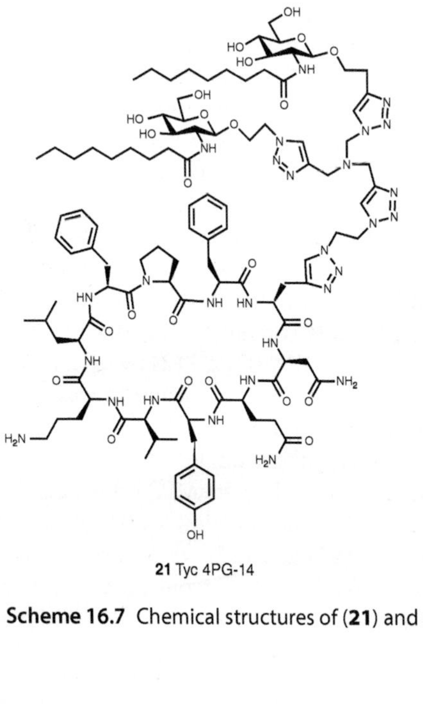

21 Tyc 4PG-14

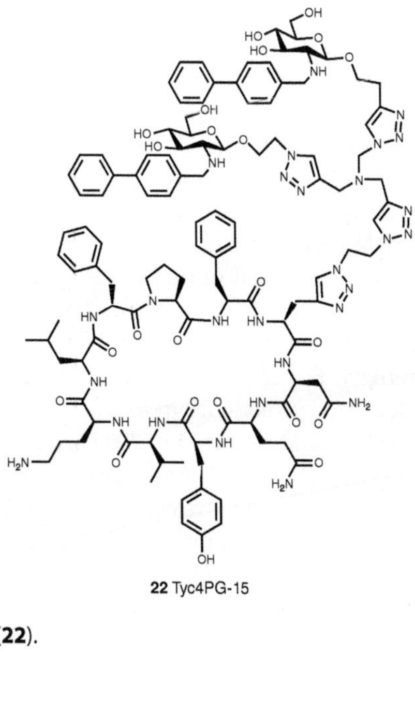

22 Tyc4PG-15

Scheme 16.7 Chemical structures of (**21**) and (**22**).

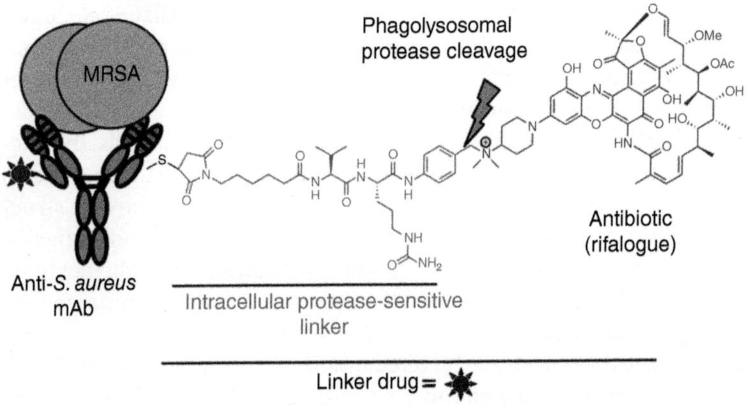

Scheme 16.8 Schematic illustration of the antibody–antibiotic conjugate. *Source:* Reproduced from Lehar et al. (2015) with permission. Copyright 2015 Nature Publishing Group.

had better pesticide effect and also had the potential to be used in curing invasive infections.

Because some drug-resistant bacteria had persisted bacterial cells, which reduced the cellular transport activity and neutralized effective antibiotics, a new antibiotic Pentobra (**23**) with bacterial cell-targeted moiety was designed and synthesized by Schmidt et al. (2014). By introducing a specific 12 amino acid sequence, the tobramycin was endowed with spontaneous permeability. All the relating biological results demonstrated that the modified molecule obtained higher antibacterial activity against both Gram-positive *S. aureus* and Gram-negative *Escherichia coli*, and it also remained noncytotoxic to eukaryotes as expected. As the mentioned resistant *S. aureus*, it has been proved that the precursor peptidoglycan terminus in vancomycin-resistant bacteria was remodeled from D-Ala-D-Ala to D-Ala-D-Lac via mutation (Perkins 1982). The single atom exchange from "NH" to "O" in the cell wall precursors of resistant bacteria resulted in the avoiding action of initial vancomycin. Based on these studies about the mutations sites of antibiotic interactions, Boger's group has done researches about antibiotic modifications for enhancing the binding activity. For countering this mutation, [Ψ[C(=NH)NH]Tpg^4] (**24**) with a complementary single atom exchange in the vancomycin core structure was synthesized and reported in 2011 (Xie et al. 2011). Although its binding ability to the natural D-Ala-D-Ala ligand is twofold less than vancomycin, while it represents a 600-fold increasing to the mutated D-Ala-D-Lac. Its MIC value against vancomycin-resistant bacteria *Enterococcus faecalis* is as low as 0.31 $\mu g\,ml^{-1}$. Furthermore, a range of durable antibiotics (**25**) and (**26**) were created, and they were described with the outstanding features that avoid many known antibiotic resistance mechanisms (Okano et al. 2017). Firstly, the binding modification was intended to provide a dual D-Ala-D-Ala/D-Ala-D-Lac binding ability, thereby directly overcoming the molecular basis of vancomycin resistance. Secondly, these side-chain structures were designed to enhance the antibacterial effect and provided additional synergy mechanisms. The introduced quaternary ammonium salt group was found to provide another binding pocket-modified vancomycin simulation with both ligands, and the added (4-chlorobiphenyl) methyl (CBP) provided another more potent antimicrobial units. The biological results indicate that drug resistance is difficult to develop in all the obtained vancomycins with antimicrobial efficacy and durability. Proved by all these mentioned examples, it is an effective strategy to modify antibiotics overcoming drug resistance through improving the drug binding ability, promoting the transfer process, and increasing the intracellular antibiotic concentration. It is encouraged that we could directly counter the evolutionary forces of drug resistance, through an individualized modification targeted for drug resistance mechanisms (Scheme 16.9).

16.3 Antibiotic Modification with Photo-Switching Units

Using optical-switching units for antibiotic modification is another innovative strategy against the threat of antibiotic resistance. Through this intelligent strategy, researches realize the regulation of antibacterial activity on-demand which expect to reduce the abuse of antibiotics under micro-environment. In 2013, Ben L. Feringa and co-workers designed and synthesized a series of quinolones derivatives (Scheme 16.10, **27**) modified with azobenzene (a photo-reversed unit) (Velema et al. 2013). Under UV light irradiation at 365 nm, these photo-responsive antibiotics exhibited different *trans* or *cis* conformations and exhibited totally different antibacterial abilities. In this way, a precise positioning control of antibacterial activity was achieved, allowing the growth of bacteria to be limited as a defined range. In further study, they designed two antibiotics FQNC (**28**) and BPOC (**29**) with orthogonally photo-triggered activity (Velema et al. 2014). Under the irradiation of different wavelengths of light, the connected photocleavable groups could be broken to release original antibiotics. In this way, they realized the selected inhibition of single pathogenic bacterial strain in the mixed bacterial samples. In 2017, to overcome the cytotoxicity and the low tissue penetration of UV light, they moved the excitation wavelength to the visible range by adjusting the substituent groups of azobenzene moiety on diaminopyrimidines (**30**) (Wegener et al. 2017). These photo-responsive antibiotics realized the *in situ* control of antibacterial activity against *E. coli* under green or violet light irradiation. Remarkably, the compound (**31**) exhibited at least eightfold activity difference before and after light irradiation at 652 nm. This work further moved photo-pharmacology into the

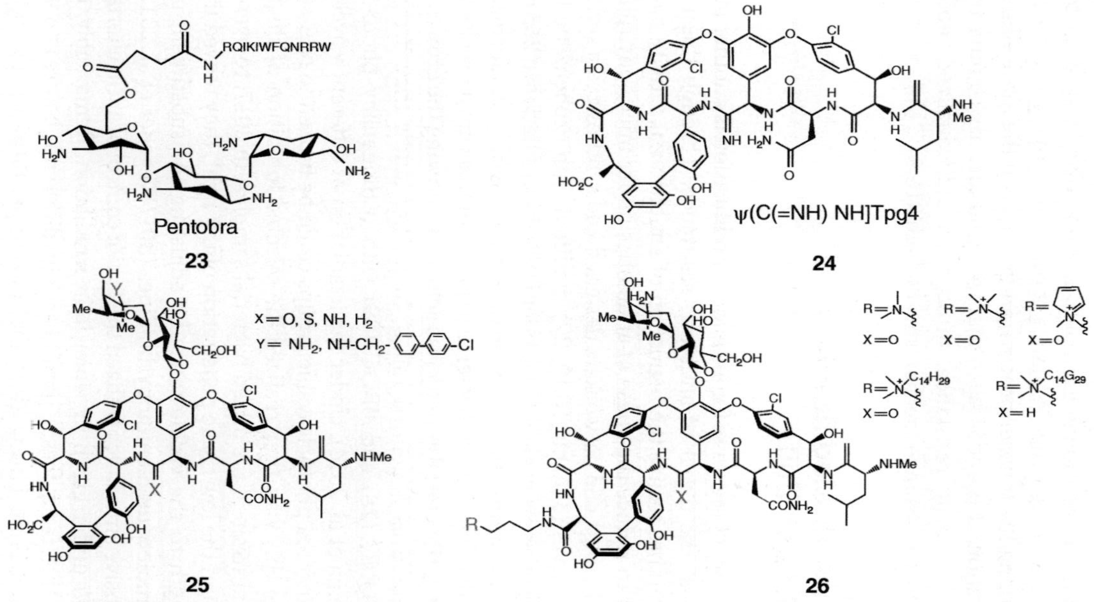

Scheme 16.9 Chemical structures of (**23**)–(**26**).

therapeutic window. Except incorporating with traditional antibiotics, photo-switching units also could be modified on new-type antibiotic–antimicrobial peptide. As the published work by Komarov's group, the diarylethene was inserted into the backbone of cyclic antimicrobial peptide Gramicidin S to construct a diarylethene-based amino acid analogue (Babii et al. 2014). Attributed to the different conformations of diarylethene under UV and visible light irradiation, the resultant products could form the "open" or "closed" status, resulting in a remarkable change in biological activity. Antibacterial regulation under light irradiation is an innovative route to explore using methods for antibiotics without the environment pressure. To develop these light-trigged antibiotics, all the photo-responsible units showed the great potential to be applied in the field of antibiotic modification. We believe that there must be some more exciting progress in the near future.

16.4 Antibiotic Modification by Supramolecular Chemistry

Supramolecular chemistry, which focuses on chemical systems beyond single molecule, is also a valuable route to modifying common antibiotics for dealing with pathogenic resistance. The self-assembly process driven by noncovalent force may endow classical drugs with some progressive properties. A promising strategy against drug-resistant bacteria is to reverse the drug resistance mechanisms, with the potential to rejuvenate large classes of previously powerful antibiotics (Zhang et al. 2008). As shown in Scheme 16.11, Structure (**32**), Patirick Couvreur and co-workers linked the penicillin G (PNG) with the hydrophobic squalene, which could form nanoparticles in aqueous environment by spontaneous assembly (Semiramoth et al. 2012). In neutral physiological environment, the synthesized SqPNG could enter cells by clathrin-dependent endocytosis. Because of the different membrane diffusion process, the new supramolecular complex could efficiently improve the concentration of PNG upon intracellular pathogens. Similarly, utilizing the unique self-assembly property of aggregation-induced emission (AIE) agents, Olof Ramstrom and co-workers synthesized a series of neoteric ciprofloxacin-based nanodrugs (Xie et al. 2017). As shown in Scheme 16.11, all these molecules (**33**)–(**36**) that ciprofloxacin bears with a perfluoroaryl ring (AIE agent) could form a propeller-shaped structure. More importantly, they are all endowed with higher bactericidal capability against both sensitive and resistant *E. coli*.

Host–guest chemistry, an important branch in supramolecular chemistry, also can be employed for improving the bactericidal activity. Utilizing the beta-cyclodextrin (β-CD) as host molecule and the antibacterial doxycycline (DOX) as guest molecule, Sinisterra et al. cleverly constructed the β-CD/DOX complex and successfully enhanced the drug activity of DOX (Suarez et al. 2014). Harshita

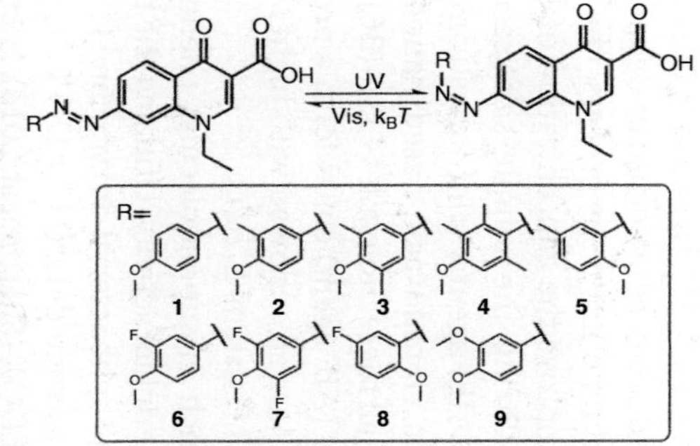

27

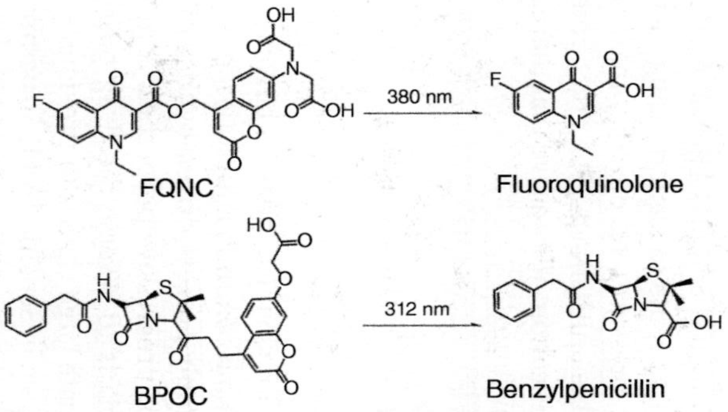

28, 29

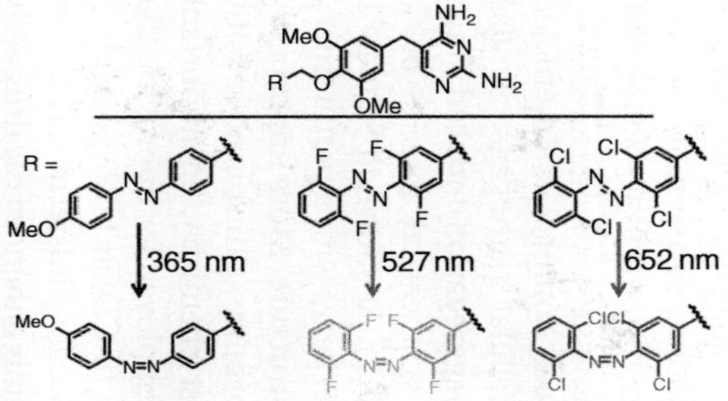

30

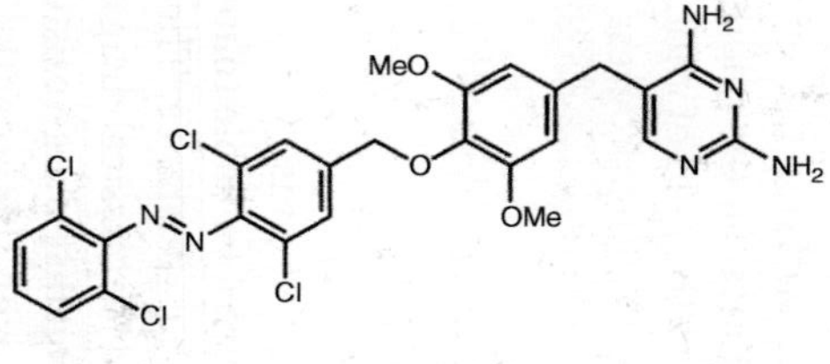

31

Scheme 16.10 Chemical structures of (**27**)–(**31**).

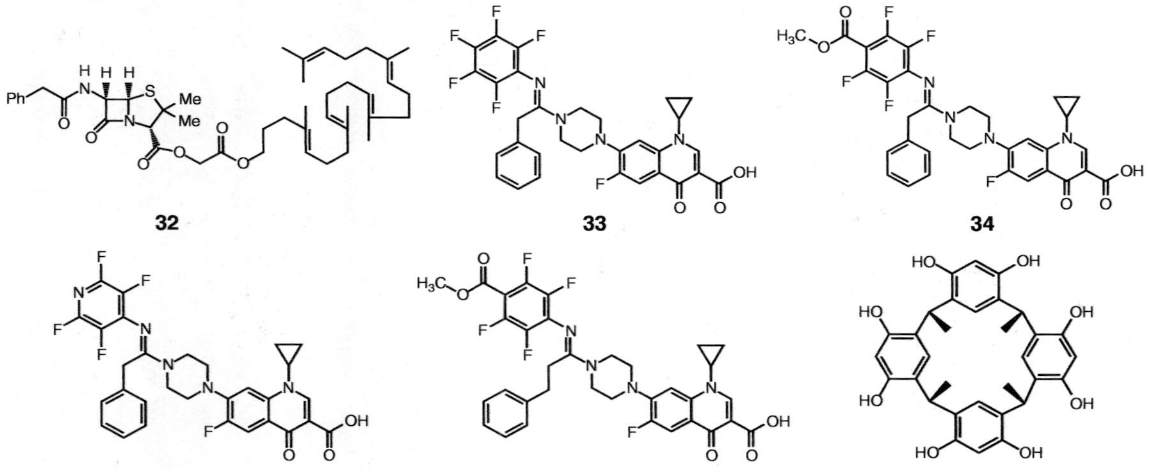

32, 33, 34, 35, 36, 37

Scheme 16.11 Chemical structures of (**32**)–(**37**).

Kumari et al. used the C-alkylresorcin[4]arene (RsC1) (**37**) as the host molecule and the antibacterial gatifloxacin as guest molecule to construct supramolecular bacteriostatic complex (Dawn et al. 2017). The antibiotic-based supramolecular drug also realized high-efficiency inhibition of Gram-negative pathogens. Because the dynamic properties of supramolecular self-assembly process only relies on the well-defined structure and properties of antibiotics and the related host molecules, this strategy does not require any chemical modification on the active site of existing antibacterial agent, which is simple, rapid, but efficient. It also should be noted that, based on the rigorous investigations in Wang's group, antibacterial regulation can also be achieved by supramolecular chemistry strategy (Bai et al. 2015, 2016, 2017b). More importantly, they further proved that the strategy could effectively and precisely treat the multidrug resistance and avoid the rapid appearance of MDR pathogens in 2017 (Bai et al. 2017a). In this work, a series of commercial germicides and cucurbit[7]uril (CB[7]) were employed for constructing supramolecular germicide switches. Their bactericidal activities were controlled reversibly by the self-assembly and disassembly manner between the host and guest molecules. As shown in Scheme 16.12, the selected germicide (dimethyl benzyl ammonium bromide [DDBAB]) killed nearly 100% pathogens in the initial treatment, and the drug-resistant pathogens occurred right after frequently treated with DDBAB in tenth generation. But the supramolecular germicides still contained highly bactericidal activity when needed. The biological analysis demonstrated that the antibacterial regulation strategy could effectively and precisely treat pathogens and avoid the rapid appearance of MDR pathogens. With all the above projects, we should be confident in the supramolecular strategy of modifying antibiotics or constructing antibiotic complexes and rejuvenating the large classes of previously powerful antibiotics.

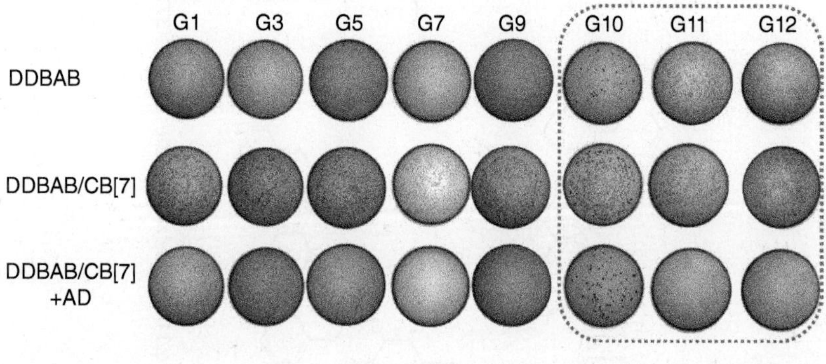

Scheme 16.12 The plate photographs of *Staphylococcus aureus* treated with DDBAB frequently and those treated with DDBAB/CB[7] complex before and after adding AD for replacement.

16.5 Antibiotic Modification by Complexed with Other Materials

With the continuous development of material science, we have acquired more biological functional materials with desired properties. It is also a unique way for handling drug resistance that combines antibiotics with these new material systems through a reasonable and ingenious method. Hao Wang et al. successfully constructed a supramolecular antibiotic delivery platform and realized the loading and transportation of vancomycin (Li et al. 2014). In this smart system, the cross-linked gelatin nanoparticles (SGNPs) were used as the core, and the uniform red blood cell (RBC) membrane is the shell. As shown in Scheme 16.13a, after loading vancomycin (Van) on SGNPs, the Van⊂SGNPs@RBC system could release drugs and efficiently killed the pathogens on demand. As shown in Scheme 16.13b, Hui Gao and Yang Ying-Wei groups co-reported

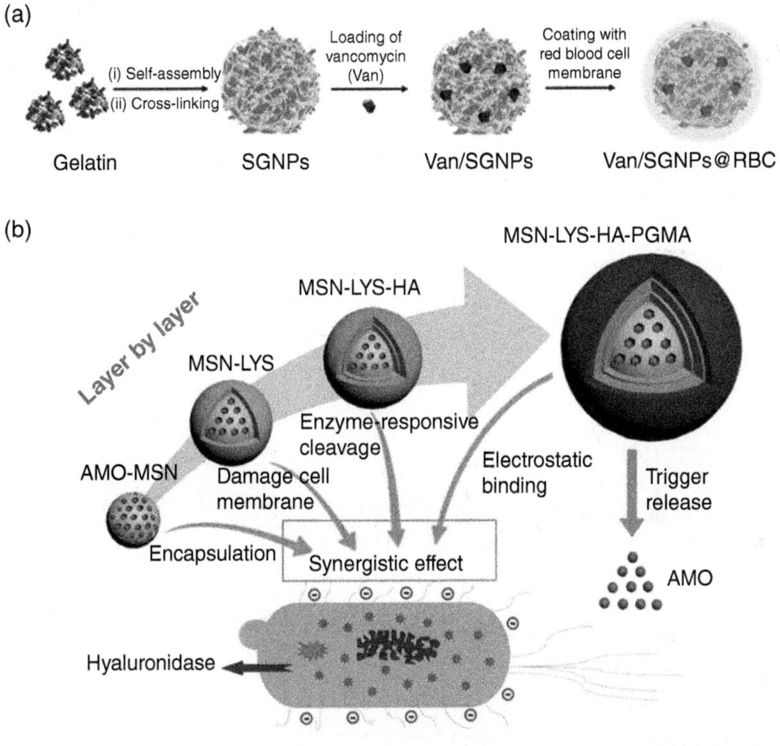

Scheme 16.13 Schematic illumination of the preparation process of (a) Van⊂SGNPs@RBC. Source: Reprinted from Bai et al. (2015) with permission. Copyright 2014, American Chemical Society. (b) MSN-LYS-HA-PGMA. Source: Reprinted from Bai et al. (2016) with permission. Copyright 2015, American Chemical Society.

an antibacterial nanomaterial utilizing mesoporous silica nanoparticle (MSN) as the porous core and lysozyme (Lys), hyaluronic acid (HA), and 1,2-ethanediamine (EDA)-modified polyglycerol methacrylate (PGMA) as the shell (Wu et al. 2015). In this study, amoxicillin (AMO) was chosen as the model drug loading at the MSN, and biocompatible layer-by-layer (LBL) technique was used for coating the hyaluronidase-responsive shell. After being treated with Gram-positive bacteria or Gram-negative bacteria, the shell could be disassembled by hyaluronidase of pathogens, with that the AMO was released to react on and kill drug-resistant bacteria.

In 2018, the concept of constructing hyperbranched polymer using antibiotics was proposed and practiced by Schoenfisch's group (Yang and Schoenfisch 2018). As shown in Scheme 16.14, they synthesized and characterized a series of hyperbranched polyaminoglycosides via respective polymerization of kanamycin, gentamicin, and neomycin. The resulting polymers with great branching degrees and antibiotic terminal groups are more effective in eradicating all selected dental pathogens. It was a promising model for the therapeutic index in future biomedical application. In addition to the above mentioned examples, traditional antibiotics can be modified with a numerous of classical or modern materials, and the collision of materials science and antibacterial drugs will enormously contribute to the development of antibacterial materials for treating clinical infections.

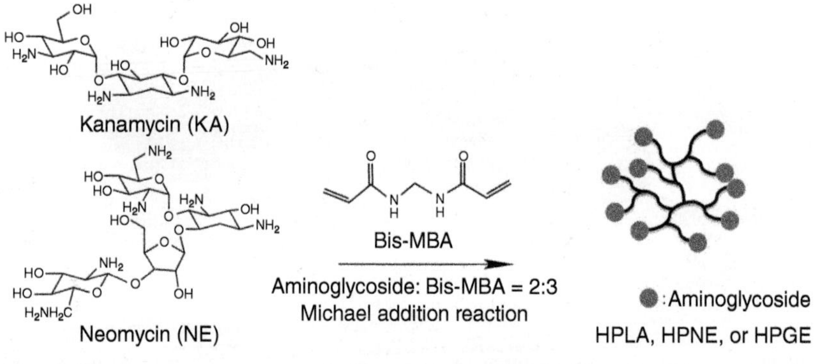

Scheme 16.14 Schematic illumination of the synthesis process of aminoglycoside-terminated hyperbranched polyaminoglycosides. *Source:* Reprinted from Bai et al. (2017b) with permission. Copyright 2018, American Chemical Society.

16.6 Conclusion

In this section, we listed five main strategies on antibiotic modification for addressing drug resistance: (i) synthesis of alternative antibiotics by traditional chemical synthesis process and screening of the effective ones, (ii) targeted modification of existing antibiotics on the basis of the known resistance mechanisms, (iii) introduction of photo-switching units for developing new antibiotics with controllable bactericidal activity, (iv) supramolecular strategies for antibiotic modification, and (v) combination of antibiotics with multiple materials with advantageous properties. It is safe to say that all these modification strategies aimed to obtain novel antibiotic derivatives with excellent bactericidal ability and all these antibacterial products are promising materials for addressing the crisis of drug resistance. All the studies shed light on combating antibiotic resistance, and it is believed that more effective/intelligent strategies to modify antibiotics will be developed to counter the antibiotic resistance.

References

Babii, O., Afonin, S., Berditsch, M. et al. (2014). Controlling biological activity with light: diarylethene-containing cyclic peptidomimetics. *Angew. Chem. Int. Ed.* 53 (13): 3392–3395.

Bai, H.T., Fu, X.C., Huang, Z.H. et al. (2017a). Supramolecular germicide switches through host-guest interactions for decelerating emergence of drug-resistant pathogens. *ChemistrySelect* 2 (26): 7940–7945.

Bai, H.T., Lv, F.T., Liu, L.B., and Wang, S. (2016). Supramolecular antibiotic switches: a potential strategy for combating drug resistance. *Chem. A Eur. J.* 22 (32): 11114–11121.

Bai, H.T., Yuan, H.X., Nie, C.Y. et al. (2015). A supramolecular antibiotic switch for antibacterial regulation. *Angew. Chem. Int. Ed.* 54 (45): 13208–13213.

Bai, H.T., Zhang, H.Y., Hu, R. et al. (2017b). Supramolecular conjugated polymer systems with controlled antibacterial activity. *Langmuir* 33 (4): 1116–1120.

Baker, W.R., Clark, J.D., Stephens, R.L., and Kim, K.H. (1988). Modification of macrolide antibiotics – synthesis of 11-deoxy-11-(carboxyamino)-6-O-methylerythromycin-a 11,12-(cyclic esters) via an intramolecular Michael reaction of O-carbamates with an alpha,beta-unsaturated ketone. *J. Org. Chem.* 53 (10): 2340–2345.

Blaskovich, M.A.T., Hansford, K.A., Butler, M.S. et al. (2018). Developments in glycopeptide antibiotics. *ACS Infect. Dis.* 4 (5): 715–735.

Bugg, T.D.H., Wright, G.D., Dutkamalen, S. et al. (1991). Molecular-basis for vancomycin resistance in *Enterococcus-faecium* Bm4147: biosynthesis of a depsipeptide peptidoglycan precursor by vancomycin resistance proteins VanH and VanA. *Biochemistry* 30 (43): 10408–10415.

Chellat, M.F., Raguz, L., and Riedl, R. (2016). Targeting antibiotic resistance. *Angew. Chem. Int. Ed.* 55 (23): 6600–6626.

Dawn, A., Chandra, H., Ade-Browne, C. et al. (2017). Multifaceted supramolecular interactions from C-methylresorcin[4]arene lead to an enhancement in in vitro antibacterial activity of gatifloxacin. *Chem. A Eur. J.* 23 (72): 18171–18179.

Fischbach, M.A. and Walsh, C.T. (2009). Antibiotics for emerging pathogens. *Science* 325 (5944): 1089–1093.

Jiang, J.W., Sun, Y., Nie, Y. et al. (2013). Synthesis and antibacterial evaluation of a novel series of 10-hydroxyl ketolide derivatives. *Bioorg. Med. Chem. Lett.* 23 (11): 3452–3457.

Jiang, Y.W., Xu, L., Fu, W. et al. (2018). Design, synthesis and biological activity of novel demethylvancomycin dimers against vancomycin-resistant *Enterococcus faecalis*. *Tetrahedron* 74 (27): 3527–3533.

Kaplan, A.H., Gilligan, P.H., and Facklam, R.R. (1988). Recovery of resistant enterococci during vancomycin prophylaxis. *J. Clin. Microbiol.* 26 (6): 1216–1218.

Kerns, R., Dong, S.D., Fukuzawa, S. et al. (2000). The role of hydrophobic substituents in the biological activity of glycopeptide antibiotics. *J. Am. Chem. Soc.* 122 (50): 12608–12609.

Lehar, S.M., Pillow, T., Xu, M. et al. (2015). Novel antibody-antibiotic conjugate eliminates intracellular *S. aureus*. *Nature* 527 (7578): 323–328.

Li, L.L., Xu, J.H., Qi, G.B. et al. (2014). Core-shell supramolecular gelatin nanoparticles for adaptive and "on-demand" antibiotic delivery. *ACS Nano* 8 (5): 4975–4983.

Lin, H.N. and Walsh, C.T. (2004). A chemoenzymatic approach to glycopeptide antibiotics. *J. Am. Chem. Soc.* 126 (43): 13998–14003.

Mackay, J.P., Gerhard, U., Beauregard, D.A. et al. (1994a). Dissection of the contributions toward dimerization of glycopeptide antibiotics. *J. Am. Chem. Soc.* 116 (11): 4573–4580.

Mackay, J.P., Gerhard, U., Beauregard, D.A. et al. (1994b). Glycopeptide antibiotic-activity and the possible role of dimerization – a model for biological signaling. *J. Am. Chem. Soc.* 116 (11): 4581–4590.

McComas, C.C., Crowley, B.M., and Boger, D.L. (2003). Partitioning the loss in vancomycin binding affinity for D-Ala-D-Lac into lost H-bond and repulsive lone pair contributions. *J. Am. Chem. Soc.* 125 (31): 9314–9315.

Mu, Y.Q., Nodwell, M., Pace, J.L. et al. (2004). Vancomycin disulfide derivatives as antibacterial agents. *Bioorg. Med. Chem. Lett.* 14 (3): 735–738.

Okano, A., Isley, N.A., and Boger, D.L. (2017). Peripheral modifications of [Psi[CH2NH]Tpg(4)]vancomycin with added synergistic mechanisms of action provide durable and potent antibiotics. *Proc. Natl. Acad. Sci. U. S. A.* 114 (26): E5052–E5061.

Pavlovic, D., Fajdetic, A., and Mutak, S. (2010). Novel hybrids of 15-membered 8a- and 9a-azahomoerythromycin A ketolides and quinolones as potent antibacterials. *Bioorg. Med. Chem.* 18 (24): 8566–8582.

Pavlovic, D. and Mutak, S. (2011). Discovery of 4″-ether linked azithromycin-quinolone hybrid series: influence of the central linker on the antibacterial activity. *ACS Med. Chem. Lett.* 2 (5): 331–336.

Pavlovic, D., Mutak, S., Andreotti, D. et al. (2014). Synthesis and structure-activity relationships of alpha-amino-gamma-lactone ketolides: a novel class of macrolide antibiotics. *ACS Med. Chem. Lett.* 5 (10): 1133–1137.

Perkins, H.R. (1982). Vancomycin and related antibiotics. *Pharmacol. Ther.* 16 (2): 181–197.

Schmidt, N.W., Deshayes, S., Hawker, S. et al. (2014). Engineering persister-specific antibiotics with synergistic antimicrobial functions. *ACS Nano* 8 (9): 8786–8793.

Semiramoth, N., Di Meo, C., Zouhiri, F. et al. (2012). Self-assembled squalenoylated penicillin bioconjugates: an original approach for the treatment of intracellular infections. *ACS Nano* 6 (5): 3820–3831.

Sheldrick, G.M., Jones, P.G., Kennard, O. et al. (1978). Structure of vancomycin and its complex with acetyl-D-alanyl-D-alanine. *Nature* 271 (5642): 223–225.

Suarez, D.F., Consuegra, J., Trajano, V.C. et al. (2014). Structural and thermodynamic characterization of doxycycline/beta-cyclodextrin supramolecular complex and its bacterial membrane interactions. *Colloids Surf. B* 118: 194–201.

Velema, W.A., van der Berg, J.P., Hansen, M.J. et al. (2013). Optical control of antibacterial activity. *Nat. Chem.* 5 (11): 924–928.

Velema, W.A., van der Berg, J.P., Szymanski, W. et al. (2014). Orthogonal control of antibacterial activity with light. *ACS Chem. Biol.* 9 (9): 1969–1974.

Wegener, M., Hansen, M.J., Driessen, A.J.M. et al. (2017). Photocontrol of antibacterial activity: shifting from UV to red light activation. *J. Am. Chem. Soc.* 139 (49): 17979–17986.

Wei, X. and You, Q.D. (2006). A facile and scaleable synthesis of 3-O-decladinose-6-methyl-10,11-dehydrate-erythromycin-3-one-2'-acetate, an important intermediate for ketolide synthesis. *Org. Process Res. Dev.* 10 (3): 446–449.

Wright, G.D. (2005). Bacterial resistance to antibiotics: enzymatic degradation and modification. *Adv. Drug Deliv. Rev.* 57 (10): 1451–1470.

Wu, Y.H., Long, Y.B., Li, Q.L. et al. (2015). Layer-by-layer (LBL) self-assembled biohybrid nanomaterials for efficient antibacterial applications. *ACS Appl. Mater. Interfaces* 7 (31): 17255–17263.

Xie, S., Manuguri, S., Proietti, G. et al. (2017). Design and synthesis of theranostic antibiotic nanodrugs that display enhanced antibacterial activity and luminescence. *Proc. Natl. Acad. Sci. U. S. A.* 114 (32): 8464–8469.

Xie, J., Pierce, J.G., James, R.C. et al. (2011). A redesigned vancomycin engineered for dual D-Ala-D-Ala and D-Ala-D-Lac binding exhibits potent antimicrobial activity against vancomycin-resistant bacteria. *J. Am. Chem. Soc.* 133 (35): 13946–13949.

Yang, L. and Schoenfisch, M.H. (2018). Nitric oxide-releasing hyperbranched polyaminoglycosides for antibacterial therapy. *ACS Appl. Bio Mater.* 1: 1066–1073.

Yarlagadda, V., Sarkar, P., Manjunath, G.B., and Haldar, J. (2015). Lipophilic vancomycin aglycon dimer with high activity against vancomycin-resistant bacteria. *Bioorg. Med. Chem. Lett.* 25 (23): 5477–5480.

Zhang, L., Gu, F.X., Chan, J.M. et al. (2008). Nanoparticles in medicine: therapeutic applications and developments. *Clin. Pharmacol. Ther.* 83 (5): 761–769.

Zhao, Y., You, Q.D., and Shen, W.B. (2003). A novel bicyclic ketolide derivative. *Bioorg. Med. Chem. Lett.* 13 (10): 1805–1807.

17

Sensitizing Agents to Restore Antibiotic Resistance

Anton Gadelii, Karl-Omar Hassan, and Anders P. Hakansson

Division of Experimental Infection Medicine, Department of Translational Medicine, Lund University, Malmö, Sweden

17.1 Introduction

Throughout history, infectious diseases have been one of the major causes of death (Nathan 2012). During the last century, with improved sanitation and nutrition combined with the development and broad use of various vaccines and the discovery of antibiotics, we have found ways to prevent and treat infections that would previously be fatal. With the growing emergence of antibiotic resistance, this might come to an end if nothing is done. In a global report published in 2014 on antibiotic resistance surveillance, the WHO presented a bleak picture for the twenty-first century if actions were not taken immediately (WHO 2014). In 2017, they followed up this report with a list of priority pathogens, whose resistance pattern and spread, listed from critical to medium concern, warrant immediate attentions regarding research and development of new strategies to combat these infections (Tacconelli et al. 2017).

To accomplish this, we need to be able to attack both heritable resistance, which is resistance accomplished through genetic mutations or acquired through horizontal spread of genetic resistance genes, and nonheritable resistance, a phenotypic tolerance to antibiotics in microorganisms that organize into biofilms, that do not replicate, and where epigenetic and transcriptional changes leading to antibiotic tolerance is observed (D'Costa et al. 2006; Levin and Rozen 2006; Smith and Romesberg 2007; Fischbach and Walsh 2009; Fernández et al. 2011).

There are several strategies to accomplish this, which are presented in this book. The obvious option would be to restrict antibiotic use severely (antibiotic stewardship), both within and outside of the healthcare sector. Such a

Antibiotic Drug Resistance, First Edition. Edited by José-Luis Capelo-Martínez and Gilberto Igrejas.

regional or worldwide restriction would need the cooperation of healthcare workers, policy makers, and other sectors, a solution that, however effective and worthwhile, seems hard to put into practice (LaCross et al. 2013). Discovery of novel antibiotics based on new scaffolds and molecules needs to be pursued further. However, this has been a challenge as the pharmaceutical industry do not have enough incentive to develop costly new antibiotics (Projan 2003; Projan and Shlaes 2004), and today only 2 of the 17 antibiotics currently undergoing phase I, II, or III studies are molecules based on new scaffolds (Taneja and Kaur 2016; Wright 2016). Producing new antibiotics is a slow and expensive process, and, more importantly, resistance emerges almost as soon as the new antibiotics are introduced in the clinic (Andersson and Hughes 2014; Ventola 2015).

In this chapter, we will discuss sensitizers or adjuvants as a different approach to combat antibiotic resistance. A sensitizer is a molecule that itself has no bactericidal or bacteriostatic activity but that affects bacteria in a way that makes them more sensitive to antibiotics (Kalan and Wright 2011; Bernal et al. 2013; Gill et al. 2015; Wright 2016). These are avenues with which we can repurpose the use of the current arsenal of safe antibiotics in order to make them effective again. However, these strategies may also decrease the dosing of antibiotic needed for effective treatment and therefore decrease the general antibiotic use. Sensitizing strategies are presented in three categories, starting with ways that are used or being developed to directly target antimicrobial resistance mechanisms (Section 17.2). The second category contains strategies that increase antibiotic sensitivity by indirectly affecting resistance mechanisms (Section 17.3), and the third category includes strategies in which stimulating host immunity synergizes to make antibiotics more effective against drug-resistant organisms (Section 17.4). These strategies are depicted in Figure 17.1.

17.2 Sensitizing Strategies Directly Targeting Resistance Mechanisms

17.2.1 Inhibition of β-Lactamases

Resistance mechanisms predate the introduction of antibiotic use (D'Costa et al. 2011; Bhandari et al. 2016). In this regard, β-lactamases are ancient, and it is estimated that they evolved already around 2.4 billion years ago (Finley et al. 2013). These enzymes degrade many of our most commonly used β-lactam antibiotics, such as penicillins, cephalosporins, and carbapenems, and are widespread in bacterial populations, where they spread between bacteria through horizontal gene transfer (Wright 2016). They are arranged in two groups: the first group of enzymes uses an active site serine

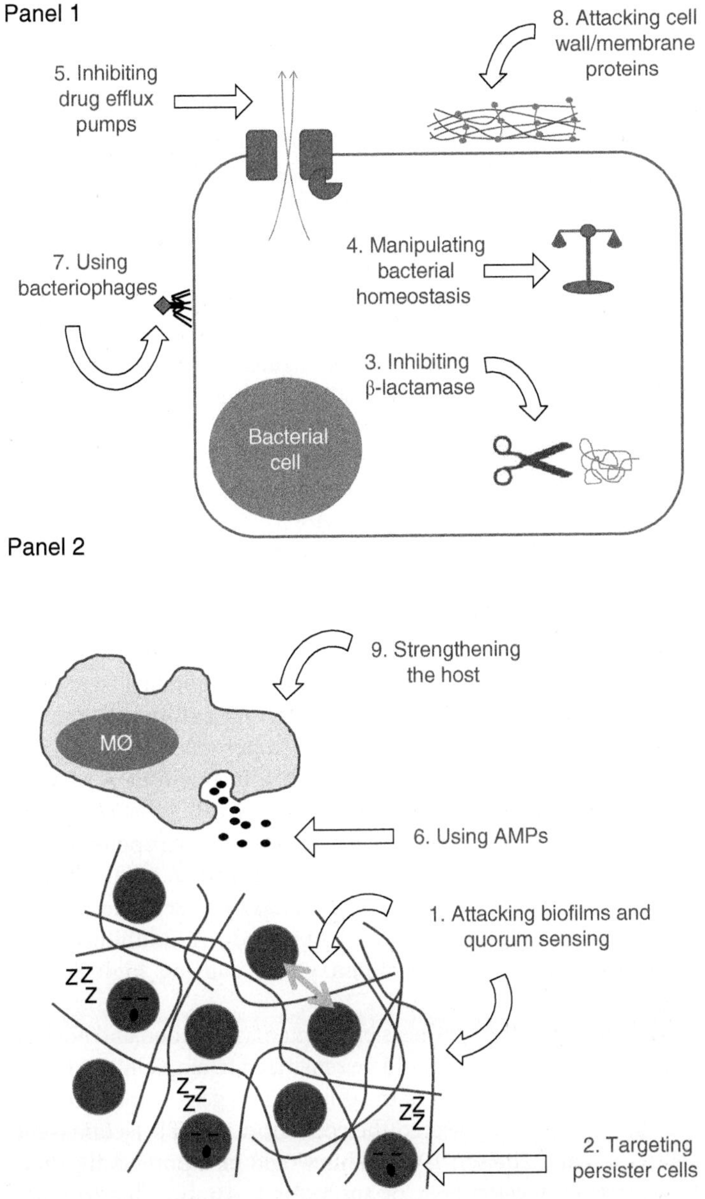

Figure 17.1 Strategies to sensitize bacteria to antibiotics. A simple sketch illustrating different methods to accomplish antibiotic sensitization. Panel 1 shows a lone bacterial cell. Panel 2 shows a biofilm with persister cells and a host macrophage.

to capture and hydrolyze their substrate, and the second group requires an active site Zn^{2+} atom for activity (metallo-β-lactamases [MBL]) (Bush and Macielag 2010). Inhibitors of β-lactamases have been the most successful sensitizers of antibiotics in the clinic thus far. In the 1970s, clavulanic acid was discovered as an irreversible inhibitor of serine β-lactamases and was the first sensitizer approved for clinical use in combination with an antibiotic. This method has greatly extended the lifetime of β-lactam antibiotics (Drawz and Bonomo 2010). Over the years, resistant alleles of β-lactamases have evolved, and extended-spectrum β-lactamases (ESBLs) have emerged, leading to a lower efficacy of clavulanic acid. Recently, novel classes of β-lactamase inhibitors, including tazobactam, sulbactam, and avibactam, have been approved for clinical use and are effective against some of the emerging ESBLs, but not all (Bassetti and Righi 2015; Bush 2015; Papp-Wallace and Bonomo 2016).

One enzyme that has remained elusive is the MBL. The emerging New Delhi MBL that quickly has become broadly distributed in *Enterobacteriaceae*, *Pseudomonas aeruginosa*, and *Acinetobacter baumannii* strains (Dortet et al. 2014) threatens to make all β-lactam antibiotics ineffective and is of such concern globally that they are listed as critical priority pathogens by the WHO in their recent report (Tacconelli et al. 2017). Many different compounds have been investigated, but no one has yet been approved for clinical use (McGeary et al. 2017). Promising compounds able to inhibit MBL is biphenyl tetrazoles, which were shown to work synergistically with β-lactam antibiotics, such as penicillin and imipenem, against MBL-expressing *Bacteroides fragilis* (Toney et al. 1998). Another interesting candidate is aspergillomarasmine A (AMA), a natural substance found in fungi. Mice infected with MBL *Klebsiella pneumoniae* receiving a combination therapy with AMA and meropenem were treated effectively, while mice treated with either substance alone showed high mortality (King et al. 2014). However, AMA has been screened earlier as an ACE inhibitor, which illustrates a problem with MBL inhibitors since MBL has similarities to our own metalloenzymes, causing high risk of adverse effects.

Another approach is to stop the synthesis of β-lactamases. Good and Nielsen showed that resistant *Escherichia coli* could be sensitized to ampicillin using an antisense nucleotide to the start codon of the β-lactamase gene, which blocked the translation of the mRNA and lowered the concentration of β-lactamase in the bacterial cell (Good and Nielsen 1998). This would be a potentially effective way of counteracting development of antibiotic resistance due to many mechanisms since the base sequence can be changed to match any gene. However, effective administration of antisense nucleotides is still challenging and not effective *in vivo*. In general, the successful use of clavulanic acid for the past several decades points to these kinds of inhibitors as a promising strategy for the future.

17.2.2 Drug Efflux Pump Inhibitors (EPIs)

Bacterial membranes contain proteins specialized for transport of a variety of molecules that require energy for transport. Extrusion of antibiotics out of the bacteria through efflux pumps is one of the most important multidrug resistance mechanism in Gram-negative bacteria, resulting in a decrease in intracellular concentration of antibiotic inside the bacterium (Mahamoud et al. 2007; Martins et al. 2008; Nikaido 2009). There are five types of efflux pumps that are driven either by ATP or by ion gradients over the membrane, working as antiporters, symporters, or uniporters (for more details on transport mechanisms, see Lorch et al. (2005) and Mahamoud et al. (2007)). The most common resistance mechanisms derive from two ion-gradient-driven groups of transporters: the major facilitator superfamily (MFS) and the resistance-nodulation-division (RND) family of transporters (Mahamoud et al. 2007). The MFS transporters are found in both Gram-positive and Gram-negative organisms and have a rather narrow spectrum of cargo allowed through the transporter, whereas RND transporters are exclusive to Gram-negative organisms and have a very wide substrate spectrum.

Inhibition of drug efflux in bacteria has the major advantage that it increases the intracellular concentration of antibiotics and thus provides higher antibiotic efficacy. Three approaches to inhibit efflux are being explored: (i) changing the structure of current antibiotics to reduce their efflux, (ii) affecting the permeability of the cell envelope to increase permeability and uptake of antibiotics in combination with decreasing the ability for escape, and (iii) direct changes to the pump activity or function.

There are no current antibiotics that, by modification of side groups or structure, have shown reduced efflux. Increased uptake of antibiotics has been seen during treatment with polymyxins and especially their peptides (Tsubery et al. 2000, 2005). This approach has shown efficacy both *in vitro* and *in vivo* against Gram-negative bacteria (Tsubery et al. 2005). Other strategies that permeabilize the membrane or affect the energetics that drive the efflux pumps are the addition of CCCP, a cyanide-based molecule that reduces membrane energy flux (Pages and Amaral 2009) and high concentration of transport inhibitors, such as reserpine and verapamil. These compound are, however, extremely toxic and cannot be used in clinical practice (Li and Nikaido 2004; Garvey and Piddock 2008). Berberin, a derivative of plant compounds, shows activity against efflux pumps but suffers from problems with low production yield, stability, and most importantly toxicity (Tegos et al. 2002). Since many efflux pumps are proton antiporters, the disruption of the proton gradient leads to efflux pump dysfunction. One example is oligo-acyl-lysyls (OAKs) that when added to bacterial cultures depolarized bacterial membranes, disrupted the proton motive force, and made bacteria increasingly sensitive to intracellular antibiotics such as erythromycin (Goldberg et al. 2013). The study was also able to show that neutropenic mice infected with *E. coli* showed synergistic effects between OAKs and erythromycin, without any visible toxicity (Goldberg et al. 2013).

The majority of work has been placed on the third group of inhibitors, those that directly inhibit pump function (Lomovskaya and Bostian 2006; Mahamoud et al. 2007, 2016; Pages et al. 2008; Pages and Amaral 2009). A large number of efflux pump inhibitors (EPIs) that directly affect pump function have been produced and are divided in four major classes. In the first group, peptidomimetics produced from the study of efflux pump function have shown synergistic effects with a variety of antibiotics against Gram-negative bacteria (Lomovskaya and Bostian 2006; Mahamoud et al. 2007 and references therein). The best characterized molecule is phenylalanine arginine beta-naphtylamide (PAβN) that sensitizes *P. aeruginosa* to levofloxacin (Lomovskaya et al. 2001) as well as other Gram-negative organisms against several classes of antibiotics (Lomovskaya and Bostian 2006). This approach suffers from problems with high toxicity, making it impossible for these compounds to be used in the clinic thus far.

A second class of compounds that have been gaining attention is quinoline derivatives that show broad-spectrum sensitization of *Enterobacter* spp. and *K. pneumoniae* to various antibiotics (Chevalier et al. 2004). These compounds have less activity against some classes of transporters, especially those expressed by *P. aeruginosa* (Mahamoud et al. 2007), and their toxicity to human cells is a major hurdle in their development.

The third class of inhibitor containing acylpiperidines and arylpiperazines and their derivatives have showed promising effects on efflux pump inhibition in both Gram-positive (*Staphylococcus aureus*) and Gram-negative (*E. coli*) bacteria, for example, reversing antibiotic resistance of *E. coli* to linezolid (Thorarensen et al. 2001). It is unclear how they work, however, as treatment with these EPIs did not produce significant increase in intracellular antibiotic concentration, which would be expected of an EPI. Although they work effectively on *Enterobacteriaceae* and *A. baumanii* (two of the critical priority pathogens on the WHO list (Tacconelli et al. 2017)), they had poor activity against the third critical pathogen *P. aeruginosa*.

Finally, other compounds have also been discovered. Dwivedi et al., showed that urosolic acid derivatives can inhibit efflux pump activity both through inducing a reduced expression of efflux pump genes and by inhibition of the efflux pump ATPase itself (Dwivedi et al. 2014). This approach sensitized *E. coli* to nalidixin and tetracycline *in vitro*. Another study shows that a group of compounds already used for other indications (chlorpromazine, amitryptiline, and transchlorprothixene) also have EPI activity and can sensitize several bacteria to antibiotics (Kristiansen et al. 2010). These substances have already been used clinically for years, proving their safety, but their alternative use may not be without obstacles due to higher concentrations needed for sensitization efficacy, with potential adverse effects. Many of the substances described so far are synthetic, but there are also natural substances under investigation. Essential oils from thyme and rosemary have been shown to possess sensitizing properties by inhibition of efflux pumps (Fadli et al. 2011).

Efflux pumps are a powerful strategy bacteria use to counter the effects of many different types of antibiotics. Even though research has been ongoing for some time, it has been difficult to find EPIs that either do not have strong adverse effects or inherent cytotoxicity (Mahamoud et al. 2007), and so far no drugs have reached the market (Mahmood et al. 2016; Spengler et al. 2017).

17.3 Sensitizing Strategies Circumventing Resistance Mechanisms

Established strategies to make bacteria susceptible to antibiotics often include direct activity against resistance mechanisms, as described above in Section 17.2. However, it is also possible to sensitize bacteria by indirect means by disrupting their basic homeostasis or their resistance growth phenotype. Examples of such strategies will be presented in this section.

17.3.1 Manipulating Bacterial Homeostasis

One strategy to accomplish sensitization of antibiotics is to target basic homeostasis of bacteria. Since many species of bacteria share proteins or enzymes that are important for bacterial homeostasis, this strategy has a lot of potential, as successful sensitization of one organism might prove effective against many other organisms. As mentioned under efflux pump inhibitors, one strategy that may have potential is to disrupt the membrane potential and/or proton motive force. This will not only close-circuit the driving force for efflux pumps but will have a general effect on bacterial membrane homeostasis and signaling. Besides the polycation, polymyxin B nonapeptide (PMBN), which was shown to sensitize enteric bacteria to several antibiotics (Vaara and Vaara 1983; Tsubery et al. 2000, 2005), high concentrations of cations have also been shown to be associated with the sensitivity of the fungus *Schizosaccharomyces pombe* to antibiotics (Alao et al. 2015).

Another interesting sensitizing compound is human alpha-lactalbumin made lethal to tumor cells (HAMLET), a protein–lipid complex from human milk, that is acting on the membrane to induce depolarization and specific ion fluxes. Marks et al., showed that this protein–lipid complex can sensitize bacteria to antibiotics (Marks et al. 2012, 2013). Although HAMLET has a direct antibacterial effect against a narrow spectrum of bacterial species, it only has nonlethal membrane effects on the majority of bacteria (Hakansson et al. 2000, 2011). HAMLET was first shown to sensitize *Streptococcus pneumoniae* to several classes of antibiotics (penicillin, macrolides, and aminoglycosides) in the presence of sublethal concentrations of HAMLET (Marks et al. 2012). The sensitizing activity was shown through a decreased MIC value for penicillin and erythromycin-resistant pneumococci that are listed on the WHO priority pathogen list. Furthermore, HAMLET provided a strong sensitizing effect both against biofilms and in an *in vivo* nasopharyngeal colonization model, where bacterial eradication was observed only after combination treatment with HAMLET and antibiotics. In a similar fashion, HAMLET was also shown to sensitize the priority pathogens *A. baumannii* and *S. aureus* to a broad range of antibiotics both *in vitro* and for *S. aureus* also *in vivo* (Marks et al. 2012, 2013). As HAMLET is a natural compound obtained from human milk, the risk of cytotoxicity and other adverse effects is low. Another promising quality of HAMLET is its wide range of activity – a result from potentially targeting highly conserved pathways in bacteria that bacteria cannot survive without. This could make HAMLET applicable as an adjuvant in treatment of many different clinically important and difficult to treat multidrug-resistant infectious diseases.

Additional compounds that affect membrane potential and proton motive force have been described. In a screen of non-antibiotic drugs, Ejim et al., found that loperamide, used as over the counter medication against diarrhea, showed a sensitizing activity for minocycline against *P. aeruginosa* (Ejim et al. 2011). Loperamide dissipated the electrical component the proton motive force in the bacteria, leading to an increased pH gradient over the inner membrane that increased the uptake of antibiotics into the cells. In a different study, compounds were identified that dissipated the proton motive force in methicillin-resistant *S. aureus* (MRSA) (Farha et al. 2013). By mixing compounds that affected the electrical gradient and the pH gradient over the membrane, a synergistic effect was obtained that lowered the concentrations of the drugs to a level where toxicity to human cell were not detected and only antibiotic sensitization was observed (Farha et al. 2013). Finally, in a study by Balakrishna et al., acyl polyamines were shown to bind to Gram-negative bacteria and increase their membrane permeability (Balakrishna et al. 2006). Unfortunately, these polyamines are potentially cytotoxic, making some of them unfit for clinical use. However, albumin in blood lowers this cytotoxicity, making polyamines interesting for future research.

In conclusion, attacking the membrane potential and the proton motive force to both affect the energetics of the bacteria and increase permeability of antibiotics over the membrane, although wrought with some potential toxicity problems, appears to be a strategy that can be well tolerated and effective clinically in the near future.

17.3.2 Cell Wall/Membrane Proteins

Many antibiotics (β-lactam antibiotics, glycopeptides, etc.) used today interfere with cell wall synthesis and concentrations of intracellular antibiotics in bacteria depend on cell wall permeability and transport. Also, several proteins in the cell wall are important to bacterial antibiotic resistance, e.g. penicillin-binding proteins (Nikolaidis et al. 2014). Different approaches that interact with cell wall integrity such as destabilization of the bacterial membrane, increase of cell wall permeability, and modification of cell wall proteins could be promising ways to sensitize bacteria to antibiotics.

Some novel strategies have been proposed that can disturb the cell wall structure and improve the sensitivity to antibiotics. For example, Taylor et al., in a screen of 30 000 small molecules, found an inhibitor of MreB, a regulator of actin filament homologues in bacteria (Taylor et al. 2012). This inhibition made *E. coli* sensitive to novobiocin, an antibiotic that usually has poor effect on Gram-negative bacteria. Ernst et al., examined MrpF, a membrane protein in *S. aureus* that induced resistance to daptomycin through modification of phosphatidylglycerol (PG) (Ernst et al. 2015). They discovered that MrpF harbors a flippase activity that translocates the modified PG from the cytoplasm to the outer membrane, changing the electrical charge of the membrane to become more positive on the outside. The increased surface charge leads to repulsion of daptomycin, causing resistance. Although the data are preliminary, the authors see an opportunity to target MrpF and other flippases as a new way of sensitizing bacteria to antibiotics in the future.

Chiosis and Boneca have described a way to overcome vancomycin resistance in enterococci due to structural changes in the cell wall peptidoglycan (Chiosis and Boneca 2001). Peptidoglycan precursors in the membrane of resistant strains contain D-Ala-D-lactate instead of D-Ala-D-Ala, which lowers the affinity of vancomycin by a factor of over a thousand. To combat this, Chiosis and Boneca found small molecules with cleaving activity specifically of the D-Ala-D-lactate moiety, which could increase the proportion of D-Ala-D-Ala in the cell wall, thus re-sensitizing the bacteria to vancomycin. Another study has shown that vancomycin-resistant enterococci (VRE) could be re-sensitized to vancomycin by treatment with flavonoids (Liu et al. 2001). The mechanism by which this sensitization works might be through the ability of flavonoids to inhibit D-Ala-D-Lac peptide production during cell wall formation.

Another interesting study that sensitized bacteria to antibiotics that target the cell wall involved the ClpXP protease in bacteria (McGillivray et al. 2012). This protease performs controlled proteolysis that regulate protein turnover in many bacterial species. Inhibition of this protease with a compound they called F2, showed an increased susceptibility to the antimicrobial peptide (AMP) cathelicidin and to antibiotics in both *S. aureus* and *Bacillus anthracis*, but did not work for ciprofloxacin or erythromycin, suggesting that ClpXP inhibition only improves the efficacy of antibiotics that target the cell wall or cell membrane. They also found that F2 increased the antimicrobial activity of human whole blood, possibly demonstrating AMP sensitization under physiological conditions (McGillivray et al. 2012). It has been long known that thioridazine (TDZ), an old neuroleptic, can sensitize MRSA to β-lactam antibiotics, but there is no clear mechanism why. The hypothesis has been that TDZ works as an EPI. Recently, Thorsing et al., presented evidence against this hypothesis by showing that TDZ interferes with cell wall synthesis in MRSA (Thorsing et al. 2013). Bacteria treated with TDZ exhibited a thickened and irregular cell wall with decreased intracellular concentrations of amino acids. The authors suggest that TDZ interferes with the formation of pentaglycine branches in the peptidoglycan that sensitize MRSA to antibiotics targeting the cell wall, exemplified in the study by a synergistic effect between TDZ and dicloxacillin (Thorsing et al. 2013). TDZ, being an old neuroleptic, might be associated with similar problems as the antipsychotic drugs mentioned above, with problematic side effects.

Nonetheless, the use of cell wall-interacting sensitizers feels like a smart approach since many important and safe antibiotics use related mechanisms. Sensitizers affecting cell wall integrity could also lead to treatment advantages associated with effect on membrane proteins (such as transporters) and increased permeability, which could increase concentrations of intracellular antibiotics.

17.3.3 Biofilms and Quorum Sensing

Biofilms are complex microbial communities encased in a self-produced matrix of extracellular substances, in which the bacteria have an altered phenotype compared with planktonically growing bacteria (Donlan and Costerton 2002; Lewis 2008). One of these phenotypes is an up to thousand times increased resistance to antimicrobial factors, including antibiotics, which in this context refers to an increased tolerance rather than an inherent resistance due to mutations or acquisition of resistance genes through horizontal transfer (Donlan and Costerton 2002; Yankaskas et al. 2004; Slinger et al. 2006). Additionally, biofilms are extremely common, and recent estimates say that they are present in 80% of bacterial infections (Lewis 2007; Wolcott and Ehrlich 2008). This makes interfering with biofilm formation or structure a promising approach to increase antibiotic resistance. As this topic will be discussed in detail in Chapter 9 in this book, we will only present the general aspects of the strategies being pursued in this chapter.

Biofilm tolerance is probably due to a combination of factors including the protective matrix and the layers of bacteria that are difficult for antimicrobial compounds to penetrate. Additionally, evidence is mounting that biofilm bacteria reduce their metabolism, resulting in slower growth and creating what has been called persister cells, which will be described in more detail in the next section (Fux et al. 2005; Lewis 2008). These cells are less sensitive to most antibiotics, as antibiotics tend to target metabolic processes in the bacterial cells (Fux et al. 2005; Li and Zhang 2007). To target resistance in biofilms, a few different strategies have been pursued. The most explored avenue is to interfere with inter-bacterial communication or so-called quorum sensing (QS) involving three separate systems of molecules: acylhomoserine lactone (AHL) QS system restricted to Gram-negative bacteria, the autoinducing peptide (AIP) system in Gram-positive bacteria, and the autoinducer-2 (AI-2) system present in both types of bacteria (Waters and Bassler 2005; Brackman et al. 2012; Brackman and Coenye 2015; Singh et al. 2017). Four strategies have been investigated to inhibit QS and thus block biofilm formation.

The first is to target signal synthesis (Hentzer and Givskov 2003). This has been successfully done for AHL production and resulted in clearance of bacterial infections, including *P. aeruginosa* lung infection *in vivo* (Parsek et al. 1999; Hoffmann et al. 2007). It has also been done for AI-2 *in vitro* although no evaluation of biofilms has been performed so far (Shen et al. 2006; Han and Lu 2009).

The second strategy, quorum quenching (QQ), is a broad term for disrupting QS signaling based on degradation or inactivation of the signaling molecules. This has been accomplished by enzymatic degradation of AHL (Reimmann et al. 2002; Chun et al. 2004; Yang et al. 2005), but as far as we are aware, no specific quenchers have been described for AIP or AI-2.

A third strategy, which sometimes is included in the concept QQ, is inhibition of signaling based on signal analogues. Both synthetic and natural molecules have been found for each of the three systems that affect biofilm formation and show increased clearance of infection *in vivo*. For more specific information see Rasmussen et al. (2005), Ren et al. (2005), Ishida et al. (2007), Morohoshi et al. (2007), Brackman et al. (2011), Jakobsen et al. (2012), Cirioni et al. (2013), Roy et al. (2013, 2018), Aggarwal et al. (2015), and Kuo et al. (2015).

Finally, the fourth strategy of QQ is to block signal transduction. One group of molecules that show big promise is furanones (originally from cinnamon) that in various studies have blocked signal transduction and affected biofilm formation *in vitro* as well as in a variety of animal models of infection with *P. aeruginosa, E. coli, Bacillus subtilis, Streptococcus* spp., and other organisms to increase the sensitivity to antibiotics (Hentzer and Givskov 2003; Ren et al. 2004; Christensen et al. 2012; He et al. 2012; Starkey et al. 2014). Although promising results have been shown with these compounds, toxicity is currently limiting their use. As QS is involved in several types of bacterial group behavior, not only biofilm formation, this means that similar strategies may be used to inhibit virulence, competence, and conjugation that are involved in pathogenesis and spread of antibiotic resistance.

Additionally, biofilms can be targeted in other ways. Based on the stabilizing aspect of the DNA-binding protein integration host factor (IHF) on extracellular DNA in biofilms, Brandstetter et al., used anti-IHF antibodies in combination with amoxicillin to eradicate *Haemophilus influenzae* biofilms (Brandstetter et al. 2013). Also, other approaches to disperse biofilms have been proposed. Studies using D-amino acids and norspermidine showed effective dispersal of bacterial biofilms *in vitro* (Kolodkin-Gal et al. 2010, 2012). As D-amino acids and norspermidine are molecules produced by many bacteria, they may contribute to the regulation of biofilm formation *in vivo*.

17.3.4 Persister Cells

Under certain conditions, bacteria can enter a state of lowered metabolism, which enhances their antibiotic resistance, as most antibiotics require active metabolism to function effectively. These cells have been designated persister cells (Lewis 2007). The resistance pattern is not the result of genetic mutations, but instead a change in phenotype controlled by gene expression, and therefore not heritable (Spoering et al. 2006; Hansen et al. 2008; Conlon et al. 2016). What causes cells to enter this persister state is not completely known, but antibiotic treatment may be a trigger (McNamara and Proctor 2000), and genes, such as *glpD*, *ygfA*, and *yigB* in *E. coli* (Spoering et al. 2006; Hansen et al. 2008), toxin–antitoxin mechanisms (Balaban et al. 2004, 2013), and ATP depletion in *S. aureus* (Conlon et al. 2016) are thought to be involved mechanistically.

Interfering with the formation of persister cells and "waking" them up from their dormant phenotype are approaches that are being explored to avoid chronic infections. A study of persister cells of *P. aeruginosa* in CF patients showed that overexpression of PvrR, a regulatory protein involved in both biofilm formation and phenotype switching to a persister type, resulted in a reduced proportion of the resistant *P. aeruginosa* (Drenkard and Ausubel 2002). Using a different approach, Pan et al., used furanones to revert *P. aeruginosa* persister cells back to metabolically active and antibiotic-sensitive organisms (Pan et al. 2012, 2013). Interestingly, they also showed that persister cells could be reactivated by glucose, having a similar effect as furanone but probably through a different mechanism. This might mimic the natural behavior of bacteria, where they revert to persister cells in times of low nutrition and activate when conditions improve. In a similar study aimed to eliminate persister cells, Allison et al., showed that sensitization of the dormant bacteria using a metabolic stimulus could increase both *E. coli* and *S. aureus* persister cell sensitivity to antibiotics (Allison et al. 2011).

It seems clear that to combat chronic infections effectively, we need to take persister cells into serious consideration. The first steps are of course to identify proteins and genes involved in phenotypic switching, but the more difficult step is to find ways to disrupt this pathway in a clinical setting.

17.3.5 Targeting Nonessential Genes/Proteins

In a screening of the complete Keio collection that includes individual knockout mutations in all nonessential genes of an *E. coli* strains, with 22 different antibiotics, 283 strains were identified to have increased sensitivity to at least one antibiotic (Liu et al. 2010). This strategy used in other organisms could identify potential targets that determine antibiotic resistance and that can be used to sensitize bacteria to antibiotics in the future. In a similar approach, Clatworthy et al., has proposed to target virulence factors to sensitize bacteria rather than identifying agents that are bactericidal (Clatworthy et al. 2007).

17.3.6 Bacteriophages

Bacteriophages are viruses with the capability of infecting and killing bacteria, which has been used clinically for a long time to treat both systemic and local bacterial infections in humans (Abedon et al. 2011). However, phages have also shown promise in sensitizing bacteria to antibiotics.

A study by Fischetti and co-workers showed synergy between the pneumococcal phage lytic enzyme Cpl-1 and antibiotics in clearing pneumococcal infection with antibiotic-resistant strains (Djurkovic et al. 2005). Comeau et al., described the phenomenon phage-antibiotic synergy (PAS) and showed that sublethal doses of antibiotics significantly increased the production of phage in *E. coli*, resulting in increased bacterial killing (Comeau et al. 2007). Similarly, low penicillin concentrations increased the lytic ability of phages in *Streptococcus thermophiles* (Verhue 1978), suggesting that bacteriophages and antibiotics can synergize both for bacteria that are sensitive and more importantly for bacteria that are resistant to antibiotics.

A unique and interesting approach was presented in a study by Yosef et al. (2015). By creating two phages, one lysogenic phage transducing CRISPR sequences that specifically targeted antibiotic-resistant plasmids into the bacteria in combination with a lytic phage that contains the same target sequence, treatment resulted in that infected bacteria became resistant to the lytic phage but sensitized to antibiotics, as the resistance plasmids were eliminated (Yosef et al. 2015).

17.4 Using and Strengthening the Human Immune System Against Resistant Bacteria

17.4.1 Strengthening Host Immune System Function

Antibiotics alone are never enough to treat an infection; they only act to tip the scale in favor of the immune system in an already ongoing battle. The studies described above all focus on the bacterial part of the treatment, i.e. how to modulate and overcome bacterial features that normally interfere with antibacterial treatment. But another interesting strategy of improving infection treatment of antibiotic-resistant organisms is to strengthen the host response. This can be done in several ways, such as using antibodies to target bacterial resistance mechanisms, stimulating bacterial eradication inside macrophages, enhancing bacterial killing by neutrophils, or reducing the immune response to minimize tissue damage.

TLR5 agonists together with antibiotics in a mouse pneumococcal infection model resulted in less risk of dissemination, increased number of early neutrophils, and improved antibiotic efficacy (Porte et al. 2015). The advantage of this strategy is an increased clearance of bacteria and the potential decreased dosing of antibiotics that may decrease the emergence of antibiotic resistance and dysbiosis of the microflora, normally adverse effects from antibiotic treatment. In a separate study screening for immunostimulatory molecules, streptazolin was identified and shown to activate phospho-inositol-3 kinases that resulted in activation of NF-κB, which increased bacterial killing by macrophages (Perry et al. 2015). There are additional studies that have been performed to investigate immunomodulation as a strategy to improve antibacterial treatment (Hancock et al. 2012). Many of those rely on molecules that activate toll-like receptors or NODs. Although promising as sensitizing options to eradicate antibiotic-resistant bacteria, immunostimulation needs to be tightly regulated not to result in immune dysregulation with adverse effects.

Specific antibodies have become a staple of anticancer treatment, and in similar ways antibodies are being investigated for their ability to attack antimicrobial resistance mechanisms. Antibody treatment is not a new concept and was explored already in the 1930s by Avery and co-workers, who showed that anti-capsular antibodies could eradicate pneumococci from the bloodstream of mice (Avery and Dubos 1931). As mentioned above, antibodies to IHF eradicated biofilms and synergized with antibiotics to eradicate antibiotic-resistant bacteria (Brandstetter et al. 2013). In a study by Barekzi et al., injection of mice with human pooled sera into the site of infection with *E. coli* or *K. pneumoniae* produced synergistic and protective effects with antibiotics in lethal wound infections (Barekzi et al. 2002). This may be a strategy that can be used more widely (Casadevall et al. 2004).

A study by Lo et al., used a different approach and showed that AR12, a compound able to help macrophages eradicate bacteria by forcing lysosome fusion, had synergistic effects with aminoglycosides on eradicating intracellular infection with *Salmonella typhi* both *in vitro* and *in vivo* (Lo et al. 2014). Intracellular bacteria are often more difficult to treat because many antibiotics have poor penetrance into eukaryotic cells. For example, GAS infections tend to recur in patients after antibiotic treatment (Cue et al. 2000). This is not because *Streptococcus pyogenes* has acquired intrinsic antibiotic resistance but because the bacteria hide inside the cell where they stay protected from the antibiotic molecules. Cue et al., tried an approach where an antagonist (SJ755) to the fibronectin (Fn)-receptor of the epithelial cells could inhibit the internalization of *S. pyogenes* into human epithelial cells (Cue et al. 2000). Combining antibiotics with SJ755 may keep bacteria extracellular, extending bacterial eradication and ending the infection permanently. Furthermore, because many bacteria and other pathogens use integrins to adhere to or invade eukaryotic cells, this is a promising strategy transferrable to treatments of other infectious diseases.

Other approaches that have been observed is that statins, used for lowering cholesterol levels in the blood, have the additional effect of increasing the formation of phagocyte extracellular traps, resulting in increased bacterial killing (Chow et al. 2010). Statin treatment of mice caused less bacterial infections *in vivo*, especially with *S. aureus*.

17.4.2 Antimicrobial Peptides (AMPs)

Antimicrobial peptides are small molecules with broad killing activity against bacteria, fungi, viruses, and some parasites (Ganz 2003; Sang and Blecha 2008). AMPs are an important part of the innate immune defense in almost every life form, from unicellular organisms to mammals, creating a wide spectrum of natural antibiotics (Hancock and Lehrer 1998). The discovery of these peptides sparked an interest in using AMPs or synthetic analogues of AMPs as stand-alone treatments of infections, but AMPs might also have sensitizing properties.

Le et al. treated penicillin-resistant *S. pneumoniae* with a combination of synthetically designed AMPs and penicillin (Le et al. 2015). This was done as a factorial random controlled trial in mice with pneumococcal systemic infection. They showed a clear synergistic effect where the bacteria were re-sensitized. The exact mechanism of how this works is not yet fully clear, but results suggest that AMPs bind and inactivate bacterial toxins.

Since AMPs are a natural part of our immune system, they should not be very cytotoxic to human cells. This makes AMPs promising compared with many other sensitization methods, where adverse effects are problematic.

17.5 Conclusion

In this chapter, we propose that sensitization of resistant bacteria could be an important method in the combat of antibiotic resistance. We identify four categories of strategies that are promising for future research. New ways of treating resistant bacteria are crucially needed, and future research within this field will be necessary.

References

Abedon, S.T., Kuhl, S.J., Blasdel, B.G., and Kutter, E.M. (2011). Phage treatment of human infections. *Bacteriophage* 1: 66–85.

Aggarwal, C., Jimenez, J.C., Lee, H. et al. (2015). Identification of quorum-sensing inhibitors disrupting signaling between Rgg and short hydrophobic peptides in streptococci. *MBio* 6: e00393-15.

Alao, J.P., Weber, A.M., Shabro, A., and Sunnerhagen, P. (2015). Suppression of sensitivity to drugs and antibiotics by high external cation concentrations in fission yeast. *PLoS One* 10: e0119297.

Allison, K.R., Brynildsen, M.P., and Collins, J.J. (2011). Metabolite-enabled eradication of bacterial persisters by aminoglycosides. *Nature* 473: 216–220.

Andersson, D.I. and Hughes, D. (2014). Microbiological effects of sublethal levels of antibiotics. *Nat. Rev. Microbiol.* 12: 465–478.

Avery, O.T. and Dubos, R. (1931). The protective action of a specific enzyme against type III pneumococcus infection in mice. *J. Exp. Med.* 54: 73–89.

Balaban, N.Q., Gerdes, K., Lewis, K., and McKinney, J.D. (2013). A problem of persistence: still more questions than answers? *Nat. Rev. Microbiol.* 11: 587–591.

Balaban, N.Q., Merrin, J., Chait, R. et al. (2004). Bacterial persistence as a phenotypic switch. *Science* 305: 1622–1625.

Balakrishna, R., Wood, S.J., Nguyen, T.B. et al. (2006). Structural correlates of antibacterial and membrane-permeabilizing activities in acylpolyamines. *Antimicrob. Agents Chemother.* 50: 852–861.

Barekzi, N.A., Felts, A.G., Poelstra, K.A. et al. (2002). Locally delivered polyclonal antibodies potentiate intravenous antibiotic efficacy against Gram-negative infections. *Pharm. Res.* 19: 1801–1807.

Bassetti, M. and Righi, E. (2015). New antibiotics and antimicrobial combination therapy for the treatment of Gram-negative bacterial infections. *Curr. Opin. Crit. Care* 21: 402–411.

Bernal, P., Molina-Santiago, C., Daddaoua, A., and Llamas, M.A. (2013). Antibiotic adjuvants: identification and clinical use. *Microb. Biotechnol.* 6: 445–449.

Bhandari, D., Thapa, P., Thapa, K. et al. (2016). Antibiotic resistance: evolution and alternatives. *Can. J. Infect. Control* 31: 149–155.

Brackman, G. and Coenye, T. (2015). Quorum sensing inhibitors as anti-biofilm agents. *Curr. Pharm. Des.* 21: 5–11.

Brackman, G., Cos, P., Maes, L. et al. (2011). Quorum sensing inhibitors increase the susceptibility of bacterial biofilms to antibiotics in vitro and in vivo. *Antimicrob. Agents Chemother.* 55: 2655–2661.

Brackman, G., Risseeuw, M., Celen, S. et al. (2012). Synthesis and evaluation of the quorum sensing inhibitory effect of substituted triazolyldihydrofuranones. *Bioorg. Med. Chem.* 20: 4737–4743.

Brandstetter, K.A., Jurcisek, J.A., Goodman, S.D. et al. (2013). Antibodies directed against integration host factor mediate biofilm clearance from nasopore. *Laryngoscope* 123: 2626–2632.

Bush, K. (2015). A resurgence of β-lactamase inhibitor combinations effective against multidrug-resistant Gram-negative pathogens. *Int. J. Antimicrob. Agents* 46: 483–493.

Bush, K. and Macielag, M.J. (2010). New β-lactam antibiotics and β-lactamase inhibitors. *Expert Opin. Ther. Pat.* 20: 1277–1293.

Casadevall, A., Dadachova, E., and Pirofski, L.A. (2004). Passive antibody therapy for infectious diseases. *Nat. Rev. Microbiol.* 2: 695–703.

Chevalier, J., Bredin, J., Mahamoud, A. et al. (2004). Inhibitors of antibiotic efflux in resistant *Enterobacter aerogenes* and *Klebsiella pneumoniae* strains. *Antimicrob. Agents Chemother.* 48: 1043–1046.

Chiosis, G. and Boneca, I.G. (2001). Selective cleavage of D-Ala-D-Lac by small molecules: re-sensitizing resistant bacteria to vancomycin. *Science* 293: 1484–1487.

Chow, O.A., von Köckritz-Blickwede, M., Bright, A.T. et al. (2010). Statins enhance formation of phagocyte extracellular traps. *Cell Host Microbe* 8: 445–454.

Christensen, L.D., van Gennip, M., Jakobsen, T.H. et al. (2012). Synergistic antibacterial efficacy of early combination treatment with tobramycin and quorum-sensing inhibitors against *Pseudomonas aeruginosa* in an intraperitoneal foreign-body infection mouse model. *J. Antimicrob. Chemother.* 67: 1198–1206.

Chun, C.K., Ozer, E.A., Welsh, M.J. et al. (2004). Inactivation of a *Pseudomonas aeruginosa* quorum-sensing signal by human airway epithelia. *Proc. Natl. Acad. Sci. U. S. A.* 101: 3587–3590.

Cirioni, O., Mocchegiani, F., Cacciatore, I. et al. (2013). Quorum sensing inhibitor FS3-coated vascular graft enhances daptomycin efficacy in a rat model of staphylococcal infection. *Peptides* 40: 77–81.

Clatworthy, A.E., Pierson, E., and Hung, D.T. (2007). Targeting virulence: a new paradigm for antimicrobial therapy. *Nat. Chem. Biol.* 3: 541–548.

Comeau, A.M., Tétart, F., Trojet, S.N. et al. (2007). Phage-antibiotic synergy (PAS): beta-lactam and quinolone antibiotics stimulate virulent phage growth. *PLoS One* 2: e799.

Conlon, B.P., Rowe, S.E., Gandt, A.B. et al. (2016). Persister formation in *Staphylococcus aureus* is associated with ATP depletion. *Nat. Microbiol.* 1: 16051.

Cue, D., Southern, S.O., Southern, P.J. et al. (2000). A nonpeptide integrin antagonist can inhibit epithelial cell ingestion of *Streptococcus pyogenes* by blocking formation of integrin alpha 5beta 1-fibronectin-M1 protein complexes. *Proc. Natl. Acad. Sci. U. S. A.* 97: 2858–2863.

D'Costa, V.M., King, C.E., Kalan, L. et al. (2011). Antibiotic resistance is ancient. *Nature* 477: 457–461.

D'Costa, V.M., McGrann, K.M., Hughes, D.W., and Wright, G.D. (2006). Sampling the antibiotic resistome. *Science* 311: 374–377.

Djurkovic, S., Loeffler, J.M., and Fischetti, V.A. (2005). Synergistic killing of *Streptococcus pneumoniae* with the bacteriophage lytic enzyme Cpl-1 and penicillin or gentamicin depends on the level of penicillin resistance. *Antimicrob. Agents Chemother.* 49: 1225–1228.

Donlan, R.M. and Costerton, J.W. (2002). Biofilms: survival mechanisms of clinically relevant microorganisms. *Clin. Microbiol. Rev.* 15: 167–193.

Dortet, L., Poirel, L., and Nordmann, P. (2014). Worldwide dissemination of the NDM-type carbapenemases in Gram-negative bacteria. *Biomed. Res. Int.* 2014: 249856.

Drawz, S.M. and Bonomo, R.A. (2010). Three decades of beta-lactamase inhibitors. *Clin. Microbiol. Rev.* 23: 160–201.

Drenkard, E. and Ausubel, F.M. (2002). Pseudomonas biofilm formation and antibiotic resistance are linked to phenotypic variation. *Nature* 416: 740–743.

Dwivedi, G.R., Maurya, A., Yadav, D.K. et al. (2014). Drug resistance reversal potential of ursolic acid derivatives against nalidixic acid- and multidrug-resistant *Escherichia coli*. *Chem. Biol. Drug Des.* 86: 272–283.

Ejim, L., Farha, M.A., Falconer, S.B. et al. (2011). Combinations of antibiotics and nonantibiotic drugs enhance antimicrobial efficacy. *Nat. Chem. Biol.* 7: 348–350.

Ernst, C.M., Kuhn, S., Slavetinsky, C.J. et al. (2015). The lipid-modifying multiple peptide resistance factor is an oligomer consisting of distinct interacting synthase and flippase subunits. *MBio* 6: e02340-14.

Fadli, M., Chevalier, J., Saad, A. et al. (2011). Essential oils from Moroccan plants as potential chemosensitisers restoring antibiotic activity in resistant Gram-negative bacteria. *Int. J. Antimicrob. Agents* 38: 325–330.

Farha, M.A., Verschoor, C.P., Bowdish, D., and Brown, E.D. (2013). Collapsing the proton motive force to identify synergistic combinations against *Staphylococcus aureus*. *Chem. Biol.* 20: 1168–1178.

Fernández, L., Breidenstein, E.B., and Hancock, R.E. (2011). Creeping baselines and adaptive resistance to antibiotics. *Drug Resist. Updat.* 14: 1–21.

Finley, R.L., Collignon, P., Larsson, D.G. et al. (2013). The scourge of antibiotic resistance: the important role of the environment. *Clin. Infect. Dis.* 57: 704–710.

Fischbach, M.A. and Walsh, C.T. (2009). Antibiotics for emerging pathogens. *Science* 325: 1089–1093.

Fux, C.A., Costerton, J.W., Stewart, P.S., and Stoodley, P. (2005). Survival strategies of infectious biofilms. *Trends Microbiol.* 13: 34–40.

Ganz, T. (2003). Defensins: antimicrobial peptides of innate immunity. *Nat. Rev. Immunol.* 3: 710–720.

Garvey, M.I. and Piddock, L.J. (2008). The efflux pump inhibitor reserpine selects multidrug-resistant *Streptococcus pneumoniae* strains that overexpress the

ABC transporters PatA and PatB. *Antimicrob. Agents Chemother.* 52: 1677–1685.

Gill, E.E., Franco, O.L., and Hancock, R.E. (2015). Antibiotic adjuvants: diverse strategies for controlling drug-resistant pathogens. *Chem. Biol. Drug Des.* 85: 56–78.

Goldberg, K., Sarig, H., Zaknoon, F. et al. (2013). Sensitization of Gram-negative bacteria by targeting the membrane potential. *FASEB J.* 27: 3818–3826.

Good, L. and Nielsen, P.E. (1998). Antisense inhibition of gene expression in bacteria by PNA targeted to mRNA. *Nat. Biotechnol.* 16: 355–358.

Hakansson, A., Svensson, M., Mossberg, A.K. et al. (2000). A folding variant of alpha-lactalbumin with bactericidal activity against *Streptococcus pneumoniae. Mol. Microbiol.* 35: 589–600.

Hakansson, A.P., Roche-Hakansson, H., Mossberg, A.K., and Svanborg, C. (2011). Apoptosis-like death in bacteria induced by HAMLET, a human milk lipid-protein complex. *PLoS One* 6: e17717.

Han, X. and Lu, C. (2009). Biological activity and identification of a peptide inhibitor of LuxS from *Streptococcus suis* serotype 2. *FEMS Microbiol. Lett.* 294: 16–23.

Hancock, R.E. and Lehrer, R. (1998). Cationic peptides: a new source of antibiotics. *Trends Biotechnol.* 16: 82–88.

Hancock, R.E., Nijnik, A., and Philpott, D.J. (2012). Modulating immunity as a therapy for bacterial infections. *Nat. Rev. Microbiol.* 10: 243–254.

Hansen, S., Lewis, K., and Vulic, M. (2008). Role of global regulators and nucleotide metabolism in antibiotic tolerance in *Escherichia coli. Antimicrob. Agents Chemother.* 52: 2718–2726.

He, Z., Wang, Q., Hu, Y. et al. (2012). Use of the quorum sensing inhibitor furanone C-30 to interfere with biofilm formation by *Streptococcus mutans* and its luxS mutant strain. *Int. J. Antimicrob. Agents* 40: 30–35.

Hentzer, M. and Givskov, M. (2003). Pharmacological inhibition of quorum sensing for the treatment of chronic bacterial infections. *J. Clin. Invest.* 112: 1300–1307.

Hoffmann, N., Lee, B., Hentzer, M. et al. (2007). Azithromycin blocks quorum sensing and alginate polymer formation and increases the sensitivity to serum and stationary-growth-phase killing of *Pseudomonas aeruginosa* and attenuates chronic *P. aeruginosa* lung infection in Cftr(−/−) mice. *Antimicrob. Agents Chemother.* 51: 3677–3687.

Ishida, T., Ikeda, T., Takiguchi, N. et al. (2007). Inhibition of quorum sensing in *Pseudomonas aeruginosa* by N-acyl cyclopentylamides. *Appl. Environ. Microbiol.* 73: 3183–3188.

Jakobsen, T.H., van Gennip, M., Phipps, R.K. et al. (2012). Ajoene, a sulfur-rich molecule from garlic, inhibits genes controlled by quorum sensing. *Antimicrob. Agents Chemother.* 56: 2314–2325.

Kalan, L. and Wright, G.D. (2011). Antibiotic adjuvants: multicomponent anti-infective strategies. *Expert Rev. Mol. Med.* 13: e5.

King, A.M., Reid-Yu, S.A., Wang, W. et al. (2014). Aspergillomarasmine A overcomes metallo-β-lactamase antibiotic resistance. *Nature* 510: 503–506.

Kolodkin-Gal, I., Cao, S., Chai, L. et al. (2012). A self-produced trigger for biofilm disassembly that targets exopolysaccharide. *Cell* 149: 684–692.

Kolodkin-Gal, I., Romero, D., Cao, S. et al. (2010). D-amino acids trigger biofilm disassembly. *Science* 328: 627–629.

Kristiansen, J.E., Thomsen, V.F., Martins, A. et al. (2010). Non-antibiotics reverse resistance of bacteria to antibiotics. *In Vivo* 24: 751–754.

Kuo, D., Yu, G., Hoch, W. et al. (2015). Novel quorum-quenching agents promote methicillin-resistant *Staphylococcus aureus* (MRSA) wound healing and sensitize MRSA to β-lactam antibiotics. *Antimicrob. Agents Chemother.* 59: 1512–1518.

LaCross, N.C., Marrs, C.F., and Gilsdorf, J.R. (2013). Population structure in nontypeable *Haemophilus influenzae. Infect. Genet. Evol.* 14: 125–136.

Le, C.F., Yusof, M.Y., Hassan, M.A. et al. (2015). In vivo efficacy and molecular docking of designed peptide that exhibits potent antipneumococcal activity and synergises in combination with penicillin. *Sci. Rep.* 5: 11886.

Levin, B.R. and Rozen, D.E. (2006). Non-inherited antibiotic resistance. *Nat. Rev. Microbiol.* 4: 556–562.

Lewis, K. (2007). Persister cells, dormancy and infectious disease. *Nat. Rev. Microbiol.* 5: 48–56.

Lewis, K. (2008). Multidrug tolerance of biofilms and persister cells. *Curr. Top. Microbiol. Immunol.* 322: 107–131.

Li, X.Z. and Nikaido, H. (2004). Efflux-mediated drug resistance in bacteria. *Drugs* 64: 159–204.

Li, Y. and Zhang, Y. (2007). PhoU is a persistence switch involved in persister formation and tolerance to multiple antibiotics and stresses in *Escherichia coli. Antimicrob. Agents Chemother.* 51: 2092–2099.

Liu, A., Tran, L., Becket, E. et al. (2010). Antibiotic sensitivity profiles determined with an *Escherichia coli* gene knockout collection: generating an antibiotic bar code. *Antimicrob. Agents Chemother.* 54: 1393–1403.

Liu, L.X., Durham, D.G., and Richards, R.M. (2001). Vancomycin resistance reversal in enterococci by flavonoids. *J. Pharm. Pharmacol.* 53: 129–132.

Lo, J.H., Kulp, S.K., Chen, C.S., and Chiu, H.C. (2014). Sensitization of intracellular Salmonella enterica serovar Typhimurium to aminoglycosides in vitro and in vivo by a host-targeted antimicrobial agent. *Antimicrob. Agents Chemother.* 58: 7375–7382.

Lomovskaya, O. and Bostian, K.A. (2006). Practical applications and feasibility of efflux pump inhibitors in the clinic – a vision for applied use. *Biochem. Pharmacol.* 71: 910–918.

Lomovskaya, O., Warren, M.S., Lee, A. et al. (2001). Identification and characterization of inhibitors of multidrug resistance efflux pumps in *Pseudomonas aeruginosa*: novel agents for combination therapy. *Antimicrob. Agents Chemother.* 45: 105–116.

Lorch, M., Lehner, I., Siarheyeva, A. et al. (2005). NMR and fluorescence spectroscopy approaches to secondary and primary active multidrug efflux pumps. *Biochem. Soc. Trans.* 33: 873–877.

Mahamoud, A., Chevalier, J., Alibert-Franco, S. et al. (2007). Antibiotic efflux pumps in Gram-negative bacteria: the inhibitor response strategy. *J. Antimicrob. Chemother.* 59: 1223–1229.

Mahmood, H.Y., Jamshidi, S., Sutton, J.M., and Rahman, K.M. (2016). Current advances in developing inhibitors of bacterial multidrug efflux pumps. *Curr. Med. Chem.* 23: 1062–1081.

Marks, L.R., Clementi, E.A., and Hakansson, A.P. (2012). The human milk protein-lipid complex HAMLET sensitizes bacterial pathogens to traditional antimicrobial agents. *PLoS One* 7: e43514.

Marks, L.R., Clementi, E.A., and Hakansson, A.P. (2013). Sensitization of *Staphylococcus aureus* to methicillin and other antibiotics in vitro and in vivo in the presence of HAMLET. *PLoS One* 8: e63158.

Martins, M., Dastidar, S.G., Fanning, S. et al. (2008). Potential role of non-antibiotics (helper compounds) in the treatment of multidrug-resistant Gram-negative infections: mechanisms for their direct and indirect activities. *Int. J. Antimicrob. Agents* 31: 198–208.

McGeary, R.P., Tan, D.T., and Schenk, G. (2017). Progress toward inhibitors of metallo-β-lactamases. *Future Med. Chem.* 9: 673–691.

McGillivray, S.M., Tran, D.N., Ramadoss, N.S. et al. (2012). Pharmacological inhibition of the ClpXP protease increases bacterial susceptibility to host cathelicidin antimicrobial peptides and cell envelope-active antibiotics. *Antimicrob. Agents Chemother.* 56: 1854–1861.

McNamara, P.J. and Proctor, R.A. (2000). *Staphylococcus aureus* small colony variants, electron transport and persistent infections. *Int. J. Antimicrob. Agents* 14: 117–122.

Morohoshi, T., Shiono, T., Takidouchi, K. et al. (2007). Inhibition of quorum sensing in *Serratia marcescens* AS-1 by synthetic analogs of N-acylhomoserine lactone. *Appl. Environ. Microbiol.* 73: 6339–6344.

Nathan, C. (2012). Fresh approaches to anti-infective therapies. *Sci. Transl. Med.* 4: 140sr2.

Nikaido, H. (2009). Multidrug resistance in bacteria. *Annu. Rev. Biochem.* 78: 119–146.

Nikolaidis, I., Favini-Stabile, S., and Dessen, A. (2014). Resistance to antibiotics targeted to the bacterial cell wall. *Protein Sci.* 23: 243–259.

Pages, J.M. and Amaral, L. (2009). Mechanisms of drug efflux and strategies to combat them: challenging the efflux pump of Gram-negative bacteria. *Biochim. Biophys. Acta* 1794: 826–833.

Pages, J.M., James, C.E., and Winterhalter, M. (2008). The porin and the permeating antibiotic: a selective diffusion barrier in Gram-negative bacteria. *Nat. Rev. Microbiol.* 6: 893–903.

Pan, J., Bahar, A.A., Syed, H., and Ren, D. (2012). Reverting antibiotic tolerance of *Pseudomonas aeruginosa* PAO1 persister cells by (Z)-4-bromo-5-(bromomethylene)-3-methylfuran-2(5H)-one. *PLoS One* 7: e45778.

Pan, J., Song, F., and Ren, D. (2013). Controlling persister cells of *Pseudomonas aeruginosa* PDO300 by (Z)-4-bromo-5-(bromomethylene)-3-methylfuran-2(5H)-one. *Bioorg. Med. Chem. Lett.* 23: 4648–4651.

Papp-Wallace, K.M. and Bonomo, R.A. (2016). New β-lactamase inhibitors in the clinic. *Infect. Dis. Clin. N. Am.* 30: 441–464.

Parsek, M.R., Val, D.L., Hanzelka, B.L. et al. (1999). Acyl homoserine-lactone quorum-sensing signal generation. *Proc. Natl. Acad. Sci. U. S. A.* 96: 4360–4365.

Perry, J.A., Koteva, K., Verschoor, C.P. et al. (2015). A macrophage-stimulating compound from a screen of microbial natural products. *J. Antibiot. (Tokyo)* 68: 40–46.

Porte, R., Fougeron, D., Muñoz-Wolf, N. et al. (2015). A toll-like receptor 5 agonist improves the efficacy of antibiotics in treatment of primary and influenza virus-associated pneumococcal mouse infections. *Antimicrob. Agents Chemother.* 59: 6064–6072.

Projan, S.J. (2003). Why is big Pharma getting out of antibacterial drug discovery? *Curr. Opin. Microbiol.* 6: 427–430.

Projan, S.J. and Shlaes, D.M. (2004). Antibacterial drug discovery: is it all downhill from here. *Clin. Microbiol. Infect.* 10 (Suppl 4): 18–22.

Rasmussen, T.B., Bjarnsholt, T., Skindersoe, M.E. et al. (2005). Screening for quorum-sensing inhibitors (QSI) by use of a novel genetic system, the QSI selector. *J. Bacteriol.* 187: 1799–1814.

Reimmann, C., Ginet, N., Michel, L. et al. (2002). Genetically programmed autoinducer destruction reduces virulence gene expression and swarming motility in *Pseudomonas aeruginosa* PAO1. *Microbiology* 148: 923–932.

Ren, D., Bedzyk, L.A., Ye, R.W. et al. (2004). Differential gene expression shows natural brominated furanones interfere with the autoinducer-2 bacterial signaling system of *Escherichia coli*. *Biotechnol. Bioeng.* 88: 630–642.

Ren, D., Zuo, R., González Barrios, A.F. et al. (2005). Differential gene expression for investigation of *Escherichia coli* biofilm inhibition by plant extract ursolic acid. *Appl. Environ. Microbiol.* 71: 4022–4034.

Roy, R., Tiwari, M., Donelli, G., and Tiwari, V. (2018). Strategies for combating bacterial biofilms: a focus on anti-biofilm agents and their mechanisms of action. *Virulence* 9: 522–554.

Roy, V., Meyer, M.T., Smith, J.A. et al. (2013). AI-2 analogs and antibiotics: a synergistic approach to reduce bacterial biofilms. *Appl. Microbiol. Biotechnol.* 97: 2627–2638.

Sang, Y. and Blecha, F. (2008). Antimicrobial peptides and bacteriocins: alternatives to traditional antibiotics. *Anim. Health Res. Rev.* 9: 227–235.

Shen, G., Rajan, R., Zhu, J. et al. (2006). Design and synthesis of substrate and intermediate analogue inhibitors of S-ribosylhomocysteinase. *J. Med. Chem.* 49: 3003–3011.

Singh, B.N., Prateeksha, Upreti, D.K. et al. (2017). Bactericidal, quorum quenching and anti-biofilm nanofactories: a new niche for nanotechnologists. *Crit. Rev. Biotechnol.* 37: 525–540.

Slinger, R., Chan, F., Ferris, W. et al. (2006). Multiple combination antibiotic susceptibility testing of nontypeable *Haemophilus influenzae* biofilms. *Diagn. Microbiol. Infect. Dis.* 56: 247–253.

Smith, P.A. and Romesberg, F.E. (2007). Combating bacteria and drug resistance by inhibiting mechanisms of persistence and adaptation. *Nat. Chem. Biol.* 3: 549–556.

Spengler, G., Kincses, A., Gajdács, M., and Amaral, L. (2017). New roads leading to old destinations: efflux pumps as targets to reverse multidrug resistance in bacteria. *Molecules* 22: E468.

Spoering, A.L., Vulic, M., and Lewis, K. (2006). GlpD and PlsB participate in persister cell formation in *Escherichia coli*. *J. Bacteriol.* 188: 5136–5144.

Starkey, M., Lepine, F., Maura, D. et al. (2014). Identification of anti-virulence compounds that disrupt quorum-sensing regulated acute and persistent pathogenicity. *PLoS Pathog.* 10: e1004321.

Tacconelli, E., Carmeli, Y., Harbarth, S. et al. (2017). Global priority list of antibiotic-resistant bacteria to guide research, discovery, and development of new antibiotics. WHO Report 1–7.

Taneja, N. and Kaur, H. (2016). Insights into newer antimicrobial agents against Gram-negative bacteria. *Microbiol. Insights* 9: 9–19.

Taylor, P.L., Rossi, L., De Pascale, G., and Wright, G.D. (2012). A forward chemical screen identifies antibiotic adjuvants in *Escherichia coli*. *ACS Chem. Biol.* 7: 1547–1555.

Tegos, G., Stermitz, F.R., Lomovskaya, O., and Lewis, K. (2002). Multidrug pump inhibitors uncover remarkable activity of plant antimicrobials. *Antimicrob. Agents Chemother.* 46: 3133–3141.

Thorarensen, A., Presley-Bodnar, A.L., Marotti, K.R. et al. (2001). 3-Arylpiperidines as potentiators of existing antibacterial agents. *Bioorg. Med. Chem. Lett.* 11: 1903–1906.

Thorsing, M., Klitgaard, J.K., Atilano, M.L. et al. (2013). Thioridazine induces major changes in global gene expression and cell wall composition in methicillin-resistant *Staphylococcus aureus* USA300. *PLoS One* 8: e64518.

Toney, J.H., Fitzgerald, P.M., Grover-Sharma, N. et al. (1998). Antibiotic sensitization using biphenyl tetrazoles as potent inhibitors of *Bacteroides fragilis* metallo-beta-lactamase. *Chem. Biol.* 5: 185–196.

Tsubery, H., Ofek, I., Cohen, S., and Fridkin, M. (2000). Structure-function studies of polymyxin B nonapeptide: implications to sensitization of Gram-negative bacteria. *J. Med. Chem.* 43: 3085–3092.

Tsubery, H., Yaakov, H., Cohen, S. et al. (2005). Neopeptide antibiotics that function as opsonins and membrane-permeabilizing agents for Gram-negative bacteria. *Antimicrob. Agents Chemother.* 49: 3122–3128.

Vaara, M. and Vaara, T. (1983). Polycations sensitize enteric bacteria to antibiotics. *Antimicrob. Agents Chemother.* 24: 107–113.

Ventola, C.L. (2015). The antibiotic resistance crisis: part 2: management strategies and new agents. *P T* 40: 344–352.

Verhue, W.M. (1978). Interaction of bacteriophage infection and low penicillin concentrations on the performance of yogurt cultures. *Appl. Environ. Microbiol.* 35: 1145–1149.

Waters, C.M. and Bassler, B.L. (2005). Quorum sensing: cell-to-cell communication in bacteria. *Annu. Rev. Cell Dev. Biol.* 21: 319–346.

WHO (2014). *Antimicrobial Resistance: Global Report of Surveillance*. Geneva: WHO. 1 p.

Wolcott, R.D. and Ehrlich, G.D. (2008). Biofilms and chronic infections. *JAMA* 299: 2682–2684.

Wright, G.D. (2016). Antibiotic adjuvants: rescuing antibiotics from resistance. *Trends Microbiol.* 24: 862–871.

Yang, F., Wang, L.H., Wang, J. et al. (2005). Quorum quenching enzyme activity is widely conserved in the sera of mammalian species. *FEBS Lett.* 579: 3713–3717.

Yankaskas, J.R., Marshall, B.C., Sufian, B. et al. (2004). Cystic fibrosis adult care: consensus conference report. *Chest* 125: 1S–39S.

Yosef, I., Manor, M., Kiro, R., and Qimron, U. (2015). Temperate and lytic bacteriophages programmed to sensitize and kill antibiotic-resistant bacteria. *Proc. Natl. Acad. Sci. U. S. A.* 112: 7267–7272.

18

Repurposing Antibiotics to Treat Resistant Gram-Negative Pathogens

Frank Schweizer

Department of Chemistry, University of Manitoba, Winnipeg, Canada

18.1 Introduction

Antimicrobial resistance is a major threat to public health and economic growth (The United Nations General Assembly 2016). The World Health Organization (WHO) declared carbapenem-resistant *Pseudomonas aeruginosa* (*CR-PA*), CR *Acinetobacter baumannii* (*CR-ABi*), and CR *Enterobacteriaceae* as the three most critical pathogens that pose the greatest threat to human health (WHO Media 2017). These Gram-negative pathogens are also frequently multidrug resistant (MDR) (resistant to ≥3 different antibiotic classes) and have become resistant to all or almost all of our antibiotic armamentarium including carbapenems and third/fourth-generation cephalosporins – the most widely used drugs to fight these infections (Peleg and Hooper 2010). Despite significant investments into antibiotic discovery in the past, no new antibiotic class with novel mode(s) of action against Gram-negative bacteria (GNB) has been approved in half a century (Brown et al. 2014). As a result there is an urgent demand to explore new approaches to combat antimicrobial resistance with the goal to preserve or revive existing antibiotics. In recent years several approaches including anti-virulence strategies and antibiotic combination approaches as well as antibiotic–adjuvant combination approaches have been put forward with the goal to restore efficacy in antibiotics against antibiotic-resistant pathogens (Li et al. 2015; Tommasi et al. 2015; Zygurskaya et al. 2015; Brown and Wright 2016; Silver 2016; The Pew Charitable Trusts 2016; Domalaon et al. 2018c).

Antibiotic Drug Resistance, First Edition. Edited by José-Luis Capelo-Martínez and Gilberto Igrejas.

18.2 Anti-Virulence Strategy

The development of agents that are not bactericidal but indirectly inhibit the molecular pathway responsible for bacterial communication is a viable strategy to address the problem of antibiotic resistance (Williams 2014; Brannon and Hadjifrangiskou 2016; Dickey et al. 2017). Compounds of this type are perceived to exert reduced evolutionary selective pressure and lower rates of resistance development (Rampioni et al. 2014). One such example involves the blocking of bacterial quorum sensing (QS). QS is characterized by bacterial production release and group-wide detection of autoinducer molecules as a mode of bacterial communication with their neighbors (Asfahl and Schuster 2017). This network of communication is triggered by environmental factors within the microbial community, such as differences in bacterial density or the presence of environmental challenges (Lee and Zhang 2015; Grandclement et al. 2016). Once these signaling molecules are detected, cascades of physiological and metabolic changes occur by orchestrated alterations in bacterial gene expression, resulting to the secretion of biomolecules needed for biofilm formation and virulence (Papenfort and Bassler 2016). Therefore, hindering QS may result in the pathogen not being able to cause harm to the host. Several agents that block QS are in preclinical development. For example, the synthetic agent meta-bromo-thiolactone (mBTL) has been reported to curb the production of the virulence factor pyocyanin and biofilm formation in *P. aeruginosa* by affecting the regulation of Las and Rhl quorum-sensing systems (O'Loughlin et al. 2013). Moreover, *in vitro* protection of human lung epithelial cells and *in vivo* protection of *Caenorhabditis elegans* by mBTL against *P. aeruginosa* have been described (O'Loughlin et al. 2013). A follow-up report detailed the optimization of mBTL for enhanced stability as the thiolactone ring is susceptible to chemical and enzymatic hydrolysis (Miller et al. 2015). Other anti-quorum-sensing agents at the preclinical stage have also been reported (Oh et al. 2010; Alasil et al. 2015; Lidor et al. 2015; Simonetti et al. 2016). Nevertheless, this strategy remains controversial (Kalia et al. 2014; Scutera et al. 2014) especially in the light that certain clinical isolates are resistant to established anti-quorum-sensing agents (Garcia-Contreras et al. 2013).

18.3 Antibiotic Combination Strategy

Combination therapy, the concomitant use of two or more antibacterial agents, has been around for more than three decades (Hilf et al. 1989). Clinicians often prescribe two or more antibiotics concomitantly, during empirical treatment to ensure coverage of all possible bacterial pathogens and resistance profiles. Moreover, the use of multiple antibiotic agents in a therapeutic cocktail may limit the development of resistance *in vitro* in comparison with drug monotherapy. The overall expected clinical outcome for this strategy is to have lower patient mortality rates. However, combination therapy is not limited to antibiotic agents but includes therapeutic interventions that may use anti-virulence agents or helper molecules, also known as adjuvants with no or poor antibacterial activity capable of enhancing the efficacy of a primary antibiotic. In fact, it has been argued that the antibiotic–adjuvant combination approach offers a more attractive option in the treatment of drug-resistant bacterial infection than using multiple antibiotics (Wright 2016).

18.4 Antibiotic–Antibiotic Combination Approach

The use of two or more antibiotic agents that have different targets is an attractive strategy to overcome drug resistance. The hypotheses of the antibiotic–antibiotic combination approach are (i) to achieve drug synergism between each drug component in a way that enhances treatment efficacy and (ii) to simultaneously impact multiple targets in pathogens, resulting in the suppression of antibiotic resistance development and complete eradication of bacterial strains with intermediate susceptibility or resistance to one of two antibiotics (Williams 2014). The assumption is that the bacterial cell will have difficulty surviving with multiple "hits" at the same time. Clinicians sometimes employ this strategy during empirical treatment of infection, and such an approach might indeed prolong the clinical utility of antibiotics. For instance, the combination of trimethoprim and sulfamethoxazole has been in use since 1968 for the treatment of bacterial infections caused by the *Enterobacteriaceae* family and non-fermentative opportunistic pathogens (Huovinen 2001; Masters et al. 2003). Both antibiotics work together to inhibit sequential steps in bacterial folic acid synthesis, which is detrimental as most bacteria are obligate folate synthesizers while humans acquire folate through diet. The sulfonamide, sulfamethoxazole, inhibits dihydropteroate synthase that converts para-aminobenzoic acid to dihydrofolate, and trimethoprim inhibits dihydrofolate reductase that converts dihydrofolate to tetrahydrofolate (Masters et al. 2003). Trimethoprim–sulfamethoxazole is an efficacious antibiotic used to treat urinary tract and select gastrointestinal bacterial infections (Libecco and Powell 2004; McIsaac et al. 2008). Sulfamethoxazole may be replaced with the sulfonamide, sulfametrole, in some European Union countries, although both, when combined with trimethoprim, exhibit the same clinical efficacy (Livermore et al. 2014). However, the success of the trimethoprim-sulfamethoxazole combination has been affected by the dissemination of resistance mechanisms that prevent both antibiotics from eliciting their biological functions. Overexpression of multidrug efflux pumps able to expel both trimethoprim and sulfamethoxazole out of the cell and membrane modifications that limit their intracellular permeation are problematic (Huovinen 2001). Many other antibiotic–antibiotic combinations are used in empirical therapy in the clinic including those of tigecycline + gentamicin, tigecycline + colistin, and carbapenem + colistin (Falagas et al. 2014).

18.5 Antibiotic–Adjuvant Combination Approach

Arguably the most successful therapeutic strategy of the twenty-first century, the antibiotic–adjuvant approach has resulted in several drug entities in the market. The paradigm entails the use of bioactive adjuvants that augment the antibiotic efficacy of a primary antibiotic against drug-resistant pathogens. The adjuvant may possess weak to no antibacterial activity on its own but is able to either impede antibiotic resistance mechanisms or potentiate antibiotic action. An adjuvant may be an efflux pump inhibitor to prevent extrusion of drugs, a membrane permeabilizer to increase the number of molecules that penetrate the membrane, or an enzyme inhibitor to prevent degradation of drugs before reaching their targets. Moreover, inhibition of the intrinsic repair pathway of cells by preventing biofilm formation or by assisting in *in vivo* clearance of an infection by the host besides may also be useful to design adjuvants (Mansour et al. 2014; Brown 2015; Gill et al. 2015; Hancock et al. 2016; Wright 2016).

18.6 β-Lactam and β-Lactamase Inhibitor Combination

Augmentin® is a clinically used broad-spectrum antibiotic combination of amoxicillin and clavulanic acid (White et al. 2004). Clavulanic acid is a β-lactamase inhibitor that acts in synchrony with the β-lactam amoxicillin to prevent bacterial growth. These β-lactamase inhibitors, such as clavulanic acid, block the function of β-lactamases or β-lactam-hydrolyzing enzymes by forming an irreversible bond with the enzyme's functional/reactive site. Clavulanic acid by itself possesses poor intrinsic activity against pathogens, but it efficiently inhibits widespread β-lactamases such as many types of the extended-spectrum β-lactamase (ESBL) family (Drawz and Bonomo 2010). Inhibition of ESBLs is especially important as this group of β-lactamases is promiscuous and is able to hydrolyze penicillins, cephalosporins (first, second, and third generations), and monobactams (such as aztreonam) (Gniadkowski 2001; Paterson and Bonomo 2005). Augmentin was first introduced in 1981 by GlaxoSmithKline and continues its clinical usefulness even today (Ball 2007; Geddes et al. 2007). Unfortunately, their use is compromised by the global spread of bacterial β-lactamase-encoding genes. The pursuit of adjuvants that inhibit β-lactamases is therefore crucial to retain clinical effectiveness of the β-lactam class of antibiotics. The recent approval of ceftolozane–tazobactam in 2014, ceftazidime–avibactam in 2015, and meropenem–vaborbactam by FDA in 2017 for the treatment of drug-resistant Gram-positive and GNB infections is indicative of the continued interest in the development of adjuvant combination therapies that includes a β-lactam and a β-lactamase inhibitor. At least four more β-lactam-based antibiotic–adjuvant combinations are currently in clinical trials (Taneja and Kaur 2016; Butler et al. 2017). We will briefly highlight a few recent examples.

18.7 Imipenem–Cilastatin/Relebactam Triple Combination

In 1985, the combination of the carbapenem, imipenem, and the adjuvant cilastatin was approved for use in the United States under the trade name Primaxin® (Jacobs 1986). Imipenem is a broad-spectrum antibiotic that is rapidly degraded by the human renal enzyme dehydropeptidase-1, and the resulting metabolite poses potential for nephrotoxicity (Hikida et al. 1992). Thus, addition of the dehydropeptidase-1 inhibitor cilastatin to imipenem prevents imipenem's degradation and nephrotoxicity. Cilastatin also blocks megalin-mediated proximal tubule uptake of cationic antibiotics (Hori et al. 2017), further lowering the risk of kidney damage. However, the recent increase in bacterial infections caused by carbapenemase-producing organisms that inactivate imipenem calls for an improvement in this therapy. The combination of imipenem–cilastatin with the addition of the diazabicyclooctane β-lactamase inhibitor relebactam is currently in phase III clinical trial for the treatment of GNB infections (Falagas et al. 2016). The adjuvant relebactam is able to inhibit the activity of ESBL, class A (e.g. KPC) and class C (e.g. AmpC) β-lactamases, against imipenem by irreversibly blocking their functional/reactive site (Blizzard et al. 2014). The triple combination was found to be generally well tolerated in patients, with commonly reported adverse effects being nausea, vomiting, and diarrhea (Lucasti et al. 2016). Recently, a phase III randomized, double-blind, non-inferiority study of imipenem-cilastatin/relebactam in comparison to imipenem–cilastatin/colistimethate sodium for the treatment of hospital-acquired bacterial pneumonia (HABP), ventilator-associated bacterial pneumonia (VABP), complicated intra-abdominal infection (cIAI), and complicated urinary tract infection (cUTI) (https://clinicaltrials.gov/ct2/show/study/NCT02452047) was completed. Results are yet to be disclosed. Another phase III randomized, double-blind, non-inferiority study of imipenem–cilastatin/relebactam against piperacillin-tazobactam for the treatment of HABP or VABP is currently recruiting (https://clinicaltrials.gov/ct2/show/

NCT02493764). Moreover, a phase III non-randomized, open label study for the efficacy and safety of imipenem–cilastatin/relebactam for the treatment of cIAI and cUTI is currently ongoing in Japan (https://clinicaltrials.gov/ct2/show/NCT03293485). The activity of this triple combination will be useful against organisms that harbor metallo-β-lactamase such as New Delhi metallo-β-lactamase-1 (NDM-1), imipenemase (IMP), and Verona integron-encoded metallo-β-lactamase (VIM) (Blizzard et al. 2014; Falagas et al. 2016).

18.8 Aspergillomarasmine A

The adjuvant aspergillomarasmine A (AMA) was recently discovered to resuscitate the biocidal activity of the carbapenem drug, meropenem, against metallo-β-lactamase-producing organisms (King et al. 2014). The fungal metabolite AMA was first isolated in the 1960s (Haenni et al. 1965) and was later evaluated for its antihypertensive properties (Arai et al. 1993; Matsuura et al. 1993). In an antibiotic era where enzymes capable of degrading even the most powerful β-lactam (e.g. carbapenems) are abundant, it is promising to find AMA able to inhibit metallo-β-lactamases such as the NDM-1 enzyme. AMA was found to sequester zinc cations (King et al. 2014) that are essential for the hydrolytic activity of metallo-β-lactamases (Karsisiotis et al. 2014; Meini et al. 2015). In a mouse model of NDM-1-positive *Klebsiella pneumoniae* infection, a single dose of meropenem (10 $mg\,kg^{-1}$) and AMA (30 $mg\,kg^{-1}$) combination led to >95% survival after five days post-infection (King et al. 2014). Meropenem alone (10 $mg\,kg^{-1}$) or AMA alone (30 $mg\,kg^{-1}$) resulted in 0% survival (King et al. 2014). These promising results stimulate the need for an optimized dosing regimen of AMA in combination with carbapenems for the treatment of metallo-β-lactamase-producing pathogens. Currently, AMA is undergoing preclinical optimization (Albu et al. 2016; Koteva et al. 2016; Liao et al. 2016).

18.9 Intrinsic Resistance Challenges and Strategies to Overcome Them

Poor outer membrane permeability and overexpression of multidrug efflux pumps, the hallmarks of intrinsic resistance in GNB, prevent many classes of antibiotics from achieving the required intracellular or periplasmic concentrations to elicit their antibacterial effect (Li et al. 2015; Tommasi et al. 2015; Zygurskaya et al. 2015; The Pew Charitable Trusts 2016; Brown and Wright 2016; Silver 2016; Domalaon et al. 2018c). One pathogen that uses intrinsic resistance at near perfection is *P. aeruginosa*. *Pseudomonas aeruginosa* expresses more than 30 selective or "slow" porins, up to 12 multidrug efflux pumps, and different types of antibiotic-inactivating enzymes (e.g. β-lactamases, aminoglycoside-modifying enzymes, and others) that concertedly reduce the concentrations of antibiotics in the cell (Breidenstein and Hancock 2011). Over the years various strategies to enhance the intracellular concentrations of antibiotics have been investigated including the design and development of antibiotic adjuvants designed to enhance the permeability of agents through the outer membrane of GNB, leading to an increase in intracellular concentration. It should be noted that it is difficult to distinguish between a permeabilizer and an EPI as the net result is the same as increase in the intracellular concentration of the antibiotic (Li et al. 2015). Adjuvants that enhance membrane permeability and/or inhibit efflux are known, but their clinical efficacy/safety has not been demonstrated (Zabawa et al. 2016; Corbett et al. 2017). Examples of permeabilizers include polybasic agents like polymyxin B (**1**) and polymyxin E (**2**) as shown in Figure 18.1, but these agents do not qualify as adjuvants because of their high antibacterial potency. However, polymyxin-derived antibiotic adjuvants with potent outer membrane-permeabilizing properties but weak antibacterial activity can be generated from polymyxins by removal or modifications of the lipid tail as shown in compounds (**3**)–(**9**) (Figure 18.1) (Vaara and Vaara 1983; Zabawa et al. 2016; Corbett et al. 2017; Domalaon et al. 2018a). However, non-polymyxin-derived polybasic antibiotic adjuvants including cationic steroid antibiotics or ceragenins (Li et al. 1998, 1999; Ding et al. 2002), oligo-acyl-lysyls (Radzishevsky et al. 2007; Jammal et al. 2015), amphiphilic aminoglycosides (Ouberai et al. 2011), cationic lipopeptides (Domalaon et al. 2018d), and some approved drugs for non-antibiotic applications have also been reported (Figure 18.2) (Ejim et al. 2011; Taylor et al. 2012; Stokes et al. 2017).

Adjuvants that work as EPIs include dibasic aromatics (Bohnert and Kern 2005; Kern et al. 2006), dibasic dipeptides (Lomovskaya et al. 2001), polybasic amphiphilic aminoglycosides (Yang et al. 2017a, 2018), and also nonbasic compounds (Yoshida et al. 2007; Opperman et al. 2014; Aron and Opperman 2016). EPIs are believed to block the function of efflux pumps in GNB by competing with or distorting the antibiotic binding site (Pagès et al. 2005; Vargiu et al. 2014; Li et al. 2015) and/or by perturbing the OM channel or assembly of the tripartite protein complex of RND pumps (Pagès et al. 2005; Li et al. 2015). Alternatively, *uncouplers*, which target the PMF required for ATP synthesis and dissipate either the transmembrane electrical ($\Delta\Psi$) or the proton gradient (ΔpH) components of the PMF located in the IM, can also block (de-energize) RND multidrug efflux pumps (Li et al. 2015). Uncouplers were initially thought to be too cytotoxic as drugs (Li et al. 2015); however, it has now been shown that several FDA-approved drugs including clofazimine, clomiphene, bedaquiline, niclosamide, telavancin, daptomycin, and others have activity as uncouplers (Kaneti et al. 1858; Feng et al. 2015). Among them, the salicylanilide

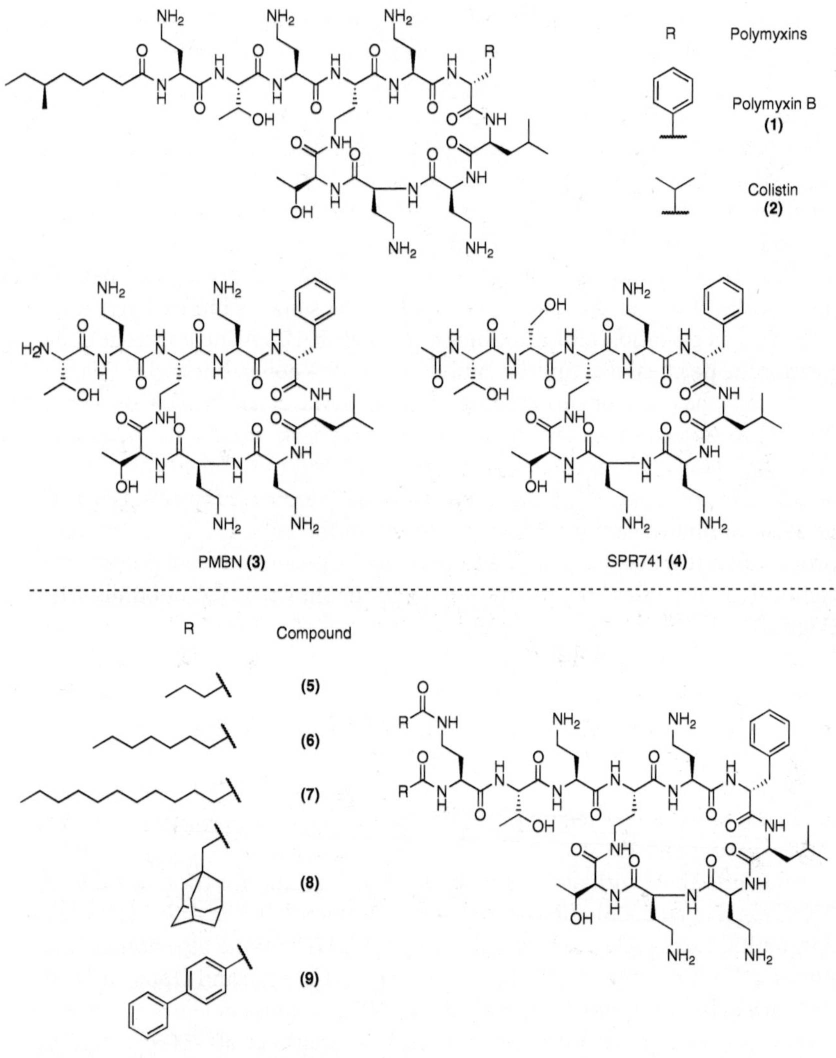

Figure 18.1 Structures of polymyxins (**1**) and (**2**) and polymyxin-derived antibiotic adjuvants (**3**)–(**9**).

niclosamide has been shown to inhibit QS in *PA* by hindering the cell's response and production of the autoinducer *N*-3-oxododecanoyl-homoserine lactone (Imperi et al. 2013). The quenching of QS by *niclosamide* phenotypically impeded bacterial motility, biofilm formation, and secretion of virulence factors. Polybasic permeabilizers and EPIs could potentially induce nephrotoxicity as was observed for certain polymyxins and aminoglycosides (Ordooei et al. 2015) or become toxic by concentrating in acidic vesicles as seen for dibasic

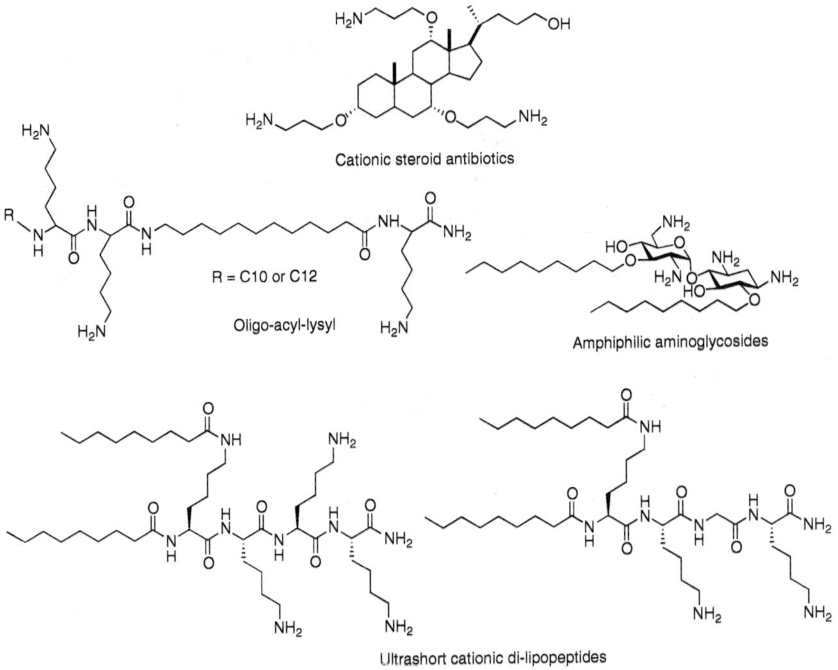

Figure 18.2 Structure of non-polymyxin-based outer membrane permeabilizers.

EPIs (Watkins et al. 2003). Some polybasic aminoglycosides can also induce ototoxicity (Lanvers-Kaminsky et al. 2017). In this case, reducing the number of amino functions can reduce nephro- and ototoxicity (Coleman et al. 2016a).

18.10 Repurposing of Hydrophobic Antibiotics with High Molecular Weight by Enhancing Outer Membrane Permeability Using Polybasic Adjuvants

Antibiotics are taken up into bacterial cells by entropy-driven diffusion across the membranes (passive transport) or by porin-mediated, energy-dependent transport (active transport). GNB possess two cellular membranes with orthogonal penetrating properties. The negatively charged lipopolysaccharide (LPS) core, stabilized by cross-linking to bivalent metal cations on the outer leaflet of the OM, forms an effective impermeable barrier for most hydrophobic agents (Silhavy et al. 2010). Compounds that are able to traverse the OM typically do so slowly, through narrow β-barrel proteins (porins) that are lined with charged amino acids (Denyer and Maillard 2002). Once inside the periplasm compounds are susceptible to multidrug efflux pumps; thus, to accumulate in GNB to a level that is sufficient for activity, compounds typically must traverse porins faster than they are pumped out (Denyer and Maillard 2002; Silhavy et al. 2010; Li et al. 2015). In addition, compounds that interfere with intracellular targets must cross the IM that is impermeable to most polar molecules. The resultant orthogonal physicochemical properties of both membranes combined with extensive efflux poses a significant challenge for traditional small molecules. Existing rules for small molecule accumulation in GNB, developed by retrospective analyses of known antibiotics, indicate that a molecular weight (MW) of less than 600 Da and *high degree of polarity* as measured by $ClogD_{7.4}$ (the predicted octanol/water distribution coefficient at pH 7.4) are optimal (Brown et al. 2014). These observations are consistent with porin architecture, *but the high polarity is inconsistent with the Lipinski's rules* (Lipinsky 2004). In addition, membrane-active antibiotics with porin-independent uptake can possess a higher MW as shown for polymyxins (MW > 1150 Da). Recently, compound accumulation rules have emerged for *Escherichia coli* (Richter et al. 2017). These rules state that compounds that are most likely to accumulate contain one or more amino functions, are amphiphilic and rigid, and have low globularity. However, these accumulation rules do not apply to intrinsically resistant organisms like PA or organisms with elevated expression of multidrug efflux pumps (Richter et al. 2017). For instance, the intrinsic resistance in *PA* is the result of the exceptional low permeability of its OM, which is 12–100 times less permeable than that of *E. coli* owing to its selective porins (Breidenstein and Hancock 2011). Influx of antibiotics and other molecules through the outer membrane can be enhanced via the self-promoted uptake (SPU) mechanism (Breidenstein and Hancock 2011). SPU is a process by which polycationic (or polybasic) molecules, preferably those containing primary amine functional groups (Richter et al. 2017), displace the divalent cations (Ca^{2+} or Mg^{2+}), which are stabilizing counterions for the phosphate groups of lipid A and the phosphorylated core sugars that presumably prevent repulsion between and among individual LPS (Figure 18.3) (Breidenstein and Hancock 2011; Silver 2016). While metal chelators such as ethylenediaminetetraacetic acid (EDTA) (at millimolar concentrations) can remove these cations and cause disaggregation of the entire outer membrane, SPU is thought to act in the region of compound/lipid A interaction, thus resulting in a localized disruption of LPS and allowing passage of the promoting molecules and other molecules (antibiotics) into the periplasm (Breidenstein and Hancock 2011; Hancock 1984; Silver 2016). Some polybasic molecules have been shown to induce SPU at low micromolar concentrations, including certain antimicrobial peptides and their mimetics, lipopeptides, polymyxins, and aminoglycosides. Molecules possessing an SPU mechanism may be of use, in combination therapy, to effectively permit cellular entry of antibiotics that otherwise poorly penetrate/permeate the outer membrane. Compounds of this

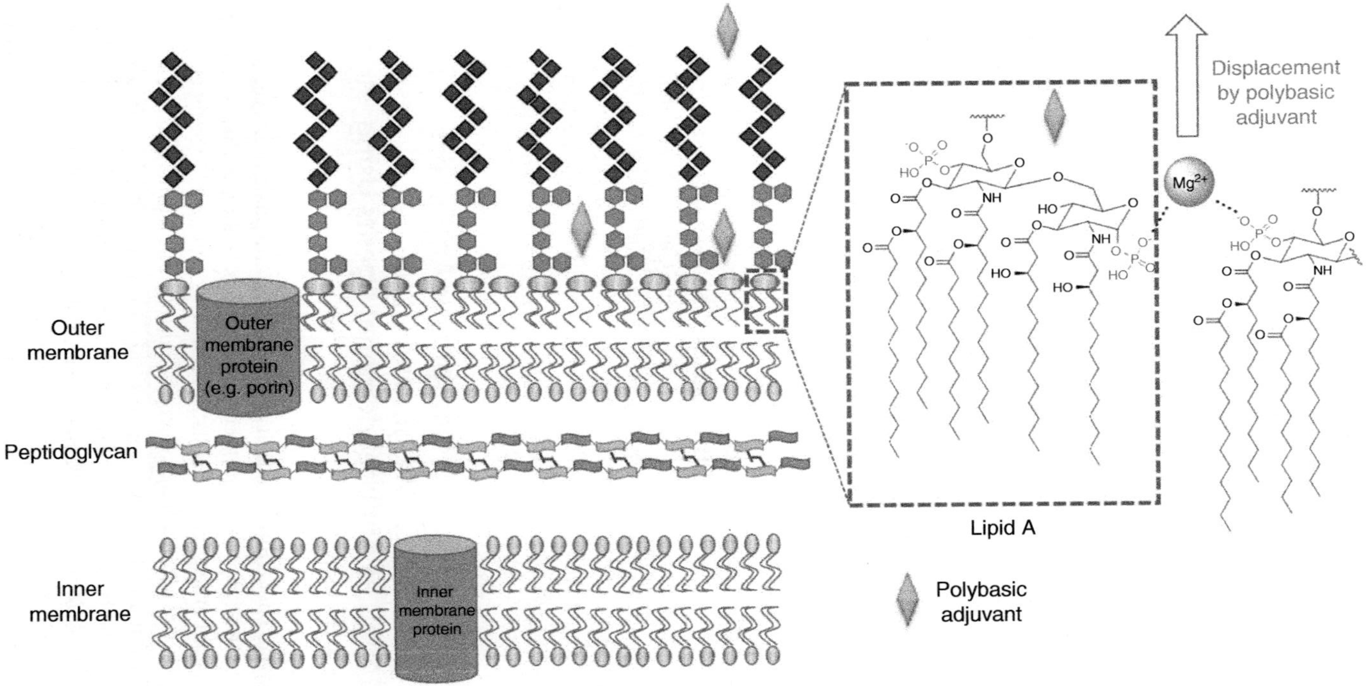

Figure 18.3 Polybasic compounds destabilize the outer membrane of Gram-negative bacteria (GNB) via replacement of divalent metal ions that stabilize the intermolecular interactions of adjacent lipid A molecules that leads to enhanced outer membrane permeability.

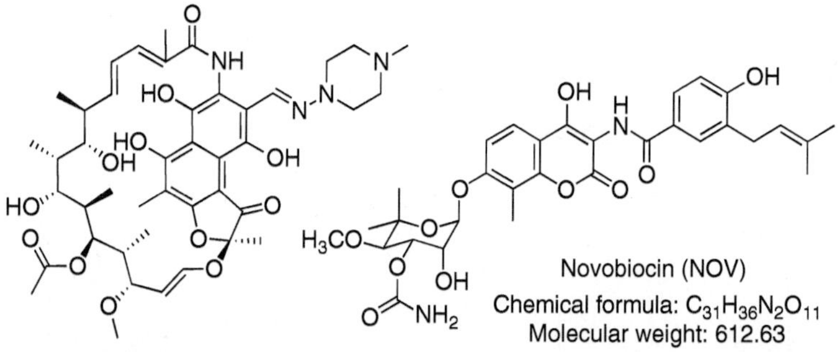

Rifampicin (RIF)
Chemical formula: $C_{43}H_{58}N_4O_{12}$
Molecular weight: 822.95

Figure 18.4 Structure of hydrophobic antibiotics with large molecular weight that display poor antipseudomonal properties.

type are vaguely defined as hydrophobic antibiotics with an MW above 600 g mol^{-1}. The polybasic molecule that enhances outer membrane uptake is called a permeabilizer or adjuvant. Several hydrophobic antibiotics with an MW above 600 Da possess selective and potent activity against Gram-positive bacilli but poor antibacterial activity against GNB. Among others these include the RNA polymerase inhibitor rifampicin (RIF) (Calvori et al. 1965), the aminocoumarin antibiotic novobiocin (NOV) that inhibits the GyrB subunit of DNA gyrase at a different site than fluoroquinolones (Biacchi and Manchester 2015) and vancomycin that inhibits the cross-linking of the peptidoglycan chains (Figure 18.4) Levine (1987).

18.11 Repurposing of Hydrophobic Antibiotics with Large Molecular Weight and Other Antibacterials as Antipseudomonal Agents Using Polybasic Adjuvants

RIF has a broad spectrum of activity including Gram-positive bacteria and certain GNB, although it is not recommended as a single therapeutic agent because of rapid emergence of resistance *in vitro* and *in vivo* (Song et al. 2008). Apart from its use against *Staphylococcus aureus* infections, RIF has been used in triple combination therapy against infections caused by *Mycobacterium tuberculosis* as well as in combination with polymyxins against CR-GNB, most notably against *A. baumannii* (Bassetti et al. 2008). The use against *A. baumannii* is based on preclinical studies indicating synergism, but this effect has not been corroborated by clinical evidence of benefit based on recent clinical trials.

Neither CLSI nor EUCAST has defined breakpoints for RIF for Gram-negative pathogens. The French Society for Microbiology has established an RIF breakpoint for *A. baumannii* based on MIC distributions (susceptible, ≤4 mg l^{-1}; intermediate, 8–16 mg l^{-1}; and resistant, ≥16 mg l^{-1}) (Bonnet et al. 2012). The intravenous administration of 300 or 600 mg of RIF over 30 minutes resulted in peak serum concentrations of 9 or 17.5 mg l^{-1}. -10 mg l^{-1} (Song et al. 2008). However, considering an 80% protein binding of RIF (Song et al. 2008) it is unlikely that unbound RIF concentrations will reach peak concentrations high enough to induce effective antibacterial killing considering the MICs of RIF against most GNB especially *P. aeruginosa*. As such, the activity of RIF most likely relies on its potential synergistic properties with other antibacterial agents or antibiotic adjuvants. A variety of polybasic adjuvants have been shown to synergize the activity of RIF against GNB. Perhaps the best known antibiotic adjuvant is pentabasic polymyxin B nonapeptide (**PMBN**) (Vaara and Vaara 1983). PMBN (**3**) is derived from the lipopeptide polymyxin B (**1**) by enzymatic hydrolysis of a lipid-diaminobutyric acid portion and possesses reduced toxicity in animals when compared with polymyxin B and E. PMBN has been shown to synergize RIF against *P. aeruginosa* PAO1 with a fractional inhibitory concentration (FIC) index of 0.01. An FIC index of ≤0.5, $0.5 < x \leq 4$, or >4 denotes for synergistic, additive, or antagonistic interaction, respectively (Odds 2003). For instance, 0.5 mg l^{-1} of PBMN reduced the MIC of RIF by 256-fold (Domalaon et al. 2018a). In our hands polymyxin B and polymyxin E did not show synergy with RIF against *P. aeruginosa*. Another polymyxin-based tricationic adjuvant is SPR741 (**4**). Similar to PMBN, SPR741 does not contain a hydrocarbon-based fatty acid chain (Figure 18.1). SPR741 is currently being developed by Evotec AG and Spero Therapeutics as an adjuvant that potentiates antibiotics against Gram-negative pathogens (Spero Therapeutics Media Centre n.d.). The recently completed randomized, quadruple-blind phase I clinical study for safety and tolerability in healthy volunteers (https://clinicaltrials.gov/ct2/show/NCT03022175) yielded favorable results for this adjuvant. SPR741 was well tolerated by healthy adult volunteers in a single dose of up to 800 mg and at doses of up to 600 mg every 8 hours for 14 days. In contrast to polymyxins, SPR741 has poor activity against Gram-negative pathogens on its own but can permeabilize the outer membrane to facilitate entry of other antibiotics into the bacterial cell (Oh et al. 2010; Lidor et al. 2015). For instance, SPR741 was reported to sensitize *Enterobacteriaceae* and *A. baumannii* but not *P. aeruginosa* to an extensive panel of antibiotics including clarithromycin, fusidic acid, and RIF (Zabawa et al. 2016; Corbett et al. 2017). These three antibiotics are not classical drugs used to treat Gram-negative bacillary infections due to intrinsic resistance, notably OM impermeability. At 2 μg ml^{-1} of SPR741, the MIC_{50} and MIC_{90} of RIF against a panel of MDR *E. coli* were 0.016 and 0.06 μg ml^{-1}, respectively (Zabawa et al. 2016; Corbett et al. 2017). RIF alone has an MIC_{50} and MIC_{90} of 16 and >128 μg ml^{-1}, respectively, against the same panel of *E. coli* strains (Oh et al. 2010; Lidor et al. 2015). At similar concentrations of SPR741, strong RIF potentiation was also described against a panel of MDR *A. baumannii* (Zabawa et al. 2016; Corbett et al. 2017). The *in vivo* efficacy of SPR741 and RIF combination was shown in murine thigh and lung infection models (Warn et al. 2016a, b). Interestingly, the characteristic nephrotoxic concerns usually associated with polymyxins (Pogue et al. 2016; Zavascki and Nation 2017) were not observed with SPR741 at a dose of 60 mg kg^{-1} day^{-1} in cynomolgus monkeys after 7 days of a 1-hour infusion thrice daily (Coleman et al. 2016b). Other polymyxin-based adjuvants include the polymyxin dilipid (**5**) with two short C-4 lipid tails but not larger lipid tails as in polymyxin analogues (**6**)–(**9**) (Domalaon et al. 2018a). Compound (**5**) is a potent synergizer of RIF against various clinical *P. aeruginosa* isolates and other GNB (Domalaon et al. 2018a).

Our research group has demonstrated that amphiphilic tobramycin (TOB)-based hybrids (**10**)–(**15**) (Gorityala et al. 2016a, b; Lyu et al. 2017; Yang et al. 2017a) are potent synergizer of RIF against wild-type and MDR *P. aeruginosa* isolates (Figure 18.5). For instance, the TOB–ciprofloxacin hybrid at 3.5 μMol reduces the MIC of RIF below 1 mg l^{-1} against a panel of MDR *P. aeruginosa* isolates (Gorityala et al. 2016a). In addition, to RIF many other non-pseudomonal agents are potentiated against *P. aeruginosa* including fluoroquinolones

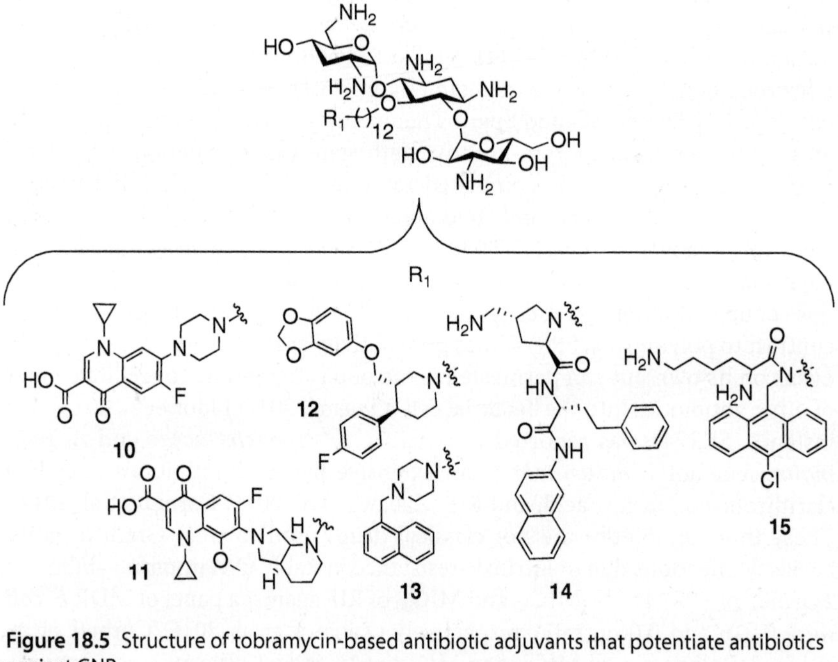

Figure 18.5 Structure of tobramycin-based antibiotic adjuvants that potentiate antibiotics against GNB.

(ciprofloxacin, moxifloxacin), tetracyclines (doxycycline, minocycline, tigecycline), NOV, chloramphenicol, fosfomycin and others except aminoglycosides, and carbapenems. Frequently, the observed effect was limited to *P. aeruginosa* except in the case of RIF where the potentiating effects were seen across most MDR GNB including *E. coli*, *K. pneumoniae*, *A. baumannii*, and *Enterobacter cloacae*. Moreover, a few polybasic antimicrobial peptides have shown to synergize RIF. For instance, the cationic antimicrobial peptides magainin II and cecropin A showed a synergistic fractional inhibitory concentration index (FICI) = 0.31 with RIF in the wild-type *P. aeruginosa* strain ATCC 27852. Moreover both peptides reduced the bacterial count in blood of wild-type and MDR *P. aeruginosa*-infected rats as well as mortality when treated with a combination of RIF and magainin II or cecropin A (Cirioni et al. 2008). Recently, short polybasic proline-rich lipopeptides (Domalaon et al. 2018d) and also ultrashort cationic di-lipopeptides (Domalaon et al. 2018b) have been shown to synergize RIF and minocycline against MDR *P. aeruginosa* isolates.

18.12 Repurposing of Antibiotics as Potent Agents Against MDR GNB

Besides polymyxin and TOB-based adjuvants, other adjuvants have been identified to enhance the antibacterial effects of antibiotics. Among them, two FDA-approved drugs pentamidine and loperamide (Figure 18.6) with no antibacterial applications stand out. The antiprotozoal drug and outer membrane permeabilizer pentamidine displayed synergy with antibiotics typically restricted to Gram-positive bacteria, yielding effective drug combinations with activity against a wide range of Gram-negative pathogens except *P. aeruginosa in vitro* and against systemic *A. baumannii* infections in mice (Stokes et al. 2017). Similarly, loperamide (an antidiarrheal agent) sensitized the tetracycline antibiotics to MDR GNB via a decrease in the electrical component of the proton motive force, resulting in increased pH gradient across the inner membrane (Ejim et al. 2011). The ΔpH would in turn increase uptake of tetracycline antibiotics, thereby overcoming intrinsic resistance. Finally, triclosan is a

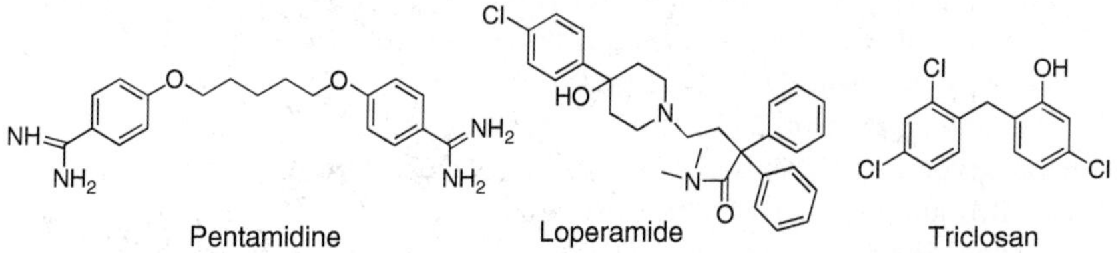

Figure 18.6 Structure of non-polymyxin or non-tobramycin-based antibiotic adjuvants that potentiate antibiotics against GNB.

broad-spectrum antimicrobial that inhibits the enoyl-acyl carrier protein reductase FabI to prevent type II fatty acid synthesis in several bacterial species (Yang et al. 2017b), which has been widely used as disinfectant. Recently, this non-antibiotic triclosan has been shown to potentiate the efficacy of TOB against *P. aeruginosa* biofilms by the eradication of bacterial biofilm. Besides, the combination of triclosan and TOB resulted in a 100-fold reduction of viable persistent cells during 8 hours and complete eradication by 24 hours (Maiden et al. 2018). Eradication of biofilm is another largely unexplored area of how an adjuvant could synergize existing antibiotics (Figure 18.6).

18.13 Outlook and Conclusions

Antibiotic resistance in GNB remains one of the most pressing global health challenges. Our inability to discover novel antibacterial agents with novel modes of action in the past 50 years forces us to look for strategies to enhance the antibacterial activity and prolong the lifespan of our existing antibiotics. Adjuvants employ multiple mechanisms to revive antibiotics. The majority of these strategies are pathogen directed and enhance the intracellular concentration of legacy antibiotics by combating bacterial resistance. In the absence of novel antibiotic classes, this approach appears to be one of the few promising avenues to cope with the global health crises.

References

Alasil, S.M., Omar, R., Ismail, S., and Yusof, M.Y. (2015). Inhibition of quorum sensing-controlled virulence factors and biofilm formation in *Pseudomonas aeruginosa* by culture extract from novel bacterial species of Paenibacillus using a rat model of chronic lung infection. *Int. J. Bacteriol.* 671562.

Albu, S.A., Koteva, K., King, A.M. et al. (2016). Total synthesis of aspergillomarasmine A and related compounds: a sulfamidate approach enables exploration of structure-activity relationships. *Angew. Chem. Int. Ed. Engl.* 2016 (55): 13259–13262.

Arai, K., Ashikawa, N., Nakakita, Y. et al. (1993). Aspergillomarasmine A and B, potent microbial inhibitors of endothelin-converting enzyme. *Biosci. Biotechnol. Biochem.* 57: 1944–1945.

Aron, Z. and Opperman, T.J. (2016). Optimization of a novel series of pyranopyridine RND efflux pump inhibitors. *Curr. Opin. Microbiol.* 33: 1–6.

Asfahl, K.L. and Schuster, M. (2017). Social interactions in bacterial cell-cell signaling. *FEMS Microbiol. Rev.* 41: 92–107.

Ball, P. (2007). Conclusions: the future of antimicrobial therapy – augmentin and beyond. *Int. J. Antimicrob. Agents* 30 (Suppl 2): S139–S141.

Bassetti, M., Repetto, E., Righi, E. et al. (2008). Colistin and rifampicin in the treatment of multidrug-resistant *Acinetobacter baumannii* infections. *J. Antimicrob. Chemother.* 61: 281–284.

Biacchi, G.S. and Manchester, J.I. (2015). A new class antibacterial – almost. Lessons in drug discovery and development: a critical analysis of more than 50 years of effort toward ATPase inhibitors of DNA gyrase and topoisomerase IV. *ACS Infect. Dis.* 1: 4–41.

Blizzard, T.A., Chen, H., Kim, S. et al. (2014). Discovery of MK-7655, a beta-lactamase inhibitor for combination with Primaxin. *Bioorg. Med. Chem. Lett.* 24: 780–785.

Bohnert, J.A. and Kern, W.V. (2005). Selected arylpiperazines are capable of reversing multidrug resistance in *Escherichia coli* overexpressing RND efflux pumps. *Antimicrob. Agents Chemother.* 49: 849–852.

Bonnet, R., Caron, F., Cavallo, J.D. et al. (2012). Comité de L-Antibiogram de las Société Française de Microbiologie – Recommandations 2012. January Edition 2012.

Brannon, J.R. and Hadjifrangiskou, M. (2016). The arsenal of pathogens and antivirulence therapeutic strategies for disarming them. *Drug Des. Devel. Ther.* 10: 1795–1806.

Breidenstein, E.B.M. and Hancock, R.E.W. (2011). *Pseudomonas aeruginosa*: all roads lead to resistance. *Trends Microbiol.* 19: 419–426.

Brown, D. (2015). Antibiotic resistance breakers: can repurposed drugs fill the antibiotic discovery void? *Nat. Rev. Drug Discov.* 14: 821–832.

Brown, D.G., May-Dracka, T.L., Gagnon, M.M., and Tommasi, R. (2014). Trends and exceptions of physical properties on antibacterial activity for Gram-positive and Gram-negative pathogens. *J. Med. Chem.* 57: 10144–10161.

Brown, E.D. and Wright, G.D. (2016). Antibacterial drug discovery in the resistance era. *Nature* 529: 336–343.

Butler, M.S., Blaskovich, M.A., and Cooper, M.A. (2017). Antibiotics in the clinical pipeline at the end of 2015. *J. Antibiot. (Tokyo)* 70: 3–24.

Calvori, C., Frontali, L., Leoni, L., and Tecce, G. (1965). Effect of rifamycin on protein synthesis. *Nature* 207 (995): 417–418.

Cirioni, O., Silvestri, C., Ghiselli, R. et al. (2008). Protective effects of the combination of α-helical antimicrobial peptides and rifampicin in three rat models of *Pseudomonas aeruginosa* infection. *J. Antimicrob. Chemother.* 62: 1332–1338.

Coleman, S., Bleavins, M., Lister, T. et al. (2016a). Assessment of SPR741 for nephrotoxicity in cynomolgus monkeys and Sprague-Dawley rats. Poster 5–23, ASM Microbe 2016.

Corbett, D., Wise, A., Langley, T. et al. (2017). Potentiation of antibiotic activity by a novel cationic peptide: potency and spectrum of activity of SPR741. *Antimicrob. Agents Chemother.* 61: e00200-17.

Denyer, S.P. and Maillard, J.Y. (2002). Cellular impermeability and uptake of biocides and antibiotics in Gram-negative bacteria. *J. Appl. Microbiol.* 92: 35S–45S.

Dickey, S.W., Cheung, G.Y.C., and Otto, M. (2017). Different drugs for bad bugs: antivirulence strategies in the age of antibiotic resistance. *Nat. Rev. Drug Discov.* 16 (7): 457–471.

Ding, B., Guan, Q., Walsh, J.P. et al. (2002). Correlation of the antibacterial activities of cationic peptide antibiotics and cationic steroid antibiotics. *J. Med. Chem.* 45: 663–669.

Domalaon, R., Berry, L., Tays, Q. et al. (2018a). Development of dilipid polymyxins: investigation in the effect if hydrophobicity through its fatty acyl component. *Bioorg. Chem.* 80: 639–648.

Domalaon, R., Brizuela, M., Eisner, B. et al. (2018b). Dilipid ultrashort cationic lipopeptides as adjuvants for chloramphenicol and other conventional antibiotics against Gram-negative bacteria. *Amino Acids* https://doi.org/10.1007/s07726-018-2673-9.

Domalaon, R., Idowu, T., Zhanel, G.G., and Schweizer, F. (2018c). Antibiotic hybrids; the next generation of agents and adjuvants against Gram-negative pathogens. *Clin. Microbiol. Rev.* 31: e00077-17.

Domalaon, R., Sanchak, Y., Koskei, L.C. et al. (2018d). Short proline-rich lipopeptide potentiates minocycline and rifampicin against multidrug-extensively drug-resistant *Pseudomanas aeruginosa*. *Antimicrob. Agents Chemother.* 62: e02374-17.

Drawz, S.M. and Bonomo, R.A. (2010). Three decades of beta-lactamase inhibitors. *Clin. Microbiol. Rev.* 23: 160–201.

Ejim, L., Farha, M.A., Falconer, S.B. et al. (2011). Combinations of antibiotics and nonantibiotics enhance antimicrobial efficacy. *Nat. Chem. Biol.* 7: 348–350.

Falagas, M.E., Lourida, P., Poulikakos, P. et al. (2014). Antibiotic treatment of infections due to carbapenem-resistant Enterobacteriaceae: systematic evaluation of the available evidence. *Antimicrob. Agents Chemother.* 58: 654–663.

Falagas, M.E., Mavroudis, A.D., and Vardakas, K.Z. (2016). The antibiotic pipeline for multi-drug resistant gram negative bacteria: what can we expect? *Expert Rev. Anti Infect. Ther.* 14: 747–763.

Feng, X., Zhu, W., Schurig-Briccio, L.A. et al. (2015). Antiinfectives targeting enzymes and the proton motive force. *Proc. Natl. Acad. Sci. U. S. A.* 112: E7073–E7082.

Garcia-Contreras, R., Martinez-Vazquez, M., Velazquez Guadarrama, N. et al. (2013). Resistance to the quorum-quenching compounds brominated furanone C-30 and 5-fluorouracil in *Pseudomonas aeruginosa* clinical isolates. *Pathog. Dis.* 68: 8–11.

Geddes, A.M., Klugman, K.P., and Rolinson, G.N. (2007). Introduction: historical perspective and development of amoxicillin/clavulanate. *Int. J. Antimicrob. Agents* 30 (Suppl 2): S109–S112.

Gill, E.E., Franco, O.L., and Hancock, R.E. (2015). Antibiotic adjuvants: diverse strategies for controlling drug-resistant pathogens. *Chem. Biol. Drug Des.* 5: 56–78.

Gniadkowski, M. (2001). Evolution and epidemiology of extended-spectrum beta-lactamases (ESBLs) and ESBL-producing microorganisms. *Clin. Microbiol. Infect.* 7: 597–608.

Gorityala, B.K., Guchhait, G., Fernando, D.M. et al. (2016a). Schweizer F adjuvants based on hybrid antibiotics overcome resistance in *Pseudomonas aeruginosa* and enhance fluoroquinolone efficacy. *Angew. Chem. Int. Ed. Engl.* 55: 555–559.

Gorityala, B.K., Guchhait, G., Goswami, S. et al. (2016b). Hybrid antibiotic overcomes resistance in *P. aeruginosa* by enhancing outer membrane penetration and reducing efflux. *J. Med. Chem.* 59: 8441–8455.

Grandclement, C., Tannieres, M., Morera, S. et al. (2016). Quorum quenching: role in nature and applied developments. *FEMS Microbiol. Rev.* 40: 86–116.

Haenni, A.L., Robert, M., Vetter, W. et al. (1965). Chemical structure of Aspergellomarasmines A and B. *Helv. Chim. Acta* 48: 729–750.

Hancock, R.E., Haney, E.F., and Gill, E.E. (2016). The immunology of host defence peptides: beyond antimicrobial activity. *Nat. Rev. Immunol.* 165: 321–334.

Hancock, R.E.W. (1984). Alterations in outer membrane permeability. *Annu. Rev. Microbiol.* 38: 237–264.

Hikida, M., Kawashima, K., Yoshida, M., and Mitsuhashi, S. (1992). Inactivation of new carbapenem antibiotics by dehydropeptidase-I from porcine and human renal cortex. *J. Antimicrob. Chemother.* 30: 129–134.

Hilf, M., Yu, V.L., Sharp, J. et al. (1989). Antibiotic therapy for *Pseudomonas aeruginosa* bacteremia: outcome correlations in a prospective study of 200 patients. *Am. J. Med.* 87: 540–546.

Hori, Y., Aoki, N., Kuwahara, S. et al. (2017). Megalin blockade with cilastatin suppresses drug-induced nephrotoxicity. *J. Am. Soc. Nephrol.* 2017 (28): 1783–1791.

Huovinen, P. (2001). Resistance to trimethoprim-sulfamethoxazole. *Clin. Infect. Dis.* 32: 1608–1614.

Imperi, F., Massai, F., Ramachandran Pillai, C. et al. (2013). New life for an old drug: the anthelmintic drug niclosamide inhibits *Pseudomonas aeruginosa* quorum sensing. *Antimicrob. Agents Chemother.* 57: 996–1005.

Jacobs, R.F. (1986). Imipenem-cilastatin: the first thienamycin antibiotic. *Pediatr. Infect. Dis.* 5: 444–448.

Jammal, J., Zaknoon, F., Kaneti, G. et al. (2015). Sensitization of Gram-negative bacteria to rifampin and OAK combinations. *Sci. Rep.* 5: 9216.

Kalia, V.C., Wood, T.K., and Kumar, P. (2014). Evolution of resistance to quorum-sensing inhibitors. *Microb. Ecol.* 68: 13–23.

Kaneti, G., Meir, O., and Mor, A. (1858). Controlling bacterial infections by inhibiting proton-dependent processes. *Biochim. Biophys. Acta* 2016: 995–1003.

Karsisiotis, A.I., Damblon, C.F., and Roberts, G.C.K. (2014). A variety of roles for versatile zinc in metallo-beta-lactamases. *Metallomics* 6: 1181–1197.

Kern, W.V., Steinke, P., Schumacher, A. et al. (2006). Effect of 1-(1-naphthylmethyl)-piperazine, a novel putative efflux pump inhibitor, on antimicrobial drug susceptibility in clinical isolates of *Escherichia coli*. *J. Antimicrob. Chemother.* 57: 339–343.

King, A.M., Reid-Yu, S.A., Wang, W. et al. (2014). Aspergillomarasmine A overcomes metallo-beta-lactamase antibiotic resistance. *Nature* 510: 503–506.

Koteva, K., King, A.M., Capretta, A., and Wright, G.D. (2016). Total synthesis and activity of the metallo-beta-lactamase inhibitor aspergillomarasmine A. *Angew. Chem. Int. Ed. Engl.* 2016 (55): 2210–2212.

Lanvers-Kaminsky, C., Zehnhoff-Dinnesen, A.A., Parfitt, R., and Ciarimboli, G. (2017). Drug-induced ototoxicity: mechanisms, pharmacogenetics, and protective strategies. *Clin. Pharmacol. Ther.* 101: 491–500.

Lee, J. and Zhang, L. (2015). The hierarchy quorum sensing network in *Pseudomonas aeruginosa*. *Protein Cell* 6: 26–41.

Levine, J.F. (1987). Vancomycin: a review. *Med. Clin. North Am.* 71: 1135–1145.

Li, C., Lewis, M.R., Gilbert, A.B. et al. (1999). Antimicrobial activities of amine- and guanidine- functionalized cholic acid derivatives. *Antimicrob. Agents Chemother.* 43: 1347–1349.

Li, C., Peters, A.S., Meredith, E.L. et al. (1998). Design and synthesis of potent sensitizers of Gram-negative bacteria based on a cholic acid scaffolding. *J. Am. Chem. Soc.* 120: 2961–2962.

Li, X.-Z., Plesiat, P., and Nikaido, H. (2015). The challenge of efflux-mediated antibiotic resistance and Gram-negative bacteria. *Clin. Microbiol. Rev.* 28: 337–418.

Liao, D., Yang, S., Wang, J. et al. (2016). Total synthesis and structural reassignment of aspergillomarasmine A. *Angew. Chem. Int. Ed. Engl.* 2016 (55): 4291–4295.

Libecco, J.A. and Powell, K.R. (2004). Trimethoprim/sulfamethoxazole: clinical update. *Pediatr. Rev.* 25: 375–380.

Lidor, O., Al-Quntar, A., Pesci, E.C., and Steinberg, D. (2015). Mechanistic analysis of a synthetic inhibitor of the *Pseudomonas aeruginosa* LasI quorum-sensing signal synthase. *Sci. Rep.* 2015 (5): 16569.

Lipinsky, C.A. (2004). Lead and drug-like compounds: the-rule-of-five revolution. *Drug Discov. Today Technol.* 1: 337–341.

Livermore, D.M., Mushtaq, S., Warner, M., and Woodford, N. (2014). Comparative in vitro activity of sulfametrole/trimethoprim and sulfamethoxazole/trimethoprim and other agents against multiresistant Gram-negative bacteria. *J. Antimicrob. Chemother.* 69: 1050–1056.

Lomovskaya, O., Warren, M.S., Lee, A. et al. (2001). Identification and characterization of inhibitors of multidrug resistance efflux pumps in

Pseudomonas aeruginosa: novel agents for combination therapy. *Antimicrob. Agents Chemother.* 45: 105–116.

Lucasti, C., Vasile, L., Sandesc, D. et al. (2016). Phase 2, dose-ranging study of relebactam with imipenem-cilastatin in subjects with complicated intra-abdominal infection. *Antimicrob. Agents Chemother.* 60: 6234–6243.

Lyu, Y., Yang, X., Goswami, S. et al. (2017). Amphiphilic tobramycin-lysine conjugates sensitize multidrug resistant Gram-negative bacteria to rifampicin and minocycline. *J. Med. Chem.* 60: 3684–3702.

Maiden, M.M., Hunt, A.M.A., Zachos, M.P. et al. (2018). Triclosan is an aminoglycoside adjuvant for eradication of *Pseudomonas aeruginosa* biolfilms. *Antimicrob. Agents Chemother.* 62: e00146-18.

Mansour, S.C., Pena, O.M., and Hancock, R.E. (2014). Host defense peptides: front-line immunomodulators. *Trends Immunol.* 35: 443–450.

Masters, P.A., O'Bryan, T.A., Zurlo, J. et al. (2003). Trimethoprim-sulfamethoxazole revisited. *Arch. Intern. Med.* 163: 402–410.

Matsuura, A., Okumura, H., Asakura, R. et al. (1993). Pharmacological profiles of aspergillomarasmines as endothelin converting enzyme inhibitors. *Jpn. J. Pharmacol.* 1993 (63): 187–193.

McIsaac, W.J., Prakash, P., and Ross, S. (2008). The management of acute uncomplicated cystitis in adult women by family physicians in Canada. *Can. J. Infect. Dis. Med. Microbiol.* 2008 (19): 287–293.

Meini, M.-R., Llarrull, L.I., and Vila, A.J. (2015). Overcoming differences: the catalytic mechanism of metallo-beta-lactamases. *FEBS Lett.* 589: 3419–3432.

Miller, L.C., O'Loughlin, C.T., Zhang, Z. et al. (2015). Development of potent inhibitors of pyocyanin production in *Pseudomonas aeruginosa*. *J. Med. Chem.* 2015 (58): 1298–1306.

Odds, F.C. (2003). Synergy, antagonism and what the checkerboard puts between them. *J. Antimicrob. Chemother.* 52: 1.

Oh, K.-B., Nam, K.-W., Ahn, H. et al. (2010). Therapeutic effect of (Z)-3-(2,5-dimethoxyphenyl)-2-(4-methoxyphenyl) acrylonitrile (DMMA) against *Staphylococcus aureus* infection in a murine model. *Biochem. Biophys. Res. Commun.* 2010 (396): 440–444.

O'Loughlin, C.T., Miller, L.C., Siryaporn, A. et al. (2013). A quorum-sensing inhibitor blocks *Pseudomonas aeruginosa* virulence and biofilm formation. *Proc. Natl. Acad. Sci. U. S. A.* 110: 17981–17986.

Opperman, T.J., Kwasny, S.M., Kim, H.S. et al. (2014). Characterization of a novel pyranopyridine inhibitor of the AcrAB efflux pump of *Escherichia coli*. *Antimicrob. Agents Chemother.* 2014 (58): 722–733.

Ordooei, J.A., Shokouhi, S., and Sahraei, Z. (2015). A review on colistin nephrotoxicity. *Eur. J. Clin. Pharmacol.* 71: 801–810.

Ouberai, M., El Garch, F., Bussiere, A. et al. (2011). The *Pseudomonas aeruginosa* membranes: a target for a new amphiphilic aminoglycoside derivative? *Biochim. Biophys. Acta* 1808: 1716–1727.

Pagès, J.M., Masi, M., and Barbe, J. (2005). Inhibitors of efflux pumps in Gram-negative bacteria. *Trends Mol. Med.* 11: 382–389.

Papenfort, K. and Bassler, B.L. (2016). Quorum sensing signal-response systems in Gram-negative bacteria. *Nat. Rev. Microbiol.* 14: 576–588.

Paterson, D.L. and Bonomo, R.A. (2005). Extended-spectrum beta-lactamases: a clinical update. *Clin. Microbiol. Rev.* 18: 657–686.

Peleg, A.Y. and Hooper, D.C. (2010). Hospital-acquired infections due to Gram-negative bacteria. *N. Engl. J. Med.* 362: 1804–1813.

Pogue, J.M., Ortwine, J.K., and Kaye, K.S. (2016). Are there any ways around the exposure-limiting nephrotoxicity of the polymyxins? *Int. J. Antimicrob. Agents* 2016 (48): 622–626.

Radzishevsky, I.S., Rotem, S., Bourdetsky, D. et al. (2007). Improved antimicrobial peptides based on acyl-lysine oligomers. *Nat. Biotechnol.* 25: 657–659.

Rampioni, G., Leoni, L., and Williams, P. (2014). The art of antibacterial warfare: deception through interference with quorum sensing-mediated communication. *Bioorg. Chem.* 2014 (55): 60–68.

Richter, M.F., Drown, B.S., Riley, A.P. et al. (2017). Predictive compound accumulation rules yield a broad-spectrum antibiotic. *Nature* 545: 299–304.

Scutera, S., Zucca, M., and Savoia, D. (2014). Novel approaches for the design and discovery of quorum-sensing inhibitors. *Expert Opin. Drug Discovery* 9: 353–366.

Silhavy, T.J., Kahne, D., and Walker, S. (2010). The bacterial cell envelope. *Cold Spring Harb. Perspect. Biol.* 2: a000414.

Silver, L.L. (2016). A gestalt approach to Gram-negative entry. *Bioorg. Med. Chem.* 24: 6379–6389.

Simonetti, O., Cirioni, O., Cacciatore, I. et al. (2016). Efficacy of the quorum sensing inhibitor FS10 alone and in combination with tigecycline in an animal model of staphylococcal infected wound. *PLoS One* 11: e0151956.

Song, J.Y., Lee, J., Heo, J.Y. et al. (2008). Colistin and rifampicin combination in the treatment of ventilator-associated pneumonia caused by carbapenem-resistant *Acinetobacter baumannii*. *Int. J. Antimicrob. Agents* 61: 417–420.

Spero Therapeutics Media Centre (n.d.). SPR206 & SPR741: IV Potentiator Platform. https://sperotherapeutics.com/pipeline/spr741-spr206-iv-potentiator-platform (accessed 23 November 2018).

Stokes, J.M., MacNair, C.R., Ilyas, B. et al. (2017). Pentamidine sensitizes Gram-negative pathogens to antibiotics and overcomes acquired colistin resistance. *Nat. Microbiol.* 2: 17028.

Taneja, N. and Kaur, H. (2016). Insights into newer antimicrobial agents against Gram-negative bacteria. *Microbiol. Insights* 9: 9–19.

Taylor, P.L., Rossi, L., De Pascale, G., and Wright, G.D. (2012). A forward chemical screen identifies antibiotic adjuvants in *Escherichia coli*. *ACS Chem. Biol.* 7: 1547–1555.

The Pew Charitable Trusts (2016). A scientific roadmap for antibiotic discovery, May 2016. http://www.pewtrusts.org/en/research-and-analysis/reports/2016/05/a-scientific-roadmap-for-antibiotic-discovery (accessed 21 August 2018).

The United Nations General Assembly (2016). Press release from 21 September 2016. http://www.un.org/pga/71/2016/09/21/press-release-hl-meeting-on-antimicrobial-resistance (accessed 11 February 2018).

Tommasi, R., Brown, D.G., Walkup, G.K. et al. (2015). ESKAPEing the labyrinth of antibacterial discovery. *Nat. Rev. Drug Discov.* 14: 529–542.

Vaara, M. and Vaara, T. (1983). Sensitization of Gram-negative bacteria to antibiotics and complement by a non-toxic oligopeptide. *Nature* 303: 526–528.

Vargiu, A.V., Ruggerone, P., Opperman, T.J. et al. (2014). Molecular mechanism of MBX2319 of *Escherichia coli* AcrB multidrug efflux pump and comparison with other inhibitors. *Antimicrob. Agents Chemother.* 58: 6224–6234.

Warn, P., Teague, J., Burgess, E. et al. (2016a). In vivo efficacy of combinations of novel antimicrobial peptide SPR741 and rifampicin in short-duration murine lung infection models of K. pneumoniae or E. cloacae infection, p. abstr P497. American Society for Microbiology Microbe 2016.

Warn, P., Thommes, P., Vaddi, S. et al. (2016b). In vivo efficacy of combinations of novel antimicrobial peptide SPR741 and rifampicin in short-duration murine thigh infection models of Gram-negative bacterial infection, p. abstr P561. American Society for Microbiology Microbe 2016.

Watkins, W.J., Landaverry, Y., Leger, R. et al. (2003). The relationship between physicochemical properties, in vitro activity and pharmacokinetic profiles of analogues of diamine-containing efflux pump inhibitors. *Bioorg. Med. Chem. Lett.* 2003 (13): 4241–4244.

White, A.R., Kaye, C., Poupard, J. et al. (2004). Augmentin (amoxicillin/clavulanate) in the treatment of community-acquired respiratory tract infection: a review of the continuing development of an innovative antimicrobial agent. *J. Antimicrob. Chemother.* 53: i3–i20.

WHO Media (2017). WHO publishes list of priority pathogens for which new antibiotics are urgently required. 27 February 2017. http://www.who.int/mediacentre/news/releases/2017/bacteria-antibiotics-needed/en (accessed 11 February 2018).

Williams, S.C.P. (2014). News feature: next-generation antibiotics. *Proc. Natl. Acad. Sci. U. S. A.* 111: 11227–11229.

Wright, G.D. (2016). Antibiotic adjuvants: rescuing antibiotics from resistance. *Trends Microbiol.* 24: 862–871.

Yang, X., Domalaon, R., Lyu, Y. et al. (2018). Tobramycin-linked efflux pump inhibitor conjugates synergize fluoroquinolones, rifampicin, fosfomycin against multidrug-resistant *Pseudomonas aeruginosa*. *J. Clin. Med.* 7: E158.

Yang, X., Goswami, S., Gorityala, B.K. et al. (2017a). A tobramycin vector enhances synergy and efficacy of efflux pump inhibitors against multidrug-resistant Gram-negative bacteria. *J. Med. Chem.* 60: 3913–3932.

Yang, X., Lu, J., Ying, M. et al. (2017b). Docking and molecular dynamics studies on triclosan derivatives binding to FabI. *J. Mol. Model.* 23 (1): 1–13. https://doi.org/10.1007/s00894-016-3192-9.

Yoshida, K., Nakayama, K., Ohtsuka, M. et al. (2007). MexAB-OprM specific efflux pump inhibitors in *Pseudomonas aeruginosa*. Part 7: highly soluble and in vivo active quaternary ammonium analogue D13-9001, a potential preclinical candidate. *Bioorg. Med. Chem.* 15: 7087–7097.

Zabawa, T.P., Zabawa, T.P., Pucci, M.J. et al. (2016). Treatment of Gram-negative bacterial infections by potentiation of antibiotics. *Curr. Opin. Microbiol.* 33: 7–12.

Zavascki, A.P. and Nation, R.L. (2017). Nephrotoxicity of polymyxins: is there any difference between colistimethate and polymyxin B? *Antimicrob. Agents Chemother.* 61: e02319-16.

Zygurskaya, H., Lopez, C.A., and Gnanakaran, S. (2015). Permeability barrier of Gram-negative cell envelopes and approaches to bypass it. *ACS Infect. Dis.* 1: 512–522.

19

Nontraditional Medicines for Treatment of Antibiotic Resistance

Ana Paula Guedes Frazzon, Michele Bertoni Mann, and Jeverson Frazzon

Federal University of Rio Grande do Sul, Porto Alegre, Brazil

19.1 Introduction

The success of the discoveries of the physician and microbiologist Alexander Fleming gave rise to "antibiotic era" brought a great relief for the medical and scientific community in the postwar period, since the treatment of bacterial infections were deemed as a historical milestone. However, less than 100 years after and, warned by the researcher himself that his discovery would have a limited time, the life span of antibiotics seems to have its days over.

However, even before the discoveries of antimicrobials, some substances were already used in the treatment of bacterial diseases. Generally, essential oils from plants have been used in the treatment of infections throughout history. Further, at the end of the nineteenth century and the start of the twentieth century, physicians and bacteriologists Emil Adolf von Behring and Shibasaburo Kitasato showed the first experimental evidence of the use of antibodies, along with the possibility of using immunoglobulins as to neutralize diphtheria toxin, a fact that has generated a great revolution in the scientific thinking at that time. Afterward, a German physician and bacteriologist, Paul Ehrlich, proposed the theory of side chain of antibody formation, quite similar to the current idea on cell surface receptors. These antibodies were called "magic bullets" due to their ability to find the target antigen through specific interactions and to neutralize and destroy the antigen. The therapeutic potential of antibodies had been unveiled.

Taking into consideration the definition of "traditional approaches" as therapeutic agents using small molecules acting directly on the target bacterial component exerting a bactericidal or bacteriostatic effect. Here, we will

Antibiotic Drug Resistance, First Edition. Edited by José-Luis Capelo-Martínez and Gilberto Igrejas.

discuss in this chapter “nontraditional approaches” to antimicrobial therapies. These therapies such as antibodies, immunomodulators, phages, essential oils, and microbiome-based therapy are less susceptible to generate resistance than traditional antibiotics treatment.

19.2 Antibodies

In spite of a considerable number of antibiotics available to treat bacterial infections, antibodies have been shown as ineffective in the fight against resistant microorganisms, especially in immunocompromised patients. With the increased microbial resistance, the pharmaceutical industry has reduced investments in development of new antibiotics. One of the alternatives to the use of antibiotics is antibodies, which regulate or induce a humoral immune response, which are promising in the fight against bacterial infections in immunocompromised subjects or in cases where there are no effective drugs. The use of passive immunotherapy, which utilizes monoclonal antibodies (mAbs), has been proposed as an alternative to the use of antibiotics against multidrug-resistant strains. mAbs may bind directly to the target antigen, acting as an opsonizing agent or neutralizing toxins and virulence factors.

19.2.1 Raxibacumab Versus *Bacillus anthracis*

Bacillus anthracis is an endospore-forming bacterium, being deemed a highly lethal bioterrorism threat. Although antibiotics can be used to effectively treat bacteremia, mortality of inhalational anthrax ranges from 45 to 80%, mainly due to pathogenesis caused by exotoxin. The anthrax toxin is a tripartite toxin containing enzyme and binding portions called (i) lethal factor (LF) and (ii) edema factor (EF). The protective antigen (PA) is the gatekeeper portion that binds to cell receptors and, therefore, binds and translocates LF and EF within the cell.

In December 2012, the Food and Drug Administration (FDA) approved raxibacumab to treat infections caused by *B. anthracis*, which is a human IgG1γ MAb produced by recombinant DNA technology in a murine cell expression system that has the ability to bind to the PA component of *B. anthracis* toxin. Raxibacumab has a molecular weight of about 146 kDa, with an equilibrium dissociation constant (Kd) of 2.78 ± 0.9 nM. For the treatment of inhalation anthrax, the antibody should be used in combination with appropriate antibacterial drugs and it is indicated for anthrax prophylaxis when alternative therapies are not available or are not appropriate.

19.2.1.1 Treatment and Mechanism of Action

Raxibacumab inhibits PA binding to its cell receptors, preventing intracellular entry of anthrax LF and EF, the enzyme toxin components responsible for pathogenic effects of anthrax toxin. This mechanism of action has no direct antibacterial activity; however, antibiotics are highly effective when given at the onset of disease onset. Thus, it was not possible to determine the added benefit over the antibiotic alone or as monotherapy treatment.

19.2.2 Bezlotoxumab Versus *Clostridium difficile*

Clostridium difficile infection (CDI) is the most common hospital-acquired infection and is associated with increased hospital costs. Although *C. difficile* spores are present in the gut innocuously, due to repeated broad-spectrum antibiotic therapy, spores germinate along with the release of exotoxins A and B, resulting in mild to severe diarrhea. Treatment options for recurrent CDI are limited and new alternatives to prevent recurrent CDI is required.

In 2016, the FDA approved the use of bezlotoxumab, an emergent mAb therapy to prevent recurrent CDI. Bezlotoxumab is a human mAb directed against *C. difficile* toxin B (TcdB). This procedure prevents the recurrent risk of CDI, collaborating for the recovery of the intestinal microbiota.

19.2.2.1 Treatment and Mechanism of Action

Antibiotic therapy is extended by adding humanized antibody bezlotoxumab so as to prevent the action of exotoxins A and B, respectively, since they provide passive immunity. Bezlotoxumab inhibits specifically the N-terminal CROP domain of TcdB region. The CROP domain is composed of four different peptide units: B1, B2, B3, and B4. Bezlotoxumab recognizes a specific epitope in TcdB and has a high affinity for that region. The GTD domain does not interact with bezlotoxumab, but seems to interact with B1, which is representative of the entire CROP domain. Bezlotoxumab interacts with B2 and B3 or with residue region overlapping between both domains. B4 does not interact with the specific CROP domain portion. The characterization of peptide B1 as a full TcdB CROP domain suggests that the antibody reacts specifically with CROP domain B2 region. This leads to the conclusion that TcdB epitope lies within the N-terminal region of CROP domain.

19.2.3 Panobacumab Versus *Pseudomonas aeruginosa*

Pseudomonas aeruginosa is an opportunistic pathogen causing infection-related mortality and morbidity in immunocompromised patients due to antibiotic resistance. Multidrug-resistant (MDR) *P. aeruginosa* may cause life-threatening conditions, such as ventilator-associated and hospital-acquired bacterial pneumonias, as well as bloodstream and urinary tract infections in susceptible patients. Thus, designing an appropriate antimicrobial therapy may be a clinical dilemma due to scarcity of effective and safe drugs that may fight MDR *P. aeruginosa* strains.

Panobacumab is a fully human IgM/κ mAb directed against *P. aeruginosa* O11 (*O*-polysaccharide lipopolysaccharide [LPS] portion). This drug, not released yet for use, is in the study phase and is being investigated for its efficacy. The serotype O11 LPS mAb panobacumab was tested in a clinically relevant cyclophosphamide-induced neutropenia murine model and in combination with meropenem in susceptible and meropenem-resistant *P. aeruginosa*-induced pneumonia. Observations showed that *P. aeruginosa*-induced pneumonia was dramatically increased in neutropenic mice compared with immunocompetent mice. First, panobacumab reduced significantly lung inflammation and improved bacterial clearance in the neutropenic host lung. Second, the combination of panobacumab and meropenem had an additive effect. Third, panobacumab retained activity in a meropenem-resistant *P. aeruginosa* strain. In conclusion, panobacumab contributes to clear *P. aeruginosa* in neutropenic hosts, as well as in combination with antibiotics in immunocompetent hosts. This suggests beneficial effects from concurrent treatment, even in immunocompromised subjects, who suffer most of morbidity and mortality of *P. aeruginosa* infections.

19.2.4 LC10 Versus *Staphylococcus aureus*

Staphylococcus aureus is one of the main causes of pneumonia in inpatients and in healthy subjects in the community. *S. aureus* pneumonia is a fatal disease, with mortality rates of 60%. Treatment of these infections is complicated by the fact that 50% of *S. aureus* isolates from pneumonia patients are methicillin-resistant *S. aureus* (MRSA), thus reducing safe and effective treatment options. To overcome this public health issue, several approaches have been developed; among these approaches is the use of mAbs directed against *S. aureus* and its virulence determinants, which can be used in prophylaxis and as an adjuvant therapy with antibiotics.

Alpha-toxin (AT) is an important virulence factor in the disease pathogenesis of *S. aureus*. Passive immunization with LC10 has increased survival and decreased the number of bacteria in lungs and kidneys of infected mice and showed protection against various clinical *S. aureus* isolates. The lungs of mice infected with *S. aureus* exhibited bacterial pneumonia, including generalized inflammation, while the lungs of mice given LC10 showed minimum inflammation and retained a healthy architecture.

The reduction in inflammation and damages in animals treated with LC10 resulted in reduced vascular protein leak and blood CO_2 levels. LC10 was also assessed for its therapeutic activity combined with vancomycin or linezolid, and the treatment resulted in a significant increase ($P < 0.05$) in survival over monotherapies. The lungs of animals treated with antibiotic plus LC10 showed less inflammatory tissue damage than those given monotherapy. This data provides information on protection mechanisms provided by AT inhibition and supports AT as a promising target for immune prophylaxis or adjuvant therapy against *S. aureus* pneumonia.

19.3 Immunomodulators

The human immune system has mechanisms preventing the entry and colonization of our body by pathogen microorganisms. However, since these bacterial pathogens escape cell defense system, these may extensively colonize the patient, resulting in severe symptoms and even death in the absence of a medical intervention. One of the possibilities to avoid it is to modulate patient's immune response as to reverse disease progression and, thus, remove pathogens from the system. Currently there is growing evidence showing that immune responses designed to diseased tissues can dramatically reverse the progression of disease. Immune therapy is a type of treatment aimed at using the immune system so that it may fight infections and other diseases, such as cancer. One of the possibilities, for instance, the use of small molecules to graft immunogenic epitopes onto target cells, is particularly attractive due to its versatility and the ability to mimic native immune responses.

A diverse range of recombinant, synthetic, and natural immunomodulatory preparations for the treatment and prophylaxis of several infections are available today. Substances such as granulocyte colony-stimulating factor, interferons, imiquimod, and bacterial-derived preparations are licensed for use in patients. Others including IL-12, chemokines, and phosphate-guanine-cytosine and synthetic oligodeoxynucleotides are being studied in clinical and/or preclinical phase.

19.3.1 Antibodies plus Polymyxins

Immunomodulators offer an attractive approach as an adjunctive therapy to control microbial diseases in the era of antibiotic resistance. Pires and colleagues developed a potent model to engage immune system components so as to induce an immune response to Gram-negative bacteria. A structure has been established for the use of small molecules to activate the recruitment of endogenous antibodies to the bacterial cell surface. Specifically, conjugates were assembled using polymyxin B (an antibiotic inherently binding to Gram-negative pathogens) and antigenic epitopes recruiting antibodies found in human serum. The compound has been shown to be effective, resulting in specific bacterial killing by the human serum. The potential of these molecules paves the way for a new type of antimicrobials.

19.3.2 Antibodies plus Vitamin D

Tuberculosis (TB) is a pandemic disease caused by *Mycobacterium tuberculosis* (MTb) responsible for millions of active lung TB cases worldwide associated with a high death rate. The prevalence of multiresistant TB and extensively drug-resistant TB converges with pandemics of HIV and diabetes, generating more problems due to the combination of diseases. Therefore, there is a need to develop new drug classes to reduce treatment duration for TB and to overcome infection from MDR. It has been shown in a series of *in vitro* studies that the active form of vitamin D, 1,25-di-hydroxyvitamin D3 or 1,25 (OH) 2D3, induces gene expression of β-defensin 2 and human cathelicidin LL-37. These peptides belong to two antimicrobial peptide classes produced in lung epithelial cells, monocytes/macrophages and neutrophils, which are able to suppress the increase of modular antimicrobial responses. Vitamin D 1,25 (OH) 2D3 also induces autophagy in macrophages/monocytes infected with MTb that may control the infection through an LL-37-dependent mechanism. Therefore, several studies showed an association between vitamin D deficiency and an increased risk of developing active TB. In total, these findings reinforced the interest in vitamin D as an adjuvant therapy for standard anti-TB treatment. Further, additional studies demonstrated a synergistic effect in the induction of LL-37 expression when vitamin D was used in combination with sodium 4-phenylbutyrate (PBA). These studies showed that a combination of PBA and vitamin D to healthy adults increased LL-37 expression and MTb intracellular death in macrophages.

19.3.3 Antibodies plus Clavanin

Clavanin A is a natural antimicrobial peptide that was isolated first from hemocytes of marine tunicate *Styela clava*. This peptide has an amino acid sequence rich in histidine, phenylalanine, and glycine residues, shown to be effective against Gram-negative and Gram-positive bacteria. Among other peptides, this peptide has the advantage of being active even at high salt concentrations and in acidic pH. To increase the potential, five nonpolar amino acids were incorporated to the N-terminal region, the hydrophobic part of clavanin A molecule, and the new engineered molecule was called clavanin-MO. Tests showed that the peptide effectively killed a panel of representative bacterial strains, including MDR clinical isolates. Its antimicrobial activity was shown in animal models, decreasing bacterial counts by six orders of magnitude and contributing to clearance of the infection. Moreover, clavanin-MO was able to modulate innate immunity through stimulating leukocyte recruitment to the site of infection site and the production of immune mediators GM-CSF, IFN-γ, and MCP-1, suppressing an excessive and potentially harmful inflammatory response by increasing anti-inflammatory cytokine synthesis, such as IL-10, and repressing pro-inflammatory cytokine levels of IL-12 and TNF-α.

Treatment with the peptide protected rats against other lethal infections caused by both Gram-negative and Gram-positive antibiotic-resistant strains. Clavanin-MO showed antimicrobial activity by killing bacteria and further helped to resolve infections through its immunomodulatory properties. Therefore, the combination of antimicrobials and immunomodulatory properties represent a new approach to treat infections from antibiotic-resistant strains.

19.3.4 Antibodies plus Reltecimod

In spite of all advantages, a few immunomodulators for use against infections have advanced to clinical stage development. A good example is Reltecimod (AB103), an immunomodulator used to treat soft tissue necrotizing infections. AB103 binds to co-stimulatory receptor CD-28, which modulates immune response. It significantly diminishes the acute inflammatory response leading to tissue and organ damage. This drug is not yet released for use and is currently in clinical development phase 3.

19.4 Potentiators of Antibiotic Activity

Resistance to antibiotics can be classified as intrinsic and acquired resistance. Intrinsic resistance is caused by the lack of target in the bacterial cell, causing some bacteria naturally not to respond to an antibiotic. However acquired resistance refers to the ability of the bacteria to acquire antibiotic resistance through mutations or acquisition of foreign DNA encoding for resistance determinants. The main mechanisms of resistance to antibiotics are (i) production of enzyme that inactivates the antibiotic by hydrolysis or transfer of a chemical group, causing steroid; (ii) overexpression of efflux pumps, which eliminate a wide variety of compounds out of the cell; (iii) reduction or elimination of cellular permeability through changes in the outer membrane permeability, which decreased the effective entry of the antibiotic; (iv) modification of the microbial target by genetic mutation or posttranslational modification of the target; and (v) biofilm formation.

These potentiators of antibiotic activity can be applied either by reversing resistance mechanisms or by sensitizing intrinsic resistant strains. The approaches used to combat MDR include antibiotic–antibiotic combinations (synergistic antimicrobial combinations – potentiators of antibiotic activity) and pairing of antibiotics with nonantibiotics (adjuvants are active molecules that in combination with antibiotics enhance the antimicrobial activity of nonantibiotics).

19.4.1 Antibiotic–Antibiotic Combinations

Antibiotic combination therapy involving the use of two, three, or even four different antibiotics – multiple antibiotic regimens – is routinely used and often is the only available treatment option for infections caused by MDR organisms. This combination therapy can be divided into (i) antibiotic that targets in different pathways, (ii) antibiotics with different targets in the same pathway, and (iii) antibiotics that target the same pathway in different ways.

19.4.1.1 Targets in Different Pathways

One example of this strategy is the combination therapy used for the treatment of MTb infections, which is a combination of four drugs, namely, isoniazid, ethambutol, and pyrazinamide, which target the cell wall biosynthesis, and rifampicin which target mRNA synthesis.

Other examples of combination therapy that inhibits targets in different pathways are those therapies used to treat MDR Gram-negative infections such as MDR *Acinetobacter* spp., which is often combination of colistin (a cationic polypeptide antibiotic) with aminoglycoside, ampicillin/sulbactam, a carbapenem, or rifampin. Colistin/tigecycline and an aminoglycoside, a carbapenem, colistin, fosfomycin, rifampin, or tigecycline have been successful against carbapenemase-producing *Enterobacteriaceae*. A combination therapy of a broad-spectrum β-lactam and an aminoglycoside or a fluoroquinolone has been used for suspected Gram-negative sepsis and severe infections caused by *Pseudomonas* spp.

19.4.1.2 Different Targets in the Same Pathway

A well-known example of combination therapy that inhibits different targets in the same pathway is the combination of sulfamethoxazole and trimethoprim, which inhibit successive steps in the folic acid biosynthetic pathway. Another example is the targeting of teichoic acid synthesis in Gram-positive bacteria by tunicamycin (drug produced by *Streptomyces lysosuperificus*), which exhibits synergism with β-lactam against *S. aureus*. Tunicamycin inhibits *N*-acetylglucosamine-1-phosphate transferase (TarO), the first enzyme in the pathway of wall teichoic acid biosynthesis. The antiplatelet drug ticlopidine (Ticlid), a nonantibiotic that is a TarO inhibitor, has the ability to potentiate the activity of the cefuroxime against MRSA strains.

19.4.1.3 Same Target in Different Ways

A very interesting example of combination antibiotic therapy that inhibits the same target in different ways is the use of streptogramins, natural cyclic peptides produced by a number of *Streptomyces* species. These peptides represent a unique class of antibacterial that has two structurally distinct macrocyclic components: group A streptogramins (macrolactones), such as virginiamycin

M_1, pristinamycin II_A, pristinamycin II_B, and pristinamycin II_B derivative (dalfopristin), and group B streptogramins (cyclic hexadepsipeptides) with virginiamycin S and pristinamycins I_A, I_B, and I_C as the principal compounds. In addition, quinupristin is a pristinamycin I_A derivative. Both macrocyclic components inhibit bacterial protein synthesis at the level of the ribosome and have bacteriostatic activity and bactericidal activity when combined.

19.4.2 Pairing of Antibiotic with Nonantibiotic

The pairing of antibiotic with non-antibiotic adjuvants can restore or enhance the activity of older antibiotic, or also extend the spectrum of activity. Nonantibiotic adjuvants can function either by (i) affecting a vital physiological bacterial function; (ii) inhibition of antibiotic resistance elements; (iii) enhancement of the uptake of the antibiotic through the bacterial membrane; (iv) inhibition of efflux pumps; and (v) changing the physiology of resistant cells (dispersal of biofilms).

19.4.2.1 Affecting a Vital Physiological Bacterial Function

Molecules that alter bacterial cell shape by blocking cytoskeleton proteins and/or peptidoglycan biosynthesis and that act synergistically with the antibiotic have been also identified. Compounds 1, A22, pivmecillinam, and echinomycin are shown to be synergistic with novobiocin, including inhibitors of the bacterial cytoskeleton protein MreB and/or cell wall biosynthesis enzymes, in *Escherichia coli.*

19.4.2.2 Inhibition of Antibiotic Resistance Elements

One of the most successful and clinically used strategies has been the combination of a β-lactam antibiotic (inhibit to cell wall biosynthesis) with a β-lactamase inhibitor adjuvant (clavulanic acid, tazobactam, and sulbactam), of which the classical example is Augmentin. The β-lactamase inhibitor enhances the action of the β-lactam antibiotic by inhibiting the β-lactamase activity, thus restoring the antibiotic activity against β-lactamase-producing pathogens. Augmentin was the best-selling antibiotic in 2001, demonstrating the effectiveness of the approach of combining an antibiotic and adjuvant in clinical settings.

Previous reports have shown that the novel bicyclic penem β-lactamase inhibitors (BLI-489) conferred activity and efficacy as an inhibitor of class A including extended-spectrum β-lactamases (ESBLs), class D, and class C β-lactamase enzymes, while the tricyclic carbapenem inhibitor LK-157 has also shown promising activity against various ESBLs.

NXL104 (avibactam) is a broad-spectrum non-β-lactam β-lactamase inhibitor that has been shown to restore cephalosporin susceptibility to a large number of ESBL-producing *E. coli* and *Klebsiella pneumoniae* strains. Avibactam in association with ceftazidime (AVYCAZ®) has demonstrated efficacy in animal models and in phase III clinical trials for the treatment of hospital-acquired bacterial pneumonia and ventilator-associated bacterial pneumonia caused by Gram-negative bacteria, such as *E. coli, K. pneumoniae, Serratia marcescens, Enterobacter cloacae, Proteus mirabilis, P. aeruginosa,* and *Haemophilus.* Like other β-lactamase inhibitors, avibactam covalently binds the enzyme but unlike β-lactam-derived inhibitors is not susceptible to hydrolysis once bound to the enzyme (which would regenerate the active enzyme and destroy the activity of the released inhibitor). The association of adjuvant cilastatin and imipenem (a β-lactam antibiotic) is another successful example. Cilastatin inhibits the action of dehydropeptidase, an enzyme that degrades the imipenem, prolonging its antibacterial effect.

However, carbapenem-hydrolyzing oxacillinases (CHDLs) and metallo-β-lactamases (MBLs) – including the New Delhi MBL (NDM-1) – are β-lactamases not inhibited by some β-lactamase inhibitor adjuvant such as clavulanic acid and other currently available inhibitors such as tazobactam. However, currently at different stages of development, there are several new β-lactamase inhibitors that are active against these classes of β-lactamases. The malonate derivative ME1071 and inhibitor cocktail BAL3036725 have shown promise in restoring the activity of β-lactam antibiotics against MBL-producing bacterial strains. BAL30367 is a triple combination of the siderophore monobactam BAL19764, a bridged monobactam that inhibits class C β-lactamases, and clavulanic acid and has shown good *in vitro* activity against MBL-producing *Enterobacteriaceae.* ME107 has been shown to inhibit the MBLs IMP-1(imipenem-resistant) and VIM-2 (Verona integron-encoded MBL) and significantly enhances the activity of the carbapenem biapenem against *P. aeruginosa.*

Non-β-lactams, such as chrysin, morin, quercetin, luteolin, hesperetin, naringenin, catechin, hyperforin, octahydrohyperforin, and many others, which are nontoxic or have low toxicity, have shown efficient antibacterial activities against both Gram-negative and Gram-positive bacteria.

19.4.2.3 Enhancement of the Uptake of the Antibiotic Through the Bacterial Membrane

Gram-negative bacteria have in their composition an outer membrane that is characterized as the first line of defense against toxic compounds. The outer membrane contains porins and LPS. It also provides a certain barrier to some antibiotics (penicillin) and digestive enzymes (lysozyme), detergents, and heavy metals.

The porins located in the outer membrane control the passage of large nutrients and other substances into the bacterium, which are proteins capable of forming channels, which allows the passive diffusion of hydrophilic solutes through the outer membrane. Potentiator molecules that increase permeability efficiency are desirable. Recently, SPR741, a cationic peptide derived from polymyxin B, was used to permeabilize the outer membrane, allowing antibiotics that would otherwise be excluded access to their targets. The combination of SPR741 and conventional antibiotics used to treat clinical *E. coli, K. pneumoniae*, and *Acinetobacter baumannii* (azithromycin, clarithromycin, erythromycin, fusidic acid, mupirocin, retapamulin, rifampin, and telithromycin) reduced the MIC of these antibiotics against these bacteria and MDR strains.

Another cationic polypeptide antibiotic, colistin – also known as polymyxin E – interferes with LPS and permeabilizes the outer membrane – and at lower concentrations this has been used as adjuvant and has shown to improve the activity of rifampicin and vancomycin against Gram-negative pathogens.

By targeting the bacterial membrane and potentiating the activity of conventional antibiotics, the Q*n*-prAP and QCybuAP macromolecules have demonstrated *in vitro* and *in vivo* experiments, a novel approach to treating Gram-negative bacterial infections (*E. coli, A. baumannii*, and *K. pneumoniae*). Antibacterial activity of conventional antibiotics, alone and in combination with the macromolecules, was also evaluated. At very low concentrations, both macromolecules disrupted the bacterial membrane of *E. coli* antibiotic-tolerant cells, indicated by reduced membrane potential and increased membrane permeability.

19.4.2.4 Inhibit Efflux Pumps

Efflux pumps are transport proteins involved in the extrusion of toxic substrates, including relevant antibiotics, from within cells into the external environment.

Molecules that prevent the efflux pump of bacteria are also desirable adjuvant. Several families of efflux pump inhibitors have been described. *In vitro* efficacies of SPR741 – a cationic peptides derived from polymyxin B – showed to potentiate antibiotics that are substrates of AcrAB-TolC efflux systems in Gram-negative bacteria.

Phe-Arg-β-naphthylamide (PAβN), also called MC-207,110, was able to inhibit MexAB-OprM, MexCD-OprJ, MexEF-OprN, and MexXYOprM (four clinically relevant pumps) in *P. aeruginosa*, as well as similar pumps in other MDR Gram-negative bacteria.

However, the most promising strategy is the use of substrate analogues that compete with the antibiotic for the pump. Reserpine, a mammalian MDR pump inhibitor, demonstrated its potential in suppressing the emergence of resistance to ciprofloxacin in *S. aureus* and *Streptococcus pneumoniae.*

19.4.2.5 Changing the Physiology of Resistant Cells

The ability to form biofilms is a universal attribute of bacteria. Bacteria within the biofilms are metabolically dormant, rendering conventional antibiotics ineffective. Adjuvants can enhance antibiotic potency by changing the physiology of resistant cells, for example, by disrupting the bacterial biofilm. Mixtures of D-amino acids have been showed to disperse biofilm of Gram-positive and Gram-negative bacteria. In addition, the combination of antibiotic and an antibiofilm exopolysaccharides is also a promising strategy to improve the antimicrobial activity of antibiotics. The *Qn*-prAP and QCybuAP molecules were able to disrupt *in vitro* the *E. coli* and *A. baumannii* biofilms. In addition, when macromolecules were associated with erythromycin, the biofilm disruption was more effective. In addition, *Qn*-prAP and QCybuAP, alone and combined with erythromycin, successfully prevented replication and killed the planktonic cells.

19.5 Bacteriophages

Bacteriophages, also called phages, are viruses that infect and kill bacteria with high efficacy. They were discovered independently by Frederick W. Twort in Great Britain (1915) and Félix d'Hérelle in France (1917). D'hérelle coined the term *bacteriophage*, meaning "bacteria eater," to describe the agent's bactericidal ability. Phages are obligate intracellular parasites that replicate using bacterial machinery. Bacteriophages attach to specific host receptor on the surface of bacteria – wherein every bacterium is likely to have their own specific phage – and inject their genetic material into the bacterial cell, thus possessing the ability to change the composition of microbial populations.

Like all viruses, phages are simple organisms that consist of a core of genetic material (DNA or RNA) surrounded by a protein capsid. The genetic material has different sizes, conformation (circular, linear, or segmented), and structure (dsDNA, ssDNA, dsRNA, or ssRNA). There are three basic structural forms of phage: an icosahedral (20-sided) head with a tail, an icosahedral head without a tail, and a filamentous form. Phages are ubiquitous in the biosphere and can be found living in higher organisms, mostly in digestive tract, vagina, respiratory and oral tract, skin, and mucosal epithelium, forming the phageome. It is believed that they have a significant role in human homeostasis, for example, gut phages are important for ensuring the balance of microbiota, preventing the dysbiosis.

19.5.1 Life Cycles of Bacteriophages

After phages attach to a bacterium and insert its genetic material into the cell, they follow one of two life cycles, lytic (virulent) or lysogenic (temperate). In the *lytic cycle*, the virus inserts the genetic material into the host cell, and requiring the machinery of the host cell to make lots of new phage particles than then destroy, or lyse the cell, releasing numerous new phages particles. On the other hand, in the *lysogenic cycle*, the phage inserts the genetic material into the host cell (either as a free plasmid or integrated into the chromosome, now called a prophage) and replicate with it as a unit without destroying the cell. Under certain stressful conditions, including the presence of some antibiotics, lysogenic phages can be induced to follow a lytic cycle.

19.5.2 Bacteriophage Therapy

Phage therapy, which refers to the utilization of phages to treat bacterial infections, has been around for almost a century. Phage therapy are considered safety, since they non-interacts with body tissues and non-target microbiota, showing minimal impact on the normal microbiota.

Félix d'Hérelle was the first to apply phages as a therapy to successfully treat child with severe dysentery caused by *Shigella dysenteriae*, and at the beginning of the twentieth century, he proposed the use of bacteriophages for the therapy of human and animal bacterial infections. In 1923, d'Hérelle and his assistant George Eliava founded the Eliava Institute. The institute played an important role in the elaboration of novel biological preparations and manufacturing products against almost all major bacterial and viral diseases, such as anthrax, rubies, TB, brucellosis, salmonellosis, dysentery, etc. The old Soviet literature indicates that phage therapy was used extensively to treat a wide range of bacterial infections in the areas of dermatology, urology, stomatology, pediatrics, and surgery. These approaches were not widely accepted in the West. After the discovery of antibiotics in the 1940s and the widespread use of penicillin during World War II, phages were reduced to being used only in Eastern countries, which had no access to antibiotics.

Due to the specificity of the phage, confers on them an advance over antibiotics and can be useful in this battle, and a wide variety of phage therapies has been used to treat. Lytic phages are preferred alternatives in phage therapies when compared with lysogenic phages for the following reasons: lytic phages will destroy their host cell, while lysogenic phages can transfer virulence and resistance genes due their life cycle.

One of the main applications of the phage therapies has been to eliminate pathogenic bacteria involved in infectious diseases in humans. Moreover, with the increase of MDR strains, there has been increased interest in alternative therapy, and phage therapy has been shown to be an effective strategy to prevent and control these strains, being employed alone or in cocktails or as supplement to conventional antibiotic therapies. Recently, *in vivo* and *in vitro* experiments showed the efficacy of the phage cocktail to control MDR *A. baumannii*. Data from Hirszfeld Institute in Poland showed promising results, with 40% rate of good response of phage therapy against several MDR bacteria.

To control the ESKAPE (*Enterococcus faecium, S. aureus, K. pneumoniae, A. baumannii, P. aeruginosa*, and *Enterobacter* species) pathogens associated to nosocomial infections around the world, the phage therapy showed to be efficacy and safety. In a murine model of *P. aeruginosa* chronic lung infections, PELP20 was able to penetrate and kill bacteria within a biofilm-associated cystic-fibrosis lung-like environment. In a mice model, intraperitoneal injection of bacteriophage efficiently rescued bacteremia caused by imipenem-resistant *Pseudomonas*. A bacteriophage in the form of eye-drop showed high efficacy against *Pseudomonas* keratitis.

Numerous virulent phages of *Staphylococcus* have been described to have the ability to kill different *S. aureus* strains – such as phage f812. Phage Sb-1 was used to successfully treat patients in the United States with diabetic foot ulcers infected with MRSA. In the food industry, phages have been employed to decontaminate food by removing *E. coli, Salmonella*, or *Listeria* and played important role in food production and cattle raising. They are used to control foodborne diseases associated with *Salmonella* (salmonellosis), *Campylobacter* (campylobacteriosis in poultry), *Listeria monocytogenes*, or *E. coli*. Phages have been used to ensure food safety since they allow removal of bacterial infections in animals and thus prevent consumption of contaminated food. Phage CEVI successfully reduced the pathogenic *E. coli* strain in sheep.

19.5.3 Phage Enzymes

Phages produce several enzymes, for example, the holin and endolysin (lysins), used by most of the lytic phage to destroy the host cell. Endolysins, which degrade the peptidoglycan in the cell wall, are a new class of antimicrobials for treatment of infections caused by MDR strains due to their rapid actions, low evidence of resistance development, and cytotoxicity to mammalians cells. This activity is particularly against Gram-positive bacteria; since the outer membrane of Gram-negative bacteria provides protection to the peptidoglycan layer. Several endolysins have showed bactericidal activity *in vitro* against *Enterococcus faecalis* and *E. faecium* including vancomycin-resistant strains. The endolysin PlySs2 from a phage infecting *Streptococcus suis* exhibited activity against vancomycin-intermediate *S. aureus*, MRSA, *Staphylococcus epidermidis*, different species of *Streptococcus*, and *Listeria* spp. The phage lysine is used to treat infection caused by *B. anthracis*. To improve efficacy against Gram-negative bacteria, researchers have been looking for endolysins with the natural ability to degrade Gram-negative bacteria; in addition the combination of lysins with other agents has been shown to destabilize the outer membrane.

Holins are commonly small proteins that elicit a variety of effects and serve various functions. Their function involves bacterial cell lysis, a process used to form open or tightly sealed channels in the host bacterial cell membrane, releasing an autolysin and ultimately, following lysis of the cell, releasing the phage progeny. The combined use of holin and endolysin may be an effective strategy to cure infections and prevent bacterial resistance to antibiotics. Holin alone, without endolysins, compromises the cell membrane, but is unable to destabilize the cell wall. Thus, peptidoglycan destruction, mediated by the holin-exported endolysin, is necessary for complete cell lysis.

19.5.4 Concerns About the Application of Phage to Treat Bacteria

There are some major concerns about phage therapy, phage selection, preparation, and storage. The emergence of bacterial resistance to phage is another concern of phage therapy. Although apparently innocuous, to defend the genome against parasitic DNA and to maintain fidelity of the genomes in stable ecologies, a mechanism known as clustered regularly interspaced short palindromic repeats (CRISPR) found in a wide range of bacteria is a crucial component of the immune systems of bacteria. If a viral infection threatens a bacterial cell, the CRISPR system can prevent the attack by destroying the genome of the invading virus. The CRISPR systems provide a type of defense in prokaryotes, conferring resistance to plasmid uptake and phage infections, and a barrier to horizontal gene transfer. This immunity depends on the presence of specific target-derived spacer sequences, the intervening repeat palindromes short and highly conserved, and nuclease activity encoded by the *cas* genes. Since the CRISPR system is reactive to the environment, it might play a critical role in the adaptation of the host to its surroundings and explain the persistence of particular bacterial strains in ecosystems where phages are present.

19.6 Therapy with Essential Oils

Essential oils are biologically active organic compounds produced as secondary metabolites in plants and are an intermediate or final product, produced by plants, being synthesized in different plant organs, such as sprouts, flowers, leaves, husks, branches, seeds, fruits, roots, wood, or barks. They are complex hydrophobic and volatile liquids containing alkaloids, phenols and polyphenols, flavonoids, quinones, tannins, coumarins, terpenes, lecithins, polypeptides, and saponins synthesized by different metabolic routes. They are obtained through water/steam distillation, hydrodistillation, cold pressing, extraction by solvents, and supercritical fluid extraction (clean, nontoxic, and non-residual technology producing high-quality products) often used in pharmaceutical, food, and cosmetic industry.

Essential oils have distinct biological properties, such as anti-inflammatory, soothing and sedative, digestive, antimicrobial, antiviral, antiparasitic, antioxidative, and cytotoxic properties. They have been used for thousands of years for their curative potential. Diseases such as respiratory and intestinal infections and wounds were previously treated with essential oils. In Eastern and Western civilizations, knowledge on the benefits and traditions were conveyed by healing masters to the apprentices. Egyptians believed that essential oils had a surprising value for medicinal, spiritual, and therapeutic treatment. These beliefs were extremely respected in ancient Egypt, and people from royalty used essential oils, cedar wood, incense, and myrrh in their burial.

In the last decades, these compounds with biological properties produced by a broad range of plants have been pointed out as an alternative to control microorganisms. While each antibiotic class inhibits a specific cell activity, essential oils have a relatively broad activity spectrum, with mechanisms of action related to membrane rupture, cell membrane permeability related to cell surface, and penetration, resulting in cell lysis and consequently bacterial death. Plant-based essential oils may act synergistically with conventional antimicrobials.

When using essential oils as a therapy, it is estimated that about 62% are used to treat bacterial, fungal, or viral infections, followed by 20% to treat dermatitis, eczema, and lupus, with the remaining 18% corresponding to cosmetic use. A number of works have been reported on antimicrobial and neuroprotective activities of dry leaves such as oregano (*Origanum vulgare*) and lavender leaves and flowers (*Lavandula officinalis*). In a study performed with *Eucalyptus globulus* fruit assessing anti-oxidative activity, essential oil inhibited the growth of pathogenic strains such as *P. aeruginosa*, *S. aureus*, *Bacillus subtilis*, *E. coli*, and *Listeria innocua* when compared with tetracycline and gentamycin antibiotics. Recent papers in which researchers assessed the antimicrobial activity of essential oils extracted from *Mentha citrata*, they checked the inhibition of eight genus of Gram-positive bacteria as well as five Gram-negative bacteria. Essential oil of *M. citrata* was chemically characterized by GC-MS. The results showed that their main components are linalyl acetate (26.69%) and D-linalyl (24%) and were shown to be promising to control resistant bacteria.

Among resistance problems, biofilm-related infections in hospital settings are currently known due to their microbial recalcitrance. Essential oils and nanoparticles have been shown to be effective in preventing microbial biofilms. Nanoparticles combined with essential oils have a significant antimicrobial potential against multiresistant pathogens due to an increase in chemical stability and solubility, decreased fast evaporation, and minimized active essential oil degradation. The application of encapsulated essential oils also supports their controlled and sustained release, which improves their bioavailability and efficacy against multiresistant pathogens (Table 19.1).

Essential oils obtained from medicinal plants represent an alternative for prospection of new drugs and are sources for inclusion of new therapies in the pharmaceutical industry. The use of essential oils and antibiotics as a combined therapy has been shown to be effective with significant decreases in antibiotic dosage so as to eradicate MDR microorganisms. Combination therapy among different antibiotic classes and essential oils may expand antimicrobial spectrum, reducing the emergence of resistant variants.

Table 19.1 Mechanism of action and dermatological use of some essential oils.

Plant of origin	Common name	Microorganism target	Mechanism of action/dermatological use
Abies balsamea	Balsam		Burns, cracks, cuts, eczema, rashes, sores, and wounds
Acacia dealbata	Mimosa		Antiseptic, general care, oily conditions, and nourisher
Allium sativum	Garlic	*Escherichia coli*	Induced leakage
Anthemis nobilis	Roman chamomile		Abscesses, acne, allergies, antiseptic, blisters, boils, burns, cleanser, cuts, dermatitis, eczema, foot blisters, general care, herpes, inflammation, insect bites and stings, nappy rash, nourisher, problematic skin, pruritus, psoriasis, rashes, rosacea, sores, sunburn, ulcers, and wounds
Cinnamon	True cinnamon	*Staphylococcus aureus*	Disruption of cell membrane
Cinnamomum camphora	Camphor (white)		Acne, burns, inflammation, oily conditions, spots, and ulcers
Coriaria nepalensis		*C. nepalensis*	Inhibition of ergosterol biosynthesis and
		Candida isolates	disruption of membrane integrity
Coriandrum sativum	Coriander		Used to prevent the growth of odor-causing bacteria
Cuminum cyminum	Cumin	*Bacillus subtilis*	
		Bacillus cereus	Changes in cytoplasm
		E. coli	
Curcuma longa	Turmeric	*Aspergillus flavus*	Inhibition of ergosterol biosynthesis
		E. coli	

(*Continued*)

Table 19.1 (Continued)

Plant of origin	Common name	Microorganism target	Mechanism of action/dermatological use
Dipterocarpus gracilis		*Proteus mirabilis*	Disruption of cell membrane
		B. cereus	
		E. coli	
Foeniculum vulgare	Fennel	*Shigella dysenteriae*	Loss of membrane integrity
Forsythia koreana		*Foodborne and other pathogenic bacteria*	Loss of membrane integrity and increased permeability
Litsea cubeba		*E. coli*	Destruction of outer and inner membrane
Mentha longifolia	Mentha	*Micrococcus luteus*	Cell wall damage
		Salmonella typhimurium	
Ocimum gratissimum	Clove basil	*Pseudomonas aeruginosa*	Permeabilized membrane
		S. aureus	
Origanum vulgare	Oregano	*S. aureus*	Permeabilized membrane
		P. aeruginosa	
		E. coli	
Piper nigrum	Black pepper	*E. coli*	Cell becomes pitted and shriveled; leakage of intercellular material
Syzygium aromaticum	Clove		Acne, antiseptic, athlete's foot, burns, cuts, cold sores, fungal infections, lupus, sores, septic ulcers, and wounds
Tagetes minuta	Tagetes		Bacterial infections, fungal infections, inflammation, and viral infections (verrucae and warts)

19.7 Microbiota-Based Therapy

Human microbiota has a unique composition of bacteria, fungi, bacteriophages, and virus and interacts dynamically in a balanced manner; however, the disruption of homeostasis may cause some diseases. The composition of these microorganisms may be beneficial and/or probiotic and pathogenic, and some metabolic activities take place jointly both by the microorganism and the host.

The inclusion of high-throughput sequence as a strategy to study microbial communities' profile, environmental factors, and gene prediction was crucial to conclude that microbiota plays an important role in our physiology, as the protection against some persisting infections, in the metabolism of drugs, vitamin synthesis, nutrition, and therapies to treat cancer. The unbalance of communities with a decrease in beneficial microorganisms and the excessive increase of pathogenic bacteria, known as dysbiosis, may cause diseases such as diabetes, obesity, cardiovascular diseases, and neurodegenerative diseases. Recent evidences suggest that dysbiosis in inpatients is attributed to several factors, including the excessive use of antibiotics. Exposure to antibiotics changes microbiota composition, and resistant opportunistic microorganisms are selected. The progressive increase of antibiotic-resistant genes with consequences such as increased morbidity and mortality leads to the quest for nontraditional therapies as alternatives for control of microorganism.

Microbiota-based therapy has been shown to be favorable treatment of gut diseases, but still faces a challenge, which is the functional understanding of a complex system such as the gut microbiota. For a better understanding of microorganism needs individuality and the interaction among communities, some studies are being performed by changing community members randomly in this dynamic system, but some responses are not being fully clarified. Due to the need of better results for problematic issues such as bacterial resistance, it would be important to take into consideration interventions such as microbiota modulation and stool microbiota transplant so as to help to decrease the use of antibiotics and reestablish a balanced microbiota.

19.7.1 Microbiota Modulation

A therapeutic strategy is microbiota modulation, which consists in changing microbiota composition by giving probiotics and prebiotics. Microbiota modulation has predictable outcomes in regard to response, without side effects and with desirable effects for the host. Symbionts have positive synergistic effects arising from probiotic and prebiotic dosing, and the strain choice is crucial for an excellent outcome.

19.7.1.1 Probiotics

Probiotics emerged in the end of nineteenth century and at the start of twentieth century and were consumed by centuries in fermented food aiming to restore beneficial microorganisms in the intestine in order to provide functional balance. Probiotics are specific bacteria that, when given at proper amounts, may confer benefits for the host, which may include prevention of gastrointestinal disorders, microbiota normalization, decreased bowel permeability, weight loss, immune function regulation, and treatment of irritable bowel syndrome, allergic diseases, and atopic dermatitis.

Recent works have shown that the administration of probiotics containing *Saccharomyces boulardii* and *Lactobacillus* spp. strains promotes the decreased incidence of CDI. *Lactobacillus salivarius* may prevent the growth of *L. monocytogenes*, and *Lactobacillus jensenii* may decrease infections caused by *Gardnerella vaginalis*, *Candida albicans*, and *E. coli*.

19.7.1.2 Prebiotics

The term prebiotic was defined in 1995 as nondigestible food components stimulating the selection of microbial populations desirable in the gut. The intake of food non-hydrolyzable in the upper gastrointestinal tract promotes the growth of certain bowel bacterial, particularly bifidobacteria and lactobacilli. Fibers are composed of poorly absorbed carbohydrates in the gastrointestinal tract, such as inulin and oligofructose. Fibers serve as a substrate for beneficial microorganisms to acidify the colon, favoring a favorable environment for a symbiotic relationship.

19.7.2 Stool Microbiota Transplant

A nonconventional alternative with an excellent potential to treat bacterial infections is stool microbiota transplant, scientifically reported for the first time in 1958. However, there are reports that Chinese medicine already prescribed stool medication for about 3000 years. Stool microbiota transplant has been used to treat patients with diarrhea related to pseudomembranous colitis caused by *C. difficile*. *Clostridium difficile* causes injuries in the colon by producing enterotoxin A, cytotoxin B, and binary toxin, causing diarrhea, fever, and leukocytosis. This bacterium has a high resistance and is often acquired by inpatients after being given antibiotics or immunosuppression periods. Studies performed had obtained results and concluded that the treatment with stool microbiota transplant for recurrent *C. difficile* colitis is more effective than oral vancomycin dosing.

In addition to treatment for bacterial infections, stool microbiota transplant is further used for a broad range of disorders, including Parkinson's disease, fibromyalgia, chronic fatigue syndrome, multiple sclerosis, obesity, insulin resistance, metabolic syndrome, and autism. The principle of stool microbiota transplant is the inoculation of bacteria in stools from healthy donors in patients with pathogenic native microorganisms. The most common stool microbiota transplant consists in the direct stool material transfer with minimum processing (first generation) from a healthy donor for a patient. In accordance with the success of outcomes obtained for the treatment of diseases caused by microorganisms, stool microbiota transplants were started from donor sample banks (second generation) and later the fully characterized microbial consortium (third generation) including culture in a laboratory.

In conclusion, nontraditional antimicrobial therapy is an alternative for the treatment and prevention of bacterial infections, slowing down the development of antibiotic resistance not only in pathogenic microorganisms but also in bacteria considered commensal, as in the case of *E. coli*. Infections are currently treated solely with antibiotics and abundantly in hospitals, and the implementation of this therapy and other types thereof will require significant changes in the pharmaceutical industry, and regulatory agencies.

Further Reading

Antibodies

Adawi, A., Bisignano, C., Genovese, T. et al. (2012). In vitro and in vivo properties of a fully human Iggl monoclonal antibody that combats multidrug resistant *Pseudomonas Aeruginosa*. *Int. J. Mol. Med.* 30 (3): 455–464.

François, B., Barraud, O., and Jafri, H.S. (2017). Antibody-based therapy to combat *Staphylococcus aureus* infections. *Clin. Microbiol. Infect.* 23 (4): 219–221.

Hua, L., Hilliard, J.J., Shi, Y. et al. (2014). Assessment of an anti-alpha-toxin monoclonal antibody for prevention and treatment of *Staphylococcus aureus*-induced pneumonia. *Antimicrob. Agents Chemother.* 58 (2): 1108–1117.

Kummerfeldt, C.E. (2014). Raxibacumab: potential role in the treatment of inhalational anthrax. *Infect. Drug Resist.* 7: 101–109.

Migone, T.-S., Mani Subramanian, G., Zhong, J. et al. (2009). Raxibacumab for the treatment of inhalational anthrax. *N. Engl. J. Med.* 361 (2): 135–144.

Navalkele, B.D. and Chopra, T. (2018). Bezlotoxumab: an emerging monoclonal antibody therapy for prevention of recurrent *Clostridium difficile* infection. *Biol. Targets Ther.* 12: 11.

Patel, A., DiGiandomenico, A., Keller, A.E. et al. (2017). An engineered bispecific DNA-encoded IGG antibody protects against *Pseudomonas Aeruginosa* in a pneumonia challenge model. *Nat. Commun.* 8 (1): 637.

Secher, T., Fas, S., Fauconnier, L. et al. (2013). The anti-*Pseudomonas Aeruginosa* antibody panobacumab is efficacious on acute pneumonia in neutropenic mice and has additive effects with meropenem. *PLoS One* 8 (9): e73396.

Immunomodulators

Feigman, M.S., Kim, S., Pidgeon, S.E. et al. (2018). Synthetic immunotherapeutics against Gram-negative pathogens. *Cell Chem. Biol.* 25 (10): 1185.e5–94.e5.

Ingersoll, M.A. and Albert, M.L. (2013). From infection to immunotherapy: host immune responses to bacteria at the bladder mucosa. *Mucosal Immunol.* 6 (6): 1041.

Mahdi, L., Mahdi, N., Musafer, H. et al. (2018). Treatment strategy by lactoperoxidase and lactoferrin combination: immunomodulatory and antibacterial activity against multidrug-resistant *Acinetobacter baumannii*. *Microb. Pathog.* 114: 147–152.

Masihi, K.N. (2001). Fighting infection using immunomodulatory agents. *Expert. Opin. Biol. Ther.* 1 (4): 641–653.

Mily, A., Rekha, R.S., Mostafa Kamal, S.M. et al. (2015). Significant effects of oral phenylbutyrate and vitamin D3 adjunctive therapy in pulmonary tuberculosis: a randomized controlled trial. *PLoS One* 10 (9): e0138340.

Silva, O.N., De La Fuente-núñez, C., Haney, E.F. et al. (2016). An anti-infective synthetic peptide with dual antimicrobial and immunomodulatory activities. *Sci. Rep.* 6: 35465.

Syn, N.L., Teng, M.W.L., Mok, T.S.K., and Soo, R.A. (2017). De-novo and acquired resistance to immune checkpoint targeting. *Lancet Oncol.* 18 (12): e731–e741.

Bacteriophage Therapy

Domingo-Calap, P. and Delgado-Martinez, J. (2018). Bacteriophages: protagonists of a post-antibiotic era. *Antibiotics (Basel)* 7 (3): 1–16.

El Haddad, L., Harb, C.P., Gebara, M.A. et al. (2019). A systematic and critical review of bacteriophage therapy against multi-drug resistant eskape organisms in humans. *Clin. Infect. Dis.* 69 (1): 167–178.

Jamal, M., Smaus, B., Andleeb, S. et al. (2018). Bacteriophages: an overview of the control strategies against multiple bacterial infections in different fields. *J. Basic Microbiol.* 59 (2): 123–133.

Jasim, H.N., Hafidh, R.R., and Abdulamir, A.S. (2018). Formation of therapeutic phage cocktail and endolysin to highly multi-drug resistant *Acinetobacter baumannii*: in vitro and in vivo study. *Iran J. Basic Med. Sci.* 21 (11): 1100–1108.

Kakasis, A. and Panitsa, G. (2019). Bacteriophage therapy as an alternative treatment for human infections. a comprehensive review. *Int. J. Antimicrob. Agents* 53 (1): 16–21.

Loc-Carrillo, C. and Abedon, S.T. (2011). Pros and cons of phage therapy. *Bacteriophage* 1 (2): 111–114.

Morozova, V.V., Vlassov, V.V., and Tikunova, N.V. (2018). Applications of bacteriophages in the treatment of localized infections in humans. *Front. Microbiol.* 9: 1696.

Paule, A., Frezza, D., and Edeas, M. (2018). Microbiota and phage therapy: future challenges in medicine. *Med. Sci. (Basel)* 6 (4): pii: E86.

Torres-Barceló, C. (2018). The disparate effects of bacteriophages on antibiotic-resistant bacteria. *Emerg. Microbes Infect.* 7: 168.

Viertel, T.M., Ritter, K., and Horz, H.P. (2014). Viruses versus bacteria-novel approaches to phage therapy as a tool against multidrug-resistant pathogens. *J. Antimicrob. Chemother.* 69 (9): 2326–2336.

Waters, E.M., Neill, D.R., Kaman, B. et al. (2017). Phage therapy is highly effective against chronic lung infections with *Pseudomonas aeruginosa*. *Thorax* 72 (7): 666–667.

Potentiation of Antibiotics Activity

Bernal, P., Molina-Santiago, C., Daddaoua, A., and Llamas, M.A. (2013). Antibiotic adjuvants: identification and clinical use. *Microb. Biotechnol.* 6 (5): 445–449.

Corbett, D., Wise, A., Langley, T. et al. (2017). Potentiation of antibiotic activity by a novel cationic peptide: potency and spectrum of activity of Spr741. *Antimicrob. Agents Chemother.* 61 (8): pii: e00200-17.

González-Bello, C. (2017). Antibiotic adjuvants – a strategy to unlock bacterial resistance to antibiotics. *Bioorg. Med. Chem. Lett.* 27 (18): 4221–4228.

Kerantzas, C.A. and Jacobs, W.R. (2017). Origins of combination therapy for tuberculosis: lessons for future antimicrobial development and application. *mBio* 8 (2): e01586-16.

Mukherjee, S.K., Mandal, R.S., Das, S., and Mukherjee, M. (2018). Effect of non-B-lactams on stable variants of inhibitor-resistant tem β-lactamase in uropathogenic *Escherichia coli*: implication for alternative therapy. *J. Appl. Microbiol.* 124 (3): 667–681.

Pieren, M. and Tigges, M. (2012). Adjuvant strategies for potentiation of antibiotics to overcome antimicrobial resistance. *Curr. Opin. Pharmacol.* 12 (5): 551–555.

Uppu, D.S.S.M., Konai, M.M., Sarkar, P. et al. (2017). Membrane-active macromolecules kill antibiotic-tolerant bacteria and potentiate antibiotics towards Gram-negative bacteria. *PLoS One* 12 (8): e0183263.

Worthington, R.J. and Melander, C. (2013). Combination approaches to combat multidrug-resistant bacteria. *Trends Biotechnol.* 31 (3): 177–184.

Therapy with Essential Oils

Ahmad, A., Khan, A., Kumar, P. et al. (2011). Antifungal activity of *Coriaria nepalensis* essential oil by disrupting ergosterol biosynthesis and membrane integrity against candida. *Yeast* 28 (8): 611–617.

Ayaz, M., Sadiq, A., Junaid, M. et al. (2017). Neuroprotective and anti-aging potentials of essential oils from aromatic and medicinal plants. *Front. Aging Neurosci.* 9: 168.

Bannour, M., Aouadhi, C., Khalfaoui, H. et al. (2016). Barks essential oil, secondary metabolites and biological activities of four organs of Tunisian Calligonum Azel Maire. *Chem. Biodivers.* 13 (11): 1527–1536.

Bhardwaji, M., BR Singh, Sinha, D.K. et al. (2016). Potential of herbal drug and antibiotic combination therapy: areview approche to treat multidrug resistant bacteria. *Pharm. Anal. Acta* 7 (11): 523.

Boire, N.A., Riedel, S., and Parrish, N.M. (2013). Essential oils and future antibiotics: new weapons against emerging 'superbugs'. *J. Anc. Dis. Prev. Rem.* 1 (2): 1000105.

Chouhan, S., Sharma, K., and Guleria, S. (2017). Antimicrobial activity of some essential oils—present status and future perspectives. *Medicines* 4 (3): 58.

Diao, W.-R., Hu, Q.-P., Zhang, H., and Xu, J.-G. (2014). Chemical composition, antibacterial activity and mechanism of action of essential oil from seeds of fennel (*Foeniculum Vulgare* Mill.). *Food Control* 35 (1): 109–116.

Farrer-Halls, G. (2005). *The Aromatherapy Bible: The Definitive Guide to Using Essential Oils*. Sterling Publishing Company.

García-Beltrán, J.M. and Esteban, M.A. (2016). Properties and applications of plants of Origanum sp. genus. *SM. J. Biol.* 2: 1006.

Gupta, P.D. and Birdi, T. (2017). Development of botanicals to combat antibiotic resistance. *J. Ayurveda Integr. Med.* 8 (4): 266–275.

Hu, Y., Zhang, J., Kong, W. et al. (2017). Mechanisms of antifungal and anti-aflatoxigenic properties of essential oil derived from turmeric (*Curcuma Longa* L.) on *Aspergillus Flavus*. *Food Chem.* 220: 1–8.

Khorshidian, N., Yousefi, M., Khanniri, E., and Mortazavian, A.M. (2018). Potential application of essential oils as antimicrobial preservatives in cheese. *Innov. Food Sci. Emerg. Technol.* 45: 62–72.

Lambert, R.J.W., Skandamis, P.N., Coote, P.J., and Nychas, G.J. (2001). A study of the minimum inhibitory concentration and mode of action of oregano essential oil, thymol and carvacrol. *J. Appl. Microbiol.* 91 (3): 453–462.

Langeveld, W.T., Veldhuizen, E.J.A., and Burt, S.A. (2014). Synergy between essential oil components and antibiotics: a review. *Crit. Rev. Microbiol.* 40 (1): 76–94.

Lawless, J. (1997). *The Complete Illustrated Guide to Aromatherapy: A Practical Approach to the Use of Essential Oils for Health and Well-being*. Element.

Orchard, A. and van Vuuren, S. (2017). Commercial essential oils as potential antimicrobials to treat skin diseases. *Evid. Based Complement. Alternat. Med.* 2017: 4517971.

Patra, J.K. and Baek, K.-H. (2016). Antibacterial activity and action mechanism of the essential oil from Enteromorpha Linza L. against foodborne pathogenic bacteria. *Molecules* 21 (3): 388.

Rehman, R., Hanif, M.A., Mushtaq, Z., and Al-Sadi, A.M. (2016). Biosynthesis of essential oils in aromatic plants: a review. *Food Rev. Intl.* 32 (2): 117–160.

Rohraff, D. and Morgan, R. (2014). The evaluation of essential oils for antimicrobial activity. Student Summer Scholars 124.

Saad, N.Y., Muller, C.D., and Lobstein, A. (2013). Major bioactivities and mechanism of action of essential oils and their components. *Flavour Fragr. J.* 28 (5): 269–279.

Said, Z.B.-O.S., Haddadi-Guemghar, H., Boulekbache-Makhlouf, L. et al. (2016). Essential oils composition, antibacterial and antioxidant activities of hydrodistillated extract of *Eucalyptus globulus* fruits. *Ind. Crop. Prod.* 89: 167–175.

Vasireddy, L., Bingle, L.E.H., and Davies, M.S. (2018). Antimicrobial activity of essential oils against multidrug-resistant clinical isolates of the burkholderia cepacia complex. *PLoS One* 13 (8): e0201835.

Wink, M. (2012). Secondary metabolites from plants inhibiting abc transporters and reversing resistance of cancer cells and microbes to cytotoxic and antimicrobial agents. *Front. Microbiol.* 3: 130.

Yap, P.S.X., Lim, S.H.E., Hu, C.P., and Yiap, B.C. (2013). Combination of essential oils and antibiotics reduce antibiotic resistance in plasmid-conferred multidrug resistant bacteria. *Phytomedicine* 20 (8-9): 710–713.

Yap, P.S.X., Yang, S.K., Lai, K.S., and Lim, S.H.E. (2017). Essential oils: the ultimate solution to antimicrobial resistance in *Escherichia coli*. In: *Escherichia Coli-Recent Advances on Physiology, Pathogenesis and Biotechnological Applications.* InTech.

Yap, P.S.X., Yiap, B.C., Ping, H.C., and Lim, S.H.E.J. (2014). Essential oils, a new horizon in combating bacterial antibiotic resistance. *Open Microbiol. J.* 8: 6–14.

Microbiota-based Therapy

Cammarota, G., Ianiro, G., Tilg, H. et al. (2017). European consensus conference on faecal microbiota transplantation in clinical practice. *Gut* 66 (4): 569–580.

Choi, H.H. and Cho, Y.S. (2016). Fecal microbiota transplantation: current applications, effectiveness, and future perspectives. *Clin. Endosc.* 49 (3): 257–265.

Cuomo, A., Maina, G., Rosso, G. et al. (2018). The microbiome: a new target for research and treatment of schizophrenia and its resistant presentations a systematic literature search and review. *Front. Pharmacol.* 9: 1040.

Galloway-Pena, J., Brumlow, C., and Shelburne, S. (2017). Impact of the microbiota on bacterial infections during cancer treatment. *Trends Microbiol.* 25 (12): 992–1004.

Gupta, S., Allen-Vercoe, E., and Petrof, E.O. (2016). Fecal microbiota transplantation: in perspective. *Ther. Adv. Gastroenterol.* 9 (2): 229–239.

Marchesi, J.R. and Ravel, J. (2015). The vocabulary of microbiome research: a proposal. *Microbiome* 3: 31.

Markowiak, P. and Slizewska, K. (2017). Effects of probiotics, prebiotics, and synbiotics on human health. *Nutrients* 9 (9).

Mintz, M., Khair, S., Grewal, S. et al. (2018). Longitudinal microbiome analysis of single donor fecal microbiota transplantation in patients with recurrent *Clostridium difficile* infection and/or ulcerative colitis. *PLoS One* 13 (1): e0190997.

Petrosino, J.F. (2018). The microbiome in precision medicine: the way forward. *Genome Med.* 10 (1): 12.

Schmidt, T.S.B., Raes, J., and Bork, P. (2018). The human gut microbiome: from association to modulation. *Cell* 172 (6): 1198–1215.

Taur, Y. and Pamer, E.G. (2016). Microbiome mediation of infections in the cancer setting. *Genome Med.* 8 (1): 40.

Tse, B.N., Adalja, A.A., Houchens, C. et al. (2017). Challenges and opportunities of nontraditional approaches to treating bacterial infections. *Clin. Infect. Dis.* 65 (3): 495–500.

Walsh, C.J., Guinane, C.M., O'Toole, P.W., and Cotter, P.D. (2014). Beneficial modulation of the gut microbiota. *FEBS Lett.* 588 (22): 4120–4130.

Wischmeyer, P.E., McDonald, D., and Knight, R. (2016). Role of the microbiome, probiotics, and 'dysbiosis therapy' in critical illness. *Curr. Opin. Crit. Care* 22 (4): 347–353.

Young, V.B. (2017). The role of the microbiome in human health and disease: an introduction for clinicians. *BMJ* 356: j831.

20

Therapeutic Options for Treatment of Infections by Pathogenic Biofilms

Bruna de Oliveira Costa[1], *Osmar Nascimento Silva*[1], *and Octávio Luiz Franco*[1,2,3]

[1] *S-Inova Biotech, Programa de Pós-Graduação em Biotecnologia, Universidade Católica Dom Bosco, Campo Grande, MS, Brazil*
[2] *Centro de Análises Proteômicas e Bioquímicas, Programa de Pós-Graduação em Ciências Genômicas e Biotecnologia, Universidade Católica de Brasília, Brasília, DF, Brazil*
[3] *Faculdade de Medicina, Programa de Pós-Graduação em Patologia Molecular, Universidade de Brasília, Brasília, DF, Brazil*

20.1 Introduction

In the current context, there is an increase in the number of human infections associated to resistant microorganisms. Many of such infections have been associated with microorganisms that grow in biofilms (Donlan 2001a; Høiby et al. 2011; Bjarnsholt 2013). Biofilms can be defined as a set of microorganisms (including pathogens) associated in suspension or as a community of microorganisms adhered to a biotic or abiotic surface surrounded by a self-secreted polymer matrix composed of polysaccharides, mineral crystals, proteins, and extracellular DNA (Bjarnsholt et al. 2013). Biofilm matrix facilitates the development of a resistant and highly hydrated structure that allows the pathogenic microorganisms to survive the stress of the local microenvironment, favoring the dissemination and consequently the resurgence of infections (Kumar et al. 2017).

Microorganisms may form biofilms in any part of the body such as the oral cavity (teeth and gums) (Chandki et al. 2011), urogenital tract (Delcaru et al. 2016), upper (Pintucci et al. 2010) and lower (Pirrone et al. 2016) respiratory tract, and wounds (Percival et al. 2015), as well as in medical devices (Donlan 2001b) – catheters (Trautner and Darouiche 2004), endotracheal tubes (Souza et al. 2014), implants (Arciola et al. 2018), contact lenses (Kackar et al. 2017), pacemakers (Santos et al. 2011), heart valves (Litzler et al. 2007),

Antibiotic Drug Resistance, First Edition. Edited by José-Luis Capelo-Martínez and Gilberto Igrejas.

artificial joints (Gbejuade et al. 2015), and prosthesis (Silverstein and Donatucci 2003). The most common biofilm-forming pathogens are *Candida albicans, Enterococcus faecalis, Escherichia coli, Klebsiella pneumoniae, Mycobacterium* sp., *Staphylococcus aureus, Staphylococcus epidermidis, Streptococcus viridans, Proteus mirabilis,* and *Pseudomonas aeruginosa* (Donlan 2001b; Kumar et al. 2017).

Biofilm-related diseases are particularly chronic, slowly developing infections but are characterized as persistent and progressive due to the ability to resist the elements of the host organism's defense system and the effects of the available antimicrobial agents (Lewis 2007). It is believed that the chronic character of biofilm infections could be associated with a subpopulation of cells located inside them, known as "persistent" (Maisonneuve and Gerdes 2014). These cells have the ability to survive the prolonged action of antimicrobials, which allows their dispersion to other organ systems, acting as sources for new infections (Lewis 2012; Maisonneuve and Gerdes 2014).

The peculiarities described about biofilms represent a critical and alarming picture for the clinical community and for the medical industry, and considerable efforts are being made to identify new approaches that can establish the basis of anti-biofilm therapies that are efficiently better than existing treatments (Markowska et al. 2013). This chapter reports the current options for the treatment of infections caused by pathogenic biofilms, such as conventional antibiotics, as well as the alternatives developed for future anti-biofilm treatments based on scientific perspective.

Note: "Antibiotics" were here defined as pharmaceutically formulated and medically administered substances". Furthermore "Antimicrobials" was defined here as a class of substances that may or may not be regulated as drugs.

20.2 Antibiotic Therapy for the Treatment of Pathogenic Biofilms

20.2.1 Monotherapy

As described earlier, treatment options to combat pathogenic biofilm infections are limited. This can be attributed to the insufficient delivery of the desired concentration of the treatment agent to the target microbial cells within the biofilm since the microorganisms organized in biofilms have an architecture and an environment that provide for the increase of microbial resistance to the conventional antibiotics (Wu et al. 2015). In addition, the limited treatment of infections associated with pathogenic biofilms can also be attributed to the fact that antibiotics currently available for clinical treatment have been developed to combat infections caused by planktonic microorganisms rather than biofilms (Bjarnsholt et al. 2013; Hancock 2015).

In this context, in clinical practice, the most efficient conventional treatment for the control of chronic infections associated with pathogenic biofilms has been the use of antibiotic monotherapy in precocious, aggressive, and intensive way, that is, the administration of the antibiotic before the establishment infection for a long period (prophylaxis) (Wu et al. 2015). Antimicrobial prophylaxis in clinical practice can be defined as the use of antibiotics in patients who do not show signs or symptoms of infection, with the aim of preventing their onset. It is based on the idea that if the antibiotics can kill the bacteria and/or prevent their growth in established infections, they can also do it in the blood or in specific sites, avoiding the installation of an infectious process (Høiby et al. 2015).

In clinical practice, it is recommended to use high doses of the antibiotic vancomycin (1 mg.mL^{-1}), for Gram-positive bacteria, and gentamicin (2 mg l^{-1}), for Gram-negative bacteria, to prevent infections by pathogenic biofilms in catheters (Wu et al. 2015). As for orthopedic devices, there is evidence that prophylaxis with antibiotics such as gentamicin, tobramycin, and vancomycin may reduce the incidence of biofilm infections associated with the prosthesis (Johannsson et al. 2010; Marschall et al. 2013). In the case of colonization by *P. aeruginosa* at cystic fibrosis patients' lungs, intermittent antibiotic therapy is essential to prevent the establishment of a biofilm infection (Høiby et al. 2015).

In surgical procedures prophylaxis has also been widely used, for example, after heart surgery, collagen implants containing gentamicin have been used to close the sternum (sternotomy) of patients, in order to reduce the incidence of biofilm infection (Mishra et al. 2014). In the case of chronic wounds, debridement is recommended, with subsequent administration of an antibiotic, either by irrigation or instillation to prevent the formation of the pathogenic biofilm (Caputo et al. 2008).

Although the practice of prophylaxis has been progressively used, especially in high-risk patients, some points are still questionable as to the durability of its effect and especially its potential for favoring the resistance of microorganisms to the antibiotics available in the market (Lynch and Robertson 2008). In this sense, combined antibiotic therapy has been widely disseminated as an alternative to prevent or delay the onset of resistance.

20.2.2 Antibiotic Combination Therapy

In the treatment of biofilm-related infections, the combination therapy of antibiotics acquires great relevance. As the pathogenic biofilms present different structural areas and different metabolic states, the combination of some antibiotics appears as an effective alternative to fight the microbial cells of the interior of the biofilm located in the different structural layers with different metabolisms (Ciofu et al. 2017). For example, the combination of some antibiotic agents that target metabolically active strains such as ciprofloxacin, tetracycline, or β-lactams with other antibiotics that preferentially kill biofilm cells with low metabolic activity, such as colistin, provides a rational option to establishment combination therapy (Pamp et al. 2008; Ciofu et al. 2017).

In this perspective, studies have demonstrated efficient combinations for the treatment of infections caused by pathogenic biofilms. For example, at *in vitro* assays using pharmacokinetic/pharmacodynamic models of mature (susceptible and methicillin-resistant) *S. aureus* biofilms, Ruiz and co-workers (2010) demonstrated that the combination of the antibiotics daptomycin or moxifloxacin with clarithromycin was significantly effective against mature biofilms, whereas in isolation, those same antibiotics tested in high concentrations were not able to demonstrate a bactericidal activity against staphylococcal biofilms (Parra-Ruiz et al. 2010). Similarly, from an *in vitro* pharmacokinetic/pharmacodynamic model, the combination of linezolid antibiotics with daptomycin was superior to each agent alone, suggesting another therapeutic option for staphylococcal biofilms (Parra-Ruiz et al. 2012). In another study, the *in vitro* anti-biofilm activity of the antibiotics clarithromycin, cefazolin, and vancomycin isolated and combined against *S. aureus* biofilms formed on titanium devices was evaluated (Fujimura et al. 2008). As a result, individual antibiotics were not able to eradicate *S. aureus* biofilm; in contrast, the combination of clarithromycin with cefazolin or vancomycin was clearly efficient (Fujimura et al. 2008). Therefore, combined therapy seems to be the most efficient way to achieve the pathogenic biofilm eradication (Jacqueline and Caillon 2014).

Regarding other biofilm configurations, it has been found that antibiotic combinations may represent an ideal anti-biofilm strategy for use in patients with cystic fibrosis (Heijerman et al. 2009). In this sense, the combination of the antibiotics tobramycin and colistin proved to be more efficient than the respective antibiotics individually tested to *in vitro* kill *P. aeruginosa* biofilm cells. Moreover, the combination significantly reduced the cell counts of *P. aeruginosa* in a lung infection model in rats and in patients with cystic fibrosis (Herrmann et al. 2010). Likewise, the unique combination of broad-spectrum antibiotics fosfomycin and tobramycin in inhalation has been shown to be effective against *P. aeruginosa*, demonstrating that it is also a therapeutic potential for patients with cystic fibrosis (Trapnell et al. 2012).

Furthermore, based on the concept of combining antibiotics targeting different metabolic states of the biofilm cell subpopulations, the combination of ciprofloxacin with colistin produced promising results for *in vitro* biofilm eradication of *P. aeruginosa*, since biofilm cells exhibiting low metabolic activity were killed by colistin and ciprofloxacin was able to specifically kill the subpopulation of metabolically active biofilm cells (Pamp et al. 2008).

Different studies suggest that concurrent combination therapy may be more effective than antibiotic monotherapy to combat pathogenic biofilms.

20.3 New Findings for the Treatment of Pathogenic Biofilms

Although combined antibiotic therapy is a promising strategy for replacing the use of antibiotics alone against pathogenic biofilms, new treatment options are required due to the biofilm resistance multifactorial nature to antibiotics available on the market. In this regard, antimicrobial peptides (AMPs) (natural or synthetic), bacteriophages, and nanotechnology have been increasingly recognized as promising strategies for the development of unusual anti-biofilm treatments (Grassi et al. 2017).

20.3.1 AMPs Applied to Treatment Pathogenic Biofilms

AMPs comprise a heterogeneous group of evolutionarily conserved molecules, which consist of important natural effectors of the innate immune system of uni/multicellular organisms (Mookherjee and Hancock 2007; Moreno et al. 2017). Structurally, AMPs can be differentiated into four main classes: α-helical, β-sheet, extended, and mixed, with the first two classes being the most recurrent in nature (Chung and Khanum 2017).

The AMPs with anti-biofilm activity, as well as the AMPs that fight against microorganisms in the planktonic state, are small in size (12–50 amino acid residues, with 2–9 basic residues of arginine and lysine) and are cationic and amphipathic (Fuente-Núñez et al. 2014a; Sharma et al. 2016).

AMPs have been proposed as potential anti-biofilm agents because they present a broad spectrum of biological activity and low specificity of their molecular target (López-Meza et al. 2015; Batoni et al. 2016b), present high potential to reach metabolically inactive cells, and have a low propensity to induce resistance mechanisms, characteristics that make them strong candidates for pharmacological applications (Grassi et al. 2017).

In the perspective of AMPs with anti-biofilm activity, Di Luca and co-workers(2015) developed a database called *BaAMPs, Biofilm-active AMPs* database (http://www.baamps.it), which aims to make available the peptide sequences and experimental data of AMPs specifically tested against biofilms. This database aims to provide the scientific community with an open-access platform for consultation and for assistance in the development of AMPs aimed at this activity (Di Luca et al. 2015).

AMPs have been isolated from numerous organisms, such as single-celled microorganisms, plants, invertebrates, and chordates (Mookherjee and Hancock 2007). Among the most well-characterized AMPs with anti-biofilm activity, we can highlight magainin II, α-helical AMP originally isolated from the skin of the African toad *Xenopus laevis* (Kim et al. 2018); cathelicidin LL-37, α-helical AMP of human origin (Jacobsen and Jenssen 2012); and defensin-1, peptide produced by bees and by them added to honey (Sojka et al. 2016).

In addition, Table 20.1 includes examples of natural AMPs that may exert an anti-biofilm activity.

Although numerous studies claim the potent anti-biofilm activity of natural AMPs, even in the case of biophimes formed by multiresistant pathogens, there are some limitations to their future clinical use, such as high production costs, high toxicity to mammalian cells, immunomodulatory effects undesirable, and rapid degradation by proteases (Fjell et al. 2011; Strempel et al. 2015). Therefore, efforts are required for the development of synthetic AMP anti-biofilm with improved properties (Strempel et al. 2015).

20.3.1.1 Synthetic Anti-Biofilm Peptides

Synthetic anti-biofilm peptides can be defined as molecules developed through *in cerebro* rational design methods, for example, physicochemical methods and methods based on a model sequence, or by computer-assisted design methods – evolutionary methods and *de novo* methods – (Diller et al. 2015) that have a broad spectrum of activity and are indicated as a favorable alternative to conventional antimicrobials for the treatment of biofilm infections (Jorge et al. 2012; Fuente-Núñez et al. 2016; Pletzer and Hancock 2016).

Synthetic AMPs are peptides that resemble natural AMPs in relation to the structure, such as reduced size, presence of cationic amino acids, and a high proportion of hydrophobic residues, but they differ in origin and activity (Hancock and Sahl 2006; Fuente-Núñez et al. 2012). As for the activity, the synthetic AMPs may present anti-biofilm action at concentrations that do not affect planktonic growth, or they may be able to fight only biofilms, thus demonstrating a selective activity (Fuente-Núñez et al. 2012, 2014b).

Another relevant approach involving this type of peptide has been the combination of these molecules with conventional antimicrobials. Given that, synthetic AMPs may enhance the antimicrobial action to prevent biofilm formation and also eradicate mature biofilms (Reffuveille et al. 2014; Fuente-Núñez et al. 2015; Ribeiro et al. 2015). In addition, this approach may decrease the effective concentration of the active molecules and broaden their action spectrum, minimizing the possible toxic side effects and selective pressure in the development of resistance often exercised by monotherapy (Fuente-Núñez et al. 2016; Walkenhorst 2016). Table 20.2 contains synthetic peptides with anti-biofilm activity.

20.3.1.2 Mechanism of Action

Different mechanisms of anti-biofilm activity have been proposed to the AMPs (natural or synthetic) (Batoni et al. 2016b). These molecules are capable of combating biofilms at different stages of formation and maintenance (Figure 20.1) (Segev-Zarko et al. 2015).

Among them, AMPs can prevent biofilm formation by (i) preventing biofilm maturation, targeting the first colonizers on the surface; (ii) inhibiting the

Table 20.1 Natural peptides with anti-biofilm activity.

Natural AMP	Source	Anti-biofilm activity	Pathogens	References
RIP	Bacteria	Inhibition	*Staphylococcus aureus*	Cirioni et al. (2007)
HsAFP1	Plant	Inhibition	*Candida albicans*	Vriens et al. (2015)
*Tn*AFP1	Plant	Inhibition	*Candida tropicalis*	Mandal et al. (2011)
Clavanin A	Tunicate	Inhibition/ eradication	Methicillin-resistant *S. aureus* (MRSA), *Escherichia coli, Klebsiella pneumoniae* KPC, *C. albicans, Aspergillus fumigatus, Alternaria* sp., and *Fusarium* sp.	Silva et al. (2016a, b) and Mandal et al. (2017)
Coprisin	Insect	Inhibition	*Enterococcus faecium, S. aureus, Streptococcus mutans, E. coli* O-157, *E. coli*, and *Pseudomonas aeruginosa*	Hwang et al. (2013)
Chrysophsin-1	Fish	Eradication	*S. mutans*	Wang et al. (2012)
NRC-16	Fish	Inhibition	*S. aureus* and *P. aeruginosa*	Gopal et al. (2013)
Pleurocidin	Fish	Inhibition/ eradication	*S. aureus, E. faecium, E. coli, E. coli* O-157, *P. aeruginosa,* and *S. mutans*	Tao et al. (2011) and Choi and Lee (2012)
Citropin 1.1	Amphibian	Inhibition	*S. aureus*	Cirioni et al. (2006)
Magainin I	Amphibian	Inhibition	*Cryptococcus neoformans*	Martinez and Casadevall (2006)
SMAP-29	Sheep	Inhibition	*Burkholderia thailandensis*	Blower et al. (2015)
Bactenecin	Cow	Inhibition	*Burkholderia pseudomallei*	Madhongsa et al. (2013)
BMAP-27	Cow	Inhibition/ eradication	*S. aureus, P. aeruginosa,* and *Stenotrophomonas maltophilia*	Pompilio et al. (2011)
Indolicidin	Cow	Inhibition	MRSA	Mataraci and Dosler (2012)
Hepcidin 20	Human	Inhibition/ eradication	*Staphylococcus epidermidis*	Brancatisano et al. (2014)
Histatin 5	Human	Inhibition	*C. albicans*	Pusateri et al. (2009)
Lactoferrampin	Human	Inhibition	*P. aeruginosa*	Xu et al. (2010)

Table 20.2 Examples of synthetic peptides with anti-biofilm properties.

Synthetic peptide	Precursor	Anti-biofilm activity	Pathogens	References
AS10	Cathelicidin-related antimicrobial peptide (CRAMP)	Inhibition	*Candida albicans*	De Brucker et al. (2014)
Battacin	Lipopeptides	Inhibition/ eradication	*Pseudomonas aeruginosa, Staphylococcus aureus,* and *Pseudomonas syringae pv. actinidiae*	De Zoysa et al. (2015)
Bac8c	Bactenecin	Eradication	*Streptococcus mutans*	Ding et al. (2014)
CAMA	Cecropin A and Melittin-A	Inhibition/ eradication	MRSA	Mataraci and Dosler (2012)
Chromofungin	Chromogranin A	Inhibition	*Neurospora crassa*	Etienne et al. (2005)
Clavanin MO	Clavanin A	Eradication	*Klebsiella pneumoniae* KPC	Silva et al. (2016a)
[CYC2]OSIP108	OSIP108	Inhibition	*C. albicans*	Delattin et al. (2014)
D-ATRA-1A	NA-CATH	Inhibition	*Burkholderia thailandensis*	Blower et al. (2015)
Dhvar4	Histatin 5	Inhibition	*C. albicans*	Prijck et al. (2010)
DI-MB-LF11-322	LF11	Eradication	*P. aeruginosa*	Sánchez-Gómez et al. (2015)
DJK5 and DJK6	Synthetic analogue of active anti-biofilm peptides	Inhibition/ eradication	*P. aeruginosa, Escherichia coli, K. pneumoniae, Acinetobacter baumannii, Salmonella entérica,* and *K. pneumoniae* KpC	Fuente-Núñez et al. (2015) and Ribeiro et al. (2015)
DRGN-1	VK25	Eradication	*P. aeruginosa* and *S. aureus*	Chung et al. (2017)
EcDBS1R5	*E. coli*	Eradication	*P. aeruginosa*	Cardoso et al. (2018)

Giovanni 2	Pg-AMP1	Inhibition/ eradication	*E. coli, S. aureus, K. pneumoniae*, and *C. albicans*	Porto et al. (2018)
GL13K	BPIFA2	Inhibition/ eradication	*P. aeruginosa, Streptococcus gordonii, and Porphyromonas gingivalis*	Hirt and Gorr (2013), Holmberg et al. (2013), and Chen et al. (2014)
IDR-1018	Bactenecin	Inhibition/ eradication	*P. aeruginosa, E. coli, A. baumannii, K. pneumoniae,* MRSA, *Salmonella typhimurium,* and *Burkholderia cenocepacia*	Mansour et al. (2015)
IMB-2	*S. mutans* ComC signaling peptide (CSP)	Eradication	*S. mutans*	Mai et al. (2011)
KSL-W	KSL	Inhibition/ eradication	*C. albicans*	Theberge et al. (2013)
KT2 and RT2	Luecrocin I	Inhibition	*E. coli*	Anunthawan et al. (2015)
Lys-a1	Hy-A1	Inhibition	*S. mutans, Streptococcus oralis, Streptococcus sanguinis, Streptococcus parasanguinis, Streptococcus salivarius,* and *Streptococcus sobrinus*	Silva et al. (2013)
Melimine	Protamine and melittin	Inhibition	*P. aeruginosa* and *S. aureus*	Chen et al. (2012)
Myxinidin2 and myxinidin3	Myxinidin	Inhibition	*P. aeruginosa, S. aureus, and Listeria monocytogenes*	Han et al. (2016)
NRC-16	Pleurocidin	Inhibition	*P. aeruginosa*	Gopal et al. (2013)

(*Continued*)

Table 20.2 (Continued)

Synthetic peptide	Precursor	Anti-biofilm activity	Pathogens	References
Pa-MAP 1.9	*Pa*-MAP	Inhibition/ eradication	*K. pneumoniae* and *E. coli*	Cardoso et al. (2016)
P60.4Ac	LL-37	Inhibition/ eradication	Methicillin-resistant *S. aureus*	Haisma et al. (2014)
P10	P60.4Ac	Inhibition/ eradication	MRSA	Haisma et al. (2014)
P15-CSP	P15 and CSP	Inhibition	*S. mutans*	Li et al. (2015)
P318	CRAMP	Inhibition	*C. albicans*	De Brucker et al. (2014)
SB056-1	SB056	Inhibition	*Staphylococcus epidermidis*	Batoni et al. (2016a)
1037	LL-37	Inhibition	*P. aeruginosa*, *B. cenocepacia*, and *L. monocytogenes*	Fuente-Núñez et al. (2012)

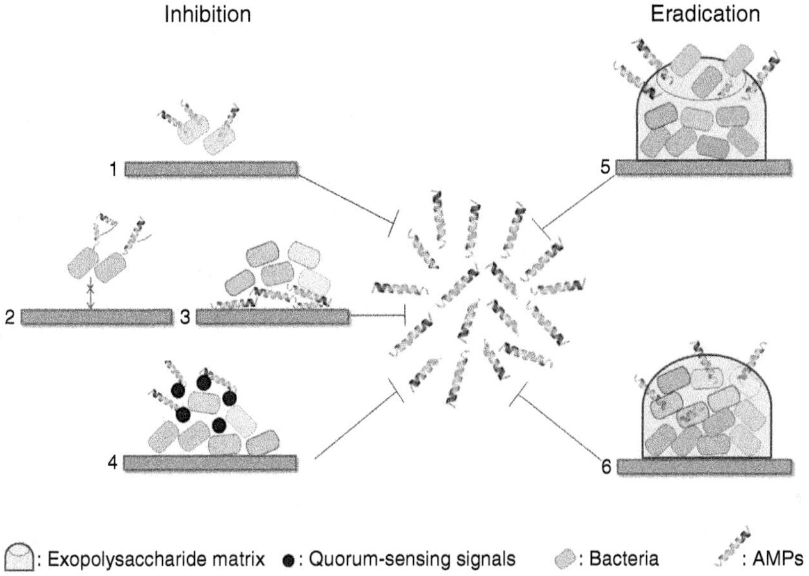

Figure 20.1 Mechanisms of anti-biofilm action of AMPs. (1) Inhibition of biofilm formation by targeting cells in the planktonic state. (2) Inhibition of adhesion on biotic surface reaching the regulatory pathways responsible for the production of adhesion defects, such as pili and flagellum. (3) Inhibition of adhesion by abiotic surface coating. (4) Inhibition of quorum sensing. (5) Eradication of mature biofilm by disruption of the extracellular matrix. (6) Eradication of the mature biofilm through the death of the adhered cells and incorporated into the extracellular matrix. *(See insert for color representation of the figure.)*

initial adhesion through the abiotic surface coating; (iii) inhibiting the adhesion on a biotic surface, reaching the regulatory pathways responsible for the production of adhesion structures; or (iv) binding to the molecules involved in *quorum sensing*, inhibiting bacterial communication (Jorge et al. 2012; Di Luca et al. 2014; Fuente-Núñez et al. 2016).

At the molecular level, studies also demonstrate that peptides can prevent the formation of pathogenic biofilms by disrupting nucleotide signaling, important signaling molecules that allow microorganisms to control the formation and maintenance of biofilm (Pletzer et al. 2016). For example, it has been described that AMPs can bind and degrade intracellular nucleotides (p)ppGpp (guanosine tetra- and pentaphosphate), thus preventing the intracellular accumulation of this molecule and consequently inhibiting the formation of pathogenic biofilms (Fuente-Núñez et al. 2015). The AMPs can also bind to the nucleotide c-di-GMP (bis-(3′,5′)-cyclic dimeric guanosine monophosphate) and prevent the expression of extracellular matrix components, also inhibiting biofilm formation (Foletti et al. 2018). In addition, they can also inhibit quorum-sensing sensors such as trp RNA-binding attenuation protein (TRAP) (sensor protein) that regulates the expression of multiple genes important for biofilm formation (Balaban et al. 2007).

In combating mature (established) biofilms, AMPs may promote eradication of the microbial community by dispersing the cells within the biofilm after degradation of the extracellular matrix or may promote eradication by killing the cells associated with the biofilm after penetration into the extracellular matrix (Jorge et al. 2012; Di Luca et al. 2014).

20.3.2 Bacteriophage Therapy Anti-Biofilm

Bacteriophages or phages are obligate parasites that propagate in bacterial hosts and can be classified according to their life cycles, lytic or lysogenic (Maciejewska et al. 2018). For infection of a bacterial cell to occur, a phage will adhere to the surface of the host cell and then inject the viral DNA into the bacterium. Subsequently, the replication strategy will depend on whether the phage is virulent or tempered (Salmond and Fineran 2015).

Virulent phages are only able to replicate via the lytic cycle, a process involving the production of viral progeny and their release from lysis of the infected cell. In contrast, the temperate phages may enter the lytic cycle or form an association with the host, called lysogeny. During lysogeny, the viral genome is termed the prophagy and has its replication in conjunction with the host DNA. Already under stress conditions, the prophages can leave the lysogenic state and produce more virion, with subsequent release from lysis of the host cell, which results in bacterial death (Young 2013; Roach and Donovan 2015; Salmond and Fineran 2015).

Phages have been used for the treatment of bacterial infections for more than 50 years; however the appearance of strains resistant to multiple antibiotics has made this type of therapy revitalized (Wu et al. 2015). In addition, the occurrence of pathogenic bacterial strains capable of forming biofilms drew more attention to the investigation of this type of therapy, thus emerging as a valuable alternative for the treatment of bacterial infections, mainly those caused by biofilm (Hughes and Webber 2017).

In the treatment of infections associated with pathogenic biofilms, phages have unique properties that offer some advantages: they are highly specific and nontoxic, do not affect the normal microbiota, and have the capacity to improve conventional treatment (Yang et al. 2012; Casey et al. 2018).

Several experimental and clinical studies have demonstrated the effectiveness of the use of phages as well as proteins derived from phages, especially enzymes, in the fight against pathogenic biofilms (Szafrański et al. 2017). In this context, Sutherland and co-workers (2004) evidenced in their work that some phages can carry on their surface very specific enzymes that promote the degradation of the bacterial polysaccharides, promoting the destruction of the integrity of the biofilms. Son and co-workers (2010) characterized in their study a cell wall degradation enzyme, SAL-2 (derived from the SAP-2 phage), which demonstrated positive activity in the removal of biofilms from the pathogenic bacterium *S. aureus*. In the study by Fu and co-workers(2010), the pretreatment of phage hydrogel-coated silicone catheters has been shown to be effective in preventing bacterial fixation and in initial formation of the pathogenic *P. aeruginosa* biofilm in an *in vitro* model system.

Hanlon and co-workers (2001) suggest in their work that phages may cause a reduction in the viscosity of the exopolysaccharide of the *P. aeruginosa* biofilm, in addition to causing a reduction in the number of cells within the biofilm. Rahman and co-workers (2011) when investigating the potential use of the phage in combination with conventional antimicrobial agents observed structural changes in the *S. aureus* biofilm matrix as well as a substantial decrease in the number of bacteria and concluded that this mixture of phage and an antimicrobial agent is an efficient alternative in the removal of a pathogenic bacterial biofilm.

In the context of mechanisms for biofilm degradation by phages, Harper and co-workers (2014) described four different types in their review (Figure 20.2). (i) The phage infects the host bacterial cells and may replicate therein, resulting in the increase and release of infective bacterial progeny from the biofilm. By spreading through the biofilm, by eliminating the bacteria that produce exopolysaccharide (EPS) matrix material, phages can progressively remove biofilm and decrease the regeneration potential. (ii) The phage can carry or express depolymerizing enzymes that are capable of degrading biofilm EPS. (iii) The phages can induce the expression of depolymerizing enzymes in host bacteria, thereby promoting the degradation of EPS from within the host genome. (iv) The phage can infect persistent cells and remain within them until they react to produce a productive and lytic infection (destruction of host cells). Note: Phages are not able to replicate and destroy persistent (inactive) cells.

Although countless studies claim the effectiveness of natural phages in combating pathogenic bacterial biofilms, some limitations are reported in their use. From the commercial point of view, natural phages have complex processes of production and quality control, have unpredictable social acceptance, have high costs of clinical trials, and are not patentable (Harper et al. 2014; Szafrański et al. 2017). This prevents them from being commercially viable in the industry, especially for large pharmaceutical corporations, thus preventing the development of infrastructure that can provide new treatments to patients (Sagona et al. 2016). Another limitation is the possibility of inducing a mammalian immune response when it enters the body. In addition, phage-induced lysis of bacteria releases toxic or pro-inflammatory products, so phage may indirectly induce an inflammatory response (Szafrański et al. 2017). In this context, the research community starts to explore solutions to overcome such limitations, and the projection of new phages appears as an alternative (Harper et al. 2014).

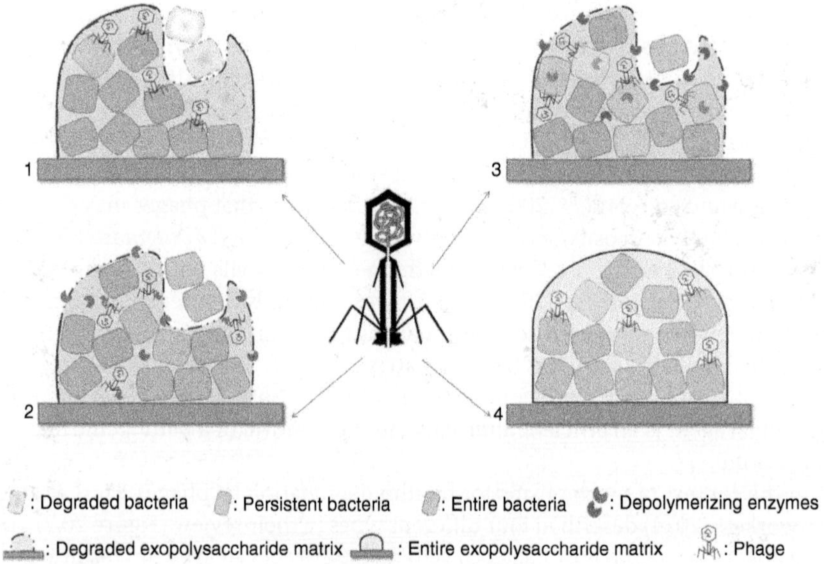

Figure 20.2 Mechanisms for biofilm degradation by phages. (1) Infection of the bacteria by the phage, with subsequent elimination of the infected bacteria and consequent degradation of EPS. (2) Degradation of EPS by depolymerization enzymes carried by the phage. (3) Degradation of EPS by depolymerization enzymes expressed by bacteria infected by the phage. (4) Infection of persistent bacteria by phage. *(See insert for color representation of the figure.)*

In this perspective, the improvement of synthetic biology allowed the development of new phages by rational design, modification, and construction of recombinant phages – which have genetically modified DNA and/or undergo genetic recombination – allowing the amplification of their innate phenotypes (Sagona et al. 2016).

Phages developed by the use of synthetic biology may be marketable; noninflammatory, from the modification of phage coat protein; and/or non-lytic, from the projections to be toxic, but not lytic (Abedon 2011; Yan et al. 2014). In addition, phages can also be developed to enhance/broaden their spectrum of activity.

In another work, Lu and Collins (2007) designed two phages to express a biofilm-degrading enzyme, so that it was possible to simultaneously attack the bacterial cells within the biofilm and the biofilm matrix. As a result it has showed the efficacy of biofilm removal by the two enzyme phages and promoted a removal significantly higher when compared to the nonenzymatic natural phage treatment (Lu and Collins 2007). Gallet and co-workers (2009) developed three phages, each with a different adsorption rate to determine their abilities to complete the settlement–production–emigration cycle.

The results demonstrated that the high rate of adsorption is beneficial for colonization but it impairs production and emigration. Therefore, the low adsorption rate is more advantageous under biofilms (Gallet et al. 2009).

In the study by Kelly et al., a cocktail of phages modified to break down biofilms was developed to prevent the formation of *S. aureus* pathogenic biofilm, as well as to reduce the density of this same established biofilm. The results demonstrated that the cocktail was capable of inhibiting biofilm formation and reducing the density of the established biofilm of the aforementioned bacterium in a time-dependent manner (Kelly et al. 2012). In another study, Tinoco and co-workers(2016) modified genes of the tempered phage φE11, by genetic engineering, whose purpose was to render it incapable of lysogeny and to extend the range of host strains within the target bacterial species. From these modifications, the efficacy of the same was evaluated in the degradation of two strains of *E. faecalis* (sensitive and resistant to vancomycin). As a result, the modified phage demonstrated a significant reduction in the biofilm biomass of *E. faecalis* from both strains tested (Tinoco et al. 2016).

In summary, the development of new phages appears to bring even more advantages to the bacteriophage therapy associated with the combat of pathogenic biofilms.

20.3.3 Nanotechnology Applied to the Treatment of Pathogenic Biofilms

Over the last decade, different approaches based on nanoparticles, with specific physicochemical properties for antimicrobial activities, have been described in several studies (Kim 2016), mainly in relation to the inhibition, control, or eradication of infections related to pathogenic biofilms, either on biotic or abiotic surfaces as medical devices (Ramasamy and Lee 2016).

Pathogenic biofilm infections, as mentioned, have limitations in treatment with conventional antimicrobial agents. In this sense, nanoparticles as drug carriers, specifically antimicrobials, appear as an advantageous strategy, since they may be able to protect antimicrobial agents from enzymatic inactivation (Han et al. 2017). They may also be able to prevent electrostatic binding to biofilm matrix components, such as DNA or polysaccharides, and may release antimicrobial agents locally in an efficient, controlled, and safe manner for the reduction of systemic side effects, as well as optimization of antimicrobial efficacy (Han et al. 2017). Thus, nanoparticles as carriers can improve the bioavailability and the targeted delivery of antimicrobials (De Jong and Borm 2008). In this context, many biocompatible and biodegradable nanoparticles have been used as carriers of antimicrobials against pathogenic biofilms (Kim 2016), including liposomal nanoparticles (Zhang et al. 2010).

Liposomes consist of small spherical vesicles composed of a phospholipid bilayer that may be produced from nontoxic phospholipids (e.g. cholesterol), and, depending on their composition, properties as stiffness or fluidity, and bilayer loading can be regulated (Santos Ramos et al. 2018). As carriers, liposomal nanoparticles may have the lipophilic treatment agents incorporated into their phospholipid bilayer, whereas the hydrophobic agents may be encapsulated in their aqueous core (Forier et al. 2014). As an example, in the study by Mugabe and co-workers(2005), it was demonstrated that the encapsulation of the antimicrobial gentamicin may result in a significantly higher antimicrobial activity against *P. aeruginosa* biofilms when compared with treatment with free gentamicin.

In another study, Li and co-workers(2013) developed liposomal nanoparticles loaded with the antimicrobial daptomycin that were able to efficiently inhibit the growth of pathogenic *S. aureus* biofilm when compared with free daptomycin. In the study developed by Meers and co-workers(2008), the antimicrobial effect of the antimicrobial amikacin encapsulated in liposomes against the pathogenic bacterium *P. aeruginosa* was investigated in *in vitro* experiments. As a result, liposomal nanoparticles with amikacin demonstrated significantly greater efficacy about amikacin in the free form (Meers et al. 2008). In summary, studies have shown that antimicrobials incorporated into liposomes demonstrate improved bioactivity relative to the administration of free antimicrobial agents.

As for the use of nanoparticles as active agents against pathogenic biofilms, metallic nanoparticles such as silver, copper, gold, titanium, and zinc are highlighted. This is because they are presented as alternatives to conventional antimicrobials because they do not significantly increase the risk of resistance development (Ramasamy and Lee 2016).

Among all metallic nanoparticles, silver has been one of the most used because it shows low levels of toxicity to animal cells and high levels of toxicity to pathogenic microorganisms (Sousa et al. 2011). This toxicity can be explained by the interaction of silver ions with the internal structures of microorganisms (Sondi and Salopek-Sondi 2004; Choi et al. 2010).

In addition, specifically against biofilms silver nanoparticles have been shown to be able to prevent bacterial adhesion on surfaces and subsequent biofilm formation (Klueh et al. 2000) and also be able to destabilize the biofilm matrix, compromising the intermolecular forces (Chaw et al. 2005). The use of nanotechnology has been proposed as one of the most promising possibilities for biological applications. For example, in the study of Kalishwaralal and co-workers(2010), the *in vitro* anti-biofilm capacity of silver nanoparticles against *P. aeruginosa* and *S. epidermidis*, important agents that cause keratitis, especially those related to the incorrect use of contact lenses, was evaluated. As a result, the silver nanoparticles exhibited a significant inhibition in the formation of the biofilms of these bacteria, besides demonstrating an efficient elimination of mature (established) biofilms of these same pathogenic bacteria (Kalishwaralal et al. 2010). Therefore, these results demonstrate that treatment with silver nanoparticles may emerge as futuristic applications for biofilm-based human ocular problems (Kalishwaralal et al. 2010). In another study, Martinez-Gutierrez and co-workers (2013) also demonstrated that silver nanoparticles effectively prevent the formation of biofilms and kill bacteria in established biofilms, which further reinforces the use of this therapy to prevent and combat infections related to pathogenic biofilms.

In a more recent study, Gillet and co-workers(2018) explored the efficacy of silver nanoparticle-based films against *E. coli* biofilms. As a result, films were able to act as a silver source sufficient to prevent unwanted initial formation of these biofilms while maintaining toxicity at low levels and therefore could be suggested as potential candidates for use in the medical industry (Gillett et al. 2018).

In the context of silver nanoparticles as active agents against pathogenic biofilms on abiotic surfaces, specifically medical devices, the literature brings several studies. For example, the study by Roe and co-workers(2008) carries tests with plastic catheters coated with silver nanoparticles, which were able to efficiently inhibit the formation of the biofilms of the bacteria *E. coli* and *S. aureus* and the fungus *C. albicans*. Moreover, in *in vivo* experiments these coated catheters provided safe use in animals, since they were not toxic (Roe et al. 2008). In another study, Secinti and co-workers(2011) with coated titanium implants with silver in *in vivo* experiments tested the toxicity and ability of these implants to inhibit the formation of *S. aureus* biofilm. As a result, the coated implants did not induce toxic effects and were able to inhibit the biofilm formation of these bacteria (Secinti et al. 2011).

In summary, despite the extensive investigation of silver nanoparticles as potential alternatives for treatment against pathogenic biofilms, either as active biofilm agents or as a coating for medical devices, there is still no in-depth knowledge about the interaction of these nanoparticles with the body human. Therefore more studies are required, since the biosafety of silver nanoparticles for humans is currently uncertain, since in the literature the studies focus on biodistribution and *in vivo* toxicity with rabbits, rats, and mice (Markowska et al. 2013).

20.4 Conclusion and Future Directions

As previously described, pathogenic biofilms are the main contributors to chronic infection. Therefore, the need for new therapies to reduce the impact of these infections becomes even more urgent. Thus, the study of biofilm and strategies to prevent and combat them consists of one of the most important fields of research today. In this sense, advances in biotechnology have allowed the opening of new therapeutic horizons and therapies with AMPs; bacteriophages and nanoparticles are currently the most promising for anti-biofilm treatments. All these approaches have the promise and are capable of effectively preventing or combating pathogenic biofilms in model systems. In addition, other approaches involving the use of weak organic acids, photoinactivation, quorum-sensing inhibitors, and single nucleotide inhibitors represent promising therapeutic niches. However, further studies to assess the safety of these therapeutic options are necessary for the fight against microbial resistance. In addition, the combined use of the agents described with antibiotics or other compounds could increase the success in eliminating microorganisms in biofilms, making such compounds incorporated into clinical use as effective anti-biofilm treatments.

It is also worth mentioning that in addition to advances in anti-biofilm therapy, improvements in the diagnosis of biofilm infections are required, as early diagnosis of this type of infection currently is a great challenge. Generally, pathogenic biofilms cause indolent infections that may not be detected by routine diagnostic procedures until the infection shows clinical signs, which makes it even more difficult to treat. Therefore, new techniques of microbiology should be introduced as efficient diagnostic complements in hospital routines.

References

Abedon, S.T. (2011). Lysis from without. *Bacteriophage* 1 (1): 46–49.

Anunthawan, T., de la Fuente-Núñez, C., Hancock, R.E.W., and Klaynongsruang, S. (2015). Cationic amphipathic peptides KT2 and RT2 are taken up into bacterial cells and kill planktonic and biofilm bacteria. *Biochimica et Biophysica Acta* 1848 (6): 1352–1358.

Arciola, C.R., Campoccia, D., and Montanaro, L. (2018). Implant infections: adhesion, biofilm formation and immune evasion. *Nature Reviews Microbiology* 16 (7): 397–409.

Balaban, N., Cirioni, O., Giacometti, A. et al. (2007). Treatment of *Staphylococcus aureus* biofilm infection by the quorum-sensing inhibitor RIP. *Antimicrobial Agents and Chemotherapy* 51 (6): 2226–2229.

Batoni, G., Casu, M., Giuliani, A. et al. (2016a). Rational modification of a dendrimeric peptide with antimicrobial activity: consequences on membrane-binding and biological properties. *Amino Acids* 48 (3): 887–900.

Batoni, G., Maisetta, G., and Esin, S. (2016b). Antimicrobial peptides and their interaction with biofilms of medically relevant bacteria. *Biochimica et Biophysica Acta* 1858 (5): 1044–1060.

Bjarnsholt, T. (2013). The role of bacterial biofilms in chronic infections. *Acta Pathologica, Microbiologica et Immunologica Scandinavica* 121: 1–58.

Bjarnsholt, T., Ciofu, O., Molin, S. et al. (2013). Applying insights from biofilm biology to drug development: can a new approach be developed? *Nature Reviews. Drug Discovery* 12 (10): 791–808.

Blower, R.J., Barksdale, S.M., and van Hoek, M.L. (2015). Snake cathelicidin NA-CATH and smaller helical antimicrobial peptides are effective against *Burkholderia thailandensis*. *PLOS Neglected Tropical Diseases* 9 (7): e0003862.

Brancatisano, F.L., Maisetta, G., Di Luca, M. et al. (2014). Inhibitory effect of the human liver-derived antimicrobial peptide hepcidin 20 on biofilms of polysaccharide intercellular adhesin (PIA)-positive and PIA-negative strains of *Staphylococcus epidermidis*. *Biofouling* 30 (4): 435–446.

Caputo, W.J., Beggs, D.J., DeFede, J.L. et al. (2008). A prospective randomised controlled clinical trial comparing hydrosurgery debridement with conventional surgical debridement in lower extremity ulcers. *International Wound Journal* 5 (2): 288–294.

Cardoso, M.H., Cândido, E.S., Chan, L.Y. et al. (2018). A computationally designed peptide derived from *Escherichia coli* as a potential drug template for antibacterial and antibiofilm therapies. *ACS Infectious Diseases* 4 (12): 1727–1736.

Cardoso, M.H., Ribeiro, S.M., Nolasco, D.O. et al. (2016). A polyalanine peptide derived from polar fish with anti-infectious activities. *Scientific Reports* 6: 21385.

Casey, E., van Sinderen, D., and Mahony, J. (2018). In vitro characteristics of phages to guide "real life" phage therapy suitability. *Viruses* 10 (4): pii: E163.

Chandki, R., Banthia, P., and Banthia, R. (2011). Biofilms: a microbial home. *Journal of Indian Society of Periodontology* 15 (2): 111–114.

Chaw, K.C., Manimaran, M., and Tay, F.E.H. (2005). Role of silver ions in destabilization of intermolecular adhesion forces measured by atomic force microscopy in *Staphylococcus epidermidis* biofilms. *Antimicrobial Agents and Chemotherapy* 49 (12): 4853–4859.

Chen, R., Willcox, M.D.P., Cole, N. et al. (2012). Characterization of chemoselective surface attachment of the cationic peptide melimine and its effects on antimicrobial activity. *Acta Biomaterialia* 8 (12): 4371–4379.

Chen, X., Hirt, H., Li, Y. et al. (2014). Antimicrobial GL13K peptide coatings killed and ruptured the wall of *Streptococcus gordonii* and prevented formation and growth of biofilms. *PLoS One* 9 (11): e111579.

Choi, H. and Lee, D.G. (2012). Antimicrobial peptide pleurocidin synergizes with antibiotics through hydroxyl radical formation and membrane damage, and exerts antibiofilm activity. *Biochimica et Biophysica Acta* 1820 (12): 1831–1838.

Choi, O., Yu, C.-P., Esteban Fernández, G., and Hu, Z. (2010). Interactions of nanosilver with *Escherichia coli* cells in planktonic and biofilm cultures. *Water Research* 44 (20): 6095–6103.

Chung, E.M.C., Dean, S.N., Propst, C.N. et al. (2017). Komodo dragon-inspired synthetic peptide DRGN-1 promotes wound-healing of a mixed-biofilm infected wound. *NPJ Biofilms and Microbiomes* 3: 9.

Chung, P.Y. and Khanum, R. (2017). Antimicrobial peptides as potential anti-biofilm agents against multidrug-resistant bacteria. *Journal of Microbiology, Immunology, and Infection* 50 (4): 405–410.

Ciofu, O., Rojo-Molinero, E., Macià, M.D., and Oliver, A. (2017). Antibiotic treatment of biofilm infections. *APMIS (Acta Pathologica, Microbiologica et Immunologica Scandinavica)* 125 (4): 304–319.

Cirioni, O., Ghiselli, R., Minardi, D. et al. (2007). RNAIII-inhibiting peptide affects biofilm formation in a rat model of staphylococcal ureteral stent infection. *Antimicrobial Agents and Chemotherapy* 51 (12): 4518–4520.

Cirioni, O., Giacometti, A., Ghiselli, R. et al. (2006). Citropin 1.1-treated central venous catheters improve the efficacy of hydrophobic antibiotics in the treatment of experimental staphylococcal catheter-related infection. *Peptides* 27 (6): 1210–1216.

De Brucker, K., Delattin, N., Robijns, S. et al. (2014). Derivatives of the mouse cathelicidin-related antimicrobial peptide (CRAMP) inhibit fungal and bacterial biofilm formation. *Antimicrobial Agents and Chemotherapy* 58 (9): 5395–5404.

De Jong, W.H. and Borm, P.J. (2008). Drug delivery and nanoparticles: applications and hazards. *International Journal of Nanomedicine* 3 (2): 133–149.

De Zoysa, G.H., Cameron, A.J., Hegde, V.V. et al. (2015). Antimicrobial peptides with potential for biofilm eradication: synthesis and structure activity relationship studies of battacin peptides. *Journal of Medicinal Chemistry* 58 (2): 625–639.

Delattin, N., De Brucker, K., Craik, D.J. et al. (2014). Plant-derived decapeptide OSIP108 interferes with *Candida albicans* biofilm formation without affecting cell viability. *Antimicrobial Agents and Chemotherapy* 58 (5): 2647–2656.

Delcaru, C., Alexandru, I., Podgoreanu, P. et al. (2016). Microbial biofilms in urinary tract infections and prostatitis: etiology, pathogenicity, and combating strategies. *Pathogens* 5 (4): E65.

Di Luca, M., Maccari, G., Maisetta, G., and Batoni, G. (2015). BaAMPs: the database of biofilm-active antimicrobial peptides. *Biofouling* 31 (2): 193–199.

Di Luca, M., Maccari, G., and Nifosì, R. (2014). Treatment of microbial biofilms in the post-antibiotic era: prophylactic and therapeutic use of antimicrobial peptides and their design by bioinformatics tools. *Pathogens and Disease* 70 (3): 257–270.

Diller, D.J., Swanson, J., Bayden, A.S. et al. (2015). Rational, computer-enabled peptide drug design: principles, methods, applications and future directions. *Future Medicinal Chemistry* 7 (16): 2173–2193.

Ding, Y., Wang, W., Fan, M. et al. (2014). Antimicrobial and anti-biofilm effect of Bac8c on major bacteria associated with dental caries and *Streptococcus mutans* biofilms. *Peptides* 52: 61–67.

Donlan, R.M. (2001a). Biofilm formation: a clinically relevant microbiological process. *Clinical Infectious Diseases* 33 (8): 1387–1392.

Donlan, R.M. (2001b). Biofilms and device-associated infections. *Emerging Infectious Diseases* 7 (2): 277–281.

Etienne, O., Gasnier, C., Taddei, C. et al. (2005). Antifungal coating by biofunctionalized polyelectrolyte multilayered films. *Biomaterials* 26 (33): 6704–6712.

Fjell, C.D., Hiss, J.A., Hancock, R.E.W., and Schneider, G. (2011). Designing antimicrobial peptides: form follows function. *Nature Reviews. Drug Discovery* 11 (1): 37–51.

Foletti, C., Kramer, R.A., Mauser, H. et al. (2018). Functionalized proline-rich peptides bind the bacterial second messenger c-di-GMP. *Angewandte Chemie (International Ed. in English)* 57 (26): 7729–7733.

Forier, K., Raemdonck, K., De Smedt, S.C. et al. (2014). Lipid and polymer nanoparticles for drug delivery to bacterial biofilms. *Journal of Controlled Release: Official Journal of the Controlled Release Society* 190: 607–623.

Fu, W., Forster, T., Mayer, O. et al. (2010). Bacteriophage cocktail for the prevention of biofilm formation by *Pseudomonas aeruginosa* on catheters in an in vitro model system. *Antimicrobial Agents and Chemotherapy* 54 (1): 397–404.

Fuente-Núñez, C., Cardoso, M.H., de Cândido, E., S. et al. (2016). Synthetic antibiofilm peptides. *Biochimica et Biophysica Acta* 1858 (5): 1061–1069.

Fuente-Núñez, C., Korolik, V., Bains, M. et al. (2012). Inhibition of bacterial biofilm formation and swarming motility by a small synthetic cationic peptide. *Antimicrobial Agents and Chemotherapy* 56 (5): 2696–2704.

Fuente-Núñez, C., Mansour, S.C., Wang, Z. et al. (2014a). Anti-biofilm and immunomodulatory activities of peptides that inhibit biofilms formed by pathogens isolated from cystic fibrosis patients. *Antibiotics (Basel, Switzerland)* 3 (4): 509–526.

Fuente-Núñez, C., Reffuveille, F., Haney, E.F. et al. (2014b). Broad-spectrum anti-biofilm peptide that targets a cellular stress response. *PLoS Pathogens* 10 (5): e1004152.

Fuente-Núñez, C., Reffuveille, F., Mansour, S.C. et al. (2015). D-enantiomeric peptides that eradicate wild-type and multidrug-resistant biofilms and protect against lethal *Pseudomonas aeruginosa* infections. *Chemistry and Biology* 22 (2): 196–205.

Fujimura, S., Sato, T., Mikami, T. et al. (2008). Combined efficacy of clarithromycin plus cefazolin or vancomycin against *Staphylococcus aureus* biofilms formed on titanium medical devices. *International Journal of Antimicrobial Agents* 32 (6): 481–484.

Gallet, R., Shao, Y., and Wang, I.-N. (2009). High adsorption rate is detrimental to bacteriophage fitness in a biofilm-like environment. *BMC Evolutionary Biology* 9 (1): 241.

Gbejuade, H.O., Lovering, A.M., and Webb, J.C. (2015). The role of microbial biofilms in prosthetic joint infections. *Acta Orthopaedica* 86 (2): 147–158.

Gillett, A.R., Baxter, S.N., Hodgson, S.D. et al. (2018). Using sub-micron silver-nanoparticle based films to counter biofilm formation by Gram-negative bacteria. *Applied Surface Science* 442: 288–297.

Gopal, R., Lee, J.H., Kim, Y.G. et al. (2013). Anti-microbial, anti-biofilm activities and cell selectivity of the NRC-16 peptide derived from witch flounder, *Glyptocephalus cynoglossus*. *Marine Drugs* 11 (6): 1836–1852.

Grassi, L., Maisetta, G., Esin, S., and Batoni, G. (2017). Combination strategies to enhance the efficacy of antimicrobial peptides against bacterial biofilms. *Frontiers in Microbiology* 8: E2409.

Haisma, E.M., de Breij, A., Chan, H. et al. (2014). LL-37-derived peptides eradicate multidrug-resistant *Staphylococcus aureus* from thermally wounded human skin equivalents. *Antimicrobial Agents and Chemotherapy* 58 (8): 4411–4419.

Han, C., Romero, N., Fischer, S. et al. (2017). Recent developments in the use of nanoparticles for treatment of biofilms. *Nanotechnology Reviews* 6 (5): 383–404.

Han, H.M., Gopal, R., and Park, Y. (2016). Design and membrane-disruption mechanism of charge-enriched AMPs exhibiting cell selectivity, high-salt resistance, and anti-biofilm properties. *Amino Acids* 48 (2): 505–522.

Hancock, R.E.W. (2015). Rethinking the antibiotic discovery paradigm. *eBioMedicine* 2 (7): 629–630.

Hancock, R.E.W. and Sahl, H.-G. (2006). Antimicrobial and host-defense peptides as new anti-infective therapeutic strategies. *Nature Biotechnology* 24 (12): 1551–1557.

Hanlon, G.W., Denyer, S.P., Olliff, C.J., and Ibrahim, L.J. (2001). Reduction in exopolysaccharide viscosity as an aid to bacteriophage penetration through *Pseudomonas aeruginosa* biofilms. *Applied and Environmental Microbiology* 67 (6): 2746–2753.

Harper, D.R., Parracho, H.M.R.T., Walker, J. et al. (2014). Bacteriophages and biofilms. *Antibiotics* 3 (3): 270–284.

Heijerman, H., Westerman, E., Conway, S. et al. (2009). Inhaled medication and inhalation devices for lung disease in patients with cystic fibrosis: a European consensus. *Journal of Cystic Fibrosis: Official Journal of the European Cystic Fibrosis Society* 8 (5): 295–315.

Herrmann, G., Yang, L., Wu, H. et al. (2010). Colistin-tobramycin combinations are superior to monotherapy concerning the killing of biofilm *Pseudomonas aeruginosa*. *The Journal of Infectious Diseases* 202 (10): 1585–1592.

Hirt, H. and Gorr, S.-U. (2013). Antimicrobial peptide GL13K is effective in reducing biofilms of *Pseudomonas aeruginosa*. *Antimicrobial Agents and Chemotherapy* 57 (10): 4903–4910.

Høiby, N., Bjarnsholt, T., Moser, C. et al. (2015). ESCMID * guideline for the diagnosis and treatment of biofilm infections 2014. *Clinical Microbiology and Infection* 21: 1–25.

Høiby, N., Ciofu, O., Johansen, H.K. et al. (2011). The clinical impact of bacterial biofilms. *International Journal of Oral Science* 3 (2): 55–65.

Holmberg, K.V., Abdolhosseini, M., Li, Y. et al. (2013). Bio-inspired stable antimicrobial peptide coatings for dental applications. *Acta Biomaterialia* 9 (9): 8224–8231.

Hughes, G. and Webber, M.A. (2017). Novel approaches to the treatment of bacterial biofilm infections. *British Journal of Pharmacology* 174 (14): 2237–2246.

Hwang, I., Hwang, J.-S., Hwang, J.H. et al. (2013). Synergistic effect and antibiofilm activity between the antimicrobial peptide coprisin and conventional antibiotics against opportunistic bacteria. *Current Microbiology* 66 (1): 56–60.

Jacobsen, A.S. and Jenssen, H. (2012). Human cathelicidin LL-37 prevents bacterial biofilm formation. *Future Medicinal Chemistry* 4 (12): 1587–1599.

Jacqueline, C. and Caillon, J. (2014). Impact of bacterial biofilm on the treatment of prosthetic joint infections. *The Journal of Antimicrobial Chemotherapy* 69 (Suppl 1): i37–i40.

Johannsson, B., Taylor, J., Clark, C.R. et al. (2010). Treatment approaches to prosthetic joint infections: results of an Emerging Infections Network survey. *Diagnostic Microbiology and Infectious Disease* 66 (1): 16–23.

Jorge, P., Lourenço, A., and Pereira, M.O. (2012). New trends in peptide-based anti-biofilm strategies: a review of recent achievements and bioinformatic approaches. *Biofouling* 28 (10): 1033–1061.

Kackar, S., Suman, E., and Kotian, M.S. (2017). Bacterial and fungal biofilm formation on contact lenses and their susceptibility to lens care solutions. *Indian Journal of Medical Microbiology* 35 (1): 80–84.

Kalishwaralal, K., BarathManiKanth, S., Pandian, S.R.K. et al. (2010). Silver nanoparticles impede the biofilm formation by *Pseudomonas aeruginosa* and *Staphylococcus epidermidis*. *Colloids and Surfaces, B: Biointerfaces* 79 (2): 340–344.

Kelly, D., McAuliffe, O., Ross, R.P., and Coffey, A. (2012). Prevention of *Staphylococcus aureus* biofilm formation and reduction in established biofilm density using a combination of phage K and modified derivatives. *Letters in Applied Microbiology* 54 (4): 286–291.

Kim, M.-H. (2016). Nanoparticle-based therapies for wound biofilm infection: opportunities and challenges. *IEEE Transactions on NanoBioscience* 15 (3): 294–304.

Kim, M.K., Kang, N., Ko, S.J. et al. (2018). Antibacterial and antibiofilm activity and mode of action of magainin 2 against drug-resistant *Acinetobacter baumannii*. *International Journal of Molecular Sciences* 19 (10): E3041.

Klueh, U., Wagner, V., Kelly, S. et al. (2000). Efficacy of silver-coated fabric to prevent bacterial colonization and subsequent device-based biofilm formation. *Journal of Biomedical Materials Research* 53 (6): 621–631.

Kumar, A., Alam, A., Rani, M. et al. (2017). Biofilms: survival and defense strategy for pathogens. *International Journal of Medical Microbiology* 307 (8): 481–489.

Lewis, K. (2007). Persister cells, dormancy and infectious disease. *Nature Reviews Microbiology* 5 (1): 48–56.

Lewis, K. (2012). Persister cells: molecular mechanisms related to antibiotic tolerance. *Handbook of Experimental Pharmacology* 211: 121–133.

Li, C., Zhang, X., Huang, X. et al. (2013). Preparation and characterization of flexible nanoliposomes loaded with daptomycin, a novel antibiotic, for topical skin therapy. *International Journal of Nanomedicine* 8: 1285–1292.

Li, X., Contreras-Garcia, A., LoVetri, K. et al. (2015). Fusion peptide P15-CSP shows antibiofilm activity and pro-osteogenic activity when deposited as a coating on hydrophilic but not hydrophobic surfaces. *Journal of Biomedical Materials Research Part A* 103 (12): 3736–3746.

Litzler, P.-Y., Benard, L., Barbier-Frebourg, N. et al. (2007). Biofilm formation on pyrolytic carbon heart valves: influence of surface free energy, roughness, and bacterial species. *The Journal of Thoracic and Cardiovascular Surgery* 134 (4): 1025–1032.

López-Meza, J.E., Ochoa-Zarzosa, A., Barboza-Corona, J.E., and Bideshi, D.K. (2015). Antimicrobial peptides: current and potential applications in biomedical therapies. *BioMed Research International* 2015: E367243.

Lu, T.K. and Collins, J.J. (2007). Dispersing biofilms with engineered enzymatic bacteriophage. *Proceedings of the National Academy of Sciences of the United States of America* 104 (27): 11197–11202.

Lynch, A.S. and Robertson, G.T. (2008). Bacterial and fungal biofilm infections. *Annual Review of Medicine* 59: 415–428.

Maciejewska, B., Olszak, T., and Drulis-Kawa, Z. (2018). Applications of bacteriophages versus phage enzymes to combat and cure bacterial infections: an ambitious and also a realistic application? *Applied Microbiology and Biotechnology* 102 (6): 2563–2581.

Madhongsa, K., Pasan, S., Phophetleb, O. et al. (2013). Antimicrobial action of the cyclic peptide bactenecin on *Burkholderia pseudomallei* correlates with efficient membrane permeabilization. *PLoS Neglected Tropical Diseases* 7 (6): e2267.

Mai, J., Tian, X.-L., Gallant, J.W. et al. (2011). A novel target-specific, salt-resistant antimicrobial peptide against the cariogenic pathogen *Streptococcus mutans*. *Antimicrobial Agents and Chemotherapy* 55 (11): 5205–5213.

Maisonneuve, E. and Gerdes, K. (2014). Molecular mechanisms underlying bacterial persisters. *Cell* 157 (3): 539–548.

Mandal, S.M., Khan, J., Mahata, D. et al. (2017). A self-assembled clavanin A-coated amniotic membrane scaffold for the prevention of biofilm formation by ocular surface fungal pathogens. *Biofouling* 33 (10): 881–891.

Mandal, S.M., Migliolo, L., Franco, O.L., and Ghosh, A.K. (2011). Identification of an antifungal peptide from *Trapa natans* fruits with inhibitory effects on *Candida tropicalis* biofilm formation. *Peptides* 32 (8): 1741–1747.

Mansour, S.C., de la Fuente-Núñez, C., and Hancock, R.E.W. (2015). Peptide IDR-1018: modulating the immune system and targeting bacterial biofilms to treat antibiotic-resistant bacterial infections. *Journal of Peptide Science* 21 (5): 323–329.

Markowska, K., Grudniak, A.M., and Wolska, K.I. (2013). Silver nanoparticles as an alternative strategy against bacterial biofilms. *Acta Biochimica Polonica* 60 (4): 523–530.

Marschall, J., Lane, M.A., Beekmann, S.E. et al. (2013). Current management of prosthetic joint infections in adults: results of an Emerging Infections Network survey. *International Journal of Antimicrobial Agents* 41 (3): 272–277.

Martinez, L.R. and Casadevall, A. (2006). *Cryptococcus neoformans* cells in biofilms are less susceptible than planktonic cells to antimicrobial molecules produced by the innate immune system. *Infection and Immunity* 74 (11): 6118–6123.

Martinez-Gutierrez, F., Boegli, L., Agostinho, A. et al. (2013). Anti-biofilm activity of silver nanoparticles against different microorganisms. *Biofouling* 29 (6): 651–660.

Mataraci, E. and Dosler, S. (2012). In vitro activities of antibiotics and antimicrobial cationic peptides alone and in combination against methicillin-resistant *Staphylococcus aureus* biofilms. *Antimicrobial Agents and Chemotherapy* 56 (12): 6366–6371.

Meers, P., Neville, M., Malinin, V. et al. (2008). Biofilm penetration, triggered release and in vivo activity of inhaled liposomal amikacin in chronic *Pseudomonas aeruginosa* lung infections. *The Journal of Antimicrobial Chemotherapy* 61 (4): 859–868.

Mishra, P.K., Ashoub, A., Salhiyyah, K. et al. (2014). Role of topical application of gentamicin containing collagen implants in cardiac surgery. *Journal of Cardiothoracic Surgery* 9: 122.

Mookherjee, N. and Hancock, R.E.W. (2007). Cationic host defence peptides: innate immune regulatory peptides as a novel approach for treating infections. *Cellular and Molecular Life Sciences* 64 (7–8): 922–933.

Moreno, M.G., Lombardi, L., and Di Luca, M. (2017). Antimicrobial peptides for the control of biofilm formation. *Current Topics in Medicinal Chemistry* 17 (17): 1965–1986.

Mugabe, C., Azghani, A.O., and Omri, A. (2005). Liposome-mediated gentamicin delivery: development and activity against resistant strains of *Pseudomonas aeruginosa* isolated from cystic fibrosis patients. *The Journal of Antimicrobial Chemotherapy* 55 (2): 269–271.

Pamp, S.J., Gjermansen, M., Johansen, H.K., and Tolker-Nielsen, T. (2008). Tolerance to the antimicrobial peptide colistin in *Pseudomonas aeruginosa* biofilms is linked to metabolically active cells, and depends on the pmr and mexAB-oprM genes. *Molecular Microbiology* 68 (1): 223–240.

Parra-Ruiz, J., Bravo-Molina, A., Peña-Monje, A., and Hernández-Quero, J. (2012). Activity of linezolid and high-dose daptomycin, alone or in combination, in an in vitro model of *Staphylococcus aureus* biofilm. *The Journal of Antimicrobial Chemotherapy* 67 (11): 2682–2685.

Parra-Ruiz, J., Vidaillac, C., Rose, W.E., and Rybak, M.J. (2010). Activities of high-dose daptomycin, vancomycin, and moxifloxacin alone or in combination with clarithromycin or rifampin in a novel in vitro model of *Staphylococcus aureus* biofilm. *Antimicrobial Agents and Chemotherapy* 54 (10): 4329–4334.

Percival, S.L., McCarty, S.M., and Lipsky, B. (2015). Biofilms and wounds: an overview of the evidence. *Advances in Wound Care* 4 (7): 373–381.

Pintucci, J.P., Corno, S., and Garotta, M. (2010). Biofilms and infections of the upper respiratory tract. *European Review for Medical and Pharmacological Sciences* 14 (8): 683–690.

Pirrone, M., Pinciroli, R., and Berra, L. (2016). Microbiome, biofilms, and pneumonia in the ICU. *Current Opinion in Infectious Diseases* 29 (2): 160–166.

Pletzer, D., Coleman, S.R., and Hancock, R.E. (2016). Anti-biofilm peptides as a new weapon in antimicrobial warfare. *Current Opinion in Microbiology* 33: 35–40.

Pletzer, D. and Hancock, R.E.W. (2016). Antibiofilm peptides: potential as broad-spectrum agents. *Journal of Bacteriology* 198 (19): 2572–2578.

Pompilio, A., Scocchi, M., Pomponio, S. et al. (2011). Antibacterial and anti-biofilm effects of cathelicidin peptides against pathogens isolated from cystic fibrosis patients. *Peptides* 32 (9): 1807–1814.

Porto, W.F., Irazazabal, L., Alves, E.S.F. et al. (2018). In silico optimization of a guava antimicrobial peptide enables combinatorial exploration for peptide design. *Nature Communications* 9 (1): 1490.

Prijck, K., Smet, N., Rymarczyk-Machal, M. et al. (2010). *Candida albicans* biofilm formation on peptide functionalized polydimethylsiloxane. *Biofouling* 26 (3): 269–275.

Pusateri, C.R., Monaco, E.A., and Edgerton, M. (2009). Sensitivity of *Candida albicans* biofilm cells grown on denture acrylic to antifungal proteins and chlorhexidine. *Archives of Oral Biology* 54 (6): 588–594.

Rahman, M., Kim, S., Kim, S.M. et al. (2011). Characterization of induced *Staphylococcus aureus* bacteriophage SAP-26 and its anti-biofilm activity with rifampicin. *Biofouling* 27 (10): 1087–1093.

Ramasamy, M. and Lee, J. (2016). Recent nanotechnology approaches for prevention and treatment of biofilm-associated infections on medical devices. *BioMed Research International* 2016: 1851242.

Reffuveille, F., Fuente-Núñez, C., Mansour, S., and Hancock, R.E.W. (2014). A broad-spectrum antibiofilm peptide enhances antibiotic action against bacterial biofilms. *Antimicrobial Agents and Chemotherapy* 58 (9): 5363–5371.

Ribeiro, S.M., de la Fuente-Núñez, C., Baquir, B. et al. (2015). Antibiofilm peptides increase the susceptibility of carbapenemase-producing *Klebsiella pneumoniae*

clinical isolates to β-lactam antibiotics. *Antimicrobial Agents and Chemotherapy* 59 (7): 3906–3912.
Roach, D.R. and Donovan, D.M. (2015). Antimicrobial bacteriophage-derived proteins and therapeutic applications. *Bacteriophage* 5 (3): e1062590.
Roe, D., Karandikar, B., Bonn-Savage, N. et al. (2008). Antimicrobial surface functionalization of plastic catheters by silver nanoparticles. *The Journal of Antimicrobial Chemotherapy* 61 (4): 869–876.
Sagona, A.P., Grigonyte, A.M., MacDonald, P.R., and Jaramillo, A. (2016). Genetically modified bacteriophages. *Integrative Biology: Quantitative Biosciences from Nano to Macro* 8 (4): 465–474.
Salmond, G.P.C. and Fineran, P.C. (2015). A century of the phage: past, present and future. *Nature Reviews Microbiology* 13 (12): 777–786.
Sánchez-Gómez, S., Ferrer-Espada, R., Stewart, P.S. et al. (2015). Antimicrobial activity of synthetic cationic peptides and lipopeptides derived from human lactoferricin against *Pseudomonas aeruginosa* planktonic cultures and biofilms. *BMC Microbiology* 15: 137.
Santos, A.P.A., Watanabe, E., and de Andrade, D. (2011). Biofilm on artificial pacemaker: fiction or reality? *Arquivos Brasileiros de Cardiologia* 97 (5): e113–e120.
Santos Ramos, M.A., Silva, P.B., Spósito, L. et al. (2018). Nanotechnology-based drug delivery systems for control of microbial biofilms: a review. *International Journal of Nanomedicine* 13: 1179–1213.
Secinti, K.D., Özalp, H., Attar, A., and Sargon, M.F. (2011). Nanoparticle silver ion coatings inhibit biofilm formation on titanium implants. *Journal of Clinical Neuroscience* 18 (3): 391–395.
Segev-Zarko, L., Saar-Dover, R., Brumfeld, V. et al. (2015). Mechanisms of biofilm inhibition and degradation by antimicrobial peptides. *The Biochemical Journal* 468 (2): 259–270.
Sharma, A., Gupta, P., Kumar, R., and Bhardwaj, A. (2016). dPABBs: a novel in silico approach for predicting and designing anti-biofilm peptides. *Scientific Reports* 6: 21839.
Silva, B.R., Freitas, V.A.A., Carneiro, V.A. et al. (2013). Antimicrobial activity of the synthetic peptide Lys-a1 against oral streptococci. *Peptides* 42: 78–83.
Silva, O.N., Alves, E.S.F., de la Fuente-Núñez, C. et al. (2016b). Structural studies of a lipid-binding peptide from tunicate hemocytes with anti-biofilm activity. *Scientific Reports* 6: 27128.
Silva, O.N., de la Fuente-Núñez, C., Haney, E.F. et al. (2016a). An anti-infective synthetic peptide with dual antimicrobial and immunomodulatory activities. *Scientific Reports* 6: 35465.
Silverstein, A. and Donatucci, C.F. (2003). Bacterial biofilms and implantable prosthetic devices. *International Journal of Impotence Research* 15: S150–S154.
Sojka, M., Valachova, I., Bucekova, M., and Majtan, J. (2016). Antibiofilm efficacy of honey and bee-derived defensin-1 on multispecies wound biofilm. *Journal of Medical Microbiology* 65 (4): 337–344.

Son, J.-S., Lee, S.-J., Jun, S.Y. et al. (2010). Antibacterial and biofilm removal activity of a podoviridae *Staphylococcus aureus* bacteriophage SAP-2 and a derived recombinant cell-wall-degrading enzyme. *Applied Microbiology and Biotechnology* 86 (5): 1439–1449.

Sondi, I. and Salopek-Sondi, B. (2004). Silver nanoparticles as antimicrobial agent: a case study on *E. coli* as a model for Gram-negative bacteria. *Journal of Colloid and Interface Science* 275 (1): 177–182.

Sousa, C., Botelho, C., and Oliveira, R. (2011). Nanotechnology applied to medical biofilms control. In: *Science Against Microbial Pathogens: Communicating Current Research and Technological Advances* (ed. A. Mendez-Vilas), 878–888. Badajoz: Formatex Research Center.

Souza, P.R., Andrade, D., Cabral, D.B., and Watanabe, E. (2014). Endotracheal tube biofilm and ventilator-associated pneumonia with mechanical ventilation. *Microscopy Research and Technique* 77 (4): 305–312.

Strempel, N., Strehmel, J., and Overhage, J. (2015). Potential application of antimicrobial peptides in the treatment of bacterial biofilm infections. *Current Pharmaceutical Design* 21 (1): 67–84.

Sutherland, I.W., Hughes, K.A., Skillman, L.C., and Tait, K. (2004). The interaction of phage and biofilms. *FEMS Microbiology Letters* 232 (1): 1–6.

Szafrański, S.P., Winkel, A., and Stiesch, M. (2017). The use of bacteriophages to biocontrol oral biofilms. *Journal of Biotechnology* 250: 29–44.

Tao, R., Tong, Z., Lin, Y. et al. (2011). Antimicrobial and antibiofilm activity of pleurocidin against cariogenic microorganisms. *Peptides* 32 (8): 1748–1754.

Theberge, S., Semlali, A., Alamri, A. et al. (2013). *C. albicans* growth, transition, biofilm formation, and gene expression modulation by antimicrobial decapeptide KSL-W. *BMC Microbiology* 13: 246.

Tinoco, J.M., Buttaro, B., Zhang, H. et al. (2016). Effect of a genetically engineered bacteriophage on *Enterococcus faecalis* biofilms. *Archives of Oral Biology* 71: 80–86.

Trapnell, B.C., McColley, S.A., Kissner, D.G. et al. (2012). Fosfomycin/tobramycin for inhalation in patients with cystic fibrosis with pseudomonas airway infection. *American Journal of Respiratory and Critical Care Medicine* 185 (2): 171–178.

Trautner, B.W. and Darouiche, R.O. (2004). Role of biofilm in catheter-associated urinary tract infection. *American Journal of Infection Control* 32 (3): 177–183.

Vriens, K., Cools, T.L., Harvey, P.J. et al. (2015). Synergistic activity of the plant defensin HsAFP1 and caspofungin against *Candida albicans* biofilms and planktonic cultures. *PLoS One* 10 (8): e0132701.

Walkenhorst, W.F. (2016). Using adjuvants and environmental factors to modulate the activity of antimicrobial peptides. *Biochimica et Biophysica Acta* 1858 (5): 926–935.

Wang, W., Tao, R., Tong, Z. et al. (2012). Effect of a novel antimicrobial peptide chrysophsin-1 on oral pathogens and *Streptococcus mutans* biofilms. *Peptides* 33 (2): 212–219.

Wu, H., Moser, C., Wang, H.-Z. et al. (2015). Strategies for combating bacterial biofilm infections. *International Journal of Oral Science* 7 (1): 1–7.

Xu, G., Xiong, W., Hu, Q. et al. (2010). Lactoferrin-derived peptides and Lactoferricin chimera inhibit virulence factor production and biofilm formation in *Pseudomonas aeruginosa*. *Journal of Applied Microbiology* 109 (4): 1311–1318.

Yan, J., Mao, J., Mao, J., and Xie, J. (2014). Bacteriophage polysaccharide depolymerases and biomedical applications. *BioDrugs: Clinical Immunotherapeutics, Biopharmaceuticals and Gene Therapy* 28 (3): 265–274.

Yang, L., Liu, Y., Wu, H. et al. (2012). Combating biofilms. *FEMS Immunology and Medical Microbiology* 65 (2): 146–157.

Young, R. (2013). Phage lysis: do we have the hole story yet? *Current Opinion in Microbiology* 16 (6): 790–797.

Zhang, L., Pornpattananangku, D., Hu, C.-M.J., and Huang, C.-M. (2010). Development of nanoparticles for antimicrobial drug delivery. *Current Medicinal Chemistry* 17 (6): 585–594.

Part V

Strategies to Prevent the Spread of AR

21

Rapid Analytical Methods to Identify Antibiotic-Resistant Bacteria

John B. Sutherland[1], *Fatemeh Rafii*[1], *Jackson O. Lay, Jr.*[2], *and Anna J. Williams*[1]

[1] *National Center for Toxicological Research, U. S. Food and Drug Administration, Jefferson, AR, USA*
[2] *Department of Chemistry and Biochemistry, University of Arkansas, Fayetteville, AR, USA*

21.1 Introduction

Using the appropriate antibiotic for patients with severe bacterial infections is critically important for the recovery and health of the patients (Marquet et al. 2015). To prevent the spread of antibiotic-resistant bacteria in a hospital due to the excessive use of broad-spectrum antibiotics, *antimicrobial stewardship programs* have been developed to help physicians select the most appropriate antibiotic for a patient as rapidly as possible (Minejima and Wong-Beringer 2016). Because of the appearance of new variants of resistance genes and the spread of antibiotic resistance among clinically important bacteria, there is a need for rapid identification of not only the species of pathogenic bacteria but also the antibiotic sensitivity (Tenover 2010; Kothari et al. 2014; Endimiani and Jacobs 2016). Since all of the standard identification methods require incubation times of at least one day and sometimes more than two weeks for bacterial growth (Sutherland and Rafii 2006; Preez et al. 2017), various rapid methods for detecting antimicrobial susceptibility have been developed to facilitate more timely use of the appropriate antibiotic (Frickmann et al. 2014). Rapid methods allow shorter stays in the hospital, better outcomes for the patients, and decreased costs (Tenover 2010; Kothari et al. 2014; Minejima and Wong-Beringer 2016). Some of the rapid methods developed for the clinical laboratory are also being adapted for veterinary and environmental microbiology.

It is the purpose of this chapter to provide a descriptive introduction to various rapid methods for determining bacterial antimicrobial susceptibility, but not to make clinical recommendations or compare commercial products.

Antibiotic Drug Resistance, First Edition. Edited by José-Luis Capelo-Martínez and Gilberto Igrejas.

Obviously, any rapid test for antibiotic resistance that is not started as soon as possible, and whose results are not implemented quickly, is no better for the patient than a standard method (Diekema and Pfaller 2013). Current official recommendations by the Food and Drug Administration (FDA) and other regulatory agencies on any methods mentioned here must be obtained from the agency websites and publications.

21.2 Standard Methods for Antibiotic Sensitivity Testing

Among the clinically important bacteria that require not only immediate identification but also *antibiotic sensitivity testing* are species of *Staphylococcus*, *Enterococcus*, *Escherichia*, *Pseudomonas*, *Klebsiella*, *Enterobacter*, *Acinetobacter*, and others (Weiner et al. 2016). After purification, a bacterium can be identified by cultural, serological, or genetic methods (Sutherland and Rafii 2006), and the antibiotic sensitivity can be determined (Pulido et al. 2013). The standard methods for characterization of antimicrobial susceptibility in clinical isolates of bacteria show phenotypic resistance by determining cell growth in the presence of different concentrations of antibiotics. Agar dilution, broth microdilution, disk diffusion, and gradient diffusion (Etest) are some of the methods currently used to determine antibiotic sensitivity of pure cultures (Levy Hara et al. 2013; van Belkum and Dunne 2013). These methods follow standardized techniques accepted by regulatory agencies and standards institutes (Leclercq et al. 2013; Levy Hara et al. 2013; CLSI 2017). The minimum concentration of an antibiotic needed to prevent the growth of a target bacterium, referred to as the *minimum inhibitory concentration (MIC)*, indicates the level of resistance. The determination of the MIC is important for the treatment of an infectious disease because it predicts the effectiveness of various doses of an antibiotic against a particular strain. Although the standard cultural methods correctly identify antibiotic-resistant strains of bacteria, they still require time for the isolation of pure cultures before testing for resistance (Osei Sekyere et al. 2015) and then additional time for detecting bacterial growth with different concentrations of antibiotics. These time requirements often delay the use of the most effective drug for critically ill patients.

Most *biochemical methods* for detecting resistant bacteria rely on phenotypes that are only seen with pure cultures. Several commercial biochemical tests are available to detect specific kinds of resistance. Chromogenic agars for the detection of methicillin-resistant *Staphylococcus aureus* (MRSA), containing inhibitors against other bacteria, indicate the growth of MRSA by a pink or mauve color and can be used directly to screen clinical samples (Wolk et al. 2009; Hernandez et al. 2016). Gram-negative bacteria suspected of possible resistance to carbapenems can be tested by the modified Hodge test (Amjad et al. 2011). This test, which uses a lawn of *Escherichia coli* on a Mueller–Hinton agar plate with an ertapenem or meropenem disk in the center, with the test bacterium and the positive and negative controls streaked from the disk to the outside, takes 16–24 hours to show resistance (Amjad et al. 2011). A modified double-disk synergy test for Gram-negative bacteria producing extended-spectrum β-lactamases, for resistance to penicillin and related antibiotics, uses four disks containing different cephalosporins and one disk containing amoxicillin with clavulanic acid (Kaur et al. 2013). A spectrophotometric method for carbapenemase activity in the *Enterobacteriaceae* requires growth of the cells for 18 hours, centrifugation, sonication, an additional centrifugation, and then measurement of imipenem hydrolysis by monitoring the absorbance at 297 nm (Bernabeu et al. 2012).

In addition to standard methods performed manually, commercial *automated culture systems* are used to detect antimicrobial resistance of bacteria (Pulido et al. 2013). They generally require pure cultures for reliable identification and detection of resistance, although some now allow the direct use of clinical samples in certain cases (Gherardi et al. 2012). Bacterial identification and an antimicrobial susceptibility test (AST) are usually performed together by inoculating the panels provided by the manufacturer with the pure culture. For a typical AST, a panel of reagent wells containing dried antimicrobial agents is inoculated with the bacterial culture and incubated in the instrument (Snyder et al. 2008). The instrument monitors the growth in the presence of each of several antimicrobial agents, using optical density, colorimetric methods, or fluorescent dyes, and compares it with the growth in control wells without antimicrobial agents. The time required for susceptibility results depends on the growth rate of the bacteria; for instance, it has been reported as 4–16 hours for pure cultures (Snyder et al. 2008) and as 6–13 hours for positive blood cultures (Gherardi et al. 2012). The results of automated susceptibility tests generally match the results obtained by conventional sensitivity tests, although confirmation of resistance by another method is recommended in some cases (Snyder et al. 2008).

21.3 Rapid Cultural Methods

Because of the importance of timely detection of antimicrobial resistance for successful treatment of bacterial infections, new methods are continually being developed to shorten the time required to detect antimicrobial susceptibility. The following are some of the proposed methods, which generally require pure bacterial cultures for testing and therefore already require several hours to allow growth.

Fluorescence microscopy has been used to measure phenotypic responses of bacterial cultures to antibiotics, either by detecting the effects of antibiotics on cell growth or by detecting cell lysis in agarose microgels (Tamayo et al. 2009; Santiso et al. 2011). Cell lysis-based methods can rapidly detect bacteria that are phenotypically resistant to antibiotics and measure their level of sensitivity (Pulido et al. 2013). The bacteria are incubated with different concentrations of the antibiotics to be tested and immobilized in microgels on microscope slides. Assays can be tailored to detect resistance to antimicrobial agents that affect bacterial cell walls or to those that damage DNA. For detecting resistance to antibiotics affecting peptidoglycans in cell walls, a lysis solution is selected that will lyse only the bacteria whose cell walls have been damaged (Santiso et al. 2011). Following removal of the lysis solution, the bacteria in the microgels are stained using a DNA-staining fluorescent dye, such as SYBR Gold. Resistant cells are not lysed and show normal DNA in the nucleoid. In contrast, the susceptible bacteria are lysed, resulting in spreading of the DNA (Santiso et al. 2011). Other cell lysis methods have been used to determine the sensitivity of bacteria in agarose microgels to ciprofloxacin and other fluoroquinolones that damage the DNA, inducing double-stranded breaks and causing DNA fragmentation (Tamayo et al. 2009). After exposure to ciprofloxacin, the cells are embedded in a microgel and treated with a stronger lysis solution that destroys protein and removes the cell wall and membrane. The lysis solution is removed, and the bacteria are stained with a fluorescent dye and visualized (Tamayo et al. 2009). In antibiotic-resistant cells, the DNA appears intact in the nucleoid, but in sensitive cells, the DNA becomes fragmented and appears to have diffused out of the nucleoid. Quantitative fluorescence microscopy with bacterial cytological profiling has been used for identifying β-lactam-resistant (including methicillin) and daptomycin-resistant strains of *S. aureus* within one to two hours by measuring changes in cellular structure induced by antibiotic exposure (Quach et al. 2016). The increased length of sensitive Gram-negative cells induced by ceftazidime has been measured by fluorescence microscopy, even without cell lysis, by staining with SYBR Gold, followed by digital image analysis (Otero et al. 2017).

A miniaturized *lab-on-a-chip* can be fabricated from various materials and substituted for agar as a support to grow bacteria for rapid testing. For instance, porous aluminum oxide ceramic chips have been used to detect trimethoprim-resistant strains of the *Enterobacteriaceae* in two to three hours and rifampicin/isoniazid-resistant strains of the slow-growing *Mycobacterium* spp. in three days (Ingham et al. 2006, 2008). Antibiotic-sensitive bacteria can be detected in tiny compartments on a chip by comparing their growth with and without antibiotics after only one hour (Besant et al. 2015). The reduction of a dye, resazurin, shows growth, and differential pulse voltammetry distinguishes the oxidized and reduced forms of the dye. Another lab-on-a-chip, which uses a silicon dioxide-coated electrical sensor with interdigitated electrodes and an antibody against *S. aureus*, shows the growth of bacteria resistant to flucloxacillin in two hours (Abeyrathne et al. 2016). Most lab-on-chips designed for growing cells use *microfluidics techniques*, which can measure the growth of MRSA and other drug-resistant bacteria in the presence of antibiotics (Pulido et al. 2013; Campbell et al. 2016; Malmberg et al. 2016). A microfluidics agarose channel system, which allows rapid comparison of the growth of bacteria with and without antibiotics, has been used to determine the MICs of sepsis-causing bacteria by time-lapse imaging with further data processing (Choi et al. 2013). In this method, isolated bacteria are added to the microfluidics channels of chips treated with different concentrations of antimicrobial agents and incubated under optimum conditions for growth (Choi et al. 2013). A different microfluidics device has been used to measure resistance of pure cultures of *Pseudomonas aeruginosa* to several drugs; the MIC of an antibiotic can be determined in three hours (Matsumoto et al. 2016). Other microfluidics applications include the production in droplets of small numbers of cells that can be tested for resistance to a variety of antibiotics by continuous monitoring of the individual droplets (Kaminski et al. 2016; Keays et al. 2016). In a variation of the microfluidics method, antimicrobial gradients may be used to detect resistance by analyzing images from time-lapse photomicrography. Differences in grayscale intensity represent the effect of increasing concentrations of antibiotics (Malmberg et al. 2016).

Raman spectroscopic analysis has been used to compare spectra of bacteria grown with high and low concentrations of antibiotics. For instance, it has been used to distinguish strains of *E. coli* and *Klebsiella pneumoniae* that produce extended-spectrum β-lactamases and carbapenemases from those that do not, as well as to distinguish outbreak strains from unrelated resistant strains (Willemse-Erix et al. 2012). Sensitivity to vancomycin of cultures of *Enterococcus faecalis* and *Enterococcus faecium*, two leading nosocomial pathogens for which only a few antibiotics are generally effective (Sood et al. 2008), has been demonstrated in less than four hours by Raman spectroscopic analysis (Schröder et al. 2015). The sensitivity of other bacteria isolated from blood cultures to different antibiotics has been measured by this method after incubation for five hours (Dekter et al. 2017).

Forward laser light scatter (FLLS), an emerging rapid electro-optical technology, uses lasers for detecting strains of bacteria resistant to antibiotics (Hayden et al. 2016). The bacteria are inoculated in the presence of different concentrations of several antimicrobial agents. A laser is used to measure both the optical density and the light scattering of the particles in a liquid over time to estimate any increases in cell density. The images are captured, mathematically processed, and calibrated against a baseline to calculate the growth rate and generation time from multiple measurements (Hayden et al. 2016).

21.4 Rapid Serological Methods

Few rapid serological methods are available for detecting antibiotic resistance. Commercially available latex agglutination and immunochromatographic assays for the detection of MRSA use latex particles and nitrocellulose membranes, respectively, which are coated with an antibody against protein PBP2a, encoded by the methicillin resistance gene *mecA*. In a *latex agglutination test* for MRSA, the antibody binds to PBP2a, resulting in the agglutination of MRSA cells with the latex particles (Cavassini et al. 1999; Alipour et al. 2014). In an *immunochromatographic assay* for MRSA, the appearance of both a sample line and a control line on a nitrocellulose membrane indicates the interaction of PBP2a from MRSA cells with an antibody fixed on the membrane (Trienski et al. 2013).

Commercial kits are available that use *polymerase chain reaction (PCR)* amplification (see next paragraph) with an *enzyme-linked immunosorbent assay* (*ELISA*); they have been used to detect *mecA* in blood cultures containing *S. aureus* (Wellinghausen et al. 2004) or different carbapenemase resistance genes in cultures of the *Enterobacteriaceae* (Ambretti et al. 2013). Another ELISA method detects the growth of *Helicobacter pylori* in cultures from biopsy fragments in the presence of clarithromycin or metronidazole in 24 hours (Perna and Vaira 2010).

21.5 Rapid Molecular (Genetic) Methods

PCR methods for the amplification of DNA sequences have been used to detect specific antimicrobial resistance genes or identify pathogens in clinical samples. In the standard PCR method (van Pelt-Verkuil et al. 2008), samples containing DNA are added to a reagent tube containing forward and reverse primers specific for the gene to be amplified, a deoxyribonucleotide triphosphate mix, and Taq polymerase. Tubes containing the reaction mixtures are inserted into the thermal block of a thermocycler, which is programmed to raise and lower the temperature during each cycle. The DNA is denatured first and then the gene located between the forward and reverse primers is synthesized by Taq polymerase. Since the DNA is amplified exponentially, there should be enough copies of the gene after 25–30 cycles to be detected. The amplified DNA is either visualized on an agarose gel after staining or else detected using labeled probes. If the genetic basis of resistance is known, PCR can be used to detect the genes whose expression may decrease the susceptibility of bacteria to particular antibiotics, such as the *mecA* gene for methicillin resistance in MRSA (Cuny and Witte 2005). Because resistance to antibiotics depends not only on the gene but also on the conditions for its expression, the amplification of a resistance gene may not indicate phenotypic resistance (Pulido et al. 2013). PCR methods are only useful for detecting resistance genes whose sequence is known. Resistance due to a new mechanism or a new gene variant, other than those that the test was designed for, will be missed.

In *quantitative PCR (qPCR)*, also known as real-time PCR, the amplification of a gene for antibiotic resistance is shown by a fluorescent dye (Maurin 2012). Typically, a fluorescent dye that binds to double-stranded DNA, like SYBR Green, is used to detect the amplified gene. The quantity of gene expression is measured by the fold increase compared to a reference gene. For instance, qPCR has been used to detect the expression of the *mecA* and *mecC* genes (methicillin resistance) and the SA442 gene (*S. aureus*) in clinical samples (Nijhuis et al. 2014). The quantity of DNA can also be measured using molecular beacons, which are oligonucleotide probes that fluoresce when they are hybridized with complementary DNA (cDNA). Their fluorescence is quenched when they are free and unbound. Resistance in *Streptococcus pneumoniae* to five different β-lactam antibiotics has been detected by qPCR in 90 minutes using selected primers and molecular beacons to show the effect of antibiotics on growth (Chiba et al. 2012). qPCR has also been used to measure bacterial growth in the presence of different concentrations of an antibiotic (Martín-Peña et al. 2013). When primers for a highly conserved region of a reference gene present in all strains of a species are used for qPCR amplification, the accumulation of product in six hours will reflect the amount of growth in the presence of different concentrations of the antibiotic (Martín-Peña et al. 2013). This method can differentiate resistant bacteria without prior knowledge of the mechanism of resistance, but it still requires a pure culture. qPCR has also been used to measure the amount of 16S ribosomal DNA present in bacteria grown with different amounts of various antibiotics in 24 hours (Waldeisen et al. 2011).

Amplification to determine gene expression may be performed by using either conventional or quantitative *reverse transcriptase PCR (RT-PCR)*. In RT-PCR, the transcription of a gene, such as *ampC* for β-lactam resistance in *E. coli* (Corvec et al. 2003), is estimated by measuring the quantity of RNA corresponding to that gene. First, using reverse transcriptase and primers specific to the gene, cDNA that corresponds to the RNA is synthesized (Aarthi et al. 2013). The cDNA produced is then amplified by PCR with Taq polymerase using the same primers and deoxyribonucleotide triphosphates. The quantities of the amplified genes in conventional RT-PCR are compared visually on agarose gels. In quantitative RT-PCR, they are compared by analyzing the data from the thermocycler, which can measure the quantity of DNA continuously by the amount of SYBR Green bound to it. The quantity of gene expression is measured by the increase in transcription compared with that in the control. Quantitative RT-PCR has been used to detect MRSA from clinical samples in six hours (Wada et al. 2010).

Digital PCR, in which the DNA from the bacterial chromosome is divided among a large number of compartments and amplified, has the potential to detect small numbers of cells that show resistance to different antimicrobial agents because of known mutations in the target genes (Pholwat et al. 2013). It has recently been proposed as a rapid method for determining antibiotic sensitivity of bacteria causing urinary tract infections (Schoepp et al. 2016).

Multiplex PCR allows the simultaneous detection of several genes at once, using pairs of primers that require the same amplification conditions (Woodford 2010). Additional primers may be provided for the identification of bacterial species in addition to the primers for the resistance genes. The amplified DNA is either visualized by electrophoresis on an agarose gel and staining or detected by hybridization to probes labeled with different fluorescent dyes or molecular beacons. Using multiplex qPCR with two different molecular beacons, resistance to methicillin and vancomycin has been detected simultaneously in *S. aureus* (Sinsimer et al. 2005). Multiplex qPCR has also been used for detecting several other genes involved in multiple antibiotic resistance (Khan et al. 2011a; Chavda et al. 2016).

Automated systems using PCR or qPCR are commercially available and can detect the major bacterial genes that confer antibiotic resistance, in some cases even from clinical samples like positive blood cultures, in less than two hours. They may automatically purify the sample, concentrate the DNA, and amplify and detect the target genes (Boehme et al. 2011; Dalpke et al. 2016). A kit provided by the supplier includes materials needed for DNA extraction, gene amplification, and detection of one or more resistance genes. The samples are added to the reagents and inserted into the instrument. Nucleic acid purification, amplification, and target sequence detection are performed automatically. The quantity of amplified DNA is measured in qPCR using target-specific hybridization probes with fluorescent dyes. To decrease the probability of false-positive results, some systems perform nested PCR, in which a second set of primers is included to amplify the internal fragments generated by the first set of primers. Automated assays have been used to detect the vancomycin resistance genes *vanA* and *vanB* in *Enterococcus* spp. (Marner et al. 2011; Dalpke et al. 2016). Methods for the automatic detection of rifampicin resistance in *Mycobacterium tuberculosis* by qPCR have also been developed (Boehme et al. 2011). In blood cultures, the genes specific for *S. aureus* and for methicillin resistance have been detected by automated multiplex qPCR in less than two hours (Gröbner et al. 2009).

Fluorescence in situ hybridization (FISH) is a method of detecting specific sequences of DNA or RNA by using oligonucleotide probes labeled with a dye that can be visualized by fluorescence microscopy (Trebesius et al. 2000; Moosavian et al. 2007; Rohde et al. 2016). It is commonly used to identify bacterial species using a probe to detect ribosomal RNA, but it may also be able to detect antibiotic resistance genes by using probes complementary to sequences that include known mutations responsible for resistance (Trebesius et al. 2000). The bacteria on a microscope slide are treated with a fixative so the cell wall and membrane will become permeable to the dye-labeled probe. The labeled probe, under appropriate hybridization conditions, binds to complementary genes in the cells. After washing, the samples are visualized under a fluorescence microscope (Rohde et al. 2016). The presence of point mutations in the 23S rRNA genes of clarithromycin-resistant strains of *H. pylori* allows the FISH technique to distinguish resistant and sensitive strains of this bacterium in two to four hours (Moosavian et al. 2007; Yilmaz and Demiray 2007). FISH has also been used with blood cultures to detect *K. pneumoniae* strains that produce extended-spectrum β-lactamases (Palasubramaniam et al. 2008) and with *E. coli* cultures to detect the messenger RNA corresponding to the TEM genes for β-lactamase, which confer ampicillin resistance (Rohde et al. 2016).

A variation of FISH, *peptide nucleic acid fluorescent in situ hybridization (PNA-FISH)*, uses a small DNA probe (13–18 base pairs) that has an uncharged polyamide backbone replacing the phosphate backbone of the DNA (Fazli et al. 2014). The uncharged nature and small size of the probes make them more resistant to nucleases and proteases and enhance their penetration through the bacterial cell wall. PNA-FISH has been used to distinguish clarithromycin-resistant strains of *H. pylori* (Cerqueira et al. 2013). Another FISH variation is *immunofluorescence*, in which a fluorescent dye-labeled antibody has been employed to detect β-lactamase-producing bacterial strains. The bacteria were fixed on a glass slide and allowed to react with an antibody against β-lactamase. Following the removal of unbound primary antibody by washing, the samples were treated with a secondary fluorescent dye-labeled antibody. After the unbound secondary antibody was washed off, the cells were visualized using a fluorescence microscope and checked for the fluorescent secondary antibody (Rohde et al. 2016).

Microarray methods use different oligonucleotides attached to multiple spots on a solid support chip to detect many different sequences with a single assay (Naas et al. 2011; Pulido et al. 2013). Oligonucleotides complementary to known antibiotic resistance genes are attached in specific locations to detect these genes (Cohen Stuart et al. 2010). The DNA extracted from bacteria is subjected to PCR with labeled primers that specifically amplify the resistance genes to generate labeled DNA fragments (Strauss et al. 2015). The labeled DNA fragments are then allowed to hybridize to the oligonucleotides on the solid support that correspond to the resistance genes (Cohen Stuart et al. 2010). If the labeled DNA contains a sequence complementary to one of the known resistance genes, it will hybridize to that spot. After non-hybridized DNA has been removed, the solid supports are washed and processed to allow detection of labeled DNA, and the resistance genes in the bacterial strain are identified. Large numbers of genes for resistance can be detected by the microarray simultaneously, as in *Salmonella enterica* strains with multiple drug resistance (Zou et al. 2009). Commercially available microarray systems have been shown to detect genes for different β-lactamases and carbapenemases in clinical and reference strains of bacteria with high sensitivity and specificity (Bogaerts et al. 2011; Naas et al. 2011). Microarrays have been used for analysis of resistance genes in some impure samples, such as positive blood cultures (Buchan et al. 2013; Han et al. 2015). However, microarray results do not show whether bacteria are phenotypically resistant, nor do they show MICs (Khan et al. 2011b). Microarrays cannot recognize resistant strains that use novel mechanisms of resistance other than the genes included in the assay (Pulido et al. 2013).

In *whole genome sequencing (WGS)*, the whole bacterial sequence is screened for known antimicrobial resistance genes using publicly available websites for comparison (e.g. www.genomicepidemiology.org). Because of the availability of technology to rapidly sequence the whole bacterial genome economically and bioinformatics tools that facilitate the analysis of data, resistance genes can be detected by WGS in a relatively short time (Dunne et al. 2017). Experiments performed to detect resistance genes from bacteria that have been determined by antibiotic susceptibility testing to be phenotypically resistant have shown correlations in many cases, but not all, between data obtained by both methods. Although the use of WGS for routine clinical testing and detection of resistant strains is not currently practical, this method has epidemiological applications for surveillance and tracking of outbreaks (Osei Sekyere et al. 2015). As with most genetics-based detection techniques, this method shows the presence of resistance genes but not their expression, and it does not identify unknown resistance mechanisms.

Loop-mediated isothermal amplification (LAMP) uses a single temperature for target gene amplification rather than the temperature cycling used for PCR (Mori et al. 2004; Tomita et al. 2008; Mu et al. 2016). A set of four especially designed primers is used to amplify a target gene. The DNA polymerase used for amplification uses the different primers to synthesize DNA, displaces the single-stranded DNA downstream of the synthesis point during amplification, and generates multiple copies of the selected genes. The copies of newly synthesized DNA fragments are connected via stem loops (Fu et al. 2011). The LAMP assay has high specificity and produces a large number of DNA copies in a short time. The amplification can be followed by measuring fluorescence or turbidity (magnesium pyrophosphate is produced in large amounts), and the products may be analyzed by gel electrophoresis (Mori et al. 2004; García-Fernández et al. 2015; Mu et al. 2016). To avoid amplification of nontarget genes and formation of primer dimers, it is important to design specific primers using LAMP primer designing software, such as PrimerExplorer V4 (http://primerexplorer.jp/e). Commercial kits are available that have been shown to be efficient for the detection of resistance genes from phenotypically resistant bacterial strains. For instance, in *S. aureus*, the genes for resistance to methicillin and those for resistance to macrolides and streptogramin have been amplified by this method (Misawa et al. 2007; Mu et al. 2016). A LAMP-based assay has been automated to detect the β-lactamase genes that confer resistance to cephalosporins and carbapenems in Gram-negative bacteria in 15 minutes (García-Fernández et al. 2015). Similar to other target amplification and detection methods, the LAMP method provides information about the presence or absence of specific resistance genes rather than whether they are involved in resistance to a particular drug.

21.6 Mass Spectrometric Methods

Mass spectrometry (MS) has both high sensitivity and high specificity. Applications of MS for the detection of antimicrobial resistance in bacteria generally require growing pure cultures first to simplify the MS profile (Lay and Liyanage 2006). MS methods are based on the measurement of a discrete and reproducible set of molecular biomarkers in these cultures that may be correlated with specific attributes (Lay 2001). These biomarkers may be nucleic acids or lipids, but are most often proteins or peptides. Generally, each class of biomarkers requires a different method (Drissner et al. 2016). Bacterial species identity, virulence, and resistance can all be probed at the molecular level and addressed in terms of chemical profiles. Identification of bacteria based on the detection of specific chemical species by MS has been referred to as MS-based chemotaxonomy (David et al. 2008). To the extent that antibiotic-resistant bacterial strains can be distinguished from their sensitive counterparts, chemotaxonomy can be applied to address resistance properties as well as identity. Some of the MS methods that have been developed thus far allow the rapid determination of antibiotic resistance in previously identified bacteria, but not in real unknowns.

The prospect of rapidly and directly detecting identification markers from whole bacterial cells by *pyrolysis MS* with little or no sample treatment was investigated first by Anhalt and Fenselau (1975). Pyrolysis MS methods were adapted for clinical applications (Morgan et al. 2006) and initially were successful under controlled experimental conditions, but the small spectral differences and problems with spectral reproducibility hindered the development of libraries for routine microbial identification. A recent improvement in pyrolysis MS, called plasma jet ionization, may allow better evaluation of small differences between bacterial cultures (Alusta et al. 2015). *Electrospray ionization (ESI) MS* (Yamashita and Fenn 1984) was developed to allow the characterization of nonvolatile but biochemically important compounds, such as proteins. The development of *matrix-assisted laser desorption/ionization time-of-flight (MALDI-TOF) MS* of proteins (Karas and Hillenkamp 1988) made possible the first practical method for rapid identification of intact bacteria (Lay 2001; Lay and Liyanage 2006; Singhal et al. 2015).

Specific protein profiles can be generated by analyzing intact bacteria by MALDI-TOF MS (Holland et al. 1996). Although MALDI MS had already been applied to bacterial protein fractions, this method successfully detected identity-specific biomarkers directly from intact bacteria. This technique demonstrated remarkable success for rapid identification because it reproducibly sampled a specific protein fraction in a mass region where time-of-flight mass spectrometers perform well (5–20 kDa), making possible the development of large digital libraries to which unknown spectra could be compared. There have been a few reports of successful applications using MALDI-TOF MS to show antibiotic resistance; for example, the detection of hydrolyzed β-lactams shows the effect of β-lactamases (Sparbier et al. 2012). Cell surface proteins have been used to differentiate MRSA from methicillin-susceptible *S. aureus* (Edwards-Jones et al. 2000). Likewise, with known strains of the *Enterobacteriaceae*, it was possible to differentiate isolates that were resistant from those that were susceptible to imipenem/avibactam (Oviaño and Bou 2017). However, the last two examples do not reflect a general capability. By themselves, the proteins observed in whole-cell MALDI-TOF MS do not reveal antibiotic susceptibility. Therefore, this approach has had to be greatly modified to enhance resistance detection while maintaining its utility for identification (Hrabak et al. 2016).

Generally speaking, MALDI-TOF MS is not a quantitative technique. The signal level depends on the number of laser shots fired, the "quality" of the spot the laser is striking, and the homogeneity of the sample. It has been proposed (Albrethsen 2007; Lange et al. 2014) to use an internal standard in MALDI experiments to estimate the quantity of protein present. If cells were cultured on media with and without antibiotics, the quantity of protein could be used to assess susceptibility, and the overall MS profile could be used for identification. This was demonstrated with a large number of *Klebsiella* strains, which gave results comparable to the Etest (Lange et al. 2014), but this approach has proved difficult in other cases.

A simpler method for measuring protein levels in cells incubated with and without antibiotics makes use of *stable isotope tags*, such as ^{13}C or ^{15}N (Stump et al. 2003). Growing bacteria incorporate the isotope profile of the medium into their macromolecules, including proteins. If replicate cultures are inoculated with and without an antibiotic present, any changes in the isotope profile may reflect the ability of the cells to grow in the presence of the antibiotic (Demirev et al. 2013). This has been demonstrated only with *E. coli* so far (Demirev 2016), but is in principle a universal approach that could be used for identification and resistance determination. Changes in stable isotope profiles can also be monitored in proteins of bacteriophages infecting a target bacterium and used to show bacterial growth and phage replication in the presence of an antibiotic (Rees and Barr 2017).

Stable isotope labeling by amino acids in cell cultures (SILAC) is an alternative method, developed for quantitative proteomics, that has been applied successfully to detect resistance to various antibiotics in strains of *S. aureus* and *P. aeruginosa* (Sparbier et al. 2013; Jung et al. 2014). SILAC incorporates mass tags only into specific amino acids, such as lysine. To the extent that the tagged amino acids are selectively incorporated into biomolecules, changes to the isotope profile will be observed in the affected molecules. The labeling is limited in distribution and amount, so the changes are not easy to detect. While SILAC may produce less profound changes in the isotope profile, a significant advantage is that labeling can be done with easily obtainable isotope-labeled amino acids (Demirev 2016).

Metal oxide laser ionization (MOLI) MS, using cerium oxide as the catalyst, has also been proposed as a method for bacterial identification (Saichek et al. 2016). The use of *in situ* hydrolysis for applications involving pyrolysis MS of bacteria was pioneered by Basile et al. (1998); later a whole-cell MALDI-TOF MS method was modified to allow measurement of free fatty acids released from bacterial lipids (Voorhees et al. 2013). Laser ionization of extracted bacterial lipids on a CeO_2 surface (MALDI target) resulted in the release of free fatty acids for MS analysis (Cox et al. 2015). This technique now has been extended to include concurrent detection of antibiotic resistance (Saichek et al. 2016). Fatty acid profiles were used not only to identify *Staphylococcus* isolates but also to differentiate β-lactam-resistant and susceptible strains. Lipid profile differences reflected antibiotic susceptibility as well as bacterial species, but the resistance-related changes to the fatty acid profile were not profound. A combination of fuzzy rule building with expert system classification and self-optimizing partial least squares discriminant analysis was needed for data interpretation (Saichek et al. 2016). This approach clearly complements the whole-cell MALDI-TOF MS method, but the generality of the detection of antibiotic resistance as shown by fatty acids has not been proven.

Other MS methods that have shown feasibility for rapid detection of antibiotic resistance include surface-enhanced laser desorption ionization (SELDI)-TOFMS(Shahetal.2011),high-performanceliquidchromatography–electrospray ionization (LC-ESI) MS (Grundt et al. 2012; Schelli et al. 2017), and membrane electrospray ionization (MESI) MS (Fan et al. 2016).

Genomics-based MS methods that allow both identification and the concurrent determination of susceptibility show great promise. However, any omics approach will be much more complex than whole-cell MALDI-TOF MS (Lichtenwalter et al. 2000). While the complexity is similar for genomics and proteomics, proteins are generally more amenable to characterization by MS than DNA and RNA. Moreover, proteins more directly indicate phenotypic properties, such as resistance, whereas genomics addresses a potential that may or may not be expressed. Because of the difficulty of probing the genome by MS, hybrid approaches have been developed that use MALDI-TOF MS for species identification in conjunction with qPCR for detection of resistance genes (Clerc et al. 2014; Chan et al. 2015).

The direct detection of nucleic acids is also possible by combining PCR with electrospray ionization MS (PCR/ESI MS) (Wolk et al. 2012; Perreten et al. 2013). The sample, split into fractions, is mixed with one or more pairs of primers that may be either broadly or highly target specific. By selecting both highly conserved and highly variable regions, it is possible to select primers for genes associated with antibiotic resistance as well as species identity. PCR/ESI MS has demonstrated high sensitivity and specificity for the detection of antibiotic resistance-related genes in bacteria whose identity had been determined beforehand by other means (Endimiani et al. 2010). Genomics-based MS has had more limited success in the identification of unknown species. Whole-cell MALDI-TOF MS can determine a species based on a digital library, but the genomics-based MS approach requires both an MS library and oligonucleotide primers for PCR.

Proteomics-based MS methods generally require isolation, separation, and processing of the bacterial proteins before analysis. Proteomics approaches include both top-down and bottom-up methods (Drissner et al. 2016). Top-down methods use extracted proteins and analyze them at the protein level (Santos et al. 2015). Bottom-up methods generally use enzymatic digestion to produce a large pool of peptides to be separated and analyzed by LC-MS/MS. The ionization method used is typically ESI, which is interfaced to HPLC, nano-HPLC, ultra-performance LC (UPLC), or capillary electrophoresis (CE) separations. Sometimes proteins are partially fractionated beforehand using techniques such as gel electrophoresis (Drissner et al. 2016).

Analysis of the entire protein pool without fractionation is a type of bottom-up proteomics called *shotgun proteomics* (Semanjski and Macek 2016). Shotgun proteomics is attractive conceptually because it can, in principle, take the entire proteome, digest it with a protease, and then separate and identify all of the peptide fragments by MS/MS and database searching. Shotgun proteomics has been able both to identify clinical isolates as *A. baumannii* and to detect β-lactam resistance (Chang et al. 2013). The peptides were produced by microwave-assisted digestion, and then the sequences generated by MS/MS were compared with sequences in a custom database, including both β-lactamases and species-specific reference proteins, to identify matches between expected and experimental amino acid sequences. These experiments took only about six hours (Chang et al. 2013). CE-based separation of digested fragments (CE-ESI MS/MS) has given comparable results for detection of carbapenem resistance in Gram-negative bacteria (Fleurbaaij et al. 2014). However, promising results from small libraries or data sets do not prove global applicability because the reliability of matches decreases as the number of proteins in the database being searched increases. The next step in the development of shotgun proteomics will be to determine how effective it is using large comprehensive databases and real unknowns.

Currently, the most promising MS approach to simultaneous bacterial identification and antibiotic resistance seems to be the use of whole-cell MALDI-TOF MS for identification with a stable isotope tag to monitor resistance (Demirev et al. 2013; Sparbier et al. 2013; Jung et al. 2014). Experiments can be run in parallel, using controls with no antibiotics and no isotopes and treated samples with antibiotics and isotope tags, to address both identification and resistance. Concurrent administration of the isotope tag with an antibiotic should clearly establish the extent of resistance, and the whole-cell MALDI-TOF MS profile should have specificity for identification.

21.7 Flow Cytometric Methods

Flow cytometry (FC) combines sensitivity, specificity, and speed for high-throughput detection and analysis of bacteria (Davey and Kell 1996; Williams et al. 2015). It has afforded scientists the ability to detect differences among many cells in a short time. A flow cytometer measures and analyzes the physical characteristics of cells as they flow in a stream through a laser beam (BD Biosciences 2002). It rapidly measures the relative size, intensity of any fluorescence, and internal complexity of each particle (BD Biosciences 2002). FC can analyze bacterial cells quantitatively and qualitatively in liquids (Álvarez-Barrientos et al. 2000; Ambriz-Aviña et al. 2014; Bari and Kawasaki 2014) with statistically significant results (Hahne et al. 2006). Some flow cytometers now have the capability to process 100 000 cells each second (May 2016).

A flow cytometer consists of three primary parts, the fluidics, the optics, and the electronics (BD Biosciences 2002; Ambriz-Aviña et al. 2014). Flow cytometers were designed to analyze larger eukaryotic cells, but they now have the capability to analyze prokaryotic cells as well (Steen 2000). The suitability of this technique for bacterial analyses depends on the ability to discriminate between the background noise or debris and the small bacterial cells (Buzatu et al. 2014). To eliminate irrelevant signals, a series of small gates can exclude nontarget background events. Fluorescent dyes targeting the bacterium of interest, as well as major signal processing, also help to eliminate false signals (Wilkes et al. 2012).

The effects of antimicrobial compounds on bacteria have been studied using FC since 1982 (Steen et al. 1982; Ambriz-Aviña et al. 2014). Flow cytometric methods can be used to measure the antibiotic susceptibilities of bacteria, especially by determining viability and growth. Changes in bacterial morphology due to antibiotics can also be measured with FC using fluorescent dyes (Renggli et al. 2013). Some dyes cross membranes in dead or injured cells, but not in live ones, and can help distinguish between the different states (Mortimer et al. 2000; Jarzembowski et al. 2010; Shrestha et al. 2011).

Several examples of the rapid detection of antibiotic-resistant bacteria by FC have been demonstrated. For instance, FC has been used to investigate the heterogeneity of penicillin resistance in strains of *E. faecalis* and MRSA using a fluorescent penicillin, BOCILLIN FL, which binds more strongly to sensitive cells (Jarzembowski et al. 2008). Vancomycin-resistant strains of *E. faecalis*, carrying the *vanA* gene, have also been detected rapidly and sensitively using FC with a fluorescent vancomycin and propidium iodide (Jarzembowski et al. 2010). FC has been reported to distinguish between MRSA and methicillin-susceptible *S. aureus* after only two hours of incubation with oxacillin. MRSA cells displayed fluorescence intensity and side scatter attributes different from sensitive cells (Shrestha et al. 2011). Strains of *E. coli* and *K. pneumoniae* have been analyzed for resistance to the fluoroquinolones ciprofloxacin and levofloxacin by tagging cells with SYBR Green (Nogueira 2013). The bacteria were grown in the presence of various concentrations of the fluoroquinolones, and FC was used to measure differences in the populations of cells that were treated differently (Nogueira 2013). A rapid flow cytometric assay for antibiotic resistance was tested in 67 bacterial strains with mezlocillin, oxacillin, piperacillin/tazobactam, cefuroxime, cefazolin, ciprofloxacin, and gentamicin (Nuding and Zabel 2013). The anionic dye $DiBAC_4(3)$, which accumulates in the cytoplasm of depolarized bacterial cells (Plášek and Sigler 1996), binds to intracellular proteins and membranes and can be detected by a bright green fluorescence. Susceptible strains of bacteria, but not resistant strains, show this fluorescence in the presence of effective antibiotics (Nuding and Zabel 2013).

Because FC methods can rapidly distinguish live cells from dead cells, they have the potential to determine the effects of specific antimicrobial agents on bacterial survival and rapidly detect the resistant strains of bacteria, regardless of their mechanism of resistance (Shrestha et al. 2011; Ambriz-Aviña et al. 2014).

21.8 Conclusions

A variety of rapid methods for the detection of antibiotic resistance in pathogens causing common bacterial infections are being developed to reduce the length of time that empirical prescription of antibiotics is needed before the most effective antibiotic can be administered (Table 21.1). These rapid methods should not only have immediate benefits for the patient but also should reduce the proliferation of bacteria resistant to multiple antibiotics.

Table 21.1 A summary of rapid methods for analysis of antibiotic-resistant bacteria, with some of their advantages and drawbacks.

Rapid method	Advantages	Disadvantages
Fluorescence microscopy using fluorescent dyes to detect cell growth or lysis	Detects morphological changes in bacteria exposed to antibiotics; shows approximate minimum inhibitory concentration (MIC)	Has been tested successfully only with pure cultures
Lab-on-a-chip, including microfluidics	Rapid detection of resistant strains, giving approximate MIC values regardless of the mechanism of resistance	Requires pure cultures, expensive instrumentation, and highly trained personnel
Raman spectroscopic analysis	Analyzes cells directly; may distinguish resistant from sensitive strains by spectral changes	Requires pure strains of resistant and sensitive bacteria for standards; does not show MIC
Forward laser light scatter (FLLS)	Electro-optical technology that measures bacterial growth in response to antibiotics; may detect MIC	Rapidity of test varies with bacterial strains; requires standardization of growth measurement for each strain
Latex agglutination and immunochromatography	Rapidly detect resistant strains by analysis of proteins	May give false negatives; do not show MIC
Enzyme-linked immunosorbent assay (ELISA)	Can be used to detect either resistance genes or proteins; may be combined with PCR	Does not show MIC
Polymerase chain reaction (PCR)	Detects known genes for resistance in short time from clinical samples without need for bacterial isolation	Variant resistance genes are not detected; the genes detected are not always correlated with phenotypic resistance; does not show MIC
Quantitative PCR (qPCR, real-time PCR)	Knowledge of resistance genes not required; measures phenotypic resistance by detecting growth indirectly	Requires pure cultures
Reverse transcriptase PCR (RT-PCR)	Shows the transcription of resistance genes	May give false positives

(*Continued*)

Table 21.1 (Continued)

Rapid method	Advantages	Disadvantages
Digital PCR	Shows bacterial chromosome segregation; may detect multiple types of resistance in mixed populations	DNA shearing may interfere; does not detect resistant strains with unknown mechanisms of resistance
Multiplex PCR	Simultaneously detects resistance to several antibiotics	May give false positives; resistance genes not always correlated with phenotypic resistance; does not show MIC
Automated PCR and qPCR systems	Require little sample handling	May require culturing for confirmation
Fluorescence *in situ* hybridization (FISH)	Detects either DNA or RNA using fluorescent probes	Detects sequences only of known genes
Peptide nucleic acid FISH (PNA-FISH)	Diffuses through cell wall and detects RNA	Autofluorescence may interfere; does not show MIC
Immunofluorescence	Detects production of resistance proteins	Requires specific antibodies; does not show MIC
Microarray methods	Can detect a broad range of resistance genes with a single assay	Detect genes that may not be expressed; results may not be correlated with phenotypic resistance; do not show MIC
Whole genome sequencing (WGS)	Detects all known resistance genes, which can be correlated with phenotypic resistance in some cases	Standard data analysis platform not yet available; may not be clinically feasible because of high costs and turnaround time
Loop-mediated isothermal amplification (LAMP)	Highly specific for target genes	Does not show the phenotypic response to antibiotics
Pyrolysis mass spectrometry (MS)	Rapid; requires little sample preparation	Spectra not easily reproducible
Electrospray ionization (ESI) MS	May detect some resistance proteins	Data analysis is complex
Matrix-assisted laser desorption/ionization time-of-flight (MALDI-TOF) MS	Rapid, automated detection of proteins, even in intact cells	Instruments are expensive (as in all MS methods); results are not quantitative

Table 21.1 (Continued)

Rapid method	Advantages	Disadvantages
MS with stable isotope tags	Detects growth of cells in presence of antibiotic	Spectra are complex
MS with stable isotope labeling by amino acids in cell cultures (SILAC)	Labeled amino acids are readily available	Changes produced in mass spectra are small
Metal oxide laser ionization (MOLI) MS	Fatty acid profiles may show antibiotic susceptibility	Data analysis is complex and difficult
Genomics-based MS methods	May be combined with qPCR to detect resistance genes	DNA and RNA are not analyzed by MS as easily as proteins
Proteomics-based MS methods	May detect proteins involved in antibiotic resistance	Proteins must be isolated and processed before analysis
Shotgun proteomics	Protease-digested proteins are analyzed by MS/MS	Comprehensive databases not yet developed
Flow cytometry (FC)	Rapid detection of live and dead cells using fluorescent dyes	Pure cultures must be exposed to antibiotics before assay; requires many pipetting and washing steps; instruments are large and expensive

Acknowledgments

We thank Drs. S. A. Khan, K. Sung, P. S. Alusta, and W. Zou for their comments on the manuscript and Dr. C. E. Cerniglia for his encouragement and support.

The views presented in this chapter do not necessarily reflect those of either the US FDA or the University of Arkansas.

References

Aarthi, P., Bagyalakshmi, R., Therese, K.L., and Madhavan, H.N. (2013). Development of a novel reverse transcriptase polymerase chain reaction to determine the Gram reaction and viability of bacteria in clinical specimens. *Microbiol. Res.* 168 (8): 497–503. https://doi.org/10.1016/j.micres.2013.03.005.

Abeyrathne, C.D., Huynh, D.H., Mcintire, T.W. et al. (2016). Lab on a chip sensor for rapid detection and antibiotic resistance determination of *Staphylococcus aureus*. *Analyst* 141 (6): 1922–1929. https://doi.org/10.1039/c5an02301g.

Albrethsen, J. (2007). Reproducibility in protein profiling by MALDI-TOF mass spectrometry. *Clin. Chem.* 53 (5): 852–858. https://doi.org/10.1373/clinchem.2006.082644.

Alipour, F., Ahmadi, M., and Javadi, S. (2014). Evaluation of different methods to detect methicillin resistance in *Staphylococcus aureus* (MRSA). *J. Infect. Public Health* 7 (3): 186–191. https://doi.org/10.1016/j.jiph.2014.01.007.

Alusta, P., Buzatu, D., Williams, A. et al. (2015). Instrumental improvements and sample preparations that enable reproducible, reliable acquisition of mass spectra from whole bacterial cells. *Rapid Commun. Mass Spectrom.* 29 (21): 1961–1968. https://doi.org/10.1002/rcm.7299.

Álvarez-Barrientos, A., Arroyo, J., Cantón, R. et al. (2000). Applications of flow cytometry to clinical microbiology. *Clin. Microbiol. Rev.* 13 (2): 167–195. https://doi.org/10.1128/CMR.13.2.167-195.2000.

Ambretti, S., Gaibani, P., Berlingeri, A. et al. (2013). Evaluation of phenotypic and genotypic approaches for the detection of class A and class B carbapenemases in *Enterobacteriaceae*. *Microb. Drug Resist.* 19 (3): 212–215. https://doi.org/10.1089/mdr.2012.0165.

Ambriz-Aviña, V., Contreras-Garduño, J.A., and Pedraza-Reyes, M. (2014). Applications of flow cytometry to characterize bacterial physiological responses. *Biomed. Res. Int.* 2014: 461941. https://doi.org/10.1155/2014/461941.

Amjad, A., Mirza, I.A., Abbasi, S.A. et al. (2011). Modified Hodge test: a simple and effective test for detection of carbapenemase production. *Iran J. Microbiol.* 3 (4): 189–193.

Anhalt, J.P. and Fenselau, C. (1975). Identification of bacteria using mass spectrometry. *Anal. Chem.* 47 (2): 219–225. https://doi.org/10.1021/ac60352a007.

Bari, M.L. and Kawasaki, S. (2014). Rapid methods for food hygiene inspection. In: Encyclopedia of Food Microbiology, 2e, vol. 3 (ed. C.A. Batt and M.L. Tortorello), 269–279. Amsterdam: Elsevier.

Basile, F., Beverly, M.B., Voorhees, K.J., and Hadfield, T.L. (1998). Pathogenic bacteria: their detection and differentiation by rapid lipid profiling with pyrolysis mass spectrometry. *Trends Anal. Chem.* 17 (2): 95–109. https://doi.org/10.1016/S0165-9936(97)00103-9.

BD Biosciences (2002). Introduction to Flow Cytometry: A Learning Guide. San Jose, CA: Becton, Dickinson and Company. 58p.

van Belkum, A. and Dunne, W.M. Jr. (2013). Next-generation antimicrobial susceptibility testing. *J. Clin. Microbiol.* 51 (7): 2018–2024. https://doi.org/10.1128/JCM.00313-13.

Bernabeu, S., Poirel, L., and Nordmann, P. (2012). Spectrophotometry-based detection of carbapenemase producers among *Enterobacteriaceae*. *Diagn. Microbiol. Infect. Dis.* 74 (1): 88–90. https://doi.org/10.1016/j.diagmicrobio.2012.05.021.

Besant, J.D., Sargent, E.H., and Kelley, S.O. (2015). Rapid electrochemical phenotypic profiling of antibiotic-resistant bacteria. *Lab Chip* 15 (13): 2799–2807. https://doi.org/10.1039/c5lc00375j.

Boehme, C.C., Nicol, M.P., Nabeta, P. et al. (2011). Feasibility, diagnostic accuracy, and effectiveness of decentralised use of the Xpert MTB/RIF test for diagnosis of tuberculosis and multidrug resistance: a multicentre implementation study. *Lancet* 377 (9776): 1495–1505. https://doi.org/10.1016/S0140-6736(11)60438-8.

Bogaerts, P., Hujer, A.M., Naas, T. et al. (2011). Multicenter evaluation of a new DNA microarray for rapid detection of clinically relevant *bla* genes from β-lactam-resistant Gram-negative bacteria. *Antimicrob. Agents Chemother.* 55 (9): 4457–4460. https://doi.org/10.1128/AAC.00353-11.

Buchan, B.W., Ginocchio, C.C., Manii, R. et al. (2013). Multiplex identification of Gram-positive bacteria and resistance determinants directly from positive blood culture broths: evaluation of an automated microarray-based nucleic acid test. *PLoS Med.* 10 (7): e1001478. https://doi.org/10.1371/journal.pmed.1001478.

Buzatu, D.A., Moskal, T.J., Williams, A.J. et al. (2014). An integrated flow cytometry-based system for real-time, high sensitivity bacterial detection and identification. *PLoS One* 9 (4): e94254. https://doi.org/10.1371/journal.pone.0094254.

Campbell, J., McBeth, C., Kalashnikov, M. et al. (2016). Microfluidic advances in phenotypic antibiotic susceptibility testing. *Biomed. Microdevices* 18 (6): 103. https://doi.org/10.1007/s10544-016-0121-8.

Cavassini, M., Wenger, A., Jaton, K. et al. (1999). Evaluation of MRSA-Screen, a simple anti-PBP 2a slide latex agglutination kit, for rapid detection of methicillin resistance in *Staphylococcus aureus*. *J. Clin. Microbiol.* 37 (5): 1591–1594.

Cerqueira, L., Fernandes, R.M., Ferreira, R.M. et al. (2013). Validation of a fluorescence *in situ* hybridization method using peptide nucleic acid probes for detection of *Helicobacter pylori* clarithromycin resistance in gastric biopsy specimens. *J. Clin. Microbiol.* 51 (6): 1887–1893. https://doi.org/10.1128/JCM.00302-13.

Chan, W.-S., Chan, T.-M., Lai, T.-W. et al. (2015). Complementary use of MALDI-TOF MS and real-time PCR-melt curve analysis for rapid identification of methicillin-resistant staphylococci and VRE. *J. Antimicrob. Chemother.* 70 (2): 441–447. https://doi.org/10.1093/jac/dku411.

Chang, C.-J., Lin, J.-H., Chang, K.-C. et al. (2013). Diagnosis of β-lactam resistance in *Acinetobacter baumannii* using shotgun proteomics and LC-nano-electrospray ionization ion trap mass spectrometry. *Anal. Chem.* 85 (5): 2802–2808. https://doi.org/10.1021/ac303326a.

Chavda, K.D., Satlin, M.J., Chen, L. et al. (2016). Evaluation of a multiplex PCR assay to rapidly detect *Enterobacteriaceae* with a broad range of β-lactamases directly from perianal swabs. *Antimicrob. Agents Chemother.* 60 (11): 6957–6961. https://doi.org/10.1128/AAC.01458-16.

Chiba, N., Morozumi, M., and Ubukata, K. (2012). Application of the real-time PCR method for genotypic identification of β-lactam resistance in isolates from invasive pneumococcal diseases. *Microb. Drug Resist.* 18 (2): 149–156. https://doi.org/10.1089/mdr.2011.0102.

Choi, J., Jung, Y.-G., Kim, J. et al. (2013). Rapid antibiotic susceptibility testing by tracking single cell growth in a microfluidic agarose channel system. *Lab Chip* 13 (2): 280–287. https://doi.org/10.1039/c2lc41055a.

Clerc, O., Prod'hom, G., Senn, L. et al. (2014). Matrix-assisted laser desorption ionization time-of-flight mass spectrometry and PCR-based rapid diagnosis of *Staphylococcus aureus* bacteraemia. *Clin. Microbiol. Infect.* 20 (4): 355–360. https://doi.org/10.1111/1469-0691.12329.

Clinical Laboratory Standards Institute (CLSI) (2017). Performance Standards for Antimicrobial Susceptibility Testing, 27e. Wayne, PA: CLSI.

Cohen Stuart, J., Dierikx, C., Al Naiemi, N. et al. (2010). Rapid detection of TEM, SHV and CTX-M extended-spectrum β-lactamases in *Enterobacteriaceae* using ligation-mediated amplification with microarray analysis. *J. Antimicrob. Chemother.* 65 (7): 1377–1381. https://doi.org/10.1093/jac/dkq146.

Corvec, S., Caroff, N., Espaze, E. et al. (2003). Comparison of two RT-PCR methods for quantifying *ampC* specific transcripts in *Escherichia coli* strains. *FEMS Microbiol. Lett.* 228 (2): 187–191. https://doi.org/10.1016/S0378-1097(03)00757-2.

Cox, C.R., Jensen, K.R., Saichek, N.R., and Voorhees, K.J. (2015). Strain-level bacterial identification by CeO_2-catalyzed MALDI-TOF MS fatty acid analysis and comparison to commercial protein-based methods. *Sci. Rep.* 5: 10470. https://doi.org/10.1038/srep10470.

Cuny, C. and Witte, W. (2005). PCR for the identification of methicillin-resistant *Staphylococcus aureus* (MRSA) strains using a single primer pair specific for SCC*mec* elements and the neighbouring chromosome-borne *orfX*. *Clin. Microbiol. Infect.* 11 (10): 834–837. https://doi.org/10.1111/j.1469-0691.2005.01236.x.

Dalpke, A.H., Hofko, M., and Zimmermann, S. (2016). Development of a real-time PCR protocol requiring minimal handling for detection of vancomycin-resistant enterococci with the fully automated BD Max system. *J. Clin. Microbiol.* 54 (9): 2321–2329. https://doi.org/10.1128/JCM.00768-16.

Davey, H.M. and Kell, D.B. (1996). Flow cytometry and cell sorting of heterogeneous microbial populations: the importance of single-cell analyses. *Microbiol. Rev.* 60 (4): 641–696.

David, F., Tienpont, B., and Sandra, P. (2008). Chemotaxonomy of bacteria by comprehensive GC and GC-MS in electron impact and chemical ionisation mode. *J. Sep. Sci.* 31 (19): 3395–3403. https://doi.org/10.1002/jssc.200800215.

Dekter, H.E., Orelio, C.C., Morsink, M.C. et al. (2017). Antimicrobial susceptibility testing of Gram-positive and -negative bacterial isolates directly from spiked blood culture media with Raman spectroscopy. *Eur. J. Clin. Microbiol. Infect. Dis.* 36 (1): 81–89. https://doi.org/10.1007/s10096-016-2773-y.

Demirev, P. (2016). Stable-isotope-based strategies for rapid determination of drug resistance by mass spectrometry. In: Applications of Mass Spectrometry in Microbiology (ed. P. Demirev and T.R. Sandrin), 317–326. Springer International Publishing.

Demirev, P.A., Hagan, N.S., Antoine, M.D. et al. (2013). Establishing drug resistance in microorganisms by mass spectrometry. *J. Am. Soc. Mass Spectrom.* 24 (8): 1194–1201. https://doi.org/10.1007/s13361-013-0609-x.

Diekema, D.J. and Pfaller, M.A. (2013). Rapid detection of antibiotic-resistant organism carriage for infection prevention. *Clin. Infect. Dis.* 56 (11): 1614–1620. https://doi.org/10.1093/cid/cit038.

Drissner, D., Brunisholz, R., Schlapbach, R., and Gekenidis, M.-T. (2016). Rapid profiling of human pathogenic bacteria and antibiotic resistance employing specific tryptic peptides as biomarkers. In: Applications of Mass Spectrometry in Microbiology (ed. P. Demirev and T.R. Sandrin), 275–303. Springer International Publishing.

Dunne, W.M. Jr., Jaillard, M., Rochas, O., and Van Belkum, A. (2017). Microbial genomics and antimicrobial susceptibility testing. *Expert Rev. Mol. Diagn.* 17 (3): 257–269. https://doi.org/10.1080/14737159.2017.1283220.

Edwards-Jones, V., Claydon, M.A., Evason, D.J. et al. (2000). Rapid discrimination between methicillin-sensitive and methicillin-resistant *Staphylococcus aureus* by intact cell mass spectrometry. *J. Med. Microbiol.* 49 (3): 295–300. https://doi.org/10.1099/0022-1317-49-3-295.

Endimiani, A., Hujer, K.M., Hujer, A.M. et al. (2010). Rapid identification of bla_{KPC}-possessing *Enterobacteriaceae* by PCR/electrospray ionization-mass spectrometry. *J. Antimicrob. Chemother.* 65 (8): 1833–1834. https://doi.org/10.1093/jac/dkq207.

Endimiani, A. and Jacobs, M.R. (2016). The changing role of the clinical microbiology laboratory in defining resistance in Gram-negatives. *Infect. Dis. Clin. North Am.* 30 (2): 323–345. https://doi.org/10.1016/j.idc.2016.02.002.

Fan, L., Ke, M., Yuan, M. et al. (2016). Rapid determination of bacterial aminoglycoside resistance in environmental samples using membrane electrospray ionization mass spectrometry. *Rapid Commun. Mass Spectrom.* 30 (Suppl 1): 202–207. https://doi.org/10.1002/rcm.7648.

Fazli, M., Bjarnsholt, T., Høiby, N. et al. (2014). PNA-based fluorescence *in situ* hybridization for identification of bacteria in clinical samples. *Methods Mol. Biol.* 1211: 261–271. https://doi.org/10.1007/978-1-4939-1459-3_21.

Fleurbaaij, F., Heemskerk, A.A.M., Russcher, A. et al. (2014). Capillary-electrophoresis mass spectrometry for the detection of carbapenemases in (multi-)drug-resistant Gram-negative bacteria. *Anal. Chem.* 86 (18): 9154–9161. https://doi.org/10.1021/ac502049p.

Frickmann, H., Masanta, W.O., and Zautner, A.E. (2014). Emerging rapid resistance testing methods for clinical microbiology laboratories and their potential impact on patient management. *Biomed. Res. Int.* 2014: 375681. https://doi.org/10.1155/2014/375681.

Fu, S., Qu, G., Guo, S. et al. (2011). Applications of loop-mediated isothermal DNA amplification. *Appl. Biochem. Biotechnol.* 163 (7): 845–850. https://doi.org/10.1007/s12010-010-9088-8.

García-Fernández, S., Morosini, M.-I., Marco, F. et al. (2015). Evaluation of the eazyplex® SuperBug CRE system for rapid detection of carbapenemases and ESBLs in clinical *Enterobacteriaceae* isolates recovered at two Spanish hospitals. *J. Antimicrob. Chemother.* 70 (4): 1047–1050. https://doi.org/10.1093/jac/dku476.

Gherardi, G., Angeletti, S., Panitti, M. et al. (2012). Comparative evaluation of the Vitek-2 Compact and Phoenix systems for rapid identification and antibiotic susceptibility testing directly from blood cultures of Gram-negative and Gram-positive isolates. *Diagn. Microbiol. Infect. Dis.* 72 (1): 20–31. https://doi.org/10.1016/j.diagmicrobio.2011.09.015.

Gröbner, S., Dion, M., Plante, M., and Kempf, V.A.J. (2009). Evaluation of the BD GeneOhm StaphSR assay for detection of methicillin-resistant and methicillin-susceptible *Staphylococcus aureus* isolates from spiked positive blood culture bottles. *J. Clin. Microbiol.* 47 (6): 1689–1694. https://doi.org/10.1128/JCM.02179-08.

Grundt, A., Findeisen, P., Miethke, T. et al. (2012). Rapid detection of ampicillin resistance in *Escherichia coli* by quantitative mass spectrometry. *J. Clin. Microbiol.* 50 (5): 1727–1729. https://doi.org/10.1128/JCM.00047-12.

Hahne, F., Arlt, D., Sauermann, M. et al. (2006). Statistical methods and software for the analysis of highthroughput reverse genetic assays using flow cytometry readouts. *Genome Biol.* 7 (8): R77. https://doi.org/10.1186/gb-2006-7-8-R77.

Han, E., Park, D.-J., Kim, Y. et al. (2015). Rapid detection of Gram-negative bacteria and their drug resistance genes from positive blood cultures using an automated microarray assay. *Diagn. Microbiol. Infect. Dis.* 81 (3): 153–157. https://doi.org/10.1016/j.diagmicrobio.2014.10.009.

Hayden, R.T., Clinton, L.K., Hewitt, C. et al. (2016). Rapid antimicrobial susceptibility testing using forward laser light scatter technology. *J. Clin. Microbiol.* 54 (11): 2701–2706. https://doi.org/10.1128/JCM.01475-16.

Hernandez, D.R., Newton, D.W., Ledeboer, N.A. et al. (2016). Multicenter evaluation of MRSA*Select* II chromogenic agar for identification of methicillin-resistant *Staphylococcus aureus* from wound and nasal specimens. *J. Clin. Microbiol.* 54 (2): 305–311. https://doi.org/10.1128/JCM.02410-15.

Holland, R.D., Wilkes, J.G., Rafii, F. et al. (1996). Rapid identification of intact whole bacteria based on spectral patterns using matrix-assisted laser desorption/ionization with time-of-flight mass spectrometry. *Rapid Commun. Mass Spectrom.* 10 (10): 1227–1232. https://doi.org/10.1002/(SICI)1097-0231(19960731)10:10<1227::AID-RCM659>3.0.CO;2-6.

Hrabak, J., Havlicek, V., and Papagiannitsis, C.C. (2016). Detection of β-lactamases and their activity using MALDI-TOF MS. In: Applications of Mass Spectrometry in Microbiology (ed. P. Demirev and T.R. Sandrin), 305–316. Springer International Publishing.

Ingham, C.J., Ayad, A.B., Nolsen, K., and Mulder, B. (2008). Rapid drug susceptibility testing of mycobacteria by culture on a highly porous ceramic support. *Int. J. Tuberc. Lung Dis.* 12 (6): 645–650.

Ingham, C.J., van den Ende, M., Wever, P.C., and Schneeberger, P.M. (2006). Rapid antibiotic sensitivity testing and trimethoprim-mediated filamentation of clinical isolates of the *Enterobacteriaceae* assayed on a novel porous culture support. *J. Med. Microbiol.* 55 (11): 1511–1519. https://doi.org/10.1099/jmm.0.46585-0.

Jarzembowski, T., Jóźwik, A., Wiśniewska, K., and Witkowski, J. (2010). Flow cytometry approach study of *Enterococcus faecalis* vancomycin resistance by detection of Vancomycin@FL binding to the bacterial cells. *Curr. Microbiol.* 61 (5): 407–410. https://doi.org/10.1007/s00284-010-9628-z.

Jarzembowski, T., Wiśniewska, K., Józwik, A. et al. (2008). Flow cytometry as a rapid test for detection of penicillin resistance directly in bacterial cells in *Enterococcus faecalis* and *Staphylococcus aureus*. *Curr. Microbiol.* 57 (2): 167–169. https://doi.org/10.1007/s00284-008-9179-8.

Jung, J.S., Eberl, T., Sparbier, K. et al. (2014). Rapid detection of antibiotic resistance based on mass spectrometry and stable isotopes. *Eur. J. Clin. Microbiol. Infect. Dis.* 33 (6): 949–955. https://doi.org/10.1007/s10096-013-2031-5.

Kaminski, T.S., Scheler, O., and Garstecki, P. (2016). Droplet microfluidics for microbiology: techniques, applications and challenges. *Lab Chip* 16 (12): 2168–2187. https://doi.org/10.1039/c6lc00367b.

Karas, M. and Hillenkamp, F. (1988). Laser desorption ionization of proteins with molecular masses exceeding 10,000 daltons. *Anal. Chem.* 60 (20): 2299–2301. https://doi.org/10.1021/ac00171a028.

Kaur, J., Chopra, S., Sheevani, and Mahajan, G. (2013). Modified double disc synergy test to detect ESBL production in urinary isolates of *Escherichia coli* and *Klebsiella pneumoniae*. *J. Clin. Diagn. Res.* 7 (2): 229–233. https://doi.org/10.7860/JCDR/2013/4619.2734.

Keays, M.C., O'Brien, M., Hussain, A. et al. (2016). Rapid identification of antibiotic resistance using droplet microfluidics. *Bioengineered* 7 (2): 79–87. https://doi.org/10.1080/21655979.2016.1156824.

Khan, S.A., Sung, K., and Nawaz, M.S. (2011a). Detection of *aacA-aphD*, *qacEδ1*, *marA*, *floR*, and *tetA* genes from multidrug-resistant bacteria: comparative analysis of real-time multiplex PCR assays using EvaGreen® and SYBR®

Green I dyes. *Mol. Cell. Probes* 25 (2–3): 78–86. https://doi.org/10.1016/j.mcp.2011.01.004.

Khan, S.A., Sung, K., and Nawaz, M.S. (2011b). A transcriptomic expression array, PCR and disk diffusion analysis of antimicrobial resistance genes in multidrug-resistant bacteria. *Agric. Food Anal. Bacteriol.* 1 (2): 123–139.

Kothari, A., Morgan, M., and Haake, D.A. (2014). Emerging technologies for rapid identification of bloodstream pathogens. *Clin. Infect. Dis.* 59 (2): 272–278. https://doi.org/10.1093/cid/ciu292.

Lange, C., Schubert, S., Jung, J. et al. (2014). Quantitative matrix-assisted laser desorption ionization-time of flight mass spectrometry for rapid resistance detection. *J. Clin. Microbiol.* 52 (12): 4155–4162. https://doi.org/10.1128/JCM.01872-14.

Lay, J.O. Jr. (2001). MALDI-TOF mass spectrometry of bacteria. *Mass Spectrom. Rev.* 20 (4): 172–194. https://doi.org/10.1002/mas.10003.

Lay, J.O. Jr. and Liyanage, R. (2006). MALDI-TOF mass spectrometry of intact bacteria. In: Identification of Microorganisms by Mass Spectrometry, vol. 169 (ed. C.L. Wilkins and J.O. Lay Jr.), 125–152. Hoboken, NJ: Wiley.

Leclercq, R., Cantón, R., Brown, D.F.J. et al. (2013). EUCAST expert rules in antimicrobial susceptibility testing. *Clin. Microbiol. Infect.* 19 (2): 141–160. https://doi.org/10.1111/j.1469-0691.2011.03703.x.

Levy Hara, G., Gould, I., Endimiani, A. et al. (2013). Detection, treatment, and prevention of carbapenemase-producing *Enterobacteriaceae*: recommendations from an international working group. *J. Chemother.* 25 (3): 129–140. https://doi.org/10.1179/1973947812Y.0000000062.

Lichtenwalter, K.G., Apffel, A., Bai, J. et al. (2000). Approaches to functional genomics: potential of matrix-assisted laser desorption ionization – time of flight mass spectrometry combined with separation methods for the analysis of DNA in biological samples. *J. Chromatogr. B Biomed. Sci. Appl.* 745 (1): 231–241. https://doi.org/10.1016/S0378-4347(00)00131-6.

Malmberg, C., Yuen, P., Spaak, J. et al. (2016). A novel microfluidic assay for rapid phenotypic antibiotic susceptibility testing of bacteria detected in clinical blood cultures. *PLoS One* 11 (12): e0167356. https://doi.org/10.1371/journal.pone.0167356.

Marner, E.S., Wolk, D.M., Carr, J. et al. (2011). Diagnostic accuracy of the Cepheid GeneXpert *vanA/vanB* assay ver. 1.0 to detect the *vanA* and *vanB* vancomycin resistance genes in *Enterococcus* from perianal specimens. *Diagn. Microbiol. Infect. Dis.* 69 (4): 382–389. https://doi.org/10.1016/j.diagmicrobio.2010.11.005.

Marquet, K., Liesenborgs, A., Bergs, J. et al. (2015). Incidence and outcome of inappropriate in-hospital empiric antibiotics for severe infection: a systematic review and meta-analysis. *Crit. Care* 19: 63. https://doi.org/10.1186/s13054-015-0795-y.

Martín-Peña, R., Domínguez-Herrera, J., Pachón, J., and McConnell, M.J. (2013). Rapid detection of antibiotic resistance in *Acinetobacter baumannii* using quantitative real-time PCR. *J. Antimicrob. Chemother.* 68 (7): 1572–1575. https://doi.org/10.1093/jac/dkt057.

Matsumoto, Y., Sakakihara, S., Grushnikov, A. et al. (2016). A microfluidic channel method for rapid drug-susceptibility testing of *Pseudomonas aeruginosa*. *PLoS One* 11 (2): e0148797. https://doi.org/10.1371/journal.pone.0148797.

Maurin, M. (2012). Real-time PCR as a diagnostic tool for bacterial diseases. *Expert Rev. Mol. Diagn.* 12 (7): 731–754. https://doi.org/10.1586/erm.12.53.

May, M. (2016). Flow cytometry: exploring the microbiome and more with faster and more accurate platforms. *Am. Lab.* 48 (8): 12–13.

Minejima, E. and Wong-Beringer, A. (2016). Implementation of rapid diagnostics with antimicrobial stewardship. *Expert Rev. Anti Infect. Ther.* 14 (11): 1065–1075. https://doi.org/10.1080/14787210.2016.1233814.

Misawa, Y., Yoshida, A., Saito, R. et al. (2007). Application of loop-mediated isothermal amplification technique to rapid and direct detection of methicillin-resistant *Staphylococcus aureus* (MRSA) in blood cultures. *J. Infect. Chemother.* 13 (3): 134–140. https://doi.org/10.1007/s10156-007-0508-9.

Moosavian, M., Tajbakhsh, S., and Samarbaf-Zadeh, A.R. (2007). Rapid detection of clarithromycin-resistant *Helicobacter pylori* in patients with dyspepsia by fluorescent *in situ* hybridization (FISH) compared with the E-test. *Ann. Saudi Med.* 27 (2): 84–88.

Morgan, S.L., Watt, B.E., and Galipo, R.C. (2006). Characterization of microorganisms by pyrolysis-GC, pyrolysis GC/MS, and pyrolysis-MS. In: Applied Pyrolysis Handbook, 2e (ed. T.P. Wampler), 201–232. Boca Raton, FL: CRC Press.

Mori, Y., Kitao, M., Tomita, N., and Notomi, T. (2004). Real-time turbidimetry of LAMP reaction for quantifying template DNA. *J. Biochem. Biophys. Methods* 59 (2): 145–157. https://doi.org/10.1016/j.jbbm.2003.12.005.

Mortimer, F.C., Mason, D.J., and Gant, V.A. (2000). Flow cytometric monitoring of antibiotic-induced injury in *Escherichia coli* using cell-impermeant fluorescent probes. *Antimicrob. Agents Chemother.* 44 (3): 676–681. https://doi.org/10.1128/AAC.44.3.676-681.2000.

Mu, X.-Q., Liu, B.-B., Hui, E. et al. (2016). A rapid loop-mediated isothermal amplification (LAMP) method for detection of the macrolide-streptogramin type B resistance gene *msrA* in *Staphylococcus aureus*. *J. Glob. Antimicrob. Resist.* 7: 53–58. https://doi.org/10.1016/j.jgar.2016.07.006.

Naas, T., Cuzon, G., Bogaerts, P. et al. (2011). Evaluation of a DNA microarray (Check-MDR CT102) for rapid detection of TEM, SHV, and CTX-M extended-spectrum β-lactamases and of KPC, OXA-48, VIM, IMP, and NDM-1 carbapenemases. *J. Clin. Microbiol.* 49 (4): 1608–1613. https://doi.org/10.1128/JCM.02607-10.

Nijhuis, R.H.T., van Maarseveen, N.M., van Hannen, E.J. et al. (2014). A rapid and high-throughput screening approach for methicillin-resistant *Staphylococcus aureus* based on the combination of two different real-time PCR assays. *J. Clin. Microbiol.* 52 (8): 2861–2867. https://doi.org/10.1128/JCM.00808-14.

Nogueira, R.F. (2013). Detection of fluoroquinolone resistance and efflux pumps activity by flow cytometry. MS thesis. Universidade Católica Portuguesa, Porto.

Nuding, S. and Zabel, L.T. (2013). Detection, identification, and susceptibility testing of bacteria by flow cytometry. *J. Bacteriol. Parasitol.* S5-005. https://doi.org/10.4172/2155-9597.S5-005.

Osei Sekyere, J., Govinden, U., and Essack, S.Y. (2015). Review of established and innovative detection methods for carbapenemase-producing Gram-negative bacteria. *J. Appl. Microbiol.* 119 (5): 1219–1233. https://doi.org/10.1111/jam.12918.

Otero, F., Santiso, R., Tamayo, M. et al. (2017). Rapid detection of antibiotic resistance in Gram-negative bacteria through assessment of changes in cellular morphology. *Microb. Drug Resist.* 23 (2): 157–162. https://doi.org/10.1089/mdr.2016.0023.

Oviaño, M. and Bou, G. (2017). Imipenem-avibactam: a novel combination for the rapid detection of carbapenemase activity in *Enterobacteriaceae* and *Acinetobacter baumannii* by matrix-assisted laser desorption ionization-time of flight mass spectrometry. *Diagn. Microbiol. Infect. Dis.* 87 (2): 129–132. https://doi.org/10.1016/j.diagmicrobio.2016.10.016.

Palasubramaniam, S., Muniandy, S., and Navaratnam, P. (2008). Rapid detection of ESBL-producing *Klebsiella pneumoniae* in blood cultures by fluorescent *in-situ* hybridization. *J. Microbiol. Methods* 72 (1): 107–109. https://doi.org/10.1016/j.mimet.2007.10.008.

van Pelt-Verkuil, E., van Belkum, A., and Hays, J.P. (2008). Principles and Technical Aspects of PCR Amplification. Springer.

Perna, F. and Vaira, D. (2010). A new 24 h ELISA culture based method for *Helicobacter pylori* chemosusceptibility. *J. Clin. Pathol.* 63 (7): 648–651. https://doi.org/10.1136/jcp.2010.076844.

Perreten, V., Endimiani, A., Thomann, A. et al. (2013). Evaluation of PCR electrospray-ionization mass spectrometry for rapid molecular diagnosis of bovine mastitis. *J. Dairy Sci.* 96 (6): 3611–3620. https://doi.org/10.3168/jds.2012-6124.

Pholwat, S., Stroup, S., Foongladda, S., and Houpt, E. (2013). Digital PCR to detect and quantify heteroresistance in drug resistant *Mycobacterium tuberculosis*. *PLoS One* 8 (2): e57238. https://doi.org/10.1371/journal.pone.0057238.

Plášek, J. and Sigler, K. (1996). Slow fluorescent indicators of membrane potential: a survey of different approaches to probe response analysis. *J. Photochem. Photobiol. B Biol.* 33 (2): 101–124.

Preez, I.D., Luies, L., and Loots, D.T. (2017). Metabolomics biomarkers for tuberculosis diagnostics: current status and future objectives. *Biomark. Med.* 11 (2): 179–194. https://doi.org/10.2217/bmm-2016-0287.

Pulido, M.R., García-Quintanilla, M., Martín-Peña, R. et al. (2013). Progress on the development of rapid methods for antimicrobial susceptibility testing. *J. Antimicrob. Chemother.* 68 (12): 2710–2717. https://doi.org/10.1093/jac/dkt253.

Quach, D.T., Sakoulas, G., Nizet, V. et al. (2016). Bacterial cytological profiling (BCP) as a rapid and accurate antimicrobial susceptibility testing method *for Staphylococcus aureus*. *EBioMedicine* 4: 95–103. https://doi.org/10.1016/j.ebiom.2016.01.020.

Rees, J.C. and Barr, J.R. (2017). Detection of methicillin-resistant *Staphylococcus aureus* using phage amplification combined with matrix-assisted laser desorption/ionization mass spectrometry. *Anal. Bioanal. Chem.* 409: 1379–1386. https://doi.org/10.1007/s00216-016-0070-3.

Renggli, S., Keck, W., Jenal, U., and Ritz, D. (2013). Role of autofluorescence in flow cytometric analysis of *Escherichia coli* treated with bactericidal antibiotics. *J. Bacteriol.* 195 (18): 4067–4073. https://doi.org/10.1128/JB.00393-13.

Rohde, A., Hammerl, J.A., and Al Dahouk, S. (2016). Rapid screening for antibiotic resistance elements on the RNA transcript, protein and enzymatic activity level. *Ann. Clin. Microbiol. Antimicrob.* 15 (1): 55. https://doi.org/10.1186/s12941-016-0167-8.

Saichek, N.R., Cox, C.R., Kim, S. et al. (2016). Strain-level *Staphylococcus* differentiation by CeO_2-metal oxide laser ionization mass spectrometry fatty acid profiling. *BMC Microbiol.* 16: 72. https://doi.org/10.1186/s12866-016-0658-y.

Santiso, R., Tamayo, M., Gosálvez, J. et al. (2011). A rapid *in situ* procedure for determination of bacterial susceptibility or resistance to antibiotics that inhibit peptidoglycan biosynthesis. *BMC Microbiol.* 11: 191. https://doi.org/10.1186/1471-2180-11-191.

Santos, T., Capelo, J.L., Santos, H.M. et al. (2015). Use of MALDI-TOF mass spectrometry fingerprinting to characterize *Enterococcus* spp. and *Escherichia coli* isolates. *J. Proteomics* 127: 321–331. https://doi.org/10.1016/j.jprot.2015.02.017.

Schelli, K., Rutowski, J., Roubidoux, J., and Zhu, J. (2017). *Staphylococcus aureus* methicillin resistance detected by HPLC-MS/MS targeted metabolic profiling. *J. Chromatogr. B* 1047 (SI): 124–130. https://doi.org/10.1016/j.jchromb.2016.05.052.

Schoepp, N.G., Khorosheva, E.M., Schlappi, T.S. et al. (2016). Digital quantification of DNA replication and chromosome segregation enables determination of antimicrobial susceptibility after only 15 minutes of antibiotic exposure. *Angew. Chem. Int. Ed.* 55 (33): 9557–9561. https://doi.org/10.1002/anie.201602763.

Schröder, U.-C., Beleites, C., Assmann, C. et al. (2015). Detection of vancomycin resistances in enterococci within 3½ hours. *Sci. Rep.* 5: 8217. https://doi.org/10.1038/srep08217.

Semanjski, M. and Macek, B. (2016). Shotgun proteomics of bacterial pathogens: advances, challenges and clinical implications. *Expert Rev. Proteomics* 13 (2): 139–156. https://doi.org/10.1586/14789450.2016.1132168.

Shah, H.N., Rajakaruna, L., Ball, G. et al. (2011). Tracing the transition of methicillin resistance in sub-populations of *Staphylococcus aureus*, using SELDI-TOF mass spectrometry and artificial neural network analysis. *Syst. Appl. Microbiol.* 34 (1): 81–86. https://doi.org/10.1016/j.syapm.2010.11.002.

Shrestha, N.K., Scalera, N.M., Wilson, D.A., and Procop, G.W. (2011). Rapid differentiation of methicillin-resistant and methicillin-susceptible *Staphylococcus aureus* by flow cytometry after brief antibiotic exposure. *J. Clin. Microbiol.* 49 (6): 2116–2120. https://doi.org/10.1128/JCM.02548-10.

Singhal, N., Kumar, M., Kanaujia, P.K., and Virdi, J.S. (2015). MALDI-TOF mass spectrometry: an emerging technology for microbial identification and diagnosis. *Front. Microbiol.* 6: 791. https://doi.org/10.3389/fmicb.2015.00791.

Sinsimer, D., Leekha, S., Park, S. et al. (2005). Use of a multiplex molecular beacon platform for rapid detection of methicillin and vancomycin resistance in *Staphylococcus aureus*. *J. Clin. Microbiol.* 43 (9): 4585–4591. https://doi.org/10.1128/JCM.43.9.4585-4591.2005.

Snyder, J.W., Munier, G.K., and Johnson, C.L. (2008). Direct comparison of the BD Phoenix system with the MicroScan WalkAway system for identification and antimicrobial susceptibility testing of *Enterobacteriaceae* and nonfermentative Gram-negative organisms. *J. Clin. Microbiol.* 46 (7): 2327–2333. https://doi.org/10.1128/JCM.00075-08.

Sood, S., Malhotra, M., Das, B.K., and Kapil, A. (2008). Enterococcal infections and antimicrobial resistance. *Indian J. Med. Res.* 128 (2): 111–121.

Sparbier, K., Lange, C., Jung, J. et al. (2013). MALDI biotyper-based rapid resistance detection by stable-isotope labeling. *J. Clin. Microbiol.* 51 (11): 3741–3748. https://doi.org/10.1128/JCM.01536-13.

Sparbier, K., Schubert, S., Weller, U. et al. (2012). Matrix-assisted laser desorption ionization-time of flight mass spectrometry-based functional assay for rapid detection of resistance against β-lactam antibiotics. *J. Clin. Microbiol.* 50 (3): 927–937. https://doi.org/10.1128/JCM.05737-11.

Steen, H.B. (2000). Flow cytometry of bacteria: glimpses from the past with a view to the future. *J. Microbiol. Methods* 42 (1): 65–74. https://doi.org/10.1016/S0167-7012(00)00177-9.

Steen, H.B., Boye, E., Skarstad, K. et al. (1982). Applications of flow cytometry on bacteria: cell cycle kinetics, drug effects, and quantitation of antibody binding. *Cytometry* 2 (4): 249–257. https://doi.org/10.1002/cyto.990020409.

Strauss, C., Endimiani, A., and Perreten, V. (2015). A novel universal DNA labeling and amplification system for rapid microarray-based detection of 117 antibiotic resistance genes in Gram-positive bacteria. *J. Microbiol. Methods* 108: 25–30. https://doi.org/10.1016/j.mimet.2014.11.006.

Stump, M.J., Jones, J.J., Fleming, R.C. et al. (2003). Use of double-depleted ^{13}C and ^{15}N culture media for analysis of whole cell bacteria by MALDI time-of-flight and Fourier transform mass spectrometry. *J. Am. Soc. Mass Spectrom.* 14 (11): 1306–1314. https://doi.org/10.1016/S1044-0305(03)00577-4.

Sutherland, J.B. and Rafii, F. (2006). Cultural, serological, and genetic methods for identification of bacteria. In: Identification of Microorganisms by Mass Spectrometry, vol. 169 (ed. C.L. Wilkins and J.O. Lay Jr.), 1–21. Hoboken, NJ: Wiley.

Tamayo, M., Santiso, R., Gosálvez, J. et al. (2009). Rapid assessment of the effect of ciprofloxacin on chromosomal DNA from *Escherichia coli* using an *in situ* DNA fragmentation assay. *BMC Microbiol.* 9: 69. https://doi.org/10.1186/1471-2180-9-69.

Tenover, F.C. (2010). Potential impact of rapid diagnostic tests on improving antimicrobial use. *Ann. N. Y. Acad. Sci.* 1213: 70–80. https://doi.org/10.1111/j.1749-6632.2010.05827.x.

Tomita, N., Mori, Y., Kanda, H., and Notomi, T. (2008). Loop-mediated isothermal amplification (LAMP) of gene sequences and simple visual detection of products. *Nat. Protoc.* 3 (5): 877–882. https://doi.org/10.1038/nprot.2008.57.

Trebesius, K., Panthel, K., Strobel, S. et al. (2000). Rapid and specific detection of *Helicobacter pylori* macrolide resistance in gastric tissue by fluorescent *in situ* hybridisation. *Gut* 46 (5): 608–614. https://doi.org/10.1136/gut.46.5.608.

Trienski, T.L., Barrett, H.L., Pasquale, T.R. et al. (2013). Evaluation and use of a rapid *Staphylococcus aureus* assay by an antimicrobial stewardship program. *Am. J. Health Syst. Pharm.* 70 (21): 1908–1912. https://doi.org/10.2146/ajhp130118.

Voorhees, K.J., Jensen, K.R., McAlpin, C.R. et al. (2013). Modified MALDI MS fatty acid profiling for bacterial identification. *J. Mass Spectrom.* 48 (7): 850–855. https://doi.org/10.1002/jms.3215.

Wada, M., Lkhagvadorj, E., Bian, L. et al. (2010). Quantitative reverse transcription-PCR assay for the rapid detection of methicillin-resistant *Staphylococcus aureus*. *J. Appl. Microbiol.* 108 (3): 779–788. https://doi.org/10.1111/j.1365-2672.2009.04476.x.

Waldeisen, J.R., Wang, T., Mitra, D., and Lee, L.P. (2011). A real-time PCR antibiogram for drug-resistant sepsis. *PLoS One* 6 (12): e28528. https://doi.org/10.1371/journal.pone.0028528.

Weiner, L.M., Webb, A.K., Limbago, B. et al. (2016). Antimicrobial-resistant pathogens associated with healthcare-associated infections: summary of data reported to the National Healthcare Safety Network at the Centers for Disease Control and Prevention, 2011-2014. *Infect. Control Hosp. Epidemiol.* 37 (11): 1288–1301. https://doi.org/10.1017/ice.2016.174.

Wellinghausen, N., Wirths, B., Essig, A., and Wassill, L. (2004). Evaluation of the Hyplex BloodScreen Multiplex PCR-enzyme-linked immunosorbent assay system for direct identification of Gram-positive cocci and Gram-negative bacilli from positive blood cultures. *J. Clin. Microbiol.* 42 (7): 3147–3152. https://doi.org/10.1128/JCM.42.7.3147-3152.2004.

Wilkes, J.G., Tucker, R.K., Montgomery, J.A. et al. (2012). Reduction of food matrix interference by a combination of sample preparation and multi-dimensional gating techniques to facilitate rapid, high sensitivity analysis for *Escherichia coli* serotype O157 by flow cytometry. *Food Microbiol.* 30 (1): 281–288. https://doi.org/10.1016/j.fm.2011.11.002.

Willemse-Erix, D., Bakker-Schut, T., Slagboom-Bax, F. et al. (2012). Rapid typing of extended-spectrum β-lactamase- and carbapenemase-producing *Escherichia coli* and *Klebsiella pneumoniae* isolates by use of SpectraCell RA. *J. Clin. Microbiol.* 50 (4): 1370–1375. https://doi.org/10.1128/JCM.05423-11.

Williams, A.J., Cooper, W.M., Summage-West, C.V. et al. (2015). Level 2 validation of a flow cytometric method for detection of *Escherichia coli* O157:H7 in raw spinach. *Int. J. Food Microbiol.* 215: 1–6. https://doi.org/10.1016/j.ijfoodmicro.2015.08.011.

Wolk, D.M., Kaleta, E.J., and Wysocki, V.H. (2012). PCR-electrospray ionization mass spectrometry: the potential to change infectious disease diagnostics in clinical and public health laboratories. *J. Mol. Diagn.* 14 (4): 295–304. https://doi.org/10.1016/j.jmoldx.2012.02.005.

Wolk, D.M., Marx, J.L., Dominguez, L. et al. (2009). Comparison of MRSA*Select* Agar, CHROMagar Methicillin-Resistant *Staphylococcus aureus* (MRSA) Medium, and Xpert MRSA PCR for detection of MRSA in nares: diagnostic accuracy for surveillance samples with various bacterial densities. *J. Clin. Microbiol.* 47 (12): 3933–3936. https://doi.org/10.1128/JCM.00601-09.

Woodford, N. (2010). Rapid characterization of β-lactamases by multiplex PCR. *Methods Mol. Biol.* 642: 181–192. https://doi.org/10.1007/978-1-60327-279-7_14.

Yamashita, M. and Fenn, J.B. (1984). Electrospray ion source: another variation on the free-jet theme. *J. Phys. Chem.* 88 (20): 4451–4459. https://doi.org/10.1021/j150664a002.

Yilmaz, O. and Demiray, E. (2007). Clinical role and importance of fluorescence *in situ* hybridization method in diagnosis of *H. pylori* infection and determination of clarithromycin resistance in *H. pylori* eradication therapy. *World J. Gastroenterol.* 13 (5): 671–675.

Zou, W., Frye, J.G., Chang, C.-W. et al. (2009). Microarray analysis of antimicrobial resistance genes in *Salmonella enterica* from preharvest poultry environment. *J. Appl. Microbiol.* 107 (3): 906–914. https://doi.org/10.1111/j.1365-2672.2009.04270.x.

22

Effective Methods for Disinfection and Sterilization

Lucía Fernández, Diana Gutiérrez, Beatriz Martínez, Ana Rodríguez, and Pilar García

Instituto de Productos Lácteos de Asturias (IPLA-CSIC), Villaviciosa, Spain

22.1 Introduction

Worldwide concern about the rise in antibiotic resistance has fostered the search for new approaches to prevent and treat human microbial infections. Likewise, notable efforts have been made to identify potential contamination routes in a wide range of premises, including healthcare facilities like hospitals and primary care centers as well as restaurants and schools. Undoubtedly, hospital-acquired infections represent a major cause of mortality in developed countries. In the United States, about 1 in 25 hospitalized patients has an infection linked to hospital care (CDC 2017). Similarly, in the European Union, approximately 4 100 000 patients are estimated to develop a hospital-related infection each year, of which at least 37 000 will die as a direct consequence of these infections (ECDC 2017). Given the worrying nature of these reports, the health authorities are aware that effective action must be taken to decrease the morbidity and mortality rates associated with nosocomial infections. One of the key strategies to achieve this goal involves the improvement of routine cleaning and disinfection procedures in healthcare facilities. This would reduce microbial contamination in the hospital environment, including air and surfaces, thereby limiting the risk of infection. Indeed, it is well known that contaminated surfaces are a potential source for the transmission of pathogens, including noroviruses and dangerous bacteria such as *Clostridium difficile*, methicillin-resistant *Staphylococcus aureus* (MRSA), vancomycin-resistant *Enterococcus* (VRE), *Acinetobacter baumannii*, and *Pseudomonas aeruginosa* (Weber et al. 2010). These microbes reach healthcare facilities through two main sources, patients and staff, and then contaminate surfaces and

Antibiotic Drug Resistance, First Edition. Edited by José-Luis Capelo-Martínez and Gilberto Igrejas.

equipment. For instance, there is a risk of cross-contamination by direct contact between the hands of healthcare personnel and the patients (Stiefel et al. 2011). Meanwhile, infected patients are a source of pathogenic microorganisms that are shed through their skin and may contaminate their clothes, bedding, and other environmental surfaces (Donskey 2010). Moreover, hospital patients can be asymptomatic carriers of opportunistic pathogens (Riggs et al. 2007). The risk of transmission to other patients is further enhanced by the ability of some bacteria to survive in the clinical environment for months (Kramer et al. 2006). Therefore, proper disinfection and sterilization of medical devices and surgical instruments are essential for preventing the transmission of pathogens to patients.

Despite the clear need to reduce microbial surface contamination, there is still controversy regarding the best cleaning and disinfection methods. Routine cleaning in hospitals generally follows predetermined policies. This task is usually carried out manually, and the intensity of surface cleaning varies depending on the clinical risk of each specific area. Of note, there is no evidence to date that surface disinfection results in lower hospital infection rates than cleaning with detergent only (Dettenkofer et al. 2004). Another concern associated with the widespread use of disinfectants is the possible co-selection of antibiotic-resistant bacteria (Gnanadhas et al. 2013). Indeed, biocides are routinely utilized in many fields besides medicine, including the food industry, veterinary medicine, agriculture, and water treatment. Repeated exposure to these non-antibiotic antimicrobial agents may select bacteria with lower susceptibility to one or more antibiotics. This may be the result of diverse molecular mechanisms such as bacterial adaptation, selection of genetic determinants, mobilization of genetic material, increased mutation, or changes in the structure of the bacterial community (Wales and Davies 2015).

Considering all these factors, it is quite clear that surface disinfection programs should be carefully designed to obtain maximum results with minimum adverse effects. To do that, it is first necessary to pinpoint the major shortcomings of traditional cleaning and disinfection practices. At the same time, the authorities should encourage efforts to develop novel technologies that can achieve proper disinfection without posing a risk to human health or the environment (e.g. natural compounds and antimicrobial surfaces). This chapter is intended to summarize key issues regarding the role of effective disinfection and sterilization to prevent transmission of pathogenic bacteria within the hospital environment as well as to provide an overview of the different mechanisms involved in bacterial resistance to disinfectants. Within this context, we will compare traditional and new techniques of disinfection and sterilization and outline future challenges to ensure a proper balance between efficacy and resistance development. Last but not least, we will briefly discuss guidelines, recommendations, and regulations regarding the use of disinfectant products.

22.2 Disinfection and Sterilization: Methods and Factors Involved in Their Efficacy

Sterilization is a process that destroys or eliminates all forms of microbial life, including bacterial spores, while disinfection is intended to eliminate many or all pathogenic microorganisms, except bacterial spores. It must also be noted that each disinfectant has a distinct activity spectrum; therefore, it is important to select the most appropriate product depending on the target microbes. Additionally, proper surface cleaning is an essential prerequisite to maximize the efficacy of both sterilization and disinfection. Cleaning is regarded as the removal of visible organic and inorganic particles from objects and surfaces and is often accomplished manually or mechanically using water with detergents or enzymatic products (Rutala and Weber 2008).

In hospitals and healthcare facilities, the utilization of different disinfection methods to reduce the risk for the patients is mainly based on a classification scheme proposed by Spaulding (1957) (Figure 22.1). This classification defines the minimum level of disinfection required depending on the infection risk associated with using a particular medical device. Thus, instruments and items for patient care are categorized as critical, semicritical, and noncritical (McDonnell and Burke 2011). For example, critical devices are those in contact with a sterile area of the body and should be sterilized (e.g. cardiac catheters). In contrast, semicritical devices interact with mucous membranes or non-intact skin, and, therefore, a high level of disinfection (complete elimination of all microorganisms, except a small number of bacterial spores) is

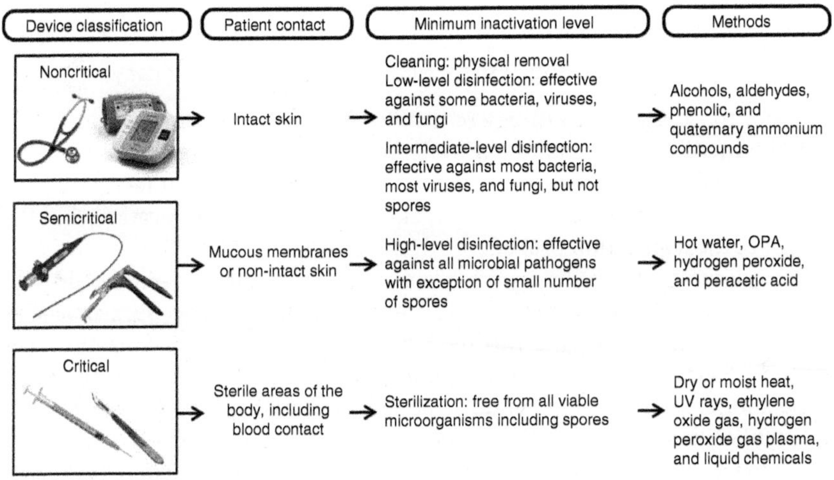

Figure 22.1 Methods of disinfection/sterilization according to the Spaulding's medical devices classification.

recommended. Finally, noncritical devices only come into contact with intact skin, and standard disinfection would suffice.

22.2.1 Methods of Sterilization and Disinfection

Both sterilization and disinfection can be achieved by physical or chemical methods (Figure 22.1). Sterilization by dry or moist heat is an adequate method for materials able to withstand high temperatures. Dry heat sterilization can be carried out by different methods such as treatment in a hot air oven, incineration, and infrared rays. Moist heat sterilization is usually performed by using an autoclave with temperatures above 100 °C and a pressure of 15 psi. Although most surgical devices are made of heat-stable material, some modern instruments require alternative sterilization techniques due to the presence of thermolabile materials used for their manufacture. The most common sterilizing agents used for heat-sensitive materials in healthcare facilities are UV rays, ethylene oxide (EtO) gas, hydrogen peroxide gas plasma, and liquid chemicals.

Disinfectant products can contain one or more active substances. In healthcare settings, objects are usually disinfected by liquid chemicals or wet pasteurization. For instance, a high level of disinfection can be achieved by using hot water, glutaraldehyde, ortho-phthalaldehyde (OPA), hydrogen peroxide, and peracetic acid. In contrast, other compounds like alcohols (ethyl alcohol, isopropyl alcohol, and methyl alcohol), aldehydes (formaldehyde, glutaraldehyde), phenolic compounds, and quaternary ammonium compounds can achieve an intermediate level of disinfection, eliminating non-enveloped viruses and some mycobacteria.

22.2.2 Factors Influencing Disinfection and Sterilization Efficacy

In order to improve the efficacy of disinfection and sterilization procedures, it is essential to understand the different factors that have an impact on the success of a given method. Perhaps, the three most relevant factors are the biological traits of the target microorganisms, their physical location, and the design of the application procedure.

Regarding the target microorganisms, there are three main features to be considered: (i) their intrinsic resistance to a given procedure or agent, (ii) the microbial load, and (iii) their location and spatial organization. For example, susceptibility to disinfection and sterilization processes varies depending on the developmental stage (e.g. bacterial spores are much more resistant than vegetative cells) or on the type of microorganism and species (Leggett et al. 2012). Moreover, high contamination levels generally require longer and/or more intense treatments, hence the importance of previous cleaning. The location and organization of microorganisms may imply a greater or lesser exposure to the disinfectant agent. For instance, the extracellular matrix in bacterial biofilms may hinder diffusion of disinfectants and reduce exposure of the embedded cells (Sanchez-Vizuete et al. 2015). Similarly, organic and inorganic solids can interfere with the antimicrobial activity of disinfectants or sterilization systems by protecting the microorganisms or inactivating the antimicrobial agent.

It is well known that the efficacy of a given method for disinfection or sterilization depends on the procedure design, the potency to kill microorganisms, concentration (for chemical treatments), and exposure time. In relation with the procedure design, standard cleaning and disinfection practices are also influenced by the personnel, and notable differences can be observed depending on individual performance (Boyce et al. 2010). The type of surface, contact time, inappropriate concentration, or contamination of the disinfectant may also contribute to reducing the effectiveness of the treatment (Boyce et al. 2016). Additionally, physical and chemical factors (temperature, pH, relative humidity, and ionic strength) may have an effect on the activity of most disinfectants.

There are several methods to assess the efficacy of disinfection procedures such as ATP bioluminescence assays (Boyce et al. 2009) or microbiological screening (White et al. 2008). In fact, these techniques have been incorporated into the disinfection routine of some hospitals (Dancer et al. 2008; Lewis et al. 2008). Data obtained from these screenings can then be compared with data from hospital-acquired infections to determine the success of different cleaning procedures. However, these monitoring systems require the use of microbiological standards. Dancer (2004) proposed two different possible standards to evaluate hospital cleaning. The first standard is the presence of some microorganisms such as *S. aureus*, including MRSA, *C. difficile*, VRE, and various Gram-negative bacilli. The second alternative is a quantitative evaluation of aerobic colony counts on frequent-hand-touch surfaces ($<5\,\text{cfu cm}^2$). Regarding the ATP bioluminescence assays, standard levels range from 25 to 500 relative light units (RLU) for surfaces of $10–100\,\text{cm}^2$ (Mulvey et al. 2011). However, it should be taken into account that, independently of the results obtained in these assessment techniques, the risk of infection also depends on the probability that a patient comes in contact with the contaminated surface.

22.3 Resistance to Disinfectants

22.3.1 Molecular Mechanisms of Biocide Resistance

A major setback for successful surface decontamination is the presence of disinfectant-resistant microorganisms. Resistance mechanisms may be inherent to a given microbe (intrinsic), acquired through horizontal transfer or mutation, or a result of microbial adaptations to a given environmental condition (adaptive). All three types of resistance can be mediated by a wide variety of molecular mechanisms. For example, the cell envelope provides a first layer of protection from the entrance of toxic agents, although the level of selectivity differs depending on the bacterial species. At one end of the spectrum, we find spores, which are the most resistant forms of microbial life known to date. As a result, biocide concentrations that may be effective against *Clostridium* or *Bacillus* vegetative cells do not kill spores formed by these bacteria (Leggett et al. 2012). Mycobacteria are also notably resistant thanks to their complex cell wall (Russell 1996; Frenzel et al. 2011). Likewise, the outer membrane of Gram-negative bacteria is an effective permeability barrier to the uptake of antibiotics and disinfectants, hence their higher intrinsic antimicrobial resistance compared with that of Gram-positive bacteria (Russell 1997). Besides the intrinsic permeability of a given microbe, stable or transient changes to the structure or composition of the cell envelope can have an effect on the resistance of certain strains. For instance, mutations that change the number or topology of porins in Gram-negative bacteria (Figure 22.2a) or mycobacteria may lead to a lesser sensitivity to some biocides (Heinzel 1998; Frenzel et al. 2011). Also, changes that alter the net charge of the cell surface may confer antimicrobial resistance by affecting electrostatic interactions with some compounds (Bruinsma et al. 2006). Besides limiting the entry of disinfectants into bacterial cells, microbes can also expel out toxic compounds by the so-called efflux pumps (Figure 22.2a). There are multiple examples of efflux pumps involved in resistance to disinfectants, such as QacA, MexCD, and AcrAB in *S. aureus*, *P. aeruginosa*, and *Escherichia coli*, respectively (Rouch et al. 1990; Ma et al. 1993; Fraud et al. 2008). Microbes can also produce proteins with disinfectant-degrading activity like catalase (Figure 22.2a), an enzyme that breaks down hydrogen peroxide, thereby protecting the bacterial cells (Stewart et al. 2000).

22.3.2 Biofilms

Biofilm formation can be considered a type of adaptive resistance, in which bacterial cells respond to certain environmental cues by attaching to a surface and multiplying while surrounding themselves with an extracellular matrix. This matrix may consist of different proportions of polysaccharides, DNA, and proteins, among other molecules (Flemming and Wingender 2010). It is well known that biofilms can withstand concentrations of antimicrobials that would easily eradicate a planktonic cell population (Costerton et al. 1999). Nonetheless, the transient nature of this resistance is evidenced by the fact that biofilm cells subcultured in a liquid medium display planktonic resistance levels (de la Fuente-Núñez et al. 2013). Nevertheless, it is worth noting that biofilms exhibit increased mutation (Driffield et al. 2008) and horizontal transfer rates

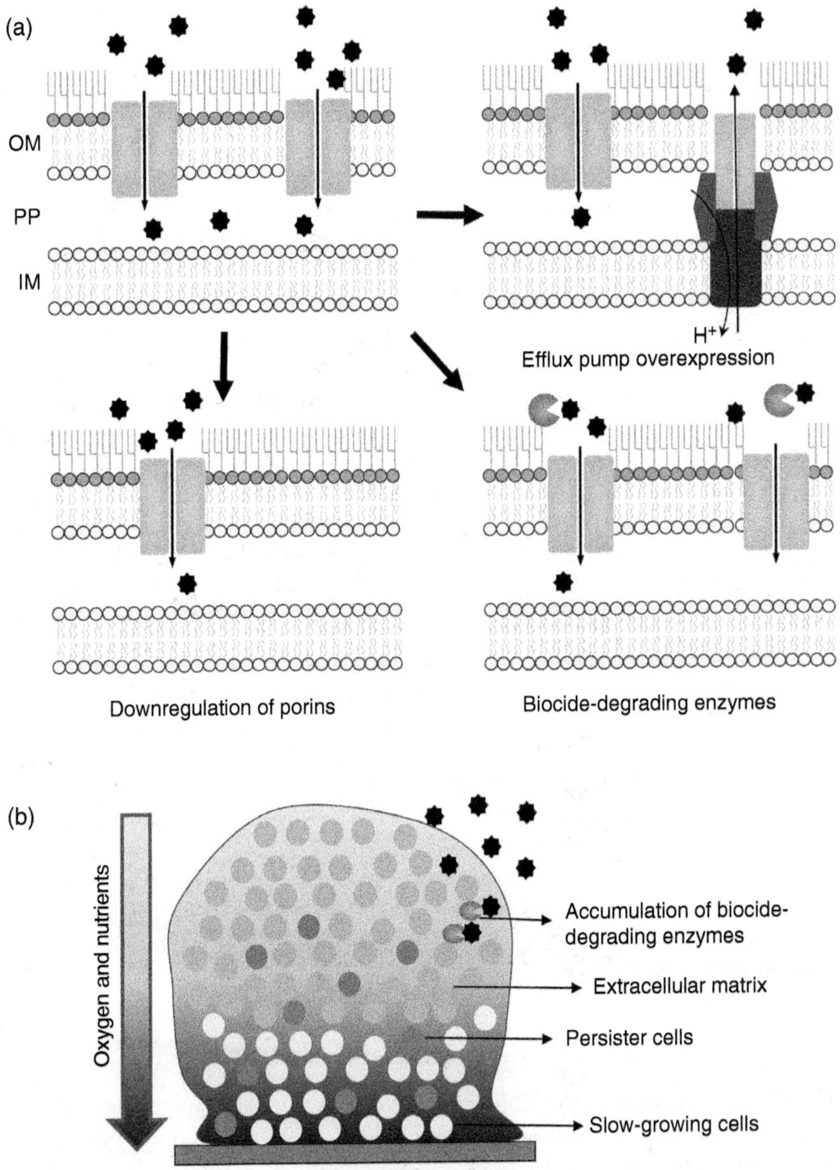

Figure 22.2 Mechanisms of adaptive resistance to disinfectants. (a) Examples of adaptations that confer decreased susceptibility to biocidal compounds in a Gram-negative bacterium: increased expression of degrading enzymes or efflux pumps and decreased expression of porin-encoding genes. (b) Examples of factors that increase resistance to disinfectants of bacterial biofilms. OM, outer membrane; PP, periplasm; IM, inner membrane. *(See insert for color representation of the figure.)*

(Cook et al. 2011; Savage et al. 2013), which may result in a higher prevalence of acquired resistance. Resistance in bacterial biofilms is a very complex phenomenon involving multiple mechanisms, many of which are directly related to the structure and composition of these communities. For example, the extracellular matrix may restrict the access of disinfectants to the cells (Figure 22.2b) (Kumon et al. 1994; Singh et al. 2010). Indeed, despite the existence of water-filled channels throughout the biofilm, interactions, such as electrostatic ones, between the matrix polymers and biocidal compounds may block their antimicrobial action (Lambert and Johnston 2001). The extracellular matrix is also an environment that permits the accumulation of disinfectant-degrading enzymes (Figure 22.2b) (Stewart et al. 2000). Another factor that contributes to biofilm resistance is the heterogeneous nature of the bacterial population. Thus, biofilms typically display nutrient and oxygen gradients that result in deep areas of lesser growth and anaerobic conditions together with areas of active growth near the surface (Figure 22.2b). These conditions also trigger stress responses in certain areas of the biofilm (Figure 22.2b). As a result of all of these factors, cells in different areas of the biofilm will display distinct biocide susceptibility profiles (Saby et al. 1999; Sabev et al. 2006). Additionally, the environmental conditions in biofilms can also modulate the expression of other resistance determinants. An example is the upregulation of some efflux pumps by low oxygen conditions, which occur in deep biofilm layers, increasing the ability of cells to expel antimicrobials (Schaible et al. 2012). Additionally, biofilms have a greater proportion of the so-called persister cells (Figure 22.2b), which exhibit tolerance to many toxic compounds (Lewis 2008). Finally, another trait that adds even more complexity to multicellular communities is the potential coexistence of multiple species in close proximity forming mixed-species biofilms (Sanchez-Vizuete et al. 2015). This phenomenon may limit the efficacy of antimicrobial treatments because the presence of different microorganisms can sometimes be more resilient to disinfectant treatments than the single-species biofilms formed by the same microbes (Sanchez-Vizuete et al. 2015).

22.3.3 Cross-Resistance Between Antibiotics and Disinfectants

The danger of promoting antibiotic resistance due to widespread use of disinfectants should not be underestimated. Indeed, there are several examples of cross-resistance in literature. For example, Loughlin et al. reported the acquisition of adaptive resistance to benzalkonium chloride and antibiotics in *P. aeruginosa* grown in subinhibitory concentrations of the biocide (Loughlin et al. 2002). We now have more information about the mechanisms involved in cross-resistance. For instance, some efflux pumps are involved in resistance to both disinfectants and antibiotics. Moreover, there is evidence that exposure to some biocides can upregulate the expression of efflux systems in pathogenic bacteria. Thus, subinhibitory levels of chlorhexidine increased the expression of MexCD in *P. aeruginosa* via the global regulator AlgU (Fraud et al. 2008). MexCD-OprJ is a pump that mediates resistance to chloramphenicol, fluoroquinolones, and tetracycline (Poole et al. 1996). Likewise, the SmeDEF efflux pump of *Stenotrophomonas maltophilia* is induced upon exposure to triclosan due to binding of the biocide to repressor SmeT (Hernandez et al. 2011). SmeDEF upregulation leads to resistance to erythromycin, fluoroquinolones, and tetracycline (Alonso and Martinez 2000; Zhang et al. 2001). Another example of adaptive antibiotic resistance in response to sublethal doses of disinfectants is the downregulation of *E. coli* outer membrane porins OmpA, OmpF, and OmpT following exposure to benzalkonium chloride (Bore et al. 2007). This confers resistance to this biocide as well as to several antibiotics, including chloramphenicol, ciprofloxacin, and ampicillin (Bore et al. 2007). Cross-resistance between disinfectants and antibiotics can also occur through induction of the SOS response by different compounds like hydrogen peroxide or fluoroquinolones (Nunoshiba et al. 1991).

22.4 New Technologies as Alternatives to Classical Disinfectants

The need to improve disinfection procedures in healthcare environments has led to the development of new technologies, which include the use of new chemical antimicrobial compounds and physical methods for bacterial destruction. Overall, these technologies are very efficient against the main nosocomial pathogens. However, their implementation as part of routine disinfection protocols still requires more studies to accurately evaluate the costs versus benefits of these techniques.

22.4.1 Chemical and Physical Disinfectants

22.4.1.1 Hydrogen Peroxide and Hydrogen Peroxide-Based Solutions

New disinfectants containing hydrogen peroxide have shown a high ability to remove contaminating bacteria, including multiresistant strains (Rutala et al. 2012). A combination between peracetic acid and hydrogen peroxide proved to be efficient even against spores (Deshpande et al. 2014). Some advantages of these compounds is that they can also be applied as vapors and aerosols and that they are effective against *Mycobacterium tuberculosis*, MRSA, viruses, spore formers, VRE, and Gram-negative bacilli, including *Acinetobacter* spp. (Falagas et al. 2011). The main drawbacks of this system are the risk of accidental human exposure, the corrosion of some materials, the economic cost, and the long cycle times needed to perform the disinfection procedure.

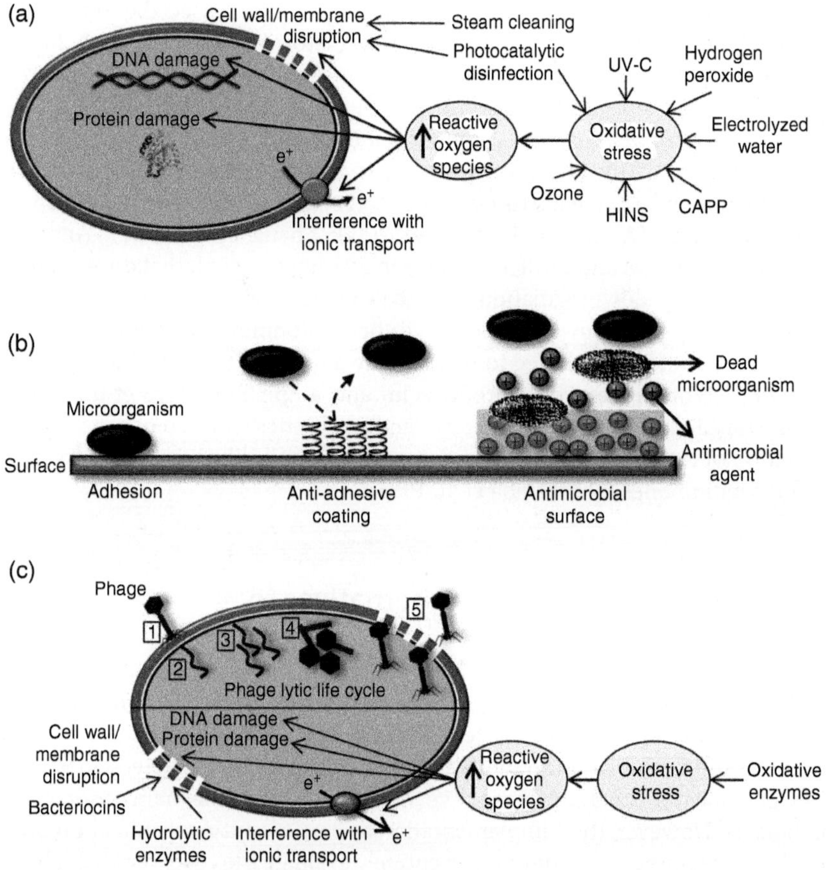

Figure 22.3 Schematic representation of the mode of action of new antimicrobial alternatives. (a) Chemical/physical disinfectants affecting different parameters in the microorganism (generating oxidative stress and the subsequent effects or provoking cell wall/membrane disruption). HINS, High-Intensity Narrow-Spectrum Light; CAPP, Cold-Air Atmospheric Pressure Plasma. (b) Coating of surfaces with anti-adhesive compounds or antimicrobial agents that prevent adhesion and kill the microorganisms. (c) Mode of action of biological disinfectants. Bacteriocins and hydrolytic enzymes degrade the cell wall structure. Oxidative enzymes result in an oxidative stress unleashing dead of the microorganism. Bacteriophages that kill the bacteria in the last step of their life cycle. (1) Adsorption. (2) DNA injection. (3) DNA replication. (4) Synthesis of new phage particles. (5) Release of viral progeny and cell lysis.

22.4.1.2 Electrolyzed Water

Electrolyzed water initially showed promising results (Fertelli et al. 2013), as the procedure reduced the number of aerobic colony counts, but more recent studies showed a further increase in staphylococci cell counts several hours after the disinfection procedure (Stewart et al. 2014). The main advantages of this technique are lack of toxicity and short application time, allowing electrolyzed water to be used in near-patient surfaces.

22.4.1.3 Cold-Air Atmospheric Pressure Plasma

Cold-air atmospheric pressure plasma is produced by the excitation of gas with electrical discharges at room temperature and atmospheric pressure. This process generates reactive oxygen species with antimicrobial activity (Cahill et al. 2014). The efficacy of this technique against spores is still under evaluation (Claro et al. 2015).

22.4.1.4 Steam Cleaning

Steam cleaning is highly effective against several pathogens, including Gram-negative bacilli like *P. aeruginosa* (Tanner 2009). However, this system has some drawbacks such as incompatibility with electrical appliances, average duration of the procedure, the presence of residual water, and the risk of exacerbating breathing problems.

22.4.1.5 Ozone

Ozone is a disinfectant with high oxidative properties and, therefore, with strong antibacterial activity against many pathogens but not bacterial spores (Sharma and Hudson 2008; Doan et al. 2012). The main advantage of ozone is its low cost but it also has some disadvantages like toxicity and corrosiveness for metals.

22.4.1.6 Ultraviolet Light Irradiation (UV-C)

Ultraviolet light irradiation (UV-C) with a wavelength of 254 nm exhibits germicidal activity and can be useful for the decontamination of surfaces, instruments, and air in healthcare facilities (Nernandzic et al. 2010). However, there are some issues to be considered such as cost, time of exposure, light intensity, and loss of effectiveness when there are obstacles between the light and the target.

22.4.1.7 High-Intensity Narrow-Spectrum (HINS) Light

High-intensity narrow-spectrum (HINS) light is a new system for disinfection using visible violet light at 405 nm (Maclean et al. 2014). This technology has a lower activity than UV-C, but a major advantage is that it can be used in rooms occupied by patients. Nonetheless, further investigation is still necessary.

22.4.1.8 Photocatalytic Disinfection

Photocatalytic disinfection uses UV-activated titanium dioxide to oxidize volatile organic compounds. It is interesting for applications against airborne microorganisms such as *S. aureus* and *C. difficile* (Cram et al. 2004).

22.4.2 Antimicrobial Surfaces

The use of antimicrobial surfaces to prevent nosocomial infections is currently under investigation. To manufacture these surfaces, materials are coated with both antimicrobial and anti-adhesive compounds to keep them germ-free. Consequently, treated surfaces will pose a lesser infection risk for patients. Microbial adhesion is generally prevented by covering the surface with a layer of compounds that creates a hydrophilic environment and hinders hydrophobic interactions between bacterial cells and the surface. This can be achieved using polyethylene glycol (PEG), diamond-like carbon (DLC), and zwitterionic head groups (Park et al. 1998; Cheng et al. 2007). Additionally, these surfaces are coated with heavy metals that exhibit antimicrobial properties. The most common metals used are silver and copper. Although the antibacterial efficacy of silver-coated surfaces in hospitals has not been assessed extensively yet, there are multiple examples that support the antimicrobial activity of this metal in catheters and other medical devices (Chaloupka et al. 2010). Regarding the use of copper-coated surfaces in hospitals, there are several studies that have shown good results (Casey et al. 2010). However, a potential problem related to widespread use of metal-coated surfaces would be the development of resistance to copper and/or silver (Percival et al. 2005; Dupont et al. 2011).

Antimicrobial surfaces can also be engineered by coating with other disinfectants such as triclosan and quaternary ammonium compounds. The main drawback of these surfaces is the release of these compounds into the environment, which could promote resistance acquisition. Finally, there are some new antimicrobials, like polycationic peptides (polyethyleneimines [PEIs]), that can be used to coat surfaces, but more research is still required to rule out toxic effects and to confirm their stability (Klivanov 2007).

22.4.3 Biological Disinfectants

22.4.3.1 Bacteriophages or Phages

Bacteriophages or phages are viruses that exclusively attack bacteria. Their antimicrobial potential is well known, and multiple applications are currently under investigation, the most notable being as an antimicrobial agent (O'Flaherty et al. 2009). Regarding the prevention of bacterial contamination, bacteriophages have been successfully attached to different materials used to manufacture medical devices like catheters (Curtin and Donlan 2006) and biodegradable polymers (Jikia et al. 2005) (Markoishvili et al. 2002). The main drawbacks of bacteriophage-containing materials are the high specificity of phages for their target host and the potential development of bacterial resistance. Both problems can be solved by using mixtures of phages with different host specificities. Thus, mixtures of phages have a wider host range and drastically reduce the frequency to find bacteria resistant to all the phages. In addition, disinfectants containing free bacteriophages can be useful for the removal of bacterial biofilms (Gutierrez et al. 2016). Although numerous studies confirm the promising results of phage preparations, no legal framework has been developed yet to allow their use in clinical environments.

22.4.3.2 Enzymes

Disinfectants containing enzymes with hydrolytic or oxidative activities can be used to remove pathogenic microorganisms. For example, laccases, haloperoxidases, and perhydrolases generate biocidal oxidants and peracetic acid. In addition, these enzymes can be incorporated in nanoparticles made of silica, gold, and/or carbon nanotubes, which may significantly increase their stability (Grover et al. 2013). More recently, bacteriophage-encoded enzymes have also been proposed as promising disinfectants, especially to facilitate the removal of bacterial biofilms. Thus, phage endolysins have displayed *in vitro* anti-biofilm activity against major clinical pathogens (Shen et al. 2013; Gutiérrez et al. 2014). Moreover, some phage exopolysaccharide depolymerase proteins have the ability to degrade extracellular material from bacterial biofilms (Gutierrez et al. 2015; Pires et al. 2016).

22.4.3.3 Bacteriocins

Many bacteria synthesize antimicrobial peptides that inhibit growth of putative competitors. Specific attention has been paid to bacteriocins synthesized by nonpathogenic lactic acid bacteria because they are active against several Gram-positive pathogens of clinical relevance, including antibiotic-resistant bacteria. Moreover, some bacteriocins like nisin have a long history of safe use as food biopreservatives that strongly support their potential use in clinical applications (Shin et al. 2016). The antimicrobial activity of bacteriocins against biofilms has been recently investigated, and nisin has been proved to inhibit the growth of MRSA biofilms on medical devices (Okuda et al. 2013).

22.5 Current Legislation

Disinfectants are used to kill living organisms and, not surprisingly, often pose a risk of toxicity to humans or the environment. For this reason, their commercialization and application must be strictly controlled. To accomplish this goal, most countries require the authorization or at least notification of a disinfectant product before it can be placed on the market. Besides safety, regulatory frameworks also aim to ensure efficacy of the products.

In the European Union, disinfectants are regulated under the Biocidal Products Regulation 528/12 (BPR, http://eur-lex.europa.eu/legal-content/EN/

TXT/PDF/?uri=CELEX:32012R0528&from=EN), which entered into force on 17 July 2012 and replaced the Biocidal Products Directive 98/8/EC (BPD, http://eur-lex.europa.eu/legal-content/EN/TXT/PDF/?uri=CELEX:31998L0008&from=en). According to BPR, biocidal products can be classified based on their use in 22 different product types (PT). Disinfectants correspond to five different PTs, namely, PT1 (human hygiene biocidal products), PT2 (private area and public health area disinfectants and other biocidal products), PT3 (veterinary hygiene biocidal products), PT4 (food and feed area disinfectants), and PT5 (drinking water disinfectants). It must be noted that cleaning products without a biocidal claim do not fall under this regulation. Likewise, products used for disinfecting medical instruments and medical devices are regulated under the Medical Device Directive 93/42/EEC (http://eur-lex.europa.eu/LexUriServ/LexUriServ.do?uri=CONSLEG:1993L0042:20071011:en:PDF). Conversely, articles that have been treated with biocidal substances are considered biocidal products, and their commercialization requires authorization under BPR. The new EU regulatory framework aims to comply with the rules and requirements set by different international organizations, including the World Trade Organization (WTO), the United Nations and the Globally Harmonized System (GHS) for Classification and Labelling of Chemicals, and the Organisation for Economic Co-operation and Development (OECD). Additionally, the European Chemicals Agency (ECHA) oversees that risk assessments for biocides and chemicals regulated under REACH (Registration, Evaluation, Authorisation and Restriction of Chemicals) are performed according to the same principles.

In the United States, disinfectants are regulated by the Environment Protection Agency (EPA) and the Food and Drug Administration (FDA). More specifically, the EPA regulates the registration of products intended for disinfecting household and clinical contact surfaces under the Federal Insecticide, Fungicide, and Rodenticide Act (FIFRA) of 1947 (amended in 1972, 1988, 1996, and 2012; https://www.agriculture.senate.gov/imo/media/doc/FIFRA.pdf). Conversely, the FDA is responsible for the commercialization of liquid chemical sterilants/high-level disinfectants used for disinfection of clinical devices under the authority of the Medical Devices Amendment to the Food, Drug, and Cosmetic Act of 1976 (https://www.fda.gov/regulatoryinformation/lawsenforcedbyfda/federalfooddrugandcosmeticactfdcact/default.htm), which has also been subsequently amended. Interestingly, the EPA and the FDA have different criteria for the classification of disinfectants. Thus, the FDA follows the same classification and terminology suggested by the Centers for Disease Control and Prevention (CDC) into high-, intermediate-, and low-level disinfectants, depending on their potency. Conversely, the EPA distinguishes between limited, general, and hospital disinfectants.

The European Union and the United States are just two examples, but, as mentioned previously, most countries have some similar legislation. A major aspect of all these regulations is the need for the manufacturers to provide the competent authorities with proof regarding the safety and efficacy of their products. Indeed, an important aspect of regulations about disinfectants concerns product labeling, especially in terms of defining the applications and instructions for the appropriate use of the product. The suitability of a product against its target organisms constitutes the so-called label claims and must always be backed up by efficacy tests that should be ideally performed according to internationally recognized guidelines, such as those published by the OECD, the European Committee for Standardisation (CEN), or the International Organization for Standardization (ISO). However, in-company standard operating procedures (SOPs) can also be valid if there is no available standard method. Based on the different studies concerning efficacy, toxicity, and environmental risks, the company will compile the instructions that will ensure a safe and effective use of the product. Indeed, it is very important to select the right disinfectant, as well as the right concentration and contact time for the intended application. During the authorization process, manufacturers are also frequently required to provide information regarding the potential development of resistance to the active substance present in the disinfectant.

Overall, it is quite clear that regulations controlling the use of disinfectants are essential to guarantee both human and environmental safety. However, regulatory burdens should not pose a major obstacle for placing new products on the market. This is especially important given that the development of novel active compounds may be the only way to overcome the antimicrobial resistance threat.

22.6 Conclusions

Nowadays, it is widely accepted that proper cleaning and disinfection of healthcare facilities is essential to control nosocomial infections. However, the increase in antibiotic resistance, somehow favored by the selection of disinfectant-resistant bacteria, is making us reconsider the efficacy of current prophylactic measures to control infection diseases. In this context, the adoption of new technologies to remove microorganisms from clinical environments is welcome, as long as they meet the requirements of safety, efficacy, low rate of bacterial resistance selection, and cost. To this end, a proper understanding of pathogen environmental reservoirs, routes of transmission, and especially disinfectant resistance mechanisms is needed.

References

Alonso, A. and Martinez, J.L. (2000). Cloning and characterization of SmeDEF, a novel multidrug efflux pump from *Stenotrophomonas maltophilia*. *Antimicrob. Agents Chemother.* 44: 3079–3086.

Bore, E., Hebraud, M., Chafsey, I. et al. (2007). Adapted tolerance to benzalkonium chloride in *Escherichia coli* K-12 studied by transcriptome and proteome analyses. *Microbiology* 153: 935–946.

Boyce, J.M., Havill, N.L., Dumigan, D.G. et al. (2009). Monitoring the effectiveness of hospital cleaning practices by use of an adenosine triphosphate bioluminescence assay. *Infect. Control Hosp. Epidemiol.* 30: 678–684.

Boyce, J.M., Havill, N.L., Lipka, A. et al. (2010). Variations in hospital daily cleaning practices. *Infect. Control Hosp. Epidemiol.* 31: 99–101.

Boyce, J.M., Sullivan, L., Booker, A., and Baker, J. (2016). Quaternary ammonium disinfectant issues encountered in an environmental services department. *Infect. Control Hosp. Epidemiol.* 37: 340–342.

Bruinsma, G.M., Rustema-Abbing, M., van der Mei, H.C. et al. (2006). Resistance to a polyquaternium-1 lens care solution and isoelectric points of *Pseudomonas aeruginosa* strains. *J. Antimicrob. Chemother.* 57: 764–766.

Cahill, O.J., Claro, T., O'Connor, N. et al. (2014). Cold air plasma to decontaminate inanimate surfaces of the hospital environment. *Appl. Environ. Microbiol.* 80: 2004–2010.

Casey, A.L., Adams, D., Karpanen, T.J. et al. (2010). Role of copper in reducing hospital environment contamination. *J. Hosp. Infect.* 74: 72–77.

CDC (2017). Healthcare-associated infections. https://www.cdc.gov/winnablebattles/healthcareassociatedinfections (accessed 26 April 2017).

Chaloupka, K., Malam, Y., and Seifalian, A.M. (2010). Nanosilver as a new generation of nanoproduct in biomedical applications. *Trends Biotechnol.* 28: 580–588.

Cheng, G., Zhang, Z., Chen, S. et al. (2007). Inhibition of bacterial adhesion and biofilm formation on zwitterionic surfaces. *Biomaterials* 28: 4192–4199.

Claro, T., Cahill, O.J., O'Connor, N. et al. (2015). Cold-air atmospheric pressure plasma against *Clostridium difficile* spores: a potential alternative for the decontamination of hospital inanimate surfaces. *Infect. Control Hosp. Epidemiol.* 36: 742–744.

Cook, L., Chatterjee, A., Barnes, A. et al. (2011). Biofilm growth alters regulation of conjugation by a bacterial pheromone. *Mol. Microbiol.* 81: 1499–1510.

Costerton, J.W., Stewart, P.S., and Greenberg, E.P. (1999). Bacterial biofilms: a common cause of persistent infections. *Science* 284: 1318–1322.

Cram, N., Shipman, N., and Quarles, J.M. (2004). Reducing airborne microbes in the surgical operating theater and other clinical settings: a study utilizing the AiroCide System. *J. Clin. Eng.* 29 (2): 79–88.

Curtin, J.J. and Donlan, R.M. (2006). Using bacteriophages to reduce formation of catheter-associated biofilms by *Staphylococcus epidermidis*. *Antimicrob. Agents Chemother.* 50: 1268–1275.

Dancer, S.J. (2004). How do we assess hospital cleaning? A proposal for microbiological standards for surface hygiene in hospitals. *J. Hosp. Infect.* 56: 10–15.

Dancer, S.J., White, L., and Robertson, C. (2008). Monitoring environmental cleanliness on two surgical wards. *Int. J. Environ. Health Res.* 18: 357–364.

de la Fuente-Núñez, C., Reffuveille, F., Fernández, L., and Hancock, R.E.W. (2013). Bacterial biofilm development as a multicellular adaptation: antibiotic resistance and new therapeutic strategies. *Curr. Opin. Microbiol.* 16: 580–589.

Deshpande, A., Mana, T.S., Cadnum, J.L. et al. (2014). Evaluation of a sporicidal peracetic acid/hydrogen peroxide-based daily disinfectant cleaner. *Infect. Control Hosp. Epidemiol.* 35: 1414–1416.

Dettenkofer, M., Wenzler, S., Amthor, S. et al. (2004). Does disinfection of environmental surfaces influence nosocomial infection rates? A systematic review. *Am. J. Infect. Control* 32: 84–89.

Doan, L., Forrest, H., Fakis, A. et al. (2012). Clinical and cost effectiveness of eight disinfection methods for terminal disinfection of hospital isolation rooms contaminated with *Clostridium difficile* 027. *J. Hosp. Infect.* 82: 114–121.

Donskey, C.J. (2010). Preventing transmission of *Clostridium difficile*: is the answer blowing in the wind? *Clin. Infect. Dis.* 50: 1458–1461.

Driffield, K., Miller, K., Bostock, J.M. et al. (2008). Increased mutability of *Pseudomonas aeruginosa* in biofilms. *J. Antimicrob. Chemother.* 61: 1053–1056.

Dupont, C.L., Grass, G., and Rensing, C. (2011). Copper toxicity and the origin of bacterial resistance – new insights and applications. *Metallomics* 3: 1109–1118.

ECDC (2017). Healthcare-associated infections. http://ecdc.europa.eu/en/healthtopics/Healthcare-associated_infections/Pages/index.aspx (accessed 26 April 2017).

Falagas, M.E., Thomaidis, P.C., Kotsantis, I.K. et al. (2011). Airborne hydrogen peroxide for disinfection of the hospital environment and infection control: a systematic review. *J. Hosp. Infect.* 78: 171–177.

Fertelli, D., Cadnum, J.L., Nerandzic, M.M. et al. (2013). Effectiveness of an electrochemically activated saline solution for disinfection of hospital equipment. *Infect. Control Hosp. Epidemiol.* 34: 543–544.

Flemming, H.C. and Wingender, J. (2010). The biofilm matrix. *Nat. Rev. Microbiol.* 8: 623–633.

Fraud, S., Campigotto, A.J., Chen, Z., and Poole, K. (2008). MexCD-OprJ multidrug efflux system of *Pseudomonas aeruginosa*: involvement in chlorhexidine resistance and induction by membrane-damaging agents dependent upon the AlgU stress response sigma factor. *Antimicrob. Agents Chemother.* 52: 4478–4482.

Frenzel, E., Schmidt, S., Niederweis, M., and Steinhauer, K. (2011). Importance of porins for biocide efficacy against *Mycobacterium smegmatis*. *Appl. Environ. Microbiol.* 77: 3068–3073.

Gnanadhas, D.P., Marathe, S.A., and Chakravortty, D. (2013). Biocides – resistance, cross-resistance mechanisms and assessment. *Expert Opin. Investig. Drugs* 22: 191–206.

Grover, N., Dinu, C.Z., Kane, R.S., and Dordick, J.S. (2013). Enzyme-based formulations for decontamination: current state and perspectives. *Appl. Microbiol. Biotechnol.* 97: 3293–3300.

Gutierrez, D., Briers, Y., Rodriguez-Rubio, L. et al. (2015). Role of the pre-neck appendage protein (Dpo7) from phage vB_SepiS-phiIPLA7 as an anti-biofilm agent in staphylococcal species. *Front. Microbiol.* 6: 1315.

Gutierrez, D., Rodriguez-Rubio, L., Martinez, B. et al. (2016). Bacteriophages as weapons against bacterial biofilms in the Food Industry. *Front. Microbiol.* 7: 825.

Gutiérrez, D., Ruas-Madiedo, P., Martínez, B. et al. (2014). Effective removal of staphylococcal biofilms by the endolysin LysH5. *PLoS One* 9: e107307.

Heinzel, M. (1998). Phenomena of biocide resistance in microorganisms. *Int. Biodeterior. Biodegrad.* 41: 225–234.

Hernandez, A., Ruiz, F.M., Romero, A., and Martinez, J.L. (2011). The binding of triclosan to SmeT, the repressor of the multidrug efflux pump SmeDEF, induces antibiotic resistance in *Stenotrophomonas maltophilia*. *PLoS Pathog.* 7: e1002103.

Jikia, D., Chkhaidze, N., Imedashvili, E. et al. (2005). The use of a novel biodegradable preparation capable of the sustained release of bacteriophages and ciprofloxacin, in the complex treatment of multidrug-resistant *Staphylococcus aureus*-infected local radiation injuries caused by exposure to Sr90. *Clin. Exp. Dermatol.* 30: 23–26.

Klivanov, A.M. (2007). Permanently microbicidal materials coatings. *J. Mater. Chem.* 17: 2479–2482.

Kramer, A., Schwebke, I., and Kampf, G. (2006). How long do nosocomial pathogens persist on inanimate surfaces? A systematic review. *BMC Infect. Dis.* 6: 130.

Kumon, H., Tomochika, K., Matunaga, T. et al. (1994). A sandwich cup method for the penetration assay of antimicrobial agents through *Pseudomonas* exopolysaccharides. *Microbiol. Immunol.* 38: 615–619.

Lambert, R.J. and Johnston, M.D. (2001). The effect of interfering substances on the disinfection process: a mathematical model. *J. Appl. Microbiol.* 91: 548–555.

Leggett, M.J., McDonnell, G., Denyer, S.P. et al. (2012). Bacterial spore structures and their protective role in biocide resistance. *J. Appl. Microbiol.* 113: 485–498.

Lewis, K. (2008). Multidrug tolerance of biofilms and persister cells. *Curr. Top. Microbiol. Immunol.* 322: 107–131.

Lewis, T., Griffith, C., Gallo, M., and Weinbren, M. (2008). A modified ATP benchmark for evaluating the cleaning of some hospital environmental surfaces. *J. Hosp. Infect.* 69: 156–163.

Loughlin, M.F., Jones, M.V., and Lambert, P.A. (2002). *Pseudomonas aeruginosa* cells adapted to benzalkonium chloride show resistance to other membrane-active agents but not to clinically relevant antibiotics. *J. Antimicrob. Chemother.* 49: 631–639.

Ma, D., Cook, D.N., Alberti, M. et al. (1993). Molecular cloning and characterization of acrA and acrE genes of *Escherichia coli*. *J. Bacteriol.* 175: 6299–6313.

Maclean, M., McKenzie, K., Anderson, J.G. et al. (2014). 405 nm light technology for the inactivation of pathogens and its potential role for environmental disinfection and infection control. *J. Hosp. Infect.* 88: 1–11.

Markoishvili, K., Tsitlanadze, G., Katsarava, R. et al. (2002). A novel sustained-release matrix based on biodegradable poly(ester amide)s and impregnated with bacteriophages and an antibiotic shows promise in management of infected venous stasis ulcers and other poorly healing wounds. *Int. J. Dermatol.* 41: 453–458.

McDonnell, G. and Burke, P. (2011). Disinfection: is it time to reconsider Spaulding? *J. Hosp. Infect.* 78: 163–170.

Mulvey, D., Redding, P., Robertson, C. et al. (2011). Finding a benchmark for monitoring hospital cleanliness. *J. Hosp. Infect.* 77: 25–30.

Nernandzic, M.M., Cadnum, J.L., Pultz, M.J., and Donskey, C.J. (2010). Evaluation of an automated ultraviolet radiation device for decontamination of *Clostridium difficile* and other healthcare-associated pathogens in hospital rooms. *BMC Infect. Dis.* 10 (197).

Nunoshiba, T., Hashimoto, M., and Nishioka, H. (1991). Cross-adaptive response in *Escherichia coli* caused by pretreatment with H2O2 against formaldehyde and other aldehyde compounds. *Mutat. Res.* 255: 265–271.

O'Flaherty, S., Ross, R.P., and Coffey, A. (2009). Bacteriophage and their lysins for elimination of infectious bacteria. *FEMS Microbiol. Rev.* 33: 801–819.

Okuda, K., Zendo, T., Sugimoto, S. et al. (2013). Effects of bacteriocins on methicillin-resistant *Staphylococcus aureus* biofilm. *Antimicrob. Agents Chemother.* 57: 5572–5579.

Park, K.D., Kim, Y.S., Han, D.K. et al. (1998). Bacterial adhesion on PEG modified polyurethane surfaces. *Biomaterials* 19: 851–859.

Percival, S.L., Bowler, P.G., and Russell, D. (2005). Bacterial resistance to silver in wound care. *J. Hosp. Infect.* 60: 1–7.

Pires, D.P., Oliveira, H., Melo, L.D. et al. (2016). Bacteriophage-encoded depolymerases: their diversity and biotechnological applications. *Appl. Microbiol. Biotechnol.* 100: 2141–2151.

Poole, K., Tetro, K., Zhao, Q. et al. (1996). Expression of the multidrug resistance operon mexA-mexB-oprM in *Pseudomonas aeruginosa*: mexR encodes a regulator of operon expression. *Antimicrob. Agents Chemother.* 40: 2021–2028.

Riggs, M.M., Sethi, A.K., Zabarsky, T.F. et al. (2007). Asymptomatic carriers are a potential source for transmission of epidemic and nonepidemic *Clostridium difficile* strains among long-term care facility residents. *Clin. Infect. Dis.* 45: 992–998.

Rouch, D.A., Cram, D.S., DiBerardino, D. et al. (1990). Efflux-mediated antiseptic resistance gene qacA from *Staphylococcus aureus*: common ancestry with tetracycline- and sugar-transport proteins. *Mol. Microbiol.* 4: 2051–2062.

Russell, A.D. (1996). Activity of biocides against mycobacteria. *Soc. Appl. Bacteriol. Symp. Ser.* 25: 87S–101S.

Russell, A.D. (1997). Plasmids and bacterial resistance to biocides. *J. Appl. Microbiol.* 83: 155–165.

Rutala, W.A., Gergen, M.F., and Weber, D.J. (2012). Efficacy of improved hydrogen peroxide against important healthcare-associated pathogens. *Infect. Control Hosp. Epidemiol.* 33: 1159–1161.

Rutala, W.A. and Weber, D.J. (2008). *Healthcare Infection Control Practices Advisory Committee: Guideline for Disinfection and Sterilization in Healthcare Facilities*, 158p. Atlanta, GA: Centers for Disease Control and Prevention (CDC).

Sabev, H.A., Robson, G.D., and Handley, P.S. (2006). Influence of starvation, surface attachment and biofilm growth on the biocide susceptibility of the biodeteriogenic yeast *Aureobasidium pullulans*. *J. Appl. Microbiol.* 101: 319–330.

Saby, S., Leroy, P., and Block, J.C. (1999). *Escherichia coli* resistance to chlorine and glutathione synthesis in response to oxygenation and starvation. *Appl. Environ. Microbiol.* 65: 5600–5603.

Sanchez-Vizuete, P., Orgaz, B., Aymerich, S. et al. (2015). Pathogens protection against the action of disinfectants in multispecies biofilms. *Front. Microbiol.* 6: 705.

Savage, V.J., Chopra, I., and O'Neill, A.J. (2013). *Staphylococcus aureus* biofilms promote horizontal transfer of antibiotic resistance. *Antimicrob. Agents Chemother.* 57: 1968–1970.

Schaible, B., Taylor, C.T., and Schaffer, K. (2012). Hypoxia increases antibiotic resistance in *Pseudomonas aeruginosa* through altering the composition of multidrug efflux pumps. *Antimicrob. Agents Chemother.* 56: 2114–2118.

Sharma, M. and Hudson, J.B. (2008). Ozone gas is an effective and practical antibacterial agent. *Am. J. Infect. Control* 36: 559–563.

Shen, Y., Koller, T., Kreikemeyer, B., and Nelson, D.C. (2013). Rapid degradation of *Streptococcus pyogenes* biofilms by PlyC, a bacteriophage-encoded endolysin. *J. Antimicrob. Chemother.* 68: 1818–1824.

Shin, J.M., Gwak, J.W., Kamarajan, P. et al. (2016). Biomedical applications of nisin. *J. Appl. Microbiol.* 120: 1449–1465.

Singh, R., Ray, P., Das, A., and Sharma, M. (2010). Penetration of antibiotics through *Staphylococcus aureus* and *Staphylococcus epidermidis* biofilms. *J. Antimicrob. Chemother.* 65: 1955–1958.

Spaulding, E.H. (1957). Chemical disinfection and antisepsis in the hospital. *J. Hosp. Res.* 9: 5–31.

Stewart, M., Bogusz, A., Hunter, J. et al. (2014). Evaluating use of neutral electrolyzed water for cleaning near-patient surfaces. *Infect. Control Hosp. Epidemiol.* 35: 1505–1510.

Stewart, P.S., Roe, F., Rayner, J. et al. (2000). Effect of catalase on hydrogen peroxide penetration into *Pseudomonas aeruginosa* biofilms. *Appl. Environ. Microbiol.* 66: 836–838.

Stiefel, U., Cadnum, J.L., Eckstein, B.C. et al. (2011). Contamination of hands with methicillin-resistant *Staphylococcus aureus* after contact with environmental surfaces and after contact with the skin of colonized patients. *Infect. Control Hosp. Epidemiol.* 32: 185–187.

Tanner, B.D. (2009). Reduction in infection risk through treatment of microbially contaminated surfaces with a novel, portable, saturated steam vapor disinfection system. *Am. J. Infect. Control* 37: 20–27.

Wales, A.D. and Davies, R.H. (2015). Co-selection of resistance to antibiotics, biocides and heavy metals, and its relevance to foodborne pathogens. *Antibiotics (Basel)* 4: 567–604.

Weber, D.J., Rutala, W.A., Miller, M.B. et al. (2010). Role of hospital surfaces in the transmission of emerging health care-associated pathogens: norovirus, *Clostridium difficile*, and *Acinetobacter* species. *Am. J. Infect. Control* 38: S25–S33.

White, L.F., Dancer, S.J., Robertson, C., and McDonald, J. (2008). Are hygiene standards useful in assessing infection risk? *Am. J. Infect. Control* 36: 381–384.

Zhang, L., Li, X.Z., and Poole, K. (2001). SmeDEF multidrug efflux pump contributes to intrinsic multidrug resistance in *Stenotrophomonas maltophilia*. *Antimicrob. Agents Chemother.* 45: 3497–3503.

23

Strategies to Prevent the Spread of Antibiotic Resistance

Understanding the Role of Antibiotics in Nature and Their Rational Use

Rustam Aminov[1,2]

[1] *School of Medicine, Medical Sciences and Nutrition, University of Aberdeen, Aberdeen, UK*
[2] *Institute of Fundamental Medicine and Biology, Kazan Federal University, Kazan, Russia*

23.1 Introduction

The history of antimicrobial drug discovery includes more than 15 classes of antimicrobials that became a cornerstone of microbial infection control and management and saved many lives (Aminov 2017). They indeed became one of the most successful forms of therapy in clinical medicine. However, their broad and often indiscriminate use in human and veterinary medicine and in agriculture resulted in the global antimicrobial resistance problem (Aminov 2010). A recent report by the Centers for Disease Control and Prevention (CDC), for example, stated: "Each year in the United States, at least 2 million people become infected with bacteria that are resistant to antibiotics and at least 23,000 die" (CDC Report 2013). According to the European Medicines Agency (EMA) estimates, infections by multidrug-resistant bacteria cause 25 000 deaths and healthcare expenditures and productivity losses approximately €1.5 billion in the EU every year (EMA 2015). Especially worrying are the trends toward the emergence of bacterial pathogens that are resistant toward all known drugs, so-called extensively or totally drug-resistant bacteria. For example, extensively drug-resistant strains of *Mycobacterium tuberculosis*, which are resistant to all tested drugs, have been already isolated in several countries around the globe (Migliori et al. 2007; Velayati et al. 2009; Udwadia et al. 2012; Klopper et al. 2013; Raviglione and Sulis 2016; Petersen et al. 2017). The worldwide number of deaths due to antibiotic-resistant infections may be estimated on the scale of many hundreds of thousands (O'Neill 2014). If no immediate actions are taken, the estimated death toll due to the antimicrobial

Antibiotic Drug Resistance, First Edition. Edited by José-Luis Capelo-Martínez and Gilberto Igrejas.

resistance will reach 10 million by the year 2050, thus surpassing, for instance, the mortality rate of cancer (O'Neill 2014).

Traditionally the problem of antibiotic resistance among pathogens has been viewed from the clinical microbiology perspective, e.g. as mainly associated with the therapeutic use and overuse/misuse of antibiotics in humans and animals. While this is certainly one of the contributing factors (Goossens et al. 2005), the problem must be contemplated in its entirety and from a broader evolutionary and ecological prospective (Aminov and Mackie 2007; Aminov 2009). In a Darwinian selection sense, the dilemma is very simple: more is the use of antibiotics, greater are the chances for selection of antibiotic resistance. Unlike the vertical inheritance of selected traits in macroorganisms, however, the main mechanisms of antibiotic resistance among bacteria are acquired horizontally. Because of this, antibiotic resistance genes selected in one ecological compartment can be transferred to other compartments aided by the extensive horizontal gene transfer (HGT) mechanisms operating in the microbial world (Aminov 2011, 2012).

23.2 Agriculture as the Largest Consumer of Antimicrobials

Several independent assessments suggest that a considerable proportion of antimicrobials produced are used in food animals for nontherapeutic purposes such as prophylactic, metaphylactic, and growth promoting (Landers et al. 2012; Krishnasamy et al. 2015). For example, an estimated half of the 210 000 tons of antibiotics produced in China are supposedly used in food animals, while the corresponding proportion in the United States may reach up to 80% (Collignon and Voss 2015). The proportion analogous to the latter can also be seen in Canada. According to the Canadian Integrated Program for Antimicrobial Resistance Surveillance (CIPARS), among antimicrobials distributed for use in Canada, 79% were intended for use in production animals (food animals and horses), 20% were intended for people, and less than 1% was intended for use in companion animals (CIPARS 2013). In Europe, about 70% of all antimicrobials are administered to animals raised for food production. According to a recent joint report compiled together by several European agencies (European Centre for Disease Prevention and Control [ECDC], European Food Safety Authority [EFSA], and EMA), among a total of 11 381.83 tons of antimicrobials consumed in 26 EU/EEA countries in 2012, only 3 399.8 tons was consumed by humans, while the amount fed to animals was more than twofold greater, at 7 982.0 tons or 70% of the total (ECDC, EFSA, and EMA 2015). From these data we may conclude, therefore, that the contemporary animal production systems in developed countries consume approximately 70–80% of antimicrobials produced. This model of animal production is closely followed by the developing countries to reach the same level of efficiency to meet the demands of growing population and lifestyle change.

The tendencies in the development of agricultural practices in animal industry suggest that the worldwide use of antimicrobials in food animals will grow rapidly, by at least 67% from 2010 to 2030 (Van Boeckel et al. 2015). Even a bigger growth of antimicrobial consumption by livestock is predicted for the developing economies. For example, in the BRICS countries, referring to Brazil, Russia, India, China, and South Africa, the quantity of antimicrobials ingested by livestock is predicted to double by 2030. The total amount of antibiotics produced for this purpose is predicted to reach 105 596 ± 3 605 tons in 2030 from 63 151 ± 1 560 tons in 2010 (Van Boeckel et al. 2015). About a third of this growth will be due to the change of farming practices, from small-scale and extensive family-owned farms to large-scale intensive farming operations (concentrated animal feeding operations [CAFOs]), that routinely use antimicrobials at subtherapeutic concentrations for prophylactic, metaphylactic, and growth-promoting purposes (Van Boeckel et al. 2015). This increase will be certainly followed by the corresponding increase in antimicrobial resistance, both in humans and in animals.

23.3 Antimicrobials and Antimicrobial Resistance

There is a considerable body of evidence suggesting an epidemiologic link between the antibiotic use in food animals and antibiotic resistance in humans (Landers et al. 2012). In a previously mentioned joint ECDC/EFSA/EMA report (ECDC, EFSA, and EMA 2015), an attempt to estimate associations between consumption of antimicrobials in humans and food-producing animals and antimicrobial resistance in bacteria from humans and food-producing animals has been implemented. This analysis suggested that the overall antimicrobial consumption, expressed in milligrams per kilogram of estimated biomass, was higher in animals than in humans. In both humans and animals, there were positive correlations between the consumption of certain antimicrobials and the corresponding resistance in bacteria. In some instances, however, there was a positive correlation between the consumption of certain antimicrobials in animals and resistance in bacteria from humans (ECDC, EFSA, and EMA 2015). Thus, as noted earlier (Landers et al. 2012), antimicrobial resistance in human-associated bacteria may be also affected by antimicrobial selection in other ecological compartments such as food-producing animals.

The number of antibiotic classes currently used in humans and animals is more than 15 (Aminov 2017). It is virtually impossible to overview all of them in terms of elaborating strategies to limit the dissemination of the corresponding resistances. Thus, the approach chosen here is to focus on one antibiotic class, tetracyclines, and the corresponding resistance genes. This will help to perform comprehensive analysis and discussion of the antibiotic discovery and usage, the corresponding antibiotic resistance mechanisms that emerged, and evolutionary and ecological aspects of antibiotics/resistance mechanisms. The general concepts derived from these analyses would be apparently applicable to other classes of antibiotics and the corresponding antibiotic resistance genes.

23.4 First-Generation Tetracyclines: Discovery and Usage

Chlortetracycline (Aureomycin®) was the first antibiotic of this class discovered by the researchers at Lederle Laboratories Division of American Cyanamid Company (Duggar 1948). Another antibiotic of this class, oxytetracycline (Terramycin®), was discovered in a collaborative drug discovery project between Pfizer and Harvard University (Hochstein et al. 1953). But even before these discoveries, there are indications that tetracyclines have been used well before the modern antibiotic era (Bassett et al. 1980; Hummert and Van Gerven 1982; Cook et al. 1989; Nelson et al. 2010). The tetracycline traces have been found on several occasions as incorporated into human skeletal remains, presumably because of diet containing the biomass of antibiotic-producing strains.

The clinical evaluation of the newly discovered chlortetracycline was almost immediately examined for the treatment of various human infections, and it was found to be equal in potency to penicillin (Wright and Schreiber 1949). Since then tetracyclines have been extensively used in the therapy of many human infections due to their efficiency and low incidences of side effects.

Another property of chlortetracycline that has been discovered accidentally at Lederle Laboratories Division of American Cyanamid Company is the growth-promoting effects of Aureomycin on animals (Duggar 1948; Aminov 2017). Following the confirmation of this effect, the company started producing animal feed with this additive, and many other companies and countries worldwide followed this example. Since then the antibiotics of this class became the drugs of choice for the use in animal production. In the United States, for example, tetracyclines comprise a large proportion of all antibiotics used in food animals; the quantity sold to the industry in 2013 reached an astounding 6514779 kg of active ingredient (FDA 2015). Proportionally, tetracyclines accounted for 71% of all antibiotics sold for the use in food-producing animals in the United States. The main route of tetracycline consumption by the livestock is via medicated feed, which accounted for most domestic sales and distribution of medically important antimicrobials approved for the use in food-producing animals. Tetracyclines were also the leading antibiotics to be administered via drinking water (FDA 2015). In China tetracyclines are also among the most prevalent antibiotic classes applied in animal production: an estimated 16 336 823 kg is used annually for swine production alone (Krishnasamy et al. 2015). Cumulative data for 26 EU/EEA countries (expressed in milligrams per kilogram of estimated biomass of animals) also exposed tetracyclines as the most common antibiotic class used in animal production in Europe (CIPARS 2013).

Besides, the nonclinical use of tetracyclines also includes aquaculture and horticulture (Chopra and Roberts 2001). Thus, the production and use of tetracyclines for nontherapeutic purposes are truly overwhelming. Probably because of the large-scale and extensive use of tetracyclines in agriculture, resistance against them is widespread, which made the first-generation tetracyclines essentially unusable for the therapy of human infectious diseases.

23.5 Tetracycline Resistance Mechanisms

Presumably because of the extensive use of tetracyclines in human medicine, food animals, aquaculture, and horticulture, the emergence and dissemination of tetracycline resistance among pathogens were very rapid. The two key mechanisms of resistance to tetracycline consist of target modification due to the production of alternative elongation factors, which prevent the binding of tetracyclines to the ribosomes, and the efflux of tetracyclines from the cell (Roberts 1996; Chopra and Roberts 2001; Roberts 2005). Production of ribosomal protection proteins is common among both Gram-positive and Gram-negative bacteria. The tetracycline efflux pumps differ in their structure depending on the cell wall architecture and thus are specific either for Gram-positive or Gram-negative bacteria. The less common mechanisms of resistance to tetracyclines are the enzymatic degradation of tetracyclines (Speer et al. 1991) and resistances with unclear mechanisms (Nonaka and Suzuki 2002; Kazimierczak et al. 2009). The former mechanism, which is encoded by the *tet*(X) gene, however, may have a potential to compromise the efficiency of therapy by recently introduced third-generation tetracyclines such as tigecycline (Aminov 2013a). This resistance mechanism seems to be rapidly penetrating potentially pathogenic microbiota due to the agricultural and clinical use of older tetracyclines.

23.6 Phylogeny of Tetracycline Resistance Genes

The clear majority of tetracycline resistance genes are located on mobile genetic elements (MGEs) and can be acquired by a variety of microbiota via HGT processes. But what are the original reservoirs of antibiotic resistance genes from which they are continuously mobilized? If the source of these genes are the producers of antibiotics because they should have protection mechanisms against the antibiotics they synthetize? It is indeed a popular view that the main contributors of antibiotic resistance genes for dissemination are the producers of antibiotics, which protect themselves this way from the lethal action of antibiotics produced (Wright 2012). To respond this question, we performed several phylogenetic analyses with antibiotic resistance genes including those from antibiotic producers. Our phylogenetic analyses of genes conferring tetracycline resistance via the ribosomal protection mechanism, for example, consistently suggested their ancient evolutionary history (Aminov et al. 2001; Aminov and Mackie 2007). They belong to the genes encoding translation elongation factors (EF-Tu, EF-Ts, EF-G, and EF-P) but diverged from the group very early in evolution forming a monophyletic branch. The next bifurcation, which was also a fairly early event, led to the separation of genes residing in antibiotic producers such as *Streptomyces lividans* and *Streptomyces rimosus* and in the rest of nonproducing microbiota including environmental, commensal, and pathogenic. According to these analyses, there are no indications of recent horizontal transfer of the *tet* genes from antibiotic producers to other microbiota following the widespread use of tetracyclines in humans and agriculture.

Some earlier DNA–DNA hybridization-based studies, however, have implied that there is a certain level of homology between the *tet* genes in mycobacteria and tetracycline-producing strains of *Streptomyces* (Pang et al. 1994; Roberts 1996). What could be detected in tetracycline-resistant mycobacteria, however, is the *tet*(M) gene (Rossi-Fedele et al. 2006), which is not present in the streptomycetes. Besides, when a large collection of environmental and clinical mycobacterial strains was checked by PCR for the presence of *otr*(A) (which is present in *S. rimosus*), no positive signal was detectable (Kyselková et al. 2012). According to this and some earlier reports (Aínsa et al. 1998; De Rossi et al. 1998), tetracycline resistance in mycobacteria is predominantly encoded by the efflux pump genes, *tet*(V) and *tap*. Thus, there is no indication of acquisition of tetracycline resistance genes from antibiotic producers by pathogens due to the recent extensive tetracycline use. What is shared between *M. tuberculosis* and *Streptomyces*, though, is a putative transcriptional activator, *whiB7*, which confers multidrug resistance, including resistance to tetracycline (Morris et al. 2005). The gene, however, is not related phylogenetically to the *tet* gene family discussed here, and it was presumably acquired by the mycobacteria long time ago, when their ancestors shared the soil ecological niche with the streptomycetes.

Our other phylogenetic analyses of genes and gene clusters conferring resistance against tetracyclines, macrolides, glycopeptides, and quinolones (Aminov et al. 2001; Aminov et al. 2002; Aminov and Mackie 2007; Koike et al. 2010) essentially supported the view that there is no indication of antibiotic resistance genes from antibiotic producers being mobilized into other nonproducing bacteria. Other phylogenetic analyses involving, for example, β-lactamases also showed the ancient independent evolution of the corresponding genes in antibiotic-producing and nonproducing microbiota (Ogawara 1993; Hall and Barlow 2003). Thus, the original antibiotic resistance gene pool was supposedly more widespread and existed in many other bacteria other than antibiotic producers.

23.7 Second-Generation Tetracyclines

The rapid emergence and dissemination of tetracycline resistance among the bacterial pathogens incited the development of the second-generation tetracyclines such as minocycline, which became available in 1966 (Redin 1966), and doxycycline, which became available in 1967 (Corey 2013). Essential for the development of these semisynthetic tetracyclines was the establishment of the chemical structure of natural tetracyclines (Nelson and Levy 2011). They are still used to treat many different infectious diseases, such as urinary and intestinal tract infections, respiratory infections, skin infections, acne, gonorrhea, tick fever, chlamydia, eye infections, periodontitis, and others. Besides the antibacterial effects, they also display other potent activities directed toward the eukaryotic cell targets (Aminov 2013b). In particular, minocycline displays strong anti-inflammatory, neuroprotective, anti-proteolytic, and anti-apoptotic properties, as well as inhibiting angiogenesis and metastatic growth (Garrido-Mesa et al. 2013). In addition, it displays antioxidant activity, inhibits several enzyme activities, and modulates immune cell activation and proliferation (Aminov 2013b). Thus, the applications of the second-generation tetracyclines could be potentially extended to other than antimicrobial use.

23.8 Third-Generation Tetracyclines

The first representative of the third-generation of tetracyclines, tigecycline (the minocycline derivative 9-tert-butylglycylamido-minocycline, GAR-936), was approved for clinical use by the FDA in 2005. In preclinical studies, tigecycline displayed good activity against tetracycline-resistant strains carrying genes encoding tetracycline efflux determinants (*tet*(A), *tet*(B), *tet*(C), *tet*(D), and *tet*(K)) as well as a gene encoding a ribosomal protection protein (*tet*(M)) (Petersen et al. 1999). Other studies confirmed its efficiency against clinical isolates of *Acinetobacter* spp. including *Acinetobacter baumannii* (Henwood et al. 2002), nontuberculous mycobacteria (Wallace et al. 2002), *Enterococcus* spp. including vancomycin-resistant enterococci (Mercier et al. 2002; Lefort et al. 2003; Nannini et al. 2003), *Staphylococcus aureus* including methicillin-resistant (MRSA) and glycopeptide-intermediate resistant strains (Mercier et al.

2002; Petersen et al. 2002), *Legionella pneumophila* (Edelstein et al. 2003), *Stenotrophomonas maltophilia* (Betriu et al. 2002), and a number of other clinical isolates including multidrug-resistant ones (Abbanat et al. 2003; Milatovic et al. 2003; Fritsche et al. 2004; Bouchillon et al. 2005; Sader et al. 2005; Borbone et al. 2008; Nørskov-Lauritsen et al. 2009; Dowzicky and Chmelařová 2011; Balode et al. 2013; Morfin-Otero et al. 2015; Marco and Dowzicky 2016; Tärnberg et al. 2016; Stefani and Dowzicky 2016; Giammanco et al. 2017). The antibiotic displayed outstanding therapeutic response in animal infection models as well as in clinical trials aimed at treating intra-abdominal and skin and soft tissue infections (Hawkey and Finch 2007).

23.9 Resistance to Third-Generation Tetracyclines

While clinical efficiency of tigecycline remains satisfactory including infections resistant to the first- and second-generation tetracyclines (Bertrand and Dowzicky 2012; Hawser et al. 2012; Mayne and Dowzicky 2012; Giammanco et al. 2017), our efforts should be directed toward preserving its efficiency and lifespan by elaborating strategies to prevent the emergence of the corresponding resistance. The presently known mechanisms of resistance in clinical isolates are mainly associated with a nonspecific efflux of the drug from the cell such as mediated via resistance-nodulation-division (RND) family of efflux pumps. In *Pseudomonas aeruginosa*, for example, the removal of tigecycline from the cell is driven by the RND family of efflux pumps, in particular the MexXY-OprM complex (Dean et al. 2003). In the absence of MexXY-OprM, however, two other efflux pumps, MexAB-OprM and MexCD-OprJ, are involved in the drug efflux. In *Proteus mirabilis*, the efflux of tigecycline is mediated by another representative of the RND family of efflux pumps, encoded by the *acrAB* homologue of *Escherichia coli*, the AcrAB-TolC system (Visalli et al. 2003; Ruzin et al. 2005). Reduced susceptibility of clinical isolates of *A. baumannii* toward tigecycline is also due to the elevated expression of the RND family of efflux pumps such as AdeABC and AdeIJK (Peleg et al. 2007; Damier-Piolle et al. 2008). Overexpression of another RND pump, AdeFGH, in *A. baumannii* confers resistance to multiple drugs, including tigecycline (Coyne et al. 2010). Reduced tigecycline susceptibility in clinical isolates of *Enterobacter cloacae* and *E. coli* is also due to the elevated expression of another efflux pump in the RND family, AcrAB (Keeney et al. 2007, 2008). In clinical *Klebsiella pneumoniae* strains, mutations in the regulatory *ramR* gene may also lead to the overexpression of RamA, which is a positive regulator of the AcrAB efflux system, thus resulting in the concomitant resistance to multiple antibiotics, including tigecycline (Hentschke et al. 2010). The similar mechanisms of tigecycline resistance may emerge because of therapy by other antibiotics. Ciprofloxacin therapy of *E. cloacae* infection, for example, may select for cross-resistance to tigecycline as a result of RamA-mediated AcrAB upregulation (Hornsey et al. 2010).

Thus, the increase of minimum inhibitory concentrations (MICs) toward tigecycline in clinical isolates is mainly associated with the increased nonspecific efflux of the drug, largely due to the overexpression of the RND family of efflux pumps, which, in turn, most likely due to mutations in the regulatory regions that drive their expression (Peleg et al. 2007). Low expression or the lack of expression of global regulators such as *ramR*, *marR*, or *soxR* also contributes to the elevated expression of the RND family of efflux pumps, resulting in the resistance to many drugs including tigecycline (Hentschke et al. 2010; Li et al. 2015a). These efflux pumps are structurally complex and chromosomally located and are subjected to complex regulation (Piddock 2006), which make these mechanisms as unlikely candidates for the acquired tigecycline resistance. Thus, the probability of horizontal dissemination of these mechanisms of tigecycline resistance via HGT remains low, although the clonal dissemination cannot be ruled out. Besides, mutants with the overexpression of nonspecific drug efflux pumps may be co-selected by structurally unrelated antibiotics and thus confer cross-resistance to multiple antibiotics, which may compromise antibiotic therapy. The worrying trend, however, is that despite being reserved as a drug of last resort, the use of tigecycline is steadily increasing (Huttner et al. 2012). This will lead, expectedly, to the higher occurrence of tigecycline-resistant clones in clinical settings. Moreover, the initial low-level resistance conferred by the overexpression of less specific drug efflux pumps may precede and facilitate the development of high-level drug resistance mediated by other mechanisms (Singh et al. 2012).

23.10 Other Potential Resistance Mechanisms Toward Third-Generation Tetracyclines

Because of the extremely broad use of tetracyclines in the past and present, the corresponding resistance genes are widespread in many ecological compartments including the environment (Chee-Sanford et al. 2001; Aminov and Mackie 2007; Chee-Sanford et al. 2009). And many of these tetracycline resistance genes are located on MGEs (Aminov 2011, 2012). Thus, there are preexisting conditions with the tetracycline resistance gene pool located on MGEs that may serve as a starting point for the emergence of strains with a lesser susceptibility to tigecycline. For example, during the early stages of drug development, it has been discovered that mutations in the interdomain loop region of *tet*A(A) may increase the efflux of minocycline and glycylcyclines (Tuckman et al. 2000). More recently, it has been shown that the increased expression of two tetracycline resistance genes, *tet*(L) (encoding efflux of the drug) and *tet*(M) (encoding ribosomal protection protein), confers tigecycline resistance in clinical isolates of *Enterococcus faecium* (Fiedler et al. 2016). A systematic approach to evaluate the potential of tetracycline resistance genes to evolve tigecycline resistance was attempted with the mutants of *tet*(A), *tet*(K), *tet*(M), and *tet*(X) (Linkevicius et al. 2015). All mutants selected were capable to decrease the susceptibility of *E. coli* host toward tigecycline. The authors predicted that *tet*(X) could become the most problematic since mutations can improve the weak intrinsic activity of *Tet*(X) against tigecycline without collateral loss of activity toward other tetracyclines (Linkevicius et al. 2015). Thus, the existing pool of resistance genes conferring resistance to the earlier tetracycline generations may serve as building blocks for the emergence of resistance to newer tetracyclines such as glycylcyclines.

23.11 Evolutionary Aspect of *tet*(X)

Another aspect that makes *tet*(X) even more problematic is the dynamic of its dissemination. These ecological aspects have been discussed by us earlier (Aminov 2009, 2013a). In our initial phylogenetic analyses, we established that TetX belongs to flavoprotein monooxygenases that hydroxylate specific regions in organic molecules and present in many metabolic pathways of the Bacteria, Archaea, and Eukarya (Harayama et al. 1992). Based on amino acid sequence and three-dimensional structure, they are distributed into six classes (A–F) (van Berkel et al. 2006). TetX belongs to the class A monooxygenases, which are involved in the microbial degradation of aromatic compounds by *ortho* or *para* hydroxylation of the aromatic ring (Moonen et al. 2002). Our phylogenetic analysis of class A monooxygenases uncovered the high diversity of these enzymes as well as their incongruence with the phylogeny that is based on standard markers such as the 16S rRNA gene (Aminov 2009). The latter may suggest extensive HGT processes during the evolution of class of flavoprotein monooxygenases. Another prominent feature of the genes encoding flavoprotein monooxygenases is frequent duplication events that accompanied their evolution. These genes are present in genomes of many bacterial species, with a wide-ranging ecological presence, including terrestrial, aquatic, and intestinal ecosystems, while some of these bacteria are opportunistic pathogens. These enzymes perform a broad range of biochemical reactions, which, interestingly, also include modification of various aromatic polyketide antibiotics other than tetracyclines such as rifampin (Andersen et al. 1997), mithramycin (Prado et al. 1999), griseorhodin (Li and Piel 2002), chromomycin (Menendez et al. 2004), and auricin (Novakova et al. 2005). Thus, TetX belongs to the class of enzymes that are ubiquitous in environmental microbiomes and perform many biochemical functions including modification of many antibiotics as concomitant functions.

More detailed phylogenetic analysis (Aminov 2009) suggested that the ancestral clade of flavoprotein monooxygenase genes, from which the presently known *tet*(X) had diverged, is localized within the genome of a fish pathogen, *Flavobacterium psychrophilum* (Duchaud et al. 2007). At the time of our original analysis, the nucleotide sequences highly similar (≥99%) to *tet*(X) formed a close-fitting cluster consisting of genes found in commensal and environmental bacteria such as *Bacteroides* spp. (Speer et al. 1991; Whittle et al. 2001) and environmental *Sphingobacterium* sp. (Ghosh et al. 2009) as well as in metagenomic libraries constructed from the human gut microbiome (Kurokawa et al. 2007) and the microbiome of biological phosphorus removal sludge (García Martín et al. 2006). Although the tigecycline-resistant phenotype is experimentally confirmed only for the *tet*(X) gene from Tn*4351* (Moore et al. 2005), the high similarity of nucleotide sequences in the clade (≥99%) is indicative of shared functionality. For more distant sequences, the phenotypic expression of minocycline resistance is confirmed for an opportunistic pathogen *P. aeruginosa* (GenBank accession number AB097942), and tetracycline resistance for an environmental plasmid (GenBank accession number FJ012881), when cloned into *E. coli*.

23.12 Ecological Aspects of *tet*(X)

No penetration of *tet*(X) into the clinical microbiota was detected at the time of the first analysis, and the gene was mainly encountered in commensal and environmental bacteria (Aminov 2009). Interesting is the case with the *Bacteroides* where it has been originally detected (Speer et al. 1991). These bacteria are anaerobic, but the corresponding tetracycline-degrading activity of the flavin-dependent monooxygenase requires the presence of oxygen. Thus, the gene cannot be expressed phenotypically in *Bacteroides*, and, therefore, it was not selected as conferring tetracycline resistance. Probably it is a result of co-selection by aminoglycosides and macrolides due to the genetic linkage with *addS* and *ermF* in the *Bacteroides* transposon (Whittle et al. 2001). In aerobic bacteria, it could be potentially selected directly as conferring resistance against the tetracyclines used.

At the later dates, *tet*(X) has been detected in several other ecosystems including human gut bacteria (de Vries et al. 2011), intestinal *Bacteroides* strains (Bartha et al. 2011), sewage treatment plants (Zhang and Zhang 2011), and an oxytetracycline production wastewater treatment system (Liu et al. 2012). A recent screening of soil functional metagenomes for tetracycline resistance has suggested that the occurrence of genes conferring high-level tetracycline resistance by enzymatic inactivation in natural ecosystems may be greater than previously anticipated (Forsberg et al. 2016). Some of the purified flavoenzymes degraded tetracyclines consistent with the corresponding activity of Tet(X), while others degraded tetracyclines through potentially different oxidation mechanisms.

Our follow-up analysis of the occurrence of *tet*(X) revealed its penetration into the pathogens of veterinary and human clinical importance (Aminov 2013a). In particular, Leski et al. (2013) reported about the occurrence of multidrug-resistant and *tet*(X)-containing clinical bacterial isolates in a hospital in Sierra Leone. All the *tet*(X)-positive isolates, 21% of total, were from urinary tract infections. Taxonomically, these were identified as *E. cloacae, Comamonas testosteroni, E. coli, K. pneumoniae, Delftia acidovorans, Enterobacter* sp., and other members of the Enterobacteriaceae and Pseudomonadaceae families (Leski et al. 2013). Our phylogenetic analysis confirmed, with a 100% bootstrap support, a close relationship of *tet*(X) sequences from Enterobacteriaceae bacterium SL1 and *Delftia* sp. SL20 with the known *tet*(X) genes, given the high similarity of sequences within the clade that exceeds 99% (Aminov 2013a). This high similarity at the nucleotide level suggests the scenario of horizontal exchange of *tet*(X) among the environmental, commensal, and clinical ecological compartments.

In this resource-limited country, tigecycline (Tygacil®, Pfizer Inc.) was not available in the hospital where *tet*(X)-positive samples were collected nor via the independent pharmacies and hospital dispensaries operating in the area (Leski et al. 2013). But 87% of pharmacies in the area sell the first and second generation of tetracyclines without prescription. The authors indicated that potentially this continuous selective pressure by older tetracyclines may have led to the spread of *tet*(X) into the pathogens. Given the multidrug-resistant phenotype of these pathogens, another possibility is the co-selection of multidrug resistance cassettes within MGEs. The authors specified the presence of MGEs in some isolates, and most of the *tet*(X)-harboring strains (10 out of 11) displayed the multidrug-resistant phenotype.

In agriculture, the acquisition of *tet*(X) by a pathogen of veterinary importance can be demonstrated for *Riemerella anatipestifer*, a causative agent of septicemia anserum exsudativa (Segers et al. 1993). The disease causes major economic losses in duck production (Ryll et al. 2001; Sarver et al. 2005), and it may additionally affect other domesticated birds (Sandhu and Rimler 1997; Hess et al. 2013). The strain of *R. anatipestifer* with resistance to multiple drugs such as amikacin, ampicillin, chloramphenicol, gentamicin, nalidixic acid, tetracycline, and trimethoprim/sulfamethoxazole was isolated in 2005 from waterfowl in Taiwan (Chen et al. 2010). The strain carries a plasmid, pRA0511, which, in addition to two chloramphenicol acetyltransferases and a multidrug ABC transporter permease/ATPase, encodes TetX as well. In our phylogenetic analysis, the gene has been confidently grouped into the same clade, which includes *tet*(X) sequences of environmental, commensal, and clinical origin (Aminov 2009). The *tet*(X) gene can also be found in the genomic sequences of at least five *R. anatipestifer* strains (Yuan et al. 2011; Wang et al. 2015b; Song et al. 2016; GenBank accession numbers CP003787 and CP004020). At the same time, in some other strains of *R. anatipestifer*, for which genomic data are available (Mavromatis et al. 2011; Zhou et al. 2011; Wang et al. 2012; Yuan et al. 2013; Wang et al. 2015a), no *tet*(X) can be found. Unfortunately, no antibiotic usage regimens have been reported for the sites of strain isolation, so it is difficult to make a conclusion about the possible involvement of antibiotic selection on dissemination of *tet*(X) at different sites.

The two prominent antimicrobials used for poultry in China are coccidiostats (3 407 220 kg) and arsenicals (2 879 624 kg), with the distant third place occupied by tetracyclines (613 120 kg) (Krishnasamy et al. 2015). The role of coccidiostats and arsenicals on the emergence of antibiotic resistance such as TetX is not clear, but we have to keep in mind that under the umbrella of coccidiostats, there are three groups of drugs: ionophores, sulfonamides, and chemical compounds. Sulfonamides are the oldest mass-produced antimicrobials still in use (Aminov 2017), and the corresponding resistance is widespread including *R. anatipestifer* RCAD0122 (Song et al. 2016). Sulfonamide resistance genes encoded by sulfonamide-resistant dihydropteroate synthetases are co-localized with integrases and numerous other antibiotic resistance genes on multidrug-resistant integrons thus facilitating the broad dissemination of multiple drug resistance genes via the co-selection by sulfonamides (Huovinen et al. 1995). Components of resistance to arsenic compounds such as genes encoding arsenate reductase gene (two copies) or ArsR family transcriptional regulator (three copies) can be also found (Wang et al. 2015b). Thus, the selection of coccidiostats and arsenicals may be behind the co-selection of other drug resistances, including TetX. In general, *R. anatipestifer* strains are multidrug resistant and carry resistances to heavy metals, which brings another possibility for co-selection.

Especially high occurrences of *tet*(X) have been found in a recent investigation of resistomes in two low-income human habitats in Latin America (Pehrsson et al. 2016). In these communities, many antibiotics can be purchased without prescription at the local pharmacies. Consistently with this, at all WWTP influents, chloramphenicol, ciprofloxacin, tetracycline, trimethoprim, and sulfamethoxazole can be detected, while erythromycin was present in only 36% of influent samples. The fecal resistomes of these communities harbor a greater number of antibiotic-resistant proteins per person compared with traditional hunter-gatherer, rural agriculturalist, or industrialized communities. Interestingly the genetic context of *tet*(X) in these two low-income human habitats in Latin America was identical in human and animal feces, latrines, and sewage influent (Pehrsson et al. 2016), suggesting the circulation of the same *tet*(X)-carrying MGE in the local ecological compartments. It is highly unlikely that tigecycline is used in these communities and, similarly to the situation in Africa (Leski et al. 2013), the prime suspects for the selection of *tet*(X) are the older tetracyclines that are widely available without prescription in these communities.

The use of the third-generation tetracyclines such tigecycline in animals is not authorized, and it can be used only under exceptional circumstances in companion animals since *the Cascade* rule allows the use of human-approved drugs in certain situations (EMA 2013). The use of tigecycline in clinical medicine also remains limited because it is an antibiotic of last resort and reserved for treatment of infections that are not susceptible to other antibiotics (although its use is continuously increasing (Huttner et al. 2012)). The observed extensive dissemination of *tet*(X) in agriculture and low-income communities, however, is driven by other factors than the direct selection by glycylcyclines. Since the gene confers cross-resistance to tetracyclines of all generations, the most likely selective pressure for its widespread dissemination is the extensive use of older tetracyclines in agriculture and its unrestricted availability in low-income communities. Co-selection by other antimicrobials such as coccidiostats, arsenicals (and other heavy metals), and other antibiotics may also play a role in the dissemination.

Thus, the examples given above, which are derived from the history of tetracycline antibiotic use and the consequent emergence of resistance, allow us to synopsize some general concepts that are apparently applicable to other classes of antibiotics and the corresponding antibiotic resistance genes. These concepts could be applicable for designing the corresponding strategies to limit the spread of antibiotic resistance. But first we need to reveal the roles played by antibiotics in nature. This background evidence is crucial for understanding what is happening when the antibiotic producers are taken from the environmental context, optimized for a large-scale antibiotic production, and the resulting antibiotics are extensively used for the purposes unrelated to their role in nature.

23.13 Antibiotics and Antibiotic Resistance as Integral Parts of Microbial Diversity

The common perception of antibiotics, which is based on clinical practice of using them for clearing bacterial infectious agents, categorizes antibiotics as killing chemical compounds. In this role, antibiotics perfectly fit the idea of a weaponry used by some bacteria against others in competition for limited resources. These aspects have been studied in detail and resulted in characterization of almost all antibiotics for the MICs *in vitro* and their pharmacokinetic/pharmacodynamic (PK/PD) properties *in vivo*. The antibiotic concentrations in natural ecosystems are generally low, and they have been consistently measured only in few occasions involving the symbiotic relationships, with the antibiotic production by symbionts to defend their hosts from fungal pathogens

(Gil-Turnes et al. 1989; Oh et al. 2009). In these cases, the ecosystem diversity is low so that the biomass of antibiotic producers (and concomitantly the amount of antibiotic produced) may reach substantially higher values than under the normal circumstances of highly diverse ecosystems.

The human use of antibiotics coerces the view that, in a Darwinian natural selection sense, the production of antibiotics by certain organisms was selected as a successful trait for elimination of competitors to ensure their own reproductive success. Is it indeed a sole role played by antibiotic producers and antibiotics in the natural ecosystems? Indeed, in certain instances such as a symbiotic relationship mentioned above, this could be a case. There are many examples, especially in the rhizosphere environment, of such relationships that involve the production of antimicrobial substances by symbionts to protect their host, which, in turn, provides them by favorable conditions for reproduction. For example, fluorescent pseudomonads in the rhizosphere may suppress soilborne pathogens via the production of diffusible or volatile antibiotics such as phenazines, phloroglucinols, pyoluteorin, pyrrolnitrin, cyclic lipopeptides, and hydrogen cyanide (Haas and Défago 2005). Production of nonribosomal peptides by Gammaproteobacteria may also have the same plant disease-suppressing activity (Mendes et al. 2011).

Several lines of evidence emerging in recent years suggest, however, that antibiotic concentrations in mostly oligotrophic natural environments may not reach the levels capable of killing competitors. The operating concentrations though may play signaling and regulatory roles in microbial communities (Davies et al. 2006; Linares et al. 2006; Yim et al. 2006; Martínez 2008; Aminov 2009; Bernier and Surette 2013; Andersson and Hughes 2014). Thus, the range of activities is more extensive, from the lethal to modulating actions. Interestingly, the modulating activities of antibiotics are extended to other domains of life, beyond the Bacteria (Aminov 2013b).

Logically, if resistance to antibiotics evolved as a mechanism of protection against the lethal action of antibiotics, then the genes of antibiotic resistance should be detectable as a gradient originating from the places of extensive use of antibiotics (used at concentrations higher than in natural unaffected ecosystems) with further release into the environment, which is indeed the case (Chee-Sanford et al. 2001; Koike et al. 2007; Baquero et al. 2008; Chee-Sanford et al. 2009; Koike et al. 2010). However, there are many cases of occurrence of antibiotic resistance genes in apparently antibiotic-free environments. In the metagenomic libraries generated from "cold-seep" sediments and a remote Alaskan soil, for example, the presence of diverse β-lactamase genes at both locations has been demonstrated (Song et al. 2005; Allen et al. 2009). Calculations based on the total amount of environmental DNA sampled from Alaskan and the number of β-lactam-resistant clones encountered suggest that about 5% of the average-sized bacterial genomes in this apparently antibiotic-free environment may harbor functional β-lactamase genes. At the phenotype level, more than 60% of the *Enterobacteriaceae* isolates from the freshwater environment unaffected by human activity could be multidrug resistant (Lima-Bittencourt et al. 2007). Isolation of functional antibiotic resistance genes from the environments that have been isolated from any human influence (D'Costa et al. 2011; Bhullar et al. 2012) suggests that these genes have an ancient evolutionary history. These experimental observations are confirmatory to the phylogenetic analyses of antibiotic resistance genes performed by us and others (Ogawara 1993; Aminov et al. 2001, 2002; Hall and Barlow 2003; Aminov and Mackie 2007, Aminov 2009, 2013a; Koike et al. 2010) that suggested the ancient evolutionary history of these classes of genes. The question then is, what is the functional role of antibiotics and antibiotic resistance in microbial (and possibly not only microbial) ecosystems?

23.14 The Role of Antibiotics in Natural Ecosystems

The occurrence of antibiotics in the environment is usually associated with anthropogenic factors such as agricultural activities, human settlements of various size, hospitals, and pharmaceutical industry (Yang and Carlson 2003; Kim and Carlson 2006; Mackie et al. 2006; Pei et al. 2006; Baquero et al. 2008; Walraven and Laane 2009; Watkinson et al. 2009; Manzetti and Ghisi 2014; aus der Beek et al. 2016). The question, however, is: what are the natural concentrations of antibiotics generated by the antibiotic producers *in situ*? In some soils, for example, in Jordan's red soils, the *in situ* concentration of antibiotics may reach therapeutic levels, so the soil applications have been used in the past and used presently as well (as an inexpensive alternative to present-day pharmaceutical products) for treating skin infections (Falkinham et al. 2009). This type of red soil in another geographical location also seems harbor a high diversity of actinomycetes that are producing a large number of secondary metabolites, including macrolides, polyethers, diketopiperazines, and siderophores (Guo et al. 2015). Interestingly, compared with the less affected recreational environments, residential environments seem to drive the changes in the composition of soil microbiota toward the increase of antibiotic-producing microorganisms (Woappi et al. 2013). The authors hypothesize that this may indicate a coevolutionary biosynthesis of novel antibiotics driven by the increase of bioactive microbiota in residential environments.

While antibiotic resistance genes can be encountered in the environments apparently unaffected by human activities, the concentrations of antibiotics in these environments are very low, usually below the detection limit. In fact, this type of environments served as controls for measuring the impact of human activity on natural ecosystems (Yang and Carlson 2003; Kim and Carlson 2006; Pei et al. 2006). These concentrations are not lethal but may be involved into the signaling, sensing, and regulatory processes in the microbial world (Davies et al. 2006; Linares et al. 2006; Yim et al. 2006; Martínez 2008; Aminov 2009; Bernier and Surette 2013; Andersson and Hughes 2014; LeRoux et al. 2015). These potential roles of antibiotics were studied mainly under laboratory conditions using low antibiotic concentrations and under the umbrella of hormesis hypothesis (Davies et al. 2006; Fajardo and Martínez 2008). In this concept, the effects of antibiotics are dose dependent: low-concentration antibiotics affect the regulation of specific sets of genes in target bacteria, while the increasingly higher concentrations provoke stress responses, with the even higher concentrations being lethal. It has been suggested that the main role of low-concentration antibiotics in nature is regulatory in contrast to the lethal effects of high-dose antibiotics used by humans for infectious disease treatment (Aminov 2009). Another relevant question is whether the drugs at different concentrations interact with the same cellular targets.

23.15 Low-Dose Antibiotics: Phenotypic Effects

Recently, the effects of low-dose antibiotics have been summarized by us in a research topic (Nosanchuk et al. 2014). Probably the best-studied model is opportunistic pathogen *P. aeruginosa*, infection with which is often fatal for cystic fibrosis patients. It has been shown during the early days of investigating the effects of low-dose antibiotics that the subinhibitory concentrations of a macrolide antibiotic, azithromycin, suppress its pathogenic properties such as alginate overproduction and biofilm formation *in vitro* (Ichimiya et al. 1996). The subinhibitory concentrations of other antibiotic classes such as ceftazidime (β-lactam) and ciprofloxacin (quinolone) also appeared to be effective in suppressing the quorum-sensing-regulated virulence factors of this bacterium (Skindersoe et al. 2008). According to the authors, the potential mechanism for this effect is the interaction of these two drugs with the cellular membrane, which affects its permeability and subsequently the flux of the quorum-sensing molecule *N*-3-oxo-dodecanoyl-L-homoserine lactone. It is important to note here that these two antibiotics at therapeutic concentrations have differential targets, the bacterial cell wall for ceftazidime and bacterial topoisomerases for ciprofloxacin.

Interestingly, however, the antibiotics of the same classes as described above, with the same model bacterium (Skindersoe et al. 2008), may demonstrate the opposite effects as well. Exposure of *P. aeruginosa* to the subinhibitory concentrations of another β-lactam, imipenem, for example, results in the elevated alginate production and in the increased volume of biofilm (Bagge et al. 2004). Another quinolone drug, norfloxacin, at subinhibitory concentrations induces the formation of biofilm by *P. aeruginosa* (Linares et al. 2006). These discrepancies suggest that the degree of freedom in interaction with cellular targets is higher at subtherapeutic concentration. Thus, the antibiotics of the same class, which share the same molecular targets to execute their lethal activity, may have differential targets when present at low concentrations.

The notion of differential cellular targets for the killing and regulatory activities of antibiotics is further supported by the effect of subinhibitory concentrations of an aminoglycoside antibiotic, tobramycin, which induces biofilm formation in *P. aeruginosa* and *E. coli* (Hoffman et al. 2005). In *P. aeruginosa*, molecular target for subinhibitory tobramycin is not the ribosome as for therapeutic aminoglycosides but an aminoglycoside response regulator. This is an inner-membrane phosphodiesterase, the substrate of which cyclic di-guanosine monophosphate is a bacterial second messenger that regulates cell surface adhesiveness. The dual targeting was also demonstrated for the subinhibitory concentrations of the macrolide–lincosamide–streptogramin antibiotics, which interact with the ribosomes and modulate transcription (Tsui et al. 2004).

It was noticed in many experiments with subinhibitory antibiotics that their effects could be pleiotropic affecting transcription/translation of many genes. Cells of *P. aeruginosa* subjected to subinhibitory concentrations of colistin, for example, show the altered expression of 30 genes involved in virulence and bacterial colonization (Cummins et al. 2009). Interestingly, effects also included the upregulation of the quinolone signal biosynthesis genes, thus inducing the central component of the quorum-sensing network in this bacterium. In *P. aeruginosa* this quorum-sensing signaling system is involved in pathogenesis, but genomic analyses demonstrated that this system is also present in various *Burkholderia* species as well as in a soil bacterium, *Pseudomonas putida* (Dubern and Diggle 2008). It is fascinating to learn about the enormous metabolic diversity of the microbial world using the quinolones as an example. The first representative of this class, nalidixic acid, was discovered during an attempt of chloroquine synthesis in the early 1960s and was found as possessing antibacterial properties (Wentland 1993). Only significantly later the presence and biological functions of the quinolone compounds as signaling molecules have been uncovered in the microbial world. This is yet another supporting evidence that the effects of the secondary metabolites are concentration dependent, ranging from regulation to stress and to killing. Another aspect of low-dose antibiotics concerns their effects on genetic processes taking place in microbial populations, and this will be discussed in the next section.

23.16 Low-Dose Antibiotics: Genetic Effects

The very first experiments to evaluate the effect of subinhibitory antibiotics on gene transfer demonstrated that HGT frequencies are enhanced under these conditions. The frequency of transfer of tetracycline resistance plasmids in *S. aureus*, for example, is enhanced 100- to 1000-fold by exposure of the culture to subinhibitory β-lactams (Barr et al. 1986). If a *Bacteroides* strain is pre-grown in the presence of low-concentration tetracycline, the mobilization of a resident non-conjugative plasmid by chromosomally encoded tetracycline conjugal elements is also accelerated (Valentine et al. 1988). Interestingly, the exposure of donor *Bacteroides* to low-dose tetracycline seems to be essential for the excision and conjugal transfer of conjugative transposons of the CTnDOT family (Stevens et al. 1993; Whittle et al. 2002). Otherwise virtually no HGT takes place. The similar stimulatory effects on conjugation frequency by subinhibitory tetracycline were reported for Tn*925* and Tn*916* transposons (Torres et al. 1991; Showsh and Andrews 1992).

The stimulatory effect of subinhibitory antibiotics on HGT is not only an *in vitro* phenomenon but is also reproducible *in vivo*. When gnotobiotic mice are given low-dose tetracycline in drinking water, the transfer frequency of Tn*1545* conjugative transposon from *Enterococcus faecalis* to *Listeria monocytogenes* in the gut is increased by about 10-fold (Doucet-Populaire et al. 1991). In gnotobiotic rats, selection for the resistant phenotype was a major factor in producing greater numbers of Tn*916* transconjugants in the presence of tetracycline (Bahl et al. 2004). In aquatic systems, exposure to low-level oxytetracycline may lead to the MGE-mediated dissemination of antibiotic resistance genes (Knapp et al. 2008).

The antibiotic concentrations, which are optimal and maximally conducive for horizontal dissemination of MGEs, are very low. It has been established experimentally that these are 150 times below the MIC (Jutkina et al. 2016). But what are the mechanistic explanations for this phenomenon imposed by low-dose antibiotics? There are several potential mechanisms involved. In a model with translation attenuation by tetracycline, the attenuation leads to the increased production of RteA and RteB and activation of *rteC* transcription. RteC then activates transcription of excision genes of CTnDOT and other mobile elements (Moon et al. 2005). At low concentration of tetracycline, the excision is not associated with growth phase and can occur at all phases (Song et al. 2009).

A less specific mechanism of HGT stimulation by antibiotics involves stress response genes. As discussed in the previous section, subinhibitory antibiotics may induce many genes including stress response genes, and some DNA-damaging antibiotics may also induce the SOS response. Thus, mitomycin C and antibiotics such as fluoroquinolones and dihydrofolate reductase inhibitors may increase the rate of HGT more than 300-fold (Beaber et al. 2004). This may result in a co-selection of other antibiotic resistance genes that are genetically linked in an MGE (Hastings et al. 2004). Other classes of antibiotics, which are not interfering with DNA metabolism such as β-lactam antibiotics, may also induce the SOS response, in particular in staphylococci (Miller et al. 2004). Thus, β-lactams may increase the rate of horizontal transfer of the staphylococcal virulence genes (Ubeda et al. 2005; Maiques et al. 2006). There are also examples of phage-mediated gene transfers that are induced by various antibiotics (Allen et al. 2011; Bearson et al. 2014; Bearson and Brunelle 2015). Horizontal transfer of MGEs, which are abundantly present in many bacterial genomes and which encode a variety of properties besides antibiotic resistance and virulence, may also be controlled via the SOS response and environmental signals (Auchtung et al. 2005; Bose et al. 2008). Some recent findings, however, suggest that the contribution of antibiotics to the promotion of HGT may be overestimated (Lopatkin et al. 2016).

23.17 Regulation of Antibiotic Synthesis in Antibiotic Producers

In the previous two sections, the effects of low-dose antibiotics were discussed from the perspective of organisms targeted by antibiotics. But what about the antibiotic producers themselves? How is the biosynthesis of these molecules that evidently have regulatory effect on the other microbiota regulated in antibiotic producers? In this regard, the regulation of antibiotic synthesis by antibiotic producers would be the best example to assess the role of these secondary metabolites in the microbial ecosystems. The absolute majority of naturally occurring antibiotics, which served as the foundation for the present-day antibiotic production, are synthesized by *Streptomyces* spp. (Weber et al. 2003; Aminov 2017). Optimization of strains for antibiotic production includes many cycles of mutagenesis, with the selection of most productive clones. During this process, the strains are selected, in which many regulatory circuits involved in fine-tuning of antibiotic synthesis are knocked out to increase the yield and inactivate the negative feedback mechanisms. For the purpose of this chapter, however, we are mainly interested in uncovering the mechanisms that operate in the original wild-type strains.

One of the pioneering studies of antibiotic synthesis regulation in the representatives of *Streptomyces* has resulted in identification of a γ-butyrolactone or A-factor, which induces differentiation and antibiotic production in *Streptomyces griseus* (Khokhlov et al. 1967). Only four decades later a biosynthetic pathway for γ-butyrolactone has been proposed (Kato et al. 2007). This group of quorum-sensing molecules, which includes 6-keto, (6*R*)-hydroxy, and (6*S*)-hydroxy types (Nishida et al. 2007), shares many properties with the best-studied quorum-sensing model that of *Vibrio fischeri* (Fuqua et al. 1996). The quorum-sensing systems are very common in the microbial world, and their function is to maintain communication not only with the closest relatives but also on a wider scale, extending up to the inter-kingdom lines of communication (Shiner et al. 2005). Like the LuxI/LuxR system in Gram-negative bacteria, the γ-butyrolactone signaling system of the streptomycetes consists of a γ-butyrolactone synthase, AfsA, and a receptor protein, ArpA (Nishida et al.

2007). The γ-butyrolactones are steadily accrued in the media during growth, and once the critical concentrations are reached, they interact with the DNA-binding cytoplasmic receptor proteins (ArpA and its homologues) to release these repressors to allow transcription from the set of target genes. These genes encode specific transcriptional factors, which express the genes involved in morphological differentiation and the synthesis of secondary metabolites such as antibiotics (Horinouchi 2007).

There are interesting similarities in the occurrence of antibiotic synthesis and antibiotic resistance compared with γ-butyrolactone synthases and receptors. In particular, the well-established proportion of a limited number of antibiotic producers and a much wider presence of antibiotic resistance has its parallels with γ-butyrolactone synthases and receptors. Only 10 or 11 putative γ-butyrolactone synthase genes were detected in the genomic sequences, all of which are representatives of *Streptomyces*, while 37–42 putative γ-butyrolactone receptor genes can be found in genomes of bacteria belonging not only to the streptomycetes but also to *Kitasatospora*, *Brevibacterium*, *Saccharopolyspora*, *Mycobacterium*, *Rhodococcus*, *Anabaena*, *Nocardia*, and *Nostoc* (Takano 2006; Nishida et al. 2007). Thus, the behavior of microbial community members may be orchestrated via a small number of γ-butyrolactone producers, with a much larger and diverse audience of signal receivers. Consistent with this, the evolution of γ-butyrolactone synthases and its receptors was not congruent (Nishida et al. 2007). Besides, the ancestral receptors initially functioned as regulatory DNA-binding proteins and only later in evolution acquired the γ-butyrolactone-sensing capability (Nishida et al. 2007). As this example demonstrates, the well-known quorum-sensing signaling network may affect the whole community mediating the processes such as differentiation, synthesis of secondary metabolites such as antibiotics, and probably other cellular processes in the community members (Nodwell 2014). If this similarity of the quorum-sensing network, with a lesser number of signal producers and a much bigger audience of signal receivers, to a limited number of antibiotic producers and a much wider presence of antibiotic resistance is just a coincidence?

Functional redundancy is an essential attribute of many microbial ecosystems to provide the stability and resilience against potential disruption of community functions. The intra- and interspecies signaling systems operating in microbial communities also provide the level of redundancy, via multiple regulatory networks, to ensure that the signals are received and executed. Quorum sensing represents one of such signaling networks, and its role in intra- and interspecies communication, as well as in regulation of many aspects of metabolism, virulence, physiology, competence, motility, symbiosis, and other functions, is well established. Other examples of regulatory networks include two-component systems and diverse sensor proteins to perceive and react to a variety of environmental clues. If antibiotics in microbial communities may also serve as one of the languages of communication? We discussed before the regulatory role of signaling via γ-butyrolactone for antibiotic synthesis. It appeared that, similarly to this signaling, antibiotics can also coordinate antibiotic biosynthesis via "pseudo"-γ-butyrolactone receptors (Xu et al. 2010). Indeed, γ-butyrolactones and antibiotics are recognized as signaling molecules playing fundamental roles in intra- and interspecies communications (Li et al. 2015b). This suggests that certain regulatory circuits in microbial communities are replicated, with the combination of quorum sensing and antibiotic signaling. Supporting this view is the evolutionary relationship and biological relevance found between the regulatory systems of quorum sensing and multidrug resistance (Xu 2016).

The redundancy of signaling networks built into microbial communities could be explained by the need to guarantee that a message is conveyed and received, even under the continuously changing environmental conditions. Also, the role could be in amplification of a weak or rapidly decaying signal to make it stronger and less specific so it can be sensed by other community members, which are less capable of perceiving and deciphering the environmental cues. Or attenuate the signal if it is too intense. The acyl-homoserine lactone quorum signal, for example, decays very rapidly in many soil types (Wang and Leadbetter 2005). The function of antibiotic resistance via antibiotic degradation/modification could be then similar to the attenuation of signal intensity via quorum quenching in the quorum-sensing communication (Dong et al. 2001). For example, removal of the initial quorum-sensing signal affects antibiotic susceptibility of *Streptococcus anginosus*, which makes it more sensitive to antibiotics (Ahmed et al. 2007). The negative feedback loop of a secondary signaling system, antibiotics, may then suppress the primary quorum-sensing signaling network (Tateda et al. 2004; Skindersoe et al. 2008), thus providing fine-tuning between the two signaling networks.

23.18 Convergent Evolution of Antibiotics as Signaling Molecules

Regulatory effects of antibiotics in antibiotic-producing and nonproducing bacteria discussed above strongly support the view for the signaling role played by antibiotics in natural microbial communities. If antibiotics evolved as the language for intra- and interspecies communication in microbial communities, then we should expect another supporting argument for this, at the evolutionary level. What is expected is the pattern of convergent evolution in biosynthesis of antibiotics – in other words, the evolvement of same type of "language" (signaling antibiotics) to be "understood" by taxonomically divergent bacteria, which, nevertheless, may employ different biosynthetic pathways to generate these signaling compounds. The antibiotic discovery programs have been confined, almost solely, to soil ecosystems, and the "dialects" we are mostly familiar with are those of the soil streptomycetes (Aminov 2017). But if antibiotics are used as a language of communication, then it must be "spoken" by a much broader diversity of microbiota by generating structurally similar compounds. The inclusion of other members of microbial communities and methods of testing for antibiotic production may uncover the genuine magnitude of the environmental antibiome language. For example, now we know that the β-lactams represent an abundant group of antibiotics and they are produced by a wide range of microorganisms (Aminov 2017), not confined to the representatives of the originally discovered *Penicillium* (Fleming 1929). Similarly, aminoglycosides and thiopeptides are produced by taxonomically divergent actinomycetes and *Bacillus* spp. (Flatt and Mahmud 2007; Brown et al. 2009; Liao et al. 2009) and cephalosporins – by fungi, actinomycetes, and Gammaproteobacteria (Liras et al. 2008; Liras and Martin 2006). Because of space limitations, only the carbapenem group of β-lactams, one of the most therapeutically important β-lactam antibiotics currently in use (Nicolau 2008), will be discussed in the context of how this signaling molecule has evolved in several lineages by convergent evolution to serve as a signaling regulatory molecule in a variety of microbiota.

23.19 Carbapenems: Convergent Evolution and Regulation in Different Bacteria

Initially, the biosynthesis of carbapenem was discovered in *Streptomyces cattleya* (Kahan et al. 1972). Later, the synthesis of structurally similar compounds was found to be performed by other *Streptomyces* species as well as by Gram-negative bacteria belonging to *Serratia* and *Erwinia* (Parker et al. 1982). Carbapenem compounds are also produced by a luminescent entomopathogenic bacterium *Photorhabdus luminescens* (Derzelle et al. 2002). Compared with the classical β-lactam biosynthesis pathway for penicillins, cephamycins, and cephalosporins, carbapenems are synthesized via a different pathway (Williamson et al. 1985). The genes involved in the synthesis of carbapenems are organized into clusters (McGowan et al. 1997; Cox et al. 1998; Derzelle et al. 2002; Núñez et al. 2003), and while the enzymes of the biosynthetic pathways share some similarities among the Gram-positive and Gram-negative producers, they are noticeably divergent (Coulthurst et al. 2005). These enzymes probably evolved from the primary metabolic enzymes in corresponding antibiotic producers and represent an example of convergent evolution.

These examples of convergent evolution can be seen from a different perspective, that is, as a convergence to kill (Fischbach 2009). Once again, we must emphasize here that achieving killing concentrations of antibiotics synthesized by a producer is a rather artificial phenomenon. What is happening with a pure culture on a Petri dish, with no nutrient limitation, is drastically different from the environmental context of complex and diverse microbiota of natural ecosystems. On a Petri dish, the concentration of secondary metabolites may reach the levels that are never encountered in highly populated and resource-limited natural ecological niches. The main counterargument here is that if even it is the case in natural ecosystems, then antibiotic concentrations should reach killing levels, akin to at least the MIC values. We are not aware about any supporting evidence for such antibiotic concentrations in natural microbial ecosystems. Besides, the killing effect must be always targeted and, therefore, specific against certain ecosystem inhabitants to eliminate potential competitors for a certain resource. Otherwise collateral damage by a high-dose and broad-range antibiotic may lead to indiscriminate elimination of ecosystem inhabitants, with a complete destruction of an ecological niche.

Regulatory aspects of carbapenem biosynthesis in Gram-negative bacteria also represent an argument against the interpretation of the antibiotic role as a mechanism for indiscriminate killing of competitors. Its biosynthesis is finely tuned and highly dependent on environmental cues, in particular on quorum-sensing molecules. Although the chemical structure of carbapenems from streptomycetes and Gram-negative bacteria is highly similar, the regulation of biosynthesis is governed by quorum-sensing signals that are specific for a given group of bacteria. In *Pectobacterium carotovorum* (formerly *Erwinia carotovora*), it is regulated by a classical autoinducer, *N*-(3-oxohexanoyl)-L-homoserine lactone (Bainton et al. 1992). In *Serratia*, despite a lesser dependence on antibiotic production from the growth phase and cell density, it is still under a quorum-sensing control in this group of bacteria as well (Thomson et al. 2000). Moreover, several other signals are also integrated into the regulatory circuits governing the carbapenem biosynthesis in *Serratia* species (Slater et al. 2003; Fineran et al. 2005). The extreme complexity of this regulation reflects the fine-tuning mechanisms adjusting the level of antibiotic production in response to the multiple signals perceived by these bacteria from the environment. Presumably, the killer role of the antibiotic would be regulated by less complex mechanisms, essentially limited to the achieving of MIC values that are sufficient to suppress/eliminate competitors. On the contrary, the highly variable antibiotic concentrations maintained by the multiple regulatory mechanisms suggest that the antibiotic itself serves as a signaling molecule for other regulatory circuits.

Another aspect of the factual concentration of carbapenems in natural ecosystems can be discussed regarding the occurrence of apparently silent gene clusters encoding its biosynthesis. Intriguingly, cryptic carbapenem antibiotic production genes are widespread in *P. carotovorum* (formerly known as *E. carotovora*), but the expression can be forced under laboratory conditions with the use of multiple copies of the apparently mutant transcriptional activator (Holden et al. 1998). The reason for selection of mutation causing the silencing of this biosynthetic cluster is not known, but, once again, we must consider a factual role played by this antibiotic in natural ecosystems. The antibiotic concentrations achieved via this transcriptional activator may indeed unable to reach the killing concentrations, but the concentrations produced could be sufficient to serve as a signaling molecule, although these concentrations may be too low to be detected in the MIC or instrumental assays. Also, given the complexity of regulation by a variety of mechanisms, laboratory conditions such as media composition may not provide all the cues necessary for the high-level expression of this cluster. In general, there are many biosynthetic pathways in natural ecosystems that are seemingly silent from the human prospective. This silent potential, however, can be activated and expressed for human purposes if necessary (Bergmann et al. 2010; Brakhage and Schroeckh 2011; Reen et al. 2015).

Much less is known about the regulation of carbapenem biosynthesis in the first producer discovered, *S. cattleya*, but, regarding the regulation of biosynthesis of other antibiotics, there are indications that there is an additional low-level crosstalk between the thienamycin and cephamycin C pathways in this bacterium (Rodríguez et al. 2008). In another cephamycin C-producing bacterium, *S. clavuligerus*, antibiotic synthesis is regulated at the primary regulatory level by a well-known quorum-sensing molecule, γ-butyrolactone (Liras et al. 2008). Therefore, in the case of carbapenems, producer cells respond to the initial quorum-sensing signals and other environmental clues such as *N*-acetylglucosamine (Rigali et al. 2008) and nutrient depletion (Hesketh et al. 2007; Lian et al. 2008), possibly integrating them, by the second-level signaling, via antibiotics.

The second-level signaling by low-dose antibiotics was discussed above in the corresponding section. Regarding the secondary-level signaling by carbapenems, the representative of this class of antibiotics, imipenem, at subinhibitory concentrations, induces global gene expression changes, including β-lactamase and alginate production, in *P. aeruginosa* biofilms (Bagge et al. 2004). In Gram-positive bacteria such as *E. faecium* and *M. tuberculosis*, low-dose carbapenems are strong inhibitors of L,D-transpeptidases, which are involved in the formation of 3→3 peptidoglycan cross-links and in the bypass of the 4→3 cross-links formed by the D,D-transpeptidase activity of penicillin-binding proteins (Mainardi et al. 2007; Lavollay et al. 2008). In *E. coli*, the L,D-transpeptidase homologue is involved in attachment of the Braun lipoprotein to peptidoglycan (Magnet et al. 2007). The lack of the lipoprotein in *E. coli* leads to sensitivity to EDTA, cationic dyes, and detergents, but no vital cellular functions are affected (Hirota et al. 1977). The question is, how other antibiotic signals are perceived, and what kind of phenotypic and genotypic responses they may evoke in other systems?

23.20 Antibiotics and Antibiotic Resistance: Environmental and Anthropogenic Contexts

In this section, a typical chain of events starting from a discovery of antibiotic to the extensive use, to the emergence resistance, and to the continuous race against resistance is illustrated using tetracycline antibiotics as an example. The fate of other antibiotics is very similar (Aminov 2017), and the main lesson learned here is that the appearance and dissemination of antibiotic resistance is just an issue of time, although it could be considerably affected by the usage practices involved. It becomes increasingly clear, however, that the current global problem of antibiotic resistance cannot be efficiently managed within the narrow remits of clinical microbiology but should include other aspects such as ecological, evolutionary, agricultural, and economical. Antibiotic producers that are used by humans for industrial production of antibiotics have been acquired from natural reservoirs, where they are involved in production of signaling molecules, antibiotics, which regulate various functions of the environmental microbiota. Thus, an extensive summary of the role played by antibiotic producers/antibiotics/antibiotic resistance is also given in the chapter. Economical background for nontherapeutic use of antibiotics in agriculture is also given.

In brief, the path of antibiotic usage by humans can be condensed to a few principal steps. First, the antibiotics, which serve as signaling molecules at low concentrations in natural ecosystems, are selected for killing activity at high concentrations. Then they are extensively used in clinical medicine, veterinary, agriculture, and other applications, thus creating hot spots of high antibiotic concentrations. In these hot spots, the naturally occurring antibiotic resistance genes are selected and amplified. At this stage, antibiotic resistance genes are integrated into the normal bacterial metabolism via reducing the fitness cost associated with their carriage. At this stage, antibiotic resistance becomes very resilient against eradication, even in the absence of antibiotic selective pressure. The pool of the antibiotic resistance genes amplified at hot spots is then released, together with the concomitant antibiotics, into other ecological compartments. These are further disseminated to even more distant ecological compartments, including pathogens, via extensive horizontal gene exchange mechanisms.

What can be done to reduce the size and "temperature" of the hot spots? The biggest consumer of antibiotics is agriculture, where a considerable proportion of antibiotics is used for nontherapeutic purposes (Koike et al. 2017). Analysis of the current strategies implemented in the development of agricultural practices forecasts a rapid growth in antibiotic consumption by this sector (Van Boeckel et al. 2015). This hasty surge of antibiotic consumption by agriculture is driven by economic incentives offered by large-scale intensive farming operations, the practice that is increasingly adapted by and becomes prevalent in many countries. This global trend requires other assessments beyond the economic considerations within the industry. The practice that results in the release of large quantities of antibiotics and antibiotic-resistant bacteria into other ecological compartments contributes significantly to the problem of antibiotic resistance we are facing today. The cost of the resulting resistance, which necessitates the development of new antibiotics, requires extended and costly therapy of infectious diseases and results in the higher morbidity and mortality rates, however, being taken out of the industry's economic equations.

Calls to constraint/stop the nonmedical use of antibiotics, specifically in agriculture, are issued from time to time, especially recently (Tollefson and Miller 2000; Heymann 2006; Maron et al. 2013; Meek et al. 2015; WHO 2015; Aitken et al. 2016; Jørgensen et al. 2016; UN 2016; Ludvigsson et al. 2017). The earliest attempts to ease the antibiotic selection pressure in agriculture have been made by the Scandinavian countries, in particular by Sweden, which prohibited the use of growth-promoting antibiotics in food animals as early as 1986. The measures involved the withdrawal of antibiotic growth promoters and implementation of optimal disease prevention management programs with the proper use of antimicrobials in food animal production (Bengtsson and Wierup 2006). During 1992–2008, these efforts resulted in the reduction of antimicrobial consumption by >50% and in improved productivity (Aarestrup et al. 2010). In other EU countries, specific antibiotics in feedstuffs were banned before 1 January 2006; after that date, all the growth-promoting antibiotics were deleted from the Community Register of authorized feed additives (EPC 2005; Castanon 2007). Reversal of resistance after the release of antimicrobial selective pressure, however, is not straightforward, and resistance may persist at low, but detectable, levels for many years in the absence of the corresponding drugs (Johnsen et al. 2009; Bortolaia et al. 2015). Although the occurrence of antibiotic resistance genes may be significantly reduced, they are still encountered in the absence of antibiotic selection (Kazimierczak et al. 2009; Koike et al. 2017). During a long-term selection by antibiotics, not only resistance mechanisms are selected but also compensatory mechanisms that ameliorate the fitness costs associated with resistance (Hernando-Amado et al. 2017). In this regard, a complete reversal to a susceptible phenotype is unlikely.

23.21 Conclusions

Agricultural use of antibiotics as growth promotors as discussed above in the previous section is only one contributing factor to the emergence and dissemination of antibiotic resistance. There are many other aspects of antibiotic use that must be tackled to decrease the effects of antibiotic/antibiotic resistance "hot spots" and limit the spread of antibiotic resistance:

- Efficient antibiotic stewardship, communication, education, and training.
- Global minimal standard for antibiotic use and legislative measures and their enforcement.
- Optimization of PK/PD-based antibiotic regimens to decrease the exposure to antibiotics and the quantities used.
- Investments in hygiene, sanitation, and infection control such as vaccination.
- Improved knowledge regarding the dynamics of antimicrobials and resistances in various ecosystems for evidence-based decision-making.
- Drug development and conservation and development of antibiotic alternatives and diagnostic tools.
- Development of livestock production systems that are less reliant on antibiotics.
- Preclusion of overlapping in antibiotic classes used for animals and humans.
- Technological solutions for secondary treatment of WWTPs to prevent the release of antibiotics and antibiotic-resistant bacteria from agriculture, community, hospitals, pharmaceutical industry, and other hot spots.

Conflict of Interest

The author declares that the research was conducted in the absence of any commercial or financial relationships that could be construed as a potential conflict of interest.

References

Aarestrup, F.M., Jensen, V.F., Emborg, H.D. et al. (2010). Changes in the use of antimicrobials and the effects on productivity of swine farms in Denmark. *Am. J. Vet. Res.* 71 (7): 726–733. https://doi.org/10.2460/ajvr.71.7.726.

Abbanat, D., Macielag, M., and Bush, K. (2003). Novel antibacterial agents for the treatment of serious Gram-positive infections. *Expert Opin. Investig. Drugs* 12: 379–399.

Ahmed, N.A., Petersen, F.C., and Scheie, A.A. (2007). AI-2 quorum sensing affects antibiotic susceptibility in *Streptococcus anginosus. J. Antimicrob. Chemother.* 60: 49–53.

Aínsa, J.A., Blokpoel, M.C., Otal, I. et al. (1998). Molecular cloning and characterization of Tap, a putative multidrug efflux pump present in *Mycobacterium fortuitum* and *Mycobacterium tuberculosis. J. Bacteriol.* 180: 5836–5843.

Aitken, S.L., Dilworth, T.J., Heil, E.L., and Nailor, M.D. (2016). Agricultural applications for antimicrobials. A danger to human health: an official position statement of the society of infectious diseases pharmacists. *Pharmacotherapy* 36 (4): 422–432. https://doi.org/10.1002/phar.1737.

Allen, H.K., Looft, T., Bayles, D.O. et al. (2011). Antibiotics in feed induce prophages in swine fecal microbiomes. *MBio* 2 (6): e00260-11. https://doi.org/10.1128/mBio.00260-11.

Allen, H.K., Moe, L.A., Rodbumrer, J. et al. (2009). Functional metagenomics reveals diverse β-lactamases in a remote Alaskan soil. *ISME J.* 3: 243–251.

Aminov, R.I. (2009). The role of antibiotics and antibiotic resistance in nature. *Environ. Microbiol.* 11: 2970–2988. https://doi.org/10.1111/j.1462-2920.2009.01972.x.

Aminov, R.I. (2010). A brief history of the antibiotic era: lessons learned and challenges for the future. *Front. Microbiol.* 1: 134. https://doi.org/10.3389/fmicb.2010.00134.

Aminov, R.I. (2011). Horizontal gene exchange in environmental microbiota. *Front. Microbiol.* 2: 158. https://doi.org/10.3389/fmicb.2011.00158.

Aminov, R.I. (2012). The extent and regulation of lateral gene transfer in natural microbial ecosystems. In: *Horizontal Gene Transfer in Microorganisms* (ed. M.P. Francino), Chapter 6. Norwich: Horizon Scientific Press.

Aminov, R.I. (2013a). Evolution in action: dissemination of *tet*(X) into pathogenic microbiota. *Front. Microbiol.* 4: 192. https://doi.org/10.3389/fmicb.2013.00192.

Aminov, R.I. (2013b). Biotic acts of antibiotics. *Front. Microbiol.* 4: 241. https://doi.org/10.3389/fmicb.2013.00241.

Aminov, R.I. (2017). History of antimicrobial drug discovery: major classes and health impact. *Biochem. Pharmacol.* 133: 4–19. https://doi.org/10.1016/j.bcp.2016.10.001.

Aminov, R.I., Chee-Sanford, J.C., Garrigues, N. et al. (2002). Development, validation, and application of primers for detection of tetracycline resistance genes encoding tetracycline efflux pumps in Gram-negative bacteria. *Appl. Environ. Microbiol.* 68: 1786–1793.

Aminov, R.I., Garrigues-Jeanjean, N., and Mackie, R.I. (2001). Molecular ecology of tetracycline resistance: development and validation of primers for detection of tetracycline resistance genes encoding ribosomal protection proteins. *Appl. Environ. Microbiol.* 67: 22–32.

Aminov, R.I. and Mackie, R.I. (2007). Evolution and ecology of antibiotic resistance genes. *FEMS Microbiol. Lett.* 271: 147–161.

Andersen, S.J., Quan, S., Gowan, B., and Dabbs, E.R. (1997). Monooxygenase-like sequence of a *Rhodococcus equi* gene conferring increased resistance to rifampin by inactivating this antibiotic. *Antimicrob. Agents Chemother.* 41: 218–221.

Andersson, D.I. and Hughes, D. (2014). Microbiological effects of sublethal levels of antibiotics. *Nat. Rev. Microbiol.* 12 (7): 465–478. https://doi.org/10.1038/nrmicro3270.

Auchtung, J.M., Lee, C.A., Monson, R.E. et al. (2005). Regulation of a *Bacillus subtilis* mobile genetic element by intercellular signaling and the global DNA damage response. *Proc. Natl. Acad. Sci. U. S. A.* 102: 12554–12559.

aus der Beek, T., Weber, F.A., Bergmann, A. et al. (2016). Pharmaceuticals in the environment: global occurrences and perspectives. *Environ. Toxicol. Chem.* 35 (4): 823–835. https://doi.org/10.1002/etc.3339.

Bagge, N., Schuster, M., Hentzer, M. et al. (2004). *Pseudomonas aeruginosa* biofilms exposed to imipenem exhibit changes in global gene expression and beta-lactamase and alginate production. *Antimicrob. Agents Chemother.* 48: 1175–1187.

Bahl, M.I., Sorensen, S.J., Hansen, L.H., and Licht, T.R. (2004). Effect of tetracycline on transfer and establishment of the tetracycline-inducible conjugative transposon Tn*916* in the guts of gnotobiotic rats. *Appl. Environ. Microbiol.* 70: 758–764.

Bainton, N.J., Stead, P., Chhabra, S.R. et al. (1992). N-(3-oxohexanoyl)-L-homoserine lactone regulates carbapenem antibiotic production in *Erwinia carotovora*. *Biochem. J.* 288: 997–1004.

Balode, A., Punda-Polić, V., and Dowzicky, M.J. (2013). Antimicrobial susceptibility of Gram-negative and Gram-positive bacteria collected from countries in Eastern Europe: results from the Tigecycline Evaluation and Surveillance Trial (T.E.S.T.) 2004-2010. *Int. J. Antimicrob. Agents* 41 (6): 527–535. https://doi.org/10.1016/j.ijantimicag.2013.02.022.

Baquero, F., Martínez, J.L., and Cantón, R. (2008). Antibiotics and antibiotic resistance in water environments. *Curr. Opin. Biotechnol.* 19: 260–265.

Barr, V., Barr, K., Millar, M.R., and Lacey, R.W. (1986). Beta-lactam antibiotics increase the frequency of plasmid transfer in *Staphylococcus aureus*. *J. Antimicrob. Chemother.* 17: 409–413.

Bartha, N.A., Sóki, J., Urbán, E., and Nagy, E. (2011). Investigation of the prevalence of tetQ, tetX and tetX1 genes in *Bacteroides* strains with elevated tigecycline minimum inhibitory concentrations. *Int. J. Antimicrob. Agents* 38: 522–525.

Bassett, E.J., Keith, M.S., Armelagos, G.J. et al. (1980). Tetracycline-labelled human bone from ancient Sudanese Nubia (A.D. 350). *Science* 209: 1532–1534.

Beaber, J.W., Hochhut, B., and Waldor, M.K. (2004). SOS response promotes horizontal dissemination of antibiotic resistance genes. *Nature* 427: 72–74.

Bearson, B.L., Allen, H.K., Brunelle, B.W. et al. (2014). The agricultural antibiotic carbadox induces phage-mediated gene transfer in *Salmonella*. *Front. Microbiol.* 5: 52. https://doi.org/10.3389/fmicb.2014.00052.

Bearson, B.L. and Brunelle, B.W. (2015). Fluoroquinolone induction of phage-mediated gene transfer in multidrug-resistant *Salmonella*. *Int. J. Antimicrob. Agents* 46 (2): 201–204. https://doi.org/10.1016/j.ijantimicag.2015.04.008.

Bengtsson, B. and Wierup, M. (2006). Antimicrobial resistance in Scandinavia after ban of antimicrobial growth promoters. *Anim. Biotechnol.* 17: 147–156.

Bergmann, S., Funk, A.N., Scherlach, K. et al. (2010). Activation of a silent fungal polyketide biosynthesis pathway through regulatory cross talk with a cryptic nonribosomal peptide synthetase gene cluster. *Appl. Environ. Microbiol.* 76 (24): 8143–8149. https://doi.org/10.1128/AEM.00683-10.

Bernier, S.P. and Surette, M.G. (2013). Concentration-dependent activity of antibiotics in natural environments. *Front. Microbiol.* 4: 20. https://doi.org/10.3389/fmicb.2013.00020.

Bertrand, X. and Dowzicky, M.J. (2012). Antimicrobial susceptibility among Gram-negative isolates collected from intensive care units in North America, Europe, the Asia-Pacific Rim, Latin America, the Middle East, and Africa between 2004 and 2009 as part of the Tigecycline Evaluation and Surveillance Trial. *Clin. Ther.* 34: 124–137.

Betriu, C., Rodriguez-Avial, I., Sánchez, B.A. et al. (2002). Comparative *in vitro* activities of tigecycline (GAR-936) and other antimicrobial agents against *Stenotrophomonas maltophilia*. *J. Antimicrob. Chemother.* 50: 758–759.

Bhullar, K., Waglechner, N., Pawlowski, A. et al. (2012). Antibiotic resistance is prevalent in an isolated cave microbiome. *PLoS One* 7 (4): e34953. https://doi.org/10.1371/journal.pone.0034953.

Borbone, S., Lupo, A., Mezzatesta, M.L. et al. (2008). Evaluation of the in vitro activity of tigecycline against multiresistant Gram-positive cocci containing tetracycline resistance determinants. *Int. J. Antimicrob. Agents* 31 (3): 209–215.

Bortolaia, V., Mander, M., Jensen, L.B. et al. (2015). Persistence of vancomycin resistance in multiple clones of *Enterococcus faecium* isolated from Danish broilers 15 years after the ban of avoparcin. *Antimicrob. Agents Chemother.* 59 (5): 2926–2929. https://doi.org/10.1128/AAC.05072-14.

Bose, B., Auchtung, J.M., Lee, C.A., and Grossman, A.D. (2008). A conserved anti-repressor controls horizontal gene transfer by proteolysis. *Mol. Microbiol.* 70: 570–582.

Bouchillon, S.K., Hoban, D.J., Johnson, B.M. et al. (2005). *In vitro* activity of tigecycline against 3989 Gram-negative and Gram-positive clinical isolates from the United States Tigecycline Evaluation and Surveillance Trial (TEST Program; 2004). *Diagn. Microbiol. Infect. Dis.* 52 (3): 173–179.

Brakhage, A.A. and Schroeckh, V. (2011). Fungal secondary metabolites: strategies to activate silent gene clusters. *Fungal Genet. Biol.* 48 (1): 15–22. https://doi.org/10.1016/j.fgb.2010.04.004.

Brown, L.C.W., Acker, M.G., Clardy, J. et al. (2009). Thirteen posttranslational modifications convert a 14-residue peptide into the antibiotic thiocillin. *Proc. Natl. Acad. Sci. U. S. A.* 106: 2549–2553.

Castanon, J.I. (2007). History of the use of antibiotic as growth promoters in European poultry feeds. *Poult. Sci.* 86: 2466–2471. https://doi.org/10.3382/ps.2007-00249.

CDC. 2013. *Antibiotic Resistance Threats in the United States*. Atlanta, GA: Centres for Disease Control and Prevention, U.S. Department of Health and

Human Services. http://www.cdc.gov/drugresistance/pdf/ar-threats-2013-508.pdf (accessed 23 April 2019).

Chee-Sanford, J., Aminov, R.I., Garrigues, N. et al. (2001). Occurrence and diversity of tetracycline resistance genes in lagoons and groundwater underlying two swine production facilities. *Appl. Environ. Microbiol.* 67: 1494–1502.

Chee-Sanford, J.C., Mackie, R.I., Koike, S. et al. (2009). Fate and transport of antibiotic residues and antibiotic resistance genes following land application of manure waste. *J. Environ. Qual.* 38: 1086–1108.

Chen, Y.P., Tsao, M.Y., Lee, S.H. et al. (2010). Prevalence and molecular characterization of chloramphenicol resistance in *Riemerella anatipestifer* isolated from ducks and geese in Taiwan. *Avian Pathol.* 39: 333–338.

Chopra, I. and Roberts, M. (2001). Tetracycline antibiotics: mode of action, applications, molecular biology, and epidemiology of bacterial resistance. *Microbiol. Mol. Biol. Rev.* 65: 232–260.

CIPARS (2013). Canadian Integrated Program for Antimicrobial Resistance Surveillance (CIPARS). http://www.phac-aspc.gc.ca/cipars-picra/2013/annu-report-rapport-eng.php (accessed 23 April 2019).

Collignon, P. and Voss, A. (2015). China, what antibiotics and what volumes are used in food production animals? *Antimicrob. Resist. Infect. Control* 4: 16. https://doi.org/10.1186/s13756-015-0056-5.

Cook, M., Molto, E., and Anderson, C. (1989). Fluorochrome labelling in Roman period skeletons from Dakhleh Oasis, Egypt. *Am. J. Phys. Anthropol.* 80: 137–143. https://doi.org/10.1002/ajpa.1330800202.

Corey, E.J. (2013). *Drug Discovery Practices, Processes, and Perspectives*, 406. Hoboken, NJ: Wiley.

Coulthurst, S.J., Barnard, A.M., and Salmond, G.P. (2005). Regulation and biosynthesis of carbapenem antibiotics in bacteria. *Nat. Rev. Microbiol.* 3: 295–306.

Cox, A.R., Thomson, N.R., Bycroft, B. et al. (1998). A pheromone-independent CarR protein controls carbapenem antibiotic synthesis in the opportunistic human pathogen *Serratia marcescens*. *Microbiology* 144: 201–209.

Coyne, S., Rosenfeld, N., Lambert, T. et al. (2010). Overexpression of resistance-nodulation-cell division pump AdeFGH confers multidrug resistance in *Acinetobacter baumannii*. *Antimicrob. Agents Chemother.* 54 (10): 4389–4393. https://doi.org/10.1128/AAC.00155-10.

Cummins, J., Reen, F.J., Baysse, C. et al. (2009). Subinhibitory concentrations of the cationic antimicrobial peptide colistin induce the pseudomonas quinolone signal in *Pseudomonas aeruginosa*. *Microbiology* 155(Pt 9: 2826–2837. https://doi.org/10.1099/mic.0.025643-0.

Damier-Piolle, L., Magnet, S., Brémont, S. et al. (2008). AdeIJK, a resistance-nodulation-cell division pump effluxing multiple antibiotics in *Acinetobacter baumannii*. *Antimicrob. Agents Chemother.* 52: 557–562.

Davies, J., Spiegelman, G.B., and Yim, G. (2006). The world of subinhibitory antibiotic concentrations. *Curr. Opin. Microbiol.* 9: 445–453.

D'Costa, V.M., King, C.E., Kalan, L. et al. (2011). Antibiotic resistance is ancient. *Nature* 477 (7365): 457–461. https://doi.org/10.1038/nature10388.

De Rossi, E., Blokpoel, M.C., Cantoni, R. et al. (1998). Molecular cloning and functional analysis of a novel tetracycline resistance determinant, *tet*(V), from *Mycobacterium smegmatis*. *Antimicrob. Agents Chemother.* 42 (8): 1931–1937.

de Vries, L.E., Vallès, Y., Agersø, Y. et al. (2011). The gut as reservoir of antibiotic resistance: microbial diversity of tetracycline resistance in mother and infant. *PLoS One* 6 (6): e21644.

Dean, C.R., Visalli, M.A., Projan, S.J. et al. (2003). Efflux-mediated resistance to tigecycline (GAR-936) in *Pseudomonas aeruginosa* PAO1. *Antimicrob. Agents Chemother.* 47: 972–978.

Derzelle, S., Duchaud, E., Kunst, F. et al. (2002). Identification, characterization, and regulation of a cluster of genes involved in carbapenem biosynthesis in *Photorhabdus luminescens*. *Appl. Environ. Microbiol.* 68: 3780–3789.

Dong, Y.H., Wang, L.H., Xu, J.L. et al. (2001). Quenching quorum-sensing-dependent bacterial infection by an N-acyl homoserine lactonase. *Nature* 411: 813–817.

Doucet-Populaire, F., Trieu-Cuot, P., Dosbaa, I. et al. (1991). Inducible transfer of conjugative transposon Tn1545 from *Enterococcus faecalis* to *Listeria monocytogenes* in the digestive tracts of gnotobiotic mice. *Antimicrob. Agents Chemother.* 35: 185–187.

Dowzicky, M.J. and Chmelařová, E. (2011). Global in vitro activity of tigecycline and linezolid against Gram-positive organisms collected between 2004 and 2009. *Int. J. Antimicrob. Agents* 37 (6): 562–566. https://doi.org/10.1016/j.ijantimicag.2011.02.004.

Dubern, J.F. and Diggle, S.P. (2008). Quorum sensing by 2-alkyl-4-quinolones in *Pseudomonas aeruginosa* and other bacterial species. *Mol. BioSyst.* 4 (9): 882–888. https://doi.org/10.1039/b803796p.

Duchaud, E., Boussaha, M., Loux, V. et al. (2007). Complete genome sequence of the fish pathogen *Flavobacterium psychrophilum*. *Nat. Biotechnol.* 25: 763–769.

Duggar, B.M. (1948). Aureomycin: a product of the continuing search for new antibiotics. *Ann. N. Y. Acad. Sci.* 51: 177–181. https://doi.org/10.1111/j.1749-6632.1948.tb27262.x.

Edelstein, P.H., Weiss, W.J., and Edelstein, M.A.C. (2003). Activities of tigecycline (GAR-936) against *Legionella pneumophila in vitro* and in guinea pigs with *L. pneumophila* pneumonia. *Antimicrob. Agents Chemother.* 47: 533–540.

EMA (2013). Use of glycylcyclines in animals in the European Union: development of resistance and possible impact on human and animal health. http://www.ema.europa.eu/docs/en_GB/document_library/Report/2013/07/WC500146814.pdf (accessed 23 April 2019).

EMA (2015). Antimicrobial resistance. http://www.ema.europa.eu/ema/index.jsp?curl=pages/special_topics/general/general_content_000439.jsp (accessed 23 April 2019).

EPC (2005). Ban on Antibiotics as Growth Promoters in Animal Feed Enters into Effect. European Commission – IP/05/1687 (22 December 2005). http://europa.eu/rapid/press-release_IP-05-1687_en.htm (accessed 23 April 2019).

European Centre for Disease Prevention and Control (ECDC), European Food Safety Authority (EFSA) and European Medicines Agency (EMA) (2015). ECDC/EFSA/EMA first joint report on the integrated analysis of the consumption of antimicrobial agents and occurrence of antimicrobial resistance in bacteria from humans and food-producing animals. Stockholm/Parma/London: ECDC/EFSA/EMA, 2015. *EFSA J.* 13 (1): 4006, 114 pp. doi:https://doi.org/10.2903/j.efsa.2015.4006.

Fajardo, A. and Martínez, J.L. (2008). Antibiotics as signals that trigger specific bacterial responses. *Curr. Opin. Microbiol.* 11: 161–167.

Falkinham, J.O. 3rd, Wall, T.E., Tanner, J.R. et al. (2009). Proliferation of antibiotic-producing bacteria and concomitant antibiotic production as the basis for the antibiotic activity of Jordan's red soils. *Appl. Environ. Microbiol.* 75: 2735–2741.

FDA (2015). Summary Report on Antimicrobials Sold or Distributed for Use in Food-Producing Animals. https://docs.google.com/viewer?url=http%3A%2F%2Fwww.fda.gov%2Fdownloads%2FForIndustry%2FUserFees%2FAnimalDrugUserFeeActADUFA%2FUCM440584.pdf (accessed 23 April 2019).

Fiedler, S., Bender, J.K., Klare, I. et al. (2016). Tigecycline resistance in clinical isolates of *Enterococcus faecium* is mediated by an upregulation of plasmid-encoded tetracycline determinants *tet*(L) and *tet*(M). *J. Antimicrob. Chemother.* 71 (4): 871–881. https://doi.org/10.1093/jac/dkv420.

Fineran, P.C., Slater, H., Everson, L. et al. (2005). Biosynthesis of tripyrrole and beta-lactam secondary metabolites in *Serratia*: integration of quorum sensing with multiple new regulatory components in the control of prodigiosin and carbapenem antibiotic production. *Mol. Microbiol.* 56: 1495–1517.

Fischbach, M.A. (2009). Antibiotics from microbes: converging to kill. *Curr. Opin. Microbiol.* 12 (5): 520–527. https://doi.org/10.1016/j.mib.2009.07.002.

Flatt, P.M. and Mahmud, T. (2007). Biosynthesis of aminocyclitol-aminoglycoside antibiotics and related compounds. *Nat. Prod. Rep.* 24: 358–392.

Fleming, A. (1929). On the antibacterial action of cultures of a *Penicillium*, with special reference to their use in the isolation of *B. influenzae*. *Brit. J. Exp. Path.* 10: 226–236.

Forsberg, K.J., Patel, S., Wencewicz, T.A., and Dantas, G. (2016). The tetracycline destructases: a novel family of tetracycline-inactivating enzymes. *Chem. Biol.* 22 (7): 888–897. https://doi.org/10.1016/j.chembiol.2015.05.017.

Fritsche, T.R., Kirby, J.T., and Jones, R.N. (2004). *In vitro* activity of tigecycline (GAR-936) tested against 11,859 recent clinical isolates associated with community-acquired respiratory tract and Gram-positive cutaneous infections. *Diagn. Microbiol. Infect. Dis.* 49 (3): 201–209.

Fuqua, C., Winans, S.C., and Greenberg, E.P. (1996). Census and consensus in bacterial ecosystems: the LuxR-LuxI family of quorum-sensing transcriptional regulators. *Annu. Rev. Microbiol.* 50: 727–751.

García Martín, H., Ivanova, N., Kunin, V. et al. (2006). Metagenomic analysis of two enhanced biological phosphorus removal (EBPR) sludge communities. *Nat. Biotechnol.* 24: 1263–1269.

Garrido-Mesa, N., Zarzuelo, A., and Gálvez, J. (2013). Minocycline: far beyond an antibiotic. *Br. J. Pharmacol.* 169: 337–352. https://doi.org/10.1111/bph.12139.

Ghosh, S., Sadowsky, M.J., Roberts, M.C. et al. (2009). *Sphingobacterium* sp. strain PM2-P1-29 harbours a functional *tet*(X) gene encoding for the degradation of tetracycline. *J. Appl. Microbiol.* 106: 1336–1342.

Giammanco, A., Calà, C., Fasciana, T., and Dowzicky, M.J. (2017). Global assessment of the activity of tigecycline against multidrug-resistant Gram-negative pathogens between 2004 and 2014 as part of the tigecycline evaluation and surveillance trial. *mSphere* 2 (1): pii: e00310-16. https://doi.org/10.1128/mSphere.00310-16.

Gil-Turnes, M.S., Hay, M.E., and Fenical, W. (1989). Symbiotic marine bacteria chemically defend crustacean embryos from a pathogenic fungus. *Science* 246: 116–118. https://doi.org/10.1126/science.2781297.

Goossens, H., Ferech, M., Vander Stichele, R. et al. (2005). Outpatient antibiotic use in Europe and association with resistance: a cross-national database study. *Lancet* 365: 579–587. https://doi.org/10.1016/S0140-6736(05)17907-0.

Guo, X., Liu, N., Li, X. et al. (2015). Red soils harbor diverse culturable actinomycetes that are promising sources of novel secondary metabolites. *Appl. Environ. Microbiol.* 81 (9): 3086–3103. https://doi.org/10.1128/AEM.03859-14.

Haas, D. and Défago, G. (2005). Biological control of soil-borne pathogens by fluorescent pseudomonads. *Nat. Rev. Microbiol.* 3: 307–319.

Hall, B.G. and Barlow, M. (2003). Structure-based phylogenies of the serine beta-lactamases. *J. Mol. Evol.* 57 (3): 255–260.

Harayama, S., Kok, M., and Neidle, E.L. (1992). Functional and evolutionary relationships among diverse oxygenases. *Annu. Rev. Microbiol.* 46: 565–601.

Hastings, P.J., Rosenberg, S.M., and Slack, A. (2004). Antibiotic-induced lateral transfer of antibiotic resistance. *Trends Microbiol.* 12: 401–404.

Hawkey, P. and Finch, R. (2007). Tigecycline: *in-vitro* performance as a predictor of clinical efficacy. *Clin. Microbiol. Infect.* 13 (4): 354–362.

Hawser, S.P., Bouchillon, S.K., Hackel, M. et al. (2012). Trending 7 years of *in vitro* activity of tigecycline and comparators against Gram-positive and Gram-negative pathogens from the Asia-Pacific region: Tigecycline Evaluation Surveillance Trial (TEST) 2004-2010. *Int. J. Antimicrob. Agents* 39 (6): 490–495.

Hentschke, M., Wolters, M., Sobottka, I. et al. (2010). *ramR* mutations in clinical isolates of *Klebsiella pneumoniae* with reduced susceptibility to tigecycline. *Antimicrob. Agents Chemother.* 54 (6): 2720–2723. https://doi.org/10.1128/AAC.00085-10.

Henwood, C.J., Gatward, T., Warner, M. et al. (2002). Antibiotic resistance among clinical isolates of *Acinetobacter* in the UK, and *in vitro* evaluation of tigecycline (GAR-936). *J. Antimicrob. Chemother.* 49: 479–487.

Hernando-Amado, S., Sanz-García, F., Blanco, P., and Martínez, J.L. (2017). Fitness costs associated with the acquisition of antibiotic resistance. *Essays Biochem.* 61 (1): 37–48. https://doi.org/10.1042/EBC20160057.

Hesketh, A., Chen, W.J., Ryding, J. et al. (2007). The global role of ppGpp synthesis in morphological differentiation and antibiotic production in *Streptomyces coelicolor* A3(2). *Genome Biol.* 8: R161.

Hess, C., Enichlmayr, H., Jandreski-Cvetkovic, D. et al. (2013). *Riemerella anatipestifer* outbreaks in commercial goose flocks and identification of isolates by MALDI-TOF mass spectrometry. *Avian Pathol.* 42: 151–156.

Heymann, D.L. (2006). Resistance to anti-infective drugs and the threat to public health. *Cell* 124 (4): 671–675.

Hirota, Y., Suzuki, H., Nishimura, Y., and Yasuda, S. (1977). On the process of cellular division in *Escherichia coli*: a mutant of *E. coli* lacking a murein-lipoprotein. *Proc. Natl. Acad. Sci. U. S. A.* 74: 1417–1420.

Hochstein, F.A., Stephens, C.R., Conover, L.H. et al. (1953). The structure of terramycin. *J. Am. Chem. Soc.* 75 (22): 5455–5475. https://doi.org/10.1021/ja01118a001.

Hoffman, L.R., D'Argenio, D.A., MacCoss, M.J. et al. (2005). Aminoglycoside antibiotics induce bacterial biofilm formation. *Nature* 436: 1171–1175.

Holden, M.T., McGowan, S.J., Bycroft, B.W. et al. (1998). Cryptic carbapenem antibiotic production genes are widespread in *Erwinia carotovora*: facile trans activation by the carR transcriptional regulator. *Microbiology* 144: 1495–1508.

Horinouchi, S. (2007). Mining and polishing of the treasure trove in the bacterial genus *Streptomyces*. *Biosci. Biotechnol. Biochem.* 71: 283–299.

Hornsey, M., Ellington, M.J., Doumith, M. et al. (2010). Emergence of AcrAB-mediated tigecycline resistance in a clinical isolate of *Enterobacter cloacae* during ciprofloxacin treatment. *Int. J. Antimicrob. Agents* 35 (5): 478–481. https://doi.org/10.1016/j.ijantimicag.2010.01.011.

Hummert, J. and Van Gerven, D. (1982). Tetracycline-labeled human bone from a medieval population in Nubia's Batn El Hajar (550-1450 A.D.). *Hum. Biol.* 54 (2): 355–363.

Huovinen, P., Sundstrom, L., Swedberg, G., and Skold, O. (1995). Trimethoprim and sulfonamide resistance. *Antimicrob. Agents Chemother.* 39: 279–289. https://doi.org/10.1128/aac.39.2.279.

Huttner, B., Jones, M., Rubin, M.A. et al. (2012). Drugs of last resort? The use of polymyxins and tigecycline at US Veterans Affairs medical centers, 2005–2010. *PLoS One* 7 (5): e36649.

Ichimiya, T., Takeoka, K., Hiramatsu, K. et al. (1996). The influence of azithromycin on the biofilm formation of *Pseudomonas aeruginosa in vitro*. *Chemotherapy* 42: 186–191.

Johnsen, P.J., Townsend, J.P., Bøhn, T. et al. (2009). Factors affecting the reversal of antimicrobial-drug resistance. *Lancet Infect. Dis.* 9 (6): 357–364. https://doi.org/10.1016/S1473-3099(09)70105-7.

Jørgensen, P.S., Wernli, D., Carroll, S.P. et al. (2016). Use antimicrobials wisely. *Nature* 537 (7619): 159–161. https://doi.org/10.1038/537159a.

Jutkina, J., Rutgersson, C., Flach, C.F., and Larsson, D.G. (2016). An assay for determining minimal concentrations of antibiotics that drive horizontal transfer of resistance. *Sci. Total Environ.* 548–549: 131–138. https://doi.org/10.1016/j.scitotenv.2016.01.044.

Kahan, J.S., Kahan, F.M., Goegelman, R. et al. (1972). Thienamycin, a new beta-lactam antibiotic. I. Discovery, taxonomy, isolation and physical properties. *J. Antibiot. (Tokyo)* 32: 1–12.

Kato, J.Y., Funa, N., Watanabe, H. et al. (2007). Biosynthesis of gamma-butyrolactone autoregulators that switch on secondary metabolism and morphological development in *Streptomyces*. *Proc. Natl. Acad. Sci. U. S. A.* 104: 2378–2383.

Kazimierczak, K.A., Scott, K.P., Kelly, D., and Aminov, R.I. (2009). Tetracycline resistome of the organic pig gut. *Appl. Environ. Microbiol.* 75: 1717–1722.

Keeney, D., Ruzin, A., and Bradford, P.A. (2007). RamA, a transcriptional regulator, and AcrAB, an RND-type efflux pump, are associated with decreased susceptibility to tigecycline in *Enterobacter cloacae*. *Microb. Drug Resist.* 13: 1–6.

Keeney, D., Ruzin, A., McAleese, F. et al. (2008). MarA-mediated overexpression of the AcrAB efflux pump results in decreased susceptibility to tigecycline in *Escherichia coli*. *J. Antimicrob. Chemother.* 61: 46–53.

Khokhlov, A.S., Tovarova, I.I., Borisova, L.N. et al. (1967). The A-factor, responsible for streptomycin biosynthesis by mutant strains of *Actinomyces streptomycini*. *Dokl. Akad. Nauk SSSR* 177: 232–235.

Kim, S.C. and Carlson, K. (2006). Occurrence of ionophore antibiotics in water and sediments of a mixed-landscape watershed. *Water Res.* 40: 2549–2560.

Klopper, M., Warren, R.M., Hayes, C. et al. (2013). Emergence and spread of extensively and totally drug-resistant tuberculosis, South Africa. *Emerg. Infect. Dis.* 19 (3): 449–455. https://doi.org/10.3201/EID1903.120246.

Knapp, C.W., Engemann, C.A., Hanson, M.L. et al. (2008). Indirect evidence of transposon-mediated selection of antibiotic resistance genes in aquatic systems at low-level oxytetracycline exposures. *Environ. Sci. Technol.* 42: 5348–5353.

Koike, S., Aminov, R.I., Yannarell, A.C. et al. (2010). Molecular ecology of macrolide-lincosamide-streptogramin B methylases in waste lagoons and subsurface waters associated with swine production. *Microb. Ecol.* 59: 487–498.

Koike, S., Krapac, I.G., Oliver, H.D. et al. (2007). Monitoring and source tracking of tetracycline resistance genes in lagoons and groundwater adjacent to swine production facilities over a three-year period. *Appl. Environ. Microbiol.* 73: 4813–4823.

Koike, S., Mackie, R., and Aminov, R.I. (2017). Agricultural use of antibiotics and antibiotic resistance. In: *Antibiotic Resistance Genes in Natural Environments and Long-Term Effects* (ed. S. Mirete and M.P. Pérez). Hauppauge, NY: Nova Science Publishers, Chapter 8.

Krishnasamy, V., Otte, J., and Silbergeld, E. (2015). Antimicrobial use in Chinese swine and broiler poultry production. *Antimicrob. Resist. Infect. Control* 4: 17. https://doi.org/10.1186/s13756-015-0050-y.

Kurokawa, K., Itoh, T., Kuwahara, T. et al. (2007). Comparative metagenomics revealed commonly enriched gene sets in human gut microbiomes. *DNA Res.* 14: 169–181.

Kyselková, M., Chroňáková, A., Volná, L. et al. (2012). Tetracycline resistance and presence of tetracycline resistance determinants *tet*(V) and *tap* in rapidly growing mycobacteria from agricultural soils and clinical isolates. *Microbes Environ.* 27 (4): 413–422.

Landers, T.F., Cohen, B., Wittum, T.E., and Larson, E.L. (2012). A review of antibiotic use in food animals: perspective, policy, and potential. *Public Health Rep.* 127 (1): 4–22.

Lavollay, M., Arthur, M., Fourgeaud, M. et al. (2008). The peptidoglycan of stationary-phase *Mycobacterium tuberculosis* predominantly contains cross-links generated by L,D-transpeptidation. *J. Bacteriol.* 190: 4360–4366.

Lefort, A., Lafaurie, M., Massias, L. et al. (2003). Activity and diffusion of tigecycline (GAR-936) in experimental enterococcal endocarditis. *Antimicrob. Agents Chemother.* 47: 216–222.

LeRoux, M., Peterson, S.B., and Mougous, J.D. (2015). Bacterial danger sensing. *J. Mol. Biol.* 427 (23): 3744–3753. https://doi.org/10.1016/j.jmb.2015.09.018.

Leski, T.A., Bangura, U., Jimmy, D.H. et al. (2013). Multidrug-resistant *tet*(X)-containing hospital isolates in Sierra Leone. *Int. J. Antimicrob. Agents* 42 (1): 83–86. https://doi.org/10.1016/j.ijantimicag.2013.04.014.

Li, A. and Piel, J. (2002). A gene cluster from a marine *Streptomyces* encoding the biosynthesis of the aromatic spiroketal polyketide griseorhodin A. *Chem. Biol.* 9: 1017–10126.

Li, H., Wang, X., Zhang, Y. et al. (2015a). The role of RND efflux pump and global regulators in tigecycline resistance in clinical *Acinetobacter baumannii* isolates. *Future Microbiol.* 10 (3): 337–346. https://doi.org/10.2217/fmb.15.7.

Li, X., Wang, J., Li, S. et al. (2015b). ScbR- and ScbR2-mediated signal transduction networks coordinate complex physiological responses in *Streptomyces coelicolor*. *Sci. Rep.* 5: 14831. https://doi.org/10.1038/srep14831.

Lian, W., Jayapal, K.P., Charaniya, S. et al. (2008). Genome-wide transcriptome analysis reveals that a pleiotropic antibiotic regulator, AfsS, modulates nutritional stress response in *Streptomyces coelicolor* A3(2). *BMC Genomics* 29 (9): 56.

Liao, R., Duan, L., Lei, C. et al. (2009). Thiopeptide biosynthesis featuring ribosomally synthesized precursor peptides and conserved posttranslational modifications. *Chem. Biol.* 16: 141–147.

Lima-Bittencourt, C.I., Cursino, L., Gonçalves-Dornelas, H. et al. (2007). Multiple antimicrobial resistance in Enterobacteriaceae isolates from pristine freshwater. *Genet. Mol. Res.* 6: 510–521.

Linares, J.F., Gustafsson, I., Baquero, F., and Martínez, J.L. (2006). Antibiotics as intermicrobial signaling agents instead of weapons. *Proc. Natl. Acad. Sci. U. S. A.* 103: 19484–19489.

Linkevicius, M., Sandegren, L., and Andersson, D.I. (2015). Potential of tetracycline resistance proteins to evolve tigecycline resistance. *Antimicrob. Agents Chemother.* 60 (2): 789–796. https://doi.org/10.1128/AAC.02465-15.

Liras, P., Gomez-Escribano, J.P., and Santamarta, I. (2008). Regulatory mechanisms controlling antibiotic production in *Streptomyces clavuligerus*. *J. Ind. Microbiol. Biotechnol.* 35: 667–676.

Liras, P. and Martin, J.F. (2006). Gene clusters for beta-lactam antibiotics and control of their expression: why have clusters evolved, and from where did they originate? *Int. Microbiol.* 9: 9–19.

Liu, M., Zhang, Y., Yang, M. et al. (2012). Abundance and distribution of tetracycline resistance genes and mobile elements in an oxytetracycline production wastewater treatment system. *Environ. Sci. Technol.* 46: 7551–7557.

Lopatkin, A.J., Huang, S., Smith, R.P. et al. (2016). Antibiotics as a selective driver for conjugation dynamics. *Nat. Microbiol.* 1 (6): 16044. https://doi.org/10.1038/nmicrobiol.2016.44.

Ludvigsson, J.F., Hadjipanayis, A., Del Torso, S. et al. (2017). Appropriate use of antibiotics is vital for public health. *Acta Paediatr.* 106 (5): 691. https://doi.org/10.1111/apa.13772.

Mackie, R.I., Koike, S., Krapac, I. et al. (2006). Tetracycline residues and tetracycline resistance genes in groundwater impacted by swine production facilities. *Anim. Biotechnol.* 17: 157–176.

Magnet, S., Bellais, S., Dubost, L. et al. (2007). Identification of the L,D-transpeptidases responsible for attachment of the Braun lipoprotein to *Escherichia coli* peptidoglycan. *J. Bacteriol.* 189: 3927–3931.

Mainardi, J.L., Hugonnet, J.E., Rusconi, F. et al. (2007). Unexpected inhibition of peptidoglycan LD-transpeptidase from *Enterococcus faecium* by the β-lactam imipenem. *J. Biol. Chem.* 282: 30414–30422.

Maiques, E., Ubeda, C., Campoy, S. et al. (2006). β-lactam antibiotics induce the SOS response and horizontal transfer of virulence factors in *Staphylococcus aureus*. *J. Bacteriol.* 188: 2726–2729.

Manzetti, S. and Ghisi, R. (2014). The environmental release and fate of antibiotics. *Mar. Pollut. Bull.* 79 (1-2): 7–15. https://doi.org/10.1016/j.marpolbul.2014.01.005.

Marco, F. and Dowzicky, M.J. (2016). Antimicrobial susceptibility among important pathogens collected as part of the Tigecycline Evaluation and Surveillance Trial (T.E.S.T.) in Spain, 2004-2014. *J. Glob. Antimicrob. Resist.* 6: 50–56. https://doi.org/10.1016/j.jgar.2016.02.005.

Maron, D.F., Smith, T.J., and Nachman, K.E. (2013). Restrictions on antimicrobial use in food animal production: an international regulatory and economic survey. *Glob. Health* 9: 48. https://doi.org/10.1186/1744-8603-9-48.

Martínez, J.L. (2008). Antibiotics and antibiotic resistance genes in natural environments. *Science* 321: 365–367.

Mavromatis, K., Lu, M., Misra, M. et al. (2011). Complete genome sequence of *Riemerella anatipestifer* type strain (ATCC 11845). *Stand. Genomic Sci.* 4: 144–153.

Mayne, D. and Dowzicky, M.J. (2012). *In vitro* activity of tigecycline and comparators against organisms associated with intra-abdominal infections collected as part of TEST (2004-2009). *Diagn. Microbiol. Infect. Dis.* 74 (2): 151–157.

McGowan, S.J., Sebaihia, M., O'Leary, S. et al. (1997). Analysis of the carbapenem gene cluster of *Erwinia carotovora*: definition of the antibiotic biosynthetic genes and evidence for a novel β-lactam resistance mechanism. *Mol. Microbiol.* 26: 545–556.

Meek, R.W., Vyas, H., and Piddock, L.J. (2015). Nonmedical uses of antibiotics: time to restrict their use? *PLoS Biol.* 13 (10): e1002266. https://doi.org/10.1371/journal.pbio.1002266.

Mendes, R., Kruijt, M., de Bruijn, I. et al. (2011). Deciphering the rhizosphere microbiome for disease-suppressive bacteria. *Science* 332: 1097–1100. https://doi.org/10.1126/science.1203980.

Menendez, N., Nur-e-Alam, M., Brana, A.F. et al. (2004). Biosynthesis of the antitumor chromomycin A3 in *Streptomyces griseus*: analysis of the gene cluster and rational design of novel chromomycin analogs. *Chem. Biol.* (1): 21–32.

Mercier, R.C., Kennedy, C., and Meadows, C. (2002). Antimicrobial activity of tigecycline (GAR-936) against *Enterococcus faecium* and *Staphylococcus aureus* used alone and in combination. *Pharmacotherapy* 22: 1517–1523.

Migliori, G.B., De Iaco, G., Besozzi, G. et al. (2007). First tuberculosis cases in Italy resistant to all tested drugs. *Euro Surveill.* 12 (5): E070517.1.

Milatovic, D., Schmitz, F.J., Verhoef, J., and Fluit, A.C. (2003). Activities of the glycylcycline tigecycline (GAR-936) against 1,924 recent European clinical bacterial isolates. *Antimicrob. Agents Chemother.* 47: 400–404.

Miller, C., Thomsen, L.E., Gaggero, C. et al. (2004). SOS response induction by beta-lactams and bacterial defense against antibiotic lethality. *Science* 305: 1629–1631.

Moon, K., Shoemaker, N.B., Gardner, J.F., and Salyers, A.A. (2005). Regulation of excision genes of the *Bacteroides* conjugative transposon, CTnDOT. *J. Bacteriol.* 186: 5732–5741.

Moonen, M.J.H., Fraaije, M.W., Rietjens, Y.M.C.M. et al. (2002). Flavoenzyme-catalyzed oxygenations and oxidations of phenolic compounds. *Adv. Synth. Catal.* 344: 1–13.

Moore, I.F., Hughes, D.W., and Wright, G.D. (2005). Tigecycline is modified by the flavin-dependent monooxygenase TetX. *Biochemistry* 44: 11829–111835.

Morfin-Otero, R., Noriega, E.R., and Dowzicky, M.J. (2015). Antimicrobial susceptibility trends among Gram-positive and -negative clinical isolates collected between 2005 and 2012 in Mexico: results from the Tigecycline Evaluation and Surveillance Trial. *Ann. Clin. Microbiol. Antimicrob.* 14: 53. https://doi.org/10.1186/s12941-015-0116-y.

Morris, R.P., Nguyen, L., Gatfield, J. et al. (2005). Ancestral antibiotic resistance in *Mycobacterium tuberculosis*. *Proc. Natl. Acad. Sci. U. S. A.* 102 (34): 12200–12205.

Nannini, E.C., Pai, S.R., Singh, K.V., and Murray, B.E. (2003). Activity of tigecycline (GAR-936), a novel glycylcycline, against enterococci in the mouse peritonitis model. *Antimicrob. Agents Chemother.* 47: 529–532.

Nelson, M.L., Dinardo, A., Hochberg, J., and Armelagos, G.J. (2010). Brief communication: mass spectroscopic characterization of tetracycline in the skeletal remains of an ancient population from Sudanese Nubia 350-550 CE. *Am. J. Phys. Anthropol.* 143: 151–154. https://doi.org/10.1002/ajpa.21340.

Nelson, M.L. and Levy, S.B. (2011). The history of the tetracyclines. *Ann. N. Y. Acad. Sci.* 1241: 17–32. https://doi.org/10.1111/j.1749-6632.2011.06354.x.

Nicolau, D.P. (2008). Carbapenems: a potent class of antibiotics. *Expert. Opin. Pharmacother.* 9: 23–37.

Nishida, H., Ohnishi, Y., Beppu, T., and Horinouchi, S. (2007). Evolution of gamma-butyrolactone synthases and receptors in *Streptomyces*. *Environ. Microbiol.* 9: 1986–1994.

Nodwell, J.R. (2014). Are you talking to me? A possible role for γ-butyrolactones in interspecies signalling. *Mol. Microbiol.* 94 (3): 483–485. https://doi.org/10.1111/mmi.12787.

Nonaka, L. and Suzuki, S. (2002). New Mg2+-dependent oxytetracycline resistance determinant tet 34 in *Vibrio* isolates from marine fish intestinal contents. *Antimicrob. Agents Chemother.* 46 (5): 1550–1552.

Nørskov-Lauritsen, N., Marchandin, H., and Dowzicky, M.J. (2009). Antimicrobial susceptibility of tigecycline and comparators against bacterial isolates collected as part of the TEST study in Europe (2004-2007). *Int. J. Antimicrob. Agents* 34 (2): 121–130. https://doi.org/10.1016/j.ijantimicag.2009.02.003.

Nosanchuk, J.D., Lin, J., Hunter, R.P., and Aminov, R.I. (2014). Low-dose antibiotics: current status and outlook for the future. *Front. Microbiol.* 5 (478): https://doi.org/10.3389/fmicb.2014.00478.

Novakova, R., Homerova, D., Feckova, L., and Kormanec, J. (2005). Characterization of a regulatory gene essential for the production of the angucycline-like polyketide antibiotic auricin in *Streptomyces aureofaciens* CCM 3239. *Microbiology* 151: 2693–2706.

Núñez, L., Méndez, C., Braña, A. et al. (2003). The biosynthetic gene cluster for the β-lactam carbapenem thienamycin in *Streptomyces cattleya*. *Chem. Biol.* 10: 301–311.

Ogawara, H. (1993). Phylogenetic tree and sequence similarity of beta-lactamases. *Mol. Phylogenet. Evol.* 2 (2): 97–111.

Oh, D.-C., Poulsen, M., Currie, C.R., and Clardy, J. (2009). Dentigerumycin: a bacterial mediator of an ant-fungus symbiosis. *Nat. Chem. Biol.* 5: 391–393. https://doi.org/10.1038/nchembio.159.

O'Neill, J. (Chair). 2014. The Review on Antimicrobial Resistance. https://amr-review.org/sites/default/files/AMR%20Review%20Paper%20-%20Tackling%20a%20crisis%20for%20the%20health%20and%20wealth%20of%20nations_1.pdf (accessed 23 April 2019).

Pang, Y., Brown, B.A., Steingrube, V.A. et al. (1994). Tetracycline resistance determinants in *Mycobacterium* and *Streptomyces* species. *Antimicrob. Agents Chemother.* 38: 1408–1412.

Parker, W.L., Rathnum, M.L., Wells, J.S. Jr. et al. (1982). SQ 27,860, a simple carbapenem produced by species of *Serratia* and *Erwinia*. *J. Antibiot. (Tokyo)* 35: 653–660.

Pehrsson, E.C., Tsukayama, P., Patel, S. et al. (2016). Interconnected microbiomes and resistomes in low-income human habitats. *Nature* 533 (7602): 212–216. https://doi.org/10.1038/nature17672.

Pei, R., Kim, S.C., Carlson, K.H., and Pruden, A. (2006). Effect of river landscape on the sediment concentrations of antibiotics and corresponding antibiotic resistance genes (ARG). *Water Res.* 40: 2427–2435.

Peleg, A.Y., Adams, J., and Paterson, D.L. (2007). Tigecycline efflux as a mechanism for nonsusceptibility in *Acinetobacter baumannii*. *Antimicrob. Agents Chemother.* 51: 2065–2069.

Petersen, E., Maeurer, M., Marais, B. et al. (2017). World TB day 2017: advances, challenges and opportunities in the "End-TB" era. *Int. J. Infect. Dis.* 56: 1–5. https://doi.org/10.1016/j.ijid.2017.02.012.

Petersen, P.J., Bradford, P.A., Weiss, W.J. et al. (2002). *In vitro* and *in vivo* activities of tigecycline (GAR-936), daptomycin, and comparative antimicrobial agents against glycopeptide-intermediate *Staphylococcus aureus* and other resistant Gram-positive pathogens. *Antimicrob. Agents Chemother.* 46: 2595–2601.

Petersen, P.J., Jacobus, N.V., Weiss, W.J. et al. (1999). *In vitro* and *in vivo* antibacterial activities of a novel glycylcycline, the 9-t-butylglycylamido derivative of minocycline (GAR-936). *Antimicrob. Agents Chemother.* 43: 738–744.

Piddock, L.J. (2006). Clinically relevant chromosomally encoded multidrug resistance efflux pumps in bacteria. *Clin. Microbiol. Rev.* 19 (2): 382–402.

Prado, L., Fernandez, E., Weissbach, U. et al. (1999). Oxidative cleavage of premithramycin B is one of the last steps in the biosynthesis of the antitumor drug mithramycin. *Chem. Biol.* 6: 19–30.

Raviglione, M. and Sulis, G. (2016). Tuberculosis 2015: burden, challenges and strategy for control and elimination. *Infect Dis Rep.* 8 (2): 6570. https://doi.org/10.4081/idr.2016.6570.

Redin, G.S. (1966). Antibacterial activity in mice of minocycline, a new tetracycline. *Antimicrob. Agents Chemother.* 6: 371–376.

Reen, F.J., Romano, S., Dobson, A.D., and O'Gara, F. (2015). The sound of silence: activating silent biosynthetic gene clusters in marine microorganisms. *Mar. Drugs* 13 (8): 4754–4783. https://doi.org/10.3390/md13084754.

Rigali, S., Titgemeyer, F., Barends, S. et al. (2008). Feast or famine: the global regulator DasR links nutrient stress to antibiotic production by *Streptomyces*. *EMBO Rep.* 9: 670–675.

Roberts, M.C. (1996). Tetracycline resistance determinants: mechanisms of action, regulation of expression, genetic mobility, and distribution. *FEMS Microbiol. Rev.* 19: 1–24.

Roberts, M.C. (2005). Update on acquired tetracycline resistance genes. *FEMS Microbiol. Lett.* 245: 195–203.

Rodríguez, M., Núñez, L.E., Braña, A.F. et al. (2008). Identification of transcriptional activators for thienamycin and cephamycin C biosynthetic genes within the thienamycin gene cluster from *Streptomyces cattleya*. *Mol. Microbiol.* 69: 633–645.

Rossi-Fedele, G., Scott, W., Spratt, D. et al. (2006). Incidence and behaviour of Tn916-like elements within tetracycline-resistant bacteria isolated from root canals. *Oral Microbiol. Immunol.* 21: 218–222.

Ruzin, A., Visalli, M.A., Keeney, D., and Bradford, P.A. (2005). Influence of transcriptional activator RamA on expression of multidrug efflux pump AcrAB and tigecycline susceptibility in *Klebsiella pneumoniae*. *Antimicrob. Agents Chemother.* 49: 1017–1022.

Ryll, M., Christensen, H., Bisgaard, M. et al. (2001). Studies on the prevalence of *Riemerella anatipestifer* in the upper respiratory tract of clinically healthy ducklings and characterization of untypable strains. *J. Vet. Med.* 48: 537–546.

Sader, H.S., Jones, R.N., Stilwell, M.G. et al. (2005). Tigecycline activity tested against 26,474 bloodstream infection isolates: a collection from 6 continents. *Diagn. Microbiol. Infect. Dis.* 52 (3): 181–186.

Sandhu, T.S. and Rimler, R.B. (1997). *Riemerella Anatipestifer* infection. In: *Diseases of Poultry*, 10e (ed. B.W. Calnek, H.J. Barnes, H.J. Beard, et al.), 161–166. Ames, IA: Iowa State University Press.

Sarver, C.F., Morishita, T.Y., and Nersessian, B. (2005). The effect of route of inoculation and challenge dosage on *Riemerella anatipestifer* infection in Pekin ducks (*Anas platyrhynchos*). *Avian Dis.* 49: 104–107.

Segers, P., Mannheim, W., Vancanneyt, M. et al. (1993). *Riemerella anatipestifer* gen. nov., comb. nov., the causative agent of septicemia anserum exsudativa, and its phylogenetic affiliation within the Flavobacterium-Cytophaga rRNA homology group. *Int. J. Syst. Bacteriol.* 43: 768–776.

Shiner, E.K., Rumbaugh, K.P., and Williams, S.C. (2005). Inter-kingdom signaling: deciphering the language of acyl homoserine lactones. *FEMS Microbiol. Rev.* 29: 935–947.

Showsh, S.A. and Andrews, R.E. Jr. (1992). Tetracycline enhances Tn*916*-mediated conjugal transfer. *Plasmid* 28: 213–224.

Singh, R., Swick, M.C., Ledesma, K.R. et al. (2012). Temporal interplay between efflux pumps and target mutations in development of antibiotic resistance in *Escherichia coli*. *Antimicrob. Agents Chemother.* 56 (4): 1680–1685. https://doi.org/10.1128/AAC.05693-11.

Skindersoe, M.E., Alhede, M., Phipps, R. et al. (2008). Effects of antibiotics on quorum sensing in *Pseudomonas aeruginosa*. *Antimicrob. Agents Chemother.* 52: 3648–3663.

Slater, H., Crow, M., Everson, L., and Salmond, G.P. (2003). Phosphate availability regulates biosynthesis of two antibiotics, prodigiosin and carbapenem, in *Serratia* via both quorum-sensing-dependent and -independent pathways. *Mol. Microbiol.* 47: 303–320.

Song, B., Wang, G.R., Shoemaker, N.B., and Salyers, A.A. (2009). An unexpected effect of tetracycline concentration: growth phase-associated excision of the Bacteroides mobilizable transposon NBU1. *J Bacteriol.* 191: 1078–1082.

Song, J.S., Jeon, J.H., Lee, J.H. et al. (2005). Molecular characterization of TEM-type beta-lactamases identified in cold-seep sediments of Edison Seamount (south of Lihir Island, Papua New Guinea). *J. Microbiol.* 43: 172–178.

Song, X.-H., Zhou, W.-S., Wang, J.-B. et al. (2016). Genome sequence of *Riemerella anatipestifer* strain RCAD0122, a multidrug-resistant isolate from ducks. *Genome Announc.* 4 (3): e00332-16. https://doi.org/10.1128/genomeA.00332-16.

Speer, B.S., Bedzyk, L., and Salyers, A.A. (1991). Evidence that a novel tetracycline resistance gene found on two *Bacteroides* transposons encodes an NADP-requiring oxidoreductase. *J. Bacteriol.* 173 (1): 176–183.

Stefani, S. and Dowzicky, M.J. (2016). Assessment of the activity of tigecycline against Gram-positive and Gram-negative organisms collected from Italy between 2012 and 2014, as part of the Tigecycline Evaluation and Surveillance Trial (T.E.S.T.). *Pharmaceuticals (Basel)* 9 (4): pii: E74.

Stevens, A.M., Shoemaker, N.B., Li, L.Y., and Salyers, A.A. (1993). Tetracycline regulation of genes on *Bacteroides* conjugative transposons. *J. Bacteriol.* 175: 6134–6141.

Takano, E. (2006). Gamma-butyrolactones: streptomyces signalling molecules regulating antibiotic production and differentiation. *Curr. Opin. Microbiol.* 9: 287–294.

Tärnberg, M., Nilsson, L.E., and Dowzicky, M.J. (2016). Antimicrobial activity against a global collection of skin and skin structure pathogens: results from the Tigecycline Evaluation and Surveillance Trial (T.E.S.T.), 2010-2014. *Int. J. Infect. Dis.* 49: 141–148. https://doi.org/10.1016/j.ijid.2016.06.016.

Tateda, K., Standiford, T.J., Pechere, J.C., and Yamaguchi, K. (2004). Regulatory effects of macrolides on bacterial virulence: potential role as quorum-sensing inhibitors. *Curr. Pharm. Des.* 10 (25): 3055–3065.

Thomson, N.R., Crow, M.A., McGowan, S.J. et al. (2000). Biosynthesis of carbapenem antibiotic and prodigiosin pigment in *Serratia* is under quorum sensing control. *Mol. Microbiol.* 36: 539–556.

Tollefson, L. and Miller, M.A. (2000). Antibiotic use in food animals: controlling the human health impact. *J. AOAC Int.* 83 (2): 245–254.

Torres, O.R., Korman, R.Z., Zahler, S.A., and Dunny, G.M. (1991). The conjugative transposon Tn925: enhancement of conjugal transfer by tetracycline in *Enterococcus faecalis* and mobilization of chromosomal genes in *Bacillus subtilis* and *E. faecalis*. *Mol. Gen. Genet.* 225: 395–400.

Tsui, W.H., Yim, G., Wang, H.H. et al. (2004). Dual effects of MLS antibiotics: transcriptional modulation and interactions on the ribosome. *Chem. Biol.* 11: 1307–1316.

Tuckman, M., Petersen, P.J., and Projan, S.J. (2000). Mutations in the interdomain loop region of the tetA(A) tetracycline resistance gene increase efflux of minocycline and glycylcyclines. *Microbiol. Drug Res.* 6: 277–282.

Ubeda, C., Maiques, E., Knecht, E. et al. (2005). Antibiotic-induced SOS response promotes horizontal dissemination of pathogenicity island-encoded virulence factors in staphylococci. *Mol. Microbiol.* 56: 836–844.

Udwadia, Z.F., Amale, R.A., Ajbani, K.K., and Rodrigues, C. (2012). Totally drug-resistant tuberculosis in India. *Clin. Infect. Dis.* 54 (4): 579–581. https://doi.org/10.1093/cid/cir889.

UN (2016). United Nations: PRESS RELEASE: High-Level Meeting on Antimicrobial Resistance, 2016. http://www.un.org/pga/71/2016/09/21/press-release-hl-meeting-on-antimicrobial-resistance (accessed 23 April 2019).

Valentine, P.J., Shoemaker, N.B., and Salyers, A.A. (1988). Mobilization of *Bacteroides* plasmids by *Bacteroides* conjugal elements. *J. Bacteriol.* 170: 1319–1324.

van Berkel, W.J., Kamerbeek, N.M., and Fraaije, M.W. (2006). Flavoprotein monooxygenases, a diverse class of oxidative biocatalysts. *J. Biotechnol.* 124: 670–689.

Van Boeckel, T.P., Brower, C., Gilbert, M. et al. (2015). Global trends in antimicrobial use in food animals. *Proc. Natl. Acad. Sci. U. S. A.* 112 (18): 5649–5654. https://doi.org/10.1073/pnas.1503141112.

Velayati, A.A., Masjedi, M.R., Farnia, P. et al. (2009). Emergence of new forms of totally drug-resistant tuberculosis bacilli: super extensively drug-resistant tuberculosis or totally drug-resistant strains in Iran. *Chest* 136 (2): 420–425. https://doi.org/10.1378/chest.08-2427.

Visalli, M.A., Murphy, E., Projan, S.J., and Bradford, P.A. (2003). AcrAB multidrug efflux pump is associated with reduced levels of susceptibility to tigecycline (GAR-936) in *Proteus mirabilis*. *Antimicrob. Agents Chemother.* 47: 665–669.

Wallace, R.J. Jr., Brown-Elliott, B.A., Crist, C.J. et al. (2002). Comparison of the *in vitro* activity of the glycylcycline tigecycline (formerly GAR-936) with those of tetracycline, minocycline, and doxycycline against isolates of nontuberculous mycobacteria. *Antimicrob. Agents Chemother.* 46: 3164–3167.

Walraven, N. and Laane, R.W. (2009). Assessing the discharge of pharmaceuticals along the Dutch coast of the North Sea. *Rev. Environ. Contam. Toxicol.* 199: 1–18.

Wang, X., Ding, C., Han, X. et al. (2015a). Complete genome sequence of *Riemerella anatipestifer* serotype 1 strain CH3. *Genome Announc.* 3 (1): pii: e01594-14. https://doi.org/10.1128/genomeA.01594-14.

Wang, X., Ding, C., Wang, S. et al. (2015b). Whole-genome sequence analysis and genome-wide virulence gene identification of *Riemerella anatipestifer* strain Yb2. *Appl. Environ. Microbiol.* 81 (15): 5093–5102. https://doi.org/10.1128/AEM.00828-15.

Wang, X., Zhu, D., Wang, M. et al. (2012). Complete genome sequence of *Riemerella anatipestifer* reference strain. *J. Bacteriol.* 194: 3270–3271.

Wang, Y.J. and Leadbetter, J.R. (2005). Rapid acyl-homoserine lactone quorum signal biodegradation in diverse soils. *Appl. Environ. Microbiol.* 71: 1291–1299.

Watkinson, A.J., Murby, E.J., Kolpin, D.W., and Costanzo, S.D. (2009). The occurrence of antibiotics in an urban watershed: from wastewater to drinking water. *Sci. Total Environ.* 407 (8): 2711–2723. https://doi.org/10.1016/j.scitotenv.2008.11.059.

Weber, T., Welzel, K., Pelzer, S. et al. (2003). Exploiting the genetic potential of polyketide producing streptomycetes. *J. Biotechnol.* 106: 221–232.

Wentland, M.P. (1993). In memoriam: George Y. Lesher, Ph.D. In: *Quinolone Antimicrobial Agents*, 2e (ed. D.C. Hooper and J.S. Wolfson), XIII–XIV. Washington, DC: American Society for Microbiology.

Whittle, G., Hund, B.D., Shoemaker, N.B., and Salyers, A.A. (2001). Characterization of the 13-kilobase ermF region of the Bacteroides conjugative transposon CTnDOT. *Appl. Environ. Microbiol.* 67: 3488–3495.

Whittle, G., Shoemaker, N.B., and Salyers, A.A. (2002). Identification of two genes involved in the modulation of conjugal transfer of the *Bacteroides* conjugative transposon CTnDOT. *J. Bacteriol.* 184: 3839–3847.

WHO (2015). Global Action Plan on Antimicrobial Resistance. http://www.wpro.who.int/entity/drug_resistance/resources/global_action_plan_eng.pdf (accessed 23 April 2019).

Williamson, J.M., Inamine, E., Wilson, K.E. et al. (1985). Biosynthesis of the beta-lactam antibiotic, thienamycin, by *Streptomyces cattleya*. *J. Biol. Chem.* 260: 4637–4647.

Woappi, Y., Gabani, P., and Singh, O.V. (2013). Emergence of antibiotic-producing microorganisms in residential versus recreational microenvironments. *Br. Microbiol. Res. J.* 3 (3): 280–294.

Wright, G.D. (2012). The origins of antibiotic resistance. *Handb. Exp. Pharmacol.* 211: 13–30. https://doi.org/10.1007/978-3-642-28951-4_2.

Wright, L.T. and Schreiber, H. (1949). The clinical value of Aureomycin: a review of current literature and some unpublished data. *J. Natl. Med. Assoc.* 41: 195–201.

Xu, G., Wang, J., Wang, L. et al. (2010). "Pseudo" gamma-butyrolactone receptors respond to antibiotic signals to coordinate antibiotic biosynthesis. *J. Biol. Chem.* 285 (35): 27440–27448. https://doi.org/10.1074/jbc.M110.143081.

Xu, G.M. (2016). Relationships between the regulatory systems of quorum sensing and multidrug resistance. *Front. Microbiol.* 7: 958. https://doi.org/10.3389/fmicb.2016.00958.

Yang, S. and Carlson, K. (2003). Evolution of antibiotic occurrence in a river through pristine, urban and agricultural landscapes. *Water Res.* 37: 4645–4656.

Yim, G., Wang, H.H., and Davies, J. (2006). The truth about antibiotics. *Int. J. Med. Microbiol.* 296: 163–170.

Yuan, J., Li, L., Sun, M. et al. (2013). Genome sequence of avirulent *Riemerella anatipestifer* strain RA-SG. *Genome Announc.* 1 (2): e0021812.

Yuan, J., Liu, W., Sun, M. et al. (2011). Complete genome sequence of the pathogenic bacterium *Riemerella anatipestifer* strain RA-GD. *J. Bacteriol.* 193: 2896–2897.

Zhang, X.X. and Zhang, T. (2011). Occurrence, abundance, and diversity of tetracycline resistance genes in 15 sewage treatment plants across China and other global locations. *Environ. Sci. Technol.* 45: 2598–2604.

Zhou, Z., Peng, X., Xiao, Y. et al. (2011). Genome sequence of poultry pathogen *Riemerella anatipestifer* strain RA-YM. *J. Bacteriol.* 193: 1284–1285.

Part VI

Public Policy

24

Strategies to Reduce or Eliminate Resistant Pathogens in the Environment

Johan Bengtsson-Palme[1,2,3] and Stefanie Heß[4]

[1] *Department of Infectious Diseases, Institute of Biomedicine, The Sahlgrenska Academy, University of Gothenburg, Gothenburg, Sweden*
[2] *Centre for Antibiotic Resistance Research (CARe), University of Gothenburg, Gothenburg, Sweden*
[3] *Wisconsin Institute of Discovery, University of Wisconsin-Madison, Madison, WI, USA*
[4] *Department of Microbiology, University of Helsinki, Helsinki, Finland*

24.1 Introduction

As antibiotic resistance has increasingly limited treatment options for bacterial infections over the last decades, the need to understand how this development takes place in order to curb it has been gradually recognized. As clinical settings are those where antibiotic-resistant bacteria cause the most severe and obvious problems, this was naturally the first area to be investigated for development and transmission of resistant bacteria. More recently, it has become increasingly clear that to understand what drives the rapid antibiotic resistance development, we have to look beyond clinical and societal situations and start contemplating how external environments, including agriculture and aquaculture, have fueled the rise of resistant pathogens in the past and how they might contribute to future resistance trends (Allen et al. 2010; Wright 2010; Finley et al. 2013; Bengtsson-Palme et al. 2018b). Recent research has started to shed light on the different roles that the environment plays in resistance propagation (Larsson et al. 2018). There are three central processes in which the environment is an important actor: (i) in dissemination of resistant bacteria between humans (and potentially animal hosts), (ii) as a reservoir or intermediate habitat of resistant bacteria and resistance genes, and (iii) in the evolution of new resistance factors – both as venue for selection for resistance genes and as a vast source of genetic diversity from which novel resistance factors can be captured (Bengtsson-Palme et al. (2018b).

Antibiotic Drug Resistance, First Edition. Edited by José-Luis Capelo-Martínez and Gilberto Igrejas.

It would be desirable to control these processes and eliminate resistant bacteria from the environment in order to reduce risks to human health. However, due to the multifaceted types of processes in which the environment is involved, achieving this goal is virtually impossible. Yet, as will be described in this chapter, several options exist to reduce risks, with different degrees of feasibility (Ashbolt et al. 2013; Pruden et al. 2013; Bengtsson-Palme 2019). It should also be pointed out that antibiotic resistance is a natural phenomenon occurring in most environmental settings (D'Costa et al. 2006, 2011; Allen et al. 2010; Van Goethem et al. 2018). Indeed, many of the clinically important resistance genes are thought to have originated in environmental bacteria, followed by subsequent transfer to pathogens in the face of anthropogenic antibiotic selection pressures (Davies and Davies 2010; Forsberg et al. 2012; Wright 2012). The ability for bacteria to exchange resistance genes with each other through horizontal gene transfer introduces a fundamental conundrum for management: resistance genes may be transferred from environmental bacteria to human pathogens over and over again, essentially erasing progress made toward eradicating resistant pathogens. This limits our mitigation options and effectively leaves us with two options to minimize risks to human health: to prevent the release of resistant pathogens and antibiotic substances into the environment and to avoid proliferation and transmission of resistant bacteria originating from environmental settings to humans.

24.2 Sources of Resistant Bacteria in the Environment

As humans have colonized virtually every corner of the world, resistant pathogens can enter the environment in a myriad of ways. However, most of those pathways would likely result in very small numbers of resistant bacteria being released. For example, humans are likely to shed small amounts of potentially pathogenic bacteria when doing recreational activities such as swimming. It would be impossible to prevent all these smaller environmental releases of human-associated bacteria, and thus strategies to reduce the occurrence of resistant pathogens into the environment need to consider where the released quantities would be relatively high. There are comparatively few settings in which the environment is exposed to pathogenic bacteria at a large scale (Harwood et al. 2014). The most obvious such exposure scenario is through sewage, either by direct releases of untreated sewage or indirectly through treated wastewater from sewage treatment plants. The sludge from those treatment plants is sometimes applied to agricultural fields as fertilizer, providing another potential exposure route. Along the same lines, waste disposal sites, such as landfills, provide another route into the environment for human pathogens (Pruden et al. 2013).

Pathogens may not only disperse into the environment from humans themselves, however. In many cases, they can be opportunistic and have the environment or other animals as their main habitat, even though they have the potential to infect, for example, immunocompromised humans. Selection for resistance in such bacteria may occur in agri- or aquacultural settings, and the resistant opportunistic pathogens could subsequently make it to humans, either directly, through the food production chain or via the environment. Such transmission of opportunistic pathogens carrying resistance genes has the potential to play an underappreciated role in resistance transmission from the environment to humans, but this is an area we currently know very little about (Bengtsson-Palme et al. 2018b).

Finally, resistant pathogens may also arise in the environment. For example, under extensive antibiotic selection pressure resulting from pollution associated with pharmaceutical production, opportunistic pathogens may attain novel resistance factors from environmental bacteria through horizontal gene transfer – or could acquire resistance mutations – rendering them resistant to antibiotics (Johnning et al. 2013; Marathe et al. 2013). These *de novo* selected resistant bacteria, which may be opportunistic human pathogens, can subsequently spread to the human population from the environment. If their resistance was acquired through gene transfer, these genes can then be shared with other human-associated bacteria, including pathogens, in the human microbiome (Porse et al. 2017). While the risks for *de novo* resistance selection are greatest when antibiotics levels are above the minimum inhibitory concentrations (MICs), there is a considerable range of sublethal concentrations that also select for resistance enrichment (Gullberg et al. 2011; Bengtsson-Palme and Larsson 2016; Lundström et al. 2016). Whether these concentrations are also high enough to drive resistance emergence and induce gene transfer is yet unknown, but there are indications for that the latter happens at least in laboratory settings (Jutkina et al. 2016). This could be relevant in sewage treatment plants where exposure to levels of antibiotics below the MIC is often chronic (Östman et al. 2017).

24.3 Sewage and Wastewater

24.3.1 Sewage Treatment Plants

Sewage treatment plants have long been recognized as potential "hot spots" for antibiotic resistance dissemination (Michael et al. 2013; Gao et al. 2015; Guo et al. 2017). To a large extent, they have received this great deal of interest because they constitute a unique setting where human-associated bacteria are mixed with environmental bacteria in conditions where a cocktail of different chemicals – including antibiotics – is present, with plentiful of nutrients and with rich opportunities for the close contacts between bacteria that are required for horizontal gene transfer. In addition, sewage treatment processes have not been designed specifically with eliminating antibiotic resistance in mind, although one aim of sewage treatment is generally to reduce the numbers of potentially pathogenic bacteria in the effluent. This goal naturally also applies to the resistant fraction of the same bacteria, effectively keeping down the number of resistant bacteria released, as long as treatment is effective. Overall, modern sewage treatment processes are very effective in reducing the number of bacteria (resistant as well as non-resistant) from influent to effluent, reaching around 97–99.5% reduction efficiency (Bengtsson-Palme et al. 2016; Flach et al. 2018). At the same time, the numbers of released bacteria are far from zero, indicating that this may still be a relevant route for resistant pathogens to reach the environment.

It is also notable that there is an ongoing discussion on the risks to human health of using municipal sewage sludge as a fertilizer in agriculture. While many factors are important in this context, including heavy metal content and the presence of persistent pollutants, the transfer of resistant bacteria via sludge has also been debated as a possibility. A key issue here is to ensure that the sludge is properly disinfected so as not to spread live resistant bacteria on agricultural fields. However, most countries that have adopted the use of sludge for fertilization also have policies in place to ensure disinfection, often by heat or application of hydrochloric acid (Vinnerås 2013). Therefore, the probability of spreading live pathogens (resistant or not) through this practice is rather slim. That said, the disinfection treatment will not remove DNA from the sludge, meaning that resistance genes may be spread onto agricultural soils as extracellular DNA. This means that those genes may be taken up by competent bacteria in the soil, which may then gain resistance to various antibiotics. This scenario becomes more likely if antibiotic residues are also present in the applied sludge, increasing the selective pressure for resistance acquisition. To what extent this scenario takes place in the environment is this far unknown.

Another (similar) concern regards the reuse of treated wastewater for irrigation, which is common in countries with low freshwater availability. This practice has been shown to spread resistance bacteria to farmland and irrigated plants, although the degree to which this happens differs between studies (Gatica and Cytryn 2013; Al-Jassim et al. 2015; Luprano et al. 2016; Christou et al. 2017; Pruden et al. 2018). The levels of resistant bacteria in treated wastewater reused for irrigation can be in the range of 10^5 and 10^6 per milliliter (Negreanu et al. 2012). A further concern in this context is that application of treated wastewater to crops may provide a direct transmission route back to the human population through the food supply chain (Bengtsson-Palme 2017). Although the risk that this would have consequences for human health is minute in each individual case, the sheer scale of industrialized agriculture makes this a fairly likely scenario to occur somewhere in the world.

24.3.2 Non-Treated Sewage

Although much attention has been paid to sewage treatment plants (Rizzo et al. 2013; Lüddeke et al. 2015; Zhang et al. 2015; Bengtsson-Palme et al. 2016; Hess et al. 2016), the abundance of resistant bacteria (as well as bacteria in general) is orders of magnitude lower in the treated effluent than in raw sewage (Bengtsson-Palme et al. 2016; Karkman et al. 2017; Flach et al. 2018). Large proportions of the sewage produced undergo treatment, but even in the most developed countries, small fractions are directly released into the environment, usually into small water recipients. The impact of such individual point sources of raw sewage is extremely hard to quantify. At the same time, in many less developed countries, the majority of sewage is released into the environment untreated. As raw sewage typically contains 10^6–10^7 cultivable bacteria per milliliter (Flach et al. 2018), these untreated releases are a growing pollution problem locally, but also has potentially global consequences in terms of easing the propagation of antibiotic-resistant bacteria (Graham et al. 2014). In addition, raw sewage does enable not only the spread of resistant bacteria between humans but also of pathogens in general.

It is also worthwhile to consider that many countries with poor sewage treatment infrastructure are located in regions with higher year-round temperatures. Since temperature is a key factor limiting bacterial growth, particularly for human-associated bacteria, as well as horizontal gene transfer (Ehlers and Bouwer 1999), higher temperatures mean more favorable conditions for human pathogens to survive and proliferate outside of the human body. Thus, proper sewage treatment may be even more urgent in warmer countries. Another aggravating factor is that countries with poor sewage treatment also generally have worse infrastructure for treatment of drinking water (WHO/UNICEF 2017). Therefore, three unfortunate factors for resistance development often come together, increasing risks for resistance transmission locally. That local effect may have global consequences due to the relatively quick spread of resistant bacteria around the world due to travel and trade (European Food Safety Authority & European Centre for Disease Prevention and Control 2013; Molton et al. 2013; Rolain 2013; Wintersdorff et al. 2014; Angelin et al. 2015; Bengtsson-Palme et al. 2015).

24.3.3 Industrial Wastewater Effluents

Besides the concerns regarding municipal sewage effluents, wastewater releases from facilities treating industrial wastewater (or industrial waste and wastewater being directly dumped or released into the environment) are also a potential source of both resistant bacteria and antibiotics, as well as other selective agents. Particularly, facilities treating wastewater from antibiotic production have repeatedly been shown to release large amounts of antibiotics into their recipients (Larsson et al. 2007; Li et al. 2008; Larsson 2014; Gothwal and Thatikonda 2017). In addition, these treatment plants seem to be breeding grounds for resistant bacteria that subsequently are released into the environment (Johnning et al. 2013; Marathe et al. 2013). The antibiotic residues released from production do in turn affect local bacterial communities and select for high levels of antibiotic resistance (Li et al. 2008; Kristiansson et al. 2011; Bengtsson-Palme et al. 2014b; González-Plaza et al. 2017). Much data also points toward that such conditions select for a general enrichment of mobile genetic elements that can be shared between bacteria, and thereby enrich resistance genes against a whole range of antibiotics, and consequently also render bacteria generally more multiresistant (Bengtsson-Palme et al. 2014b; Pal et al. 2016; Bengtsson-Palme 2018). Even if this selection does not happen directly in human pathogens, opportunistic pathogens with extensive resistance phenotypes have been isolated from such environments (Johnning et al. 2013; Marathe et al. 2013). Given the enrichment of genes responsible for moving genes between bacteria in the same environments, it is not far-fetched to imagine that such resistance traits may be transferred to human pathogens after dispersal back to a human host (Bengtsson-Palme et al. 2018b).

24.3.4 Environmental Antibiotic Resistance is a Poverty Problem

Worldwide, 2.4 billion people still do not have access to improved sanitation, and nearly one billion people still practice open defecation (United Nations World Water Assessment Programme [WWAP] 2017) (Figure 24.1). This results in a major health risk for the local population if human pathogenic microorganisms enter the environment and potentially acquire antibiotic resistance genes from environmental bacteria via horizontal gene transfer. In addition, resistance genes end up in the environment and might be incorporated by and spread within the allochthonous microbial community and, under optimal conditions, later be taken up by human pathogenic species. The permanent and direct contact of people with untreated wastewater additionally increases the risk of infection with antibiotic-resistant human pathogens.

If one projects the locations of wastewater treatment plants and the respective applied treatment technologies onto a world map, the absence of a well-functioning wastewater treatment system overlays with low income, i.e. environmental antibiotic resistance is largely a poverty problem (Figure 24.1). Whereas in high-income countries about 70% of the municipal and industrial wastewater is currently treated, only 8% of the wastewater generated in developing countries undergoes treatment in any way. Globally, 80% of the total wastewater is directly discharged into receiving water bodies. Thinking further, it becomes clear that the lack of a functioning sanitary system in poor countries has effects on the global scale. It is possible to fly around the world within less than two days, and due to globalization, air traffic becomes increasingly important. As a result,

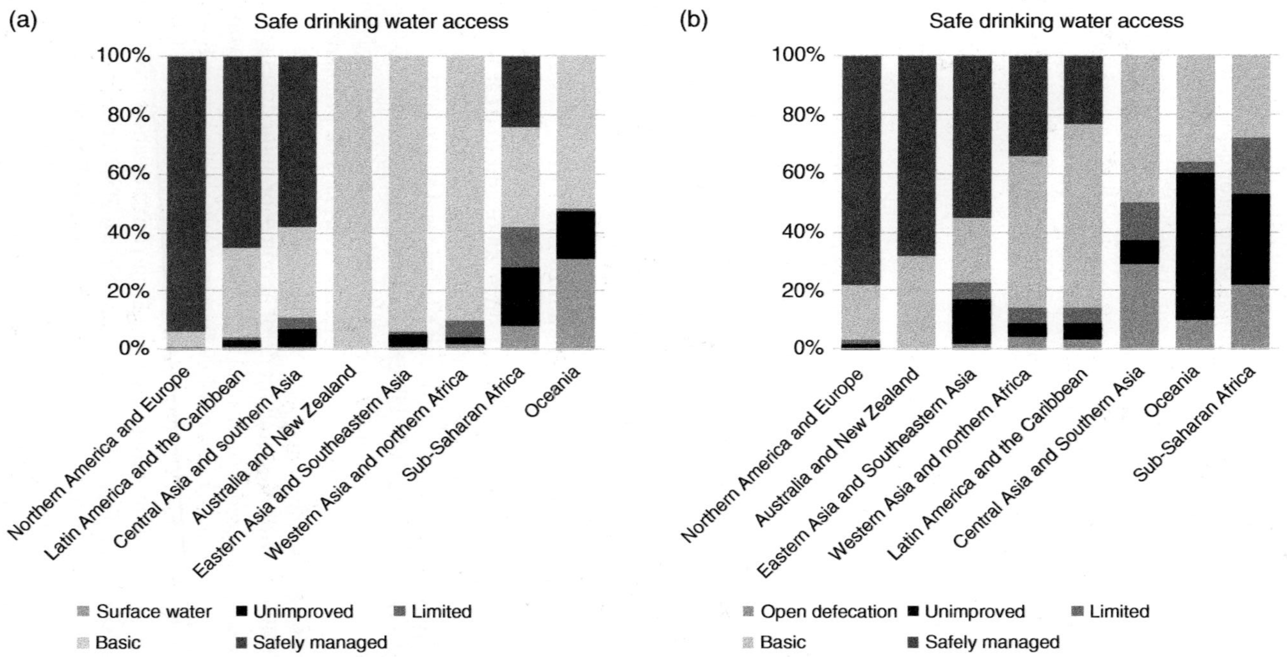

Figure 24.1 Access to safe drinking water (a) and sanitation services (b) differs substantially between different regions of the world, with northern America and Europe generally having access to the best and sub-Saharan Africa and Oceania the poorest infrastructure. *Source:* Data from WHO/UNICEF (2017).

antibiotic resistance genes and clinically relevant pathogens are distributed worldwide within very short periods of time (Bengtsson-Palme et al. 2015).

Efforts are made to build and maintain low-cost, low-energy, low-maintenance, decentralized wastewater treatment systems in developing countries. Ponds and retention soil filters/constructed wetlands seem promising and could be suitable for this purpose (Kivaisi 2001). The retention of *Escherichia coli* by passing a retention soil filter can even be comparable with that of a wastewater treatment plant equipped with a tertiary treatment step (mean retention of 2.7 log units; Scheurer et al. 2015). However, it has to be taken into account that many of these studies were carried out in the temperate zone and that the transferability of the results due to different climatic and edaphic conditions has to be proven.

24.4 Agriculture

24.4.1 Intensive, Large-Scale Animal Husbandry

In supermarkets, meat and other animal products such as eggs and milk are sold at very low prices. In order to be able to produce these products profitably, farmers breed many animals on very little space (Welfare of Livestock Regulations 1994: not less than 450 cm^2 for each laying hen where four or more hens are kept in the cage; each calve of 150 kg or more live weight must have at least 1.5 m^2; pigs with more than 110 kg live weight must have at least 1 m^2). Keeping animals close together increases the risk of disease spreading in case of infection of individual animals. To minimize this threat, antibiotics are applied as food additives. In addition, drugs are fed to promote growth (officially banned in the European Union [EU] since 2006) and to treat infectious diseases. For these purposes, mainly tetracyclines (32%), penicillins (26%), and sulfonamides are currently used (European Medicines Agency 2018). In 2014, an average of 152 mg antibiotics per kilogram biomass was consumed European in animal husbandry. In human medicine, the consumption was slightly lower with 124 $mg\,kg^{-1}$ biomass (ECDC, EFSA, and EMA 2017). The problem in this context is not only the high consumption, but that in veterinary and human medicine, antibiotics with the same mode of action – or sometimes even the same substances – are used. For example, in Australia and the EU, avoparcin has been used for years, mainly in poultry farming. Avoparcin acts in the same way as vancomycin. In human medicine, vancomycin is considered as last-line antibiotic to treat infections caused by multidrug-resistant enterococci. Since avoparcin was banned in the EU in 1997, the prevalence of vancomycin-resistant enterococci (VRE) has dropped significantly (e.g. van den Bogaard et al. 2000; Aarestrup et al. 2001). For instance, in 1997 in the Netherlands, 80% of the *Enterococcus* isolates obtained from chickens were resistant to vancomycin

(VRE). Only two years later, the percentage of VRE dropped about 50% (van den Bogaard et al. 2000).

24.4.2 Manure Application

Potential pathways for resistance dissemination do not only exist within animal husbandry, but resistant bacteria can also be spread onto crop fields via the application of manure. Typically, this fertilizer is of animal origin, but as discussed above, sewage sludge is also used as manure to some degree, although regulations around this practice differ vastly between countries (Kelessidis and Stasinakis 2012). There is a large body of literature showing that manure applied to fields contains antibiotic-resistant bacteria and resistance genes (Heuer et al. 2011; Munir and Xagoraraki 2011; Wichmann et al. 2014; Ruuskanen et al. 2016). In addition, it has been repeatedly shown that these bacteria persist in soils for extended amounts of time (Rahube et al. 2014; Riber et al. 2014; Chen et al. 2016) and that the application of manure also increases the frequency of horizontal gene transfer in soils (Jechalke et al. 2013; Ross and Topp 2015). It has even been suggested that application of manure increases the resistance frequencies of the indigenous soil bacteria (Udikovic-Kolic et al. 2014).

24.4.3 Agriculture in Developing Countries

Agriculture practices differ substantially in industrialized countries, but the span of different ways to do agriculture may be even greater in the developing parts of the world. In addition, there are very different farm sizes, ranging from essentially single household supplies to industrialized plantations managed using large numbers of generally low-paid workers. This also leads to very different scenarios for chemical use. Smaller farms may not have the resources to use virtually any chemicals, including antibiotics, for their crops or animals. At the same time, there are often limited regulations (or limited enforcement of regulations), prompting wealthier farmers to use antibiotics in excessive quantities. These differences lead to a situation where much of farmed animals and land are not exposed to any significant quantities of antibiotics (or other co-selective agents), but with specific local settings that can be subjected to exposure with excessive quantities of antimicrobial compounds. Further complicating this picture is that some antibiotics, including sulfonamides, are widely and cheaply available in many low- and middle-income countries, which has led to widespread and largely uncontrolled used in both humans and animals (Pruden et al. 2013). It is likely that this has contributed to the wide-reaching incidence of resistance genes toward these antibiotics in humans and animals, both locally and globally (He et al. 2014; Pal et al. 2016). In addition, the infrastructure for controlling and monitoring farming practices is much poorer in most developing countries, restricting the insight we have into both antibiotic use and disease outbreaks. Together, this creates a situation with a high uncertainty compared to industrialized countries, and the more complicated governmental structure also often prevents effective implementation of any mitigation strategies.

24.4.4 Aquaculture

Another food production sector where extensive amounts of antibiotics are used is in aquaculture (Cabello 2006). Infectious diseases cause losses of fish stock, and as aquaculture is globally increasing, so are the risks for diseases (Bostock et al. 2010; Cabello et al. 2016). Most antibiotic use in aquaculture is motivated by preventing stock losses, and in many parts of the world, the use of antimicrobials for these purposes is largely unregulated (FAO/OIE/WHO 2006). There are no reliable estimates of the total use of antibiotics in production of fish and other seafood, but it is likely to be counted in hundreds of thousands or even millions of tons annually (Done et al. 2015). Typically, the antibiotic is applied together with the feed, providing both a selection pressure for resistance and a more nutrient-rich environment at the same time. A wide array of antibiotics, many of which are used in human medicine, are used in aquaculture, including quinolones, sulfonamides, tetracyclines, and beta-lactams. This provides a selective environment for bacteria resistant to clinically relevant antibiotics, and there are indications of resistance being transferred between fish pathogens and human-associated bacteria (Cabello 2006; Ryu et al. 2012). Further complicating the problem is that integrated animal–fish–vegetable farming with antibiotic use is fairly common in Southeast Asia, causing direct antibiotic exposure and potentially selection for resistant bacteria (Suzuki and Hoa 2012).

Since fish farms are typically in contact with surrounding water bodies, both antibiotic residues and resistant bacteria can easily migrate out of the confined areas. In addition, the gut microbiome in the fishes will over time acquire higher degrees of resistance to the used antibiotics, as will the fish pathogens that the antibiotics are supposed to control. Many fish pathogens have relatively close phylogenetic relationships with human pathogens, making it fairly easy for them to transfer their resistance factors to human-associated bacteria. Indeed, a significant proportion of the clinically relevant mobile resistance genes circulating among pathogens today are likely to have originated from fish-associated bacteria (Rhodes et al. 2000; Lupo et al. 2012). At the same time, there seem to be relatively minor effects on the bacterial communities in sediments associated with intensive aquaculture with antibiotic use (Han et al. 2018).

24.5 *De Novo* Resistance Selection

While the vast majority of clinically relevant resistant pathogens would make their way into the environment through the paths outlined above, it is important to also recognize the role of the environment in *de novo* selection of resistant bacteria. Bacteria with antibiotic resistance mutations and other resistance determinants are likely to appear continuously in the environment, but generally cause little harm as they appear in non-human-associated species (Bengtsson-Palme et al. 2018b). Typically, there is also no selection pressure for maintaining such resistance mutations and genes, and they will therefore most often be lost from environmental populations due to the fitness costs associated with them, even if this fitness cost is exceedingly small (Andersson and Hughes 2010). That said, there are environments where antibiotics seem to be used by fungi to compete against bacteria (Bahram et al. 2018), and even if the concentrations of antibiotics likely are low in such environments, there is plenty of evidence that subinhibitory concentrations of antibiotics do select for resistance (Gullberg et al. 2011, 2014; Lundström et al. 2016; Kraupner et al. 2018). Thus, there might be a natural selection for antibiotic-resistant strains in certain environments, while in more hostile settings, where nutrient availability is lower, there is probably a long-term selection against such variants, due to the higher fitness costs of carrying them or general genome streamlining (Giovannoni et al. 2005; Bengtsson-Palme et al. 2014a, 2018b).

Furthermore, there are environments where selection for resistance is immense due to anthropogenic input of antibiotic substances. The most glaring such example is in effluents from pharmaceutical production, which have been shown to expose the receiving environments to high concentrations of antibiotics in many different parts of the world (Larsson et al. 2007; Li et al. 2008; Sim et al. 2011; Larsson 2014; Bielen et al. 2017; Gothwal and Thatikonda 2017). At these concentrations, bacteria that are not extensively resistant will be unable to survive, providing a very strong selection pressure favoring resistant strains. Since these environments have also been shown to harbor high levels of the genes enabling sharing of genetic material between bacteria (Bengtsson-Palme et al. 2014b), many of these strains are likely to be resistant to multiple antibiotics.

24.6 Relevant Risk Scenarios

The risk picture associated with antibiotic resistance in the environment is fairly complicated and multifaceted, making it possible to rank risks differently depending on what perspective one takes. It is also important to realize that while antibiotics and resistant bacteria are often released together (for example,

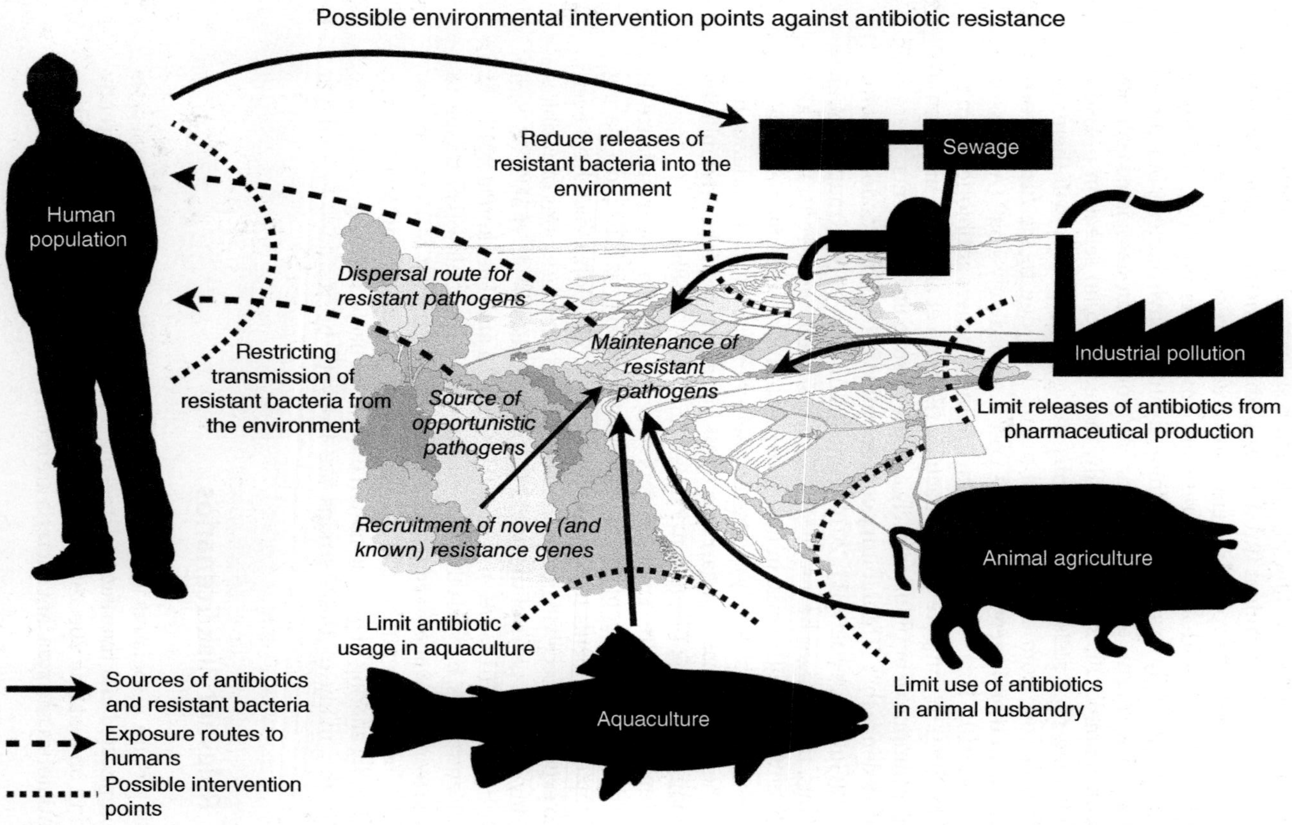

Figure 24.2 Overview of where in the environment it would be possible to intervene in the processes leading to development of antibiotic resistance in human pathogens. The outlines represent important sources of antibiotics and resistant bacteria into the environment. Solid arrows correspond to sources of antibiotics and dissemination routes for resistant bacteria, while dashed arrows represent routes through which humans are exposed to resistant bacteria from the environment. The dotted lines show possible points of intervention (bold labels).

in sewage and from agriculture), the relative levels of the two are different and the risks associated with these releases reflect these differences (Figure 24.2).

It is fairly clear that the main driver of antibiotic resistance in human pathogens is human consumption of antibiotics, both in the healthcare sector and among the general population (Chatterjee et al. 2018). At the same time, human use of antibiotics is – with some exceptions – unlikely to be directly responsible for most of the environmental exposure to antibiotics. This is because the quantities of antibiotics excreted from the human body are often fairly small and are further diluted in sewage water. Sewage treatment plants then generally reduce antibiotic levels several orders of magnitude before the substances are being released into surface waters (Lindberg et al. 2014; Östman et al. 2017). That said, untreated sewage can occasionally contain comparatively high loads of antibiotics (Zuccato et al. 2010; Lundborg and Tamhankar 2017; Östman et al. 2017), and it has also been shown that some antibiotics, such as fluoroquinolones, are very slowly degraded and seem to accumulate in sewage sludge (Lindberg et al. 2006; Marx et al. 2015; Bengtsson-Palme et al. 2016). Still, this pathway into the environment is unlikely to contribute levels of antibiotics that plausibly would select for antibiotic resistance in environmental bacteria.

In contrast, the use of antibiotics in food production is considerably more likely to contribute concentrations of antibiotics that are relevant from a resistance selection and maintenance perspective (Bengtsson-Palme 2017). There are several reasons for this, including that animal feces generally is not as thoroughly treated as human-derived sewage, that the quantities of antibiotics used in animal husbandry are fairly high, and the substantial and largely unregulated use of antimicrobial substances in aquaculture (Cabello et al. 2013; Klase et al. 2019).

Another greatly contributing source of antibiotics to the environment is releases from production facilities (Larsson 2014). There is little or no regulation of releases of active ingredients from pharmaceutical production virtually anywhere in the world, although such legislation is now being discussed at several levels, both nationally and internationally (MPA [Swedish Medical Products Agency] 2009, 2011; Laurell et al. 2014; SPHS Secretariat, UNDP Istanbul Regional Hub 2015; Government of India 2017). The lack of limits to production-related releases has created a situation where – in order to bring down prices – many pharmaceutical companies and their subcontractors are willing to accept a larger environmental impact of their products in order to secure valuable sales contracts within, e.g., procurement programs (Bengtsson-Palme et al. 2018a). The releases of active antibiotic substances from production facilities lead to very high local levels of antibiotics that are persistent over time and create selective environments for highly resistant environmental bacteria (Larsson et al. 2007; Johnning et al. 2013; González-Plaza et al. 2017; Gothwal and Thatikonda 2017). It is conceivable that these bacteria can spread their resistance factors to human-associated bacteria or opportunistic pathogens and further on to bacteria that are clinically relevant pathogens, even though the routes through which this happens can be quite complicated and are hard to elucidate (Bengtsson-Palme et al. 2018b).

While releases from agriculture, aquaculture, and pharmaceutical production are likely to be the major contributors of antibiotic substances to the environment, the sources of resistant bacteria and resistance genes are more diverse. Based on current knowledge, it can be easily calculated that thousands or millions of resistant bacteria are released from a typical sewage treatment plant every day. While these are diluted in the recipient waters, they still constitute a significant contribution of resistant bacteria, many of which are human associated. At the same time, the numbers of resistant bacteria released with raw sewage are orders of magnitude greater and are more likely to originate from the human microbiome than those in sewage treatment plant effluents, making raw sewage a vastly more acute risk in terms of disseminating resistant bacteria. There is also no strong evidence that the levels of antibiotics in sewage treatment plants would be sufficiently high to select for antibiotic-resistant bacteria (Bengtsson-Palme et al. 2016; Flach et al. 2018). Resistant bacteria are also released via agriculture, as they are often present in fairly high quantities in manure (Heuer et al. 2011; Munir and Xagoraraki 2011; Wichmann et al. 2014). Indeed, application of fertilizers containing resistant bacteria has been shown to alter the composition and enrich resistance genes in soil microbial communities (Jechalke et al. 2013; Udikovic-Kolic et al. 2014; Ross and Topp 2015). While the manure often contains antibiotic residues as well, this is likely to have a minor effect on resistance selection in the indigenous community compared with the effects of adding already resistant bacteria to the soil. The relative contribution of aquaculture in the dissemination of already resistant bacteria is less clear, as this specific subject has not received a lot of attention.

In essence the high-risk settings for environmental antibiotic resistance development and/or dissemination boil down to (i) animal husbandry, (ii) aquaculture, (iii) pharmaceutical production, and (iv) untreated sewage. There are also some settings that clearly warrant concern, but may not be of the same immediate importance, including sewage treatment plants and their effluents, manure application, and the food supply chain. The four high-risk settings mentioned above all have different problems in terms of managing risks, but in contrast to the direct ethical problem associated with reducing human access to antibiotics in order to curb resistance, they are all mainly complicated due to economic concerns. From a human health safety perspective, there is a clear overuse of antibiotics in animal husbandry and aquaculture, as it jeopardizes the long-term effectivity of drugs used to save human lives. Similarly, there are no ethical dilemmas involved in expanding treatment infrastructure in order to treat larger proportions of the world's generated sewage. Regulating release of antibiotics from pharmaceutical production should be a low-hanging fruit in this context, as it is associated with defined point sources of releases and virtually the only benefit of releasing active substances from production is to lower costs. Thus, it should be politically possible to address these high-risk settings with appropriate interventions, which will be outlined in the next section.

24.7 Management Options

As described earlier, there is a multitude of pathways for resistant bacteria (and resistance genes) to migrate to, from, and through the environment (Figure 24.2). This myriad of potential mitigation points means that no single intervention will constitute a panacea to the resistance problem. Instead, a combination of actions would need to be taken to control the flow of antibiotic resistance through the environment (Table 24.1). In addition, the urgency of available options varies in different parts of the world, and the practical feasibility of implementing certain mitigation strategies will also be largely confined by the political and economic climate.

24.7.1 Possible Interventions on the Level of Releases of Resistant Bacteria

To reduce the dissemination of resistant bacteria through the environment, it is instrumental to limit their release from human activities. As discussed earlier, one of the main routes by which resistant bacteria from the human microbiome reach the environment is via sewage. There is already an efficient means in place to lower the levels of bacteria – resistant as well as sensitive ones – in sewage: municipal wastewater treatment plants. It has been estimated that more than two billion people lack access to even basic sanitation (WHO/UNICEF 2017). This population will excrete human-associated bacteria – including those resistant to antibiotics – directly into the environment, often ending up in surface waters. There has been a large amount of resources invested in developing conventional sewage treatment technologies to even further reduce the numbers of bacteria in the treated effluent. However, building out basic sewage treatment infrastructure in countries and regions largely lacking those resources would be almost guaranteed to constitute a larger risk reduction per money spent. A major barrier to this is likely to be the political unwillingness to invest large amounts of capital, generally deriving from tax payers, in technological developments elsewhere in the world. At the same time, expanding sewage treatment in the developing world would not only reduce risks for spreading antibiotic-resistant bacteria but would also aid in preventing a large number of other diseases associated with poor water quality, including cholera and virus-borne diarrheal diseases. Such investments would

Table 24.1 Strategies to reduce the flow of antibiotics and resistant bacteria through the environment.

Possible interventions on the level of releases of resistant bacteria
• Build out sewage treatment in the developing world
• Improve treatment technologies in WWTPs
• Special treatment of hospital effluents at the source
• Limit transmission via sludge, biosolids, and other solid wastes
Restricting transmission of resistant bacteria from the environment
• Monitoring approaches should include resistance (and resistance genes)
• Improved sanitation in the developing world
• Improved food safety
• Ensuring that wastewater reuse is safe
• Better education of the general public on hygiene and transmission risks
Better agriculture practices to sustain the lifespans of antibiotics
• Using alternatives to antibiotics as growth promoters and supplements
• Maintaining good animal health through better farming practices
• Optimizing agricultural antibiotic use toward compounds not used in human medicine
• Reduced meat consumption
Limiting selection for resistance in the environment
• Put limits on industrial releases of antibiotics
• Better control of antibiotic use in agriculture
• Restrict antibiotics use in aquaculture
• Further use of vaccines in aquaculture
• Incinerate solid waste rather than depositing it on landfills
• Reduced antibiotic usage in humans

therefore have a whole range of beneficial effects aside from reducing antibiotic resistance dissemination. A compelling argument for funding sewage infrastructure problems in other countries is the fact that antibiotic-resistant bacteria rapidly spread around our globally connected world and that poor sanitation in one place therefore poses a threat to human health everywhere in the long run (van der Bij and Pitout 2012; Wilson and Chen 2012; Molton et al. 2013; Bengtsson-Palme et al. 2015).

Beyond building out treatment infrastructure where it today is lacking, there has as mentioned been much research done on advanced treatments of effluents to further reduce the numbers of resistant bacteria (and usually bacteria in general). The most promising such technologies in terms of efficiency versus cost of investment seem to be ozone treatment, activated carbon filtration, and membrane separation (Dodd et al. 2010; Dodd 2012; Riquelme Breazeal et al. 2013; Björlenius 2018). Ozone treatment is an effective method for disinfecting the effluent water and at the same time removing many harmful chemicals. However, it also creates new substances from the reaction products, some of which may be toxic or otherwise harmful. It is also a fairly energy-demanding technique and requires substantial expansions of current sewage treatment plants to be employed at full scale. Filtration with activated carbon essentially binds bacteria and chemicals to porous carbon and effectively removes those. However, it requires to be either replaced or cleaned regularly to be kept functional, and supplementing the entire sewage treatment infrastructure with this type of treatment would create a need to cheaply produce activated carbon at a very large scale. The carbon also has to be properly taken care of post-use to avoid spreading resistant bacteria from the carbon filters. Membrane separation solutions, which essentially keep bacteria within the sewage treatment plants using size filtration, have similar issues with replacement cycles and end-of-life disposal of the membranes. While all of these techniques are quite efficient in bringing the number of released bacteria down, the gains from this reduction relative to other possible interventions need to be carefully evaluated. One important consideration in this respect is whether other improvements to the general treatment process can be made. There are, for example, indications that thermophilic anaerobic sludge digestion could be better at removing resistance genes than mesophilic digestion (Diehl and LaPara 2010; Ma et al. 2011). Another consideration should be if other gains, besides antibiotic resistance reductions, are achieved with ozonation. If ozone treatment is to be used to reduce the concentrations of other harmful compounds in the effluent water, the reduction of (resistant) bacteria may come as an added benefit, essentially without increasing the costs.

There could also potentially be benefits in installing different types of highly efficient treatments at particularly relevant sources, such as hospitals. Such technologies include those mentioned above, as well as membrane bioreactors (Kovalova et al. 2012). These could remove not only resistant bacteria but also antibiotics and other drugs, which are known to be present in higher quantities in hospital wastewater (Verlicchi et al. 2012), and often are mixed with public sewage systems and treated in the same treatment plants. Targeting hospital wastewater would make sense because it is likely to eliminate higher numbers of the most relevant human pathogens before they have a chance to reach the sewage treatment plants and potentially interact with bacteria of other origins, reducing the possibility for resistance genes to be shared between bacteria. This is a fairly cost-effective approach and for which there seems to be reasonably easy to harness political support (Lienert et al. 2011).

In parts of the world, sludge from the sewage treatment process is used as fertilizer on fields, which is an appealing solution as it allows the disposal of a costly by-product from sewage treatment and at the same time is a form of resource recovery. Many antibiotics bind to sludge particles, rendering their concentrations much higher in final sludge than in the effluent water (Lindberg et al. 2006; Hörsing et al. 2011; Östman et al. 2017). Whether or not these antibiotics are still bioavailable when bound to particles is an open question, as is their release rates in soil. The effects of application of sewage-derived sludge to fields are mixed, with different researchers reporting somewhat different results (Brooks et al. 2007; Munir et al. 2011; Munir and Xagoraraki 2011). The increase and decrease of resistance genes and resistant bacteria after application seem to be dependent on many different factors, including the characteristics of the soil that the sludge is applied to.

Similarly to sludge application, manures from animals are often used as fertilizers in agriculture. Prior to application, the manure is generally stored or composted, which overall reduces the number of resistant bacteria (Storteboom et al. 2007). The specific dynamics of this process is, however, fairly complicated due to the quite different microbial ecology in different types of manure (Storteboom et al. 2007; Sharma et al. 2009; McKinney et al. 2010; Wichmann et al. 2014). Individual resistance genes are often enriched in these processes because they are present in some specific hosts that happen to increase in numbers. Still, overall resistance levels usually go down at least 10- or 100-fold in the storage or composting process. That said, the numbers are generally only reduced, and resistant *E. coli* can be detected after composting. Adding an anaerobic treatment step of the manure may be a way to reduce resistant bacteria even further (Rysz et al. 2013). Storage temperature above 50 °C also puts most human-associated bacteria at a disadvantage and could be a more cost-efficient alternative to full-scale disinfection of manure prior to application. It is also important to contain manure and other animal waste during storage. This includes avoiding spills and seepage from storage lagoons and limiting surface runoff (Pruden et al. 2013). This can partially be alleviated by improving the handling of manure and increasing the storage capacity, which allows for better management of when and how manure is added to fields. Importantly, refined containment strategies also contribute to better nutrient management and therefore improved water and soil quality. In terms of policy changes, there is a need to develop standards for acceptable concentrations of antibiotics in animal manures. Combined with standards for resistant bacteria, this could be used to monitor the levels of both selective agents and resistance prior to application on fields. This is not an easy undertaking, as it would be a further complication for farmers and would incur additional costs. Still, aside from raw sewage, manure is one of the major routes by which resistant bacteria reach the environment and therefore deserves attention and monitoring. Since manure is applied to soil used for growing food, there is also a potential spreading route back into the human population, providing a clear link from animals to humans (Bengtsson-Palme 2017), which will be further discussed below.

24.7.2 Restricting Transmission of Resistant Bacteria from the Environment

An alternative – or complementary – approach to reducing releases of resistant bacteria into the environment is to instead focus on limiting the ability of resistant bacteria to disperse back into the human population, i.e. providing the public with protective measures from resistant bacteria in the environment (Larsson et al. 2018). To some extent, such practices are already in place, for example, in the monitoring of the quality of waters used for recreational swimming, of tap water, and of food products. While these approaches almost never test for resistant bacteria specifically, they do provide a warning system for pathogenic bacteria, restricting their ability to reach and infect the human population. This, by consequence, also provides a basic protection against resistant pathogens. To make the already existing monitoring systems more useful to prevent transmission of antibiotic-resistant bacteria and resistance genes from the environment to humans, specific screening for phenotypic resistance or resistance genes would be necessary. There is not any single established protocol for monitoring resistance or resistance genes. Testing bacteria for phenotypic resistance could be carried out in a day or two, but would require somewhat elaborate screening against a panel of different antibiotics. That said, there are standard protocols in place for susceptibility testing that are used clinically and could be relatively easily transformed into an environmental monitoring setting. An alternative to phenotypic testing are methods based on PCR, which can cheaply determine the levels of a defined set of resistance genes in a matter of hours. A problem with this approach is that it is still debated what genes that are most relevant to test, with different gene sets being suggested by different authors and for different purposes (Berendonk et al. 2015; Bengtsson-Palme 2018). There is also limited baseline data for what would be typically expected in these monitoring scenarios, which would be required to be established before monitoring for resistance would be useful (Pruden et al. 2013; Angers-Loustau et al. 2018). Once the methods have been standardized and the environmental baseline of resistance and resistance genes is established, monitoring for resistance genes could provide guidance on when to close beaches, find alternative temporary sources of drinking water, and recall or provide additional advisory for food products in order to protect humans from resistant bacteria. As these practices could essentially be integrated in existing monitoring programs for water and food quality, they are likely to have a relatively high risk-reduction-to-cost ratio compared with many other interventions discussed in this chapter.

It is well known that access to clean water and other sanitation is integral to preventing the spread of infectious diseases in the general public, and this holds true for preventing the spread of resistant bacteria as well (Bain et al. 2014; Graham et al. 2014; WHO/UNICEF 2017). In countries with access to clean water and with relatively high living standards, this is generally mostly an issue of educating the public on basic hygiene and food safety (together with the monitoring approaches mentioned above). In many low-income countries, however, there is a need to build out a functional infrastructure for water treatment and supply (Figure 24.1). It is also key to quickly provide clean water and sanitary materials after large flooding events or other catastrophes in order to prevent the spread of resistant bacteria, as well as other infections. The main barriers to providing access to clean water to everyone on earth are the unequal distribution of resources and the widely varying levels of available freshwater in different regions of the world. To alleviate water scarcity, wastewater reuse is being employed in an increasing number of regions (Fatta-Kassinos et al. 2015). Wastewater reuse is not without risks from a bacterial dissemination standpoint. Importantly, providing a functional water and sanitation infrastructure globally would not only benefit in restricting the dissemination of resistant bacteria but also provide many other health benefits as well as improved standards of living.

Besides water, fresh or undercooked food is another important route by which resistant bacteria can reach the human population (European Food Safety Authority & European Centre for Disease Prevention and Control 2013; Bengtsson-Palme 2017; Oniciuc et al. 2018). Despite the fact that many food products are being monitored for microbial contamination, this surveillance is far from comprehensive, and much of the produce is likely to slip through until it has already reached consumers, as in the case of the German EHEC outbreak in 2011 (Buchholz et al. 2011; Rasko et al. 2011; Robert Koch Institute 2011). While it is definitely possible to step up the efforts to detect pathogenic and resistant bacteria in food products, the food supply chain is so diverse and complex that it is most likely impossible to extend the scale of this monitoring to completely prevent resistant bacteria to spread through food. Instead, social interventions are necessary to educate the public on the risks of handling raw foods and how to as much as possible avoid catching resistant or pathogenic bacteria from food. Here, again, access to clean water is instrumental to allow, for example, cleaning of vegetables before consumption.

24.7.3 Better Agriculture Practices to Sustain the Lifespans of Antibiotics

The use of antibiotics in animals is problematic because the substances used have large overlaps with those used in human medicine. In contrast to human medicine, the use in animal husbandry is often motivated by increasing growth or preventing diseases, which is arguably a wasteful use of a precious (and due to resistance dwindling) resource. An obvious way around this would be to use other feed additives instead of antibiotics as growth promoters. A common alternative is to use metals, such as copper, zinc, and arsenic

(Bolan et al. 2010). While this is likely to be better than antibiotics, there is a potential that such metals could co-select for antibiotic resistance because the genes providing tolerance to them often are located on the same genetic elements that carries genes encoding resistance to antibiotics (Pal et al. 2015, 2017). There are even indications that copper may be a stronger selector for some antibiotic resistance genes than the antibiotics corresponding to these genes themselves (Chee-Sanford et al. 2009). Therefore, other growth promoters are preferable than metals, if the use of growth promoters cannot be avoided altogether. Limiting these practices will be complicated, as many regions of the world completely lack regulations on antibiotic and metal use in farming (Wu et al. 2010; Zhu et al. 2013). Still, the use of antibiotics as growth promoters has been banned in the EU since 2006, showing that such actions are possible. A first obvious target for regulation should be to restrict the use of critically important antibiotics in agriculture (FAO/OIE/WHO 2004). This, however, must also be followed up with stricter monitoring and enforcement (Aarestrup 2012). With regard to the preventive use of antibiotics, a good alternative would be vaccination, and developing inexpensive (preferably orally administered) vaccines to the most common animal pathogens should be of high priority in order to reduce unnecessary use of antibiotics in animal farming. Other practices that would reduce the need to use antibiotics for animals is to not keep animals as tightly packed together and to the extent possible avoid industrializing husbandry. Ultimately, the most effective mitigation strategy for the overuse of antibiotics in animal farming would probably be to drastically reduce the global consumption of meat and other animal products such as milk, with the most drastic cuts having to occur in the richest parts of the world. This would bring a whole range of other benefits, including reduced climate impacts (Springmann et al. 2018), and likely also benefit human health, but whether a reduction of meat consumption in order to control antibiotic resistance is realistic is questionable. Lifestyle changes tend to be complicated to enforce on the population in general, and proposals such as taxes on meat have received lukewarm responses from lawmakers (Röös and Tjärnemo 2013; Säll and Gren 2015).

24.7.4 Limiting Selection for Resistance in the Environment

Thus far, we have mostly discussed interventions to the dissemination of resistant bacteria through the environment. However, as we described earlier, the environment also functions as an arena for selection of resistance and as a source of new resistance determinants. In order to curb these processes, the most important interventions involve avoiding to create environments that provide resistant bacteria with a selective advantage over sensitive ones (Bengtsson-Palme et al. 2018b; Larsson et al. 2018). To avoid creating such settings, the most important measure is to limit the concentrations of antibiotics (and other co-selective agents) in the environment. It is likely that the point source that provides the highest local concentrations of antibiotics in the environment is effluents from pharmaceutical manufacturing (Larsson 2014; Larsson et al. 2018). Therefore, it is critical to enforce limits to the levels of antibiotics that may be released into the environment. A complication in this context is that there is little knowledge of the concentrations that can provide bacteria with a selective advantage (Ågerstrand et al. 2015). We know from laboratory studies that concentrations far below those that inhibit the growth of bacteria can still select for resistant strains or species over susceptible ones (Gullberg et al. 2011; Lundström et al. 2016). However, there is no comprehensive data for what concentrations that are selective in environmental bacterial communities. In 2016, Bengtsson-Palme and Larsson published a set of 111 predicted no-effect concentrations for antibiotics (Bengtsson-Palme and Larsson 2016), and at the moment these are the best starting point for regulation available. That said, these data need to over time be supplemented with real effect data to better reflect the increasing knowledge of antibiotic effects on environmental communities (Lundström et al. 2016; Kraupner et al. 2018; Murray et al. 2018).

Since animals are exposed to high levels of antibiotics, husbandry is also a setting where selection for resistant bacteria can occur. Many opportunistic pathogens can also thrive in farmed animals, so selection for resistance in animals may provide a much more direct path to human exposure than selection in external environments. The measures that can be taken to reduce the use of antibiotics in farmed animals have already been discussed earlier. However, in terms of resistance selection, the use of antibiotics in aquaculture is probably even more important. In order to reduce or abolish the use of antibiotics in aquaculture, alternatives are needed, most importantly vaccines (Sommerset et al. 2014). Combined with improving the fish farming facilities, this has been shown to be highly effective in reducing the use of antibiotics in Norwegian aquaculture (Heuer et al. 2009). Reducing animal density and preventing wild fish from entering the pens are also effective ways of reducing infections, minimizing the need for antibiotics (Grigorakis and Rigos 2011), as is overfeeding, which enables heterotrophic bacteria to thrive. Importantly, it is to avoid complicated to achieve a quantitative risk assessment of antibiotic resistance in aquaculture, largely due to a lack of data. Monitoring of antibiotic use and resistance levels in aquaculture should be implemented, and a baseline for resistance needs to be established.

While the largest gains in terms of limiting releases of antibiotics into the environment are likely to be made in the abovementioned areas, there are also releases of antibiotics from municipal sewage treatment plants (Michael et al. 2013). The most effective intervention in this area would probably be to implement the same concentration limits as suggested for pharmaceutical production (Bengtsson-Palme and Larsson 2016) as part of environmental quality standards for antibiotics in surface water. Antibiotics are also commonly disposed in household and hospital solid wastes. In such biosolids, many antibiotics have been detected in microgram per kilogram levels or higher (Zhang and Li 2011). Landfill disposal is a suboptimal solution to biosolid waste as leachates may reach groundwater and surface water (Renou et al. 2008). A comparatively easy solution to the biosolid waste problem is incineration, which essentially eliminates the risks, but is costlier and requires proper filtration of the airborne waste products. When used as part of a larger strategy, incineration can provide a source of alternative energy. In Sweden, for example, 99% of all household waste was incinerated for energy or recycled in 2010 (Naturvårdsverket 2012), showing that this approach is feasible. Finally, a reduced use of antibiotics in humans, restricting them to when only strictly necessary, would indirectly reduce the levels of antibiotics reaching the environment. Firstly, it would reduce the excretion from human use. Secondly, it would indirectly lessen the burden of releases from antibiotic production, as the demand would go down. Lastly, it would lessen the selection of resistant bacteria in the human microbiome, limiting the release of resistant bacteria into the environment.

24.8 Final Remarks

To a large extent, the strategies for limiting antibiotic resistance dissemination and selection in the environment are already quite clear. The remaining problems largely surround how to prioritize between different strategies, which technologies that would provide the largest benefits to invest in, and – perhaps most prominently – the political willingness and feasibility to pursue these strategies. Several of the most efficient resistance prevention options involve high costs, investments in technology and infrastructure in other countries, or proposals that are likely to be rather unpopular with the general public. This poses a substantial problem for implementation and leads to less effective strategies being promoted because they are more feasible to put to effect in the short run. For example, expanded sewage treatment and clean water infrastructure in low-income countries would likely be much more effective for reducing releases of resistant bacteria into the environment than advanced treatment of sewage in high-income countries, but the latter has received more attention in Europe in this context because it can be enacted locally. Along the same lines, reduced meat consumption would contribute to lowering the use of antibiotics in animal husbandry, but taxation of meat is not a popular policy with the general public.

While development of better technologies to reduce releases of antibiotics and resistant bacteria to the environment is welcome and needed, a more urgent mitigation focus should be to implement a global action plan on antibiotic resistance in the environment. There needs to be a greater recognition that resistant bacteria do not respect borders and that sponsoring preventive efforts in other countries therefore is a means of protecting public health locally. Both regulation and mitigation need to take place on a global scale to be truly effective. It is also clear that some areas, like the releases from pharmaceutical production, should be easier to address than others. These constitute low-hanging fruits, and every day of policy inaction in regulating these sectors is a lost day against the growing resistance problem. Finally, it deserves to be highlighted that even though we have a fairly good understanding of how the environment contributes to antibiotic resistance development, we have very little knowledge of the relative contributions of the different processes involved (Bengtsson-Palme 2019). It is therefore urgent to start finding these numbers out so that we can perform quantitative risk modeling and rank the risk scenarios in a more informed manner (Larsson et al. 2018). Putting numbers, or at least ranges, to the risks also makes it easier to calculate the costs associated with policy inaction, which would serve as a political motivation to speed up the implementation of elimination and prevention strategies.

Acknowledgments

JBP is funded by the Swedish Research Council for Environment, Agricultural Sciences and Spatial Planning (FORMAS) (grant 2016-00768). SH is funded by the German Research Foundation (DFG) (grant HE8047-1).

Conflict of Interest

The authors have no conflicts of interest to declare.

References

Aarestrup, F.M. (2012). Sustainable farming: get pigs off antibiotics. *Nature* 486 (7404): 465–466.

Aarestrup, F.M. et al. (2001). Effect of abolishment of the use of antimicrobial agents for growth promotion on occurrence of antimicrobial resistance in fecal enterococci from food animals in Denmark. *Antimicrobial Agents and Chemotherapy* 45 (7): 2054–2059.

Ågerstrand, M. et al. (2015). Improving environmental risk assessment of human pharmaceuticals. *Environmental Science and Technology* 49 (9): 5336–5345.

Al-Jassim, N. et al. (2015). Removal of bacterial contaminants and antibiotic resistance genes by conventional wastewater treatment processes in Saudi Arabia: is the treated wastewater safe to reuse for agricultural irrigation? *Water Research* 73: 277–290.

Allen, H.K. et al. (2010). Call of the wild: antibiotic resistance genes in natural environments. *Nature Reviews Microbiology* 8 (4): 251–259.

Andersson, D.I. and Hughes, D. (2010). Antibiotic resistance and its cost: is it possible to reverse resistance? *Nature Reviews Microbiology* 8 (4): 260–271.

Angelin, M. et al. (2015). Risk factors for colonization with extended-spectrum beta-lactamase producing Enterobacteriaceae in healthcare students on clinical assignment abroad: a prospective study. *Travel Medicine and Infectious Disease* 13 (3): 223–229.

Angers-Loustau, A. et al. (2018). The challenges of designing a benchmark strategy for bioinformatics pipelines in the identification of antimicrobial resistance determinants using next generation sequencing technologies. *F1000Research* 7: 459.

Ashbolt, N.J. et al. (2013). Human Health Risk Assessment (HHRA) for environmental development and transfer of antibiotic resistance. *Environmental Health Perspectives* 121 (9): 993–1001.

Bahram, M. et al. (2018). Structure and function of the global topsoil microbiome. *Nature* 320: 1039.

Bain, R. et al. (2014). Fecal contamination of drinking-water in low- and middle-income countries: a systematic review and meta-analysis. *PLoS Medicine* 11 (5): e1001644.

Bengtsson-Palme, J. (2017). Antibiotic resistance in the food supply chain: where can sequencing and metagenomics aid risk assessment? *Current Opinion in Food Science* 14: 66–71.

Bengtsson-Palme, J. (2018). The diversity of uncharacterized antibiotic resistance genes can be predicted from known gene variants-but not always. *Microbiome* 6 (1): 125.

Bengtsson-Palme, J. (2019). Assessment and management of risks associated with antibiotic resistance in the environment. In: *Management of Emerging Public Health Issues and Risks*, 243–263. Elsevier.

Bengtsson-Palme, J., Alm Rosenblad, M. et al. (2014a). Metagenomics reveals that detoxification systems are underrepresented in marine bacterial communities. *BMC Genomics* 15: 749.

Bengtsson-Palme, J., Boulund, F. et al. (2014b). Shotgun metagenomics reveals a wide array of antibiotic resistance genes and mobile elements in a polluted lake in India. *Frontiers in Microbiology* 5: 648.

Bengtsson-Palme, J., Gunnarsson, L., and Larsson, D.G.J. (2018a). Can branding and price of pharmaceuticals guide informed choices towards improved pollution control during manufacturing? *Journal of Cleaner Production* 171: 137–146.

Bengtsson-Palme, J., Kristiansson, E., and Larsson, D.G.J. (2018b). Environmental factors influencing the development and spread of antibiotic resistance. *FEMS Microbiology Reviews* 42 (1): 25.

Bengtsson-Palme, J. and Larsson, D.G.J. (2016). Concentrations of antibiotics predicted to select for resistant bacteria: proposed limits for environmental regulation. *Environment International* 86: 140–149.

Bengtsson-Palme, J. et al. (2015). The human gut microbiome as a transporter of antibiotic resistance genes between continents. *Antimicrobial Agents and Chemotherapy* 59 (10): 6551–6560.

Bengtsson-Palme, J. et al. (2016). Elucidating selection processes for antibiotic resistance in sewage treatment plants using metagenomics. *The Science of the Total Environment* 572: 697–712.

Berendonk, T.U. et al. (2015). Tackling antibiotic resistance: the environmental framework. *Nature Reviews Microbiology* 13 (5): 310–317.

Bielen, A. et al. (2017). Negative environmental impacts of antibiotic-contaminated effluents from pharmaceutical industries. *Water Research* 126: 79–87.

van der Bij, A.K. and Pitout, J.D.D. (2012). The role of international travel in the worldwide spread of multiresistant Enterobacteriaceae. *Journal of Antimicrobial Chemotherapy* 67 (9): 2090–2100.

Björlenius, B. (2018). *Pharmaceuticals–Improved Removal from Municipal Wastewater and Their Occurrence in the Baltic Sea*. Stockholm: KTH Royal Institute of Technology.

van den Bogaard, A.E., Bruinsma, N., and Stobberingh, E.E. (2000). The effect of banning avoparcin on VRE carriage in The Netherlands. *Journal of Antimicrobial Chemotherapy* 46 (1): 146–148.

Bolan, N., Adriano, D., and Mahimairaja, S. (2010). Distribution and bioavailability of trace elements in livestock and poultry manure by-products. *Critical Reviews in Environmental Science and Technology* 34 (3): 291–338.

Bostock, J. et al. (2010). Aquaculture: global status and trends. *Philosophical Transactions of the Royal Society B: Biological Sciences* 365 (1554): 2897–2912.

Brooks, J.P. et al. (2007). Occurrence of antibiotic-resistant bacteria and endotoxin associated with the land application of biosolids. *Canadian Journal of Microbiology* 53 (5): 616–622.

Buchholz, U. et al. (2011). German outbreak of *Escherichia coli* O104:H4 associated with sprouts. *The New England Journal of Medicine* 365 (19): 1763–1770.

Cabello, F.C. (2006). Heavy use of prophylactic antibiotics in aquaculture: a growing problem for human and animal health and for the environment. *Environmental Microbiology* 8 (7): 1137–1144.

Cabello, F.C. et al. (2013). Antimicrobial use in aquaculture re-examined: its relevance to antimicrobial resistance and to animal and human health. *Environmental Microbiology* 15 (7): 1917–1942.

Cabello, F.C. et al. (2016). Aquaculture as yet another environmental gateway to the development and globalisation of antimicrobial resistance. *The Lancet Infectious Diseases* 16 (7): e127–e133.

Chatterjee, A. et al. (2018). Quantifying drivers of antibiotic resistance in humans: a systematic review. *The Lancet Infectious Diseases* 18 (12): e368–e378.

Chee-Sanford, J.C. et al. (2009). Fate and transport of antibiotic residues and antibiotic resistance genes following land application of manure waste. *Journal of Environmental Quality* 38 (3): 1086–1108.

Chen, Q. et al. (2016). Long-term field application of sewage sludge increases the abundance of antibiotic resistance genes in soil. *Environment International* 92-93: 1–10.

Christou, A. et al. (2017). The potential implications of reclaimed wastewater reuse for irrigation on the agricultural environment: the knowns and unknowns of the fate of antibiotics and antibiotic resistant bacteria and resistance genes – a review. *Water Research* 123: 448–467.

Davies, J. and Davies, D. (2010). Origins and evolution of antibiotic resistance. *Microbiology and Molecular Biology Reviews* 74 (3): 417–433.

D'Costa, V.M. et al. (2006). Sampling the antibiotic resistome. *Science* 311 (5759): 374–377.

D'Costa, V.M. et al. (2011). Antibiotic resistance is ancient. *Nature* 477 (7365): 457–461.

Diehl, D.L. and LaPara, T.M. (2010). Effect of temperature on the fate of genes encoding tetracycline resistance and the integrase of class 1 integrons within anaerobic and aerobic digesters treating municipal wastewater solids. *Environmental Science and Technology* 44 (23): 9128–9133.

Dodd, M.C. (2012). Potential impacts of disinfection processes on elimination and deactivation of antibiotic resistance genes during water and wastewater treatment. *Journal of Environmental Monitoring* 14 (7): 1754–1771.

Dodd, M.C. et al. (2010). Transformation of β-lactam antibacterial agents during aqueous ozonation: reaction pathways and quantitative bioassay of biologically-active oxidation products. *Environmental Science and Technology* 44 (15): 5940–5948.

Done, H.Y., Venkatesan, A.K., and Halden, R.U. (2015). Does the recent growth of aquaculture create antibiotic resistance threats different from those associated with land animal production in agriculture? *The AAPS Journal* 17 (3): 513–524.

ECDC, EFSA and EMA (2017). ECDC/EFSA/EMA second joint report on the integrated analysis of the consumption of antimicrobial agents and occurrence of antimicrobial resistance in bacteria from humans and food-producing animals – Joint Interagency Antimicrobial Consumption and Resistance Analysis (JIACRA) Report. *EFSA Journal* 15 (7): 4872, 135.

Ehlers, L.J. and Bouwer, E.J. (1999). RP4 plasmid transfer among species of pseudomonas in a biofilm reactor. *Water Science and Technology* 39 (7): 163–171.

European Food Safety Authority & European Centre for Disease Prevention and Control (2013). The European Union summary report on antimicrobial resistance in zoonotic and indicator bacteria from humans, animals and food in 2011. *EFSA Journal* 11 (5): 3196.

European Medicines Agency (2018). Sales of veterinary antimicrobial agents in 30 European countries in 2016: trends from 2010 to 2016, 15 October 2018.

FAO/OIE/WHO (2004). *Second Joint FAO/OIE/WHO Expert Workshop on Non-human Antimicrobial Usage and Antimicrobial Resistance: Management Options*. Geneva: WHO.

FAO/OIE/WHO (2006). *Antimicrobial Use in Aquaculture and Antimicrobial Resistance*. Geneva: WHO.

Fatta-Kassinos, D. et al. (2015). COST Action ES1403: new and emerging challenges and opportunities in wastewater reuse (NEREUS). *Environmental Science and Pollution Research International* 22 (9): 7183–7186.

Finley, R.L. et al. (2013). The scourge of antibiotic resistance: the important role of the environment. *Clinical Infectious Diseases* 57 (5): 704–710.

Flach, C.-F. et al. (2018). A comprehensive screening of *Escherichia coli* isolates from Scandinavia's largest sewage treatment plant indicates no selection for antibiotic resistance. *Environmental Science and Technology* 52 (19): 11419–11428.

Forsberg, K.J. et al. (2012). The shared antibiotic resistome of soil bacteria and human pathogens. *Science* 337 (6098): 1107–1111.

Gao, P. et al. (2015). Impacts of coexisting antibiotics, antibacterial residues, and heavy metals on the occurrence of erythromycin resistance genes in urban wastewater. *Applied Microbiology and Biotechnology* 99 (9): 3971–3980.

Gatica, J. and Cytryn, E. (2013). Impact of treated wastewater irrigation on antibiotic resistance in the soil microbiome. *Environmental Science and Pollution Research International* 20 (6): 3529–3538.

Giovannoni, S.J. et al. (2005). Genome streamlining in a cosmopolitan oceanic bacterium. *Science* 309 (5738): 1242–1245.

González-Plaza, J.J. et al. (2017). Functional repertoire of antibiotic resistance genes in antibiotic manufacturing effluents and receiving freshwater sediments. *Frontiers in Microbiology* 8: 2675.

Gothwal, R. and Thatikonda, S. (2017). Role of environmental pollution in prevalence of antibiotic resistant bacteria in aquatic environment of river: case of Musi river, South India. *Water and Environment Journal* 121: 993.

Government of India (2017). National Action Plan on Antimicrobial Resistance (NAP-AMR) 2017–2021.

Graham, D.W. et al. (2014). Underappreciated role of regionally poor water quality on globally increasing antibiotic resistance. *Environmental Science and Technology* 48 (20): 11746–11747.

Grigorakis, K. and Rigos, G. (2011). Aquaculture effects on environmental and public welfare: the case of Mediterranean mariculture. *Chemosphere* 85 (6): 899–919.

Gullberg, E. et al. (2011). Selection of resistant bacteria at very low antibiotic concentrations. *PLoS Pathogens* 7 (7): e1002158.

Gullberg, E. et al. (2014). Selection of a multidrug resistance plasmid by sublethal levels of antibiotics and heavy metals. *mBio* 5 (5): e01918-14.

Guo, J. et al. (2017). Metagenomic analysis reveals wastewater treatment plants as hotspots of antibiotic resistance genes and mobile genetic elements. *Water Research* 123: 468–478.

Han, Y. et al. (2018). Combined impact of fishmeal and tetracycline on resistomes in mariculture sediment. *Environmental Pollution* 242 (Pt B): 1711–1719.

Harwood, V.J. et al. (2014). Microbial source tracking markers for detection of fecal contamination in environmental waters: relationships between pathogens and human health outcomes. *FEMS Microbiology Reviews* 38 (1): 1–40.

He, L.-Y. et al. (2014). Dissemination of antibiotic resistance genes in representative broiler feedlots environments: identification of indicator ARGs and correlations with environmental variables. *Environmental Science and Technology* 48 (22): 13120–13129.

Hess, S., Lüddeke, F., and Gallert, C. (2016). Concentration of facultative pathogenic bacteria and antibiotic resistance genes during sewage treatment and in receiving rivers. *Water Science and Technology* 74 (8): 1753–1763.

Heuer, H., Schmitt, H., and Smalla, K. (2011). Antibiotic resistance gene spread due to manure application on agricultural fields. *Current Opinion in Microbiology* 14 (3): 236–243.

Heuer, O.E. et al. (2009). Human health consequences of use of antimicrobial agents in aquaculture. *Clinical Infectious Diseases* 49 (8): 1248–1253.

Hörsing, M. et al. (2011). Determination of sorption of seventy-five pharmaceuticals in sewage sludge. *Water Research* 45 (15): 4470–4482.

Jechalke, S. et al. (2013). Increased abundance and transferability of resistance genes after field application of manure from sulfadiazine-treated pigs. *Applied and Environmental Microbiology* 79 (5): 1704–1711.

Johnning, A. et al. (2013). Acquired genetic mechanisms of a multiresistant bacterium isolated from a treatment plant receiving wastewater from antibiotic production. *Applied and Environmental Microbiology* 79 (23): 7256–7263.

Jutkina, J. et al. (2016). An assay for determining minimal concentrations of antibiotics that drive horizontal transfer of resistance. *The Science of the Total Environment* 548–549: 131–138.

Karkman, A. et al. (2017). Antibiotic-resistance genes in waste water. *Trends in Microbiology* 26 (3): 220–228.

Kelessidis, A. and Stasinakis, A.S. (2012). Comparative study of the methods used for treatment and final disposal of sewage sludge in European countries. *Waste Management* 32 (6): 1186–1195.

Kivaisi, A.K. (2001). The potential for constructed wetlands for wastewater treatment and reuse in developing countries: a review. *Ecological Engineering* 16 (4): 545–560.

Klase, G. et al. (2019). The microbiome and antibiotic resistance in integrated fishfarm water: implications of environmental public health. *The Science of the Total Environment* 649: 1491–1501.

Kovalova, L. et al. (2012). Hospital wastewater treatment by membrane bioreactor: performance and efficiency for organic micropollutant elimination. *Environmental Science and Technology* 46 (3): 1536–1545.

Kraupner, N. et al. (2018). Selective concentration for ciprofloxacin resistance in *Escherichia coli* grown in complex aquatic bacterial biofilms. *Environment International* 116: 255–268.

Kristiansson, E. et al. (2011). Pyrosequencing of antibiotic-contaminated river sediments reveals high levels of resistance and gene transfer elements. *PLoS One* 6 (2): e17038.

Larsson, D.G.J. (2014). Pollution from drug manufacturing: review and perspectives. *Philosophical Transactions of the Royal Society of London Series B, Biological Sciences* 369 (1656): 20130571.

Larsson, D.G.J., de Pedro, C., and Paxeus, N. (2007). Effluent from drug manufactures contains extremely high levels of pharmaceuticals. *Journal of Hazardous Materials* 148 (3): 751–755.

Larsson, D.G.J. et al. (2018). Critical knowledge gaps and research needs related to the environmental dimensions of antibiotic resistance. *Environment International* 117: 132–138.

Laurell, M., Norhstedt, P., and Ryding, S.-O. (2014). Significance of Sustainability in Supply Chains on Behalf of SPP: A Pre-study. Swedish Environmental Management Council 12, June 30 2014.

Li, D. et al. (2008). Determination and fate of oxytetracycline and related compounds in oxytetracycline production wastewater and the receiving river. *Environmental Toxicology and Chemistry* 27 (1): 80–86.

Lienert, J. et al. (2011). Multiple-criteria decision analysis reveals high stakeholder preference to remove pharmaceuticals from hospital wastewater. *Environmental Science and Technology* 45 (9): 3848–3857.

Lindberg, R.H. et al. (2006). Behavior of fluoroquinolones and trimethoprim during mechanical, chemical, and active sludge treatment of sewage water and digestion of sludge. *Environmental Science and Technology* 40 (3): 1042–1048.

Lindberg, R.H. et al. (2014). Occurrence and behaviour of 105 active pharmaceutical ingredients in sewage waters of a municipal sewer collection system. *Water Research* 58: 221–229.

Lüddeke, F. et al. (2015). Removal of total and antibiotic resistant bacteria in advanced wastewater treatment by ozonation in combination with different filtering techniques. *Water Research* 69: 243–251.

Lundborg, C.S. and Tamhankar, A.J. (2017). Antibiotic residues in the environment of South East Asia. *BMJ* (Clinical research ed) 358: j2440.

Lundström, S.V. et al. (2016). Minimal selective concentrations of tetracycline in complex aquatic bacterial biofilms. *The Science of the Total Environment* 553: 587–595.

Lupo, A., Coyne, S., and Berendonk, T.U. (2012). Origin and evolution of antibiotic resistance: the common mechanisms of emergence and spread in water bodies. *Frontiers in Microbiology* 3: 18.

Luprano, M.L. et al. (2016). Antibiotic resistance genes fate and removal by a technological treatment solution for water reuse in agriculture. *The Science of the Total Environment* 571: 809–818.

Ma, Y. et al. (2011). Effect of various sludge digestion conditions on sulfonamide, macrolide, and tetracycline resistance genes and class I integrons. *Environmental Science and Technology* 45 (18): 7855–7861.

Marathe, N.P. et al. (2013). A treatment plant receiving waste water from multiple bulk drug manufacturers is a reservoir for highly multi-drug resistant integron-bearing bacteria. *PLoS One* 8 (10): e77310.

Marx, C. et al. (2015). Mass flow of antibiotics in a wastewater treatment plant focusing on removal variations due to operational parameters. *The Science of the Total Environment* 538: 779–788.

McKinney, C.W. et al. (2010). tet and sul Antibiotic resistance genes in livestock lagoons of various operation type, configuration, and antibiotic occurrence. *Environmental Science and Technology* 44 (16): 6102–6109.

Michael, I. et al. (2013). Urban wastewater treatment plants as hotspots for the release of antibiotics in the environment: a review. *Water Research* 47 (3): 957–995.

Molton, J.S. et al. (2013). The global spread of healthcare-associated multidrug-resistant bacteria: a perspective from Asia. *Clinical Infectious Diseases* 56 (9): 1310–1318.

MPA (Swedish Medical Products Agency) (2009). *Opportunities for Strengthening the Environmental Requirements Pertaining to the Manufacture of Medicinal Products and Active Pharmaceutical Ingredients in a National and International Context [in Swedish with English Summary]*. MPA (Swedish Medical Products Agency).

MPA (Swedish Medical Products Agency) (2011). *Underlag för att möjliggöra initieringen av en revidering av EUlagstiftningen om god tillverkningssed, GMP, med syfte att lagstiftningen även ska omfatta miljöhänsyn [in Swedish]*. MPA (Swedish Medical Products Agency).

Munir, M., Wong, K., and Xagoraraki, I. (2011). Release of antibiotic resistant bacteria and genes in the effluent and biosolids of five wastewater utilities in Michigan. *Water Research* 45 (2): 681–693.

Munir, M. and Xagoraraki, I. (2011). Levels of antibiotic resistance genes in manure, biosolids, and fertilized soil. *Journal of Environmental Quality* 40 (1): 248–255.

Murray, A.K. et al. (2018). Novel insights into selection for antibiotic resistance in complex microbial communities. *mBio* 9 (4): e00969-18.

Naturvårdsverket (2012). *Från avfallshantering till resurshushållning: Sveriges avfallsplan 2012–2017*. Stockholm: Naturvårdsverket.

Negreanu, Y. et al. (2012). Impact of treated wastewater irrigation on antibiotic resistance in agricultural soils. *Environmental Science and Technology* 46 (9): 4800–4808.

Oniciuc, E.-A. et al. (2018). Food processing as a risk factor for antimicrobial resistance spread along the food chain. *Current Opinion in Food Science* 30: 21–26.

Östman, M. et al. (2017). Screening of biocides, metals and antibiotics in Swedish sewage sludge and wastewater. *Water Research* 115: 318–328.

Pal, C. et al. (2015). Co-occurrence of resistance genes to antibiotics, biocides and metals reveals novel insights into their co-selection potential. *BMC Genomics* 16 (1): 964.

Pal, C. et al. (2016). The structure and diversity of human, animal and environmental resistomes. *Microbiome* 4 (1): 54.

Pal, C. et al. (2017). Metal resistance and its association with antibiotic resistance. *Advances in Microbial Physiology* 70: 261–313.

Porse, A. et al. (2017). Genome dynamics of *Escherichia coli* during antibiotic treatment: transfer, loss, and persistence of genetic elements in situ of the infant gut. *Frontiers in Cellular and Infection Microbiology* 7: 126.

Pruden, A. et al. (2013). Management options for reducing the release of antibiotics and antibiotic resistance genes to the environment. *Environmental Health Perspectives* 121 (8): 878–885.

Pruden, A. et al. (2018). An environmental science and engineering framework for combating antimicrobial resistance. *Environmental Engineering Science* 35 (10): 1005–1011.

Rahube, T.O. et al. (2014). Impact of fertilizing with raw or anaerobically digested sewage sludge on the abundance of antibiotic-resistant coliforms, antibiotic resistance genes, and pathogenic bacteria in soil and on vegetables at harvest. *Applied and Environmental Microbiology* 80 (22): 6898–6907.

Rasko, D.A. et al. (2011). Origins of the *E. coli* strain causing an outbreak of hemolytic-uremic syndrome in Germany. *The New England Journal of Medicine* 365 (8): 709–717.

Renou, S. et al. (2008). Landfill leachate treatment: review and opportunity. *Journal of Hazardous Materials* 150 (3): 468–493.

Rhodes, G. et al. (2000). Distribution of oxytetracycline resistance plasmids between aeromonads in hospital and aquaculture environments: implication of Tn1721 in dissemination of the tetracycline resistance determinant tetA. *Applied and Environmental Microbiology* 66 (9): 3883–3890.

Riber, L. et al. (2014). Exploring the immediate and long-term impact on bacterial communities in soil amended with animal and urban organic waste fertilizers using pyrosequencing and screening for horizontal transfer of antibiotic resistance. *FEMS Microbiology Ecology* 90 (1): 206–224.

Riquelme Breazeal, M.V. et al. (2013). Effect of wastewater colloids on membrane removal of antibiotic resistance genes. *Water Research* 47 (1): 130–140.

Rizzo, L. et al. (2013). Urban wastewater treatment plants as hotspots for antibiotic resistant bacteria and genes spread into the environment: a review. *The Science of the Total Environment* 447: 345–360.

Robert Koch Institute (2011). Final presentation of the epidemiological findings in the EHEC O104:H4 outbreak, Germany 2011. Berlin 2011.

Rolain, J.-M. (2013). Food and human gut as reservoirs of transferable antibiotic resistance encoding genes. *Frontiers in Microbiology* 4: 173.

Röös, E. and Tjärnemo, H. (2013). Challenges of carbon labelling of food products: a consumer research perspective. *British Food Journal* 113 (8): 982–996.

Ross, J. and Topp, E. (2015). Abundance of antibiotic resistance genes in bacteriophage following soil fertilization with dairy manure or municipal biosolids, and evidence for potential transduction. *Applied and Environmental Microbiology* 81 (22): 7905–7913.

Ruuskanen, M. et al. (2016). Fertilizing with animal manure disseminates antibiotic resistance genes to the farm environment. *Journal of Environmental Quality* 45 (2): 488–493.

Rysz, M. et al. (2013). Tetracycline resistance gene maintenance under varying bacterial growth rate, substrate and oxygen availability, and tetracycline concentration. *Environmental Science and Technology* 47 (13): 6995–7001.

Ryu, S.-H. et al. (2012). Antimicrobial resistance and resistance genes in *Escherichia coli* strains isolated from commercial fish and seafood. *International Journal of Food Microbiology* 152 (1–2): 14–18.

Säll, S. and Gren, I.-M. (2015). Effects of an environmental tax on meat and dairy consumption in Sweden. *Food Policy* 55: 41–53.

Scheurer, M. et al. (2015). Removal of micropollutants, facultative pathogenic and antibiotic resistant bacteria in a full-scale retention soil filter receiving combined sewer overflow. *Environmental Science Processes and Impacts* 17 (1): 186–196.

Sharma, R. et al. (2009). Selected antimicrobial resistance during composting of manure from cattle administered sub-therapeutic antimicrobials. *Journal of Environmental Quality* 38 (2): 567–575.

Sim, W.-J. et al. (2011). Occurrence and distribution of pharmaceuticals in wastewater from households, livestock farms, hospitals and pharmaceutical manufactures. *Chemosphere* 82 (2): 179–186.

Sommerset, I. et al. (2014). Vaccines for fish in aquaculture. *Expert Review of Vaccines* 4 (1): 89–101.

SPHS Secretariat, UNDP Istanbul Regional Hub (2015). *2014 Annual Report of the Informal Interagency Task Team on Sustainable Procurement in the Health Sector (SPHS)*, UNDP.

Springmann, M. et al. (2018). Options for keeping the food system within environmental limits. *Nature* 562 (7728): 519–525.

Storteboom, H.N. et al. (2007). Response of antibiotics and resistance genes to high-intensity and low-intensity manure management. *Journal of Environmental Quality* 36 (6): 1695–1703.

Suzuki, S. and Hoa, P.T.P. (2012). Distribution of quinolones, sulfonamides, tetracyclines in aquatic environment and antibiotic resistance in indochina. *Frontiers in Microbiology* 3: 67.

The Welfare of Livestock Regulations (1994). 1994 No. 2126. http://www.legislation.gov.uk/uksi/1994/2126/made. Accessed on December 21, 2018.

Udikovic-Kolic, N. et al. (2014). Bloom of resident antibiotic-resistant bacteria in soil following manure fertilization. *Proceedings of the National Academy of Sciences of the United States of America* 111 (42): 15202–15207.

Van Goethem, M.W. et al. (2018). A reservoir of "historical" antibiotic resistance genes in remote pristine Antarctic soils. *Microbiome* 6 (1): 40.

Verlicchi, P. et al. (2012). Hospital effluent: investigation of the concentrations and distribution of pharmaceuticals and environmental risk assessment. *The Science of the Total Environment* 430: 109–118.

Vinnerås, B. (2013). *Hygieniseringsteknik för säker återföring av fosfor i kretsloppet.* Stockholm: Naturevårdsverket.

WHO/UNICEF (2017). *Progress on Drinking Water, Sanitation and Hygiene: 2017 Update and SDG Baselines.* WHO.

Wichmann, F. et al. (2014). Diverse antibiotic resistance genes in dairy cow manure. *mBio* 5 (2): e01017.

Wilson, M.E. and Chen, L.H. (2012). NDM-1 and the role of travel in its dissemination. *Current Infectious Disease Reports* 14 (3): 213–226.

von Wintersdorff, C.J.H. et al. (2014). High rates of antimicrobial drug resistance gene acquisition after international travel, The Netherlands. *Emerging Infectious Diseases* 20 (4): 649–657.

Wright, G.D. (2010). Antibiotic resistance in the environment: a link to the clinic? *Current Opinion in Microbiology* 13 (5): 589–594.

Wright, G.D. (2012). The origins of antibiotic resistance. *Handbook of Experimental Pharmacology* 211: 13–30.

Wu, N. et al. (2010). Abundance and diversity of tetracycline resistance genes in soils adjacent to representative swine feedlots in China. *Environmental Science and Technology* 44 (18): 6933–6939.

WWAP (United Nations World Water Assessment Programme) (2017). *The United Nations World Water Development Report 2017. Wastewater: The Untapped Resource.* Paris: United Nations Education, Scientific and Cultural Organization.

Zhang, T. and Li, B. (2011). Occurrence, transformation, and fate of antibiotics in municipal wastewater treatment plants. *Critical Reviews in Environmental Science and Technology* 41 (11): 951–998.

Zhang, T., Yang, Y., and Pruden, A. (2015). Effect of temperature on removal of antibiotic resistance genes by anaerobic digestion of activated sludge revealed by metagenomic approach. *Applied Microbiology and Biotechnology* 99 (18): 7771–7779.

Zhu, Y.-G. et al. (2013). Diverse and abundant antibiotic resistance genes in Chinese swine farms. *Proceedings of the National Academy of Sciences of the United States of America* 110 (9): 3435–3440.

Zuccato, E. et al. (2010). Source, occurrence and fate of antibiotics in the Italian aquatic environment. *Journal of Hazardous Materials* 179 (1-3): 1042–1048.

Index

Antibiotic Drug Resistance, First Edition. Edited by José-Luis Capelo-Martínez and Gilberto Igrejas.

C

d